Structural Geology

Structural Geology

Robert J. Twiss
Eldridge M. Moores
University of California at Davis

W. H. Freeman and Company
New York

We dedicate this book
to our wives
Helen and Judy

Cover Photograph, © John Shelton 1973.
Illustrations by Dennis and Karen Tasa, unless otherwise noted.

Library of Congress Cataloging-in-Publication Data

Twiss, Robert J.
    Structural geology/Robert J. Twiss and Eldridge M. Moores.
        p.   cm.
    Includes bibliographical references and index.
    ISBN 0-7167-2252-6
    1. Geology, Structural.   I. Moores, Eldridge M., 1938–
II. Title.
QE601.T894   1992
551.8—dc20   92-4058
                                                                CIP

Printed in the United States of America

Fourth printing 1997, KP

Contents

Preface

IN WRITING this book, we have attempted to present the subject of structural geology at an introductory level appropriate for students who have had a course in physical geology and some background in introductory physics, particularly mechanics. The book provides an integrated treatment that includes classical descriptive structural geology at the scale of field observation, the theory and applications of continuum mechanics, the results of experimental rock deformation, the microscopic mechanisms and fabrics of deformed rocks, and the synthesis of structures at a regional scale. Because structural geology is taught with many different emphases and at various levels, we have designed the book to accommodate such differences.

Most students studying structural geology at this level do not have the background in mathematics and mechanics to handle a detailed continuum mechanical treatment of the subject. This material is usually introduced only in higher-level courses in physics or engineering. Although we believe firmly in the critical importance of continuum mechanics for the understanding of the deformation of the Earth's crust, it is impossible to provide a rigorous introduction to continuum mechanics as well as to teach structural geology in one course. We therefore walk a tightrope by providing an introduction to the ideas of continuum mechanics while not presupposing a working knowledge of linear algebra or vector or tensor analysis. We emphasize the intuitive and the geometric understanding of mechanical concepts so that they are sufficient to provide an appreciation for the origin of structures and a correct and solid foundation on which more advanced quantitative work can be built. In this way, we hope to demystify this aspect of the subject and encourage students to pursue it in advanced courses. More quantitative approaches can be found in the boxes.

As a result of many years of teaching this material, we have adopted a somewhat novel organization for the book. Our aim is to introduce the observations about the Earth first, followed by the relevant mechanics and experimental results that are needed to understand the observations. Thus we introduce the concepts of stress and fracture mechanics only after we have described fractures and faults as they are observed in the Earth. The descriptive information provides a context and a reason for studying the mechanics as well as a basis of knowledge for applying the theoretical and experimental results to the Earth. Similarly, we embark on a major discussion of strain only after we have covered structures that are created by ductile deformation. The relevance of strain is then clear, and its application to understanding structures rests on an established foundation of knowledge about the Earth.

We emphasize the use of models as a means of understanding our observations, while maintaining a clear distinction between what we know from observing the Earth and what we infer from interpretation. In principle, the observations remain reliable even though the models we use to interpret them can change. Models are always limited by simplifying assumptions, and it is often by challenging such assumptions that we open doorways to new and improved models. We therefore present the topic not as a fixed body of information but as an open-ended field of study where interpretations can be changed and understanding advanced.

Figure P.1 outlines the topics treated in the book, the major connections among them, and the chapter numbers and titles in which the topics are covered (the lines that leave the page at the black triangles come back on the page at the same level on the opposite side). The diagram suggests a variety of starting points and paths through the material. We divide our own course into two sections emphasizing brittle and ductile deformation, and the sequence of chapter numbers indicates the path we ourselves prefer through the material. Because our organizational preferences may not be shared by everyone, we have attempted to structure the chapters and to separate the topics so that individual instructors can easily chart their own paths. It would be possible, for example, to start with stress and strain, work into the mechanics and experimental material, and end up with the description and interpretation of the structures observed in the Earth. Alternatively, one could begin with the large-scale tectonic features of the Earth, from there describe the smaller-scale structures, and finish with the mechanics and mechanical interpretations of the structures.

We do not expect that the entire book will necessarily be covered in a single course. For colleges and universities organized on the quarter system, such as ours, it is impossible. It seems to us that all the material could be covered, however, in a fairly intensive one-semester course. For a course taught in one quarter, or for semester courses in which more time is spent on individual sections, some topics must be omitted. For any course, Chapters 1 through 15 constitute the basic core of structural geology. The sections and boxes chosen from these first chapters determine the depth to which certain topics are pursued. For example, the topics on stress and mechanics could be presented in less detail by omitting all or some combination of Sections 8.4–8.5, 9.6–9.7, 10.1, 10.3–10.5, 10.10–10.12 and the boxes in those chapters. Chapters 16 through 22 offer more advanced material. Application of concepts of strain to the analysis of structures is discussed in Chapters 16 and 17; mechanics, and the mechanisms and fabrics of ductile flow are the main topics of Chapters 18–20; and the description of the large-scale structural and tectonic features of the Earth is covered in Chapters 21 and 22. The particular emphasis preferred by individual instructors will determine the selection of topics from the later chapters.

At U.C. Davis, we do not emphasize tectonics in the structural geology course because we offer a second quarter follow-up course. The material for this course is the subject of a companion volume to this book: *Tectonics* (Moores and Twiss). In writing these two books, we have tried to show that structural geology and tectonics constitute a unified subject that is divided only by the scale of the

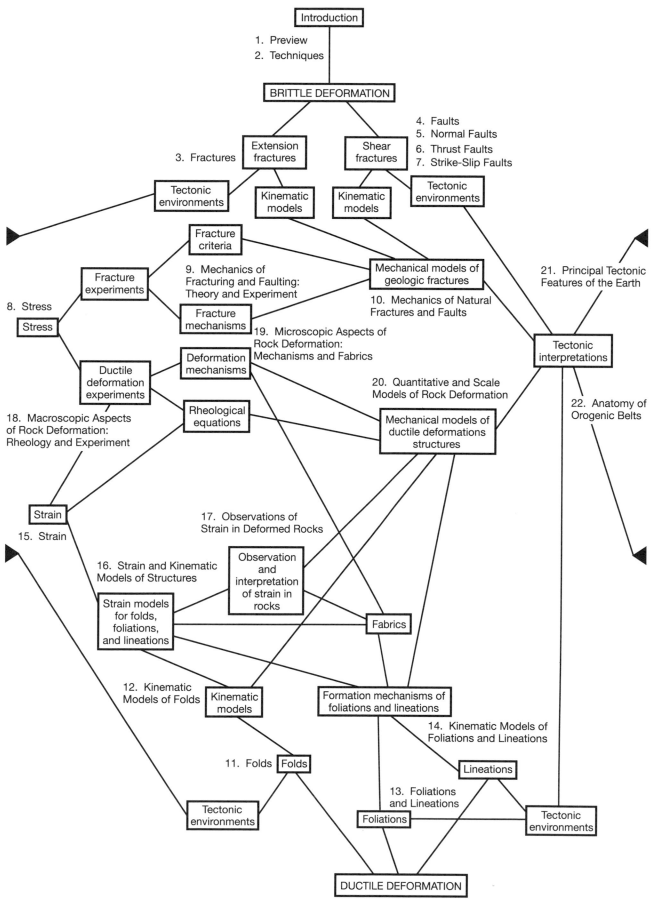

Figure P.1

features at which deformation of the Earth's crust is examined. The scale of structures, however, varies continuously so there is really no clear boundary between structural geology and tectonics. We need to integrate our understanding of deformation at all scales to understand the dynamics of the Earth as it affects the easily observable outermost parts. In *Structural Geology,* we consider structural features at scales from the submicroscopic to the regional, and in *Tectonics* structural features at scales from the regional to the global. Throughout both books, we emphasize the value of integrating information over all available scales, because we firmly believe that structural geologists and tectonicists cannot ignore each other's work.

We are eager for feedback, not only from professors who teach with the book but also from students who learn from it. We want to hear what parts of the book you do not like, what parts are still not clear, and what parts you particularly like. A manuscript such as this one never ceases evolving. Yet at some point one must commit to print and hope the worst bugs have been eliminated. With your help, however, we can keep striving for improvement.

Acknowledgments

In writing this book, we have benefited immensely from the help and suggestions of a number of colleagues and our own students who have read parts or all of the manuscript in its various stages of revision. Roy Dokka, Roy Kligfield, William MacDonald, Stephen Marshak, Peter Mattson, Cris Mawer, Sharon Mosher, Raymond Siever, Carol Simpson, and Doug Walker all improved the manuscript significantly with their comments. We are particularly grateful to William MacDonald, Sharon Mosher, and Carol Simpson for the enormous effort they invested in providing us with exceptionally detailed and helpful comments. Our own students have also helped us by their reactions to and questions about the book. For their input and their patience with early photocopied versions of the manuscript we are most grateful.

We also thank all those authors who have helped us by providing copies of their photographs and by granting us permission to use their diagrams. They are too numerous to list individually here, but they are credited in the list of sources at the end of the book. Without such generosity, this book could not have been published in its present form. We must single out John Ramsay for special thanks, however, for permitting us to use an unusually large number of figures from his publications in their original or modified form. Our reliance on his work is testimony to his leading role in the development of modern structural geology and to the seminal quality of his many published works.

Last, but not least, we wish to thank everyone at W. H. Freeman and Company who worked on the book, especially Jeremiah Lyons, Publisher, and Christine Hastings, Project Editor, who with carrot and stick have kept us going, and the late John Staples, who initiated the project.

PART I Introduction

STRUCTURAL geology and tectonics are two branches of geology that are closely related in both their subject matter and approach to the study of the evolution of the Earth. Although traditionally they are taught as courses that are distinct from other branches of geological study such as petrology, paleontology, and geophysics, structural geology and tectonics are very interdependent and overlap with these other fields. We are all trying to understand the evolution of a single planet, after all, and the pieces of the jigsaw puzzle must inevitably fit together. The practice and study of structural geology and tectonics require a familiarity with scientific method and a variety of standard techniques of data acquisition and analysis. Thus an overview of the realm of structural geology and tectonics and a discussion of some of these basic techniques are the topics we discuss in this first section of the book.

In the first chapter, after defining what we mean by structural geology and tectonics, we discuss the use of models as a technique of scientific investigation. This technique is common to most sciences today, and it is a fundamental theme of the book that provides the basis for much of the organization we use in presenting different topics. We then give a perspective of the fields of structural geology and tectonics and their position with regard to the study of Earth processes and the large-scale view of planetary bodies.

In the second chapter we present brief summaries of techniques of major importance used in structural geology and tectonics for the collection and presentation of data. We do not intend these brief discussions to substitute for a laboratory course, in which much of this information is normally covered, but simply to provide an introductory understanding of the concepts and the diagrams that we use in the book and to refresh the memories of those who are already familiar with the material.

In Chapter 2 we also introduce a number of geophysical techniques that are our primary sources of data for the structure of the deep crust and upper mantle. They enable us to view the third dimension of the solid Earth that is so central to the studies of structure and tectonics. It is important to have at least a rudimentary understanding of the meaning and significance of geophysical data, but we encourage students to go beyond these rather cursory discussions and to take separate courses in geophysics.

CHAPTER

Overview

1.1 What Are Structural Geology and Tectonics?

The Earth is a dynamic planet. The evidence is all around us. Earthquakes and volcanic eruptions regularly jar many parts of the world. Many rocks exposed at the Earth's surface reveal a continuous history of such activity, and some have been uplifted from much deeper levels in the crust where they were broken, bent, and contorted. These processes proceed in slow motion on the scale of a human lifetime, however, or even the scale of human history. The "continual" eruption of a volcano can mean that it erupts once in one or more human lifetimes. The "continual" shifting and grinding along a fault in the crust means that a major earthquake might occur once every 50 to 150 years. At the almost imperceptible rate of a few millimeters per year, high mountain ranges can be uplifted in the geologically short span of only a million years. A million years, however, is already more than two orders of magnitude longer than the whole of recorded history. These processes have been going on for hundreds of millions of years, leaving in the Earth's crust a record of constant dynamic activity.

Structural geology and tectonics are concerned with the reconstruction of the inexorable motions that have shaped the evolution of the Earth's outer layers. These terms are derived from similar roots. *Structure* comes from the Latin word *struere,* which means "to build," and *tectonics* from the Greek word *tektos,* which means "builder." The motion may be simply a **rigid body motion** that transports a body of rock from one place to another with no change in its size or shape and, therefore, no permanent imprint. Or the motion may be a **deformation** that breaks a rock or changes its shape or size and thus leaves a permanent record. The structures discussed in this book result from permanent deformation of the rocks; they provide a record of the nature of that deformation.

For example, the Earth's crust breaks along faults, and the two pieces slide past one another. Sections of continental crust break apart as oceans open, and they collide with each other as oceans close. These events result in bending and breaking of rocks in the shallow crust and in puttylike flowing of rocks at greater depth. Mountain ranges are uplifted and subsequently eroded, exposing the deeper levels of the crust. The breaking, bending, and flowing of rocks all produce permanent structures such as fractures, faults, and folds that we can use as clues to reconstruct the deformation that produced them. Even on a much smaller scale, the preferred alignment of platy and elongate mineral grains in the rocks and the submicroscopic imperfections in crystalline structure all help us trace the course of the deformation.

The fields of structural geology and tectonics both deal with motions and deformation in the Earth's crust and upper mantle. They differ in that structural geology is predominantly the study of deformation in rocks at a scale ranging from the submicroscopic to the regional, whereas tectonics is predominantly the study of the history of motions and deformation at a regional to global scale. The two realms of study are interdependent, and at the regional scale, structural geology and tectonics overlap considerably. Our interpretation of the history of large-scale motions must be consistent with the observations of deformation that has occurred in the rocks. Conversely, the origins of deformation can be understood in the context of the history of the large-scale motions that we deduce from plate tectonics.

Structural geology and tectonics have undergone a period of rapid development since the 1960s. Structural geology has changed from an almost purely descriptive discipline to an increasingly quantitative one. Application of theoretical principles of **continuum mechanics**,[1] the ability to deform rocks directly in the laboratory, and the ability to study deformed minerals at the submicroscopic level have yielded new insights into the processes of deformation and the formation of structures. The insight gained from such studies has been used in many field-based investigations vastly to improve our understanding of naturally deformed rocks.

Tectonics has undergone a revolution based largely on the formulation of the theory of plate tectonics. This theory now provides the framework for study of almost all large-scale motions and deformation affecting the Earth's crust and upper mantle. Field-based studies have taken on new meaning because plate tectonic theory has given us a new basis for the tectonic interpretation of structures and the history of regional deformation.

Geophysical data have become increasingly important to both structure and tectonics, as indicated by the number of diagrams throughout this book that present geophysical data. Seismic and gravity studies provide information on the geometry of large-scale structures at depth, which adds the critical third spatial dimension to our observations. Studies of rock magnetism and paleomagnetism, as well as seismology, provide data on past and present motions of the plates, which are essential for reconstructing global tectonic patterns.

Structure and tectonics also depend on other branches of geology, which, however, we do not emphasize in this book. Petrology and geochemistry provide data on temperature, pressure, and ages of deformation and metamorphism, which are necessary for the accurate interpretation of deformation and its tectonic significance. Sedimentology and paleontology are also important in reconstructing the patterns and ages of tectonic events.

1.2 Structure, Tectonics, and the Use of Models

All field studies in structural geology and many in tectonics rely on observations of deformed rocks at the Earth's surface. These studies generally begin with observations of features at outcrop scale—that is, a scale of a few millimeters to several meters. They may then proceed "downscale" to observations made at the microscopic or even electron-microscopic level of microns[2] or "upscale" to more regional observations at a scale of hundreds to thousands of kilometers. At the largest scale, observations are generally based on a compilation of observations from smaller scales. *None of these observations alone provides a complete view of all structural and tectonic processes.* Our understanding increases as we integrate our observations of the Earth at all scales. In addition to direct observation of the Earth, we also use the results of laboratory experiments and mathematical calculations in making our interpretations.

Our first task in trying to unravel the deformation and its history is to observe and record, carefully and systematically, the structures in the rock, including such features as lithologic contacts, fractures, faults, folds, and preferred orientations of mineral grains. In general, this process consists of determining the **geometry** of the structures. Where are the structures located in the rocks? What are their characteristics? How are they oriented in space and with respect to one another? How many times in the past have the rocks been deformed? Which structures belong to which episodes of deformation? Answering these and similar questions constitutes the initial phase of any structural investigation.

In some circumstances, determining the geometry of rock structures is an end in itself (it is important, for example, in the location of economic deposits). For understanding the processes that occur in the Earth, however, we need an explanation for the geometry. Ultimately we want to know the **kinematics** of for-

---

[1] Continuum mechanics is the study of the mechanics of **continua**, which are deformable materials whose properties vary smoothly or remain constant in any direction throughout a given volume. Rocks and many other substances, such as metals, are generally composed of grains that are very small compared with the volume being described. Thus their behavior approximates that of a continuum, and for purposes of analysis, we can ignore inherent discontinuities such as grain boundaries. This approximation enables us to describe the deformation of the rock in simple mathematical terms.

[2] One micron is one micrometer, or one-millionth (10^{-6}) meter, or one-thousandth (10^{-3}) millimeter.

mation of the structures—that is, the motions that have occurred in producing them. Beyond that, we want to understand the **mechanics** of formation—that is, the forces that were applied, how they were applied, and how the rocks reacted to those forces to form the observed structures.

In large part, we improve our understanding by making conceptual models of how structures form and then testing predictions derived from the models against observation. **Geometric models** are three-dimensional interpretations of the distribution and orientation of structures within the Earth. They are based on mapping, geophysical data, and any other information we have. We present such models as geologic maps and as vertical cross sections along a particular line through an area.

Kinematic models prescribe a specific history of motion that could have carried the system from the undeformed to the deformed state or from one configuration to another. They are not concerned with why or how the motion occurred or with what the physical properties of the system were. The model of plate tectonics is a good example of a kinematic model. We can assess the validity of such models by comparing the geometries of the motion and deformation observed in the Earth with those deduced from the model.

Mechanical models are based on our understanding of basic laws of continuum mechanics, such as the conservation of mass, momentum, angular momentum, and energy, and on our understanding of how rocks behave in response to applied forces. This latter information comes largely from laboratory experiments in which rocks are deformed under conditions that reproduce as nearly as possible the conditions within the Earth. Using mechanical models, we can calculate the theoretical deformation of a body of rock that is subjected to a given set of physical conditions such as forces, temperatures, and pressures. A model of the driving forces of plate tectonics based on the mechanics of convection in the mantle is an example of a mechanical model. Such models represent a deeper level of analysis than kinematic models, for the motions of the model are not assumed but must be a consequence of the physical and mechanical properties of the system. Thus the models are constrained not only by the geometry of the deformation, but also by the physical conditions and mechanical properties of the rocks when they deformed.

We use geometric, kinematic, and mechanical models to help us understand deformation on both the small scale and the global scale. It is important to realize, however, that even though we may be able to invent some model whose properties resemble observations of part of the Earth, such a model is not necessarily a good one. Predictions based on a model tell us only about the properties and characteristics of the *model*, not the actual conditions in the Earth. The relevance of a model for understanding the Earth, therefore, must always be tested.

Observations guide the formulation of models. The models in turn provide predictions that can be compared with reality, thereby stimulating new observations of the real world. Comparisons of what the model predicts with what we observe of the Earth constitute tests of the model's relevance. New observations that confirm the predictions support the model. To that extent, we accept it as a reasonable representation of the processes occurring in the Earth. If at any time observations contradict the predictions, we must refine the old model or reject it and devise a new one. Our understanding of structural and tectonic processes improves gradually by a continual repetition of the processes of making observations, formulating models based on those observations, deriving predictions from the models, testing the models with new observations, and fine-tuning the models accordingly. This process, in fact, is common to all science.

1.3 The Interiors of the Earth and the Other Terrestrial Bodies

Although most structural and tectonic processes discussed in this book are observed either at the surface or within the outermost layer of the Earth (the crust), the large scale at which these motions are consistent indicates that they reflect deeper, interior processes. Moreover, Earth is not alone in the solar system. In this age of space exploration, the models we make of the dynamic processes in the Earth are relevant to our understanding of other planets.

According to current models, the Earth is divided into three approximately concentric shells; these are, from the center outward, the core, the mantle, and the crust (Figure 1.1). The core is composed of very dense material believed to be predominantly an iron–nickel alloy. It includes a solid inner core and a liquid outer core. Surrounding the core is the mantle, a thick shell much lower in density than the core and composed largely of solid magnesium–iron silicates. The crust is a thin layer of relatively low-density minerals that surrounds the mantle. It is composed of igneous rocks of granitic to basaltic composition, sediments and sedimentary rocks, and their metamorphic equivalents, predominantly sodium, potassium, and calcium alumino-silicates.

The temperature of the Earth increases with depth with a gradient of approximately 30°C per kilometer in the crust and upper mantle and with considerably smaller gradients deeper within the Earth. Several

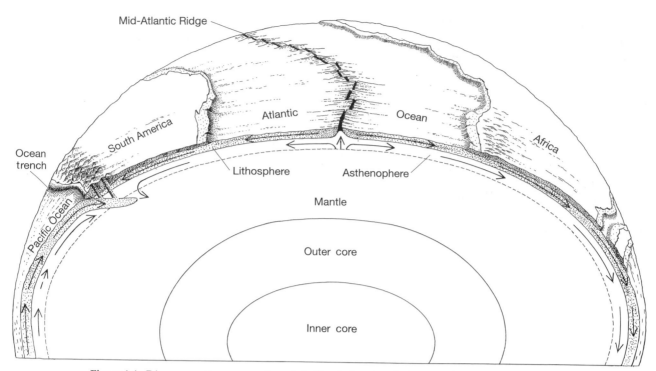

Figure 1.1 Diagrammatic cross section of the Earth showing the inner and outer core, the mantle, the lithosphere, and the crust. The spreading centers and subduction zones of plate tectonics are also indicated.

sources of heat account for this increase of temperature with depth: residual heat trapped during the original accretion of the Earth approximately 4500 million years ago, heat produced continually by the spontaneous decay of radioactive elements within the Earth, possibly latent heat of crystallization from slow solidification of the liquid outer core, and heat produced by the dissipation of tidal energy resulting from the gravitational interaction among the Earth, Moon, and Sun.

The increase of temperature with increasing depth results in a flow of heat energy toward the surface that may involve the convective cycling of material both within the liquid core and within the solid mantle. Heat convected out of the core is transferred to the lower mantle and is then carried toward the surface in a separate mantle convection system. The top of the mantle is a relatively cold, strong boundary layer called the **lithosphere,** which includes the crust and upper mantle to a depth of about 100 km. Heat escapes through the lithosphere to the surface largely by conduction and by transport in upward-migrating igneous melts.

Our studies of structural geology and tectonics focus on the motion and deformation of the lithosphere, as revealed mostly in its outer 20 to 30 km. It puts into perspective the task of understanding the dynamics of the Earth to realize that what we have available to study

is the thin rind of relatively cold material that probably rides passively on top of mantle convection currents.

The Earth is one of a class of similar objects in the solar system called the **terrestrial bodies.** They are the innermost four planets (Mercury, Venus, Earth, and Mars), Earth's Moon, and several moons of Jupiter and Saturn, including Io, Titan, and Europa. All the terrestrial bodies apparently consist of a central core of very dense material that is probably an iron–nickel alloy, a mantle of dense silicates, and a thin crust of relatively low-density silicates. The inner four planets differ in size and relative volume of core. The core is proportionately greatest for Mercury and is progressively smaller for Venus, Earth, and Mars. The core of the Moon is relatively smaller still.

Although our study of the other terrestrial bodies is in its infancy, the current state of our knowledge invites fascinating comparisons with the Earth and speculation about the origin of various similarities and differences. As our understanding of the other terrestrial bodies increases, it will provide tests for models we have devised to explain dynamic processes occurring within the Earth. These models will have to account for the presence or absence of such processes on the basis of the size, structure, and internal physical conditions of the other terrestrial bodies.

1.4 Characteristics of the Earth's Crust and Plate Tectonics

The crust of the Earth is broadly divided into continental crust of approximately granitic composition and oceanic crust of roughly basaltic composition. Land—that part of the Earth's surface above sea level—is principally continental, the exceptions being islands in the oceans. At the present time, 29.22 percent of the Earth's surface is land and 70.78 percent is oceans and seas. Continental crust makes up 34.7 percent of the total area of the Earth, and it underlies most of the land area as well as the continental shelves and continental regions covered by shallow seas, such as Hudson's Bay and the North Sea. The remaining 65.3 percent of the Earth is oceanic crust (Figure 1.2).

The distribution of the Earth's surface elevation is strongly bimodal. Most of the continental surface lies within a few hundred meters of sea level, and most of the ocean floor lies approximately 5 km below the sea surface. This distribution is evident from the two types of **hypsometric diagrams** shown in Figure 1.3. The term *hypsometric* is derived from the Greek words *hypsos*, which means "elevation," and *metron*, which means "measure." The cumulative plot (Figure 1.3A) shows the total percentage of surface above a given

elevation; the specific plot (histogram) shows the percentage of the surface within a given elevation interval. The **continental freeboard,** the difference in elevation between continent and ocean floor (Figure 1.3B), results from a number of factors, including the thickness and density differences between the continental and oceanic crusts, tectonic activity, erosion, sea level, and the ultimate strength of continental rocks, which determines their ability to maintain an unsupported slope above oceanic crust.

The characteristics of the Earth's crust are largely the direct or indirect result of motions of the lithosphere. The **theory of plate tectonics** describes these motions and accounts for most observable tectonic activity in the Earth, as well as the tectonic history recorded in the ocean basins. The theory holds that the Earth's lithosphere is divided at present into seven major and several minor **plates** that are in motion with respect to one another (Figure 1.2) and that the motion of each plate is, to a first approximation, a rigid-body motion. Deformation of the plates is concentrated primarily in belts tens to hundreds of kilometers wide along the plate boundaries. In a few regions, however, deformation extends deep into plate interiors.

The different types of boundaries between these plates include divergent boundaries, where plates move away from one another; convergent or consuming

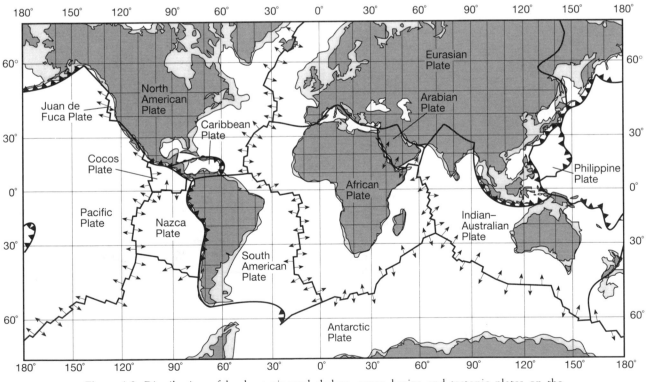

Figure 1.2 Distribution of land, continental shelves, ocean basins and tectonic plates on the surface of the Earth.

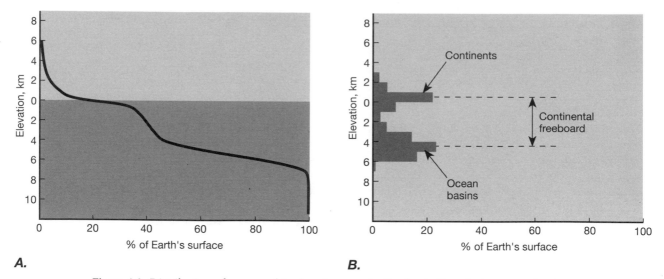

Figure 1.3 Distribution of topographic elevations on the Earth. *A.* Cumulative curve showing the percent of the Earth's surface that is above a particular elevation. *B.* Histogram showing the percent of the Earth's surface that falls within each 1-km interval of elevation. The continental freeboard is the difference between the dominant elevations of the continents and of the ocean basins.

boundaries (also called subduction zones), where plates move toward each other and one descends beneath another; and conservative or transform fault boundaries, where plates move horizontally past one another, without creation or destruction of lithosphere.

The most direct evidence for plate tectonic processes and sea floor spreading comes from the oceanic crust, where divergent motion at mid-oceanic ridges adds new material to lithospheric plates. As indicated in Figure 1.4, however, the maximum age of the oceanic crust limits this evidence to the last 180 million years—that is, to the *last 4 percent* of Earth history. Any evidence of plate tectonic processes for the preceding 96 percent of Earth history must come from the continental crust, which contains a much longer record of the Earth's activity. We must therefore learn to understand the large-scale tectonic significance of deformation in the continental crust so that we can see further back into the history of the Earth's dynamic activity.

In the geologic record, highly deformed continental rocks tend to be concentrated in long linear belts comparable to the belts of deformation associated with present plate boundaries. This observation suggests that belts of deformation in the continental crust record the existence and location of former plate boundaries. If this hypothesis is correct, and if we can learn what structural characteristics of deformation correspond to the different types of plate boundaries, we can use these structures in ancient continental rocks to infer the pattern and processes of tectonic activity. In this sense, the plate tectonic model has united the disciplines of structural geology and tectonics and made them interdependent.

The types of structures that develop in rocks during deformation (characteristically along plate boundaries) depend on the orientation and intensity of the forces applied to the rocks; on the physical conditions, such as the temperature and pressure, under which the rocks are deformed; and on the mechanical properties of the rocks, which are strongly affected by the physical conditions. At relatively low temperatures and pressures and at a high intensity of applied forces, a rock generally undergoes **brittle deformation** by loss of cohesion along discrete surfaces to form fractures and faults. At relatively high temperatures and pressures (but below the melting point) and at a relatively low intensity of applied forces, a rock commonly reacts by **ductile deformation**[3]—the flow, or coherent change in shape, of the rock in the solid crystalline state. This behavior may produce folding of stratigraphic layers, the stretching and thinning of layers, and the parallel alignment of mineral grains in the rock to form pervasive planar and linear preferred orientations.

In the belts of deformation along the plate boundaries, the different relative motions of adjacent plates, that is, whether the boundary is divergent, convergent, or conservative, largely determine the style of defor-

---

[3] We use the term *ductile* to imply coherent nonrecoverable deformation that occurs in the solid state without loss of cohesion (brittle fracturing) at the scale of crystal grains or larger. The term has broader significance in other contexts, but there is no other word that adequately describes this behavior. In particular, the term *plastic* has other specific connotations that do not accurately reflect the behavior we wish to describe. See the introduction to Part III for a more detailed discussion of these terms.

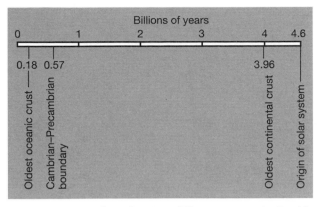

Figure 1.4 Time line showing different events in Earth's history and the ages of the oldest oceanic and continental crusts.

mation. Differences between oceanic and continental crust also affect the nature of deformation along plate boundaries.

At divergent boundaries, oceanic crust is produced by the partial melting of upwelling mantle material to form basaltic magma. Igneous intrusion and extrusion of these basalts produce the new oceanic crust. The relative motion of the plates creates structures in the crust that accommodate stretching, such as systems of normal faults near the surface and ductile thinning at deeper levels. When a divergent plate boundary develops within a continent, the associated stretching and thinning lower the mean elevation of the boundary zone, resulting in flooding of the surface by the sea. Such stretched and thinned continent commonly underlies the wide continental shelves (Figure 1.2).

Subduction zones at convergent boundaries are the places where oceanic crust is recycled back into the mantle. The plate plunges back into the interior of the Earth. Sediments on the down-going plate are partly scraped off, and partial melting of the down-going plate produces characteristic volcanic arcs on the over-riding plate. Structures at the plate boundary are predominantly systems of thrust faults. Along the volcanic arc, however, normal faults are common. If a continent is involved with either the over-riding or the down-going plate at a subduction zone, it commonly experiences shortening and thickening of its crust by means of characteristic systems of thrust faults.

At conservative or transform fault boundaries, the structures that form are typically systems of strike-slip faults or, at deeper levels, vertical zones of ductile deformation that have a subhorizontal direction of displacement.

A variety of secondary structures also develop in any of these tectonic environments, so the presence of any particular structure per se is not necessarily diagnostic of the type of boundary at which it developed.

The genesis can be inferred only after a careful study of the regional pattern of the structures and their associations.

Structural and tectonic processes profoundly influence other Earth processes as well. For example, the continents have varied in number, size, and geographic position as a result of plate tectonic processes. As plate tectonics changed the shape and distribution of continents and ocean basins, the patterns of oceanic and atmospheric circulation changed accordingly. The resulting changes in environmental conditions affected the patterns of sedimentary environments, as revealed by studies of stratigraphy and sedimentology, and the patterns of natural selection and evolution, as revealed by paleontologic studies.

Because plate boundaries are sites of major thermal anomalies in the crust and upper mantle, these areas control the occurrence and distribution of igneous and metamorphic rocks, which are studied in "hard rock" petrology. Similarly, the formation, concentration, and preservation of mineral deposits are profoundly affected by structures and their tectonic environments, as well as by the thermal anomalies at plate boundaries. As the Earth's resources are increasingly depleted, increasingly sophisticated and subtle exploration strategies are required in order to find and develop new deposits. Structural geology and tectonics are assuming a crucial role in the search for metal and hydrocarbon deposits.

1.5 Summary and Preview

In a sense, the study of Earth deformation processes is a detective exercise. As in all other branches of geology, our evidence is usually incomplete, and we must use all available paths of investigation to limit the uncertainties. Thus we study modern processes to help us understand the results of past deformations. We use indirect geophysical observation to detect structures that lie beneath the surface where we cannot see them. We make observations on all scales, from the submicroscopic to the regional, and try to integrate them into a unified model. We perform laboratory experiments to study the behavior of rocks under conditions that at least partially reproduce those found in the Earth. And we use mechanical modeling, in which we apply the principles of continuum mechanics to calculate the expected behavior of rocks under different conditions.

At the level of this book, we cannot hope to cover all these aspects in detail. Our aim, rather, is to provide a thorough basis for field observation of geologic structures and to introduce the various paths of investigation that can add valuable data to our observations and lead to deeper understanding of structural and tectonic processes. We hope also to instill an appreciation for the interdependence and essential unity of the disciplines of structural geology and tectonics.

We have arranged the topics to be covered into four major parts following this introductory Part I: Part II covers the structures typically associated with brittle deformation. Part III discusses structures formed during ductile deformation. Part IV deals with rheology, or the characteristics and mechanisms of ductile flow in rocks. And Part V discusses tectonics and the relationships among plate tectonics, crustal deformation, and the structures formed in different tectonic settings.

Our approach is first to describe the characteristics and geometry of the different types of structures that we can observe in the field. For each class of structures, we then discuss possible kinematic models that can account for the observations. After discussing the types of structures in each section, we present a more detailed analysis of their formation by introducing relevant concepts from continuum mechanics and pertinent results from laboratory experiments. We introduce the concept of stress in Part II to describe the intensity of forces so that we can explain the origin of fractures and faults in rocks. We introduce the concept of strain in Part III to describe deformation so that we can better understand the structures formed during ductile deformation.

The manner in which deformation of rock depends on the intensity of forces applied to it is determined by the relationships between the stress and the strain. These relationships are the subject of Part IV, where we also discuss the mechanisms that give rise to the flow of rocks and the characteristic microstructures that result from the operation of those mechanisms.

By applying these ideas to the observable deformation in the Earth, we can understand the conditions necessary for the formation of different structures, and this in turn helps us to determine the deformational processes and tectonic environments in which structures form. Our presentation generally follows the process of research and interpretation, which must of necessity start with the geometric analysis of the structures that exist in the rocks, and which then ideally proceeds to kinematic and mechanical interpretation of those structures.

Finally, in Part V, we look briefly at the major tectonic features of the Earth's crust with particular emphasis on orogenic belts (mountain belts). For two centuries, orogenic belts have fascinated those who study the Earth. They preserve much of the information that exists about the interaction of plates with each other through geologic history.

In a companion volume we examine in more detail the tectonic processes that are ultimately the origin of the deformation recorded by structures. We explore modern tectonic processes and describe the structures and associations of structures that develop in response to these processes.

Using the models that we construct from observations of recent tectonic processes, we look at systems of structures in ancient orogenic belts that are exposed largely in major mountain ranges of the world. By studying the structures, and by applying our understanding of how they form, we try to reconstruct the geometry, the kinematics, and the mechanics of their formation and finally to integrate this information into a tectonic interpretation consistent with the tectonic models. In this manner, we can push our reconstructions of the tectonic history of the Earth further and further back into the past, where the geologic record becomes increasingly fragmentary and obscure.

Finally, we briefly compare what we currently understand to be the tectonics of the nearby terrestrial bodies (the Moon, Mercury, Venus, and Mars) with the tectonics of the Earth.

The current theory of plate tectonics does not answer all the questions we have about the structural and tectonic evolution of the Earth. And, of course, tectonic processes have not necessarily remained the same throughout the Earth's entire history. One of the challenges of modern structural geology and tectonics is to study ancient deformation to see whether, in fact, models based on modern tectonics are appropriate, or whether the observed structural patterns and associations require different models for the various stages in the Earth's evolution. Plate tectonic theory is a major advance in our understanding, but it is itself evolving. The problems that remain today are generating provocative research questions. Answering them will lead to further advances.

We hope these books will stimulate the curiosity and ambition of a new generation of geologists to explore in greater detail the various paths of investigation we introduce and, ultimately, to create new approaches to enhance our understanding.

Additional Readings

Bally, A. W. 1980. Basins and subsidence: A summary. In *Dynamics of plate interiors,* ed. A. W. Bally, P. L. Bender, T. R. McGetchin, and R. I. Walcott, 5–20. Geodynamics Series, Vol. 1. Washington, DC: American Geophysical Union; Boulder, CO: Geological Society of America.

Siever, R., ed. 1983. The dynamic earth. *Scientific American.* September.

Uyeda, S. 1978. *The new view of the earth.* San Francisco, Freeman.

CHAPTER

Techniques of Structural Geology and Tectonics

Investigations in structural geology require familiarity with basic techniques of observation and of reporting and displaying three-dimensional formation. Thus we use standardized methods for measuring the orientations of planes and lines in space. Techniques for distinguishing the relative ages of strata in a sequence of layered sediments are crucial to interpreting their significance. Various graphical displays have proved most useful for plotting and interpreting orientational or three-dimensional data. Such displays include geologic maps and cross sections that present data in a geographic framework, as well as histograms and spherical projections that portray only orientational data without regard to geographic location. Geophysical data such as seismic reflection profiles and gravity measurements are essential to the interpretation of many large-scale structures, and a structural geologist must understand how these data are obtained and interpreted in order to take their limitations into account. Most discussions of structural geology and tectonics assume the reader is familiar with all these techniques.

Proper interpretation of geologic structures depends critically on the ability to visualize spatial relationships among various features. This ability to think in three dimensions does not necessarily come easily, but one can learn it with practice. Learning to "think in 3-D" is a major goal of laboratory courses in structural geology.

2.1 The Orientation of Structures

Many of the structures observed in outcrops are approximately planar or linear features. Planar features include bedding, fractures, fault planes, dikes, unconformities, and planar preferred orientations of micas. Linear features include grooves and streaks on a surface, intersections of two planar features, and linear preferred orientations of mineral grains. We can represent features that are not planar, such as folded surfaces and folded linear features, by measuring a series of tangent planes or lines around the structure.

Thus the **attitude** of a plane or a line—that is, its orientation in space—is fundamental to the description of structures. We specify the attitudes both of planes and of lines with two angles measured, respectively, from geographic north and from a horizontal plane. The attitude of a plane is specified by its strike and its dip, or by the trend of the dip line and the dip angle. The attitude of a linear feature is given by its trend and its plunge.

The **strike** is the horizontal angle, measured relative to geographic north, of the horizontal line in a given planar structure (Figure 2.1A). This horizontal line is the **strike line,** and it is defined by the intersection of a horizontal plane with the planar structure. It has a unique orientation for any given orientation of plane

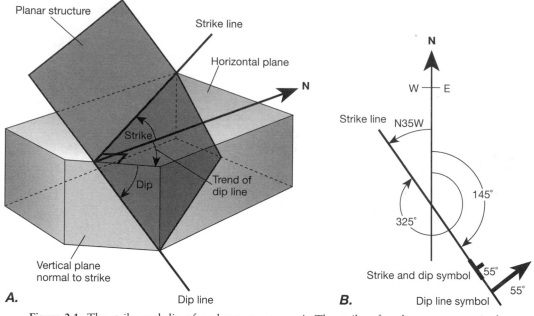

Figure 2.1 The strike and dip of a planar structure. *A.* The strike of a planar structure is the azimuth of a horizontal line in the plane and is defined by the intersection of a horizontal plane with the planar structure. The dip angle is measured between horizontal and the planar structure in a vertical plane normal to the strike line. The dip direction in this diagram is to the NE. *B.* The strike shown here is measured on a quadrant compass as N50W or on a 360° compass as either 325° or 145°. The attitude is indicated on a map by a T-shaped symbol with the stem parallel to the dip direction. The dip angle is written beside this symbol.

except a horizontal one. The **dip** is the slope of a plane defined by the dip angle and the dip direction. The **dip angle,** also referred to simply as the dip, is the angle between a horizontal plane and the planar structure, measured in a vertical plane that is perpendicular to the strike line (Figure 2.1*A*). It is the largest possible angle between the horizontal plane and the inclined plane. For a given strike, a particular value of the dip angle identifies two planes which slope in opposite directions. To distinguish between them, we specify the approximate dip direction by giving the quadrant (NE, SE, SW, NW) or the principal compass direction (N, E, S, W) of the down-dip direction.

The dip line is perpendicular to the strike line and is the direction of steepest descent on the planar structure. In some cases, particularly in British usage, the attitude of a plane is specified by the trend and plunge (defined below) of the dip line. The plunge of the dip line is the same as the dip of the plane.

For the attitude of a linear structure, the **trend** is the strike of the vertical plane in which the linear structure lies; it is unique except for an exactly vertical linear structure (Figure 2.2*A*). The **plunge** is the angle between the horizontal plane and the linear structure, measured in the vertical plane (Figure 2.2*A*). The direction of plunge, like the dip direction, can be specified by the

quadrant or by the principal compass direction of the down-plunge direction. Alternatively, it can be specified implicitly by using the convention that the trend always be measured in the down-plunge direction.

We generally use one of two conventions to record a strike or a trend. We can specify the angle as a **bearing** (the angle measured between 0° and 90° east or west of north or, in some cases, south) or as an **azimuth** (the angle between 0° and 360°, increasing clockwise from north). Measurements differing by 180° have the same orientation. Thus a northeast strike or trend could be reported as bearings N45E or S45W and as azimuths 045° or 225°, each pair differing by 180°. Using the same conventions, we could give a northwest strike or trend as bearings N45W or S45E and as azimuths 135° or 315°. It is good practice always to write the azimuth as a three-digit number, using preceding zeros where necessary, to distinguish it from dip or plunge angles which are always between 0° and 90°.

A variety of conventions are in common use for writing the attitudes of planes and lines. We generally write the strike and dip in the order strike, dip, dip direction, regardless of whether bearing or azimuth is used. Strike bearings are always reported relative to north; no distinction is made between azimuths differing by 180°. Thus N35W;55NE, 325;55NE, and 145;55NE

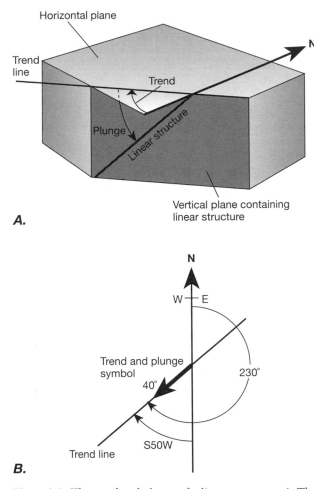

A.

B.

Figure 2.2 The trend and plunge of a linear structure. *A.* The trend is the angle between north and the strike of the vertical plane that contains the linear structure. The plunge is the angle between horizontal and the linear structure, measured in the vertical plane that contains the structure. *B.* The trend shown here is given as N20W on a quadrant compass or as either 340° or 160° on a 360° compass. If by convention the trend is recorded in the down-plunge direction, then it can be given only by S20E or 160°. On a map, linear features are plotted as an arrow parallel to the trend and pointing in the down-plunge direction. The plunge angle is written beside this arrow.

all specify the same attitude of plane (Figure 2.1B). The dip direction is always in a quadrant adjacent to that of the strike.

The most convenient convention for the attitudes of lines is to measure the trend in the down-plunge direction, in which case that direction need not be stated explicitly. We first write the plunge and then the trend (the reverse order is also used) so that the measurements 40;S50W and 40;230 both refer to the same attitude (Figure 2.2B).

We plot the attitude of a plane on a map by drafting a line parallel to strike with a short bar indicating the down-dip direction. If the trend and plunge of the dip

line is measured, we plot the attitude with an arrow pointing in the down-dip direction (Figure 2.1B). The attitude of a line is indicated by an arrow pointing in the down-plunge direction (Figure 2.2B; the arrow symbols used to indicate the attitude of a line and of the dip line of a plane should be different). In all cases, the value of the dip or plunge, as appropriate, is written beside the symbol so that the attitude can be seen at a glance.

2.2 Geologic Maps

Geologic maps are the basis of all studies of structure and tectonics. They are two-dimensional representations of an area of the Earth's surface on which are plotted a variety of data of geologic interest. These data are based on observations from many outcrops and on the judicious inference of relationships that are not directly observable. The data plotted may include the distribution of the different rock types, the location and nature of the contacts between the rock types, and the location and attitude of structural features.

Contacts are lines on the topographic surface of the Earth where the boundary surface between two different rock types intersects the topography. Contacts can be **stratigraphic,** where one unit lies depositionally upon another; **tectonic,** where the units are faulted against one another; or **intrusive,** where one unit invades another. On a geologic map, different types of contacts and the reliability of the contact location are indicated by different styles of lines. The shape of the contact on the map depends both on the geometry of the contact surface (for example, whether it is planar or folded and what its attitude is) and on the topography that the contact surface intersects. An experienced observer can determine the geometry of structures in an area, as well as the quality of information available, simply by careful inspection of a good geologic map.

All geologic maps are smaller than the area they represent. Exactly how much smaller is represented by the **scale** of the map, which is the ratio of the distance on the map to the equivalent distance on the ground. A scale of 1:25,000 (one to twenty-five thousand), for example, indicates that 1 unit of distance on the map (such as a centimeter or an inch) represents a horizontal distance of 25,000 of the same unit on the ground. Because the scale is a ratio, it applies to any desired unit of measurement. The scales of most maps used in structural and tectonic work range between 1:1000 and 1:100,000, though other scales are also used. In particular, maps of large regions such as states, provinces, countries, and continents are published at scales between 1:500,000 and 1:20,000,000.

As the scale of a map changes, the size of the features and the amount of detail that can be represented on the map also change. If a map is of a very small region (an area a few meters in dimension for example), then correspondingly small features and great detail can be portrayed. If the map represents a large region (an area hundreds of kilometers in dimension for example), then only large features and little detail can be shown.

Unfortunately, the word *scale* is used in two different and opposite ways, which can lead to confusion. With regard to geologic features, the term refers to the dimensions of the feature. Thus small-scale features have a characteristic dimension roughly in the range of centimeters to perhaps hundreds of meters. Large-scale features have a characteristic dimension of roughly hundreds of meters to thousands of kilometers. With regard to maps, however, *scale* refers to the distance on a map divided by the equivalent distance on the ground. Thus small-scale maps (such as 1:100,000) cover larger areas than large-scale maps (such as 1:1000). Confusion arises because large-scale features are portrayed on small-scale maps, and vice versa.

Geologic mapping is usually done on an accurate topographic base map. Most countries publish topographic maps at a scale between 1:25,000 and 1:50,000. If an area is very small, however, more detailed base maps must generally be prepared by using surveying or photographic techniques.

Because the Earth's surface is very nearly spherical, any planar map of the surface is a distortion of true shape. For small areas, even up to standard 1:24,000 or 1:25,000 quadrangles (approximately 17 km in a north–south direction and in midlatitudes 11 km to 15 km east to west), this distortion is minor and usually is ignored. For larger regions, however, distortions become significant. Many different types of projections of the spherical surface onto a plane are used. Each represents a compromise between minimizing the distortion of the shape of the region and minimizing the distortion of its area.

Small-scale regional maps are useful in structure and tectonics in order to portray large-scale structural features. Such maps are constructed by combining the large-scale maps on which the geology was originally mapped into a single map, usually at a smaller scale. Thus it is common to see features originally mapped at a scale of 1:24,000, for example, compiled on a map at 1:250,000, 1:1,000,000, or even 1:10,000,000 covering an entire continent. As the map scale decreases, the amount of detail must also decrease. Such compilations therefore require that many choices be made about what information to represent and what to omit, and the resulting map is highly interpretive. Often the information available from adjacent map sheets is not entirely consistent, and the compiler must contend with

problems such as different levels of detail on different maps, disagreements on the nature of map units, and the inconsistent location of contacts. In cases where such discrepancies cannot be resolved, discontinuities may appear on the smaller-scale regional map.

Other differences among geologic or tectonic maps of a particular area may be due to the particular purpose for which each map is made. A map of soil and surficial deposits would look very different from a map of the bedrock in the same area. Maps may also be compiled to emphasize various aspects of geology. A geologic map emphasizes the distribution of lithologies and their ages, whereas a tectonic map of the same area combines units of similar tectonic significance.

2.3 Cross Sections: Portrayal of Structures in Three Dimensions

A geologic map provides the basis for detailed understanding of the structural geometry of an area. The map, however, is only a two-dimensional representation of three-dimensional structures. Cross sections, on the other hand, show the variation of structure with depth, usually as it would appear on a vertical plane that cuts across the area of a geologic map. Fundamentally, cross sections are interpretations of structure at depth extrapolated from data available at the surface, but they may also be constrained by data on the regional stratigraphy or lithology, by direct observation from drill holes or mines, and by geophysical data (see Section 2.6). Without such independent constraints, cross sections are *highly interpretive,* because they are based on the assumption that the structure at depth is a simple projection of the structure observed at the surface and the attitudes and geometry do not change along the line of projection. The validity of that assumption varies a great deal, depending on the characteristics of the local geology.

Cross sections ideally are oriented normal to the dominant strike of planar structures in an area. In this orientation, the dip of those structures is accurately represented on the section. In some cases, however, if the structure is complicated with a variety of attitudes, no one orientation of cross section adequately represents all the structures, and the apparent dip of any plane whose strike is not perpendicular to the cross-section line is less than the true dip of that structure.

In order to show an undistorted view of structures at depth, we must take the vertical scale ratio equal to the horizontal scale ratio. In some cases, however, the features of interest are best shown by using **vertical exaggeration,** for which the vertical scale ratio is larger than the horizontal scale ratio. The relatively small changes in topography and in stratigraphic thickness that commonly occur over large distances can be shown clearly only with vertical exaggeration. As a result, vertically exaggerated cross sections are standard in marine geology and stratigraphy.

The habitual use of vertical exaggeration to portray certain features, however, gives a false impression of the nature of those features. Figure 2.3, for example, shows cross sections of a subduction zone and an adjacent volcanic island arc. The true scale section is shown in Figure 2.3A. Note that the topographic variation is almost impossible to see! A properly vertically exaggerated section is shown in Figure 2.3B. The vertical exaggeration necessary to emphasize the topography also exaggerates the surface slopes so that their appearance is not at all representative of actual slopes. The effect of vertical exaggeration is just as dramatic on the dip of planar features (compare the angle of the subducting lithospheric slab in Figure 2.3A with that in Figure 2.3B). Published informal cross sections often give an even more confusing view by combining vertically exaggerated topography with no vertical exaggeration below the surface, as shown in Figure 2.3C.

Vertical exaggeration makes dipping structures appear much steeper than they are. The effect is much stronger on shallowly dipping planes than on steeply dipping ones, and at high values of the vertical exaggeration, the distinction between shallow and steep true dips effectively disappears.

Another effect of vertical exaggeration is to cause beds of the same thickness but different dip to appear to differ in thickness. In Figure 2.3B the subducted lithospheric slab appears to be thinner where it is dipping than where it is horizontal, an effect caused simply by exaggeration of the vertical dimension.

Thus vertical exaggeration causes distortions of the geometry of features that seriously alter the way they look. Because much of structural geology involves visualizing the true shape of features in three dimensions, the use of vertical exaggeration with structural cross sections should be avoided whenever possible.

In areas of complex structure where a single cross section cannot be constructed to reveal the true angular

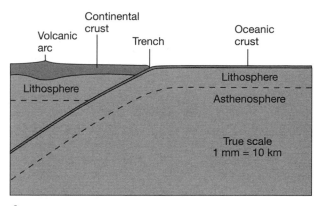

A.

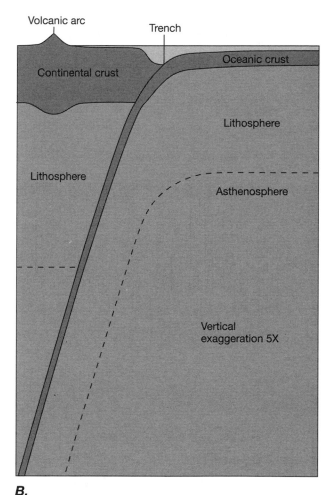

B.

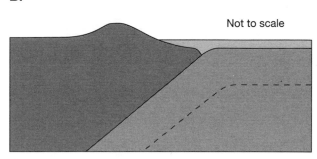

C.

Figure 2.3 (*Right*) The effect of vertical exaggeration on cross sections. A. True-scale cross section of a subduction zone in which the horizontal and vertical scales are equal (1 mm = 10 km). Note that topographic variations are almost imperceptible. B. A cross section of the same subduction zone as in part A, here shown at a vertical exaggeration of 5 ×. That is, the vertical scale is 5 times the horizontal scale. Note the exaggeration of the dip of the subducted slab. C. Schematic cross section that combines vertical exaggeration for the topography with no vertical exaggeration for structures below the surface. This mixing of vertical scales precludes the construction of an accurate cross section.

relationships of all the structures, multiple cross sections at different orientations may be used. Two constructions are regularly employed—block diagrams and fence diagrams. **Block diagrams** show a perspective diagram of a block of crust with appropriate geometric representation of the structure on the top, which is the map view, and a view of the sides, which includes two cross sections generally at high angles to each other. We have already used a number of block diagrams to illustrate structural features (see Figures 2.1*A* and 2.2*A*). Although these views of the structure on three different surfaces enable us to visualize the three-dimensional geometry, sections themselves are not accurate representations of the angular relationships because of the necessary distortion of perspective. **Fence diagrams** represent three-dimensional structures by showing a perspective view of intersecting cross sections. They are difficult to draw and require a great deal of information; thus they are relatively rare in the published structural geologic literature.

2.4 Stratigraphic Sequence Indicators

Crucial to the interpretation of a stratigraphic sequence of deformed rocks is determination of the relative ages of the different layers—that is, the "stratigraphic up" or "younging" direction in the sequence. Consider, for example, the schematic cross sections in Figure 2.4, which illustrate a series of layered rocks folded in two different geometries. Without knowledge of the stratigraphic up directions, the simplest interpretation we could make of the limited exposure in Figure 2.4*B* would be that shown in Figure 2.4*A*, which is incorrect.

Features that are useful in the determination of relative age include some primary structures in sedimentary or igneous rocks, unconformities, and an independent knowledge of the stratigraphic sequence.

Primary Structures in Sedimentary Rocks

Many structures formed during or shortly after sedimentation are useful in determining relative stratigraphic age. The most common of these structures are bottom markings, graded bedding, cross bedding, and scour-and-fill or channel structures.

Bottom markings are features formed mainly in association with turbidity currents and preserved on the underside of many sandstone beds in interlayered sandstone–shale sequences. They represent casts of the small-scale surface topography imposed on mud by an overlying sand layer, or upon which the sand layer was deposited. Bottom markings include flute casts and load casts. **Flute casts** form from the deposition of sand in spoon-shaped depressions scoured out of the underlying mud by high-velocity currents. Subsequent lithification of the sand preserves a cast of the depression on the bottom of the sandstone layer (Figure 2.5*A*, *B*). Flute casts tend to be markedly asymmetric in longitudinal cross section; they indicate the direction in which the

A.

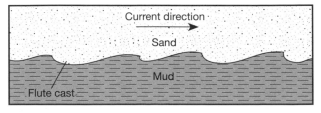

B.

Figure 2.5 Flute casts. *A.* Flute casts on the bottom of a sandstone bed in a turbidite. *B.* Cross section of flute casts, shown right side up.

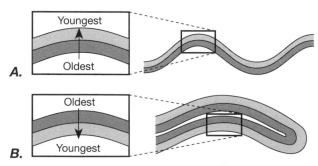

Figure 2.4 Cross sections showing "stratigraphic up" directions. *A.* A fold with beds right side up, as indicated by the arrow. The structure is simple, as shown schematically to the right. *B.* A fold with upside-down beds, as indicated by the arrow. The structure must be more complex, as shown to right.

current was moving at the time the scours were cut (Figure 2.5B).

Load casts are sandstone casts of depressions in underlying mud that form as the denser sand sinks into the soft, water-saturated mud. They appear as bulges on the bottoms of sandstone layers and, in some cases, can actually become isolated balls of sand in a mud matrix. If the mud forms sharp peaks pointing upward into the sand, and the sand forms rounded masses protruding down into the mud, the structures are described as **flame structures** which result from the down-slope shearing of the sediments. The directionality of these features serves as an indicator of the relative age of the sediment layers.

Graded bedding in sediments or sedimentary rocks is characterized by the gradual change in grain size and mineralogy within a single bed, usually from coarse and clay-poor at the bottom to fine and clay-rich at the top (see Figure 2.6). Grading occurs in a suspension of unsorted sediment because the coarsest fraction settles out faster than the finest fraction. Thus the oldest part of the layer is the coarsest, and the youngest is the finest. In some cases, however, metamorphism can actually reverse the direction of grading. Because the finest fraction has a higher surface area and is therefore more chemically reactive, that part of the layer may grow coarser crystals during the metamorphism than the originally coarser part of the layer.

Cross bedding consists of thin beds that occur at angles of as much as 20° to 30° to the principal bedding planes. These beds result from deposition of material

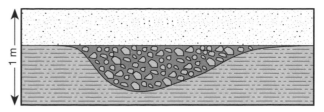

Figure 2.7 Sketch illustrating a cross section of channel or scour-and-fill structure. The depression was eroded in the lower unit (siltstone) and first conglomerate, and then sand deposited on top.

in ripples or dunes. Characteristically, the cross beds are concave upward, and they are tangent to the lower major bedding surface. Where subsequent erosion has removed the top of the bed, the crossbeds are truncated against the upper major bedding surface. In this case, they are useful as indicators of the stratigraphic up direction.

Channel structures, or **scour-and-fill structures,** are deposits that fill in stream or current channels cut by erosion into underlying sediment. Most channel deposits are conglomerates, but finer-grained sediments are also found. These structures can be identified by the characteristic truncation of layers of the underlying sediment against the side of the channel (Figure 2.7). Identifying which layer has been eroded, and which subsequently deposited, makes it possible to determine the stratigraphic up direction.

Primary Structures in Igneous Rocks

Structures that indicate the stratigraphic up direction in igneous rocks are more abundant in extrusive than in intrusive rocks, but they are not so common as in sedimentary rocks. Telltale features in extrusive rocks include flow-top breccia, pillow lava structures, and filled cavities.

Flow-top breccia develops at the upper surface of basalt flows, thereby making it possible to identify the original top of the layer (Figure 2.8). It forms by the breakup of a solidified layer during renewed movement of the underlying lava. In some cases, the breccia formed on top can roll under the flow as it advances. Thus caution is advisable in the use of this indicator.

Many volcanic rocks contain vesicles—cavities formed by solidification around gas bubbles (Figure 2.8). Vesicles may be filled by minerals precipated from solution, in which case they are called amygdules (the Greek word *amygdalon* means "almond"). In rare instances, sediment or late magma may be deposited in the vesicles. Because the cavities fill from the bottom up, partially filled amygdules provide an indicator of

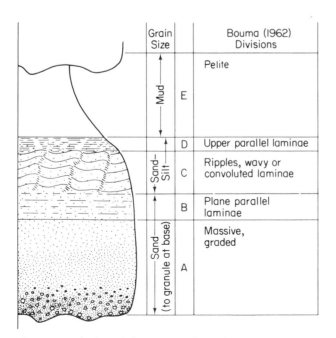

	Grain Size	Bouma (1962) Divisions
	Mud — E	Pelite
	Sand–Silt — D	Upper parallel laminae
	Sand–Silt — C	Ripples, wavy or convoluted laminae
	Sand–Silt — B	Plane parallel laminae
	Sand (to granule at base) — A	Massive, graded

Figure 2.6 Ideal graded structure of a turbidite bed, showing grading, ripples, and upper mud-rich layer (pelite).

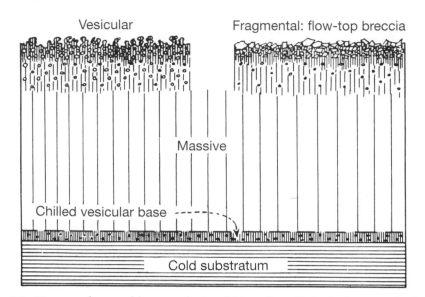

Vesicular Fragmental: flow-top breccia

Massive

Chilled vesicular base

Cold substratum

Figure 2.8 Diagram of top and bottom of subaerial lava flow. Vesicular lava may be found at the top or at the bottom of a flow. Flow-top breccia forms on top by solidification and subsequent breakup of the top during flow.

the stratigraphic up direction, as do composite amygdules if the filling sequence is known.

Pillow lavas form during extrusion of lavas under water. Lava often issues from a central vent at the top of a mound and flows downward onto the horizontal floor. Because the hot lava is quenched by the water, it flows by forming tubelike fingers that have a chilled exterior which insulates the lava within. The bottom of the top is convex upward. In cross section many tubes have a distinctive pillow-shaped upper surface and a downward pointing cusp on the lower surface (Figure 2.9), thereby indicating the original up direction.

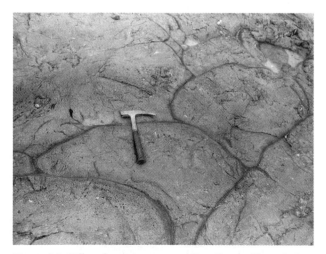

Figure 2.9 Pillow lavas in cross section. Smartville complex, Sierra Nevada, California.

Structures that reveal the original up direction are much less common in plutonic rocks than in volcanic rocks or sediments. In rare cases, however, plutonic rocks form by crystal settling at the bottom of a magma chamber, and sedimentary structures may develop that can be used to infer the stratigraphic up direction. Size grading of grains of a single mineral in a layer can be used as a stratigraphic up indicator (Figure 2.10A). The existence of gradations between two layers of different mineralogy, or **phase layering**, is more common than size grading (Figure 2.10B). It probably does not result from crystal settling, however, so cannot be used as an indicator of the stratigraphic up direction. Scour-and-fill structures are present in some layered igneous rocks. As in sedimentary rocks, they indicate the presence of currents during deposition of the rocks and are a reliable indicator of the stratigraphic up direction.

In some areas, magmas with a significant proportion of suspended crystals have intruded or extruded as sills or flows. As the molten material came to rest, the crystals settled, forming crystal accumulations toward the bottom of the sill or flow.

Unconformities

Unconformities provide a valuable means of determining relative age in a stratigraphic section. Unconformities may be disconformities, which are time gaps within a sequence of parallel layers (Figure 2.11A); angular unconformities, which are erosional surfaces that cut across older beds at an angle and are overlain by parallel beds (Figure 2.11B); or nonconformities, which are con-

A.

B.

Figure 2.10 Size and phase grading in plutonic igneous rocks. *A.* Size grading in olivine−clino-pyroxene cumulate, Duke Island ultramafic complex. Pocket knife for scale. *B.* Photograph showing repeated graded phase layering in gabbro, Vourinos ophiolite complex, northern Greece. This gradational alteration of pyroxene-rich (dark) and plagioclase-rich (light) layers may not be the result of gravity settling.

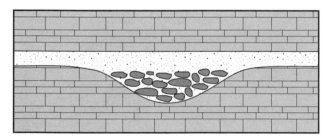

A. Disconformity

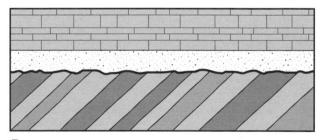

B. Angular unconformity

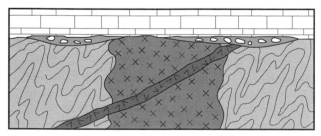

C. Nonconformity

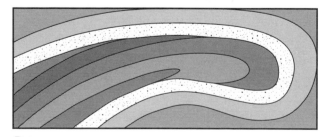

D. Folded angular unconformity

Figure 2.11 Types of unconformities. Parts *A–C* are undeformed.

tacts between sedimentary rocks and underlying igneous or metamorphic rocks (Figure 2.11C). Figure 2.11D shows how an angular unconformity might appear when folded into a large fold. Note the lack of parallelism of the beds below and above the unconformity.

Some unconformities display fossil soil horizons. If one is present, it provides a means of establishing the relative stratigraphic age. Such a horizon also makes it easy to distinguish between a sedimentary unconformity and a tectonic contact such as a fault; these can be difficult to tell apart, especially if the rocks have been deformed. Even in some metamorphosed rocks, fossil soils are sometimes preserved as a thin band of aluminum-rich metamorphic rocks sandwiched between a metasedimentary sequence and an older metamorphic or igneous terrane.

Other Indicators

Many orogenic zones include belts of rocks characterized by deformed, but little metamorphosed, fossiliferous sediments. In these rocks, it is possible to determine the stratigraphic sequence by means of biostratigraphic analysis of the sediments themselves. In highly deformed and/or metamorphosed areas, it may be necessary to infer stratigraphic relationships from the stratigraphy in less deformed or metamorphosed areas.

In some cases the rocks are unfossiliferous, the stratigraphy is unknown, and the rocks do not possess any structures that indicate the relative age. In such situations, radiometric age determinations may yield the only age information available. Sedimentary processes do not reset radiometric clocks, so sediments cannot be dated directly this way. Metamorphic or igneous events can be dated, and the age of formation of structural or tectonic features can be established if radiometrically dated igneous or metamorphic events bracket the tectonic event in time.

2.5 Graphical Presentation of Orientation Data

Often it is desirable to present orientation data in such a way that the distribution of orientations is emphasized independently of the geographic location of the data. For example, it may be useful to know whether there is a pattern of preferred orientation of beds, joints, or linear features in an area, regardless of how the orientations vary across a map. The types of diagrams most frequently used to present such information are histograms, rose diagrams, and spherical projections.

Orientation **histograms** (in Greek, *histos* means "mast" or "web," and *gram* means "line") are plots of one part of the orientation data, such as strike azimuth, against the frequency of orientations that are found within particular orientation intervals. The frequency may be plotted as a percent of all observations or as the number of observations within each interval. The plots characteristically consist of a series of rectangles, where the width of the rectangle represents the orientation interval and its height represents the frequency.

Rose diagrams are essentially histograms for which the orientation axis is transformed into a circle to give a true angular plot. The intervals of angle are plotted as pie-shaped segments of a circle in their true orientation, and the length of the radius is proportional to the frequency of that orientation. The use of the true angle conveys an intuitive sense of the orientation distribution. Rose diagrams are used for displaying such features as the direction of sediment transport and the strike of vertical joints (see, for example, Figure 2.12).

Both histograms and rose diagrams can present only one aspect of the attitude of planar or linear fea-

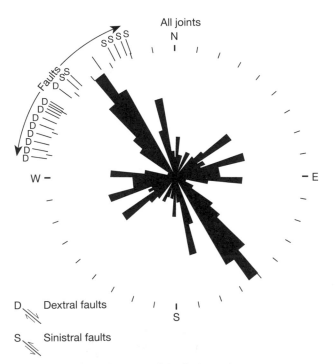

Figure 2.12 Rose diagram of the faults and joints in an area. Azimuth intervals are 5°. The length of the radius from center in any given segment is proportional to the percentage of features with an orientation in that sector. The length of the longest radius represents 40 percent of the measurements. The diagram has a twofold axis of symmetry; that is, a 180° rotation of the diagram is identical to the original diagram.

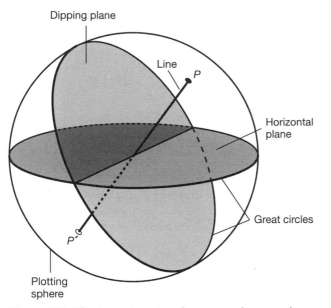

Figure 2.13 Plotting orientation data on a plotting sphere. Planar and linear features are considered to pass through the center of the sphere and to intersect its surface. Thus the attitude of a plane is defined by the great circle that is the intersection of the plane with the plotting sphere. A horizontal plane and a dipping plane are shown. The attitude of a line is defined by two points (P and P') that are the intersections of the line with the plotting sphere.

tures, such as the strike or bearing, respectively. In cases where the dip of the planar features is always essentially vertical or the plunge of linear features is always essentially horizontal, or some other constant angle, this limitation is of no consequence. If the dip or plunge of the feature is an important variable in defining its attitude, however, the best method of plotting orientation data is the spherical projection.

When orientation data are plotted on a **spherical projection**, all planes and lines are considered to pass through the center of the plotting sphere, and their attitudes are then defined by their intersections with the surface of the sphere. For planes, the intersection is a great circle; for lines, the intersection is a point (Figure 2.13). Using the full sphere is actually redundant; a hemisphere is all that is needed. Applications for structural geology and tectonics generally use the hemisphere below the horizontal plane (the lower hemisphere), whereas applications in mineralogy usually employ the upper hemisphere. In both cases, the hemisphere is projected onto a flat plane to permit the convenient graphical presentation of data.

There are a variety of methods of projecting a sphere onto a plane, although two projections are most often encountered in structural geology and tectonics. The first may be referred to as a **stereographic projection**

or as an **equal-angle projection;** the other is a **Lambert projection** or an **equal-area projection.** They differ only in the way the hemisphere is projected onto a plane called the **image plane.**

For both types of projections, the image plane for the projection is tangent to the hemisphere at T (Figure 2.14) and parallel to the plane containing the edge of the hemisphere. The equal-angle projection (Figure 2.14A) is constructed by using the highest point on the sphere opposite the image plane, the zenith point Z, as the projection point. The projection of a point on the hemisphere is defined by constructing a line from the zenith point Z through the point on the hemisphere (P or Q) to the image plane (P' or Q'). Thus any orientation of line has a unique projection on the image plane. A great circle on the hemisphere is projected by drawing

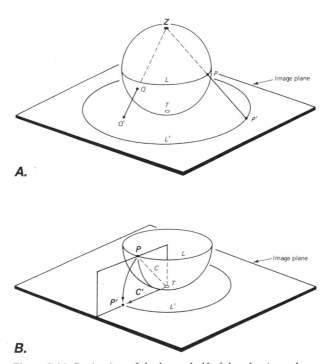

Figure 2.14 Projection of the lower half of the plotting sphere onto the image plane. A. The principle of stereographic, or equal-angle, projection. The zenith point Z is the projection point. Points P and Q on the plotting hemisphere are projected to points P' and Q' on the image plane by a line passing from Z through P and Q, respectively, to the image plane. A great circle is described on the image plane by projecting each point of the great circle on the plotting hemisphere to the image plane. B. The principle of Lambert, or equal-area, projection. The point P is projected to the image plane by constructing a cord of the sphere from T (the tangent point of the image plane) to P and rotating that cord in a vertical plane about T down to C' in the image plane. The end of the rotated cord P' is the projected point. The great circle is projected by rotating each point of the great circle on the plotting hemisphere down to the image plane in a similar manner.

lines from the zenith point through all points along that great circle. The locus of intersections of those lines with the image plane defines the projection, which again is unique for any given orientation of plane.

The advantage of this type of projection is that angles between lines on the hemisphere are not distorted by the projection. Moreover, circles on the hemisphere remain circles on the projection, although the center of the circle on the hemisphere does not project to the center of the circle on the image plane. All great circles are also arcs of circles on the image plane. Areas that are equal on the hemisphere, however, are in general not equal on the image plane.

An equal-area projection is constructed by using the point of tangency T between the hemisphere and the image plane as a center of rotation (Figure 2.14B). The projection of a point on the hemisphere is determined by constructing a chord C from T to the point P on the hemisphere and rotating this chord in a vertical plane about T down to C' in the image plane. The end of the chord P' in the image plane defines the projection of the point. Projections of great circles are constructed, in principle, by rotating all points on the great circle from the hemisphere down to the image plane in a similar manner. This projection has the advantage that any two different but equal areas on the hemisphere are also equal on the image plane. It is therefore used to present data when the statistical concentration of points is important to the interpretation, because those concentrations are not distorted by the projection. The shapes of areas on the hemisphere are not preserved by the projection, however, so angular relationships are distorted, although they can still be determined if the angles are measured along a great circle.

2.6 Geophysical Techniques

Although mapping rocks that are exposed at the surface provides good information about the three-dimensional structure near the surface, it cannot reveal the structure of areas covered by alluvium, deep soils, vegetation, or water such as lakes, seas, and oceans. Nor can surface mapping provide information about structure at great depth. Information about the shapes of major faults at depth, the presence of magma chambers at depth, the location of the crust–mantle boundary, or the thickness and nature of the lithosphere and the lower mantle can come only from the interpretation of geophysical measurements, and especially from seismic, gravity, and magnetic measurements. We review briefly the application of these aspects of geophysics to large-scale structure and tectonics because they have become essential, and because a structural geologist must at least be aware

of the techniques and their limitations. Adequate coverage of these topics, however, would require at least a separate book, and we encourage students to take appropriate courses in geophysics.

Seismic Studies

Seismic waves are oscillations of elastic deformation that propagate away from a source. Waves from large sources such as major earthquakes and nuclear explosions can be detected all around the world. Small explosions are often used as sources to investigate structure at a more local scale. Body waves, which can travel anywhere through a solid body, are of two kinds: compressional (P) waves, for which the particle motion is parallel to the direction of propagation, and shear (S) waves, for which the particle motion is normal to the direction of propagation. (The designations "P" and "S" refer to primary and secondary waves, so named because of the normal sequence of arrival of the waves as revealed on a seismogram.) P waves travel faster than S waves. They are therefore the first waves to arrive at a detector from a source and thus are the easier to recognize and measure. For this reason, and because explosions generate mostly P waves, they are the waves predominantly used to investigate the structure of the Earth. The propagation of seismic waves may be described by the seismic rays, which are lines everywhere perpendicular to the seismic wave fronts. Three types of seismic studies are particularly important in structure and tectonics: seismic refraction, seismic reflection, and first-motion studies.

Seismic refraction studies investigate the structure of the Earth by means of those seismic rays that are transmitted through boundaries at which seismic velocity changes. The change in seismic velocity refracts, or bends, the rays (see Box 2.1) so that in the Earth, where seismic velocity generally increases with depth, the ray paths tend to be concave upward. The travel time of rays from the source to different receivers is plotted against the distances of the receivers from the source. Such plots make it possible to determine the ray velocity in the deepest layer through which the ray travels. By measuring travel times for rays that penetrate to greater and greater depths, the investigator can determine the velocity structure of the Earth.

The velocities of P and S waves depend on the density and the elastic constants of the rock. Thus knowing how seismic P and S velocities vary with depth provides information about the distribution of the density and elastic properties in the Earth, and locating where changes in seismic velocity occur reveals where the rock type changes.

Although seismic refraction studies give a good "reconnaissance" view of the structure of a large area,

Box 2.1 Seismic Refraction

The time required for seismic rays to travel directly from a source to different detection stations distributed around the source is affected by the particular paths the seismic rays take, and these in turn are determined by the structure and the seismic velocity of the material along each path. If a seismic ray travels obliquely across a boundary from a low- to high-seismic-velocity material, it is refracted away from the normal to the boundary (Figure 2.1.1). If the ray travels from high- to low-velocity material, it is refracted toward the normal to the boundary.

Travel-time measurements can be interpreted to reveal the variation of seismic wave velocity with depth. The principle is illustrated in Figure 2.1.2, which shows the location of a seismic source and an array of detectors. Some of the ray paths shown stay within the crust; others travel in part through the upper mantle. A time–distance plot indicates the arrival times of those different rays at the detectors. Because the seismic velocity in the mantle is higher than that in the crust, mantle rays reach distant detectors before crustal rays. For the layered structure shown, the difference in arrival times at the different detectors reflects the speed of the rays through the deepest layer along the ray path. Thus the slopes of the two lines on the time–distance plot are the inverse of the velocities in the crust and mantle, respectively.

If a layer that has a lower seismic velocity occurs at depth between rocks that have higher seismic velocities, rays are bent toward the normal to the boundary upon entering the layer and away from the normal

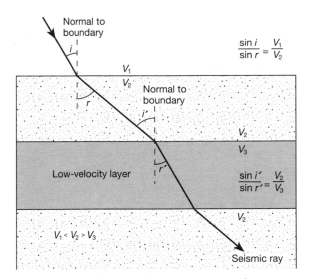

Figure 2.1.1 Refraction of seismic rays. Refraction is away from the normal to the boundary if the ray travels from a low to a high-seismic-velocity material (here V_1 to V_2 or V_3 to V_2). Refraction is toward the normal to the boundary if the ray travels from a high- to a low-seismic-velocity material (V_2 to V_3).

upon leaving (Figure 2.1.1). Seismic rays, therefore, can never reach their maximum depth in that layer, and the seismic velocity of that layer—and therefore its very existence—cannot be detected on a time–distance plot.

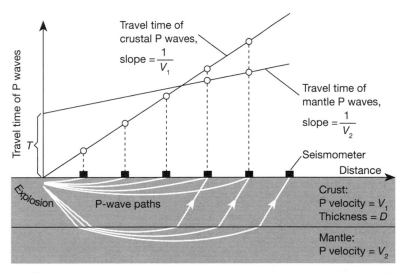

Figure 2.1.2 Illustration of the principle of seismic refraction in a two-layer structure. The diagram shows ray paths for P waves through the structure. The travel-time plot indicates the arrival times of the rays at the different detectors.

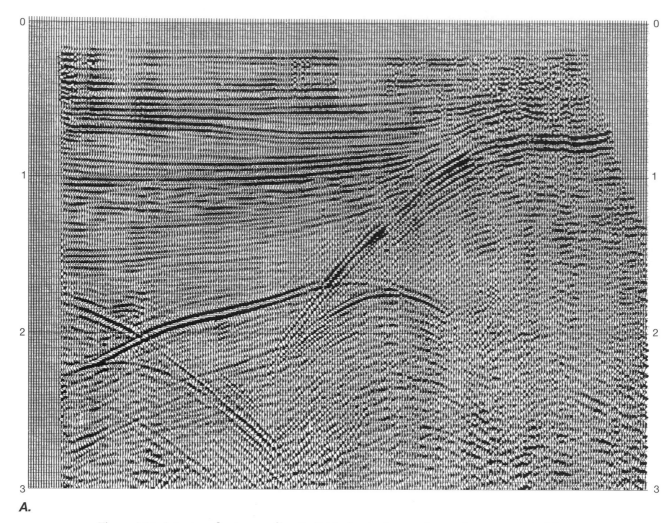

A.

Figure 2.15 Seismic reflection profiles. Individual seismic records are the wavy vertical lines plotted side by side along the distance axis. The vertical axis is the two-way travel time. Peaks in each record are shaded black to show up reflectors that can be traced from one record to the next. Good horizontal reflectors are particularly evident below about 1.7 s in the left half of the profile. *A.* Unmigrated seismic profile. *B. (Facing page)* Migrated seismic profile.

the technique has several disadvantages. The presence of low-velocity layers cannot be detected (Box 2.1). Deep structures ordinarily can be detected only at distances from the source that are greater than the depth. The properties of the Earth are averaged over large distances, so details of structure are lost. And nonhorizontal or discontinuous layers and complex structure are difficult or impossible to resolve.

In **seismic reflection** studies, reflections of P waves off internal boundaries are used to investigate the structure of the Earth. Seismic signals are recorded by as many as several hundred to several thousand geophones at a time. The resulting data are analyzed by computer (Boxes 2.2 and 2.3), and the seismograms are plotted side by side on the distance axis (Figure 2.15A). The

individual peaks on the seismogram, or events, record the two-way travel time for each reflection, which is the time required for a wave to travel from a surface point to a reflector and back to the same surface point. Peaks in each record are shaded black so that strong signals in adjacent seismograms at the same two-way travel time show up as lines that indicate a continuous reflector.

Sophisticated computerized digital processing of the seismic records, including the very important processes of stacking and migration, allow complex structures to be resolved (Figure 2.15B) and therefore yield an incomparable image of the subsurface structure. The **stacking** of seismic records is a method of enhancing the signal-to-noise ratio by adding together reflections

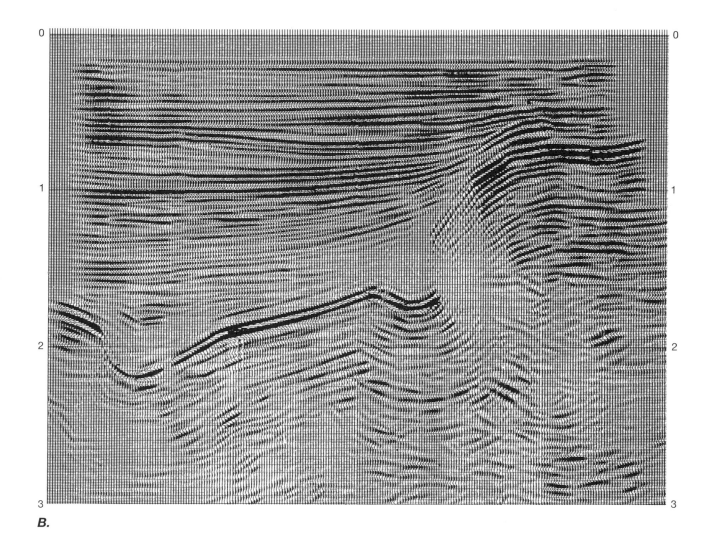

B.

that occur at different angles from the same subsurface point (Box 2.2). **Migration** is a technique that allows the true locations of reflectors to be determined. All reflection signals on a seismogram are plotted as though they were vertical reflections (Figure 2.15A). Any given reflection, however, could have come from any point on an arc around the receiver, and the process of migration corrects the seismograms to determine the actual location of the reflector (Figure 2.15B; Box 2.3).

We can illustrate the benefits of migration by comparing an unmigrated seismic record (Figure 2.16A), a migrated profile (Figure 2.16B), and the true geologic cross section (Figure 2.16C). Migration eliminates the artifacts and errors in the unmigrated record. The actual geologic section, however, can be determined only if

the two-way travel time can be converted into depth, which requires knowledge of the way velocity varies with depth. Note that if the seismic velocity changes with depth, the seismic (time) section distorts the actual vertical scale.

Despite their obvious value, the utility of reflection seismic studies is limited by their expense. Producing one seismic reflection profile can involve making hundreds of shots, each of which is recorded by hundreds to thousands of geophones.

The first arrival of a P wave can be either a compression of the material (a decrease in volume) or its opposite, a rarefaction (an increase in volume). Whether a compression or a rarefaction arrives first is revealed by the direction of first motion on the seismogram. The

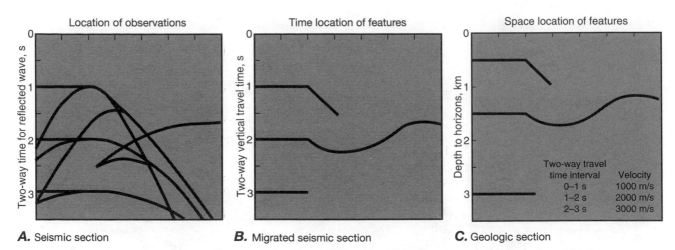

A. Seismic section **B.** Migrated seismic section **C.** Geologic section

Figure 2.16 Diagrams illustrating the effects of migration. *A.* An unmigrated seismic section with multiple intersecting curved reflections. *B.* The same section as in part *A* after migration. The ambiguities and artifacts of the unmigrated section are all removed. *C.* The corresponding geologic section. The depth scale is different from the two-way travel-time scale because seismic velocity varies with depth.

pattern of compression or rarefaction first motions that radiate out from a sudden slip event on a fault is characteristic of the orientation of the fault and the sense of slip (Box 2.4). **First-motion studies,** therefore, are used to determine the orientation of, and sense of slip on, faults at depth. Regional patterns of first motions reveal large-scale tectonic motions of the plates. Figure 2.17 shows the radiation patterns that are characteristic of the three basic types of faulting: normal, thrust, and strike-slip.

Analysis of Gravity Anomalies

Gravity measurements are perhaps the second most important geophysical technique (after seismic techniques) used in structural geology and tectonics. A gravity anomaly (from the Greek word *anomalia,* which means "irregularity or unevenness") is the difference between a measured value of the acceleration of gravity, to which certain corrections are applied, and the reference value for the particular location. The reference value is determined from an internationally accepted formula that gives the gravitational field for an elliptically symmetric Earth. Because gravity anomalies arise from differences in the density of rocks, the goal in structural geology is to relate these differences in density to structural features. If no density contrasts exist, then the structure can have no effect on the gravitational field, and gravity anomalies cannot aid in the interpretation of that structure.

The structure at depth is interpreted by matching the gravity anomaly profile observed along a linear trav-

erse with the anomaly profile calculated from an assumed model of the structure. The model is adjusted until the model anomaly profile shows a satisfactory fit to the observed anomaly profile. Although the model can never be unique, it is usually constrained by surface mapping and possibly by seismic data.

In order to calculate an anomaly, we must correct the measured value to the same reference used for the standard field. All measurements are therefore corrected to sea level as a common reference level. This altitude correction, the **free-air correction,** results in an increase in most land-based values but leaves surface observa-

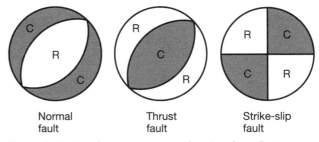

Normal fault Thrust fault Strike-slip fault

Figure 2.17 Equal-area projections showing the radiation pattern of compression first motions (C) and rarefaction first motions (R) for the three main types of faults. The fault on which the earthquake occurs is assumed to be at the center of the plotting sphere. All orientations that plot within a given sector are the orientations of rays when they leave the source that have the indicated first motion. Planes separating the sectors are nodal planes, one of which must be the fault plane. Material on each side of the fault moves toward the compression sectors, defining the sense of shear on the fault.

Box 2.2 Stacking of Seismic Records

Figure 2.2.1 illustrates the principle involved in the common depth-point stacking of seismic records. If explosions are detonated at shot points S_1, S_2, S_3, and S_4, reflections from the same point P on a horizontal subsurface boundary will be received at geophones G_1, G_2, G_3, and G_4, respectively. The same is true for all other horizontal reflectors below the point p. The corresponding shot points and geophones (S_i and G_i) are equidistant from the point p above P. If the travel times are corrected for the difference in length of the ray paths, the records can be added together, or stacked. The time-corrected signals from the reflections at P reinforce one another, and the signals from random noise tend to cancel out, thereby increasing the signal-to-noise ratio. The result is an enhanced seismogram showing the reflections as they would appear if the shot point and receiver were both at p.

In practice, data are gathered from a large linear array of shot points and geophones. Each geophone records many reflections from different depths, and the stacking is done by computer to produce enhanced seismograms at each point in the profile. Figure 2.15A is an example of a stacked seismic profile, which shows abundant horizontal reflectors at shallow depths.

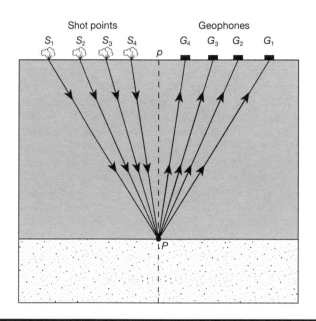

Figure 2.2.1 The principle of stacking of seismic records. Explosions are set off at shot points S_1, S_2, S_3, and S_4. The rays reflected from the same point P on a reflector at depth are received, for each of the explosions, at geophones G_1, G_2, G_3, and G_4, respectively . Adjusting the arrival times of these reflections for the different lengths of ray path allows the four records to be added together, which enhances the signal from the reflection at P and cancels out random noise. The effect is an increase in the signal-to-noise ratio.

---

tions at sea unchanged. If this is the only correction applied, the calculated anomaly is called a **free-air anomaly.**

The **Bouguer correction** is also frequently applied. It is assumed that between sea level and the altitude of the measurement is a uniform layer of continental crustal rock that represents an excess of mass piled on the surface. This assumed excess gravitational attraction is therefore subtracted from land-based measurements. At sea, it is assumed that all depths of water represent a deficiency of mass, because water is less dense than rock. The assumed deficiency in gravitational attraction is therefore added to sea-based measurements. The **Bouguer anomaly** results from application of both the free-air and Bouguer corrections.

The gravitational effects of local topography, however, differ measurably from those of a uniform layer. Thus a refinement of the simple Bouguer anomaly called the **complete Bouguer anomaly** requires a **terrain correction** to account for the local effects.

Thus, the Bouguer anomaly compares the mass of existing rocks at depth to the mass of standard continental crust whose elevation is at sea level. Bouguer anomalies are generally strongly negative over areas of high topography, indicating that there is a deficiency of mass below sea level compared to standard continental crust. They are strongly positive over ocean basins, indicating that there is an excess of mass below the ocean bottom compared to standard continental crust.

The area under a gravity anomaly profile provides a unique measure of the total excess or deficiency of mass at depth, and the shape of the profile constrains the possible distribution of the anomalous mass. The interpretation of mass distribution is not unique, however, because a given anomaly profile can be produced by a wide range of density differences and distributions. Figure 2.18A, for example, shows three symmetric bodies of the same density, each of which produces the same symmetric gravity anomaly. Figure 2.18B, C illustrates how the faulting of different density distributions affects

Box 2.3 Migration of Seismic Records

Although horizontal or very shallowly dipping reflectors are common in undeformed sedimentary basins (shallow parts of Figure 2.15), much of the structure of interest in structural and tectonic investigations is a great deal more complex (deeper parts of Figure 2.15). For example, beds with significant and variable dips and discontinuous beds (possibly truncated by a fault) are common. Such structures give rise to distortions and artifacts in seismic profiles (Figure 2.15), which must be corrected by **migration.**

We shall describe the principle of migration by using an example for which the seismic velocity of the material is constant, and the source and detector are at the same point p (Figure 2.3.1A). A particular reflection that apparently plots at P below the detector could come from a boundary that is tangent to any point on a circular arc of constant two-way travel time having radius pP around p—for example, from P'. On two adjacent reflection seismograms (Figure 2.3.1B), a reflection apparently plots at P_1 beneath p_1 and at P_2 beneath p_2. It would therefore appear that the reflector had the dip of the line P_1P_2. In fact, however, the reflector must be the common tangent to the two constant-travel-time arcs of radius p_1P_1 about p_1 and p_2P_2 about p_2. The reflections must therefore come from points P'_1 and P'_2. Thus an un-

corrected profile shows erroneous locations and dips for dipping reflectors, and the reflection points P_1 and P_2 must be migrated along their respective constant-travel-time arcs to the correct locations at P'_1 and P'_2. The higher the true dip, the greater the distortion. Vertical reflectors plot on unmigrated seismic profiles as an alignment of reflections having a 45° dip.

If the seismic source and the detector are not at the same point, the arc of constant two-way travel time becomes an ellipse, and it is further distorted if the velocity is not constant. These are complications that must be accounted for in any analysis of a real seismic record, although the principle remains the same.

Another problem occurs if a reflector is discontinuous. The end of the reflector acts as a defraction point which takes energy from any angle of incidence and radiates it in all directions as though the point were a new source (point D in Figure 2.3.2). The signals recorded by the nearby detectors—for example, at p_1, p_2, and p_3—then plot along a parabolic arc on an uncorrected seismic profile (the dotted line in Figure 2.3.2; see also Figures 2.15A, 2.16A), because the two-way travel times for the signal increase as the distance of the detector from the end of the reflector increases. The location for all possible dif-

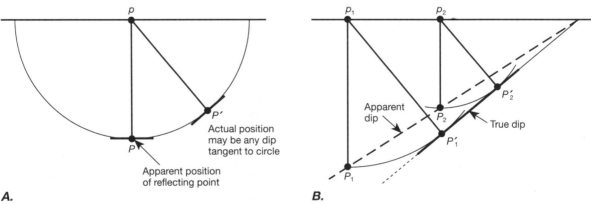

A. **B.**

Figure 2.3.1 The migration of seismic signals corrects the seismic records to give the true location and dip of reflectors. In this example, seismic velocity is considered constant, and the shot point and receiver are both located at the same point. A. A reflection received at p appears on the seismic record at a two-way travel time that plots at P. In fact, the signal could come from any reflector, such as P', that is tangent to the semicircular arc of radius pP around p. That arc is the locus of constant two-way travel time. B. Reflections detected at p_1 and p_2 plot vertically below each point at P_1 and P_2, respectively, giving the reflector the apparent dip and location of the line P_1P_2. The true location and dip of the reflector, however, must be given by the line $P_1'P_2'$, which is the common tangent to the constant two-way travel-time arcs about p_1 and p_2, respectively. Note that P_1' is the actual location of the reflector below p_2.

fraction points that could generate the signal recorded at a given receiver, however, must lie along an arc of constant two-way travel time about that receiver (the dashed arcs in Figure 2.3.2). The true location of the diffraction point is the common intersection of the arcs constructed for several detectors. Thus migrating each signal along its arc to the common point identifies the true location of the diffraction point D.

In practice, the process of migration consists of taking each individual event on a reflection seismogram, migrating it along its arc of constant two-way travel time, and adding that event to any other seismogram intersected by that arc at the point of intersection. The resulting seismic profile is then a series of seismograms, each of which consists of the original record altered by the addition of all the events that migrate to that record. With this procedure, reflecting boundaries appear as coherent traces of events across the section in their correct location, and diffracted signals sum together at the location of the diffraction point. The other additions to the different seismograms tend to cancel each other out and do not produce coherent patterns on the seismic profile.

Determining the constant two-way travel-time arc, of course, requires determining the velocity structure. The amount of computation required to migrate every event in a profile to every seismogram intersected by its constant-travel-time arc is prodigious; in practice, it can be handled only by a computer.

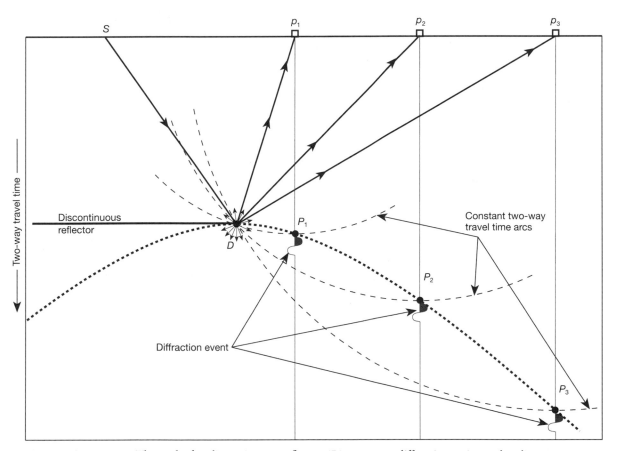

Figure 2.3.2 The end of a discontinuous reflector (D) acts as a diffraction point and radiates seismic energy in all directions for any angle of incidence. The further the receiver is from D, the later the diffracted ray arrives. Thus the diffracted energy arrives at receivers p_1, p_2, and p_3, for example, at times that fall along a parabolic arc at P_1, P_2, and P_3, respectively. The three constant-travel-time arcs constructed about the three receivers with radii p_1P_1, p_2P_2, and p_3P_3, respectively, must intersect at the location of the diffraction point. Thus we must migrate each event along its constant-travel-time arc to the common point at D in order to reconstruct its true location.

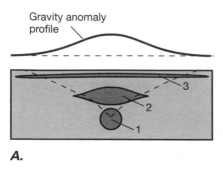

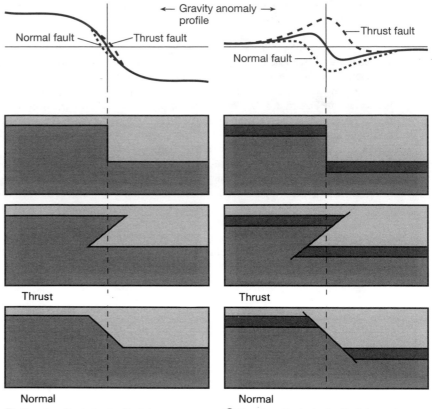

Figure 2.18 Illustration of the ambiguity and the nonuniqueness inherent in the interpretation of gravity anomalies. *A.* Three symmetrical bodies of the same density, each of which can produce the observed symmetric anomaly. *B.* The effect on gravity anomalies of vertical, thrust, and normal displacement of a dense basement overlain by less dense strata. The asymmetry of the anomaly reflects that of the underlying structure, but the distinction among the three different structures is almost negligible. *C.* The effect on gravity anomalies of vertical, thrust, and normal displacement of a dense layer within less dense layers. The three structures produce markedly different gravity anomalies.

the gravity anomaly profile. If a low-density layer overlies a thick higher-density layer and the structure is faulted (Figure 2.18*B*), the gravity anomaly profile is asymmetric, but the different geometries of faulting have only a minor effect on the anomaly shape. If the denser material is in a relatively thin layer (Figure 2.18*C*), the gravity anomaly profile is again asymmetric, though the shape is different from that in Figure 2.18*B*, and the effect of different fault geometry is significant. Thus although the anomaly shape imposes constraints on the possible structure, to be reliable, gravity models should be based on additional structural and geophysical information.

Geomagnetic Studies

A magnetic field is a vector quantity that has both magnitude and direction. For the Earth's magnetic field, the magnitude can be specified by the magnitudes of the horizontal and vertical components of the field. The orientation is specified by the declination and inclination, which are essentially the trend and plunge of the field line, though the inclination also includes the polarity, which defines whether the magnetic vector points up or down. Studies of the Earth's magnetic field include the study of magnetic anomalies and of paleomagnetism.

Magnetic anomalies are measurements of the variation of the Earth's magnetic field relative to some locally defined reference. There is no international standard reference field from which anomalies are measured, because the Earth's magnetic field is not constant and changes significantly even on a human time scale.

Regional maps of magnetic anomalies are made by using both aerial and surface measurements. The principal use of continental magnetic anomaly maps is to infer the presence of rock types and structures that are covered by other rocks, sediments, or water. In some cases, the presence of particular rock types at depth can be inferred on the basis of characteristic patterns on a magnetic anomaly map. For example, the extension of rocks of the Canadian shield beneath thrust faults of the Canadian Rocky Mountains can be inferred from the extension of the shield magnetic pattern beneath the thrust front.

Marine magnetic surveys have resulted in the well-known maps of the symmetric patterns of magnetic anomalies which have been so fundamental to the development of plate tectonic theory. When correlated with the magnetic reversal time scale, these maps can be interpreted to give a map of the age of ocean basins.

Magnetic anomalies also can be used in a manner similar to gravity anomalies to infer structure at depth, except that the magnetic anomalies are due to differences in magnetic properties of the rocks rather than in their densities. Modeling of magnetic information is more complex, because a given anomaly in total field intensity can result either from differences in intensity of magnetization of the rocks or from different orientations of the magnetic vector. Models of structure based on magnetic anomalies suffer from the same lack of uniqueness as models based on gravity anomalies, and for similar reasons.

By a variety of processes that include cystallization, cooling, sedimentation, and chemical reaction in the Earth's magnetic field, rocks can become magnetized in a direction parallel to the ambient field and can preserve that magnetism even if the rocks are rotated to new orientations. Studies of **paleomagnetism** involve measuring the orientation of the magnetic field preserved in rocks and comparing it to the orientation of the present-day field. If the original horizontal plane in the sample is known, these measurements can be interpreted to indicate the declination and inclination of the Earth's field at the time of magnetization. Gently dipping, unaltered sediments and volcanic rocks provide the most reliable paleomagnetic measurements, but more deformed or metamorphosed rocks and plutonic rocks are sometimes useful. Rocks that have been tilted since magnetization are generally assumed to have tilted about a horizontal axis, so they are restored to the original horizontal by rotation about an axis parallel to the strike of the bedding.

The earth's magnetic field is approximately symmetric about the axis of rotation, and the inclination of the field lines varies systematically with latitude from vertically down at the north pole through horizontal at the equator to vertically up at the south pole. Because this relationship is assumed to have been constant throughout geologic time, the paleo-declination determined for the sample in its original horizontal attitude indicates the amount of rotation a rock has undergone about a vertical axis, and the paleo-inclination relative to the original horizontal indicates the latitude at which the sample was magnetized. Such measurements can therefore define the changes in latitude and the rotations about a vertical axis resulting from the large-scale tectonic motions that rocks have experienced since magnetization. They can provide no information, however, on the changes in longitude associated with these motions.

Plotting the apparent paleomagnetic pole position for different time periods from a particular region provides an approximate indication of the movement of that area with respect to the Earth's geographic pole. These results are usually presented in the form of apparent polar wander maps, such as the map of paleopole positions for North America and Europe during the

Phanerozoic (Figure 2.19A). If the continents are restored to their relative positions before the opening of the Atlantic Ocean, the apparent polar wander paths coincide approximately from Silurian through Triassic, indicating the period of time the continents were joined (Figure 2.19B).

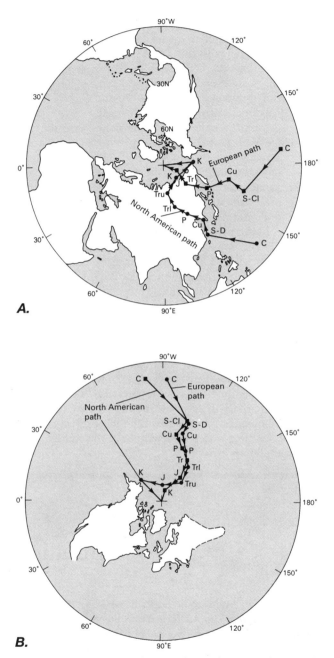

Figure 2.19 Apparent polar wander path (APW) for North America (circles) and for Europe (squares). C = Cambrian, S = Silurian, D = Devonian, Cl and Cu = lower and upper Carboniferous, P = Permian, Tr, Trl, and Tru = Triassic, lower Triassic, and upper Triassic, K = Cretaceous. A. Polar wander paths for the continents in their present positions. B. Polar wander paths for the continents before the opening of the Atlantic Ocean.

Box 2.4 First-Motion Radiation Pattern from a Faulting Event

The first-motion radiation pattern from a fault slip event can be accounted for by the two-dimensional model shown in Figure 2.4.1. The undeformed state is shown in Figure 2.4.1*A*; it is represented by two squares drawn on opposite sides of an east–west (E–W) line that represents the future location of a fault. Gradual prefaulting deformation of the rock (Figure 2.4.1*B*) deforms the squares into parallelograms, shortening the NW-oriented dimensions of the squares (such as *AD* and *CF*) and lengthening the NE-oriented dimensions (such as *BC* and *DE*). N–S and E–W dimensions remain unchanged.

An earthquake occurs when cohesion on the fault plane is lost, and sudden slip returns each square separately to its undeformed condition (Figure 2.4.1*C*). During faulting, the outer points *A, B, E,* and *F* remain stationary, while the points on the fault, *C* and *D*, separate into the respective pairs C_N and D_N, and C_S and D_S.

In this process, the NW-oriented dimensions suddenly become longer (for example, D_N moves away from *A*, and C_S moves away from *F*), creating a rarefaction for the first motion. The NE-oriented dimensions, however, suddenly become shorter (for example, C_N moves closer to *B* and D_S moves closer to *E*), creating a compression for the first motion. Again the N–S and E–W dimensions remain unchanged. Thus compressive first motions radiate outward in the NE and SW quadrants, and rarefaction first motions radiate outward in the NW and SE quadrants. The quadrants are separated by nodal planes, which are the fault plane and the plane normal to it, for dimensions do not change in these directions during faulting, and the amplitude of the seismic wave is therefore zero.

The first-motion radiation pattern of compressions and rarefactions thus enables us to identify the fault plane and the nodal plane, and it indicates the sense of slip on either plane that would generate that pattern, because slip would have to be toward the compression quadrant on either side of either nodal plane. The actual fault plane can often be identified by taking geologic considerations into account or by studying the location of aftershocks that occur along the fault plane.

The same principle works in three dimensions, and the nodal planes can be identified from first-motions by using a worldwide array of seismometers that in effect form a three-dimensional array surrounding

Figure 2.4.1 (*Facing page*) A two-dimensional model for the mechanism of first-motion radiation patterns. *A.* Undeformed state represented by squares on either side of a future fault. *B.* Deformed state before faulting. N–S and E–W dimensions of the squares are unchanged, but NE–SW dimensions (such as *BC* and *DE*) are lengthened and NW–SE dimensions (such as *AD* and *CF*) are shortened. *C.* Faulted state: sudden slip on the fault generates an earthquake. N–S and E–W dimensions of the squares are still unchanged, but NE–SW dimensions (such as BC_N and D_SE) are suddenly shortened, and NW–SE dimensions (such as AD_N and C_SF) are suddenly lengthened. Thus first motions are compressions for rays leaving the source in the quadrants marked C and are rarefactions for rays leaving the source in the quadrants marked R. The fault plane and the plane normal to it are nodal planes along which no change in dimension occurs, and the amplitude of the first motion is therefore zero.

Additional Readings

Blatt, H, G. V. Middleton, and R. C. Murray. 1980. *Origin of sedimentary rocks.* New York: Freeman.

Compton, R. R. 1962. *Field geology.* New York: Wiley.

Lindseth, R.O. 1982. *Digital processing of geophysical data: A review.* Continuing Education Program, Society of Exploration Geophysicists, Teknica Resource Development Ltd., Calgary, Alberta, Canada.

Sheriff, R. E. 1978. *A first course in geophysical exploration and interpretation.* Boston: International Human Resources Development Corporation.

the earthquake. In Chapter 9 we show that the maximum and the minimum compressive stresses lie in the rarefaction and the compression quadrants, respectively. The stress orientations are sometimes approximated by P and T axes, respectively, where the P axis bisects the nodal planes in the rarefaction quadrants, and the T axis bisects the nodal planes in the compressional quadrants (Figure 2.4.1C). It is a potential source of confusion that the minimum compressive stress, which is the deviatoric tensile stress (see Chapter 9), is located in the quadrant of compressional first arrivals.

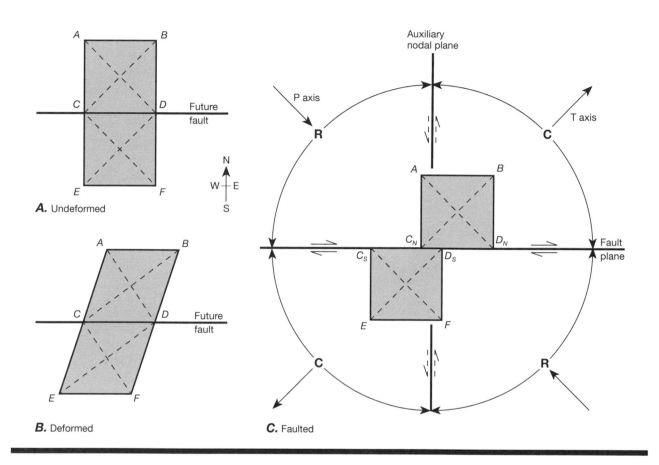

A. Undeformed

B. Deformed

C. Faulted

II Brittle Deformation

PART II of this book focuses on the structures formed in rocks predominantly by brittle deformation—that is, the breaking of rocks along well-defined fracture planes or zones. Depending on the relative motion that occurs across the fracture plane, the fractures are either extension fractures or shear fractures. We describe the general characteristics of these two types of fractures in Chapter 3. Faults are shear fractures in rocks generally at the scale of an outcrop or greater. Larger faults are commonly structures of major tectonic importance. We introduce the general characteristics of faults in Chapter 4. In Chapters 5, 6, and 7, we discuss the characteristics and tectonic significance of each of the three major types of faults: normal, thrust, and strike-slip.

Having described the structures of predominantly brittle origin that we observe in rocks, we next turn our attention to understanding how and why these structures form. Rocks break when they are subjected to an excessive amount of force; we introduce in Chapter 8 the concept of stress as a measure of the intensity of forces applied to a material. In Chapter 9 we review experimental evidence and theory about how the stress imposed on a rock is related to the types of fractures that form and to the mechanism of formation. With this background, we then return, in Chapter 10, to the interpretation of brittle structures that we find in the Earth. By understanding the mechanisms by which fractures form, and by being able to interpret the evidence we observe in the rocks, we can deduce the physical conditions that prevailed in the rock during this fracturing process, thereby opening another window on the tectonic evolution of the Earth's crust and the dynamic processes that drive that evolution.

Beyond their use in investigating the tectonic evolution of the Earth's crust, fractures are of major importance to our environment and to the continued viability of our society. First, because fractures often serve as conduits for ground water, they are the site of preferential weathering and thereby control the form of much of the Earth's topography. Indeed, some of the world's most inspiring land forms, such as Yosemite Valley in California, the Grand Canyon, the Alps, the islands of the Mediterranean, and Ayer's Rock in Australia owe much of their form to preferential erosion caused by the presence of fractures.

Furthermore, because fractures provide conduits for the migration of fluids through solid rock, they are of great significance in the migration of ground water, of hydrocarbons, and of hydrothermal and/or metamorphic fluids. Thus they are significant in the fields of hydrogeology, geothermal heat extraction, and oil and gas migration and recovery. In addition, fractures affect the location of hydrothermal mineral deposits and the integrity of nuclear waste disposal sites, which must safely contain their lethal waste for 10,000 years or more. As a consequence of this association with the world economy and the safety of future generations, understanding the characteristics of fractures and the conditions of their formation is of very real social importance.

Finally, because the cohesion of the rocks is lost across fracture surfaces, they are planes of weakness in the rock. This inherent weakness must be accounted for in the building of dams, bridge abutments, tunnels, mines, and similar engineering projects.

CHAPTER

3 Fractures and Joints

Fractures (from the Latin *fractus*, which means "broken") are surfaces along which rocks or minerals have broken; they are therefore surfaces across which the material has lost cohesion. Fractures are distinguished by the relative motion that has occurred across the fracture surface during formation. For **extension fractures, or mode I fractures,** the relative motion is perpendicular to the fracture walls (Figure 3.1*A*). For **shear fractures** the relative motion is parallel to the surface. For **mode II** shear fractures, the motion is a sliding motion perpendicular to the edge of the fracture (Figure 3.1*B*); for **mode III** shear fractures, it is a sliding motion parallel to the fracture edge (Figure 3.1*C*). A fracture that has components of displacement both parallel and perpendicular to the fracture surface is an **oblique extension fracture, or mixed mode fracture.**

Fractures are among the most common of all geologic features; hardly any outcrop of rock exists that does not have some fractures through it. Because the outcrop scale is easy to observe and is the basis of all field geology, we emphasize the descriptive characteristics of fractures at the outcrop scale.

The study of geologic history of fractures is notoriously difficult. Evidence bearing on the mode of fracture formation and the relative time of formation of different fractures is often ambiguous. As planes of weakness in the rock, fractures are subject to reactivation in later tectonic events, so some of the observable features of a fracture may be completely unrelated to the time and mode of its formation.

The investigation of fracturing comprises four general categories of observations: (1) the distribution and geometry of the fracture system, (2) the surface features of the fractures, (3) the relative timing of the formation of different fractures, and (4) the geometric relationship of fractures to other structures.

3.1 Classification of Joints and Extension Fractures

Most outcrops of rock exhibit many fractures that show very small displacement normal to their surfaces and no, or very little, displacement parallel to their surfaces. We call such fractures **joints.**[1] If there is no shear displacement, a joint is an extension fracture. A joint with very small shear displacement, however, may be an extension fracture on which shear displacement has later accumulated.

---

[1] Unfortunately, there is no universally accepted definition of the term *joint*. The definition set down here is conservative in that fractures satisfying this definition would be called joints by every other definition of the term.

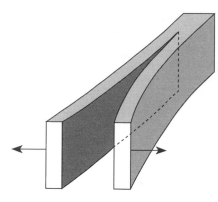
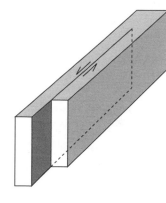
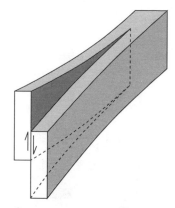

A. Extension (mode I)   **B.** Shear (mode II)   **C.** Shear (mode III)

Figure 3.1 The distinctions among the major types of fractures are based on the relative displacement of the material on opposite sides of the fracture *A.* Extension, or mode I, fractures. The relative displacement is perpendicular to the fracture. *B.* Shear fractures, mode II. Relative displacement is a sliding parallel to the fracture and perpendicular to the edge of the fracture. *C.* Shear fractures, mode III. Relative displacement is a sliding parallel to the fracture and to the edge of the fracture.

If many adjacent joints have a similar geometry, the fractures collectively are called a **joint set. Systematic joints** are characterized by roughly planar geometry, regular parallel orientations, and regular spacing (Figures 3.2 and 3.3A). **Nonsystematic joints** are curved and irregular in geometry, although they may occur in distinct sets of regional extent (Figure 3.3B). Nonsystematic joints nearly always terminate against older joints which commonly belong to a systematic set (Figure 3.3B). Because most joints we see are systematic, the term *joint* is generally used to refer to them. A **joint zone** is a quasi-continuous fracture that is composed of a series of closely associated parallel fractures and that extends much further than any of the individual fractures (Figure 3.3A, C). In practice, such a joint zone is also called simply a joint.

Most outcrops contain more than one set of joints, each with a characteristic orientation and spacing. Two or more joint sets affecting the same volume of rock constitute a **joint system**[2] (Figures 3.2 and 3.3D). If systematic joints of one set consistently terminate

---

[2] Note that the terms *joint system* and *systematic joint* have different meanings and should not be confused.

A.   **B.**

Figure 3.2 Joints. *A.* Outcrop showing a joint system made up of three distinct sets of joints. *B.* Joints of different orientations terminating against lithologic contacts.

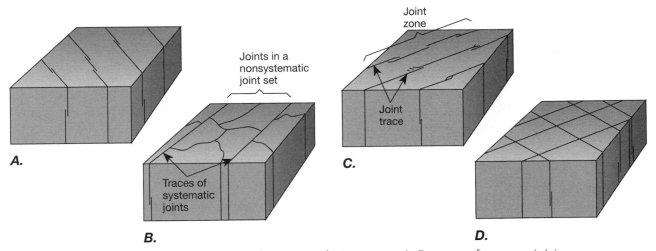

Figure 3.3 Diagrammatic views of joint sets and joint systems. *A.* Geometry of a systematic joint set. *B.* Typical pattern of nonsystematic joints and their characteristic termination against systematic joints. *C.* Joint zones forming quasi-continuous joints of much larger extent than the individual fractures. *D.* Two sets of mutually intersecting joints. Joints in each set cut joints of the other set. There is no consistent relationship whereby joints of one set terminate on joints of the other set.

against the joints of another set, they are referred to as **cross joints**. Joint sets and systems may persist over hundreds to thousands of square kilometers, each set displaying a constant or only gradually varying orien-

Figure 3.4 Fracture-controlled topography, Sinai peninsula, Egypt. Gulf of Aqaba to right about 200 km long.

tation (Figure 3.4; see also Figure 3.12). Such systems can show up as lineaments on high-altitude photographic and radar images (Figure 3.4).

In some circumstances it is useful to refer to joints in terms of their relationship to other structures. Thus strike joints and dip joints are vertical joints parallel to the strike or dip of the bedding, and bedding joints are parallel to the bedding. Joints that cut a fold or some other linear feature at high angles are also called cross joints, and those in other orientations are **oblique joints** or **diagonal joints**.

Sheet joints, sheeting, or **exfoliation joints** are curved extension fractures that are subparallel to the topography and result in a characteristic smooth, rounded topography (Figure 3.5). Sheet joints may be found in many kinds of rocks, but the characteristic topography is best displayed in plutonic rocks in mountainous regions where the joints appear to cut the rock into sheets like the layers of an onion. Many sheet joints apparently formed later than other joint sets, although in some cases they predate late phases of intrusive activity, as evidenced by dikes present along the joints.

Columnar joints are extension fractures characteristic of shallow tabular igneous intrusions, dikes or sills, or thick extrusive flows. The fractures separate the rock into roughly hexagonal or pentagonal columns (Figure 3.6), which are often oriented perpendicular to the contact of the igneous body with the surrounding rock.

Other types of extension fractures are common in deformed rocks. **Veins** are extension fractures that are filled with mineral deposits. The deposit may be massive

Figure 3.5 Sheet joints in granitic rock at Little Shuteye Pass, Sierra Nevada.

or composed of fibrous crystal grains such as quartz or calcite. The fibrous fillings can be very useful in interpreting the deformation associated with opening of the vein, as we discuss in detail in Sections 13.8 and 14.6.

Pinnate fractures, or **feather fractures,** are extension fractures that form *en echelon* arrays along brittle shear fractures (Figure 3.7). The acute angle between the extension fracture and the fault plane is a unique indicator of the sense of shear along the fault, and it points in the direction of relative motion of the block containing the fractures (see also the discussion in Section 4.3; Figure 4.16 *A, E*).

Gash fractures are extension fractures, usually mineral-filled, that may form along zones of ductile shear in the same orientation as the pinnate fractures. They are generally S- or Z-shaped, depending on the sense of shear along the zone. The photograph in Figure 3.8 shows two *en echelon* sets of gash fractures (white veins) arrayed along crossing shear zones. The Z-shaped and S-shaped veins to the left and right, respectively, show the "top down" sense of shear along each zone of fractures. The orientation of planar gash fractures can be used in the same way as feather fractures to determine the sense of shear on the associated shear zone. Extension fractures may also be associated with other structures, including folds and igneous intrusions, as described in Section 3.5.

A.   *B.*

Figure 3.6 Columnar jointing. Devil's Postpile National Monument, California. *A.* Columnar jointing in an andesitic flow. Note that the orientation of the columns varies from vertical to a shallow plunge. *B.* Cross section of the columnar joints.

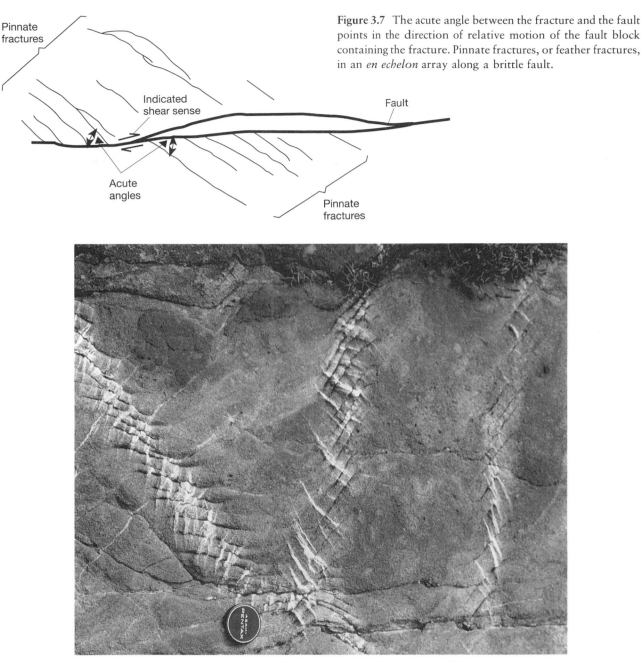

Figure 3.7 The acute angle between the fracture and the fault points in the direction of relative motion of the fault block containing the fracture. Pinnate fractures, or feather fractures, in an *en echelon* array along a brittle fault.

Pinnate fractures

Indicated shear sense

Fault

Acute angles

Pinnate fractures

Figure 3.8 Gash fractures (white veins) are extension fractures that commonly develop in a shear zone. Dark seams are solution features. Gash fractures are aligned along differently oriented planar shear zones that make an angle of approximately 50° with each other. The ends of the fractures tend to bisect the angle between these shear zones.

3.2 Geometry of Fracture Systems in Three Dimensions

In any study of the origin of fractures in rocks, we need to collect data on each fracture set that includes the orientation of the fractures, the scale of the fractures, the spacing of the fractures and their relationship to lithology and bed thickness, and the spatial pattern and distribution.

Orientation of Fractures

The orientation of fractures can help us identify fracture sets and infer the orientation and extent of the tectonic forces that produced them. In general, we collect representative orientations of all the major fracture sets in each outcrop, over a large area, and rely in part on consistency of orientation to correlate sets from one outcrop to another.

Interpretations that rely too heavily on orientation data, however, have resulted in significant misunderstanding. For example, because shear fractures commonly intersect at an angle of roughly 60°, it is often incorrectly assumed that all fractures that intersect at such an angle are shear fractures. Similarly, the consistent orientation of joints relative to other structures is often taken to be indicative of a genetic relationship. Although such interpretations cannot be ruled out a priori, they are unreliable unless corroborated by other evidence.

More than one orientation of fracture may be associated with a single fracturing event, and it is important to understand the relationships among fractures instead of blindly measuring every orientation of fracture surface available in the hope that statistical analysis will compensate for lack of careful observation. Genetically related fractures may differ in orientation as a result of segmentation and twisting of the fracture plane, curving of the plane, reorientation of the fracture into parallelism with a local planar weakness in the rock, or branching of the fracture into two or more orientations. Some fractures may be of only local extent and may even result from human activity such as excavation or blasting. Careful study is therefore required to identify the significant data.

The orientation of genetically related fractures may differ from one lithology to another; on the other hand, the fractures in various rock types may result from different events. Figure 3.9, for example, contains the same data as Figure 2.12, except that in this split rose diagram, the joints in coal are plotted in the top half, and those in the shale in the lower half. The joint patterns from the two lithologies are distinctly different. Careful investigation has led to the conclusion that the joints in the coal formed first. Gentle folding of the region then broadened the distribution of the earlier joint orientations, and finally the joints in the shale formed. Separating the joints by lithology, therefore, is important in decifering the history of joint formation.

Orientation data on fractures are conveniently collected and compared by using orientation histograms, rose diagrams, or spherical projections, all of which we discussed in Section 2.5.

Scale and Shape of Fractures

Individual fracture planes are not of indefinite extent. Field observations indicate that a joint may terminate by simply dying out (Figure 3.10A), by curving and dying out (Figure 3.10B), by branching and dying out (Figure 3.10C), by curving into a preexisting joint (Figure 3.10D, E, F), or by segmenting into an *en echelon* set of small extension fractures (Figure 3.10G). The amount of displacement across the joint decreases toward the joint

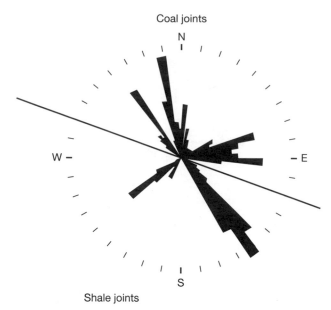

Figure 3.9 A split rose diagram showing the strikes of joints in coal in the upper half of the diagram and those in shale in the lower half. Note the difference in orientation of the joints in the two rock types.

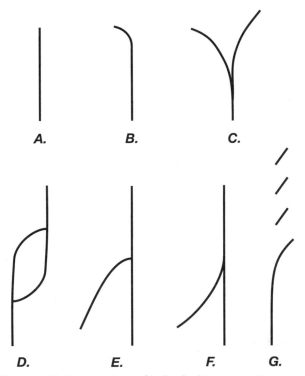

Figure 3.10 Terminations of individual joints. *A.* Dying out of a straight joint. *B.* Curving and dying out of a joint. *C.* Branching and dying out of a joint. *D.* Two overlapping joints, each curving toward a perpendicular intersection with the other. *E.* One joint curving toward a perpendicular intersection with another. *F.* One joint curving toward a parallel intersection with another. *G.* Dying out of a joint in a series of *en echelon* fractures.

termination. In a given joint set, individual joint traces or joint zones (see Figure 3.3C) range in length from a few centimeters to many meters—and even up to kilometers in the case of **master joints.** Fractures also exist at a scale down to the microscopic level; such fractures are better referred to as microfractures than as joints.

The shape of individual joints depends largely on the rock type and on its structure. In uniform rock, such as granite, argillite, or thin-bedded rocks of uniform composition, individual joint planes tend to be roughly circular to elliptical in shape, with the long axis horizontal. In sedimentary sequences involving rocks of highly different mechanical properties, such as interbedded sandstone and shale, one dimension of a joint is commonly constrained by the upper and lower contacts of the bed in which the joint forms, and the joint tends to be of much greater extent parallel to bedding than across it. Joints in individual beds of one lithology often end against beds of another lithology (Figure 3.2B).

The shape of master joints is not well known because of the difficulty of seeing the third dimension. In areas of great vertical relief, however, joints can be traced to a depth of more than 1 km.

Spacing of Fractures

The spacing of fractures in a systematic set can be measured in terms of either the average perpendicular distance between fractures or the average number of fractures found in a convenient standard distance normal to the fractures. The average spacing of joints tends to be remarkably consistent, and it depends in part on the rock type and on the thickness of the bed in which the fractures are developed (Figure 3.11). Data sets A through C are measurements made in the sandy layers of a sequence of wackes from several locations with different thicknesses of shale interbeds. Data sets D and E are from different limestones. Two features of the plot are significant. First, the spacing between adjacent joints increases with increasing bed thickness up to a maximum value independent of the thickness. Second, the maximum spacing is considerably greater for the limestones than for the wackes, which demonstrates the effect of lithology independent of the thickness of the beds.

Spatial Pattern and Distribution of Fracture Systems

The most useful method of studying the pattern and distribution of fracture sets is to plot maps of the location and orientation of the fractures. In areas of very good exposure, it may be possible to map joints indi-

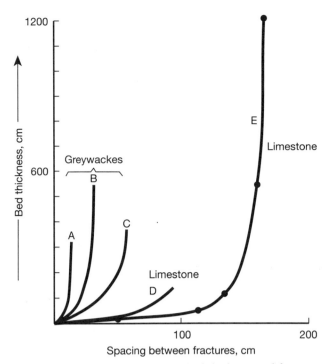

Figure 3.11 Relationship between bed thickness and fracture spacing measured in sandy layers in wackes (data sets A, B, and C) and in limestones (data sets D and E). For data set A, the shale interlayers are less than 5 cm thick, and for B and C they are greater than 5 cm thick.

vidually and to trace out the relationship of joints to one another and to lithology. On such maps, we can plot the strikes and dips of the fractures, their relationship to other local structures, and the amount and direction of shear (if any) on the fractures.

In most cases, there is neither enough exposed rock nor enough time available to permit such detailed mapping. Usually data from outcrops scattered over a large area are plotted on a map. From these data one constructs **form lines,** or **trajectories,** of the individual joint sets by projecting the strikes of joints in the same set between data points. Figure 3.12 shows such a map for an area on the Appalachian plateau. The consistency of orientation of joints over such large areas indicates that they record regional tectonic conditions.

3.3 Features of Fracture Surfaces

The features on the surfaces of a fracture can provide information critical to interpretation of the fracture's origin. Many joints display a regular pattern of subtle ridges and grooves called **hackle** that diverges from a point or a central axis. The pattern is known as **plumose**

Figure 3.12 Map of joints in the Appalachian Plateau in New York state. The lines are constructed parallel to the dominant strike of joints measured in local outcrops throughout the area.

structure or as a **hackle plume** (Figure 3.13), named for its resemblance to the shape of a feather. Plumose structure is present on joints in a variety of rock types, but it is most clearly displayed in rocks of uniform fine-grained texture, when the surface is illuminated at low angles. Figure 3.13 shows plumose structure developed in two rock types.

The characteristic features of a hackle plume are illustrated in Figure 3.14*A*. The main joint face displays the hackle plume with hackle lines that diverge from an axis. Toward the edges of the main joint face, the joint plane may segment into a series of planes that are slightly twisted from the main joint face. The twisting may increase gradually away from the axis (lower half

Figure 3.13 Plumose structure, or hackle plumes, in *A* mudstone, and *B* chalk.

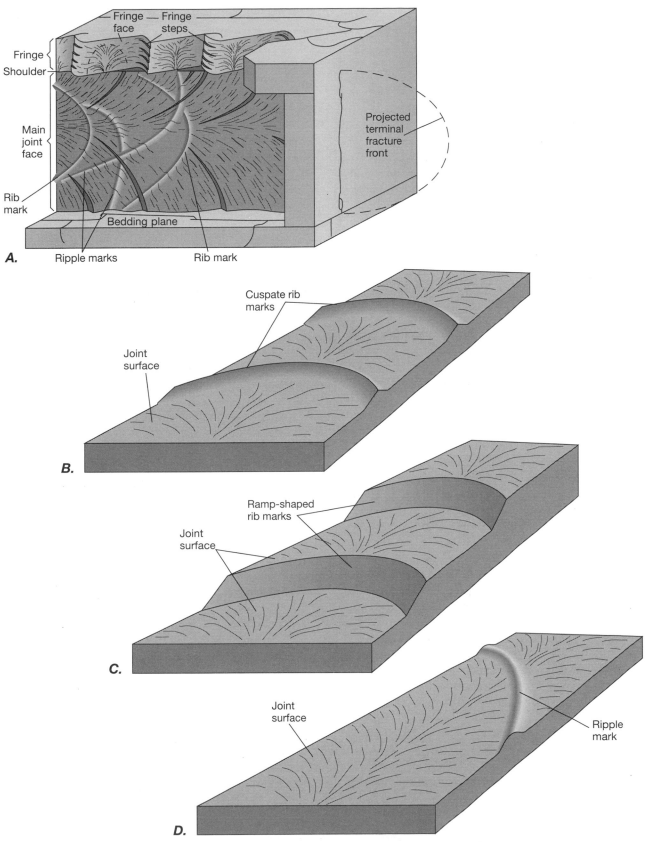

Figure 3.14 Schematic block diagrams illustrating surface markings on joint faces. *A.* Features that characteristically appear on the surface of a systematic joint. *B.* Cuspate rib marks. *C.* Ramp-shaped rib marks. *D.* Ripple marks.

of Figure 3.14A), or it may develop abruptly at a shoulder (top of Figure 3.14A) to form a hackle fringe composed of an *en echelon* set of extension fractures, or fringe faces, connected to one another by curving fringe steps. The fringe faces themselves may show second-order hackle plumes with associated fringes. The fringe is aligned along the trend of the main joint face.

Fringe faces at the edge of a joint should not be confused with pinnate fractures and gash fractures, even though they are all extension fractures that form *en echelon* arrays. Fringe faces usually make a considerably smaller angle with the main joint face than pinnate or gash fractures make with the shear surface. Moreover, in three dimensions, fringe faces are restricted to the edge of a joint surface, which commonly displays plumose structure, whereas pinnate and gash fractures occur along the entire shear fracture.

In some cases, curvilinear features called rib marks and ripple marks cross the lines of hackle on the fracture face. The rib marks either are cuspate in cross section (Figure 3.14B) or are composed of smoothly curved ramps connecting adjacent parallel surfaces of the joint face (Figure 3.14C; see also Figure 3.13D). They tend to be perpendicular to the hackle lines. The ripple marks are rounded in cross section and oblique to the hackle lines (Figure 3.14D). Hackle plumes form a variety of different patterns (compare Figure 3.13), which can characterize particular sets of joints and which reflect important differences in the fracturing process.

Plumose structure is a unique feature of brittle extension fractures that distinguishes them from shear fractures. The direction of divergence of the hackle lines is the direction in which the fracture propagated, and the lines of hackle form normal to the fracture front. When traced back along the plume axis, the hackle is usually found to radiate from a single point, which is the point of origin of the fracture. Rib marks are interpreted to be arrest lines where fracture propagation halted temporarily. Ripple marks are interpreted to form during very rapid fracture propagation, in which case they are called Walner lines. By careful study of the surfaces of joints, therefore, we can learn a great deal about where fractures initiated and how they propagated. (The use of hackle plumes is treated in greater detail in the next section. And we discuss the interpretation of joints further in Chapter 9 after we have examined the mechanics and mechanisms of fracture formation.)

In some cases a fracture displays **slickenside lineations** on its surface, indicating that shear has occurred on the fracture (see Sections 13.7, 13.8, and 14.6). Slickenside lineations may be defined by parallel sets of ridges and grooves, by light and dark streaks of fine-grained pulverized rock, or by linear mineral fibers (see Figures 13.25, 13.26). Because extension fractures commonly accumulate small amounts of shear displacement during tectonic movements subsequent to their formation, such displacements do not necessarily indicate that the fracture formed by shearing.

Joints and other fractures may have a thin deposit of mineral—such as quartz, feldspar, calcite, zeolites, chlorite, and epidote—along their surfaces. These mineral layers indicate either that the fracture was open or that fluids under pressure were able to force it open, flow along the fracture, and deposit minerals from solution. In some cases, a fracture is clearly associated with a zone of alteration in the surrounding rock, indicating diffusion of material into or out of the rock surrounding the fracture. Some joints have been affected by dissolution resulting in open **fissures**.

3.4 Timing of Fracture Formation

The interpretation of fracture sets relies on determination of the timing of their formation relative to other fracture sets and structures. Although these relationships are often ambiguous, especially for extension fractures, we can make a few generalizations.

Where more than one set of joints is developed, younger joints terminate against older joints because an extension fracture cannot propagate across a free surface such as another extension fracture. In Figure 3.3B, for example, the nonsystematic joints are clearly younger than the systematic ones. Many such terminations are at a high angle, forming T-shaped intersections, and the younger extension fracture may curve toward a high-angle intersection where it approaches an older fracture (Figure 3.10E). Low-angle intersections also occur in some joint systems (Figure 3.10F).

In many cases, however, joints form a mutually cross-cutting system, so the relative ages of the joints are ambiguous. This relationship can arise in several ways. If the first-formed joint is cemented by mineral deposits, it no longer acts as a free surface, and a later joint can cut across the older one. Subsequent dissolution of the mineral deposit leaves an ambiguous cross-cutting relationship. An early shallow joint in a layer may be cut by a later deeper joint that propagates up to one side of the older joint, continues underneath it, and propagates back to the opposite side, leaving a joint intersection for which the interpretation of relative time of formation is ambiguous (Figure 3.15A). The relative age can be decifered only if the hackle on the younger joint can be examined to determine the propagation direction of the fracture front. Two joints of the same orientation may also initiate from the same place on opposite sides of an older fracture (Figure 3.15B). Those two joints appear as a continuous joint, leaving am-

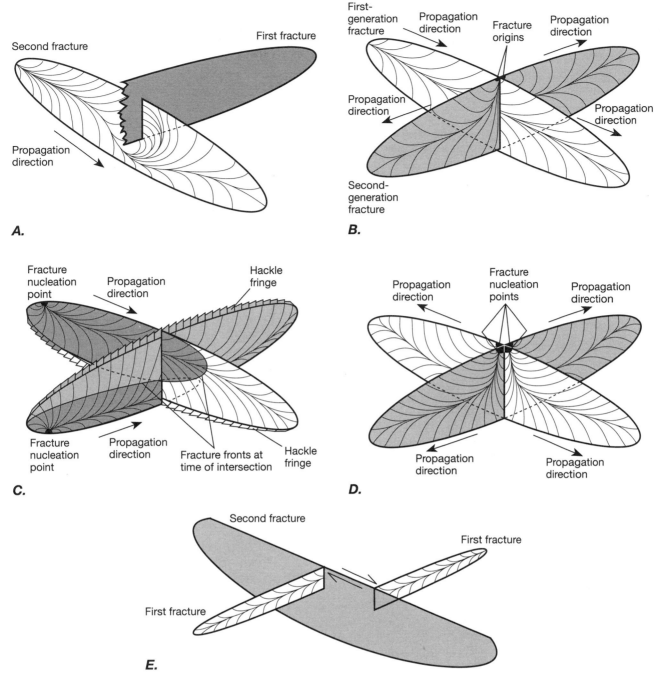

Figure 3.15 Origin of fracture intersections. Solid arrows indicate the general direction of fracture propagation. *A.* An early shallow fracture (dotted pattern) cut by a later deeper fracture. The intersection relationship would be ambiguous were it not for the distinctive hackle pattern. *B.* Two coplanar fractures originate at adjacent points on opposite sides of an earlier fracture to produce an ambiguous intersection. The hackle patterns resolve the ambiguity at the intersection and make the origins of the fractures clear. *C.* Two intersecting fractures that originated in different orientations at the top and bottom of the layer. The shaded portions of the fracture surfaces and the fracture fronts (heavy lines) indicate the geometry of the propagating fractures at the time the two fracture fronts intersected. Hackle patterns indicate that the top of fracture A and the bottom of fracture B were the earliest to form at the intersection. *D.* Three fractures, two of them coplanar, originate at the same point producing an intersection indistinguishable, from the one shown in part *B,* except for the telltale hackle pattern. *E.* Shear offset of an early fracture on a later shear fracture produces an apparent termination of the older fracture against the younger.

biguous the interpretation of the intersection with the earlier fracture.

Two joint sets could, in principle, also form during the same fracturing event. Joints could originate in one orientation at the top of a layer and propagate down, and in another orientation at the bottom of a layer and propagate up (Figure 3.15C). Their intersection would then show inconsistent relative-age relationships, because at different points along the intersection, the hackle lines would indicate that different joints had formed earliest. A similar fracture geometry would be created if the three joints formed at the same time and place (Figure 3.15D). It is possible to distinguish the histories illustrated in Figure 3.15B and D only by examining the hackle plume geometry on the joint surfaces. Inconsistent relationships can also result when shear displacement occurs on a later fracture. The offset first fracture can then appear to be a younger fracture terminating against the second fracture (Figure 3.15E). For all the cases illustrated in Figure 3.15, the hackle plume geometry is of critical importance in interpreting the fracturing history of joints.

Several structures indicate that extension fractures can form in sediments before they have consolidated into rock. When such fractures form before the deposition of overlying sediments, the open fractures may be filled by the sediment subsequently deposited on top. Mudcracks are one obvious example. If a steeply dipping fracture forms in uncompacted sediments and becomes mineralized before compaction is complete, the mineralized fracture may form a series of folds to accommodate the shortening associated with the subsequent compaction of the sediment. Extension fracturing of unconsolidated sediment in the presence of high pore-fluid pressure can result in the formation of **clastic dikes**. The opening of the fracture creates a low-pressure area into which pore fluid rushes, carrying unconsolidated material of contrasting lithology. The existence of such structures proves that some joints, at least, can form very early in the history of a rock.

Fractures that cross-cut a geologic boundary or a geologic structure clearly postdate the formation of that boundary or structure. For example, a joint set that cuts across an intrusive contact is younger than the intrusive event, and joints that maintain a constant orientation across folded layers must have formed after the folding. Fractures that are clearly affected by a geologic structure are older than that structure. A joint set that changes orientation over a fold but everywhere maintains the same angular relationship with the bedding could either predate or be synchronous with the folding, but it is not likely to be younger.

If one set of joints is consistently mineralized or has igneous rocks injected along the fractures, then it must be older than the mineralizing or intrusive event.

If a second set of joints in the same rocks is free of the mineralization or intrusion, then it probably formed after the mineralizing or intrusive event.

Applying criteria such as these has shown clearly that joints can form at any time in the history of a rock—from the earliest time, when it is still unconsolidated sediment to the latest time when the joints postdate all other structures in the rock. It is likely, therefore, that more than one mechanism produces joints. We discuss possible mechanisms in Chapter 10.

3.5 Relationship of Fractures to Other Structures

Fractures Associated with Faults

Fractures often form as subsidiary features spatially related to other structures. If such a relationship can be documented, the fractures can provide information about the origin of the associated structure.

In some cases, faults are accompanied by two sets of small-scale shear fractures at an angle of approximately 60° to each other with opposite senses of shear. These are called **conjugate shear fractures.** Figure 3.16 shows data for a system of conjugate fractures that developed in an area closely associated with a known fault. The rose diagram shown in Figure 3.16B is plotted in the vertical plane normal to the strike of the fault, and the distribution of fracture dips is plotted below the horizontal line. The orientation of the fault is also indicated on the figure. The major set of fractures is clearly parallel to the fault; the second and less well-developed set is approximately 65° from the first set.

Extension fractures associated with faulting include pinnate fractures and gash fractures, which were described in Section 3.1 (Figures 3.8 and 3.9).

Fractures Associated with Folds

Fractures often develop in rocks in association with folding. A variety of orientations, related symmetrically to the fold, have been reported. It is convenient to refer the orientations to an orthogonal system of coordinates (a, b, c) related to the fold geometry and the bedding. The b axis is parallel to the fold axis, which in most cases is a line lying in the bedding plane that has the same orientation regardless of the attitude of the folded surface (Figure 3.17). The a axis lies in the bedding plane perpendicular to the fold axis, and the c axis is everywhere perpendicular to the bedding.

Figure 3.17 is a diagrammatic illustration of the orientations of fractures that have been reported from folds. Fractures parallel to the plane of the a and c axes

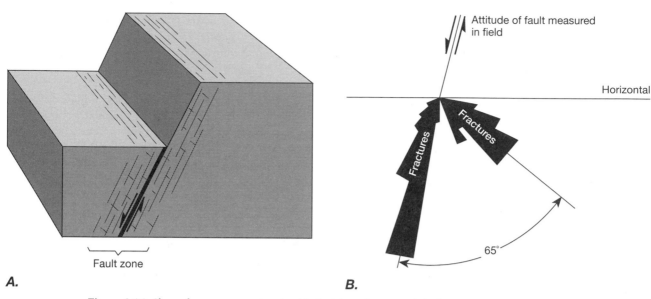

Figure 3.16 Shear fractures associated with faulting. *A.* Normal fault with dominant parallel shear fractures (long lines) and subordinate conjugate shear fractures (short lines). *B.* Rose diagram plotted in the vertical plane, showing the distribution of dips of two sets of fractures associated with a normal fault.

and to the plane of the **b** and **c** axes are called **ac fractures** and **bc fractures**, respectively. The fractures shown in sets *A*, *B*, and *D* are all perpendicular to the bedding. In sets *A* and *B*, the ac fractures and the bc fractures, respectively, bisect the acute angle between the two other fracture sets, which are **oblique fractures.** In set *D*, the bc fractures bisect the obtuse angle between the

oblique fractures. The inclined fractures in sets *C* and *E* are parallel to the fold axis **b**. Those in set *C* make a low angle with bedding, and those in set *E* make a high angle with bedding.

Fractures in sets *A* and *D* are particularly common on fold limbs. Sets *B* and *E* tend to be associated with the convex sides of a fold where the curvature is strong-

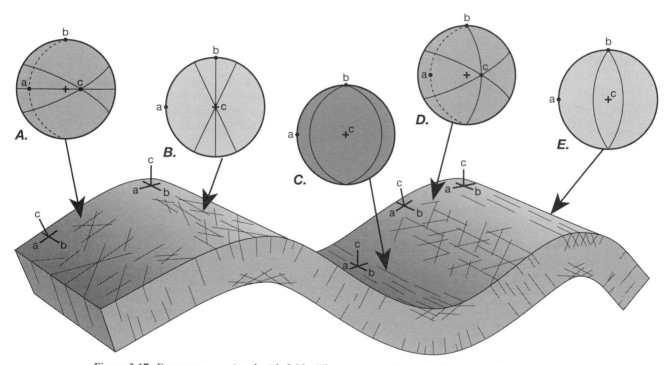

Figure 3.17 Fractures associated with folds. The stereographic projections show the orientations of the coordinate system, the bedding where it is not horizontal (dotted great circles), and the fractures (solid great circles).

est. And set C occurs on the strongly curved concave sides of folds.

It is possible that the fractures of all these orientations formed in association with folding, that the ac and bc fractures are extension fractures, and that the oblique and inclined fractures are shear fractures. That interpretation is not justified, however, simply on the basis of fracture pattern and orientation. Such fractures have been shown to predate the folding in some cases, and to postdate it in others, and as we have remarked before, the presence of shear displacement on a fracture does not necessarily mean that the fracture originated as a shear fracture. In order to make a well-documented interpretation of complex fracture systems, it is critical to describe all the characteristics of the various fracture sets that we have discussed. This includes citing specific evidence for extensional or shear displacement and fracturing, the spatial distribution of fractures, and evidence suggesting the relative sequence of formation of the fractures in different sets.

Fractures Associated with Igneous Intrusions

Fractures form an association with igneous intrusions, and some types occur *only* within the igneous rock. We have already described columnar jointing in sills and thick flows (Figure 3.6) and sheet joints or exfoliation structure in plutonic rocks (Figure 3.5). In many cases, the internal structure of plutonic rocks is related in a simple manner to the orientations of other fractures that develop. Especially near the margins of plutonic bodies, platy minerals such as mica and tabular mineral grains may be aligned parallel to one another, creating a planar structure in the rock called a **foliation**. Elongate mineral grains also may be aligned parallel to one another within the foliation, creating a linear structure called a **lineation** (see Chapter 13).

We can describe the orientations of fractures with respect to these structures by using coordinate axes (**a**, **b**, **c**) where **a** is parallel to the lineation, **b** lies in the foliation perpendicular to **a**, and **c** is perpendicular to the foliation. Joints commonly form parallel to **c** and thus perpendicular to the foliation. If they are also parallel to the lineation, they are referred to as ac joints; if perpendicular to the lineation, they are called **cross joints** or bc joints. Diagonal joints also occur, usually at an angle of about 45° to the lineation and normal to the foliation. Cross joints (bc) typically contain pegmatite dikes or hydrothermal deposits.

Additional Readings

Bahat, D., and T. Engelder. 1984. Surface morphology on cross-fold joints of the Appalachian plateau, New York and Pennsylvania. *Tectonophysics* 104: 299–313.

Beach, A. 1975. The geometry of en-echelon vein arrays. *Tectonophysics* 28: 245–263.

Engelder, T., and P. Geiser. 1980. On the use of regional joint sets as trajectories of paleostress fields during the development of the Appalachian Plateau, New York. *Journal of Geophysical Research* 85: 6319–6341.

Kulander, B. R., and S. L. Dean. 1985. Hackle plume geometry and joint propagation dynamics. In *Proceedings of the International Symposium on Fundamentals of Rock Joints, Björkliden, 15–20 Sept. 1985,* ed. Ove Stephans-son. Luleå, Sweden: Swedish Natural Science Research Council; Centek Publishers.

Kulander, B. R., C. C. Barton, and S. L. Dean. 1979. The application of fractography to core and outcrop fracture investigations. U.S. Dept. of Energy, METC/SP-79/3; National Technical Information Service, U.S. Dept. of Commerce, Springfield, Va., 22161.

Ladeira, F. L., and Price, N. J. 1981. Relationship between fracture spacing and bed thickness. *J. Struct. Geol.* 3: 179–183.

Price, N. J. 1966. *Fault and joint development in brittle and semibrittle rock.* New York: Pergamon Press.

CHAPTER 4

Introduction to Faults

A **fault** is a surface or narrow zone along which one side has moved relative to the other in a direction parallel to the surface or zone. Most faults are brittle shear fractures (Figure 4.1*A*) or zones of closely spaced shear fractures (Figure 4.1*B*), but some are narrow shear zones of ductile deformation where movement took place without loss of cohesion at the outcrop scale (Figure 4.1*C*). We generally use the term *fault* for shear fractures or zones that extend over distances of meters or larger. Features at the scale of centimeters or less are called **shear fractures,** and shear fractures at the scale of a millimeter or less are **microfaults** that may be visible only under a microscope. Faults are often structural features of first-order importance on the Earth's surface and in its interior. They affect blocks of the Earth's crust thousands or millions of square kilometers in area, and they include major plate boundaries hundreds or even thousands of kilometers long.

The word *fault* is derived from an eighteenth and nineteenth-century mining term for a surface across which coal layers were offset. Many such mining terms were transferred to geology in its early days, despite the fact that the mining lexicon was often complex and ambiguous. This century has seen a number of attempts to rationalize and systematize this terminology, although there is still no agreement on precise definitions for some words. We try to employ only those terms that are the most prevalent and the most useful in describing faults.

4.1 Types of Faults

A fault divides the rocks it cuts into two **fault blocks.** For an inclined fault, geologists have adopted the miners' terms **hanging wall** for the bottom surface of the upper fault block and **footwall** for the top surface of the lower fault block (Figures 4.1, 4.2). In a tunnel, these are the surfaces that literally *hang* overhead or lie under *foot.* The fault block above the fault is the **hanging wall block,** and the block below the fault is the **footwall block.** For a vertical fault, of course, these distinctions do not apply, and the sides of the fault are named in accordance with geographic directions: the northwest side and the southeast side, for instance.

Faults are classified in terms of the attitude of the fault surface. If a fault dip is more than 45°, it is a **high-angle fault;** if less than 45°, it is a **low-angle fault.**

We also divide faults into three categories depending on the orientation of the **relative displacement,** or **slip,** which is the net distance and direction that the hanging wall block has moved with respect to the footwall block (Figure 4.2). On **dip-slip faults,** the slip is approximately parallel to the dip of the fault surface; on **strike-slip faults,** the slip is approximately horizontal, parallel to the strike of the fault surface; and on **oblique-slip faults,** the slip is inclined obliquely on the fault surface. An oblique-slip vector can always be described as the sum of a strike-slip component and a dip-slip

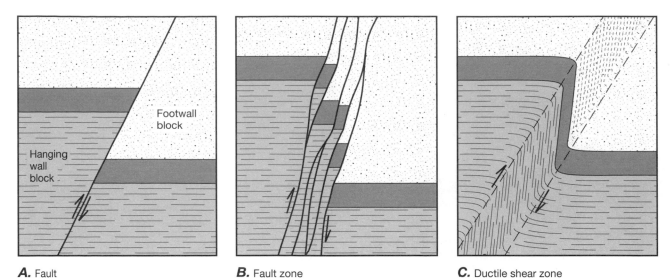

A. Fault   B. Fault zone   C. Ductile shear zone

Figure 4.1 Three styles of faulting. *A.* A single fault consists of a single shear fracture. A fault zone comprises *B.* a set of associated shear fractures or *C.* a zone of ductile shear.

component, or as the sum of a horizontal component and a vertical component. The dip-slip component may in turn be described as the sum of a vertical component and a horizontal component, which are sometimes called the **throw** and the **heave,** respectively.

We subdivide faults further in terms of the relative movement along them. Inclined dip-slip faults on which the hanging wall block moves down relative to the footwall block are **normal faults** (Figure 4.3*A*). Those on which the hanging wall block moves up relative to the footwall block are **thrust faults** (Figure 4.3*B*).[1] Vertical

---
[1] Thrust faults that dip more steeply than 45° are sometimes called reverse faults.

faults characterized by dip-slip motion cannot, of course, be classified as either normal or reverse faults, so we simply specify which side of the fault has moved up or down. Strike-slip faults are **right-lateral,** or **dextral,** if the fault block across the fault from the observer moved to the right (Figure 4.3*C*); they are **left-lateral,** or **sinistral,** if that block moved to the left (Figure 4.3*D*). Oblique-slip faults may be described according to the nature of the strike-slip and dip-slip components. Figure 4.3*E*, for example, shows sinistral normal slip, and Figure 4.3*F* shows sinistral reverse slip. For **rotational faults** the slip changes rapidly with horizontal distance along the fault (Figure 4.3*G*).

Figure 4.2 Slip vectors and slip components on a fault surface.

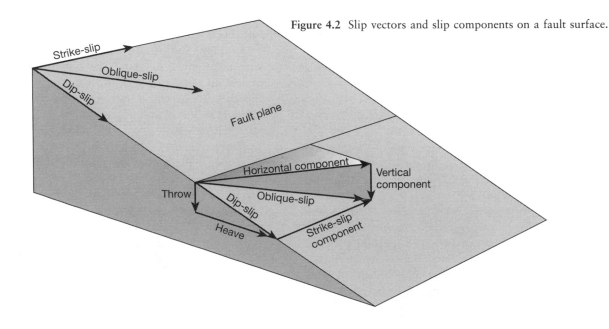

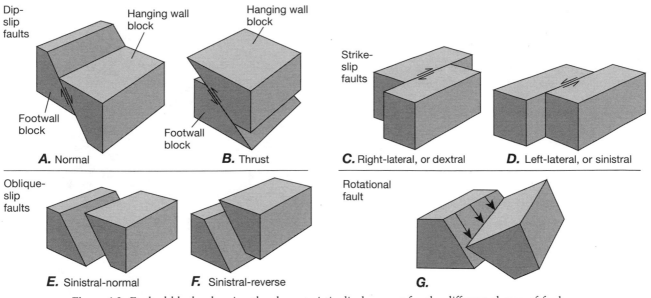

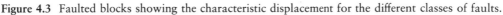

Figure 4.3 Faulted blocks showing the characteristic displacement for the different classes of faults.

4.2 Recognition of Faults

The criteria for recognizing faults can be divided into three broad categories: features intrinsic to faults themselves, effects on geologic or stratigraphic units, and effects on physiographic features. We briefly consider each of these categories.

Features Intrinsic to Faults

Faults can often be recognized by the characteristic textures and structures developed in rocks as a result of shearing (Table 4.1). These textures and structures vary with the amount and rate of shear and with the physical conditions at which the faulting occurred, including

Table 4.1 Fault Rock Terminology[a]

Cataclastic rocks				
Fabric	Texture	Name	Clasts	Matrix
Generally no preferred orientations	Cataclastic: sharp, angular fragments	Breccia series { Megabreccia	> 0.5 m	< 30%
		Breccia	1–500 mm	< 30%
		Microbreccia	< 1 mm	< 30%
		Gouge	< 0.1 mm	< 30%
		Cataclasite	Generally ≤ ~10 mm	> 30%
		Pseudotachylite		Glass, or grain size ≤ 1 μm

Mylonitic rocks				
Fabric	Texture	Name	Matrix grain size	Matrix
Foliated and lineated	Metamorphic: Interlocking grain boundaries, sutured to polygonal	Mylonitic gneiss	> 50 μm	
		Mylonite series { Protomylonite	< 50 μm	< 50%
		Mylonite	< 50 μm	50%–90%
		Ultramylonite	< 10 μm	> 90%

[a] The terminology applied to fault rocks is by no means generally agreed upon. The definitions of the different categories, and the quantitative boundaries we have placed on them, should therefore be understood as guidelines to present usage, which can vary from one geologist to another. We believe, for example, that what we have defined as mylonite would fit anyone's definition, but other geologists use *mylonite* in a broader sense, even to include what we call mylonitic gneiss.

Figure 4.4 Schematic block diagram of a portion of the earth's crust, showing the surface trace of a fault zone (i.e., its exposure on the Earth's surface) and the variation with depth of the type of fault rock within the fault zone. Incoherent cataclasites (plus pseudotachylite if dry) characterize depths above 1–4 km. Below that, coherent cataclasites (plus pseudotachylite if dry) are present at depths of up to 15 km. Mylonites are present at depths greater than 10–15 km and temperatures greater than 250–350° C.

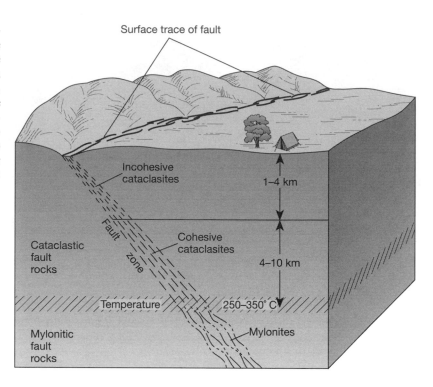

temperature and pressure which typically are a function of the depth of faulting (Figure 4.4).

Faults formed at depths less than about 10 to 15 km typically have **cataclastic rocks** present in the fault zone. This term refers in general to rocks that have been fractured into clasts or ground into powder during brittle deformation. Individual fragments are generally sharp, angular, and internally fractured. Cataclastic rocks usually lack any internal planar or linear structure. Friable cataclastic rocks are typical of faulting above depths of 1 to 4 km. Cohesive cataclasites may form at depths up to 10 to 15 km.

The terminology and classification are not universally agreed upon (see Table 4.1). We divide cataclastic rocks into four main categories. The breccia series, gouge, and cataclasite are distinguished on the basis of clast size and the percentage of matrix; whereas pseudotachylite forms a separate category based on the character of the matrix. The percentage of fine-grained matrix distinguishes rocks in the **breccia series** (less than about 30 percent; Figure 4.5A) from cataclasites (more than about 30 percent; Figure 4.5B). We subdivide the breccia series into **megabreccia** (Figure 4.5A), **breccia**, and **microbreccia**. In megabreccia and breccia, the clasts

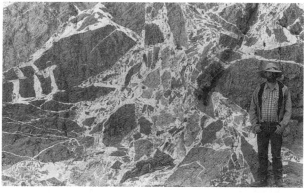

A.

Figure 4.5 Cataclastic rocks. *A.* Megabreccia composed of very large fragments of limestone (Titus Canyon, Death Valley National Monument). *B.* Cataclasite from the Whipple Mountain detachment, southeast California.

B.

Figure 4.6 Veins of dark pseudotachylite cutting a light-colored gneiss. Actual width of view shown is about 6 cm.

are predominantly rock fragments. In microbreccia the clasts are principally mineral grain fragments. **Gouge** is essentially a continuation of the breccia series to finer clast size. In outcrop, it appears as a finely ground, whitish rock powder. Remarkably, all cataclastic rocks look very much the same over a wide range of scales (compare Figure 4.5*A*, *B*). Although breccia, microbreccia, and gouge are generally noncohesive, deposi-

tion of silica during or subsequent to formation can turn them into hard, cohesive, silicified fault rock.

Cataclasites include a range of clast sizes and vary from 30 percent fine-grained matrix up to 100 percent matrix (Figure 4.5*B*). They are generally cohesive rocks.

Pseudotachylite (Figure 4.6) is a massive rock that frequently appears in microbreccias or surrounding rocks as dark veins of glassy or cryptocrystalline material. It characteristically contains a matrix of crystals less than 1 μm in diameter and/or small amounts of glass or devitrified glass cementing a mass of fractured material together. Under a petrographic microscope, the matrix appears isotropic; that is, between crossed polarizers no light is transmitted. This behavior is characteristic of glass and of extremely fine-grained material. During an earthquake under dry conditions at depths generally less than 10 to 15 km, frictional heating can be sufficient to melt small portions of the rock. The resulting material may intrude through fractures in the adjacent rock before quenching to form veins of pseudotachylite.

Cataclastic rocks occur in zones ranging from a few millimeters in thickness up to extensive zones one or more kilometers thick. In general, the greater the thickness and the smaller the grain size, the greater the amount of displacement that has accumulated on the fault.

Fault zones formed at depths exceeding about 10 to 15 km are characterized by another type of very fine-grained rock called **mylonitic rocks** (Figure 4.7). These

⊢——⊣
1 mm

Figure 4.7 Quartz mylonite showing large feldspar porphyroclast in a much finer-grained matrix of strongly recrystallized quartz grains.

rocks form only as a result of ductile deformation, which occurs in crustal rocks at temperatures generally in excess of 250°C to 350°C.[2] Mylonitic rocks have a matrix of very fine grains that are derived by reduction of grain size from the original rock. Variable amounts of relict coarse mineral grains, called **porphyroclasts,** may be present, surrounded by the fine-grained matrix. The fine grains show an interlocking grain boundary texture characteristic of metamorphic rocks; the grain boundaries themselves may be polygonal, forming 120° triple junctions, or they may be highly sutured. Mylonitic rocks exhibit a strong planar and linear internal structure, called foliation and lineation, respectively (see Chapter 13). These structures are subparallel to the fault zone.

Mylonitic rocks form as a result of the recrystallization of mineral grains during rapid ductile deformation. Their polygonal to sutured grain boundaries differ from fine-grained cataclasites, in which the grains have the sharp, angular shape characteristic of brittle fracturing.

If the grain size is reduced from the original grain size but is coarser than about 50 μm, the rock is a **mylonitic gneiss.** If the matrix grain size is less than 50 μm, the rock belongs to the **mylonite series,** which we subdivide on the basis of increasing percentage of fine-grained matrix (Table 4.1) into **protomylonite, mylonite,** and **ultramylonite.** In ultramylonites, the characteristic grain size of less than 10 μm causes the matrix to appear glassy in a hand sample.

Mylonites are generally found in ductile shear zones ranging in thickness from a fraction of a meter to several meters. Some mylonites, however, are much thicker bodies that apparently define wide shear zones. All transitional stages, from original country rock through mylonitic gneisses to ultramylonites, may be present in such a zone.

Where exposed, fault planes commonly are smooth, polished surfaces called **slickensides,**[3] which form in response to shearing on the fault planes or in the fault gouge. Fault surfaces, including slickensides, typically contain strongly oriented linear features, known as **slickenlines, slickenside lineations,** or **striations,** that are parallel to the direction of slip. These lineations are of three types: ridges and grooves, mineral streaks, and mineral fibers or **slickenfibers.** Ridges and grooves may result from scratching and gouging of the fault surface (Figure 4.8*A*), from the accumulation of

gouge behind hard protrusions or **asperities,** from the development of irregularities in the fracture surface itself forming **ridge-in-groove lineations** or **fault mullions** (Figure 4.8*B*), or from growth of slickenfibers (Figure 4.8*C*). Slickenfibers are long, single-crystal mineral fibers that grow parallel to the direction of fault displacement. They fill gaps that develop along the fault during gradual shearing (see Figure 4.15 and Sections 4.3, 13.8, and 14.6). Mineral streaks are streaks on slickensides that develop as a result of the pulverization and shearing out of mineral grains within the gouge.

Faults that develop at relatively shallow depths are **dilatant**—that is, they develop open spaces that increase the rock's volume. Such fault zones provide pathways for the flow of ground water and hydrothermal fluids. As a result, many fault zones contain secondary deposits of minerals, including calcite (see Figure 4.5*A*) and silica (quartz, opal, or chalcedony) as vein deposits or as cement for the preexisting fault gouge or breccia. Many ore deposits form by precipitation of ore minerals from hydrothermal fluids flowing along fault zones.

Effects of Faulting on Geologic or Stratigraphic Units

Displacement along faults generally places adjacent to one another rocks that do not belong together in ordinary geologic sequences. The resulting discontinuity provides some of the best evidence for the presence of the fault.

A break in an otherwise continuous geologic feature, such as sedimentary bedding, may indicate the presence of a fault. A stratigraphic discontinuity, however, may also result from an unconformity or an intrusive contact, and it is important to distinguish such features from faults. Characteristic features of unconformities include fossil soil horizons, erosional channels, basal conglomerates, depositional contacts, and the parallelism or near parallelism of the strata that overlie the unconformity (see Figure 2.11 and the discussion in Section 2.4). Distinctive features of intrusive contacts (Figure 4.9) include metamorphism in the adjacent country rocks, fragments of country rock suspended in the intrusion (**xenoliths**), and dikes or veins of igneous rock cutting the country rock adjacent to the intrusion.

The presence of **horses,** or **fault slices,** along a discontinuity is clear evidence of a fault. Horses are volumes of rock that are bounded on all sides by faults (see Figure 4.24). They are sliced from either the footwall or the hanging wall block by a branch of the fault and are displaced a significant distance from their original position. Thus they may appear markedly out of place stratigraphically. If the local stratigraphy is known, identification of the original stratigraphic position of

---

[2] We discuss in detail the structures and processes of ductile deformation in Parts III and IV of this book. We include here a brief description of some of these features because they characterize many fault zones.

[3] Confusion exists about the exact meaning of this term. Many authors use *slickenside* to refer to the lineations that occur on the fault surfaces. This usage is not consistent with the original definition, however.

A.

B.

C.

Figure 4.8 (*Left*) Lineations on fault surfaces formed during fault slip. *A.* Lineations formed by scratching and gouging of the fault surface. *B.* Ridge-in-groove lineations, or fault mullions. *C.* Serpentine slickenfiber lineations.

the rocks in a horse provides a constraint on the direction and amount of movement. In areas where horses separate two similar rock types, a horse of a different lithology may be the only observable evidence of a fault.

Repetition of strata or **omission of strata** in a known stratigraphic sequence is another possible indication of a fault. This criterion is especially important in subsurface geology, where often only drill-hole data are available. Figure 4.10 shows a diagrammatic cross section of a region of horizontal bedding with drill-hole information indicating the repetition (Figure 4.10*A*) and omission (Figure 4.10*B*) of strata. Such data are referred to as showing **repeated section** and **missing section,** respectively. If enough information is present, it is possible to map a fault in the subsurface solely on the basis of information obtained by drilling.

As in the truncation of structures, it is important to make sure that the omission of strata does not result from an unconformity and that the repetition of strata does not result from a facies change associated with alternating transgressions and regressions. The distinction between faults and facies changes can be subtle, and failure to distinguish them correctly has resulted in some spectacular geologic errors.

Bedding surfaces near faults may curve in the direction of motion of the opposite fault block. These folds are called **drag folds.** They are most likely to develop where the traces of the sedimentary layers on

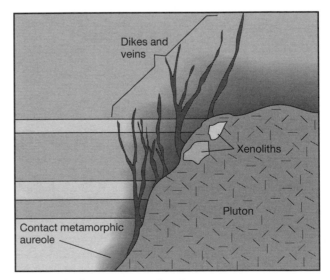

Figure 4.9 Characteristics of a plutonic intrusive contact in sedimentary rocks.

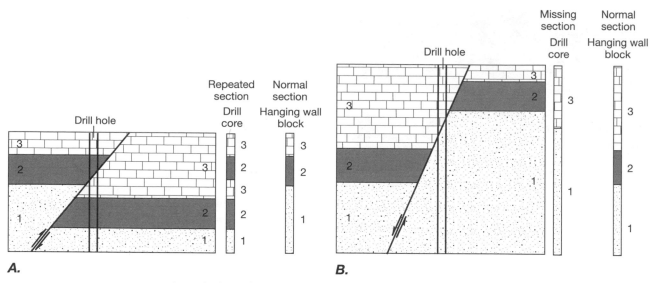

Figure 4.10 *A.* Thrust fault resulting in repeated section in a vertical drill hole. *B.* Normal fault resulting in missing section in a vertical drill hole.

the fault plane—that is, the cutoff lines—are at a high angle to the slip direction on the fault (Figure 4.11*A*, *B*). Drag folds are less likely to form if the cutoff lines are nearly parallel to the slip direction (Figure 4.11*C*).

Drag folds are especially well developed along many thrust faults (Figure 4.11*A*). Along normal faults,

rollover anticlines, which develop in the hanging wall block, are more common (Figure 4.11*D*). The direction of bending in these folds is opposite to that found in drag folds, and they reflect the deformation necessary to accommodate the hanging wall block to a curved fault surface (see Sections 5.1 and 5.4).

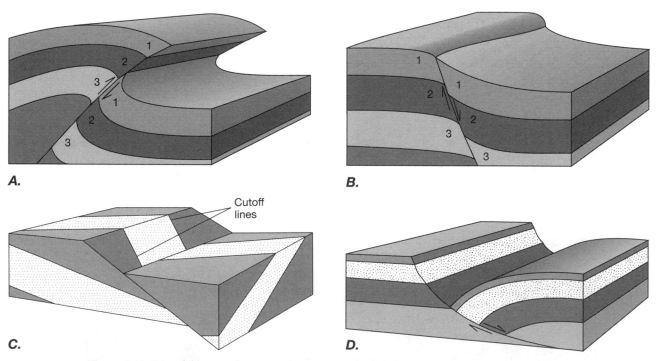

Figure 4.11 Drag folds in sedimentary layers along faults. *A.* A thrust fault, and *B.* a normal fault. *C.* If the cutoff line of the bedding makes a small angle with the displacement direction on the fault surface, the formation of drag folds is less likely. *D.* Rollover anticline in the hanging wall of a normal fault.

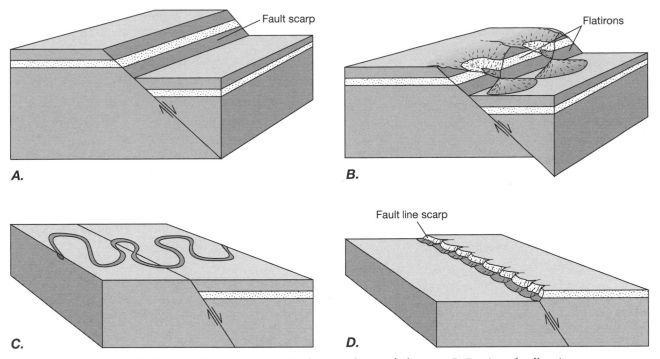

Figure 4.12 Erosion of fault scarps. *A.* Faulting produces a fault scarp. *B.* Erosion of valleys in the fault scarp produces flatirons. *C.* Erosion wears away the thin resistant layer in the topographically high footwall block and levels the topography. *D.* Erosion reaches the level of the resistant layer in the hanging wall block. More rapid erosion in the less resistant layers in the footwall block leaves a topographic step, a fault line scarp.

Physiographic Criteria for Faulting

Many active and inactive faults have pronounced effects on topography, stream channels, and ground water flow. Because these effects frequently suggest the existence of a fault, they are useful in geologic mapping.

Scarps are linear features characterized by sharp increases in the topographic slope; they suggest the presence of faults. There are three types of fault-related scarps: **Fault scarps** are continuous linear breaks in slope that result directly from displacement of topography by a fault (Figure 4.12A). **Fault line scarps** are erosional features that are characteristic of both active and inactive faults. Figure 4.12 illustrates three steps in the progressive erosion at a fault. Initially, the upthrown footwall block (Figure 4.12A) forms a fault scarp. Erosion carves valleys in the fault scarp, leading to formation of **flatirons** along the mountain front (Figure 4.12B). Eventually the upthrown block is eroded down to the same level as the downthrown hanging wall block (Figure 4.12C). Subsequent erosion exposes the thin layer in the hanging wall block, which is more resistant than the surrounding layers. Further erosion occurs most rapidly in the least resistant rocks (Figure 4.12D), leaving a scarp that, as in this case, does not necessarily indicate the sense of displacement on the fault.

Fault benches are linear topographic features characterized by an anomalous decrease in slope. They form where a fault displaces an originally smooth slope so that a strip of shallower slope results, or where erosion of the less resistant rocks in a fault zone produces a shallower slope than is supported by the surrounding, more resistant rocks. Fault benches may be associated with any of the different fault types.

Ridges, valleys, or streams may be offset along a fault. Figure 4.13 shows two stream channels that have been offset by strike-slip motion on a fault. The deflection of the stream channels may indicate the sense of slip on the fault, but if the fault displacement is sufficiently large, original stream channels may be abandoned. In this case, the "dog leg" in the channel may not correspond to the sense of displacement.

A fault surface or fault zone may act as either a conduit or a barrier for ground water, depending on the permeability of the material both in the fault and on either side of the fault. A breccia zone forms an excellent conduit for water, but a thick gouge zone containing abundant clay minerals may act as a barrier to flow. If faulting offsets an aquifer, or juxtaposes an impermeable rock, such as a plutonic or metamorphic rock, against a good aquifer, it may also significantly alter the flow of ground water. Thus fault traces are

Figure 4.13 Photograph of San Andreas fault, central California, showing streams offset along the trace of the fault.

cribed to the change in erosional resistance of the bedrock may betray a fault, but further evidence should be sought.

Determination of Fault Displacement

Complete determination of the total displacement on a fault requires knowledge of the **magnitude** and **direction** of its displacement. Some features indicate the total displacement; others permit a partial or approximate determination; still others only place a constraint on the possible displacements.

Complete Determination of Displacement

The complete determination of displacement on a fault requires identification of a particular preexisting linear feature that intersects the fault surface and is displaced by it. The **piercing points** of a linear feature are the points where it intersects the fault surface (Figure 4.14). The vector connecting the two piercing points uniquely determines the direction and magnitude of fault displacement, and the relative positions of the linear feature

often characterized by springs and by water-filled depressions called sag ponds.

A stream tends to form a consistent equilibrium profile characterized by a slope that gradually steepens toward the headwaters. Such a profile can change because of fault movements or because of the variable erodability of the bedrock. Any sharp changes in the profile or valley shape of a stream that cannot be as-

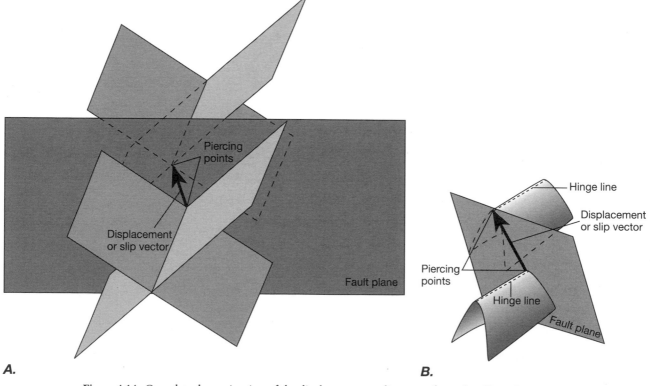

A.

B.

Figure 4.14 Complete determination of the displacement or slip vector from the offset of a unique linear feature cut by a fault. A. The intersection of two planar features. B. The hinge line of a fold.

on opposite sides of the fault give the sense of shear of the fault.

Several linear geologic features provide piercing points on fault surfaces. The intersection of two distinct planes always defines a unique line that, when cut by a fault, can be used to determine the fault displacement (Figure 4.14A). Examples include two other faults; two differently oriented veins or dikes; a bedding plane and a fault, vein, or dike; and an unconformity and a geologic contact such as a bedding plane or intrusive contact. The point of maximum curvature, or hinge line, of a fold also provides a unique line by which to determine the displacement (Figure 4.14B). Buried river channels and linear sandstone bodies (shoestring sands) are linear stratigraphic features that can be used, and cylindrical bodies such as volcanic necks and some ore deposits can serve in the same manner.

Partial Determination of Displacement from Small-Scale Structures

In many cases where a fault or ductile shear zone is identified, it is possible to determine the orientation of the displacement vector and the sense of shear, but not the magnitude of the displacement. This type of information can be obtained by examining features at the microscopic to hand sample scale.

As we have noted, slickenside lineations form parallel to the direction of displacement on a fault (see Figure 4.8), but the magnitude of the displacement is more difficult to obtain. Ridge-in-groove lineations that form during propagation of the fracture may be longer than the displacement vector on the fault. Mineral streaks may result from comminution and smearing out of mineral grains and may give a minimum estimate of the displacement magnitude, although this correlation has never been proved.

Slickenfiber lineations grow at a small angle to the shear fracture boundary such that an arrow pointing along a fiber from its point of attachment to one fault block indicates the direction of relative motion of the opposite block (Figure 4.15A). The fibers probably grow during slow aseismic movement on a fault, and the fiber growth keeps up with the displacement. Opposite ends of the same mineral fiber should therefore join points that were adjacent when the fiber started growing. Thus, for the period of time of fiber growth, the length of the fiber is a measure of the displacement magnitude on a particular fracture. Shear displacement that might have occurred before the onset of fiber growth would not, of course, be recorded. The maximum displacement magnitudes that are recorded on individual fractures by these fibers are rather small—on the order of 10 to 20 cm. Fibers much longer than this are either not formed or not preserved. In principle, the minimum total

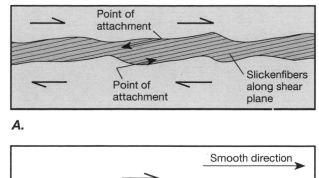

A.

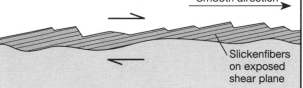

B.

Figure 4.15 Slickenfibers as indicators of shear sense and minimum displacement. A. An arrow along a slickenfiber with its base at the point of attachment to one wall of the fault points in the direction of relative slip of the opposite wall of the fracture. The length of the fiber from one wall to the other is a measure of the minimum displacement on that fault. B. The smooth, or "downstairs," direction on the stepped surface of an exposed set of slickenfibers defines the direction of relative slip of the missing block.

displacement across a fault zone should be the sum of the fiber lengths on all shear fractures in a cross section of the zone, but where displacements on the order of meters or more are involved, this is an impractical measurement to make.

Generally, slickenfibers are found on fault surfaces that have been exposed by erosion of one of the fault blocks. Because the crystals grow at a low angle to the fault surface and tend to break off either along the fibers or at a high angle to them, such a fault surface tends to have a stepped texture (Figure 4.15B). The surface feels smoothest to the hand when rubbed in the direction of relative motion of the missing block.

During brittle faulting, minor secondary fractures can develop along the fault surface at low to moderate angles to the fault. These secondary fractures may be either extension or shear fractures. In general, the extension fractures are not striated and may be filled with a secondary mineral; secondary shear fractures are striated.

Secondary fractures provide four criteria that are useful for determining the sense of shear on the fault surface.

1. As viewed on an exposed fault surface, extension fractures cut below the fault surface in the direction of movement of the missing fault block (Figure 4.16A). These extension fractures are essentially the

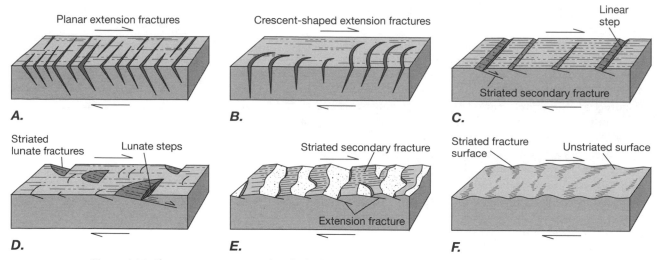

Figure 4.16 Shear sense criteria on brittle faults. Block diagrams show the relationship between secondary fractures and the sense of shear on a brittle fault. The top plane is the shear plane; relative motion is indicated by arrows. Extension fractures are unstriated and may be filled with secondary minerals. Striated fracture surfaces are shear fractures.

same as the pinnate fractures shown in a section normal to the fault in Figure 3.7 (Section 3.1). In cross section, sets of these fractures may form *en echelon* arrays of gash fractures.

2. If the extension fractures are crescent-shaped as exposed on the fault surface, they are concave in the direction of motion of the missing fault block (Figure 4.16*B*).

3. If striated secondary fractures extend beneath the main fault plane, then the fractures cut down into the surface in the direction of motion of the missing fault block (Figure 4.16*C*). Fracturing of the acute wedge of rock between the secondary shear and the fault surface produces steps in the surface that face in a direction opposite to the motion of the missing fault block. The steps may be predominantly linear (Figure 4.16*C*), or they may have a lunate morphology (Figure 4.16*D*).

4. Some striated secondary shears do not cut below the fault surface. They may alternate with unstriated secondary extension fractures that cut below the fault surface (Figure 4.16*E*), or they may be simply the faces of irregularities in the fault surface (Figure 4.16*F*). In these cases, the striated surfaces face opposite to the direction of motion of the missing fault block, and steps formed by breaking across the acute angle between the secondary shear and extension fractures also face in that direction (Figure 4.16*E*).

Because of the step faces formed in association with these secondary fractures, the fault surface generally feels smoothest to the hand when rubbed in a direction *opposite* to the sense of motion of the missing block. This criterion is opposite to that for determining shear sense on a fault surface dominated by slickenfibers. Thus one must be careful in using the smoothest direction on a fault surface to determine shear sense.

Ductile shear zones may contain a number of small-scale structures that indicate the shear sense. Platy minerals may become aligned to form a foliation (labeled S in the figure) that makes an angle of about 45° to the shear zone at its boundary and becomes roughly parallel to the zone at its center (Figure 4.17*A*). The sigmoidal pattern of the foliation defines the sense of shear as indicated in the figure.

Ductile faults also characteristically exhibit extended tube-shaped folds in layering that are called **sheath folds** (Figure 4.17*B*). The long dimension of these folds is approximately parallel to the direction of slip on the ductile fault.

Many rocks in ductile shear zones contain large crystals. Some are relict crystals, or **porphyroclasts** (after the Latin word *porphyry*, which means purple, and the Greek word *klastos* which means broken), that survived the shearing and reduction in grain size from the original rock. Others are **porphyroblasts** (after *porphyry* and the Greek word *blastos* which means growth), which are mineral grains that grow to a relatively large size in a rock during metamorphism and deformation. (The use of *porphyry* for a rock with large crystals comes from the fact that statues of Roman emperors were carved from purple volcanic rock containing large feldspar phenocrysts.)

Porphyroclasts found in mylonites may have asymmetric "tails" of very fine grains that are recrystallized from the edges of the porphyroclast itself. The sense of asymmetry of the tails defines the sense of shear in the deformed rock. We distinguish two different tail mor-

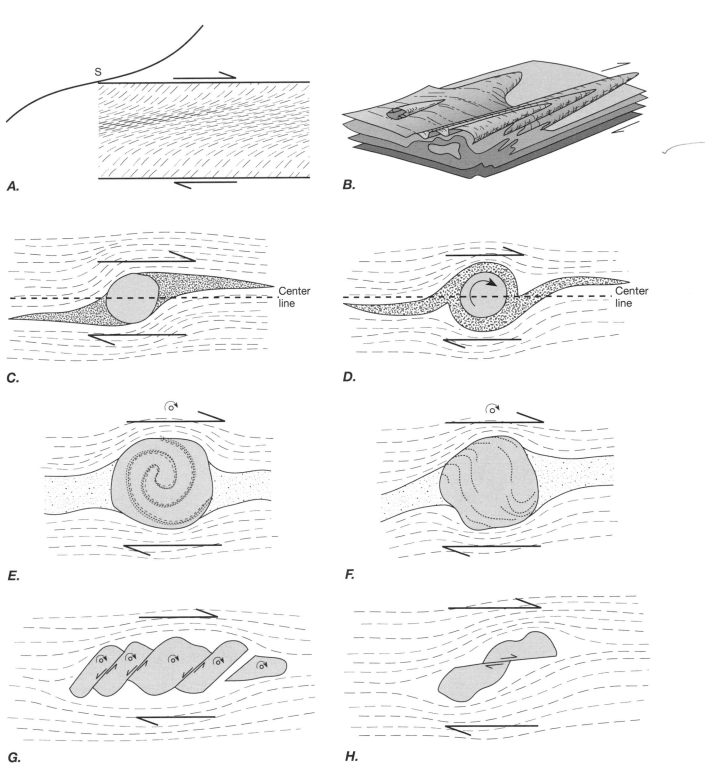

Figure 4.17 Shear sense criteria in ductile shear zones. *A.* Sense of curvature of a foliation defined by the parallel alignment of platy mineral grains. *B.* The orientation and asymmetry of sheath folds shown with the surrounding layers of rock removed. *C.* The sense of asymmetry of σ-type "tails" of recrystallized porphyroclastic material relative to the porphyroclast. *D.* δ-type "tails." *E.* The sense of rotation of a helical train of mineral grains accumulated within a porphyroblastic crystal grain such as garnet or staurolite. *F.* An inclusion train that indicates only a small amount of rotation is not, by itself, adequate evidence of shear sense. *G* and *H.* The sense of shear on fractures in mineral grains such as feldspar or mica depends on the shear sense of the fault and the angle between the fracture and the shear plane.

phologies, the σ-type and the δ-type. On σ-porphyro-clasts, the tails extend from each side of the grain in the "downstream" direction of the relative shear in the matrix, and the tails do not cross the line parallel to the foliation through the center of the grain (Figure 4.17C). The δ-type is derived from the σ-type by rotation of the porphyroclast in a sense consistent with the shear, and the tails do cross the center line (Figure 4.17D). Similarly shaped asymmetric tails may also develop from the crystallization of a mineral different from the porphyroclast in asymmetric zones called **pressure shadows**. These tails, however, are difficult to interpret unambiguously and should not be confused with the tails composed of recrystallized porphyroclast mineral.

Porphyroblasts do not deform with the rest of the rock but rotate as rigid grains during ductile deformation of the matrix. Minerals that form porphyroblasts include garnet and staurolite. As they grow during deformation, they enclose adjacent minerals, such as micas, from the matrix. Continued rotation and growth of a porphyroblast result in a helical train of inclusions that defines the sense of rotation of the grain and thereby the sense of shear in the rocks (Figure 4.17E, F). Interpretation of shear sense from inclusion trains that indicate only small amounts of rotation, however, is unreliable because such inclusion trains may actually preserve crenulations in the original foliation rather than a record of porphyroblast rotation (Figure 4.17F).

Some porphyroclastic minerals, such as mica and feldspar, tend to shear on discrete fractures or crystallographic planes to accommodate ductile deformation in the surrounding matrix. If the fractures initially make a high angle with the shear plane, then the shear sense on these planes is opposite to that in the surrounding matrix (Figure 4.17G). If, on the other hand, the fractures make a relatively low angle with the shear plane, the shear sense on the fractures is the same as it is in the matrix (Figures 4.17H and 4.7).

Partial Determination of Displacement from Large-Scale Structures

In regions where fault displacement measures tens to hundreds of kilometers, large-scale geologic features that have been offset can be used to determine the displacement direction and shear sense and to estimate the displacement magnitude of the fault. Such features include shorelines, sides of sedimentary basins, and the source and depositional site of distinctive sediments. Figure 4.18 shows a paleogeographic map of an Oligocene sedimentary basin in western California that has been cut by the San Andreas fault. The different map patterns distinguish marine from nonmarine deposits. Offset of the shoreline along the fault from A to B

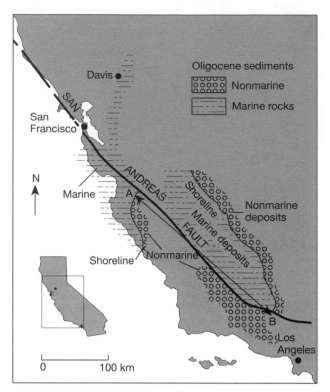

Figure 4.18 Shorelines of an Oligocene sedimentary basin offset about 300 km along the San Andreas fault, California. Points A and B were originally adjacent to each other.

suggests a right-lateral component of displacement of approximately 300 km since Oligocene time.

Isopach maps are maps showing the contours of thickness of a geologic unit. (The term *isopach* is derived from the Greek words *isos,* which means equal, and *pachos,* which means thickness.) If there is a regular variation in the thickness of a layer, fault offset of the layer (Figure 4.19A) may show up as a discontinuity on an isopach map (Figure 4.19B). Each isopach is a unique line of constant layer thickness, and matching isopachs across the discontinuity should in principle make it possible to determine the horizontal component (H) or the strike-slip component (S) of the displacement. If the data are from well logs, however, they are not very closely spaced. Thus the locations of isopachs and their intersection with the fault may be only approximate. Moreover, unless the fault strike is known, the strike-slip component of displacement cannot be determined accurately. Because isopach maps do not include elevation information, we cannot use them to determine the dip-slip and the vertical components of displacement.

Structure contour maps are maps of elevation contours on a particular geologic surface at depth, generally a stratigraphic horizon. Faults in the subsurface can be identified from discontinuities in structure contours (Figure 4.20). If the structure contours display a linear

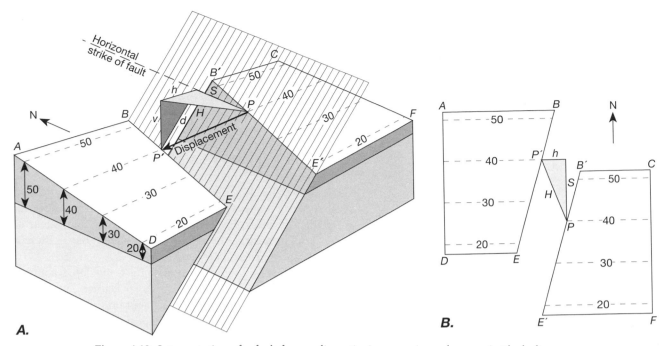

A.

B.

Figure 4.19 Interpretation of a fault from a discontinuity on an isopach map. A. Block diagram showing a bed of varying thickness cut by a fault. Contours of equal thickness (isopach lines) are drawn on the top surface of the bed. The true displacement and the strike-slip (S), dip-slip (d), horizontal (H), and vertical (v) components are shown. For ease of interpreting the three-dimensional drawing, we show a special case for which the isopachs are parallel to the strike of the layer surface and perpendicular to the strike of the fault. B. Isopach map of the structure shown in part A. The horizontal component of displacement (H) is determined by connecting the map projections of two points P and P' that mark where the same isopach on opposite sides of the fault intersects the fault surface. The strike-slip component of displacement (S) is determined by connecting the extensions of equal isopach lines with a line parallel to the strike of the fault.

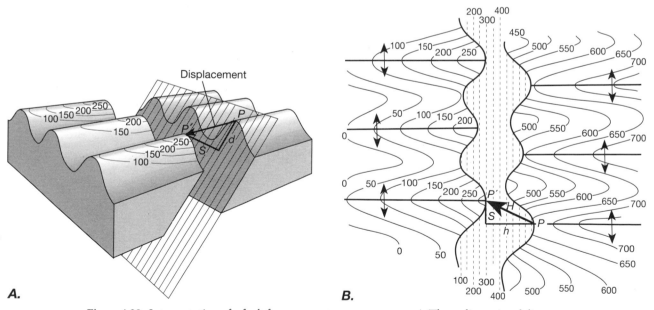

A.

B.

Figure 4.20 Interpretation of a fault from a structure contour map. A. Three-dimensional diagram showing the folded surface of a stratigraphic contact that has been cut and displaced by a fault. The contact has contours of equal elevation (structure contours) drawn on it. B. A structure contour map of the same structure shown in part A. The horizontal component of displacement (H) is determined by joining the points on the map that are the vertically projected piercing points P and P' of the fold hinge on opposite sides of the fault. The strike-slip component (S) is parallel to the fault strike, and the horizontal dip-slip component (h) is normal to the fault strike.

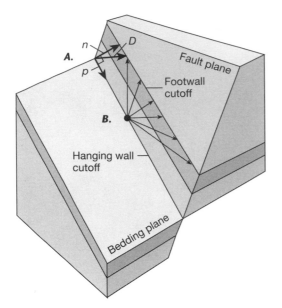

Figure 4.21 Faulted planar features are nonunique indicators of fault displacement. A. The true displacement on the fault (D) can be specified as the sum of the component normal to the cutoff line of the planar feature (n) and the component parallel to the cutoff line (p). B. Because the parallel component of displacement (p) does not produce any offset in the plane, any value of p results in the same geometry for the faulted plane, and the displacement D is therefore not uniquely defined.

structure such as a fold hinge that can be matched uniquely across the fault (Figure 4.20A), the fault displacement can be determined. The horizontal component of fault displacement (H) (Figure 4.20B) is determined by matching the map pattern of the structure contours across the fault; the vertical component is determined from the different elevations of two initially adjacent structural features. If the strike of the fault can be determined, we can use trigonometry to find 1) the strike-slip component of displacement (S) from the horizontal component (H) and 2) the actual displacement (D) from the strike-slip (S) and dip-slip (d) components (Figure 4.20A). As with isopachs, however, the determination is subject to the accuracy of the contours themselves.

Nonunique Constraints on Displacement

Frequently, the principal evidence for a fault consists of the offset of a planar structure, typically sedimentary bedding. It is very important to realize that this offset alone can never define the displacement on the fault, regardless of the appearance of the outcrop pattern. The reason for this is not difficult to understand. We express the true displacement (D) as the sum of its components in the fault plane normal (n) and parallel (p) to the

cutoff line, which is the intersection of the planar feature with the fault (Figure 4.21A). The normal component (n) is the normal distance between the matching cutoff lines on the hanging wall and the footwall, and this component can be measured easily. The component parallel to the cutoff line (p), however, produces no change in the orientations or locations of the cutoff lines and thus cannot be observed. The complete description of the displacement (D) is therefore indeterminant. Figure 4.21B shows six of the infinite number of displacement vectors that could produce the same geometry for the faulted bedding plane. Included among these six vectors are components of normal, reverse, dextral, and sinistral displacement. Each vector has the same component of displacement normal to the cutoff line (n) but a different component parallel to the cutoff.

Thus, if the only information available about a fault is the offset of parallel planar features, we cannot talk about the slip or displacement on the fault because we cannot determine it. We speak instead of the **separation,** which is the distance measured in a specified direction between the same planar feature on opposite sides of the fault (Figure 4.22). The separation enables us to determine only the component of displacement normal to the cutoff line. Any component parallel to that line is indeterminate. Illustrated in the figure are two of the common separations and the directions in which each is measured—the **strike separation,** measured parallel to the strike of the fault, and the **dip separation,** measured parallel to the dip of the fault. Other separations used in some cases include the **stratigraphic separation,** measured normal to a bedding plane; the **vertical separation,** measured in a vertical direction (it is the vertical component of both the dip and the stratigraphic separations); and the **horizontal separation,** measured in a plane normal to the strike of

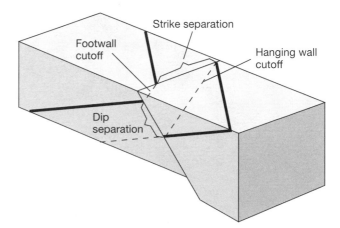

Figure 4.22 Block diagram indicating the strike separation and the dip separation of a faulted layer on a dip-slip normal fault where the footwall block has been eroded down to the same elevation as the hanging wall block.

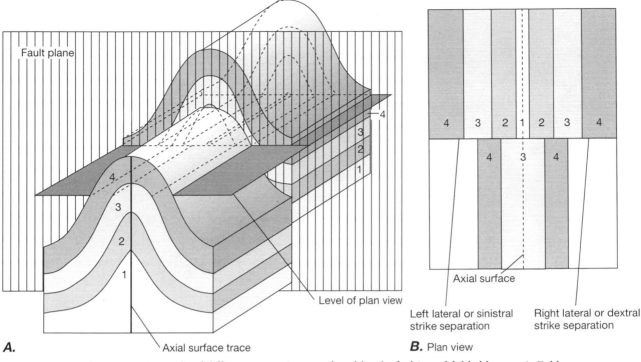

A.

Fault plane

4
3
2
1

Level of plan view

Axial surface trace

B. Plan view

Axial surface

Left lateral or sinistral
strike separation

Right lateral or dextral
strike separation

Figure 4.23 Example of different separations produced by the faulting of folded layers. *A.* Fold offset by dip-slip movement along a vertical fault. *B.* Map view at the level of the horizontal plane in part *A*, showing the opposite separations obtained on opposite sides of the fold.

the fault (it is the horizontal component of the dip separation).

On the same fault, a separation measured for one plane is different from the separation measured in the same direction for a plane of a different orientation. In fact, if the two planar features are appropriately oriented, one strike separation can be right-lateral and the other left-lateral. For example, opposite limbs of a faulted fold can be very different in separation (Figure 4.23). Similarly, opposite senses of dip separation can develop for appropriately oriented planes and displacements.

These examples emphasize the nonuniqueness of the separation and the difficulty, or impossibility, of using it to characterize a fault. If, however, the separation in a particular direction can be determined for two planes of different orientation offset by a fault, then it is possible to determine a unique magnitude and direction for the displacement vector. In effect, the intersection of the two planes defines a line that intersects the fault at unique piercing points on the hanging wall and footwall (see Figure 4.14*A*).

4.4 Faults in Three Dimensions

All faults are three-dimensional, somewhat irregular surfaces of finite extent. It is all too easy to lose track of this fact, because generally faults appear in outcrop

as lines on a relatively two-dimensional surface, and they are represented on the printed page as lines on a map (Figure 4.24*A*) or on a cross section (Figure 4.24*B*). Although such representations are certainly useful, they encourage us to ignore the three-dimensional consequences of fault geometry (Figure 4.24*C*).

An individual fault surface generally is not a flat or smoothly curved surface but instead may have **fault ramps** connecting segments of the fault (Figure 4.25*A*). A **frontal ramp** is oriented such that its intersection with the main fault surface is approximately perpendicular to the displacement direction on the fault. On strike-slip faults, frontal ramps are called **jogs** or **bends.** A **lateral ramp** is oriented such that its intersection with the main fault surface is oblique or parallel to the displacement direction on the fault. If parallel, it is also a step or **sidewall ramp;** otherwise, it is an **oblique ramp.**

In general, displacement of a fault block over a ramp induces in the block a deformation whose characteristics depend on the orientation of the fault and the displacement. A frontal ramp or jog can be extensional (Figure 4.25*B*) or contractional (Figure 4.25*C*), depending on whether the material is pulled apart across the ramp or pushed together by the dominant shear on the fault zone.

During faulting at a ramp, the location of the ramp can migrate as the fault surface cuts in discrete jumps into one fault block or the other. The result is a **fault duplex,** characterized by a stack of horses bounded by

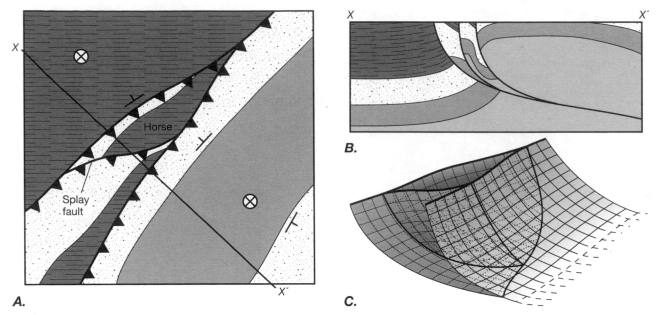

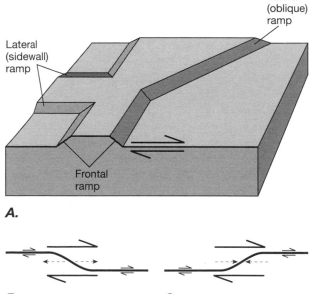

Figure 4.24 Three-dimensional representation of faults. *A.* Geologic map of branching imbricate thrust faults connected by a subsidiary splay fault which isolates a horse. *B.* Cross section along XX′ through the fault system in part *A*. *C.* Portrayal of the three-dimensional geometry of the faults in parts *A* and *B*.

Figure 4.25 Geometry of fault ramps. Any of the diagrams can be oriented arbitrarily and therefore can apply to any fault type. *A.* Schematic shape of a fault surface. *B.* An extensional frontal ramp or jog. The dashed arrows indicate that material tends to be pulled apart at the ramp. *C.* A contractional frontal ramp or jog. The dashed arrows indicate that material tends to be pushed together at the ramp.

traces of the main fault. Duplexes can be normal (Figure 4.26*B*), thrust (Figure 4.26*A*), or strike-slip (Figure 4.26*C*). We discuss the characteristics of duplex formation that are specific to the different fault types in Chapters 5 through 7.

Every fault surface, no matter what its type, must come to an end in every direction, and the end is marked by a **termination line**. A termination line must be continuous and must form a closed line about the fault surface; it cannot simply end. It has different features, depending on the geometry of the termination.

The termination of a fault at the surface of the Earth is the **fault trace** on the topographic surface (Figures 4.27 through 4.29). It may be the original boundary of the fault or the intersection of an originally deeper part of the fault with a surface of erosion. It is in essence the cutoff line of the Earth's surface on the fault.

At a brittle–fluid or brittle–ductile interface, the displacement discontinuity on a fault is easily accommodated by the flow of the fluid or ductile material. Thus the discontinuity cannot extend beyond the brittle material, and the cutoff line of the interface defines the termination of the fault.

Termination of one fault against another results in two relationships between the termination line and the displacement on the younger fault. If a younger fault terminates against an older one, its displacement vector

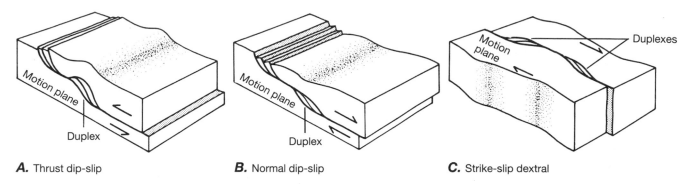

A. Thrust dip-slip   **B.** Normal dip-slip   **C.** Strike-slip dextral

Figure 4.26 Block diagram illustrating duplexes. The motion plane contains the slip direction and the line perpendicular to the fault.

must be parallel to the termination line. In Figure 4.27A, a vertical strike-slip fault terminates on a coeval horizontal fault. The termination line must be parallel to the displacement vectors on the strike-slip fault. Similarly, in Figure 4.27B, a normal fault terminates against an older vertical fault. The vertical fault initially was a strike-slip fault between block C and the unfractured pair of blocks A and B. Subsequently, however, part of that fault was reactivated as an oblique-slip fault between blocks B and C. In this case the dip-slip displacement on the younger fault must be parallel to the termination line of that fault.

If, on the other hand, a younger fault has cut and displaced an older fault, the termination line of the older fault is its cutoff line on the younger fault, and the displacement vector is not related to the termination line. For example, in Figure 4.27C, a right-lateral strike-slip fault cuts an older normal fault. The termination line of the normal fault against the strike-slip fault is not parallel to the displacement vectors on either fault.

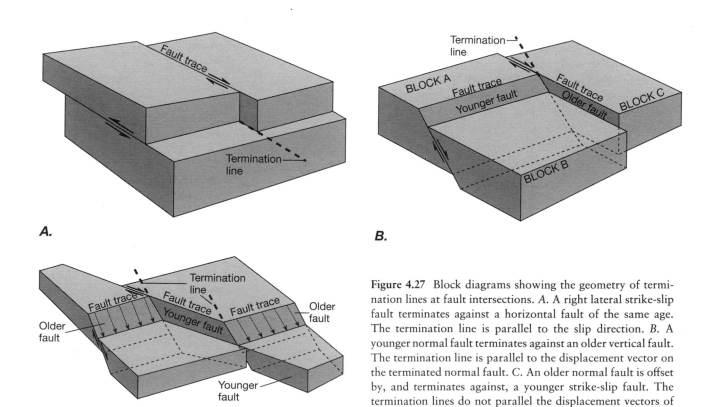

Figure 4.27 Block diagrams showing the geometry of termination lines at fault intersections. *A.* A right lateral strike-slip fault terminates against a horizontal fault of the same age. The termination line is parallel to the slip direction. *B.* A younger normal fault terminates against an older vertical fault. The termination line is parallel to the displacement vector on the terminated normal fault. *C.* An older normal fault is offset by, and terminates against, a younger strike-slip fault. The termination lines do not parallel the displacement vectors of either fault.

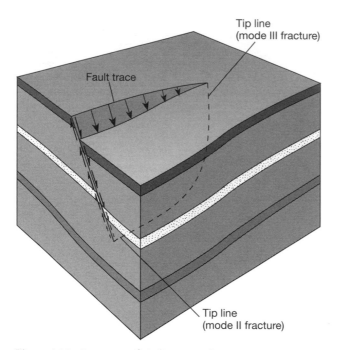

Figure 4.28 Geometry of tip lines. Displacement on a normal fault dies out along the strike and down the dip. The tip line is a continuous quasi-elliptical line. The angle it makes with the displacement vector changes around the perimeter of the fault, changing the fault from a mode III fracture (tip line parallel to displacement) to a mode II fracture (tip line perpendicular to displacement).

For very large strike-slip displacements, the observed fault intersections might resemble those in Figure 4.27B, which could result in an erroneous interpretation of the age of the faults.

A look at any geologic map of a faulted terrane demonstrates that the traces of individual faults are of limited extent. The termination line of a fault is a **tip line** where the fault displacement has decreased to the extent that it can be accommodated by coherent deformation distributed through the solid rock (Figures 4.28 and 4.29). If a fault trace ends without running into another fault, it must end at a tip line. In Figure 4.28 the tip line is parallel to the displacement vector on the fault where it intersects the top surface, and it is perpendicular to the displacement where it intersects the vertical side. Because tip lines of faults are often roughly elliptical in shape at depth, all relative orientations of displacement and tip line between these two end-member types occur on a single fault. Thus, a single fault has the characteristics of both mode II and mode III fractures (compare Figure 3.1) depending on the relative orientation of the displacement and the tip line.

Below the surface, then, a fault can be bounded on all sides by a continuous tip line that connects at both ends with the surface trace of the fault. A **blind** fault, however, does not break the surface anywhere and thus is completely surrounded by a termination line that is either a tip line or a branch line (see below). Later, possibly long after active faulting ceases, the tip line at the top of the fault surface may be eroded away to create a trace of the fault on the Earth's surface.

A **branch line** is a line of intersection where a fault surface splits into two fault surfaces of the same type, or two fault surfaces of the same type merge into one. All segments of the fault shown in Figure 4.29A are completely surrounded by a fault trace at the surface, a tip line, or a branch line, except, of course, for the trace of the fault on the vertical left side of the block, which is an artificial line on the cross section. In Figure

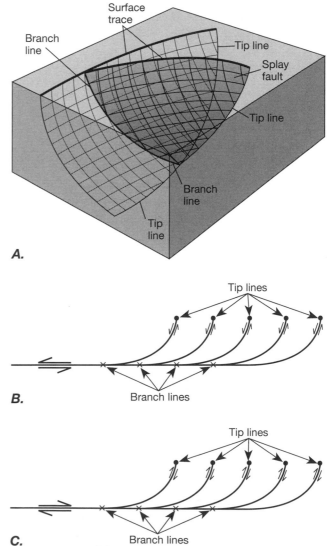

Figure 4.29 Splay faults and the geometry of branch lines. A. The three-dimensional geometry of a splay fault shows how the fault surface is completely bounded by a surface trace, a branch line, or a tip line. B. Extensional imbricate fans. C. Contractional imbricate fans.

4.24C, the horse is bounded on all edges below the surface by branch lines.

Faults of all types commonly die out in a set of **splay faults,** which are smaller, subsidiary faults that branch off from the main fault (Figure 4.29A). Where splay faults branch off from the main fault at fairly regular intervals and have comparable geometries, they form an **imbricate fan,** which can be either extensional (Figure 4.29B) or contractional (Figure 4.29C). Each splay in turn dies out at its own tip line or intersects a free surface. Where a single fault ends in a series of splay faults, the effect is to distribute the deformation at the end of the main fault over a larger volume of rock, thereby decreasing its intensity.

Two circumstances occur in which a fault can end, in a sense, without being bounded by a termination line. If the fault surface curves, the nature of the fault may change completely, but no termination line exists. Figure 4.30 illustrates a normal fault that changes orientation to become a vertical oblique-slip fault.

The other circumstance involves faults that extend deep into the crust or even into the upper mantle. With increasing depth in the Earth, the temperature and pressure rise, and if they are sufficiently high, rocks become ductile. This ability of rock to flow ultimately limits the depth to which a fault can maintain its identity as a shear zone. As we trace a fault deeper into the crust, we expect it first to change from a zone of brittle deformation into a ductile shear zone. At some depth, which we do not know well, the zone of deformation must spread out, until the nature of the fault as a discrete shear zone is lost, and the displacement is accommodated by the slow, widespread flow of the rocks. In this circumstance, the boundary of the fault is indistinct.

The depth at which a fault loses its identity is not well known, but it probably depends in part on the magnitude and rate of the displacement on the fault. Brittle fracturing is replaced by ductile flow at depths of about 10 to 20 km. Fault zones have been traced by seismic reflection techniques into the lower crust to depths of about 25 km, and in some places, slices of the upper mantle are exposed at the surface along faults, suggesting that faults have extended at least to the Moho. Subduction zones, in fact, are major thrust faults some of which can be traced hundreds of kilometers into the mantle, although we do not expect most crustal faults to extend to such depths. We have limited opportunity to observe rocks that have deformed near and below the base of the crust, however, so our knowledge is very poor.

4.5 Refinements

All our discussion about fault movement has concerned movement of one side *relative* to the other. An absolute sense of movement can rarely be obtained, because we generally do not have a reference point independent of both of the fault blocks. Thus, for example, it generally cannot be determined whether the hanging wall of a normal fault has decreased in elevation as a result of faulting, or the footwall has increased in elevation.

Faulting associated with the Alaska earthquake of 1964 near Anchorage provided one interesting exception to this statement. Measurement of motion of fault blocks relative to sea level showed that in one place where a normal fault had formed, *both* sides had moved *upward*. Figure 4.31A shows part of the area affected by the fault and gives contours of the amount of uplift that occurred during faulting. Note that the uplift contours show both sides of the Patton Bay Fault to have moved up but that the northwest side moved up farther than the southeast side. This effect is more clearly shown in Figure 4.31B, which is a plot of the amount of uplift along the line AA' that crosses the fault. This interesting result on a historical earthquake makes one cautious about assuming that the relative motion of the hanging wall is also the absolute motion.

Another exception to our general inability to determine absolute movements on faults occurs with strike-slip faults of very large displacement. In this case, we can take the geomagnetic field as an absolute reference and use paleomagnetic studies of the rocks to determine which side of a fault has changed latitude.

One of the major features used for classifying faults and interpreting their tectonic significance is the original orientation of both the fault plane and the displacement vector at the time the fault moved. The original orientation is important, because many faults are rotated into different attitudes by later tectonic activity. Thus,

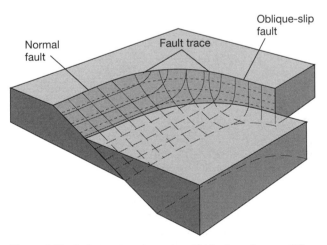

Figure 4.30 A change in orientation of a fault surface modifies a normal fault into a vertical oblique-slip fault. No termination line can be identified.

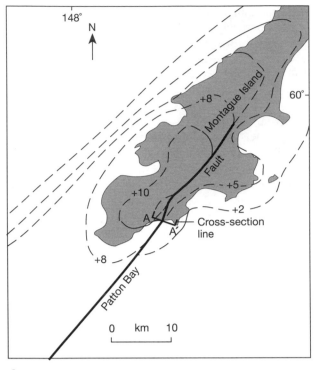

A.

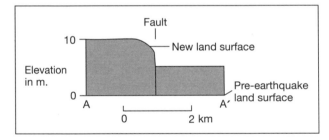

B.

Figure 4.31 Faulting of Montague Island, southern Alaska, associated with the 1964 Alaska earthquake. The map shows that the uplift contours are positive on both sides of the Patton Bay Fault. Contours are marked in meters of uplift. The cross section shows a plot of the uplift magnitudes along the line $A–A'$ that crosses the Patton Bay Fault. The fault is a normal fault resulting from uplift by different amounts on opposite sides of the fault.

for example, a thrust fault with the hanging wall up could be rotated about a horizontal axis to such an extent that it resembled a normal fault with the hanging wall down. The present attitude may or may not correspond to the original one.

Cross sections in sedimentary sequences that have not undergone significant deformation parallel to the regional strike must preserve the original cross-sectional area of the undeformed section. Such cross sections are **balanced**. The requirement that cross-sectional area be

preserved places significant constraints on possible interpretations of structure at depth, because any proposed interpretation must make it possible to restore the fault to an acceptable undeformed state characterized by continuous layers with no gaps or overlaps in the stratigraphic section. The cross section must be balanced between two **pinning points,** which are vertical reference lines chosen to pass through undeformed sections of a stratigraphic sequence. Thus the shape of these reference lines is assumed to have been unchanged by the deformation. In highly deformed regions, the pinning points can be chosen through the bottoms of flat synclines or the tops of flat anticlines, where presumably there has been no slip between beds.

If the deformation has not changed the thicknesses of units in the stratigraphic section, then balancing the section can be achieved by line balancing, which requires that the length of each contact between the pinning points be the same before and after deformation. If deformation has shortened and thickened units within the plane of the section, however, the lengths of contact lines can change, and balancing must be done by area balancing which requires that the area of each unit between pinning points be conserved during deformation.

Although the balancing requirement imposes an important constraint on the construction of cross sections of unmetamorphosed sedimentary rocks, in other geologic situations there may be significant deformation normal to any possible cross section. Under those conditions, material moves in or out of the cross-section plane, and the fundamental assumption of constant cross-sectional area, which is the basis for constructing balanced cross sections, is no longer valid. Some techniques have been developed to account roughly for this deformation, but the process requires large amounts of detailed information, and balancing cross sections under these conditions becomes an increasingly complicated and approximate procedure.

In this chapter, we have described many features common to faults in general, but we have limited our discussion to individual fault surfaces or to simple sets of faults. In the following chapters, we complete our description of faults, dealing separately with each of the major types of faults. We concentrate on those features that are unique to each fault type, including the structures associated with the different faults, the geometry of complicated fault systems, and the tectonic settings in which the different types of faults are found. Despite the differences in tectonic setting of the three major types of faults, the geometric characteristics discussed in this chapter are remarkably consistent for all fault systems, and the differences that do exist are largely attributable to the difference in orientation of the faults with respect to the Earth's surface.

Additional Readings

Fleuty, M. J. 1975. Slickensides and slickenlines. *Geol. Mag.* 112: 319–322.

Petit, J. P. 1987. Criteria for the sense of movement on fault surfaces in brittle rocks. *J. Struct. Geol.* 9: 597–608.

Sibson, R. H. 1977. Fault rocks and fault mechanisms. *J. Geol. Soc. Lond.* 133: 190–213.

Sibson, R. H. 1983. Continental fault structure and the shallow earthquake source. *J. Geol. Soc. Lond.* 140: 741–767.

Simpson, C. 1986. Determination of movement sense in mylonites. *J. Geol. Educ.* 34: 246–261.

Wise, D. U., D. E. Dunn, J. T. Engelder, P. A. Geiser, R. D. Hatcher, S. A. Kish, A. L. Odom, and S. Schamel. 1984. Fault-related rocks: Suggestions for terminology. *Geology* 12: 391–394.

Woodcock, N. J., and M. Fischer. 1986. Strike-slip duplexes. *J. Struct. Geol.* 8: 725–735.

Woodward, N. B., S. E. Boyer, and J. Suppe. 1989. Balanced geological cross-sections: An essential technique in geological research and exploration. Short Course in Geology: Volume 6. Washington, D.C.: American Geophysical Union.

CHAPTER

Normal Faults

Normal faults[1] (Figure 5.1) are inclined dip-slip faults along which the hanging wall block has moved down with respect to the footwall block. Generally, they emplace younger rocks on top of older rocks, and in a vertical section through the fault, stratigraphic section is missing. Most normal faults have steep dips of about 60°, but many have lower dips, some approaching horizontal. As a result of the hanging-wall-down motion, normal faults accommodate a lengthening, or extension, of the Earth's crust.

5.1 Characteristics of Normal Faulting

Separation and Normal Faulting

Strike and dip separations produced by normal faulting depend on the relative orientations of the fault and the stratigraphic layering. As we noted in Section 4.3, separation can be quite misleading as an indication of the

nature of a fault. For example, Figure 5.2 shows a series of block diagrams of a normal fault cutting various orientations of bedding, none of which are overturned. In all the diagrams on the left, pure dip-slip motion on the fault displaces the stratigraphy and produces a scarp on the footwall block. Each diagram on the right shows the same geometry as the diagram to its left, except that the fault scarp has been eroded away, leaving a horizontal planar surface.

In Figure 5.2A, the fault cuts horizontal beds, leaving a simple stratigraphic discontinuity. In Figure 5.2B–D, the bedding is inclined at various angles to the fault, resulting in some potentially confusing separations. In Figure 5.2B, for example, the horizontal plane shows a repetition of the stratigraphy across the fault, although stratigraphy is missing on the vertical section. This example emphasizes the fact that for a normal fault, the characteristic of *missing* stratigraphic section applies only to a vertical section through the fault. If, however, on a vertical section normal to the fault, the apparent dip of the beds is in the same direction as, but steeper than, the fault, a vertical hole through the fault reveals *repeated* stratigraphy. Moreover, if the beds are not overturned, then such a fault places older beds on top of younger.

In Figure 5.2C and D, the stratigraphic pattern is characterized by strike separation on the horizontal plane and dip separation on a vertical section. In C, the

---

[1] The terms *normal* and *reverse* applied to faults stem from eighteenth- and nineteenth-century mining. The "normal" situation in the coal mines of Britain was that the coal seams in the hanging wall block moved down on the fault relative to the same seam in the footwall block. If the hanging wall block had moved up, the "reverse" of the normal situation existed.

A.

B.

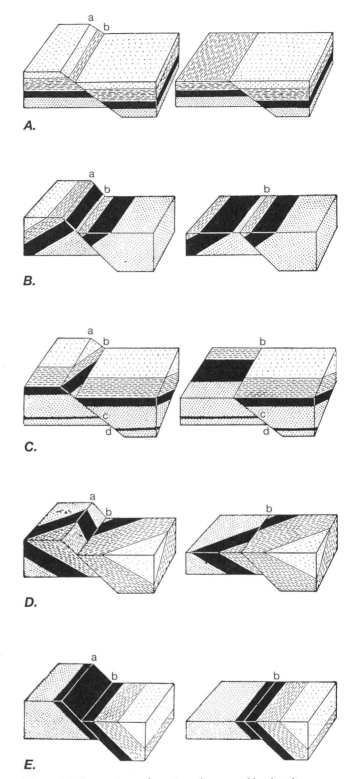

A.

B.

C.

D.

E.

Figure 5.1 Normal faults. *A.* Small-scale normal faults including examples with opposite directions of dip in beds of volcanic ash exposed in a road cut approximately 5 m high near Klamath Falls, Oregon. *B.* A normal fault bounds the east side of the Stillwater Range in Nevada, separating the rugged topography of the range from the flat valley floor. Note the sinuous trace of the fault and the flatirons, or faceted ridges, along the length of the fault.

Figure 5.2 Separations of stratigraphy created by dip-slip normal faults cutting different attitudes of bedding. In the diagrams on the left, the fault blocks have been displaced, leaving scarps on the footwall block. In the diagrams on the right, the scarp has been eroded down to a level even with the hanging wall block.

strike separation is right-lateral, and in *D* it is left-lateral, even though both faults have identical, pure dip-slip displacement.

In Figure *5.2E,* the fault and the displacement are both parallel to the bedding, leaving no separation visible on any bedding plane. In fact, for any situation, including all those shown in Figure 5.2, any component of displacement parallel to the cutoff line of the bedding on the fault surface produces no effect on the separation in any section through the fault. Thus, because of the indeterminate magnitude of this component of displacement, none of the patterns of separation in Figure 5.2 is unique to normal dip-slip motion on the fault.

A surface across which the metamorphic grade of the rocks changes abruptly from high-grade rocks below to lower-grade or unmetamorphosed rocks above may be a normal fault. The cutting out of metamorphic grades is comparable geometrically to the cutting out of stratigraphy.

Folds Associated with Normal Faults

In areas where flat-lying beds are deformed by normal faults, rollover folds in the hanging wall block are common, illustrated by the deeper strata in Figure 5.3 (compare Figure 4.11*D*). In these folds, the beds in the hanging wall block tilt down into the fault, which is opposite to the direction of tilt on drag folds (compare Figure 4.11*B*). They form on **listric normal faults,** which are concave-upward faults—that is, faults whose dip decreases with increasing depth (the Greek word *listron*

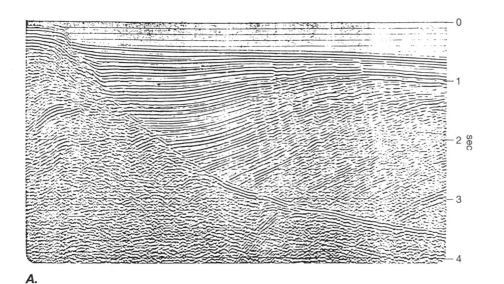

A.

B.

Figure 5.3 Seismic reflection profile of a listric normal fault. The hanging wall block is cut by a set of subsidiary "synthetic" normal faults, which are faults having the same dip direction and sense of shear as the major fault. Note the small scale drag of the layers in both blocks along the steep part of the fault and the larger-scale rollover fold in the layers of the hanging wall block along the shallowly dipping parts. *A.* Seismic reflection profile. *B.* Interpretation of beds and faults on the seismic section.

means "shovel") (see Section 5.2). As the hanging wall block slips on the fault, it deforms to maintain contact with the footwall block across the fault, thereby producing a bend in the layering (see Figure 5.17).

Drag folds tend to be smaller-scale features on normal faults than rollover folds and may be less common. In these folds, the beds in the hanging wall block tilt up against the fault, and those in the footwall block tilt down into the fault, as shown by the shallow strata in Figure 5.3, and the small-scale folds against the eastern fault of Railroad Valley in Figure 5.6. Where they can be definitely recognized, they indicate the sense of relative displacement across the fault.

Features of Fault Surfaces

Like all faults, normal faults exist at all levels in the crust. The surface features of faults vary with the shape of the fault, the depth at which movement on the fault occurred, and whether faulting was accommodated by brittle fracture with frictional sliding, or by ductile deformation.

Like all faults at shallow levels (Section 4.2), normal faults develop cataclastic rocks (see Figures 4.5 and 4.6), slickensides, and slickenside lineations (Figure 4.8) along their surfaces. Some gently dipping normal faults in regions such as the Basin and Range province of the United States are characterized by large thicknesses of breccia and megabreccia (see Figure 4.5). In many cases, such breccias develop in hanging wall blocks from the pervasive fracturing and internal faulting that accommodate brittle, layer-parallel extension above shallow normal faults. In other cases, they may be associated with large, low-angle landslides.

At deeper structural levels, normal faults develop features associated with ductile deformation, including mylonitic textures (see Figure 4.7) which may be present in shear zones tens to hundreds of meters in thickness.

Shape of the Surface Trace

The sharp planar discontinuities that we commonly use to represent normal faults, such as those in Figure 5.2, are idealized representations of the structures we actually observe in nature. The surface trace of a normal fault is generally not a straight line but instead may be a sinuous curve or a series of connected, roughly straight line segments (Figure 5.1B; see also Section 4.4 and Figure 5.10). Although surface trace irregularities may result in part from the intersection of inclined faults with irregular topography, in many places the faults themselves must be nonplanar surfaces.

Shape at Depth

Normal faults need not maintain a constant dip with increasing depth as is shown in Figure 5.3, for example. Some listric normal faults join or turn into a detachment fault at depth (Figure 5.4). A **detachment fault** is a low-angle fault that marks a major boundary between unfaulted rocks below and a hanging wall block above that is commonly deformed and faulted. Normal faults in the hanging wall block may form a set of **imbricate faults**, which are closely spaced parallel faults of the same type that either terminate against (left end of Figure 5.4) or merge with (right end of Figure 5.4) the detachment fault.

Lateral ramps on the fault show up at the surface as an irregular fault trace (Figure 5.1B), and frontal ramps may also occur (Figures 4.25A and 5.5). Thus a map or cross section alone does not provide a complete picture of the geometry of a normal fault at depth, and we must keep in mind the potential complications of the third dimension when interpreting a simple two-dimensional view.

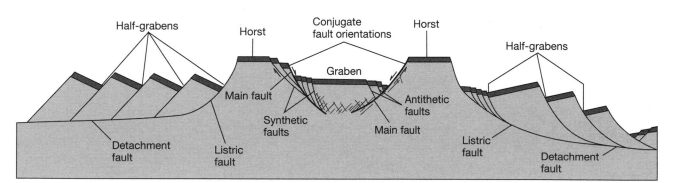

Figure 5.4 Systems of normal faults commonly are characterized by a main fault with associated subsidiary faults and by low-angle detachment faults with imbricate fault blocks in the hanging wall block.

Displacement on Normal Faults

By definition, displacement on ideal normal faults is parallel to the dip of the fault surface. If the strike of the fault varies, however, rigid movement of the hanging wall block relative to the footwall block cannot everywhere be down the dip of the fault. This fact is illustrated in Figure 4.30, where the displacement on the fault varies from pure normal dip slip to oblique slip as the fault trace curves. Thus the complex shape of real faults requires that they depart from our idealized models.

Movement on normal faults can be either nonrotational or rotational, depending on whether the orientations of the fault blocks remain constant or change as a result of the faulting. Both conditions exist in nature. If the apparent dip of the fault, measured in the direction of the displacement, does not change with depth (for example, Figures 5.2 and 4.30), and if the fault itself is not rotated during faulting, then the orientations of the fault blocks do not change during slip. Horizontal beds in the blocks remain horizontal; inclined beds maintain the same strike and dip. If, however, the dip changes with depth, then slip must result in rotation or deformation of the hanging wall block. On listric normal faults, the rotation takes place ideally about an axis parallel to the strike of the fault plane, and originally horizontal bedding ends up dipping toward the fault on which the fault block rotates. The angle between the bedding and the fault remains constant (see left and right ends of Figure 5.4).

A hanging wall block moving over a fault with ramp-flat geometry must in general deform internally. If a ramp connects two more shallowly dipping segments of the fault, slip on the fault produces a fault-ramp syncline (Figure 5.5A). If a flat connects two more steeply dipping segments of the fault, slip produces a fault-bend anticline (Figure 5.5B), which is comparable in part to a rollover anticline. These folds must parallel the associated ramp or flat, whether it intersects the main fault surface in a line perpendicular or oblique to the displacement direction. Deformation in the hanging wall block may take place by ductile bending (Figure 5.5), distributed faulting (see Figure 5.17D), or a combination of both.

5.3 Structural Associations of Normal Faults

Normal faults generally are present as systems of many associated faults. In many cases, the orientations of the faults fall into two groups, which are referred to as conjugate orientations; they have comparable dip angles but opposite dip directions and opposite senses of shear (Figure 5.4).

Commonly in such systems, some of the faults have a major amount of displacement along them and accommodate the major deformation, whereas other faults have a relatively small amount of displacement and provide the minor adjustments required for the large-scale displacements. If the smaller-scale faults are parallel to the major fault and have the same sense of shear, they are **synthetic faults**; if they are in the conjugate orientation, they are **antithetic faults**.

A **graben** is a down-dropped block bounded on both sides by conjugate normal faults (Figure 5.4) (the German *grabe* means "ditch"). A **half-graben** is a down-dropped tilted block bounded on only one side by a major normal fault. A **horst** is a relatively uplifted block bounded by two conjugate normal faults (the German *horst* means "a retreat" or "eyrie"—the nest of a bird of prey, typically built on a high cliff). These terms may refer either to the topographic feature formed by the faulting or to the structural feature of relatively down-dropped or uplifted fault blocks (Figure 5.1A). Alternating uplifted and down-dropped fault blocks are called a **horst and graben structure.** In some cases a series of half-grabens results in tilted fault blocks that also form alternating topographic highs and lows (Figure 5.4).

In regions of active normal faulting, horsts and the higher ends of tilted fault blocks provide the sediments that accumulate in the basins formed by the grabens

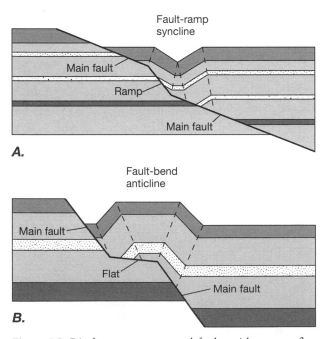

Figure 5.5 Displacement on normal faults with a ramp-flat geometry showing characteristic deformation of the hanging wall block. *A.* A fault-ramp syncline. *B.* A fault-bend anticline.

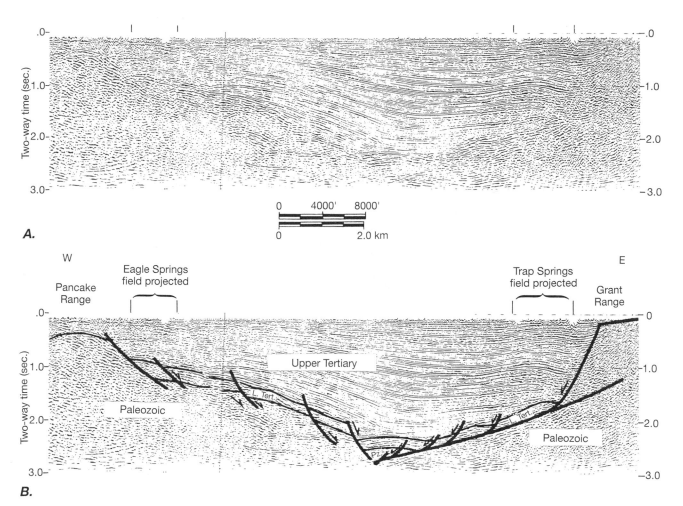

Figure 5.6 Seismic reflection profile of Railroad Valley, east central Nevada, showing typical asymmetric graben form. *A.* Seismic reflection profile. *B.* Interpretation of the faults on the seismic section.

and the lower ends of tilted blocks. As faulting continues during deposition of the basin sediments, the sediments themselves often become involved in the faulting. Study of these faults and of the age, composition, thickness, and distribution of the sediments can reveal when the major periods of uplift occurred, as well as the sequence and the time of exposure of the different rock types in the uplifted fault blocks.

Railroad Valley (Figure 5.6) in east-central Nevada illustrates the nature of an individual graben. This structure clearly is down-dropped on both sides, though more so on the east. It had a valley fill of late Tertiary and Quarternary sediments approximately 6 km thick that includes coarse to very coarse alluvial deposits shed from the surrounding ranges, playa lake sediments, and landslide deposits. These sediments record the interplay between faulting and concurrent sedimentation.

Systems of normal faults exist either on a local scale, subsidiary to other structures, or on a regional scale, where they dominate the structure. We briefly consider each scale of structure in turn.

Local Normal Faults Associated with Other Structures

Normal faults of local extent are generally associated with other structures whose geometry requires extension of crustal layers. Examples include domes, folds, cavities, and pull-apart structures on strike-slip faults.

Structural domes cut by a system of normal faults commonly result from the intrusion of bodies of salt or magma. The faults radiate from the center of the dome and may include a single major fault, one or two grabens (Figure 5.7A), or a Y-shaped set of grabens. The displacement on the faults is greatest at the center and dies out at tip lines near the margin of the dome. At depth, the faults terminate at or near the intrusive margin (Figure 5.7B). Elongate domes, described as doubly plunging anticlinal folds, often exhibit a comparable pattern of normal faults.

If a cavity forms at depth, surficial rocks commonly collapse into it along a set of concentric normal faults, forming a system of **ring faults.** Examples of such struc-

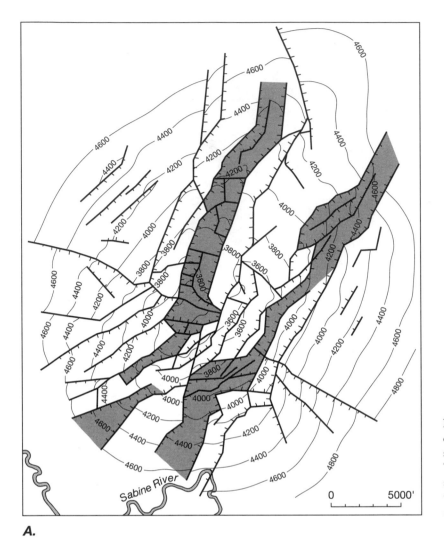

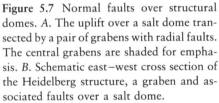

Figure 5.7 Normal faults over structural domes. *A.* The uplift over a salt dome transected by a pair of grabens with radial faults. The central grabens are shaded for emphasis. *B.* Schematic east–west cross section of the Heidelberg structure, a graben and associated faults over a salt dome.

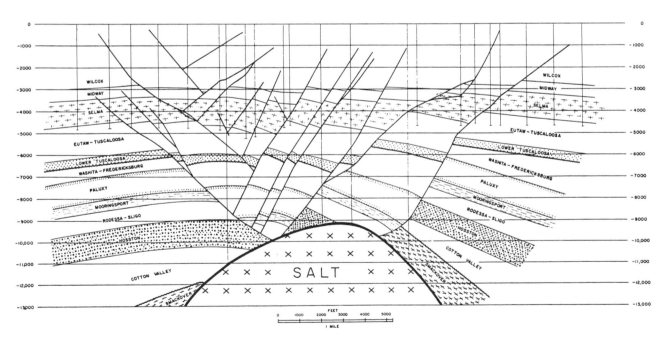

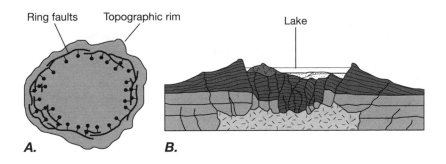

Ring faults   Topographic rim   Lake

A.   **B.**

Figure 5.8 Normal faults associated with a caldera, a volcanic collapse structure. *A.* Schematic map of ring faults around a caldera. *B.* Schematic cross section of the caldera structure at Crater Lake, Oregon.

tures include **calderas,** which form by the collapse of surficial rocks into a magma chamber emptied during an explosive eruption (Figure 5.8); **diatremes,** which are volcanic pipes explosively blasted through crustal rocks; and collapse structures, which result from the dissolution of limestone, salt, or gypsum at depth. Individual faults are not continuous around the circumference of such structures. Where one fault dies out, however, the displacement associated with the collapse is taken up by adjacent or overlapping faults, thereby forming concentric rings of discontinuous normal faults (Figure 5.8*A*). At depth, the major faults must terminate at the cavity boundary (Figure 5.8*B*).

Regional Systems of Normal Faults

Regional systems of normal faults define large, distinct structural provinces in many parts of the world (Figure 5.9). In the oceanic crust, the midoceanic ridge system constitutes an active, world-encircling extensional province characterized by normal faulting. Continental examples of such provinces include the Basin and Range province in western North America, the East African Rift region, the western Turkey–Aegean Sea region, and the Shansi graben in China, all of which are currently active. Inactive continental provinces of normal faulting include the Triassic–Jurassic–age graben system of the

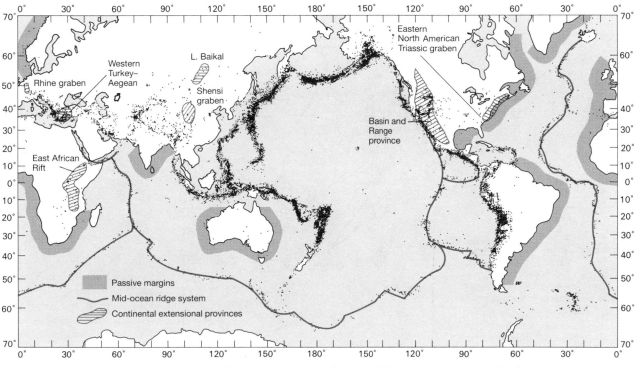

Figure 5.9 Map of the world, showing regions dominated by extensional normal faulting.

eastern United States and the Tertiary Rhine graben system in western Europe, as well as systems of normal faults in basement rocks of rifted continental margins, such as along eastern North America, western Europe, western Africa, eastern South America, southern India, and western and southern Australia. The sedimentary cover rocks of many such margins also contain systems of active normal faults not necessarily associated with the underlying basement systems. We concentrate our discussion on the Basin and Range province of North America (Figure 5.10), which provides a good example of a continental regional system of normal faults, and on the Gulf Coast province, which exemplifies the structures in the sedimentary cover rocks of an extended continental margin (see Figure 5.14).

In all of these regions, normal faults form in conjugate sets, with approximately the same strike and with dips of varying magnitude in both directions (Figures 5.6 and 5.14). The faults are always of limited length (Figure 5.10). Where one fault dies out at a tip line, regional extension is taken up by displacement on adjacent faults. Between these faults there is commonly a

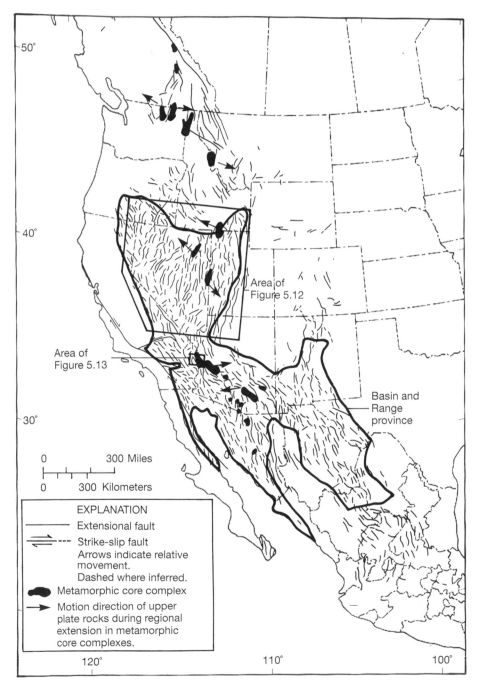

EXPLANATION

— Extensional fault

⇉ --- Strike-slip fault
Arrows indicate relative movement.
Dashed where inferred.

◼ Metamorphic core complex

→ Motion direction of upper plate rocks during regional extension in metamorphic core complexes.

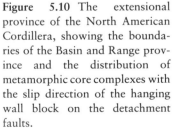

Figure 5.10 The extensional province of the North American Cordillera, showing the boundaries of the Basin and Range province and the distribution of metamorphic core complexes with the slip direction of the hanging wall block on the detachment faults.

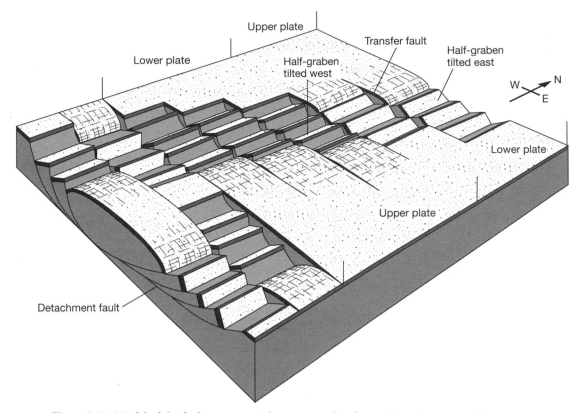

Figure 5.11 Model of the fault geometry in basement rocks of a continental extensional province. Different domains of normal faulting are separated by transfer faults. Some domains, such as the two on the left, may contain sets of oppositely dipping normal faults separated by an unfaulted block.

transfer zone within which deformation is accommodated by folding, faulting, and fracturing. In some cases, these transfer zones may be distinct strike-slip transfer faults. Transfer zones or faults may divide an extensional province into domains distinguished by different amounts of extension, different predominant orientations of faults, or different predominant directions of tilting. A schematic model of the geometry is shown in Figure 5.11.

Many rifted passive continental margins in the world originated as extensional terranes during the plate tectonic breakup of continental masses. Beneath layers of younger sediments, these margins are characterized by systems of normal faults with geometries similar to that shown in Figure 5.11.

In the Great Basin area of the Basin and Range province, several strike-slip faults have been recognized, some in part by the mapping of paleontologic associations (Figure 5.12). Both dextral and sinistral faults occur. These faults may be transfer faults of the type shown in Figure 5.11. The direction of tilting of fault blocks tends to be consistent over large areas, which suggests a structural association at depth and requires some discontinuity between major domains. The boundaries of these tilt domains therefore may be other transfer zones or faults.

In cases of extreme extension, normal faulting effectively strips off the shallower layers of rock to expose rocks that originally were deeper in the crust. This process enables us to examine rocks that were deep enough to undergo ductile faulting. There are, in the Basin and Range province, numerous regions called metamorphic core complexes (Figure 5.10) where the crust has been extended in a roughly east–west direction on major detachment faults by amounts on the order of 100 percent to 400 percent. These faults are characterized by extensive development of mylonite (see Section 4.2). As a result, the metamorphic and plutonic rocks that lie beneath the detachment faults have been brought up to the surface from depths as great as 20 km. In the Whipple Mountains of southeastern California, for example (Figure 5.13), the rocks beneath the detachment fault are extensively mylonitized and have a gently dipping foliation. The detachment fault itself contains mylonitic rocks, which in turn have been deformed by cataclasis, reflecting the change from ductile to brittle deformation as normal faulting brought the deeper rocks up toward the surface and the temperature and pressure decreased.

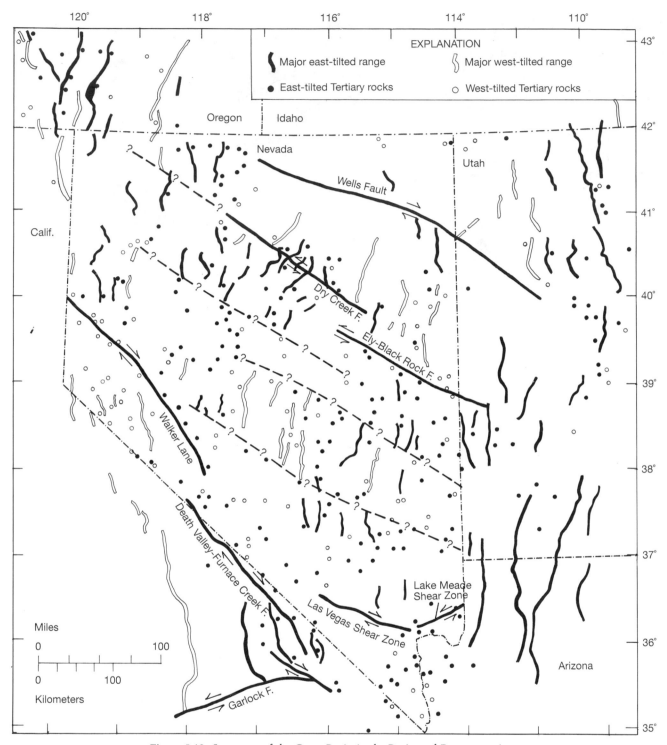

Figure 5.12 Structure of the Great Basin in the Basin and Range province of Nevada and neighboring regions, showing the tilt direction of major ranges and of Tertiary rocks. Strike-slip faults in northeastern Nevada have been identified by the offset of stratigraphic and structural trends. Hypothetical transfer faults, indicated by question marks, are suggested by the possible boundaries of tilt domains and domains of major normal faulting (cf. Figure 5.11).

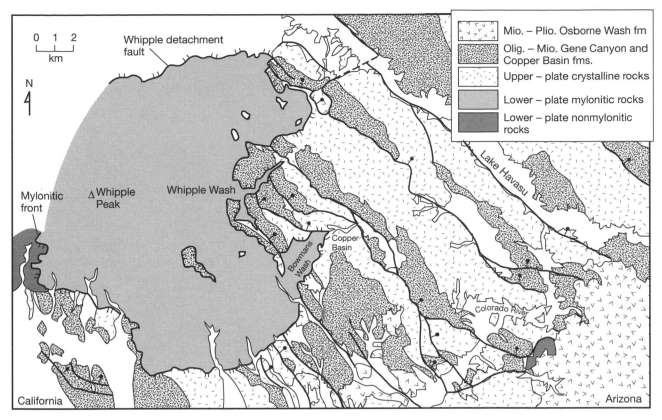

A.

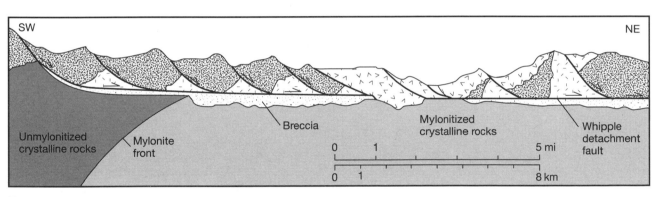

B.

C.

Figure 5.13 Faulting in the Whipple Mountain metamorphic core complex of the Basin and Range province, southeastern California. *A.* Map of the Whipple Mountains metamorphic core complex. *B.* Diagrammatic cross section through the Whipple Mountains before uplift domed the detachment fault. *C.* The Whipple Mountain detachment fault (arrow) is marked by a topographic ledge of cataclastic rocks (see Figure 4.5*B*) along which the dark-colored tilted Tertiary strata are faulted against the underlying, lighter-colored mylonitic gneisses.

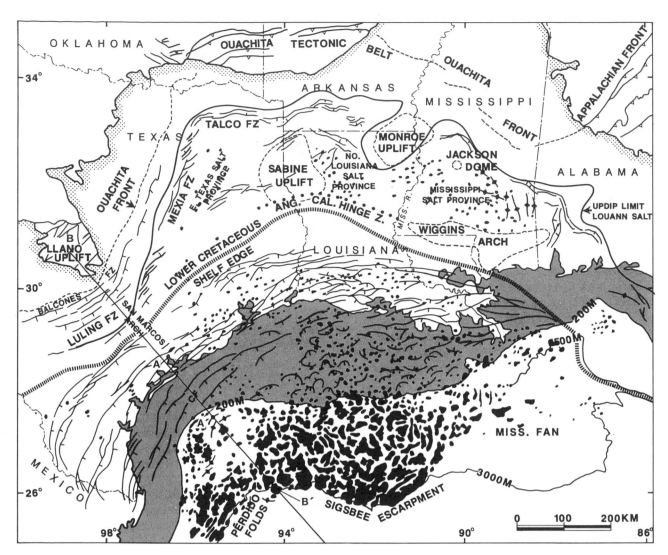

A.

Figure 5.14 Normal fault province in the Gulf Coast area. *A*. Map of the Gulf Coast region, showing major normal faults and salt structures (black). Major salt deposits occur south of the line marking the updip limit of the Louann salt, and this area which is closely associated with the province of normal faulting. The area shaded grey marks the continental shelf to a depth of 200 m. *B*. (*Facing page*) Cross section *A–A'* in part *A* across the continental shelf of southwest Texas. Much of the area is believed to be underlain by salt deposits, which are not shown because of a lack of seismic resolution. Note the growth faults and salt structure on the right. There is no vertical exaggeration. *C*. Interpretive cross section of the Gulf Coast from the Llano Uplift in the northwest to the Gulf abyssal plain in the southeast *B–B'* in part *A*. Jurassic salt is believed to underlie much of the shallow structure and to form the major detachment zone, although the structure is not known. Note the salt nappe behind the Sigsbee escarpment and the underlying Perdido fold belt. Vertical exaggeration 5 ×.

The "upper plate" rocks above the detachment fault are unmetamorphosed and have been strongly rotated on a set of imbricate listric normal faults that merge at depth with the detachment fault (Figure 5.13*B, C*).

The widespread development of core complexes in the Basin and Range province and the large amount of extension associated with them indicate that they are of major tectonic significance. Moreover, reports of similar features from the Aegean Sea and Papua–New Guinea suggest that core complexes have worldwide importance.

The northern Gulf Coast region of the United States from Texas to Alabama is an example of normal faulting along a modern continental margin. The area is char-

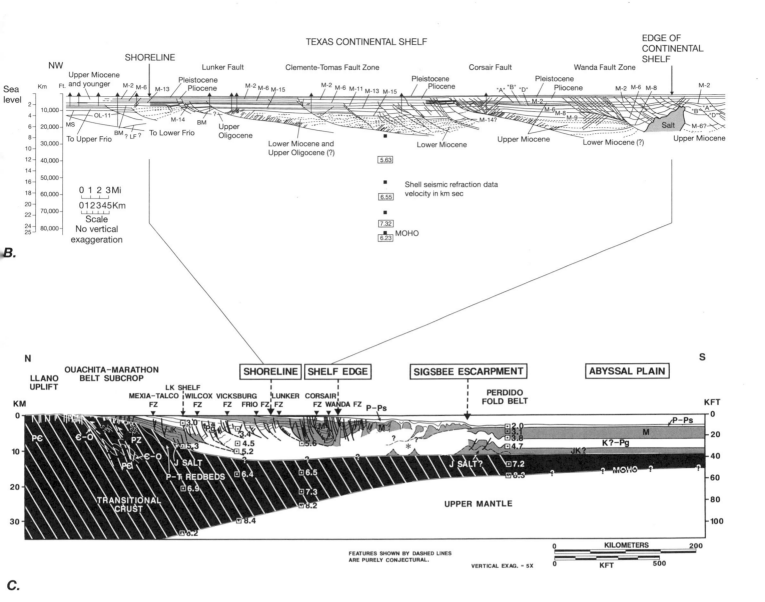

B.

C.

FEATURES SHOWN BY DASHED LINES ARE PURELY CONJECTURAL.

VERTICAL EXAG. = 5X

acterized by thick accumulations of sediment, rapid subsidence, and an arcuate system of normal faults whose extent is closely associated with the extent of major Jurassic salt deposits. Most faults dip southward (Figure 5.14*A, B, C*), although north-dipping faults also exist. Southward-dipping faults commonly show rollover anticlines, indicating that the faults are listric, with gentler dips at depth (Figure 5.14*B*).

Many normal faults along the Gulf Coast are **growth faults,** also referred to as **regional contemporaneous faults,** which are active during sedimentation. These faults characteristically have stratigraphic sequences that are thicker on the hanging wall block than the equivalent sequences on the footwall block (Figure 5.15; compare Figure 5.14*B*). This disparity in thickness

develops because as faulting continues, sediments accumulate most rapidly in the deepest parts of the basin, thereby keeping the surface of deposition approximately flat and horizontal. The down-faulted block thus accumulates a greater thickness of any particular unit. Older units show greater amounts of displacement and large tilts than younger units, because they have experienced a longer history of faulting.

Growth faults apparently form in two ways: by differential compaction of shale layers in a sandstone–shale sequence and by gravity sliding toward the basin. Coarser sands are generally deposited closer to shore than muds, but they undergo less compaction during consolidation and lithification. The greater compaction of the shale can initiate a growth fault near the boundary

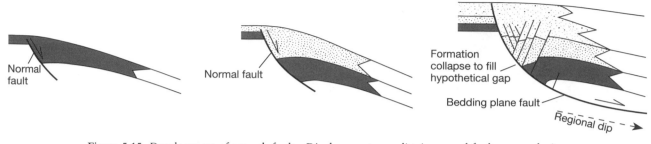

Figure 5.15 Development of growth faults. Displacement on a listric normal fault occurs during sedimentation, resulting in equivalent beds being thicker in the hanging wall block than in the footwall block. The fault passes into a bedding-plane fault at depth.

with the less compactable sands. Growth faults can also develop by formation of a detachment at the base of a sequence, usually in easily deformed shale or salt deposits. In the Gulf Coast, accumulation of large thicknesses of sediment on the continental shelf has caused the thick underlying salt deposits (originally up to 1500 to 2100 m thick) to flow toward the basin. This has created listric growth faults along the continental margin that have a detachment in the salt layers. Associated compressional structures developed in the basin, such as the huge salt nappe behind the Sigsbee escarpment and the Perdido fold belt (Figure 5.14A, C).

Similar faults are present along many rifted continental margins, such as the Atlantic Ocean margins of North America, Europe, and Africa and the Indian Ocean margins of Africa, India, and Australia. In all these regions, the fault systems are important traps for hydrocarbons and thus are of great interest to the petroleum industry.

A single set of parallel normal faults can accommodate extension in only one direction approximately perpendicular to the strike of the faults. In regions where extension occurs in two perpendicular horizontal directions, more than one orientation of normal fault is required, and a rhombic pattern of fault traces commonly develops. The angle between the faults depends on the relative magnitude of the extension in the two directions.

5.4 Kinematic Models of Normal Fault Systems

A kinematic model of any fault system is a description of the motions that have occurred on the faults in the system. A fundamental constraint on basic models of faulting is that the volume of the blocks of rock involved in the faulting must be conserved. If the deformation is two-dimensional, the cross-sectional area of each unit must remain constant, and appropriate kinematic

models and associated cross sections must be balanced.[2] Thus running the inferred fault motions backward from the present configuration must not produce overlaps of different fault blocks or large gaps between them. The model must also account for horizontal extension in the footwall block of major detachment faults.

We generally use cross sections of normal faults to display the geometry of faulting at depth. Any cross section inevitably implies some kinematic model of faulting, whether intended or not. Cross sections, however, are usually incomplete in that they do not include all fault termination lines. This incompleteness may reflect a lack of data, which makes it impossible to determine how apparent geometric problems are accommodated at depth, or it may be required by the scale of the section needed to portray important details of structure or stratigraphy. In any case, such cross sections make it easy to ignore the necessity to conserve volume. The result may be unbalanced cross sections that are geometrically impossible, that leave unresolved fundamental problems about the tectonics of an area, or that contain unintended implications about the kinematics of faulting.

Figure 5.16A, for example, shows an unbalanced cross section of a graben that is geometrically impossible. There is no way in which the motion on the two faults can be reversed to produce an originally continuous layer without large gaps (Figure 5.16B) or overlaps (Figure 5.16C). Nor does this cross section specify what happens to the fault at depth. This type of inconsistency is difficult to recognize in a cross section showing multiple intersecting faults. We note other examples of incomplete cross sections in the following discussion.

Normal fault provinces typically show tilted fault blocks (Figure 5.13B), and in some cases the rotations may approach 90°. Horizontal bedding typically is ro-

---

[2] Any component of motion out of the plane of the cross section, or volume loss due to solution, however, makes the balancing exercise unreliable (see Section 4.5).

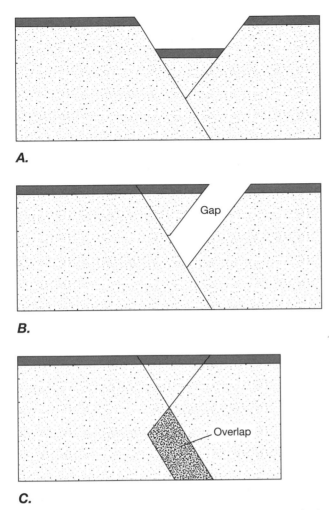

A.

B.

C.

Figure 5.16 Geometrical constraints on cross sections. *A.* A geometrically impossible cross section of a graben. Attempted reversal of the fault motion leads to *B.* major gaps or *C.* major overlap of fault blocks.

tated about an axis roughly parallel to the strike of the fault so that the beds dip toward the fault. On geometric grounds, this type of block rotation must imply either that the fault surfaces curve toward shallower dips with increasing depth or that planar faults rotated with the fault blocks during faulting. We shall discuss three kinematic models for normal faulting that result in tilted fault blocks.

Figure 5.17 illustrates some geometric problems inherent in accommodating extension on a listric normal fault. Horizontal extension of the block on a listric normal fault by an amount *d* opens a large gap between the hanging wall and footwall blocks (Figure 5.17*A*, *B*). If the bottom edge of the hanging wall block must conform to the shape of the listric fault, while at the same time keeping constant the length *L* of the surface layer and the total area of the block (Figure 5.17*C*), then the process requires internal deformation of the hanging wall block. The resulting geometry is a rollover

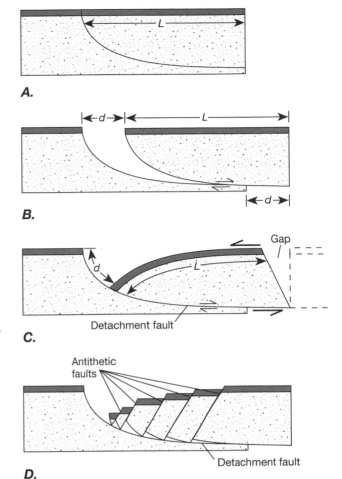

A.

B.

C.

D.

Figure 5.17 Model for the geometry of displacement on a listric normal fault accompanied by rollover folding or antithetic normal faulting. *A.* Crustal block with the incipient fault. The length *L* of the hanging wall block is kept constant. *B.* Rigid displacement of the hanging wall block a distance *d* parallel to the horizontal part of the listric normal fault results in the opening of a geologically ridiculous gap. *C.* Deformation distributed through the hanging wall block allows contact to be maintained along the fault and results in rollover folding of the layers. The length *L* remains constant, resulting in the development of another gap problem in the hanging wall block. *D.* Distributed faulting on antithetic faults in the hanging wall block reduces the gap problem along the normal fault to small misfits along the listric fault.

anticline similar to those commonly associated with listric normal faults.

One problem introduced by this model is that if there is no extension of the layers in the hanging wall block, a shearing must be distributed throughout the block, as indicated by the large arrows in Figure 5.17*C*. If the entire hanging wall block does not shear, a triangular gap must open up between the sheared and unsheared portions of the block. Neither shearing of

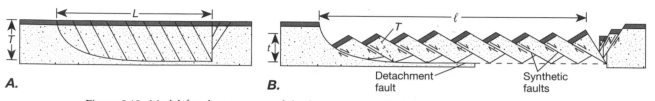

A.   **B.**

Detachment fault   Synthetic faults

Figure 5.18 Model for the geometry of displacement on a listric normal fault accompanied by synthetic normal faulting. *A.* Rotation on the listric fault can be accommodated by forming a set of synthetic faults. *L* and *T* are the original length and thickness, respectively, of the hanging wall block. *B.* After extension of the crust, the fault blocks have slipped and rotated on the system of normal faults synthetic to the listric normal fault. Here ℓ and *t* are the final length and average thickness, respectively, of the hanging wall block. The original thickness *T*, measured normal to the bedding, is shown in the second fault block from the left.

the entire hanging wall block nor the triangular gap is geologically reasonable.

The problem illustrated in Figure 5.17C can be alleviated by allowing extension parallel to the layer in the hanging wall block. A set of antithetic faults cutting the hanging wall block (Figure 5.17D) permits the block to conform fairly well to the listric fault and effectively extends the block above the curved part of the detachment. As a result, the right edge of the block remains perpendicular to the base. Greater continuity and smaller gaps under the antithetic fault blocks can be obtained if the spacing of the faults is smaller. The residual gaps are easily accommodated by local fracturing of the blocks.

A second model for slip on a listric fault requires the hanging wall block to maintain contact along the curved part of the listric fault by rotating as it slides. This mechanism can work only if the hanging wall block breaks up into a set of dominolike blocks along synthetic faults dipping in the same direction as the main fault (Figure 5.18A, B). Rotation of the fault blocks requires the synthetic fault planes to rotate as well, and the result is comparable to the collapse of a row of standing dominos. The triangular gap that opens at the right where the set of synthetic faults ends could be closed by a set of antithetic faults, as shown in Figure 5.18B. Again, the small gaps that occur below the synthetic fault blocks can be accommodated by closer spacing of the faults and by localized fracturing near the base of the fault blocks.

A third model of slip on listric normal faults requires slip of tapered fault blocks on a set of imbricate listric normal faults. As the fault blocks slip down the faults (Figure 5.19A, B), they must deform to conform to the shape of the fault. At large amounts of extension, the imbricate blocks are almost completely flattened out on the listric detachment fault, and bedding in the fault blocks is rotated to very steep dips. Rotation of the surface layer approaches a value equal to the initial dip of the fault where it cut the layer (Figure 5.19A). The

unfaulted part of the hanging wall block, however, encounters the same geometric problems in Figure 5.17.

In principle, the latter two models might be distinguishable in the field. For the "domino block" model of planar rotating normal faults (Figure 5.18B), the dip of the bedding is constant across the entire hanging wall block above the detachment fault. For the imbricate listric fault model, however, if extension has not been too great (Figure 5.19B), the dip of the bedding should increase with distance in the direction of displacement on the detachment fault.

The models we have discussed are, of course, simplified idealizations of the natural world. The presence of triangular gaps in the models, for example, results from our implicit assumption that the fault blocks behave rigidly. Although this assumption is in accordance with our intuitive experience with rock on a relatively small scale over short periods of time, on the scale of tens of kilometers and millions of years, the mechanical behavior of rocks is quite different. When the natural

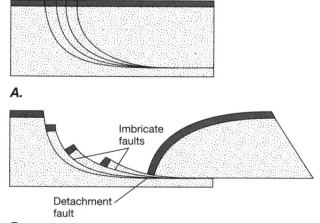

A.

Imbricate faults

Detachment fault

B.

Figure 5.19 Model for the geometry of displacement on a set of imbricate listric normal faults. *A.* Geometry of incipient imbricate listric normal faults. *B.* As the imbricate fault blocks slip down the faults, they rotate and straighten out.

behavior is scaled down to models we might make in the laboratory, the mechanical properties of rock are closer to those of sand or clay (see Sections 20.5 and 20.7). Thus the deformation in the hanging wall block required by the model in Figure 5.17C and the flattening of the imbricate fault blocks shown in Figure 5.19B are not outrageous propositions, and the small gaps that open up along the detachment fault in models such as Figures 5.17D and 5.18B could readily be accommodated by local deformation (see Figure 20.18).

The geometry of normal fault systems is generally more complex than our model cross sections imply. It is common, for example, for listric normal faults to have a ramp–flat geometry and to cut progressively into either the hanging wall block or the footwall block as faulting proceeds. In some cases, later normal faults may cross-cut earlier systems of normal faults in the same episode of extension. Figure 5.20, for example,

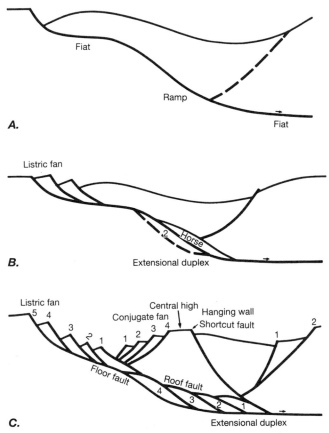

Figure 5.20 Model for the progressive development of a listric fan and an extensional duplex associated with a ramp and flat in a normal fault. A. Listric normal fault with a ramp-flat geometry. A rollover anticline and a ramp syncline are developed in the hanging wall block. B and C. Progressive propagation of the fault into the footwall block produces a listric fan near the surface and an extensional duplex at depth. Eventually other faults, including the conjugate fan, develop to accommodate the deformation of the hanging wall block.

illustrates the structure resulting from the progressive cutting of the active fault back into the footwall block, as indicated by the number sequence. An imbricate listric fan of faults develops at the surface, and with an adjacent set of conjugate faults, it defines a graben. At depth, an **extensional duplex** develops, characterized by a stack of horses that are progressively cut from the footwall block and added to the hanging wall block. The **floor fault,** which defines the bottom of the duplex, is the active fault, whereas the **roof fault,** which bounds the top of the duplex, is never active at one time as a single fault.

Although all three models in Figures 5.17 through 5.19 can account for the rotation of fault blocks, all of them ignore a major tectonic problem implied by a horizontal or low-angle detachment fault. Normal faulting inherently increases the distance between two points on opposite sides of the fault and, on the average, must thin the faulted block if cross-sectional area is conserved (compare parts A and B of Figure 5.18). Thus if extension is accommodated by normal faulting on a detachment surface, and the rocks below that surface are not extended by the same amount, then there must exist a region in the hanging wall block where an equivalent amount of shortening compensates for the extension. This compensation is presumably the relationship in the Gulf Coast region between the system of normal growth faults and the Perdido fold belt and salt nappe structures (Figure 5.14C). Alternatively, the basement below the detachment must extend by the same amount, although probably by a mechanism other than brittle faulting. Stretching and thinning of the crust must, in turn, be accommodated at depth by a flow of material in the mantle. Because the Earth is not expanding, horizontal stretching of the crust must be compensated for somewhere by crustal shortening, for example, at a subduction zone or in an orogenic belt.

Figure 5.21 shows two models that account completely for the geometry at depth of crustal normal fault systems. In Figure 5.21A, major normal faults in the upper crust join one of two symmetrically located detachment faults that become horizontal at the depth where deformation changes from brittle to ductile. The tip line for each detachment is at the same location in the middle of the faulted terrane. Below the detachments, the crust extends and thins by ductile deformation, and the extension may also be accommodated to some extent by magmatic intrusion. Below the crust, the mantle accommodates the crustal extension by ductile inflow of rock.

In Figure 5.21B, a major detachment fault extends completely through the lithosphere, changing from a brittle to a ductile fault at a depth of roughly 15 km to 20 km. Predominantly synthetic imbricate normal faulting in the hanging wall block produces an asymmetric

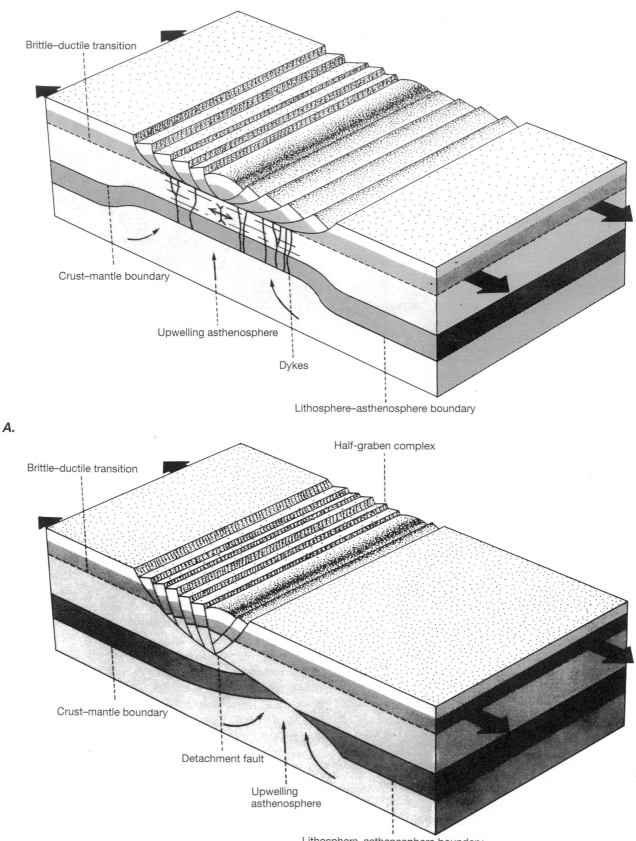

A.

Brittle–ductile transition

Crust–mantle boundary

Upwelling asthenosphere

Dykes

Lithosphere–asthenosphere boundary

Half-graben complex

Brittle–ductile transition

Crust–mantle boundary

Detachment fault

Upwelling asthenosphere

Lithosphere–asthenosphere boundary

B.

Figure 5.21 (*Left*) Complete cross sections accounting for the extension in provinces of normal faulting. *A.* The shallower crust extends by brittle normal faulting. The deeper crust extends and thins by ductile deformation. The extension is accommodated in the mantle by ductile inflow of material. Dip directions of normal faults are symmetrical about the center of the province. *B.* Extension occurs by displacement along a normal detachment fault that extends completely through the lithosphere. The brittle shallow crust extends by brittle imbricate listric normal faulting. Faulting on the detachment at depth is by ductile shear. The extension is accommodated in the mantle by ductile inflow of material. The dips of normal faults are predominantly in a direction synthetic to the detachment.

normal fault province. Thinning of the crust by faulting on the detachment fault is accommodated in the mantle by ductile inflow of material. The termination of the fault is at the base of the lithosphere, where it and the asthenosphere are both moving to the right at approximately the same rate. A large amount of extension would emplace the deep crustal mylonites of the footwall block against the brittlely faulted blocks of the hanging wall block. Pervasive ductile extension of the crust in the footwall block is not required.

Both models include all termination lines of all the faults and account for all the required tectonic motions. The actual driving force for the extension could be the same in both cases. Structural aspects of each model that could in principle be tested include the predicted symmetry or asymmetry of normal faulting across the structural province, the extent of ductile extension of the crust below the detachment, and the geometry of the Moho. The tests are not easy to make, however, and which model is better remains to be determined through field and geophysical investigations.

5.5 Determining Extension Associated with Normal Faults

In studying normal fault systems, we wish to estimate quantitatively the amount of extension in a region. We define the **extension** e as the change in length in a given direction caused by the deformation, divided by the original length. Thus in Figure 5.18, for example, $e = (\ell - L)/L$. Estimates of the extension can constrain our reconstructions of an area and help us better understand its tectonic history. We can estimate the amount of extension by examining fault geometry and by using map relationships to restore the stratigraphy to its original state.

Estimating Extension on the Basis of Fault Geometry

To evaluate the extension across a region by using fault geometry, we must make a few simplifying assumptions. We assume that the fault strike is uniform and that the change in length of the region is the sum of the horizontal extensions on all the faults (Figure 5.22A). The extension is then this change in length divided by the original distance across the region, which must be measured normal to the strike of the faults.

For example, let us take a simple cross-sectional model of planar nonrotating normal faults producing a horst and graben structure (Figure 5.22A)[3]. The segments of a particular stratigraphic layer labeled L_i meaning (L_1, L_2, L_3, and so on) when summed together equal the original length of the cross section. The segments labeled ΔL_i when summed together give the total change in length. The extension e is calculated for N segments by the formula

$$e = \frac{\sum_{i=1}^{N} \Delta L_i}{\sum_{i=1}^{N} L_i}$$

For an individual fault, the change in length ΔL is related to the dip-slip displacement d and the dip angle of the fault ϕ by

$$\Delta L = d \cos \phi$$

For a model of rotating planar normal faults (Figure 5.18A, B), if we assume that bedding is initially horizontal and that, on the average, the faults have the same orientation, spacing, and slip, we can easily derive a relationship among the extension e, the dip of the rotated bedding θ, and the dip of the rotated fault planes ϕ (Figure 5.22B).

$$e \equiv \frac{\overline{AB}_{final} - \overline{AB}_{initial}}{\overline{AB}_{initial}}$$

$$e = \frac{d}{L} \cos \phi + \cos \theta - 1$$

$$\frac{d}{L} = \frac{\sin \theta}{\sin \phi} \qquad \sin(\theta + \phi) = \sin \theta \cos \phi + \sin \phi \cos \theta$$

$$e = \frac{\sin(\theta + \phi)}{\sin \phi} - 1$$

The assumptions limit the applicability of the model, but it can be used in some cases to give a rough approximation of the extension.

---

[3] Note that this model ignores the obvious problem of the fault geometry at depth.

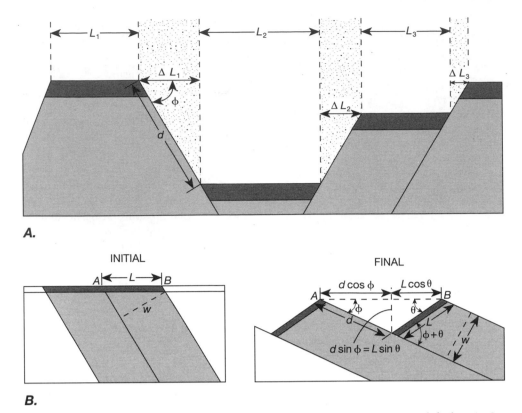

Figure 5.22 Determination of extension in a terrane faulted by planar normal faults. *A.* On nonrotating faults, the overall extension is the ratio of the sum of the extensions on each fault ΔL_i, divided by the sum of the original lengths L_i of the strata in each fault block. *B.* Geometric relationship—for equally spaced, planar, rotating normal faults above a detachment fault (Figure 5.18*B*)—among the dip of the faults ϕ, the dip of the beds θ, and the extension *e*.

For the models of listric normal fault slip shown in Figures 5.17*C* and 5.19*B*, which involve deformation of the fault blocks above the detachment, there is no simple geometric relationship between fault dip, bedding dip, and displacement.

Determining Extension from Map Relationships

In some cases, extension in a normal-faulted terrane can be determined by a **palinspastic restoration**—that is, by constructing a balanced cross section parallel to the slip direction and restoring the geology to its original configuration prior to deformation. (The term *palinspastic* is derived from two Greek words: *palin*, which means "again," and *spasmos*, which means "contraction" or "breaking.") Comparison of the restored cross-sectional length with the length after faulting, as measured in the field, makes it possible to calculate the extension. Reliable application of the method requires good data from the subsurface as well as from the surface, including firm constraints on the slip direction. These requirements severely limit the utility of the method. If erosion has removed much of the originally faulted terrane, and if large-displacement listric normal faults are prevalent, then the difficulty in knowing the original curvature of the faults also limits the reliability of the restoration.

Additional Readings

Brun, J.-P, and P. Choukroune. 1983. Normal faulting, block tilting, and décollement in a stretched crust. *Tectonics* 2: 345–356.

Coney, P. J. 1980. Cordilleran metamorphic core complexes: An overview, in M. D. Crittenden, Jr., P. J. Coney, and G. H. Davis (eds.), *Cordilleran metamorphic core com-*

plexes. *Geol. Soc. Am. Memoir* 153: 7–31.

Lister, G. S., and G. A. Davis. 1989. The origin of metamorphic core complexes and detachment faults formed during Tertiary continental extension in the northern Colorado River region, U.S.A. *J. Struct. Geol.* 11 (1/2): 65–94.

Gibbs, A. D. 1984. Structural evolution of extensional basin margins. *J. Geol. Soc. Lond.* 141: 609–620.

Lister, G. S., M. A. Etheridge, and P. A. Symonds. 1986. Detachment faulting and the evolution of passive continental margins. *Geology* 14: 246–250.

Wernicke, B., and B. C. Burchfiel. 1982. Modes of extensional tectonics. *J. Struct. Geol.* 4: 104–115.

Worrall, D. M., and S. Snelson. 1989. Evolution of the northern Gulf of Mexico, with emphasis on Cenozoic growth faulting and the role of salt, in A. W. Bally and A. R. Palmer (eds.), *The geology of North America: An overview, DNAG*. The geology of North America, Vol. 4. Boulder, Col.: Geological Society of America, pp. 97–138.

Zoback, M. L., R. E. Anderson, and G. A. Thompson. 1981. Cainozoic evolution of the state of stress and style of tectonism of the Basin-Range province of the western United States. *Phil. Trans. Roy. Soc. Lond.* A-300: 407–434.

CHAPTER

6 Thrust Faults

Thrust faults and reverse faults are dip-slip faults on which the hanging wall block has moved up relative to the footwall block (Figure 4.10A). Generally, older rocks are emplaced over younger rocks, and in a vertical section through the fault, stratigraphic section is duplicated. These faults accommodate shortening of the Earth's crust. Reverse faults have dips greater than 45°, whereas thrust faults dip less than 45°. We limit our discussion to thrust faults, because they are much more abundant than reverse faults.

Thrust faults exist at all scales. They range from small ones with extents and displacements on the order of millimeters to meters (Figure 6.1A), through major low-angle thrusts at the scale of mountain ranges that show displacements on the order of tens to hundreds of kilometers (Figure 6.1B), up to global-scale features such as convergent plate margins. The latter are enormous, complex zones of thrust faults that have total displacements as large as thousands of kilometers.

A hanging wall block above a very low-angle thrust commonly has an areal extent that is large compared with its thickness and therefore is called a **thrust sheet** or a **nappe** (the French word *nappe* means "sheet"). A thrust sheet that has moved a large distance and is thus geologically out of place is an **allochthon,** and the rocks within it are said to be **allochthonous** (from the two Greek words *allo,* which means "other," and *chthonos,* which means "ground" or "earth"). A large region of

rock that has not been moved and is close to its original location, such as the basement rocks in the footwall block of a thrust, is an **autochthon,** and the rocks within it are **autochthonous** (the Greek word *auto* means "this"). Figure 6.1B is a photo of the Keystone thrust in southern Nevada. The irregular dark/light contact is the trace of the low-angle thrust fault, which dips gently to the west (left) and is cut by irregular topography. The light rocks forming the cliff are autochthonous Jurassic Aztec sandstones, whereas the overlying dark rocks are allochthonous Paleozoic sequence that extends from lower Paleozoic at the fault to upper Paleozoic in the snow-covered peaks in the background. These rocks have moved up to 20 km on the thrust fault.

6.1 Recognition of Thrust Faults

Most thrust fault surfaces exhibit structures formed either by brittle or ductile deformation. The deformation features intrinsic to faults, as well as other features associated with them, are discussed in Section 4.2. In this section, we confine our discussion to stratigraphic characteristics that are unique to thrust faults.

Thrust faults characteristically emplace older rocks on top of younger rocks. On a vertical section through

A.

B.

Figure 6.1 The geometry and expression of thrust faults at different scales. *A.* Thrust fault cutting carbonate strata, Valley and Ridge province of the Appalachian Mountains, Tennessee. *B.* The Keystone thrust, southern Nevada.

a thrust fault, stratigraphic section is generally duplicated (Figure 6.2; compare Figures 4.10*A* and 6.1*B*).

The horizontal separation across a thrust fault can be variable, depending on the attitude of the layers before faulting, as illustrated in Figure 6.2. Each pair of block diagrams shows the results of thrust faulting of upright stratified rocks that have a particular orientation relative to the fault. The left of each pair of block diagrams shows the hanging wall block suspended over the footwall block, and the right diagram shows the same structure with the hanging wall block eroded down to the same level as the footwall block. The separations on either the horizontal surface or the vertical section do not define uniquely the displacement on the fault. In all the diagrams, however, a vertical section through the fault shows the duplication of stratigraphy with older rocks resting on top of younger rocks across the fault. Different orientations of the layers with respect

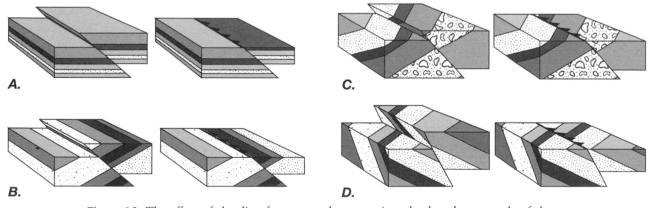

A.

B.

C.

D.

Figure 6.2 The effect of the dip of strata on the separations developed as a result of thrust faulting. The right diagram of each pair shows the hanging wall block eroded down to the same level as the footwall block. On the top surfaces (map views), *A* shows a simple discontinuity, *B* shows the cutting out of strata, and *C* and *D* show left and right lateral separations, respectively.

to the fault produce very different horizontal separations (compare, for instance, parts C and D). If, however, the layers are upright and their apparent dips on a vertical section normal to the strike of the fault are larger than the dip of the fault itself but in the same direction, the conventional relationships for a thrust fault do not hold. In particular, younger rocks are emplaced on top of older, and stratigraphic section is missing along a vertical line through the fault. Of course, in areas of complex deformation—for example, where the stratigraphy is overturned or folded—the conventional relationships also need not hold.

Several types of stratigraphic contrasts may indicate the presence of a thrust fault that has displaced the hanging wall block from substantially deeper levels or large horizontal distances from its final location.

Plutonic or high-grade metamorphic rocks are generally associated with deeper structural levels than unmetamorphosed or low-grade rocks. Thus if plutonic or high-grade metamorphic rocks overlie low-grade or unmetamorphosed sedimentary rocks, the normal structural sequence is inverted, suggesting that thrusting has occurred.

Some thrust faults separate stratigraphic sequences of essentially the same age but of markedly different sedimentary facies. Thrust faults of this nature commonly emplace allochthonous rocks of oceanic or deepwater environments, usually shales, cherts, and/or oceanic crustal rocks, on top of autochthonous shallowwater deposits such as limestones and sandstones or even rocks of continental origin. The striking discontinuity in the sedimentary environments suggests that the contact between the two sequences is a thrust fault of large displacement.

In some cases, highly deformed allochthonous rocks overlie slightly deformed or undeformed autochthonous rocks. If, for example, the rocks above and below the fault are the same stratigraphic layers, but those above are deformed by folds and those below are not, a thrust fault probably separates the two sequences.

All these criteria are only general indicators: Each pattern conceivably could form in some other fashion, for example, from a sequence of two or more episodes of deformation. The possibility of such structural complications means that we must pay careful attention to stratigraphic sequence and to evidence for sedimentary environment, conditions of igneous crystallization, and conditions of metamorphism. The geologic literature contains many examples of egregious errors committed by conscientious geologists who, when mapping a region, failed to take adequate account of these factors.

6.2 Shape and Displacement of Thrust Faults

Shape of Thrust Faults

Many map traces of thrust faults are highly irregular, a feature that results either from the intersection of a shallow-dipping fault with the irregularities of topography (Figure 6.1B) or from folding of the fault surface. In some cases, an irregular fault surface may reflect the original path cut by the fault through the stratigraphy.

At depth, thrust faults generally are listric faults that curve toward shallow or horizontal dips (see Figure 6.12). Some faults continue at a dip of roughly 30° through most of the crust (Figure 6.3). Others become steeper with depth, such as where they accommodate

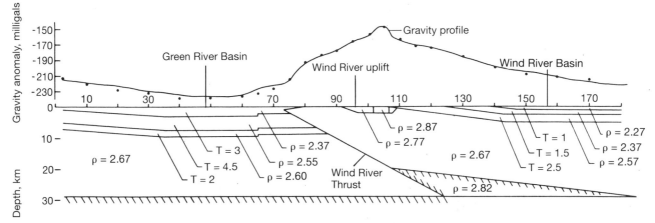

Figure 6.3 The Wind River thrust under the Wind River Mountains in Wyoming maintains an almost constant angle through the entire crust. This gravity model has been constrained by seismic reflection data. Values of T and ρ are, respectively, the thicknesses (in kilometers) and densities (in grams per cubic centimeter) of the different layers. On the gravity profile, dots are measured values of the gravity anomaly, and the solid line is the anomaly computed from the model.

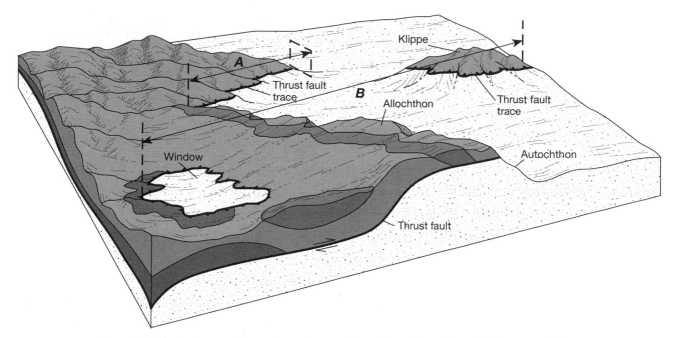

Figure 6.4 Block diagram illustrating a thrust surface, thrust sheet, thrust fault trace, window or fenster, klippe, allochthon, autochthon, and conventional thrust symbol with teeth on the hanging wall side. Minimum constraints on the displacement are given by *A* the sinuosity of the thrust fault trace and *B* the distance from the back of the window to the front of the klippe.

compression at a jog in a strike-slip fault (see Section 7.2; Figures 7.7C, 7.8B) or where they end against an upward moving intrusion (see Figure 6.9B).

Erosion of a thrust sheet that lies above a shallowly dipping fault commonly leaves an isolated remnant of the allochthon, a **klippe,** resting on the lower plate (Figure 6.4). (*Klippe* comes from a German word that means "cliff," reflecting the fact that most klippen in the Alps are bounded by cliffs.) A klippe indicates a minimum extent of the original thrust sheet. In other cases, erosion can create a hole through the thrust sheet, a **window** or **fenster** (*Fenster* is the German word for "window"), and expose an isolated area of the rocks that lie beneath the thrust (Figure 6.4). These rocks may be part of the autochthon or part of another underlying thrust sheet.

A window provides a minimum constraint on how far the thrust sheet extends out over the underlying rocks.

A low-angle thrust fault generally does not form a smooth, simple surface. The fault plane characteristically cuts through the stratigraphy in steps, alternately following flat bedding planes or easily deformed layers such as shale or evaporite beds, and then cutting up-section in the direction of displacement to form a frontal ramp (Figure 6.5; compare Figure 4.25 and Section 4.4). The fault surface thereby develops a characteristic **ramp–flat** geometry. Figure 6.5 also shows that in some places where the fault parallels the bedding, a normal stratigraphic sequence is preserved despite the presence of the fault.

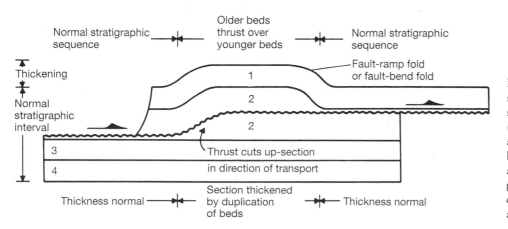

Figure 6.5 Idealized cross section of a low-angle thrust, showing the thrust surface (wavy line) cutting up-section at a frontal ramp from one horizontal glide surface to another. Changes in stratigraphic thickness and duplication of stratigraphic section are localized near the ramp.

Figure 6.6 is a composite photograph of the McConnell Thrust in Alberta, Canada, that shows such a ramp. At the left, the thrust is parallel to bedding and located within a lower Cambrian unit. To the right, the lower and middle Cambrian units of the thrust sheet are truncated against the fault. Still further right, the fault is again parallel to bedding, but here it is located within the upper Cambrian unit. The geometry is comparable to the diagram shown in Figure 6.5, the lower Cambrian corresponding roughly to layer 2.

Lateral ramps are also common features of low-angle thrust faults, as illustrated in Figure 6.7 for the Lewis thrust, which is part of the Canadian Rockies foreland fold and thrust belt in Montana (locations shown in Figure 6.11*B*). Both oblique and sidewall ramps may have dips as low as 15° to 20° (Figure 6.7*A, B*). If sidewall faults are steeply dipping, they are strike-slip faults called either **tear faults** or **transfer faults,** such as occur at the ends of the Pine Mountain thrust in the Appalachian Valley and Ridge province (Figure 6.7*C, D;* location shown in Figure 6.11*A;* see also Figure 6.13).

Displacement on Thrust Faults

Although displacement on thrust faults is typically up the dip of the fault surface, on irregularities in the fault plane such as lateral ramps, the displacement must in general be oblique slip. Ramps in the fault surface also require that the thrust sheet deform as it moves. Movement of a thrust sheet over a ramp causes a fold to develop in the thrust sheet, which is called a **fault-ramp fold** or a **fault-bend fold** (Figures 6.5, 6.6). The trend of the fold reflects the trend of the ramp below the

thrust sheet. Frontal ramps are steeper than the main fault surface and therefore produce **ramp anticlines** (Figure 6.5). On lateral ramps, if the displacement has a component up the ramp, then an anticline forms (Figure 6.8*A, B*); if displacement has a component down the ramp, then a syncline develops (Figure 6.8*A, C*).

6.3 Structural Environments of Thrust Faults

Thrust faults exist as local faults, as sets of faults subsidiary to larger structures, and as large systems involving multiple thrusts and extending over whole mountain ranges. We consider each structural environment in turn.

Local Thrust Faults

Subsidiary thrust faults form wherever the geometry of other structures requires local convergence or shortening and the rocks react brittlely.

Diapiric structures involve less dense rocks that move upward through more dense surroundings. In some cases, the diapir shoves the surrounding rocks upward and outward, and marginal thrust faults develop. One common example of such features is the thrust faults marginal to some salt domes (Figure 6.9). In plan view, the thrust faults mimic the outline of the diapir. Because the cover rocks must also be stretched by this motion, normal faults develop over the top of the dome, as described in Chapter 5.

Figure 6.6 The McConnell thrust at the Brazeau River in Alberta. Composite photograph looking north at a cliff containing the McConnell thrust. The background and the foreground to the cliff do not match up well across the joins between the frames, because the vantage points for the photographs are different. The scale is given by the thickness of the Fairholme formation, which is about 1400 ft. The McConnell thrust cuts up-section in the hanging wall block from the lower Cambrian units (center) to the upper Cambrian (right) along the cliff (compare Figure 6.5).

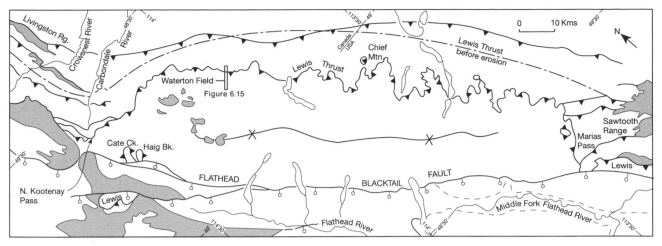

A.

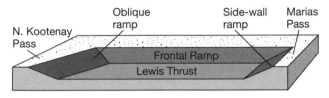

B.

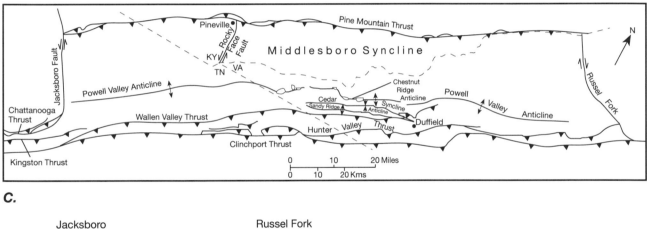

C.

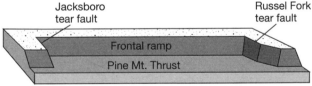

D.

Figure 6.7 Structure of low-angle thrust faults. *A.* Map of the Lewis thrust near the Canada–USA border between Alberta and Montana (see the location marked on Figure 6.11*B*). The irregular nature of the thrust is a reflection of topography on the shallowly dipping fault surface. Note the Chief Mountain klippe near the border and the Cate Creek and Haig Brook windows near North Kootenay Pass. *B.* Schematic block diagram showing the geometry of the Lewis thrust surface. Note in particular the frontal ramp that brings the fault up to the surface, the sidewall ramp near Marias Pass, and the oblique ramp near North Kootenay Pass. *C.* Map of the Pine Mountain thrust in the southern Appalachian Valley and Ridge province (see the location marked on Figure 6.11*A*). Tear faults mark the northeast and southwest ends of the Pine Mountain thrust sheet. *D.* Schematic block diagram showing the geometry of the Pine Mountain thrust surface. The tear faults bound the frontal ramp at either end.

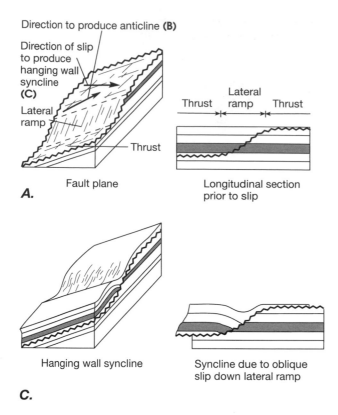

A.

Direction to produce anticline **(B)**

Direction of slip to produce hanging wall syncline **(C)**

Lateral ramp

Thrust

Fault plane

Thrust — Lateral ramp — Thrust

Longitudinal section prior to slip

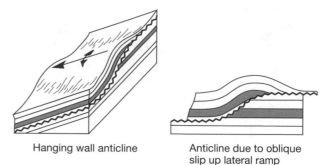

B.

Hanging wall anticline

Anticline due to oblique slip up lateral ramp

Hanging wall syncline

Syncline due to oblique slip down lateral ramp

C.

Figure 6.8 Effects of oblique displacement on sidewall ramps. The diagram on the left of each pair is a block diagram; the diagram on the right is the section shown as the right-hand face of the block. *A.* Geometry of the thrust surface and the sidewall ramp shown with the hanging wall block removed. The arrows on the block diagram show directions of displacement leading to the ramp folds shown in parts *B* and *C*. *B.* Hanging wall anticline produced by oblique slip up the sidewall ramp. *C.* Hanging wall syncline produced by oblique slip down the sidewall ramp.

Thrust faults are commonly present where bends in strike-slip faults result in compression of the rocks across the fault. We discuss this structural environment in greater detail in Chapter 7 (see Figures 7.7, 7.14).

Thrust Faults Associated with Folds

Thrust faults are associated with folds in four ways. First, as some folds develop, they reach a stage at which the sides, or limbs, of the fold cannot be rotated any closer together. Continued shortening of the layered sequence results in development of a thrust fault that generally cuts the steep or overturned limb of the fold

(Figure 6.10*A*). Second, folds may develop as a result of thrusting to accommodate the deformation above the tip line of the thrust; these are therefore called **fault-propagation folds.** As the displacement on the thrust increases, the tip line propagates through the layers and cuts the steep limb of the fold (Figures 6.10*B*, 12.27). Third, folds may develop a steep or inverted limb that becomes progressively sheared and thinned until, in effect, it is a ductile thrust fault (Figure 6.10*C*). Fourth, where thrust faults have an alternating ramp–flat geometry, movements along the faults cause fault-bend folds to form in the hanging wall block (Figures 6.5, 6.6, 6.8).

Thrust Systems

Thrust faults in **foreland fold and thrust belts,** which mark the margins of major orogenic belts, are by far the most common examples of large thrust systems. These belts consist of a set of low-angle listric thrust faults that have the same general strike and dip (Figures 6.11, 6.12). Because of the economic importance of the major reserves of hydrocarbons found in these belts, and because of their intrinsic interest as a major type of tectonic feature of the world, these systems have been the object of an enormous amount of research.

The geometry of such thrust systems is distinctive. In plan view, a foreland fold and thrust belt consists of

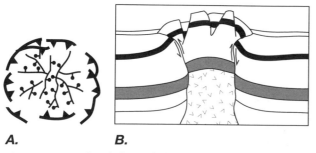

A.   **B.**

Figure 6.9 Peripheral thrust faults produced by diapiric intrusion. Normal faults in the central area accommodate extension associated with uplift. *A.* Schematic map. *B.* Schematic cross section.

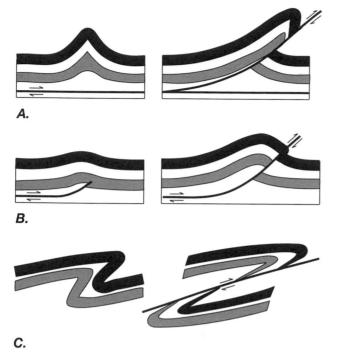

A.

B.

C.

Figure 6.10 Diagrammatic cross sections illustrating relationships between folds and thrust faults (see also Figure 6.9). *A.* Thrust fault cuts up from the décollement through the foreland limb of a fold when the fold becomes too tight to accommodate further shortening. *B.* Fold forms in association with the propagation of a thrust fault. *C.* Formation of a fold by ductile flow can result in the shearing out of one limb to form a ductile thrust fault.

a set of folds and thrusts, more or less parallel, that extends for tens or hundreds of kilometers. The area in front of the thrusts toward which the thrust sheet moved is the **foreland,** and the region behind the thrusts is the **hinterland.** Although in some places these systems are nearly straight, they are generally curved, as shown in Figure 6.11.

We describe the curvature in terms of its relationship to the direction of relative motion of the thrust sheet. In a **salient** or **virgation,** faults and folds form an arcuate belt convex toward the foreland. In a **reentrant** or **syntaxis,** the arcuate belt is concave toward the foreland. Figure 6.11 shows three examples of such thrust systems. As is evident from the figures, there are many faults and folds in these systems, and these particular systems extend for hundreds or even thousands of kilometers.

Thrust systems also display differences in elevation along strike. Relatively high areas, or **culminations,** are usually present along salients or virgations, and relatively low regions, or **depressions,** accompany reentrants or syntaxes.

In cross section, fold and thrust belts characteristically overlie an undeformed basement along a gently sloping **décollement** surface (*décollement* means "detachment" in French), also called a detachment or sole thrust (Figure 6.12). In such cases, the deformation and shortening in the thrust system are confined to the rocks above the décollement. The décollement cuts up through the stratigraphic section toward the foreland, forming a wedge-shaped thrust sheet that is thinnest near the foreland and thickens toward the hinterland. Individual thrust faults generally are listric and asymptotically join the décollement, as shown in Figure 6.12.

Most thrust faults include frontal and lateral fault ramps, as described in Section 6.2 (Figures 6.5 through 6.8; compare Figure 4.25A). As a result, fault-ramp folds are a common feature. Not all folds in thrust sheets are necessarily fault-ramp folds, however; some may form to accommodate shortening of the thrust sheet above a flat portion of the décollement.

Thrust sheets are not structurally continuous features. Rather, they are segmented by tear faults, or transfer faults, that accommodate differential displacement of different parts of the sheet or connect noncoplanar parts of the active thrust (Figure 6.13). For example, one part of the thrust sheet may shorten by faulting, and an adjacent part may shorten by folding. The discontinuity in displacement is then taken up by a tear fault (Figure 6.13A).

Many thrust systems include an **imbricate fan,** or **schuppen zone** (the German word *Schuppe* means "scales"), in which a number of individual listric thrust sheets, all dipping in the same direction, overlap like a series of roofing tiles (Figure 6.12). Faults in such imbricate systems generally are concave upward, decreasing in dip with increasing depth and distance behind the thrust front. These listric faults either cut the topographic surface or are blind and terminate upward at tip lines within the stratigraphic section. At depth, they terminate at branch lines along the basal décollement.

A thrust duplex is a system of imbricate thrust faults that branch off from a **floor thrust** below and curve upward to join a **roof thrust** at a branch line above, thereby forming a stack of horses (Figure 6.14; compare Figure 4.26A). Unlike an imbricate fan that can break through to the surface, a duplex is necessarily contained within the stratigraphic section. Like an imbricate fan, however, a duplex can develop along a frontal ramp on which the main thrust rises toward the foreland.

Duplexes exhibit a variety of forms, depending on the amount of displacement of the individual horses. Where the displacement of the horses is relatively small, they dip predominantly toward the hinterland (hinterland dipping) and form a zone of roughly constant thickness between the roof and the floor thrusts (Figure

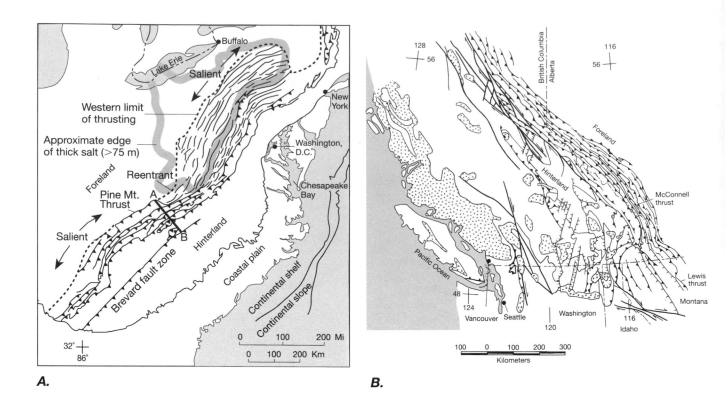

A.

B.

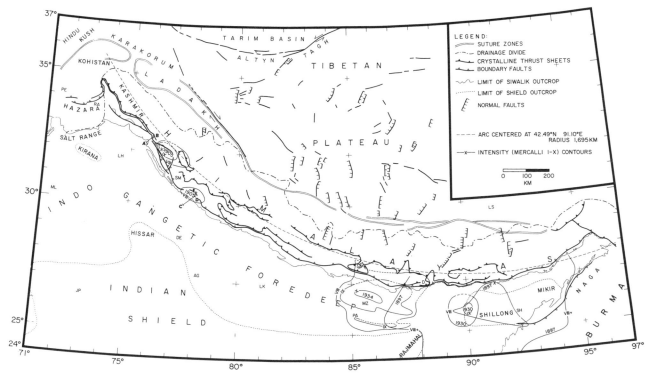

C.

Figure 6.11 Major thrust systems, showing the foreland, hinterland, salient or virgation, and reentrant or syntaxis relative to the direction of movement for each fold and thrust belt. Teeth on the thrust faults are on the side of the hanging wall. *A.* Generalized map of the Appalachians. Plain lines are fold hinges; barbed lines are thrust faults. *B.* Generalized map of the Canadian Cordillera. *C.* Generalized map of the Himalaya.

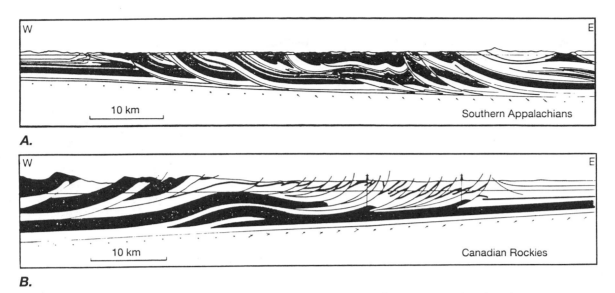

Figure 6.12 Cross sections of the major fold and thrust belts shown in parts *A* and *B* of Figure 6.11. *A.* Southern Appalachians. *B.* Canadian Cordillera.

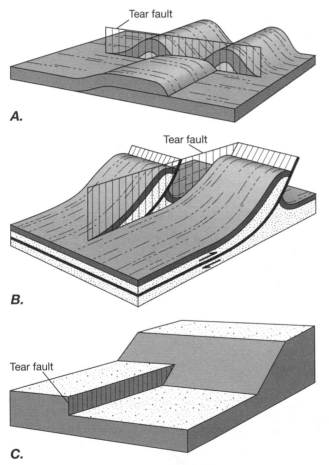

Figure 6.13 Thrust sheets segmented by tear faults. *A.* Shortening is accommodated by thrusting on one side of a tear fault and by folding on the other. *B.* Two noncoplanar imbricate thrusts are connected by a tear fault. *C.* Two segments of a thrust fault surface, each at a different structural level, are connected by a vertical sidewall ramp, or tear fault.

6.14*B*). With greater displacement, they form an antiformal stack over which the roof thrust curves through an antiform (Figure 6.14*C*). With still larger displacement, they dip predominantly toward the foreland (foreland dipping), again defining a zone of roughly constant thickness between the roof and the floor thrusts (Figure 6.14*D*).

Strata within the stack of horses generally display characteristic asymmetric anticline–syncline pairs, as shown in Figure 6.14*B, D*. Beds above and below the duplex generally parallel the roof and floor faults.

The Lewis thrust in the Waterton Field (Figure 6.7*A*) displays a more complex duplex structure (Figure 6.15*A*). There the Lewis thrust appears as the floor thrust, and the Mount Crandell thrust is the roof thrust. The duplex geometry combines elements of a hinterland-dipping duplex and an antiformal stack. Higher thrust faults in the duplex, and the horses between them, are folded over fault ramps and associated horses lower in the section, indicating that slip on the higher thrusts must have occurred before the lower ones became active. Thus formation of the thrusts progressed in time downward and toward the foreland (see Section 6.4). The Waterton Field duplex illustrates the fact that in duplexes, earlier thrusts may be folded by displacement on later faults, resulting in culminations in the earlier thrust. Erosion of such culminations subsequently can produce windows exposing lower structural levels. Restoration of this cross section (Figure 6.15*B*) is discussed in detail in Section 6.5.

Individual thrust faults, like any other structure, are not of indefinite extent; generally they are considerably shorter than the fold and thrust belt as a whole. If the shortening normal to a fold and thrust belt is

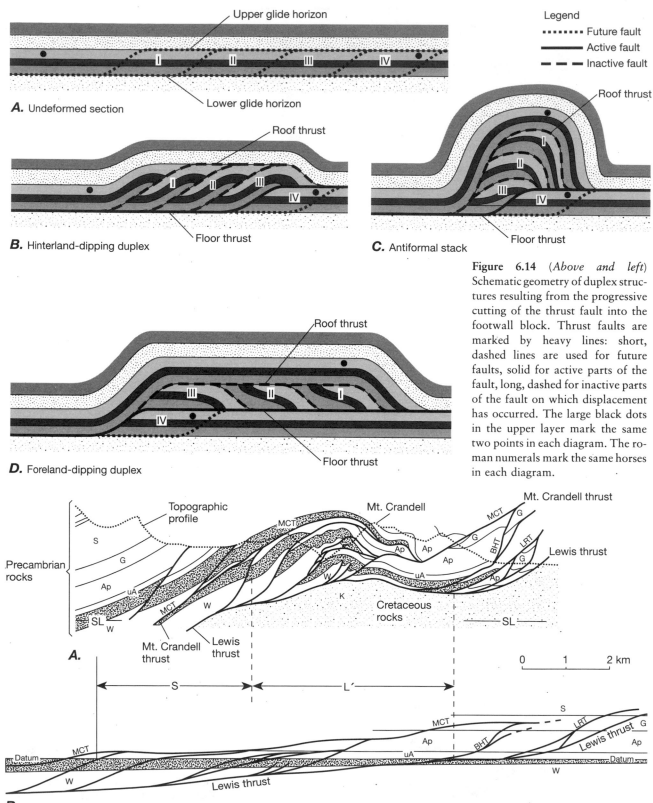

Figure 6.14 (*Above and left*) Schematic geometry of duplex structures resulting from the progressive cutting of the thrust fault into the footwall block. Thrust faults are marked by heavy lines: short, dashed lines are used for future faults, solid for active parts of the fault, long, dashed for inactive parts of the fault on which displacement has occurred. The large black dots in the upper layer mark the same two points in each diagram. The roman numerals mark the same horses in each diagram.

Legend

······· Future fault
——— Active fault
– – – Inactive fault

A. Undeformed section

B. Hinterland-dipping duplex

C. Antiformal stack

D. Foreland-dipping duplex

Figure 6.15 Cross section of duplex structure near the Waterton field in the Lewis thrust sheet near the Canada–USA border (compare Figure 6.7A). *A.* Generalized cross section showing that the Lewis thrust is the floor of the duplex where Precambrian rocks are thrust over Cretaceous siliciclastics. The Mount Crandell thrust is the roof thrust. *B.* Palinspastic balanced cross section restoring the duplex in part A to its inferred original configuration.

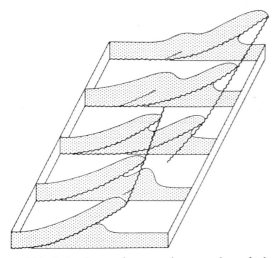

Figure 6.16 A simple transfer zone where one thrust fault dies out and the displacement is transferred through the transfer zone by folds to an *en echelon* thrust fault.

relatively constant along the belt, then where a thrust fault dies out along strike at a tip line, its displacement must be transferred to another overlapping, or *en echelon*, thrust. Figure 6.16 shows an example of the **transfer zone** between two *en echelon* thrusts that merge into the same basal décollement. As the displacement on the upper thrust decreases, shortening is taken up first by a fold in its footwall block and then by a new thrust that cuts the fold (compare Figure 6.10*A* or *B*). Finally, the upper thrust decays into a fold in the hanging wall of the lower thrust, and the displacement is progressively transferred to the lower thrust.

6.4 Kinematic Models of Thrust Fault Systems

To understand how thrust systems form, we need to know the sequence of development of the faults in duplexes and imbricate fans, that is, whether new faults develop in front of the older faults (toward the foreland) or behind (toward the hinterland).

The structure of duplexes is diagnostic of the sequence and indicates the progressive extension of the fault system toward the foreland. For example, the duplex structure illustrated in Figure 6.14*C* indicates that the youngest thrusts are those that branch from the sole fault closest to the foreland. A thrust must exist before it can be folded. If folding is a result of movement along ramps that splay off the sole fault, then displacement on the folded thrust must predate that on the underlying ramps. Moreover, a highly folded surface is not a surface of easy slip, and we expect the active part of the fault to have a simpler geometry. These relationships imply that the duplex formed as a result of the stepwise ad-

vance of the sole fault into the footwall and the progressive incorporation of the resulting horses into the hanging wall. We examine the consequences of this mechanism by looking at four idealized models for the formation of duplexes.

In Figure 6.14*A*, the sole fault in the lower glide horizon at the base of the section first cuts up across the strata on the frontal ramp to the left of block I and continues along the upper glide horizon. After the front edge of the thrust sheet has advanced part way across the top of block I, the sole fault in the lower glide horizon branches and extends into the footwall block under block I to a frontal ramp between blocks I and II. Block I becomes a horse, and the ramp and flat overlying block I cease to be active (heavy dashed line, Figure 6.14*B, C, D*). Block I is incorporated into the hanging wall, rides up over the new frontal ramp, and advances part way across the top of block II. In the process, block I and its overlying inactive thrust segment become folded. The sequence repeats itself, the sole fault branching in the lower glide horizon and advancing under block II, and the same process occurs twice more to form the final duplex.

In this model the active part of the thrust fault cuts progressively deeper into the footwall block as the horses become incorporated into the thrust sheet (the hanging wall block). The roof thrust above the duplex structure is never active as a distinct fault. It consists of segments of the upper glide horizon that are inactive by the time they become parts of the roof fault. For example, when block IV in Figure 6.14*B* is cut from the footwall block and advances up the ramp, a segment of the inactive fault above block III becomes part of the roof thrust, but by this time the segment is no longer active. Activity ceases at different times on the various segments of the roof thrust, and the segments above the youngest horses have the longest history of thrusting. In this model, the folds that develop in the horse immediately above an active ramp are slightly unfolded when the fault branches and the next horse slips up its frontal ramp.

The models in Figure 6.14*B* and *D* are constructed such that the roof thrust appears as a smooth, horizontal fault. This appearance depends entirely on how far the overriding horse advances beyond its frontal ramp; in Figure 6.14*B*, for example, it advances only a distance equal to the ramp's up-dip length. If its displacement is less or more than this amount, the roof thrust is an uneven structure. If the displacement carries the front tip of each horse just beyond the point where the next frontal ramp will emerge, the duplex develops an antiformal stack (Figure 6.14*C*). If the displacement is still greater, so that only the rear end of the youngest horse overlaps the front end of the incipient horse, a foreland dipping duplex develops (Figure 6.14*D*). This model

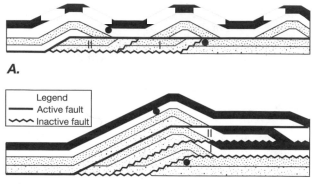

A.

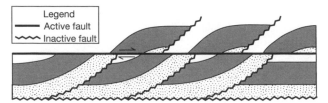

B.

Figure 6.17 Two kinematic models for the formation of duplex structure in which the duplex developed by the stepwise retreat as the thrust fault frontal ramp steps back into the hanging wall block leaving horses in the footwall block. Thrust faults are marked by heavy lines: solid for active segments of the thrust system, and wavy for inactive segments of the thrust system. *A.* The upper glide horizon remains the same with each stepwise retreat of the frontal ramp. *B.* The upper glide horizon steps up in the structure with each stepwise retreat of the frontal ramp.

Figure 6.18 A duplex structure develops if an imbricate fan is truncated by a younger thrust, which then forms the roof thrust of the duplex. Heavy lines indicate faults: solid where the fault is active, wavy where the fault is inactive.

therefore seems able to account for the geometry of duplexes found in nature. Hinterland-dipping, antiformal, and foreland-dipping duplexes constitute a continuum of structures resulting from the progressively larger displacement of each horse in the structure.

Duplexes could also develop if ramps in the thrust fault cut progressively into the hanging wall block and therefore toward the hinterland. Figure 6.17 shows the geometric consequences for two models of this mode of thrusting. With each stepwise retreat of the frontal ramp, the upper glide horizon in Figure 6.17*A* remains at the same structural level, and in Figure 6.17*B* it steps up in the structure, increasing the length of the ramp with each step. Evidence for such structures is unusual, indicating that these models do not represent common geologic processes.

In a fourth model, a duplex forms if a younger thrust fault truncates a preexisting imbricate thrust fan, thereby becoming the roof fault to the duplex in its footwall block (Figure 6.18). The major characteristics of this model are that the roof thrust is the youngest fault in the system; it is a single fault, active at the same time over its entire length; and the anticlinal parts of the horses are offset, resulting in a truncated imbricate fan in the hanging wall block.

For imbricate thrust fans, it is difficult in many cases to determine the sequence of formation of the splay faults. The kinematic models are similar to those for duplexes shown in Figures 6.14 and 6.17, but because the splay faults eventually break the surface, no roof fault forms, and a stack of imbricate thrust slices would

form regardless of whether the new faults cut into the footwall or the hanging wall block.

Thus the geometry of an imbricate thrust fan alone does not indicate whether the development of a given imbricate system progressed toward the foreland or the hinterland. Stratigraphic information, however, can provide additional evidence. In some regions where sedimentation is active, progressively younger sediments are found involved in the thrust wedges closer to the foreland. Inactive faults may be covered over by undisturbed sediments, but the same sediments in other places are cut by faults which therefore must be younger. Distinctive sediments such as conglomerates deposited toward the foreland from an active thrust sheet may subsequently be cut by the next listric fault that propagates into the footwall block. In the Idaho–Wyoming fold and thrust belt, for example, which is a southern continuation of the Canadian Rockies fold and thrust belt, such data demonstrate that the development of the imbricate fan progressed toward the foreland.

Most large fold and thrust systems appear to be dominated by the progressive cutting of fault ramps and fault splays into the footwall block and therefore toward the foreland (Figure 6.14), although examples of a progression toward the hinterland also occur at least locally, in which case they are referred to as **out-of-sequence thrusts**.

6.5 Geometry and Kinematics of Thrust Systems in the Hinterland

None of the thrust system models discussed so far deals with a complete cross section containing all fault termination lines (Section 4.4). Because thrust systems accommodate substantial shortening of the crust, and because these systems are composed of shallowly dipping faults, we must consider what happens to the continental crust below the sole fault. Moreover, to complete the model we must consider what becomes of the sole fault beneath the hinterland. We examine three models that provide a geometrically complete system.

In the first model, the sole fault could return to the surface somewhere so that the shortening along listric thrust faults in one area of the crust is balanced by extension along a system of listric normal faults in another region (Figure 6.19A). The implied pairing of a belt of shortening with a belt of lengthening may occur with shallow fault systems, such as in the sediments of the Gulf Coast (Figure 5.14C). The scale of displacement there, however, is probably much less than that observed in typical foreland fold and thrust belts, which have never been paired with an area of comparable extension.

In the second model, the basement rocks could be shortened by processes other than thrusting. The hinterland of an orogenic belt is characterized by high-grade metamorphic rocks that show abundant ductile deformation. Perhaps sole faults of the foreland fold and thrust belts terminate in a so-called **root zone** of ductile deformation within the metamorphic core. The gravitational collapse of the topographic high created by the shortening and thickening of the metamorphic core would be responsible for compression in the shallow wedge-shaped fold and thrust belt on the margin of the orogenic belt (Figure 6.19B; compare model experiment Figure 20.16B).

Still a third possible model—and an intriguing one—is that the large displacements and shortening in fold and thrust belts reflect the involvement of continental crust in the largest type of thrust system we know of on Earth, a subduction zone (Figure 6.19C). According to seismic evidence, some subduction zones are continuous down to depths of 1200 km or more and therefore have at least this much displacement on them. Where continental crust is on a down-going plate, it can be carried into a subduction zone and subducted to some extent. The ultimate sole fault to a foreland fold and thrust belt, then, may be simply a subduction zone, in which case the belt is a series of splay faults off a convergent plate boundary fault, and the driving force for thrusting is that for subduction itself.

6.6 Analysis of Displacement on Thrust Faults

On large thrust faults, the existence of piercing points (Section 4.3) on both sides of the fault is rare, and it is necessary to resort to other methods to obtain an in-

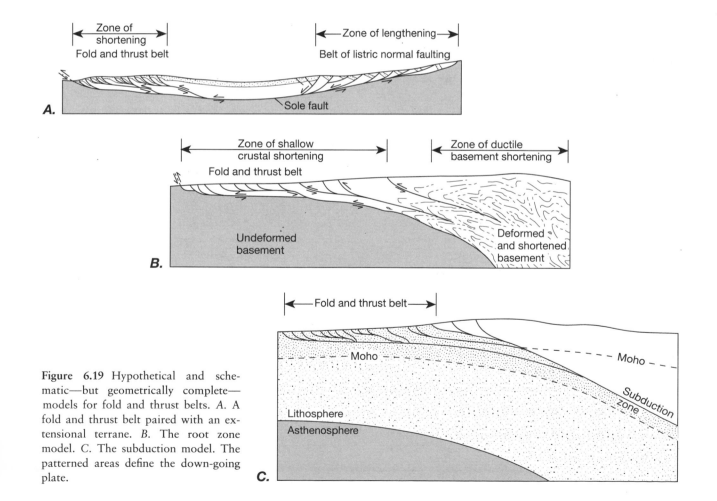

Figure 6.19 Hypothetical and schematic—but geometrically complete—models for fold and thrust belts. *A.* A fold and thrust belt paired with an extensional terrane. *B.* The root zone model. *C.* The subduction model. The patterned areas define the down-going plate.

dication of the displacement. As with other faults, the three things we need to determine are the displacement direction, the shear sense, and the amount of the displacement.

Direction and Sense of Displacement

In addition to general features of faults that indicate shear sense and relative displacement direction (Sections 4.3 and 6.1), some structures within the thrust sheet as well as the geometry of the thrust system itself can be used to constrain the thrust motion.

If, on a regional scale, the total amount of displacement on a thrust is generally the same along its length, and if erosion has not worn the sheet back unevenly, then the displacement is commonly taken to be approximately normal to the regional strike of the thrust fault or thrust system. Using this criterion, we find that the thrust sheets in the southern Appalachian Mountains (Figure 6.11A) moved to the northwest, and those in the Canadian Rockies (Figure 6.11B) moved to the northeast.

On many thrust faults, it may not be obvious in which direction the thrust sheet moved on the fault, especially on a local scale, and we must then rely on stratigraphic evidence. Although we tend to think of the hanging wall block as moving up the dip of the thrust fault, this is an unreliable assumption, because after thrusting, fault dips may be altered by folding. The slip direction of a thrust sheet is best indicated by the tendency of thrust faults to cut up-section in the direction of displacement (Figure 6.12 and 6.15).

The problem with a folded thrust fault is illustrated by the Mt. Crandell thrust shown in Figure 6.15A. In the vicinity of Mt. Crandell, the folded fault has a northeast dip, which is opposite to its dip further to the northeast and to the southwest. An inference of up-dip displacement of the thrust sheet, based only on the exposures in Mt. Crandell, would indicate top to the southwest. A restoration of the initial structure of the Mt. Crandell area (Figure 6.15B), however, shows that in fact the Mt. Crandell thrust cuts gradually but consistently up-section from southwest to northeast throughout its length, indicating that the relative displacement of the thrust sheet is toward the northeast, consistent with the regional picture.

In imbricate thrust systems, the thrusts branch up from the basal décollement in the direction of relative movement of the thrust sheet (Figures 6.12 and 6.15). In a complexly deformed region, therefore, identification of thrusts branching off the sole fault and determination of the direction in which they cut up-section provide other indications of their relative displacement.

The use of any of these criteria requires accurate data on the overall attitude of the faults, the nature of their intersections, and their relationships with the surrounding stratigraphy.

In many thrust sheets, asymmetric folds develop during thrusting. Such folds have one longer side, or limb, that dips at a relatively shallow angle and one shorter limb that dips steeply or is overturned (see Figure 6.10C, for instance). The tops of the folds can be thought of as leaning in the direction of the steep or overturned limb. This direction of leaning, the vergence of the fold, indicates the direction of relative motion of the thrust sheet. We discuss the geometry of folds in more detail in Chapter 12.

Determining the Amount of Displacement from Maps

In order to gain a better understanding of large-scale tectonic processes, as well as to predict possible sites for accumulation of economic deposits such as oil, we need to determine the amount of displacement along a thrust fault or system. Unfortunately, a map cannot provide unequivocal determinations of displacement, and it is therefore advisable to employ more than one method in order to obtain a more comprehensive understanding of the thrusting.

In some cases, the irregularities in the thrust trace, including windows and klippes, provide a minimum estimate of displacement. The desired measurement is the distance between the exposures closest to, and those farthest from, the hinterland in the movement direction. Figure 6.4 illustrates the technique. Figure 6.4A shows the displacement determinable from irregularities in the trace of the thrust fault. Using the exposure of a window and a klippe (Figure 6.4B) provides a larger lower bound for the displacement.

For the Lewis thrust (Figure 6.7A), we assume a displacement direction of N55E perpendicular to the average trend of the fault trace. The minimum possible displacement determined from the fault trace irregularities (Figure 6.4A) is approximately 12 km. Measuring parallel to the same displacement direction but using the thrust trace and the Cate Creek window, however, gives a minimum possible displacement of about 40 km.

Figure 6.20 shows the application of this analysis to two very famous windows, the Engadine and Tauern windows of the eastern Alps. In each case, the upper thrust sheets or nappes, called the Austro-Alpine nappes, overlie a lower series called the Penninic nappes, which are exposed in the windows. The distance from the rear of these windows roughly northward to the front of the main thrust trace indicates a minimum displacement of as much as 100 km for the Austro-Alpine nappes.

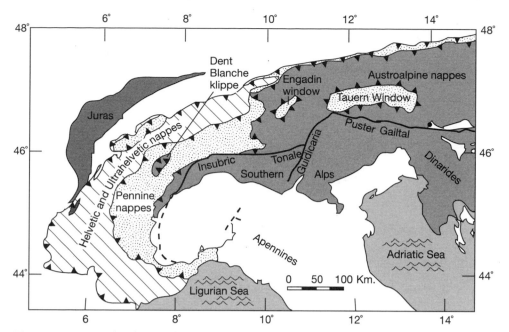

Figure 6.20 Generalized map of the Alpine region, Europe, showing three principal thrust complexes: the Helvetic (diagonal ruling), Penninic (stipple pattern), and Austro-Alpine systems (medium grey). Two windows, the Tauern and the Engadin, show the Penninic nappes underneath the Austro-Alpine nappes. The Dent Blanche klippe is an erosional outlier of the Austro-Alpine nappes on top of the Penninic nappes. To the northwest is another klippe of the Penninic nappes on top of the Helvetic nappes.

Determining the Amount of Displacement from Cross Sections

Cross sections of thrust faults can be used to determine the magnitude of the displacement if the section is parallel to the displacement direction on the fault. Figure 6.21 shows the simplest example of this situation, in which the displacement and shortening are related by the dip angle (ϕ) of the fault, and the displacement (d) is determined with a simple linear measurement.

For more complicated structures, determination of the total amount of displacement and shortening is more difficult. The cross section through the Lewis thrust system in Figure 6.15, for example, consists of a combination of imbricate and duplex faults, some of which have been folded above the younger thrusts. The original continous stratigraphic sequence appears intact at the left side of the cross section.

In such cases, we construct a balanced cross section (see Section 4.5), concentrating for this example on the area between the Lewis thrust and the Mt. Crandell thrust. The lower Altyn formation, shown as the shaded layer, is used as the reference layer because it is contained in most of the thrust wedges and horses of the thrust system. The pinning points must be to the northeast (right) of where the Lewis thrust cuts up through the stratigraphic section that is being balanced, and to

the southwest (left) of the duplex between the Mt. Crandell and Lewis thrust.

Figure 6.15*B* is the balanced palinspastic cross section, showing the undeformed stratigraphic sequence with the paths of the various thrust faults through the sequence. Two reference points at the top of the shaded lower Altyn unit in both the deformed and the palinspastic balanced cross sections show that the amount of shortening (S) caused by the thrusting amounts to

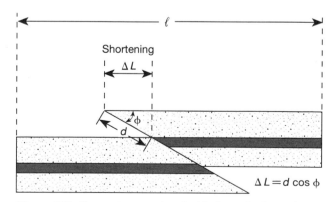

Figure 6.21 Shortening associated with thrust faulting, showing how the change in length ΔL is related to the displacement d and the dip angle of the fault ϕ for simple faults.

almost 3.5 km for a section that was originally only 8 km long. Thus this part of the section has been shortened by about 43 percent.

Across the Appalachian Valley and Ridge province from the Pine Mountain thrust to the Brevard fault (between points A and B on Figure 6.11A; see Figure 6.12A), the fold and thrust belt has accommodated roughly 280 km of shortening of the Earth's crust. "Retrodeforming" the thrust faults and taking eroded section into account reveal that the original width of the belt must have been about 435 km. A shortening of over 60 percent has occurred!

Additional Readings

Boyer, S. E., and D. Elliot. 1982. Thrust systems. *AAPG Bull.* 66: 1196–1230.

Brewer, J. A., S. B. Smithson, J. E. Oliver, S. Kaufman, and L. D. Brown. 1980. The Laramide orogeny: Evidence from COCORP deep crustal seismic profiles in the Wind River Mts., Wyoming. *Tectonophysics* 62: 165–189.

Ernst, G., 1973. Interpretive synthesis of metamorphism in the Alps. *Geol. Soc. Am. Bull.* 84: 2053–2078.

Harris, L. D., and K. C. Bayer. 1979. Sequential development of the Appalachian orogen above a master décollement—a hypothesis. *Geology* 7: 568–572.

Mitra, S. 1986. Duplex structures and imbricate thrust systems: Geometry, structural position, and hydrocarbon potential. *AAPG Bull.* 70: 1087–1112.

Price, R. A. 1981. The Cordilleran foreland thrust and fold belt in the southern Canadian Rocky Mountains, in N. J. Price (ed.), *Thrust and nappe tectonics, Geol. Soc. Lond. Sp. Pub. 9.*

CHAPTER

7 Strike-Slip Faults

Strike-slip faults are generally vertical faults that accommodate horizontal shear within the crust. Their traces on the Earth's surface may vary from straight to gently curved (Figure 7.1). Displacement on a given fault may be either right-lateral or left-lateral, and it results in no net addition or subtraction of area to the crust. In some cases, oblique strike-slip motion results from the addition of a component of horizontal contraction or extension perpendicular to the fault trace. Strike-slip faults exist on all scales in both oceanic and continental crust. In this chapter we concentrate on the geologic structures associated with continental strike-slip faults.

Tear faults are relatively small-scale, local strike-slip faults that are commonly subsidiary to other structures such as folds, thrust faults, or normal faults (see, for example, Figures 6.7C, D and 6.13). They are steeply dipping and oriented subparallel to the regional direction of displacement. They occur in the hanging wall blocks of low-angle faults and accommodate different amounts of displacement either on different parts of the fault or between the allochthon and adjacent autochthonous rocks.

The term **transfer fault** is applied to two different geometries of strike-slip faults. In extensional terranes they are parallel to the regional direction of displacement and mark domains of different normal fault geometry and displacement (see Figures 5.11, 5.12). Imbricate systems of normal faults—and possibly their detachments—terminate against such transfer faults and may have different amounts of displacement and different orientations from the normal faults in adjacent

domains. There is no clear distinction between these faults and the tear faults described above, except perhaps that transfer faults may be of larger scale and may accommodate larger amounts of slip. In strike-slip terranes, transfer faults lie at a high angle to the regional direction of displacement and connect adjacent or *en echelon* parallel strike-slip faults. They accommodate the transfer of displacement from one fault to the next, and slip on these faults is generally oblique.

Transform faults and **transcurrent faults** are major regional strike-slip fault systems that generally comprise zones of many associated faults (Figure 7.2). Transform faults are strike-slip faults that form segments of lithospheric plate boundaries (Figure 7.2A). Transcurrent faults, on the other hand, are regional-scale strike-slip faults in continental crust that are not parts of plate margins[1] (Figure 7.2B). Both types of faults may be many hundreds of kilometers long and may have accumulated relative displacements of up to several hundred kilometers.

---

[1] The specific usage of these two terms is not universally agreed on. The confusion arises in part from the fact that before the development of plate tectonics, *transcurrent fault* was used to refer to all major strike-slip faults, some of which are now recognized to be plate boundaries. Moreover, *transform fault* originally referred to faults connecting offset segments of oceanic spreading ridges. Its use has now been generalized to include all plate boundary strike-slip faults. *Wrench fault* is another term used to refer to strike-slip faults in a variety of specific senses. (We do not use this term.)

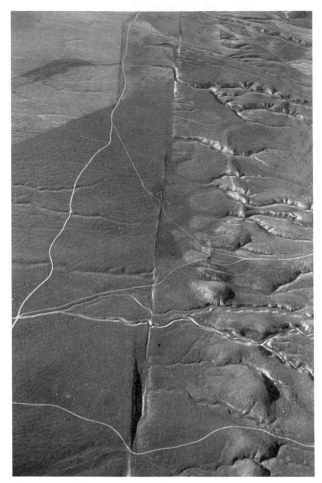

A.

B.

Figure 7.1 Photographs of strike-slip faults illustrating rectilinear fault traces. *A.* San Andreas fault, Carrizo plain, California. Aerial view looking northwest along trace of the San Andreas fault. *B.* Landsat image of Altyn Tagh fault, China showing through-going nature of the structure (see location in Figure 7.2*B*).

At outcrop or local scale, transform and transcurrent faults are indistinguishable. One must identify them on the basis of the regional plate tectonic environment and the tectonic role that each plays. For most plates, recognizing a transform boundary is straightforward. In a few situations, however, such as in Asia (Figure 7.2*B*), the distinction between transform and transcurrent faults depends in part on how small a block one chooses to accept as a "tectonic plate."

The San Andreas fault system of California (Figures 7.1*A* and 7.2*A*), is a right-lateral transform fault system 1300 km long that connects two triple junctions, one south of the Gulf of California and the other at Cape Mendocino on the north coast of California. It consists of many roughly parallel faults in a zone as much as 100 to 150 km wide. It displays along its length many of the characteristic features of strike-slip faults, and because it has been exceptionally well studied, it furnishes numerous examples of structures that we describe in the following sections.

Central and eastern Asia contains a complex system of transcurrent faults (Figures 7.1*B* and 7.2*B*), dominated by left-lateral faults in eastern Tibet and by right-lateral faults in an area extending from Lake Baikal in the northeast to the Quetta-Chaman fault in the southwest. Many workers attribute this complex system of faults to the effects of the northward-moving Indian plate impinging against the Asian crustal block, and this model accounts for many of the observed features. Several examples of characteristic strike-slip fault structures that we discuss in the following sections come from this complex.

7.1 Characteristics of Strike-Slip Faults

Most strike-slip faults are approximately planar and vertical, at least near the surface of the Earth. As a result, their fault traces tend to be straight lines on a map, even across rugged topography. Many large strike-slip faults are marked by prominent continuous topographic features on the Earth's surface that are visible even from space (Figure 7.1*B*). The topographically high side of a strike-slip fault commonly changes from one side to the other along the fault trace. The topographic

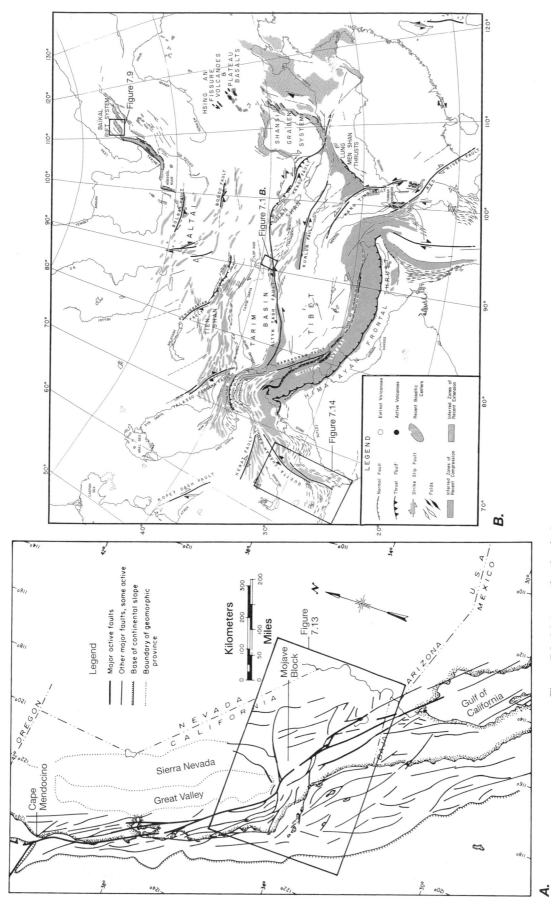

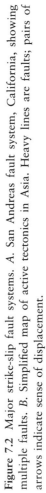

Figure 7.2 Major strike-slip fault systems. *A.* San Andreas fault system, California, showing multiple faults. *B.* Simplified map of active tectonics in Asia. Heavy lines are faults; pairs of arrows indicate sense of displacement.

expression of the fault may result from minor components of vertical slip along segments of the fault associated with a component of contraction or extension across the fault, with differences in temperature of the rocks across the fault, with juxtaposition of originally separated topographic features, or with juxtaposition of rocks that differ in resistance to erosion.

The dominantly horizontal slip on strike-slip faults produces a horizontal separation that often is used as an indication of strike-slip faulting. If the cutoff line of the bedding on the fault is parallel to the displacement, however, no separation is evident (Figure 7.3A). If beds are inclined such that their cutoff lines are oblique to the displacement, the separation on a vertical cross section of the fault can be either right-side-up (Figure 7.3B) or left-side-up (Figure 7.3C), depending on the relative orientation of the beds and the fault and on the sense of displacement on the fault. Large strike separations of a planar boundary, such as a lithologic contact, amounting to many tens or hundreds of kilometers constitute reasonable evidence for strike-slip faulting (see Figure 4.18), although small strike separations can result from other types of fault slip (see Figures 5.2 and 6.2).

Strike-slip faults display the typical features that we discussed in Chapter 4. Slickenside lineations are subhorizontal. Drag folds may form along some strike-slip faults if the bedding is favorably oriented (Section 4.2), although folds reflecting distributed deformation on either side of the fault are more common (see Figure 7.4B and the following discussion). Geomorphic features characteristic of strike-slip faults include linear erosional depressions (Figure 7.1A), sag ponds, springs, and offset streams (Figure 4.13) and topography, including

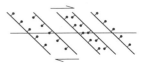

A. Subsidiary R, R′, and P shear fractures **B.** Folds

C. Thrust faults **D.** Normal faults

Figure 7.4 Structures associated with strike-slip faults and their orientations relative to the shear sense on the fault.

shutter ridges, which occur where ridge and canyon topography is transected by a strike-slip fault, and a segment of a ridge has been displaced in front of a canyon, shutting it off.

A variety of shear fractures, folds, normal faults, and thrust faults are found associated with strike-slip faults. The orientations of these structures relative to the main strike-slip fault are characteristic of the sense of shear on the fault (Figure 7.4). Subsidiary shear fractures, known as **Riedel shears** or **R shears**, develop at a small angle (roughly 10° to 20°) to the main fault in an *en echelon* array (Figure 7.4A). R shears are synthetic to the main fault, which means they are subparallel and have the same shear sense, and the acute angle formed by the traces of the R shear and the main fault points in the direction of relative motion of the block containing the R shear. Other subsidiary shears may also develop. P shears are synthetic to the main fault and are oriented symmetrically with respect to the fault from the orientation of the R shears. Conjugate Riedel shears, or R′ shears, are antithetic shear fractures that are oriented at high angles to the fault (roughly 70° to 80°) and have a shear sense opposite to that of the main fault. On a small scale, these various secondary shear fractures are responsible for some of the sense-of-shear criteria for brittle faulting that we discuss in Section 4.3 (see Figure 4.16). On a large scale, they can form a complex anastomosing network of faults that become very difficult to interpret.

Folds and thrust faults form in an *en echelon* arrangement above or beside major strike-slip faults (Figure 7.4B, C). The trend of the fold hinges and the strike of the thrust faults are oriented at 45° or less to the strike-slip fault, and the acute angle defined by the intersection of the strike-slip fault trace with the fold hinge or the thrust fault trace points in the direction of relative motion of the fault block *opposite* the one containing the fold or thrust. These structures record a component of contraction oblique to the strike-slip fault and

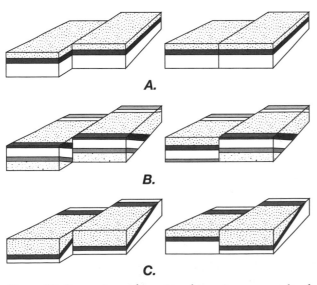

Figure 7.3 Separations of stratigraphic units as a result of left-lateral (sinistral) strike-slip faulting. Drawings on the right show vertical sections after faulting.

roughly perpendicular to the trends of the folds and the thrust faults.

Normal faults may also form *en echelon* arrays along strike-slip faults, and they are oriented at roughly 45° to the main fault and close to perpendicular to the orientations characteristic of fold hinges and thrust faults (Figure 7.4D). Thus the acute angle defined by the intersection of the traces of the strike-slip and normal faults points in the direction of relative motion of the block containing the normal fault. These structures record a component of extension that is oblique to the strike-slip fault and is perpendicular to both the normal faults and the contraction orientation recorded by folds and thrust faults.

Many of these associated structures develop as a result of the inherent geometry of strike-slip faults and the displacement along them, as we discuss in Section 7.2. Other structures reflect a distributed component of displacement along or across the fault, as we describe in Section 7.4.

7.2 Shape, Displacement, and Related Structures

Single Faults

At depth, strike-slip faults may terminate on another fault, such as a low-angle detachment, or they may continue through the crust and lose their identity at depth in a zone of ductile deformation. Earthquakes along modern strike-slip faults typically are present only down to depths of about 15 km. Below this seismic zone, aseismic shear is probably accommodated by cataclastic deformation in a transition zone—and below that by ductile deformation. A strike-slip fault terminating against a horizontal detachment is geometrically equivalent to a dip-slip fault terminating against a vertical tear fault.

Although they are characteristically vertical with straight map traces, strike-slip faults also include **bends** (or **jogs**) and **stepovers** (or **offsets**) (Figure 7.5). Bends are curved parts of a continuous fault trace that connect two noncoplanar segments of fault. Stepovers are regions where one fault ends and another *en echelon* fault of the same orientation begins. Bends and stepovers are described geometrically as being either right or left depending on whether the bend or step is to the right or to the left as one progresses along the fault. This description remains the same regardless of the sense of shear on the fault zone. Bends are geometrically equivalent to frontal ramps on dip-slip faults.

Displacement on strike-slip faults ideally is horizontal and therefore parallel to the strike of the fault. For a vertical fault, the trace of the fault on any topographic surface is straight and parallel to strike—and therefore also parallel to the ideal displacement direction. We describe a bend or stepover kinematically as **contractional** or **restraining,** if material is pushed together by the dominant fault shear (dashed arrow pairs, Figure 7.5A, D); the bend or stepover is **extensional,**

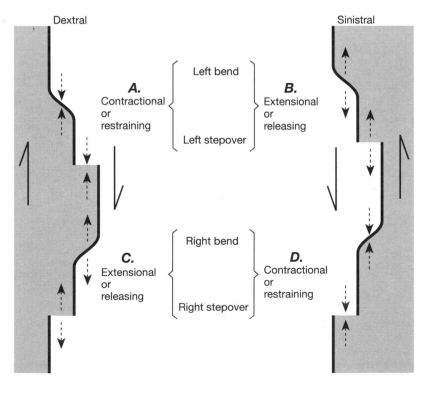

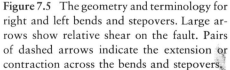

Figure 7.5 The geometry and terminology for right and left bends and stepovers. Large arrows show relative shear on the fault. Pairs of dashed arrows indicate the extension or contraction across the bends and stepovers.

releasing, or **dilatant** if material is pulled apart by the dominant shear (dashed arrow pairs, Figure 7.5B, C). On dextral faults (Figure 7.5A, C), right bends and stepovers are extensional and left bends and stepovers are contractional, whereas on sinistral faults (Figure 7.5B, D), left bends and stepovers are extensional and right bends and stepovers are contractional. Thus bends, stepovers, and broad curves in the trace of a strike-slip fault do not permit pure strike-slip motion but require some accommodating deformation.

Strike-Slip Duplexes

Displacement along strike-slip faults with bends or stepovers produces a complex zone of deformation. Commonly the result is a **strike-slip duplex,** which is a set of horizontally stacked horses bounded on both sides by segments of the main fault (see Figure 4.26C). Such a duplex may be extensional (Figure 7.6) or contractional (Figure 7.7), depending on whether it forms at an extensional or a contractional bend or stepover.

Strike-slip duplexes must differ from duplexes that form along dip-slip faults, because the different orientation of the shear plane places different constraints on the deformation. For dip-slip faults, the faulting accommodates either a thickening or a thinning of the crust, which results in a vertical displacement of the surface of the Earth, which is a free surface (Figure 4.26A, B). For strike-slip faults, however, the corresponding thick-ening or thinning would have to occur in a horizontal direction (Figure 4.26C), which is impossible because of the constraint imposed by the rest of the crust. There being no free vertical surface, the required thickening or thinning can be accommodated only by vertical motion of the free horizontal surface, and therefore slip on strike-slip duplex faults cannot be purely strike-slip but must be oblique. To accommodate this component of motion, faults in a strike-slip duplex must have a different geometry from those in dip-slip duplexes.

The oblique slip on the faults bounding the horses in an extensional strike-slip duplex must be a combination of strike-slip and normal slip (Figure 7.6C); on the faults in a contractional strike-slip duplex, it must be a combination of strike-slip and reverse slip (Figure 7.7C). The shortening associated with contractional duplexes can also be accommodated by folding subparallel to the reverse faults (see Figure 7.4). The deformation required at contractional or extensional bends provides one mechanism for producing the *en echelon* folds and the normal faults and thrust faults associated with strike-slip faults that we describe in Section 7.1 (Figure 7.4; see also Figure 7.13).

In a strike-slip duplex, the shape of the faults on the vertical section normal to the main fault trace is referred to as a **flower structure.** If the dip-slip component is normal, the faults tend to be concave up and to form a **normal,** or **negative, flower structure,** also known as **tulip structure** (Figures 7.6C, 7.8A). If the dip-slip component is reverse, the faults tend to be con-

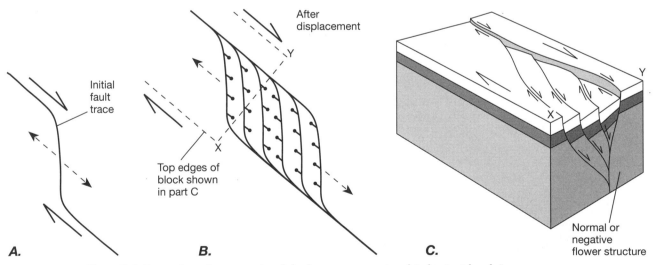

Figure 7.6 Formation of an extensional duplex at an extensional (releasing) bend. Large arrows indicate the dominant shear sense of the fault zone; small arrows indicate the sense of strike-slip and normal components of motion on the fault splays. *A.* Extensional bend on a dextral strike-slip fault. *B.* An extensional duplex developed from the bend in part *A. C.* A block diagram showing a normal, negative, flower structure in three dimensions. The block faces are vertical planes along the dashed lines in part *B.*

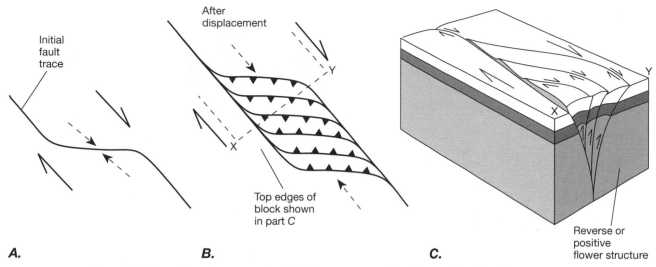

A.   **B.**   **C.**

Figure 7.7 Formation of a contractional duplex at a contractional (restraining) jog. Large arrows indicate the dominant shear sense of the fault zone; small arrows indicate the sense of strike-slip and reverse components of motion on the fault splays. *A.* Contractional bend on a dextral fault. *B.* A contractional duplex developed from the bend in part *A*. *C.* A block diagram showing reverse, or positive, flower structure in three dimensions. The block faces are vertical planes along the dashed lines in part *B*.

vex up and to form a **reverse, or positive, flower structure,** also known as **palm tree structure** (Figures 7.7*C*, 7.8*B*). All these botanical names suggest the similarity in cross-sectional form between the plants and the faults, but given the difference in the third dimension, they are not particularly apt.

In actual cases, the slip on faults in strike-slip duplexes may vary along strike from having a normal component at one end to having a thrust component at the other. Such faults, which are sometimes called **scissor faults,** accommodate the rotation of horst blocks in the duplex.

Examples of these two types of flower structures can be seen in seismic reflection profiles from the southern Andaman Sea (Figure 7.8*A*) and from the Ardmore Basin in southern Oklahoma (Figure 7.8*B*). In Figure 7.8*A*, the grabenlike offset of reflectors across the upper faults in the fault zone indicates a component of normal slip that is characteristic of normal (negative) flower structure. In Figure 7.8*B*, the major faults show a component of thrusting that is characteristic of reverse (positive) flower structure.

Displacement at extensional bends and stepovers produces topographic depressions known as **pull-apart basins,** which commonly fill with water to produce sag ponds or lakes. On a large scale, pull-apart basins are usually rhomb-shaped, fault-bounded basins several kilometers or tens of kilometers in dimension (Figure 7.9). Faulting may be accompanied by volcanic eruptions that cover the floor of the basin. Because they form topo-

graphic depressions, pull-apart basins generally accumulate large thicknesses of alluvial and/or lake deposits. With continued displacement, the basin may eventually be split by a younger segment of the fault, which separates opposite sides of the basin from each other.

Terminations

Strike-slip faults can terminate in the crust at a zone of either extensional or contractional deformation, depending on the location of the deformation zone relative to the slip vectors on the fault. Extension may be accommodated where strike-slip faults splay and turn into an imbricate fan of normal faults (Figure 7.10*A*, *B*). Similarly, contraction may be accommodated by an imbricate fan of thrust faults and/or folds (Figure 7.10*C*, *D*). Within such zones, strike-slip displacement diminishes progressively to zero along the fault.

In some cases, the fault may branch into a fan of strike-slip splay faults (also called a **horsetail splay**) that commonly curve toward the receding fault block (Figure 7.10*E*). The displacement on any individual splay is relatively small, but the sum of the displacements on all the faults in the splay equals the displacement on the main strike-slip fault. The fan thereby distributes the deformation through a large volume of crust. The geometry of a horsetail splay on a strike-slip fault is comparable to that of an imbricate fan of listric faults on a low-angle thrust fault or normal fault.

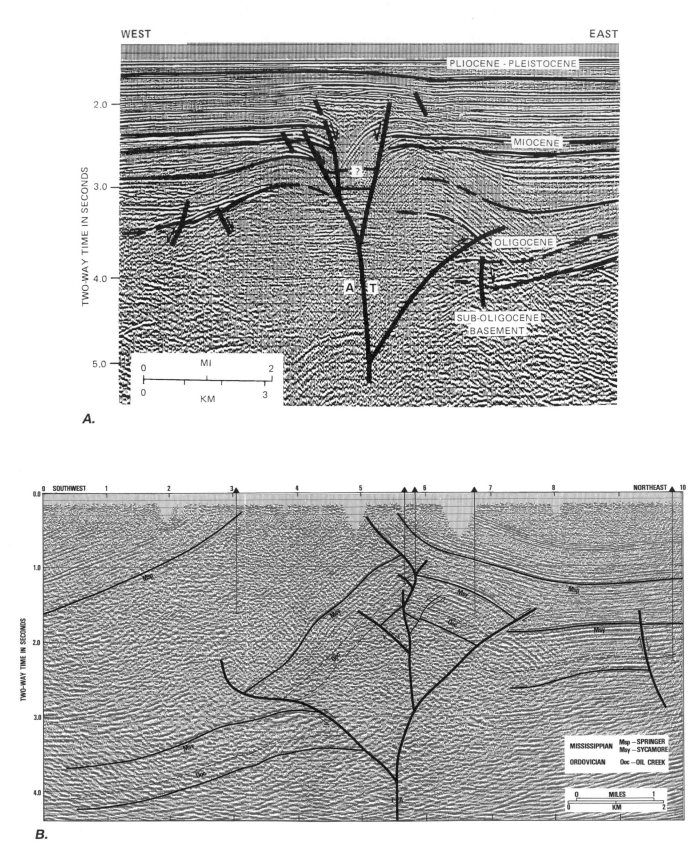

Figure 7.8A. Example of negative flower structure from an extensional duplex on a dextral strike-slip fault from the Andaman Sea between India and the Malay Peninsula. Unmigrated seismic reflection profile. *B.* Example of positive flower structure from a contractional duplex on a sinistral strike-slip fault in the Ardmore Basin, Oklahoma. Migrated seismic profile.

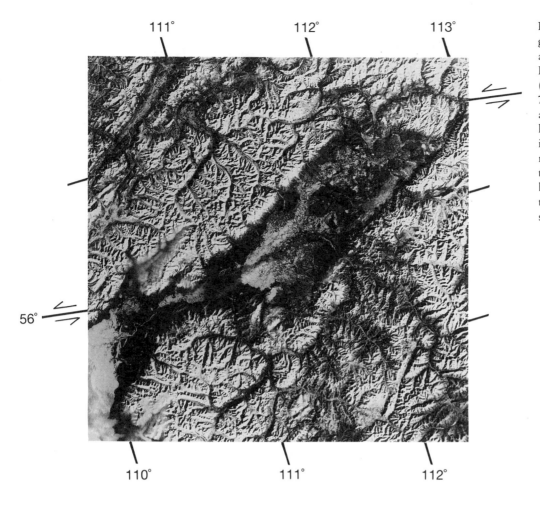

Figure 7.9 The Angara graben, a major pull-apart basin northeast of Lake Baikal in Siberia (see location in Figure 7.2B). The basin formed at a left stepover in a left-lateral strike-slip fault, indicated at the upper right and lower left of the photo. The basin is bounded by northeast-trending normal fault scarps.

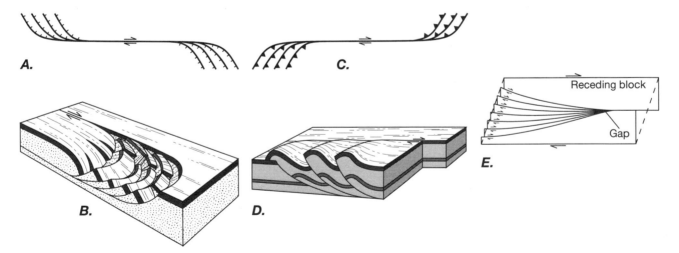

Figure 7.10 Termination of faults by the formation of imbricate fans. *A.* Geometry of extensional imbricate fans at the ends of a dextral fault. *B.* Extensional normal faulting at the termination of a dextral strike-slip fault. *C.* Geometry of contractional imbricate fans at the ends of a dextral fault. *D.* Contractional folding and thrust faulting at the termination of a dextral strike-slip fault. *E.* Geometry of a horsetail splay of strike-slip faults at the ends of a dextral strike-slip fault. The total displacement on the single fault at the right side of the block is the sum of small displacements on the individual splay faults at the left of the block. Splay faults tend to curve toward the receding block.

Transform faults terminate at major plate boundaries, where the relative slip on the fault is accommodated either by production of crust at a spreading center or by destruction of crust at a subduction zone.

7.3 Structural Associations of Strike-Slip Faults

Tear Faults

Strike-slip faults commonly are secondary structures associated with major faults and folds. Tear faults characteristically develop in regions of normal faulting (see Figures 5.11 and 5.12) and in fold and thrust sheets (see Section 6.2, Figures 6.7C, D and 6.13). They accommodate different amounts of extension or contraction in adjacent regions. The Jura Mountains of Switzerland (Figure 7.11; see Figure 6.20) are a classic example of a fold and thrust belt that is segmented by generally N-to-NW–trending tear faults. Fold hinges terminate laterally against the tear faults (compare Figure 6.13A, B), separating sections of the thrust sheet that have different magnitudes of displacement on the décollement.

Bends, Stepovers, and Duplexes

Transcurrent and transform faults never occur as simple planar faults through the crust. They are characterized by complex zones of anastomosing, parallel, or *en echelon* faults that are not perfectly straight (Figure 7.2A) and that therefore result in a variety of accommodation structures (Figure 7.4).

An excellent example of an extensional duplex occurs on the active Dasht-E Bayaz fault in northeastern Iran (Figure 7.12). The duplex is in the process of developing at a left bend on the left-lateral fault. The main trace of the fault trends obliquely through the middle of the duplex. Subsidiary faults to the east and a dense concentration of fractures to the west outline two horses in which the fracture density is much lower. The inset on Figure 7.12 shows an idealization of the duplex geometry.

An even larger left bend in the San Andreas fault system occurs in Southern California in the region where the Garlock fault intersects the San Andreas (Figures 7.2A, 7.13). Here the contraction expected at a left bend in a dextral fault is reflected by the Transverse Ranges, a block of crust that has been uplifted on east-west–trending thrust faults (Figure 7.13). Along this contrac-

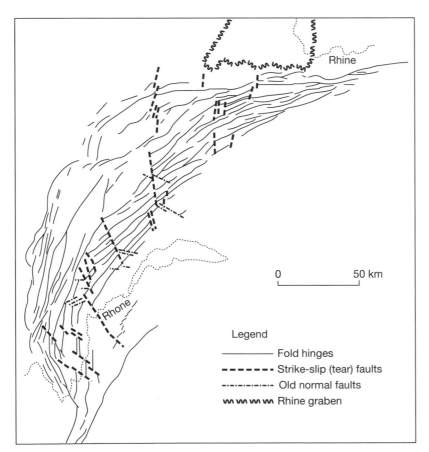

Figure 7.11 Tear faults in the Jura fold and thrust belt (see Figure 6.20 for location). The generalized map of Jura mountains shows the major fold axes tear faults, and the boundary of the Rhine graben. Note how fold axes tend to terminate against the tear faults.

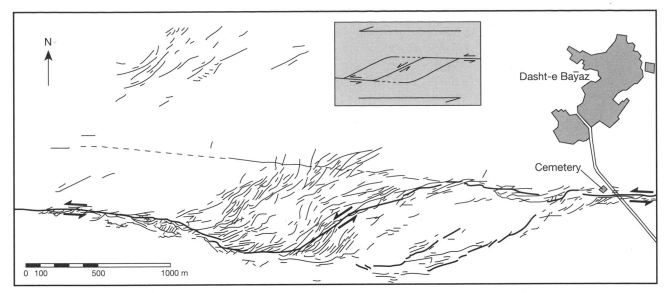

Figure 7.12 An extensional duplex developing on the active Dasht-e Baȳaz fault in northeastern Iran.

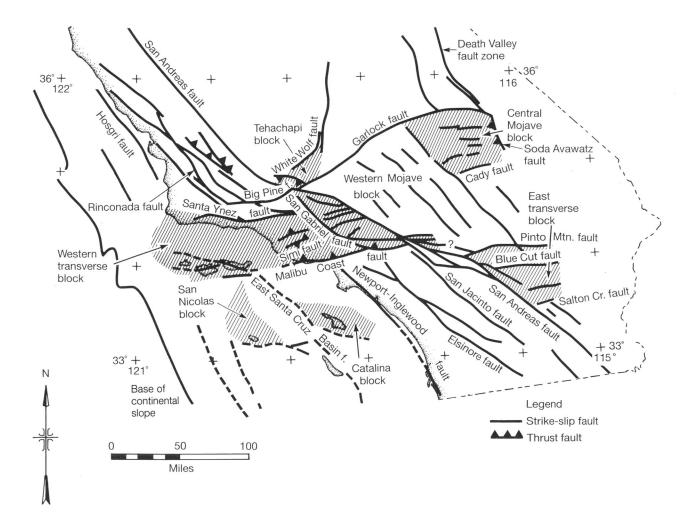

Figure 7.13 The San Andreas–Garlock fault systems in southern California.

tional bend, however, extensional basins are also present, illustrating the complex interplay of extensional and contractional structures in major strike-slip fault systems. These basins, which are filled with Neogene sediments, probably represent remnants of pull-apart basins that originally formed in extensional duplexes, some of which have been displaced from their original location.

The bend at the Transverse Ranges coincides with the intersection of the right-lateral San Andreas fault and the left-lateral Garlock and Big Pine faults. The Mojave block between the Garlock and San Andreas faults contains NW-trending dextral strike-slip faults as well as west-trending sinistral strike-slip faults, and parts of the block have experienced large amounts of roughly east–west extension. All these faults are no older than Tertiary. Understanding such a complex mosaic of faults requires an understanding of the history of each individual fault in relation to all the others—a

challenge indeed! In the following section, we discuss one kinematic model for the faulting in this region that accounts for some aspects of the geology but is only a partial solution to the complex puzzle of deformation.

Terminations

Where strike-slip faults turn at their ends into thrust faults, the fault curves around, shallows in dip, and ends up trending approximately perpendicular to the direction of movement (Figure 7.10C, D). The active Quetta-Chaman fault system of Pakistan is a good example of this structure (Figure 7.14; see the west edge of Figure 7.2B). There the left-lateral Quetta-Chaman fault system terminates southward into a series of thrust faults and folds that in fact are part of a modern convergent plate margin.

At its east end, the Garlock fault apparently turns south and becomes a thrust fault that dips westward

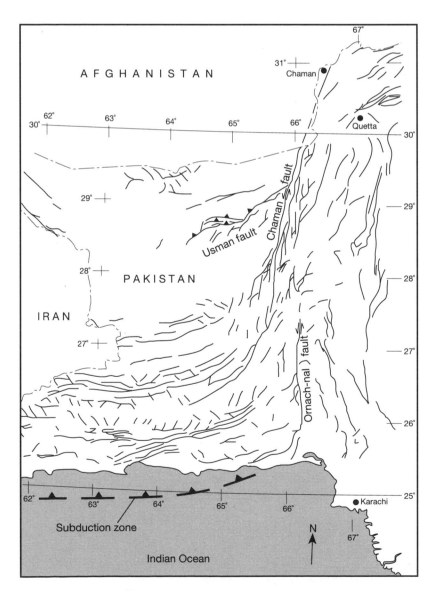

Figure 7.14 Map of faults in southern Pakistan. Major north-south–trending faults such as the Chaman and the Ornach-Nal faults are sinistral strike-slip faults that pass southward into an east-west–trending fold and thrust belt. Most of the east-west–trending faults are interpreted to be thrust faults synthetic to the subduction zone that lies in the Indian Ocean to the south. The short northeast- and northwest-trending faults in the south may be conjugate orientations of tear faults (see location in Figure 7.2B).

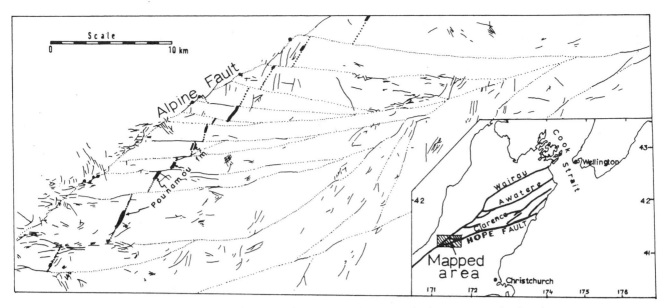

Figure 7.15 Termination of the Hope fault against the Alpine fault in New Zealand. Splaying of the Hope fault and curving of the splays toward the receding fault block are both evident. The displacement on the splays is defined by the offset of the Pounamou formation.

underneath the Soda and Avawatz Mountains (Figure 7.13). The geology here, however, is complicated by the intersection of the Garlock fault with the Death Valley fault zone.

The Hope fault, which is one strand of the Alpine fault system in New Zealand, provides a good example of the termination of a fault at a horsetail splay (Figure 7.15). The fault splays out against the Alpine fault with a relatively small amount of displacement distributed to each of the splays, as indicated by the horizontal separation of the Pounamou formation (compare Figure 7.10E).

7.4 Kinematic Models of Strike-Slip Fault Systems

As with other faults, it is useful to consider simplified kinematic models of strike-slip fault systems in order to gain insight into the complexities that can develop. In this section, we discuss models of distributed shear and of oblique strike-slip that can account for some of the folds, thrust faults, and normal faults that develop near strike-slip faults. We also discuss a model that accounts for some aspects of the regional deformation associated with the fault systems in southern California.

Many of the structures that develop near strike-slip faults can be accounted for by assuming that part of the shearing is distributed through the rock on either side of the fault. This model is illustrated in Figure 7.16, which shows two squares inscribed across a strike-slip fault (Figure 7.16A) that become separated by motion

on the fault and also deformed into parallelograms by shearing distributed on either side of the fault (Figure 7.16B). As a result of the distributed shear, one diagonal of the square becomes shorter and the other longer. This model accounts for the formation and orientation of the folds and faults in Figure 7.4. The folds and thrust faults trend perpendicular to the direction of shortening and the normal faults trend perpendicular to the direction of lengthening.

The orientation of major strike-slip faults is not necessarily exactly parallel to the direction of relative

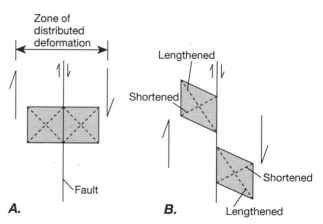

Figure 7.16 Kinematic model of a strike-slip fault where part of the shearing is distributed on either side of the fault. *A*. Before, and *B*. after shearing on and near the fault. Folds and thrust faults form at 45° to the main fault perpendicular to the direction of shortening, and normal faults form at 45° to the main fault perpendicular to the direction of lengthening (see Figure 7.4).

motion of the adjacent fault blocks. For transform faults, for example, minor changes in plate motion can result in a component of contraction or extension across the fault that can be accommodated only by the development of other structures such as folds and thrust faults or normal faults, respectively. The contractional strike-slip model for the San Andreas fault may be the explanation for the uplift of the Coast Ranges on the west side of the Central Valley of California and for a series of thrust faults and folds that are currently active along the west side of the valley.

One kinematic model of the San Andreas system is shown in Figure 7.17. It represents an effort to integrate the numerous strike-slip faults in the region into a rational pattern. The model idealizes the domains of roughly parallel faults by assuming that they are perfectly straight faults defining the boundaries of rigid fault blocks (compare Figures 7.2A and 7.13). The domains comprise either a set of right-lateral NW-trending faults or a set of left-lateral faults originally trending N but now trending ENE. The model predicts that domains dominated by NW-trending right-lateral faults should not have rotated significantly during the deformation but that domains dominated by left-lateral faults, such as the Transverse Ranges, should have rotated clockwise by as much as 80° (Figure 7.17B). Subsequent to rotation, the Transverse Ranges have been split into eastern and western blocks by slip on the San Andreas fault. Slip on the fault system produces a net N–S shortening and E–W lengthening of southern California.

The assumption of rigid fault blocks requires that numerous gaps open during shearing, especially along domain boundaries (shaded areas in Figure 7.17B). Although deformation around these gaps would certainly be complex, many of the model gaps can be correlated with deep basins filled with large thicknesses of young, locally derived sediments. In addition, paleomagnetic determinations of the orientation of the paleopole in parts of the region are consistent with the large rotations, and with progressive rotation through time, as indicated by the model. These data are consistent with the model in that they permit only small counterclockwise rotation of the right-lateral fault domains, but large clockwise rotations of up to 70° to 80° for the western Transverse Ranges block.

Despite its complexity, this controversial model is not complete. It does not account, for example, for slip along the Garlock fault, for Basin and Range extension north of the Garlock, for extension in the Mojave block, for the rise of the Transverse Ranges on thrust faults associated with the contractional duplex along the San Andreas fault, or for any nonrigid behavior of the various fault blocks. Moreover, it does not consider the problem of the fault geometry and the displacement at

depth. Nevertheless, such tentative models are useful because they provide testable predictions and focus attention on critical problems.

As with other faults, complete models of strike-slip faults must account for the termination of the faults at depth as well as along strike (see Section 7.2). For large fault systems such as the San Andreas fault, the Alpine fault of New Zealand, and the Red River and Altyn Tagh faults in Asia, displacements of several hundred kilometers have accumulated. The only crustal

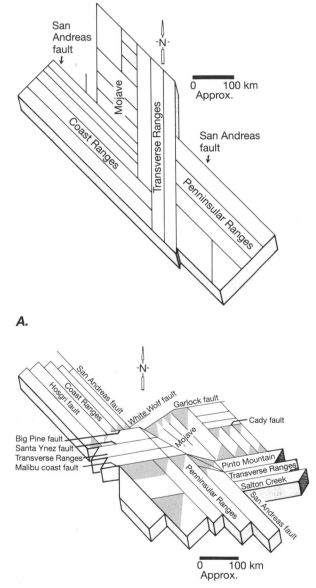

A.

B.

Figure 7.17 Rigid strike-slip fault block model for the development of the structures associated with the San Andreas fault system in southern California. A. Configuration of faults before displacement in Oligocene time. B. Present configuration, showing the right-lateral offset of the Transverse Ranges. Shaded areas indicate the location of basins that would open up as a result of sliding of the rigid blocks.

structures that seem capable of absorbing such enormous displacements are plate boundaries.

The function of the San Andreas fault as a transform fault between triple junctions in the Gulf of California and off the northwest coast of California and Oregon is well recognized. The association of the megashears in Asia with the collision of India has been proposed, but except for the Quetta-Chaman faults, the structures themselves do not serve as transform faults, and their association with plate boundaries other than the collision zone is not obvious. Accommodation of such large displacements, however, would seem to require involvement not just of crustal, but also of mantle, rocks.

7.5 Analysis of Displacement on Strike-Slip Faults

In Section 4.3 we discussed the most important methods of determining displacement on strike-slip faults. The matching of displaced geologic features on opposite sides of the fault provides the most reliable determinations. For relatively small displacements, the problem of distinguishing the separation from the displacement is an important one. The possibility is rather remote, however, that horizontal separations on the order of a hundred kilometers were produced by displacement on a fault other than a strike-slip fault. Figure 4.18 shows an example of the determination of large displacement on the San Andreas fault.

The geometric model used for strike-slip faulting in Figure 7.17 is identical to the model for rotating planar, normal faults (Figure 5.22B). One need only imagine Figure 5.22B to be a map view of strike-slip faults instead of a cross section of normal faults: θ is the angle of rotation of the blocks from the fixed boundary, ϕ is the angle between the faults and the fixed boundary, and the width of the fault block is $w = L \sin (\phi + \theta)$. Because $L \sin \theta = d \sin \phi$, the four basic parameters describing the deformation—d, w, ϕ,

and θ—are related by the equation

$$\frac{d}{w} = \left(\frac{d}{L}\right)\left(\frac{L}{w}\right) = \left[\frac{\sin \theta}{\sin \phi}\right]\left[\frac{1}{\sin (\theta + \phi)}\right]$$

which must be satisfied if the model is correct. In principle, all the parameters are measurable. We can determine the rotation θ of a crustal block by measuring the rotation of the paleomagnetic pole; we can measure the present angle ϕ of the faults from a fixed boundary, we can find the width w of a particular fault block, and we can use the displacement of geologic features on individual faults to determine the displacement d. As discussed in Section 7.4, application of this test of the model to the southern California region has confirmed the predictions of the model.

7.6 Balancing Strike-Slip Faults

In Section 4.5 we discuss the technique of balancing cross sections of dip-slip (normal and thrust) faults. The assumptions used in balancing are valid only when the deformation has been essentially two-dimensional such that no net movement of material has taken place into or out of the plane of the cross section. Vertical cross sections perpendicular to strike-slip faults do not meet this requirement, and it is generally inappropriate to attempt to balance such cross sections.

The appropriate plane for possible balancing of strike-slip faults is the plane of the map, which contains the fault slip vector. This plane is generally parallel to bedding, however, so the boundaries that are used to measure lengths and areas in balancing dip-slip faults are commonly not available for strike-slip faults. Moreover, the vertical displacement that accompanies deformation at bends and stepovers violates the strict condition of two-dimensional deformation. Thus any valid balancing of strike-slip faults would have to account for the motion normal to the plane of balancing.

Because of these difficulties, the use of balanced sections as a method of analyzing strike-slip fault zones is not commonplace.

Additional Readings

Aydin, A., and A. Nur. 1982. Evolution of pull-apart basins and their scale independence. *Tectonics* 1: 91–105.

Christie-Blick, N. and K. T. Biddle. 1985. Deformation and basin formation along strike-slip faults. In K. T. Biddle and N. Christie-Blick (ed.), Strike-slip deformation, basin formation, and sedimentation. *Society of economic pleontologists and mineralogists special publication 37*, 1–34.

Garfunkel, Z. and H. Ron. 1985. Block rotation and deformation by strike-slip faults 2: The properties of a type of macroscopic discontinuous deformation. *J. Geophys. Research* 90: 8589–8602.

Sylvester, A. G. 1988. Strike-slip faults. *Geol. Soc. Am. Bull.* 100: 1666–1703.

Woodcock, N. J., and M. Fischer. 1986. Strike-slip duplexes. *J. Struct. Geol.* 8: 725–35.

CHAPTER

8   Stress

We have described a variety of structures that can form in rocks as a result of brittle deformation. Knowing what structures exist, we are naturally inclined to ask why they exist. What caused them to form? What does their existence tell us about the processes operating in the Earth at the time they formed?

Our experience tells us that things break when too much force is applied to them. Thus we must consider what happens when forces are applied to a body of rock. In doing so, we are led to the concept of stress as a means of describing the physical state of material to which forces are applied.

8.1   Preview

The concept of stress can initially be confusing, partly because quantities that require several numbers to represent them are unfamiliar, and partly because the notation that we must use to represent these quantities is unfamiliar. In fact, however, the physical idea of stress is not difficult. Using two-dimensional geometry, we briefly introduce the physical ideas leading to the concept of stress (see Table 8.1).

We start with the idea of **force** because it is the basic concept and because we all have a physical intuition of what force is from our everyday experience of pushing and pulling on things. Force is a vector

quantity which has a magnitude (how hard the push is) and direction (which way the push is), and it is diagrammed by an arrow (*A* in Table 8.1). In a given system of coordinates (x, z) a force vector can be represented by components parallel to each of the coordinate axes (see Box 8.1).

The intensity of the force depends on the area of the surface over which the force is distributed. It is called a **traction,** and it has units of force per unit area (*B* in Table 8.1). The larger the area over which a given force is distributed, the smaller the traction on that surface. Thus the weight of a gallon of water produces a higher traction on the bottom of a tank 0.5 m on a side than on the bottom of a tank 1 m on a side. The traction is commonly represented in terms of its components perpendicular and parallel to the surface on which it acts.

In order to satisfy the requirements of equilibrium, any surface must have a pair of equal and opposite tractions acting on opposite sides of the surface. This pair of tractions defines the **surface stress,** which is commonly represented by a pair of equal and opposite components acting perpendicular to the surface and another pair acting parallel to the surface (*C* in Table 8.1).

For a given system of forces applied to a body of material, the surface stress at a given point varies with the orientation of the surface through the point. In order to know the effect at a point of all the forces acting on the body, we must be able to determine the surface stress on any plane through the point. Imagine,

Table 8.1 Development of the Concept of Stress

Diagrams		Definitions

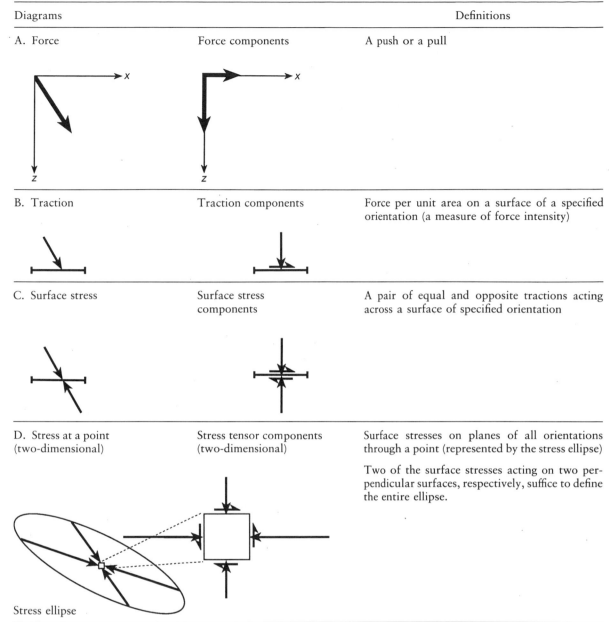

A. Force — Force components — A push or a pull

B. Traction — Traction components — Force per unit area on a surface of a specified orientation (a measure of force intensity)

C. Surface stress — Surface stress components — A pair of equal and opposite tractions acting across a surface of specified orientation

D. Stress at a point (two-dimensional) — Stress tensor components (two-dimensional) — Surface stresses on planes of all orientations through a point (represented by the stress ellipse)

Two of the surface stresses acting on two perpendicular surfaces, respectively, suffice to define the entire ellipse.

Stress ellipse

---

example, a cube that is compressed perpendicular to its faces by three vices each applying a different force. The surface stress on each pair of cube faces would be different, and each is independent of the surface stresses on the other two pairs of faces. It turns out that the surface stress on any other orientation of plane through the center of the cube can be determined from these three independent surface stresses. In fact, if we know the surface stresses on any three mutually perpendicular planes through a point, we can calculate the surface stress on any other plane through that point. The components of these three surface stresses measured perpendicular and parallel to their respective planes make up the components of the stress tensor. Thus the **stress**

tensor is a quantity that simply permits us to calculate the surface stress on a plane of any possible orientation at a given point. If we know that, we know completely what the material "feels" at that point as a result of the forces applied to the body.

In two dimensions, if we plot from a common origin the surface stresses for all the orientations of surfaces at a point, they define an ellipse (*D* in Table 8.1), which is therefore a complete representation of the two-dimensional stress tensor. The size, shape, and orientation of the ellipse are completely defined if we know the surface stresses on any pair of perpendicular planes through the point. The components of these two surface stresses are the components of the two-dimensional stress tensor (*D* in Table 8.1).

Fundamentally, that is all there is to the idea of stress. In this chapter, we further develop these concepts and the notation to express them.

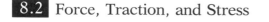 8.2 Force, Traction, and Stress

Force

Because we are concerned with vectors such as force, we review in Box 8.1 some basic properties of vectors. Forces applied to a body and originating outside the body are of two types:

1. **Body forces** act on each particle of mass, independent of the surrounding material. By far the most important body force to a structural geologist is the Earth's force of gravity. It exerts on each volume of rock a force that is proportional to the mass within that volume.

2. **Surface forces** arise either from the action of one body on another across the surface of contact between them or from the action of one part of a body on another part across an internal surface. For example, if our hand pushes on the end of a block of rock, we apply a surface force across the area of contact between our hand and the block. Moreover, across any internal surface of arbitrary orientation that divides the block in two, one side of the block applies a surface force on the other side.

For the present discussion we focus our attention on surface forces.

The Traction: A Measure of Force Intensity

A large force clearly has a greater effect on an object than a small force. But just knowing how much force is applied to a body does not give us all the information we need to determine how a deformable body will respond. For example, a thick wooden pillar might easily support the force exerted by a large mass (Figure 8.1*A*); that same force, however, would break the thin wooden leg of a table (Figure 8.1*B*). Because the type of material supporting the force is the same in both cases, we expect that the *intensity* of the force must be higher on the table leg than on the pillar, even though the magnitude of the force is the same.

The **traction** Σ is the force intensity, and it is defined by dividing the force applied **F**, by the area *A* across which it acts.[1] It therefore has physical units of force per unit area.[2] Figure 8.2*A* shows one force $\mathbf{F}^{(top)}$ acting on the top side of the surface whose area is *A*, and another $\mathbf{F}^{(bottom)}$ acting on the bottom side. Figure 8.2*B* shows the corresponding tractions $\Sigma^{(top)}$ and $\Sigma^{(bottom)}$ that act on the opposite sides of the surface. In some sources, the traction is called the stress vector. We avoid this usage because the quantity is not a true vector, as we see below, and because using the same word for both traction and stress blurs the distinction between them.

---

[1] We generally represent scalars like the area *A* in italic type; vectors like the force **F**, and tensors like the stress σ, as well as tractions and surface stresses, in boldface type.

[2] The units of the traction are the same as for the hydrostatic pressure on a surface. The two quantities differ in that hydrostatic pressure is always perpendicular to the surface on which it acts, whereas the traction in general is not.

A. **B.**

Figure 8.1 The intensity of an applied force increases as the area across which it is distributed decreases. *A.* A tyrannosaur is happily supported on a large pillar of cross-sectional area A_1. *B.* The tyrannosaur, to her dismay, is not supported by the table leg having a much smaller cross-sectional area A_2.

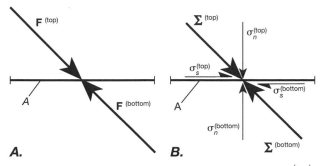

A.

B.

Figure 8.2 Force, traction, and surface stress. *A.* A force $\mathbf{F}^{(\text{top})}$ applied to the top of the surface of area *A* is balanced by an equal and opposite force $\mathbf{F}^{(\text{bottom})}$ on the bottom of the surface. *B.* The force intensity is given by the associated tractions $\mathbf{\Sigma}^{(\text{top})}$ and $\mathbf{\Sigma}^{(\text{bottom})}$ which are equal and opposite. Each traction can be expressed in terms of its normal component σ_n and its shear component σ_s. The balanced pair of tractions is the surface stress; the balanced pairs of components are the normal stress component and the shear stress component.

It is usually convenient to resolve the traction into two components, one perpendicular to the surface on which it acts and the other parallel to that surface. These components are, respectively, the **normal traction component** σ_n and the **shear traction component** σ_s (Figure 8.2*B*).

If the force is uniformly distributed over a large area *A*, and **F** represents the total force, then the traction on the whole area is given by

$$\mathbf{\Sigma} \equiv \frac{\mathbf{F}}{A} \qquad (8.1)$$

If the force is nonuniformly distributed over the area, that is, if it changes direction and magnitude across the surface, then we can define the traction only *at a point* on the surface. We represent the point as an infinitesimal area d*A* of the surface on which an infinitesimal part of the total force d**F** acts.

$$\mathbf{\Sigma} \equiv \frac{d\mathbf{F}}{dA} \qquad (8.2)$$

The magnitude and direction of the traction can then vary from point to point across the surface.

The Surface Stress

We require the surface to be in mechanical equilibrium, which means it cannot accelerate independently of the material in which it lies. For this to be true, according to Newton's second law, opposing forces must exist on opposite sides of the surface such that all the forces on the surface sum to zero (Figure 8.2*A*):

$$\mathbf{F}^{(\text{top})} + \mathbf{F}^{(\text{bottom})} = 0$$

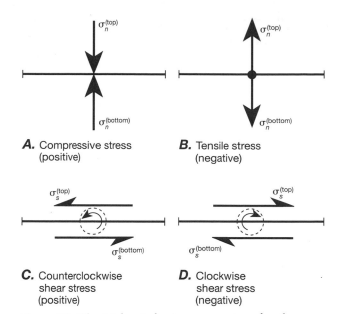

A. Compressive stress (positive) **B.** Tensile stress (negative)

C. Counterclockwise shear stress (positive) **D.** Clockwise shear stress (negative)

Figure 8.3 The Mohr circle sign conventions for the components of the stress at a point.

We can express this balance of forces in terms of the tractions by dividing the forces by the area across which they act.

$$\frac{\mathbf{F}^{(\text{top})}}{A} + \frac{\mathbf{F}^{(\text{bottom})}}{A} = 0$$

$$\mathbf{\Sigma}^{(\text{top})} + \mathbf{\Sigma}^{(\text{bottom})} = 0$$

$$\mathbf{\Sigma}^{(\text{top})} = -\mathbf{\Sigma}^{(\text{bottom})} \qquad (8.3)$$

Equation (8.3) asserts that the tractions on the top and bottom of the surface must be equal and opposite; the same relationship must therefore apply individually to the normal traction and shear traction components (Figure 8.2*B*).

$$\sigma_n^{(\text{top})} = -\sigma_n^{(\text{bottom})} \qquad \sigma_s^{(\text{top})} = -\sigma_s^{(\text{bottom})} \qquad (8.4)$$

The **surface stress,** or each of its components, consists of a pair of equal and opposite tractions, or a pair of equal and opposite traction components, acting on a surface. If the two equal and opposite normal traction components, $\sigma_n^{(\text{top})}$ and $\sigma_n^{(\text{bottom})}$, point toward each other, they define a **compressive stress** which tends to press the material together across the surface (Figure 8.3*A*). If they point away from each other, they define a **tensile stress** which tends to pull the material apart across the surface (Figure 8.3*B*). We consider that *compressive stresses are positive* and *tensile stresses are negative.*

Two equal and opposite shear traction components, $\sigma_s^{(\text{top})}$ and $\sigma_s^{(\text{bottom})}$, define a **shear stress** or a **shear couple.** The shear stress may be **clockwise** or **counterclockwise,** depending on which way a ball would turn if it were placed between the two arrows repre-

Box 8.1 What is a Vector? A Brief Review

A vector is a quantity that has both a magnitude and a direction. A scalar quantity, on the other hand, has only a magnitude. Temperature and mass density are familiar physical quantities that are scalars. These quantities are each represented by a single number that has no directional quality associated with it, such as 35°C and 2500 kg/m^3. Familiar examples of vector quantities include velocity and force. We can define a vector quantity completely only by giving its magnitude *and* the direction in which it acts: A plane travels 400 km/hr in a horizontal northeast direction. Vectors can be represented diagramatically by arrows. The length of the arrow shaft is made proportional to the magnitude of the vector, and the direction of the shaft and point indicates the direction of the vector.

Two vectors can be added using the parallelogram rule. If, for example, we wish to add two forces **V** and **W** that act on a point p, we draw the arrows representing the forces tail to tail and construct a parallelogram with the arrows defining two adjacent sides (Figure 8.1.1). The sum of the forces, called the resultant force **R**, is then the vector from the common origin to the diagonally opposite corner of the parallelogram. Thus the effect of applying the forces **V** and **W** to p is the same as if the resultant force **R** were applied to p.

In order to specify a direction, it is necessary to have some frame of reference, such as the geographic coordinates north, east, and down (which we used above to describe the velocity of the airplane). The frame of reference in three-dimensional space is commonly taken to be a mutually orthogonal system of coordinates, and we assume that its orientation is known. We label the axes x_1, x_2, and x_3 according to the *right-hand rule*. By this rule, if the fingers of the right hand are oriented to curve along the direction of rotation from positive x_1 to positive x_2, then the thumb points along positive x_3 (Figure 8.1.2). These axes are also often labeled x, y, and z, but it is more convenient to use the subscript numbers.

If we consider a vector **V** to represent, for example, a force in three-dimensional space (Figure

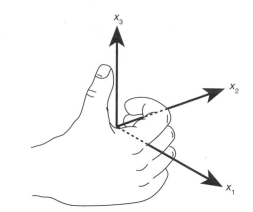

Figure 8.1.2 The right-hand rule defining the orientation of axes in a right-handed orthogonal Cartesian coordinate system.

8.1.3), using the parallelogram rule shows that it can be considered the resultant of two forces: one, **V**$_3$, parallel to the x_3 axis and the other, **W**, lying in the x_1–x_2 plane.

$$\mathbf{V} = \mathbf{W} + \mathbf{V}_3 \qquad (8.1.1)$$

Using the parallelogram rule again for **W** shows that it can be considered the resultant of two forces **V**$_1$ and **V**$_2$, which parallel the x_1 and x_2 axes, respectively.

$$\mathbf{W} = \mathbf{V}_1 + \mathbf{V}_2 \qquad (8.1.2)$$

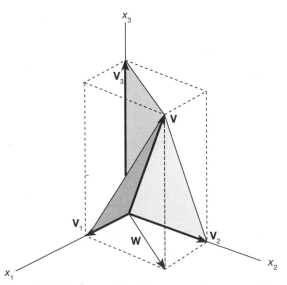

Figure 8.1.3 A vector **V** and its vector components (**V**$_1$, **V**$_2$, **V**$_3$) in three-dimensional space. **W** is the projection of **V** on the x_1–x_2 plane.

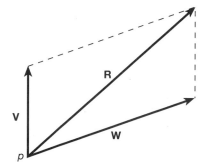
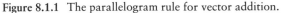

Figure 8.1.1 The parallelogram rule for vector addition.

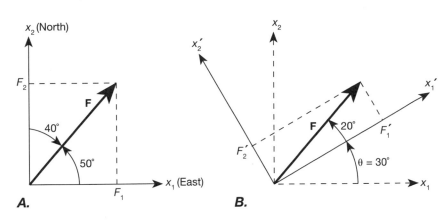

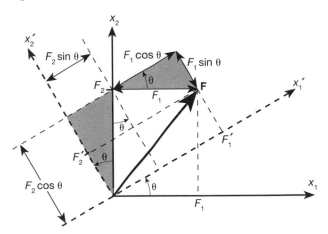

Figure 8.1.4 The dependence of the scalar components of the vector on the orientation of the coordinate system. A. Components (F_1, F_2) of $\mathbf{F}$ in the x_1–x_2 coordinate system. B. Components (F_1', F_2') of $\mathbf{F}$ in the x_1'–x_2' coordinate system.

Combining Equations (8.1.1) and (8.1.2) shows that the force $\mathbf{V}$ is the resultant of three forces, each acting parallel to one of the coordinate axes.

$$\mathbf{V} = \mathbf{V}_1 + \mathbf{V}_2 + \mathbf{V}_3 \qquad (8.1.3)$$

$\mathbf{V}_1$, $\mathbf{V}_2$, and $\mathbf{V}_3$ are the component vectors of $\mathbf{V}$. If we designate their lengths by V_1, V_2, and V_3, respectively, then these are called the scalar components, or just the components, of the vector $\mathbf{V}$ in the given coordinate system. By convention, the components are always written in order. Thus the vector $\mathbf{V}$ can be represented by an ordered array of three scalar components (V_1, V_2, V_3).

For a fixed vectorial quantity such as a given force, the values of the components representing that vector quantity depend not only on the magnitude and direction of the quantity but also on the orientation of the coordinate system in which the components are defined.

The problem is simpler to explain in two dimensions for which the reference coordinates are x_1 positive due east and x_2 positive due north (Figure 8.1.4A). If $\mathbf{F}$ is a force of 100 N (newtons) acting 40° east of north (or, equivalently, 50° north of east), then the force vector is completely defined by its components (F_1, F_2) in the x_1–x_2 coordinate system:

$$(F_1, F_2) = (64.3, 76.6) \text{ N}$$

where

$$F_1 = |\mathbf{F}| \cos 50° = (100)(0.643) = 64.3 \text{ N}$$
$$F_2 = |\mathbf{F}| \sin 50° = (100)(0.766) = 76.6 \text{ N} \qquad (8.1.4)$$

If, however, we use a coordinate system x_1'–x_2', where x_1' is 30° counterclockwise from x_1 (Figure 8.1.4B), then exactly the same force vector $\mathbf{F}$ has components given by

$$(F_1', F_2') = (94.0, 34.2) \text{ N}$$

where

$$F_1' = |\mathbf{F}| \cos 20° = (100)(0.940) = 94.0 \text{ N}$$
$$F_2' = |\mathbf{F}| \sin 20° = (100)(0.342) = 34.2 \text{ N} \qquad (8.1.5)$$

For a given vector $\mathbf{F}$, the components in different coordinate systems are systematically related. If, in Figure 8.1.5, we designate the angle between x_1 and x_1' and the angle between x_2 and x_2' by θ, then using the sides of the shaded triangles, it is not difficult to show that

$$F_1' = F_1 \cos \theta + F_2 \sin \theta \qquad (8.1.6)$$
$$F_2' = -F_1 \sin \theta + F_2 \cos \theta$$

The same situation exists in the more general three-dimensional case. Although the equations are slightly more complicated, the principle is the same: The vector $\mathbf{F}$ is the physical quantity, such as force, and it is represented by a different ordered set of components in each different coordinate system.

Because Equations (8.1.6) enable us to transform the component values from one known coordinate system to another, they are called the transformation equations. For a quantity to be a vector, its components must transform according to the rule given in these equations for two dimensions or in comparable equations for three dimensions.

Figure 8.1.5 Geometric relationships between the scalar components of the same vector in two differently oriented coordinate systems. The sides of the shaded triangles can be used to deduce the values of the components (F_1', F_2') from the components (F_1, F_2) and the angle θ.

senting the shear stress components (Figure 8.3C, D). We consider that *counterclockwise shear couples are positive* and *clockwise shear couples are negative*.[3]

Generally, we use the symbol Σ to refer to both the traction and the surface stress, and we use the symbols σ_n and σ_s to refer to the components for both quantities. It is important to realize, however, that a surface stress is defined by *pairs* of equal and opposite traction components acting across a surface. The absolute values of the traction components and of the associated surface stress components are the same; the two differ, however, in the sign convention for the com-

ponents. For example, a compressive surface stress is positive, but it is defined by one positive and one negative traction component. We will normally deal with stress components, unless it is important to consider one particular traction of the pair that defines the stress. In that case, we will identify the traction by using a superscript, such as (top) and (bottom) in Equation (8.4).

A Numerical Example

As an illustration, let us calculate the components of the surface stress acting on two different planes in the pillar supporting the tyrannosaur in Figure 8.1A. First we determine the surface stress components on a horizontal cross section of the pillar (Figures 8.1A, 8.4A). Suppose the area of the pillar cross section is $A = L \times L = 2$ m^2 and the tyrannosaur weighs $W = 80{,}000$ N, which is the magnitude of a force acting downward. The force per unit area that the upper part of the pillar exerts on the lower part, across the area A, is the traction, and the lower part exerts an equal and opposite traction on the upper part. The magnitude of the surface stress is just the same as the magnitude of the traction[4] (Figure 8.4B). We choose the signs of

[3] We refer to this sign convention as the Mohr circle sign convention, because it is used for plotting stress components as a Mohr circle, which we discuss in Section 8.3. The sign convention is not unique, however, because the same stress looks clockwise and counterclockwise when viewed from opposite directions. For this reason, it differs from the sign convention used for the components of the stress tensor. We discuss the tensor sign convention in Section 8.4, and the origin of the vexing but unavoidable difference between the Mohr circle and the tensor sign conventions in Section 8.5.

[4] A newton is the amount of force required to accelerate 1 kilogram of mass at 1 meter per second per second (1 N = 1 kg m/s^2). Appropriately enough, a force of 1 newton is approximately equal to the weight of an apple (1 N = 0.225 lb). Newtons per square meter (N/m^2), pascals (Pa), megapascals (MPa), bars (b), and kilobars (kb) are all units of stress—that is, force per unit area—related by 10^6 N/m$^2 = 10^6$ Pa $= 1$ MPa $= 10$ b $= 0.01$ kb. The pressure of 1 Pa is approximately the pressure created by grinding one apple into applesauce and spreading it in an even layer over an area of 1 m^2. It is a rather small pressure. Atmospheric pressure (14.7 lb/in^2) is approximately 10^5 Pa, 0.1 MPa, or 1 b.

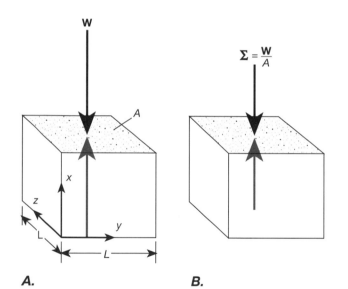

A.   **B.**

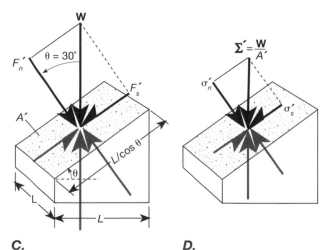

C.   **D.**

Figure 8.4 Determination of the stress components on surfaces of different orientation. A. In supporting the weight W of the tyranosaur in Figure 8.1A, the upper part of the pillar exerts a force $\mathbf{W}$ on the lower part across an arbitrary cross-section plane that has area A normal to the pillar axis. The lower part of the pillar exerts an equal and opposite force on the upper part. B. The magnitude of the surface stress Σ on the plane of area A is the force divided by the area. The force and the surface stress are normal to the surface, so there is no shear stress component. C. On a plane that is inclined at an angle $\theta = 30°$ through the pillar and has an area A', the same force $\mathbf{W}$ acts in the same direction. Components of the force normal and parallel to the plane are F_n and F_s. D. The magnitude of the surface stress Σ' is the same force $\mathbf{W}$ divided by the larger area A'. The magnitude of the normal stress and shear stress components σ'_n and σ'_s are the normal and parallel force components divided by A'.

the surface stress components according to the Mohr circle convention: compressive stress is positive.

$$\Sigma = \frac{W}{A} = \frac{80,000 \text{ N}}{2 \text{ m}^2} = 40,000 \text{ Pa} = 0.04 \text{ MPa} \quad (8.5)$$

Here we are considering only the magnitudes of the vectors and surface stresses, so we do not use boldface type.

Because the force—and therefore the surface stress—acts exactly perpendicular to the area A, the normal stress component σ_n equals the magnitude of the surface stress Σ itself, and the shear stress component σ_s is zero. Thus,

$$\sigma_n = \frac{W}{A} = \Sigma = 40,000 \text{ Pa} \qquad \sigma_s = 0 \quad (8.6)$$

Suppose, now, that we wanted to calculate the magnitude of the surface stress Σ' acting on a plane in this same column that is inclined at an angle $\theta = 30°$ to the left and has an area A' (Figure 8.4C). We have,

$$\Sigma' = \frac{W}{A'} = \frac{W}{A/\cos \theta} = \frac{80,000 \text{ N}}{2.309 \text{ m}^2} \approx 34,640 \text{ Pa}$$
$$= 0.03464 \text{ MPa} \quad (8.7)$$

where the areas A and A' are related by

$$A = LL \qquad A' = L(L/\cos \theta) = A/\cos \theta \quad (8.8)$$

Notice from Equation (8.7) that although the weight is the same, the magnitude of the surface stress on A' is smaller than that on A, because the area A' is larger than A (Figure 8.4D). The force components normal and parallel to the inclined plane, F'_n and F'_s, are

$$F'_n = W \cos \theta \qquad F'_s = W \sin \theta \quad (8.9)$$

and the normal stress and shear stress components σ'_n and σ'_s are simply the corresponding force components divided by the area across which they act:

$$\sigma'_n = \frac{F'_n}{A'} = \frac{W \cos \theta}{A'} = \Sigma' \cos \theta$$
$$= 30,000 \text{ Pa} = 0.03 \text{ MPa} \quad (8.10)$$

$$\sigma'_s = \frac{F'_s}{A'} = \frac{W \sin \theta}{A'} = \Sigma' \sin \theta$$
$$= 17,320 \text{ Pa} = 0.01732 \text{ MPa} \quad (8.11)$$

We can relate the components σ'_n and σ'_s on the surface A' to the normal stress component σ_n on the surface A. In Equations (8.10) and (8.11), we write the area A' in terms of A using Equation (8.8) and then we use Equation (8.6) to obtain

$$\sigma'_n = \frac{W \cos \theta}{A'} = \frac{W \cos \theta}{A/\cos \theta} = \sigma_n \cos^2 \theta \quad (8.12)$$

$$\sigma'_s = \frac{W \sin \theta}{A'} = \frac{W \sin \theta}{A/\cos \theta} = \sigma_n \sin \theta \cos \theta \quad (8.13)$$

This example shows that neither the traction nor the surface stress (a pair of equal and opposite tractions) is a vector quantity because they are both inseparable from the area, and thus the orientation, of the surface on which they act. The transformation equations relating the normal and shear components of the surface stress on two differently oriented planes (Equations 8.12 and 8.13) are very different from those relating the normal and tangential components of the force vector (Equations 8.9). Equations (8.12) and (8.13) include the transformation equations for the force vector (the numerators) as well as the equations accounting for the change in area with orientation (the denominators). These two effects result in products and squares of sine and cosine functions rather than just the first-order terms in these functions as in Equations (8.9).

Thus a traction has the characteristics of a vector only if we consider a surface of fixed orientation. The force vector, however, is independent of the orientation of the surface on which it acts. This difference is the most important distinction between traction, or stress, and force.

The Two-Dimensional Stress at a Point

We know the **stress** σ at a point in a body if we can determine the normal stress and shear stress components—written (σ_n, σ_s)—that act on a plane of *any* orientation passing through that point. There are, of course, an infinite number of such planes, so we need to know what minimum amount of information enables us to determine the stress components on any plane.

For the two-dimensional case, if the normal stress components on the planes are either all compressive or all tensile, the stress is particularly easy to visualize because it can be represented by an ellipse. If we plot all possible surface stresses as pairs of arrows from a common origin, the ends of the arrows fall on an ellipse called the **stress ellipse** (Figure 8.5A). States of stress are possible in which some normal stresses are compressive and some are tensile; in this situation, the stress ellipse is not defined. We concentrate on the intuitively simpler case in which the stress ellipse is a complete representation of the state of stress σ at a point in a body.

In general, the surface stresses are not perpendicular to the planes on which they act. Thus both the normal stress and the shear stress components (σ_n, σ_s) on an arbitrary surface are nonzero. The only exceptions are the surface stresses that are parallel to the major and minor axes of the ellipse (see the caption that accompanies Figure 8.6A). These two surface stresses are the **principal stresses**[5] $\hat{\sigma}_1$ and $\hat{\sigma}_3$ (Figure 8.5A). The planes on which the principal stresses act are the **principal planes,** and coordinate axes parallel to the prin-

---

[5] Here and throughout the book, we use "hats" (circumflexes) above symbols to indicate principal values or principal coordinates.

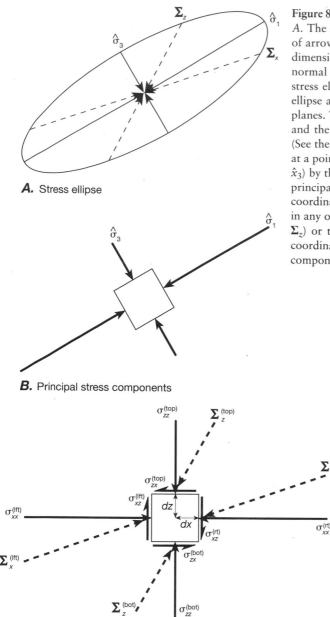

A. Stress ellipse

B. Principal stress components

C. General stress components

Figure 8.5 Representation of the state of two-dimensional stress at a point. A. The stress ellipse is the locus of all the surface stresses (plotted as pairs of arrows) that act on planes of all orientations through the point. In two dimensions, those planes are all normal to the plane of the diagram. All normal stress components must have the same sign in order to define a stress ellipse. The principal stresses are the major and minor axes of the ellipse and are the maximum and minimum of all normal stresses on the planes. The principal coordinate axes are parallel to the principal stresses, and the planes normal to the principal stresses are the principal planes. (See the legend accompanying Figure 8.6A for other details.) B. The stress at a point can be completely defined in the principal coordinate system ($\hat{x}_1$, $\hat{x}_3$) by the two principal stresses ($\hat{\sigma}_1$, $\hat{\sigma}_3$) that act on the two perpendicular principal coordinate planes. We represent the point by an infinitesimal coordinate square. C. The stress at a point can also be completely defined in any other coordinate system (x, z) by specifying the surface stresses (Σ_x, Σ_z) or their components (σ_{xx}, σ_{xz}), (σ_{zz}, σ_{zx}) on the two perpendicular coordinate planes. Superscripts identify specific tractions and traction components.

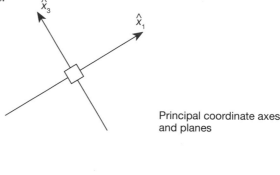

Principal coordinate axes and planes

Arbitrary coordinate axes and planes

cipal stresses are the **principal coordinates** or **principal axes**[6] $\hat{x}_1$ and $\hat{x}_3$. The principal stresses are the maximum and minimum of all the surface stresses acting on planes of any orientation through the point, and by convention we label them such that

$$\hat{\sigma}_1 \geq \hat{\sigma}_3 \qquad (8.14)$$

---

[6] The principal coordinates are labeled $\hat{x}_1$ and $\hat{x}_3$ instead of x and z so that they are directly associated with the principal stresses $\hat{\sigma}_1$ and $\hat{\sigma}_3$, respectively, to which they are parallel. The practice of distinguishing coordinate axes by different subscripts is a common one which we use specifically to label the components of the stress tensor. We describe the notation below and in more detail in Section 8.4.

The magnitudes and orientations of the principal stresses $\hat{\sigma}_1$ and $\hat{\sigma}_3$ completely define the stress ellipse—and therefore the stress $\boldsymbol{\sigma}$ at a point.

The principal stresses are perpendicular to the principal planes on which they act, so the shear stress components on the principal planes are zero. Thus the magnitudes of the principal surface stresses are completely defined by their normal stress components $\hat{\sigma}_1$ and $\hat{\sigma}_3$. The converse is also true: Any plane on which the shear stress is zero (such as plane A in Figure 8.4A) must be a principal plane, and the normal stress on that plane must be a principal stress. Because the shear stresses are zero on the principal planes, using the principal stresses is a particularly simple way to define the stress at a point.

We represent a point in a two-dimensional body as an infinitesimally small square of material. Opposite sides of the square represent the opposite sides of a plane through the point, and the perpendicular pairs of sides therefore represent two perpendicular planes through the point. Figure 8.5B shows the coordinate square in the principal coordinate system, with the principal stresses $\hat{\sigma}_1$ and $\hat{\sigma}_3$, the principal axes $\hat{x}_1$ and $\hat{x}_3$, and the principal planes, which are the sides of the square.

One surface stress does not define the complete state of stress at a point, as is evident from the fact that the length of one of the surface stresses in the stress ellipse does not define the complete shape of the ellipse. The two principal stresses completely define the shape of the stress ellipse. The surface stresses that act on *any* two perpendicular planes through the point, however, also completely define the shape of the stress ellipse.

We define general coordinates x and z perpendicular to the sides of a square of any specified orientation. We refer to planes *perpendicular* to x as x planes, and we refer to planes *perpendicular* to z as z planes. We can then label each stress component according to both the plane on which it acts and the coordinate to which it is parallel (Figure 8.5C). The components of the surface stress Σ_x acting on the x plane of the coordinate square are σ_{xx} and σ_{xz}. The first subscript x shows that both components act on the x plane; the second subscript shows that the components are parallel to the x and z coordinates, respectively. Thus σ_{xx} is the normal stress component, and σ_{xz} is the shear stress component (in Figure 8.5C, each of the tractions and traction components is labeled individually). Similarly, for the surface stress Σ_z acting on the z plane, σ_{zz} is the normal stress component, and σ_{zx} is the shear stress component.

Thus the stress $\boldsymbol{\sigma}$ at a point is completely defined either by the principal stresses $(\hat{\sigma}_1, \hat{\sigma}_3)$ and their orientations, or, in the $x-z$ coordinate system, by the surface stresses Σ_x and Σ_z or their components $(\sigma_{xx}, \sigma_{xz})$ and $(\sigma_{zx}, \sigma_{zz})$.

$$\boldsymbol{\sigma} = \begin{cases} \hat{\sigma}_1 \\ \hat{\sigma}_3 \end{cases} \quad \text{or} \quad \boldsymbol{\sigma} = \begin{cases} \Sigma_x: & (\sigma_{xx}, \sigma_{xz}) \\ \Sigma_z: & (\sigma_{zz}, \sigma_{zx}) \end{cases} \quad (8.15)$$

The only case for which one surface stress is sufficient to define the stress at a point is for a hydrostatic pressure, in which case the stress ellipse is a circle.

We require that the coordinate square be in mechanical equilibrium, which means that its acceleration parallel to each of the coordinate axes must be zero and that its angular acceleration must be zero. Thus both the forces and the moments of these forces acting on the square must sum to zero.

We know from Equation (8.4) that the normal traction and shear traction components on opposite sides

of a plane must be equal and opposite (Figure 8.2B). Accordingly, from (Figure 8.5C),

$$\sigma_{xx}^{(rt)} = -\sigma_{xx}^{(lft)} \qquad \sigma_{zz}^{(top)} = -\sigma_{zz}^{(bot)}$$
$$\sigma_{xz}^{(rt)} = -\sigma_{xz}^{(lft)} \qquad \sigma_{zx}^{(top)} = -\sigma_{zx}^{(bot)} \qquad (8.16)$$

The product of a traction component and the area on which it acts defines a force component acting on the coordinate square. Using Figure 8.5C, we sum all the force components that are parallel to the x axis, and separately we sum all force components that are parallel to the z axis. Thus we require

$$\|x: \quad \sigma_{xx}^{(rt)}A_x + \sigma_{xx}^{(lft)}A_x + \sigma_{zx}^{(top)}A_z + \sigma_{zx}^{(bot)}A_z = 0$$
$$\|z: \quad \sigma_{zz}^{(top)}A_z + \sigma_{zz}^{(bot)}A_z + \sigma_{xz}^{(rt)}A_x + \sigma_{xz}^{(lft)}A_x = 0 \qquad (8.17)$$

If we use Equations (8.16) to eliminate one of each of the traction component pairs from Equations (8.17), we obtain the identity $0 = 0$. Thus Equations (8.16) are the conditions that must be met if the forces are to sum to zero.

Taking moments of the forces about the origin—or, in essence, about the y axis—involves only the shear traction components, because the moment arms for the normal traction components are all zero. From Figure (8.5C), the infinitesimal dimensions of the square are $2dx$ and $2dz$, whereby taking all the moments and requiring their sum to be zero gives

$$\sigma_{xz}^{(rt)}A_x dx + \sigma_{xz}^{(lft)}A_x(-dx) + \sigma_{zx}^{(top)}A_z dz$$
$$+ \sigma_{zx}^{(bot)}A_z(-dz) = 0 \qquad (8.18)$$

Because $A_x = A_z$ and $dx = dz$, we can eliminate these quantities from the equation by division and, using Equation (8.16), show that the shear tractions, and therefore the shear stresses, are related, respectively, by

$$\sigma_{xz}^{(lft)} = -\sigma_{zx}^{(bot)} \qquad \text{and} \qquad \sigma_{xz} = -\sigma_{zx} \qquad (8.19)$$

Thus of the four stress components in the $x-z$ coordinate system (second equation 8.15), only three are independent: σ_{xx}, $\sigma_{xz} = -\sigma_{zx}$, and σ_{zz}.

The Three-Dimensional Stress at a Point

The description of the stress in three dimensions is a direct extrapolation of its description in two dimensions. In the simple case for which all normal stress components have the same sign, the stress at a point is represented by a stress ellipsoid (Figure 8.6A). The major, intermediate, and minor principal axes of the ellipsoid are parallel to the principal coordinate axes. They represent the maximum, intermediate, and minimum principal stresses, respectively, which we label in accordance with the convention

$$\hat{\sigma}_1 \geq \hat{\sigma}_2 \geq \hat{\sigma}_3 \qquad (8.20)$$

The principal stresses are the surface stresses acting on

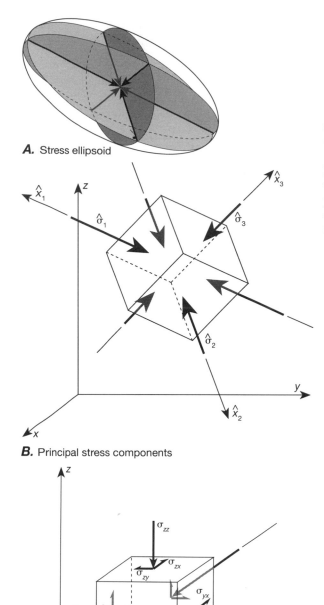

A. Stress ellipsoid

B. Principal stress components

C. General stress components

Figure 8.6 The state of three-dimensional stress at a point. *A.* The stress ellipsoid is defined by the surface stresses that act on planes of all possible orientations through a point. Shaded planes are the principal planes. Stress components are the principal stresses. For the representation of stress to be an ellipsoid, the normal components of the surface stresses must all be either compressive or tensile. The orientation of the plane on which any particular surface stress acts is not immediately obvious from the stress ellipsoid. The components of the outward unit normal vector **n** to the plane, however, are $(n_1, n_2, n_3) = (\hat{\Sigma}_1^{(n)}/\hat{\sigma}_1, \hat{\Sigma}_2^{(n)}/\hat{\sigma}_2, \hat{\Sigma}_3^{(n)}/\hat{\sigma}_3)$; where $(\hat{\Sigma}_1^{(n)}, \hat{\Sigma}_2^{(n)}$, and $\hat{\Sigma}_3^{(n)})$ are the components of the particular surface stress parallel to the three principal axes of the ellipsoid, and $\hat{\sigma}_1$, $\hat{\sigma}_2$, and $\hat{\sigma}_3$ are the principal stresses that parallel those three axes. *B.* Representation of the stress in principal coordinates. *C.* Representation of the stress in general coordinates.

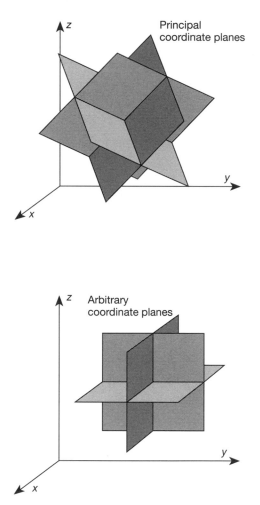

the three mutually perpendicular principal planes through a point. On the principal planes, normal stresses have extreme values and shear stresses are zero. We represent the point as an infinitesimal cube whose faces are parallel to the principal planes and perpendicular to the principal axes $\hat{x}_1$, $\hat{x}_2$, and $\hat{x}_3$ (Figure 8.6B).

The stress ellipsoid can also be defined by the surface stresses $\boldsymbol{\Sigma}_x$, $\boldsymbol{\Sigma}_y$, and $\boldsymbol{\Sigma}_z$ and their components that act on *any* three mutually perpendicular planes through

a point (Figure 8.6C). The coordinate axes x, y, and z are parallel to the mutual intersections of those three planes, and the point is represented as an infinitesimal cube whose faces are parallel to the three planes. The surface stress acting on each face of the cube has three components, one parallel to each of the coordinate axes (compare Figure 8.1.3). One component is a normal stress; the other two are shear stresses. Each stress component is composed of a pair of equal and opposite traction components that act on opposite faces of the cube. We label the components of these three surface stresses by using the same convention we used for two-dimensional stress. Each component with two identical subscripts is a normal stress; each component with two different subscripts is a shear stress.

Thus the stress ellipsoid, which defines the stress at a point, is uniquely described by the three principal stresses $\hat{\sigma}_1$, $\hat{\sigma}_2$, and $\hat{\sigma}_3$ and their orientations (Figure 8.6B) or by three surface stresses $\boldsymbol{\Sigma}_x$, $\boldsymbol{\Sigma}_y$, and $\boldsymbol{\Sigma}_z$ or their components acting on three mutually perpendicular surfaces through the point (Figure 8.6C).

$$\boldsymbol{\sigma} = \begin{cases} \hat{\sigma}_1 \\ \hat{\sigma}_2 \\ \hat{\sigma}_3 \end{cases} \quad \text{or} \quad \boldsymbol{\sigma} = \begin{cases} \boldsymbol{\Sigma}_x: & (\sigma_{xx}, \sigma_{xy}, \sigma_{xz}) \\ \boldsymbol{\Sigma}_y: & (\sigma_{yy}, \sigma_{yx}, \sigma_{yz}) \\ \boldsymbol{\Sigma}_z: & (\sigma_{zz}, \sigma_{zx}, \sigma_{zy}) \end{cases} \quad (8.21)$$

In general, nine components are required to define the three-dimensional stress at a point. Of these nine components, only six are independent however, because the moments of the forces acting on the cube taken about each coordinate axis must sum to zero, giving (compare Equations 8.18, and 8.19)

$$\sigma_{xy} = -\sigma_{yx} \qquad \sigma_{xz} = -\sigma_{zx} \qquad \sigma_{yz} = -\sigma_{zy} \quad (8.22)$$

In order for us to analyze a problem as a two-dimensional case, the plane in which the problem is analyzed must be a principal plane containing two principal stresses—for example, $\hat{\sigma}_1$ and $\hat{\sigma}_3$—and it must be perpendicular to the third principal stress—for example, $\hat{\sigma}_2$.

Stress Tensor Notation

In continuum mechanics, the stress is defined by a mathematical quantity called the **stress tensor,** which we discuss in Section 8.4 (see Box 8.2). The stress tensor components have the same numerical values as the Mohr circle stress components, but the signs of the components are determined by a different convention, and in particular, the signs of the shear stress components may be different. To distinguish the components of the stress tensor from the Mohr circle stress components discussed above, we label the three orthogonal coordinate axes x_1, x_2, and x_3 instead of x, y, and z, and we label the components of the stress tensor with numerical subscripts. The first subscript is the number of the coordinate axis that is perpendicular to the plane on which the stress component acts, and the second subscript is the number of the coordinate axis that is parallel to the stress component.[7] For example, the stress component σ_{13}, is a shear stress component that acts on the x_1 plane

(first subscript) and is parallel to the x_3 axis (second subscript). The component σ_{11} is a normal stress component that acts on the x_1 plane (first subscript) and is parallel to the x_1 axis (second subscript). As before, the normal stress components have two identical subscripts, and the shear stress components have two different subscripts.

The components of the stress tensor are written in a specific order to form a **matrix**. The surface stresses that act on the three coordinate surfaces are written in a column, in order of increasing subscript from top to bottom. The components for each of those surface stresses are then written in a row, the rows being in the same order as the surface stresses. Thus the first subscript is the same in each row, and it increases in each column from top to bottom; the second subscript increases in each row from left to right. The components of the three-dimensional stress tensor are written in principal coordinates $\hat{x}_1$, $\hat{x}_2$, and $\hat{x}_3$ or in general coordinates x_1, x_2, and x_3, respectively, as

$$\boldsymbol{\sigma} = \begin{bmatrix} \hat{\boldsymbol{\Sigma}}_1 \\ \hat{\boldsymbol{\Sigma}}_2 \\ \hat{\boldsymbol{\Sigma}}_3 \end{bmatrix} = \begin{bmatrix} \hat{\sigma}_1 & 0 & 0 \\ 0 & \hat{\sigma}_2 & 0 \\ 0 & 0 & \hat{\sigma}_3 \end{bmatrix} \quad \text{or}$$

PRINCIPAL DIAGONAL

$$(8.23)$$

$$\boldsymbol{\sigma} = \begin{bmatrix} \boldsymbol{\Sigma}_1 \\ \boldsymbol{\Sigma}_2 \\ \boldsymbol{\Sigma}_3 \end{bmatrix} = \begin{bmatrix} \sigma_{11} & \sigma_{12} & \sigma_{13} \\ \sigma_{21} & \sigma_{22} & \sigma_{23} \\ \sigma_{31} & \sigma_{32} & \sigma_{33} \end{bmatrix}$$

PRINCIPAL DIAGONAL

The normal stress components appear along the principal diagonal of the matrix, and the shear stress components appear in the off-diagonal positions. With the tensor sign convention, Equation (8.22) becomes

$$\sigma_{12} = \sigma_{21} \qquad \sigma_{13} = \sigma_{31} \qquad \sigma_{23} = \sigma_{32} \quad (8.24)$$

The three relationships in Equation (8.24) define the **symmetry of the stress tensor,** a term that refers to the equality of the shear stress components that occur, in the matrix, in symmetric positions relative to the principal diagonal, as in Equation (8.23). Note that with this sign convention, the symmetrically related shear stress components are *equal,* not opposite (compare Equation 8.22), even though one is a clockwise and the other a counterclockwise shear stress (Figure 8.6C). The notation for the principal stresses is unchanged, but the matrix shows explicitly that all the shear stress components associated with the principal stresses are zero.

In two dimensions, we have only two coordinate directions, which we generally take either to be $\hat{x}_1$ and $\hat{x}_3$ or to be x_1 and x_3, in which case the x_1–x_3 plane must be perpendicular to the intermediate principal axis $\hat{x}_2$. Thus the state of two-dimensional stress is specified only by the two surface stresses that act on the two

---

[7] Using two different notations to distinguish between the Mohr circle sign convention and the tensor sign convention for the stress components is a convenience we adopt for this book; it is not a distinction that is generally observed.

coordinate surfaces, and each surface stress has only two components. The matrix representing the two-dimensional stress at a point therefore has only four components.

$$\sigma = \begin{bmatrix} \hat{\Sigma}_1 \\ \hat{\Sigma}_3 \end{bmatrix} = \begin{bmatrix} \hat{\sigma}_1 & 0 \\ 0 & \hat{\sigma}_3 \end{bmatrix} \quad \text{or}$$

$$\sigma = \begin{bmatrix} \Sigma_1 \\ \Sigma_3 \end{bmatrix} = \begin{bmatrix} \sigma_{11} & \sigma_{13} \\ \sigma_{31} & \sigma_{33} \end{bmatrix} \qquad (8.25)$$

Here again the matrix is symmetric because $\sigma_{13} = \sigma_{31}$.

We use the stress tensor notation throughout this book in applications of stress to the study of deformation in the Earth. Table 8.2 presents a convenient reference to the notation we use for discussing stress.

For simplicity, we restrict our discussion in this section to two-dimensional stress. The extension to three-dimensional stress is discussed in Box 8.4.

The stress ellipse indicates that the normal stress and shear stress components on a plane must change progressively with the orientation of the plane. The relationship between the orientation of the plane and the values of normal stress and shear stress on the plane is difficult to extract from the stress ellipse (see the legend that accompanies Figure 8.6A). That relationship is remarkably simple, however, when the stress is plotted

Table 8.2 Notation for Stress[a]

Σ	Traction or surface stress acting on a planar surface of specified orientation.
σ_n, σ_s	Normal and shear components respectively for both the surface stress and the traction.
$\Sigma_x, \Sigma_y, \Sigma_z$ Σ_k	Surface stresses or tractions acting on the coordinate surfaces that are normal, respectively, to the coordinate axes x, y, z (or to axes x_k, where k takes the values 1, 2, and 3).
σ	The stress at a point: a second-rank symmetric tensor quantity.
$\hat{\sigma}_k$	Principal stresses (maximum, intermediate, and minimum for $k = 1, 2,$ and 3, respectively), which are normal stress components acting on coordinate planes in the principal coordinate system $\hat{x}_k$. The shear stress components on these planes are zero. Because these values are the lengths of the principal axes of the stress ellipsoid, they define the stress σ at a point.
$\sigma_{xx}, \sigma_{xy}, \sigma_{xz}$ $\sigma_{yx}, \sigma_{yy}, \sigma_{yz}$ $\sigma_{zx}, \sigma_{zy}, \sigma_{zz}$	Mohr circle stress components defining the stress σ at a point in the (x, y, z) coordinate system. Each row contains the components, one of the surface stresses, or tractions, Σ_x, Σ_y, Σ_z, respectively. The first subscript is the axis normal to the coordinate plane on which the component acts; the second subscript is the axis parallel to the stress component. In defining σ, compressive normal stress and counterclockwise shear stress components are positive, tensile normal stress and clockwise shear stress components are negative.
$\sigma_{k\ell}$	Components the stress σ at a point. These components are the same as the components for the three surface stresses or tractions Σ_k that act on the three coordinate surfaces, referred to the x_k coordinate system. For each value of subscript $k = 1, 2,$ and 3, subscript ℓ takes on the values 1, 2, 3, which indicate the three components of each surface stress; x_k is normal to the coordinate surface on which a component acts, and x_ℓ is parallel to the direction of the component. Normal components have $k = \ell$; shear components have $k \neq \ell$. These components differ from the Mohr circle stress components only in the sign convention. For the geologic tensor sign convention, tensor components have the same sign as the traction components that act on the *negative* side of the coordinate surface.
$_D\sigma$	The differential stress. A positive scalar quantity equal to the difference between the maximum and minimum principal stresses.
$\bar{\sigma}_n$	The mean normal stress: the average of the normal stress components of the stress tensor in any coordinate system. It is a scalar invariant of the stress tensor.
$_\Delta\sigma_{k\ell}$	The deviatoric stress components, equal to the stress tensor components with the mean stress subtracted from each of the normal stress components.
$_E\sigma_{k\ell}$	The effective stress components, equal to the components of the stress tensor with the pore fluid pressure subtracted from each of the normal stress components.

[a] Boldface type, either with or without subscripts, indicates vectors and tensors; normal type with subscripts indicates scalar components of vectors and tensors. We use the same notation for the traction and its components as for the stress at a point and its components, even though the stress is actually defined by the pair of equal and opposite tractions acting on opposite sides of the surface. The sign of a component may be different, depending on whether it is a traction component or a component of the stress, although its absolute value is the same. Where the distinction is important, the context makes the intent clear. In some cases, however, we specify a particular traction by using additional superscripts whose meaning is self-evident.

on a **Mohr diagram**[8] for which the horizontal axis is the value of the normal stress σ_n, and the vertical axis is the value of the shear stress σ_s.

For a given stress, we can show (see Section 8.5) that on the Mohr diagram, the normal stress and shear stress components on planes of all possible orientations through a point plot on a circle called the **Mohr circle**. The center of the circle lies on the normal stress axis. As before, compressive normal stresses and counterclockwise shear couples are considered positive. Characteristics of the Mohr circle show clearly how the stress at a point is related to the surface stresses on planes through the point. We number these characteristics to provide a convenient means of referencing them in subsequent sections.

1) *The Mohr Diagram*

(i) The diagram has axes that are values of stress. It is therefore very important to distinguish the Mohr diagram from a diagram of physical space, whose axes are spatial coordinates. It is always necessary to draw a separate diagram of physical space, along with the Mohr diagram, and to transfer data carefully from one diagram to the other (Figure 8.7).

---

[8] Named after Christian Otto Mohr (1835–1918), a German professor of mechanics and civil engineering.

(ii) The Mohr circle is a complete representation of the stress at a point, because the normal stress and shear stress components of the surface stress on planes of all possible orientations through the point are included on the circle. Each point on the circle represents the surface stress on a different plane.

2) *Principal Stresses*

(i) The maximum and minimum normal stresses have values defined by the intersection of the Mohr circle with the σ_n axis (Figure 8.7B). Note that these two points are the only surface stresses on the Mohr circle for which the shear stress is zero.

3) *Surface Stress and the Orientation of Planes*

(i) The orientation of a plane in physical space is defined relative to known coordinate axes by the orientation of its *normal* **n**, not the orientation of the plane itself (Figure 8.7A). For example, the angle θ in physical space (Figure 8.7A) is measured between $\hat{x}_1$ and the normal **n** to a plane P. θ is also the angle between the normal stress components on the $\hat{x}_1$ coordinate plane (σ_1) and on the plane P $(\sigma_n^{(P)})$ because $\hat{\sigma}_1$ is parallel to x_1 and $\sigma_n^{(P)}$ is parallel to n.

(ii) Angles measured in physical space are doubled when plotted on the Mohr diagram. Angles are measured in the same sense on the Mohr diagram as in

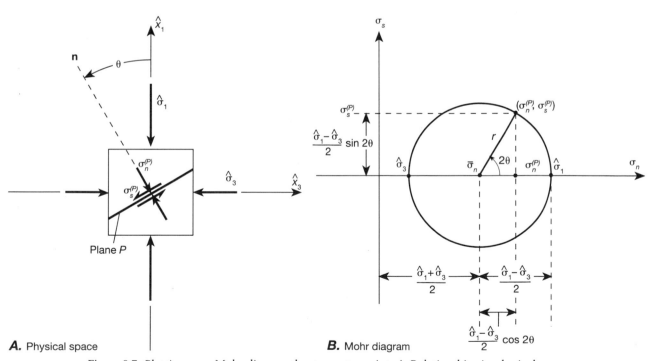

A. Physical space   **B.** Mohr diagram

Figure 8.7 Plotting on a Mohr diagram the stress at a point. A. Relationships in physical space among the stress components, the principal coordinate axes, and the plane P with its normal **n**. B. Stress at a point represented on a Mohr diagram by the Mohr circle. Superscripts (P) identify stress components acting on a plane P.

physical space. As θ takes on values from $0°$ to $180°$, the angle plotted on the Mohr diagram 2θ takes on values between $0°$ and $360°$, and the entire Mohr circle is swept out (Figure 8.7B). All planes have two normals which are $180°$ apart. Therefore in physical space, the angles $180° \leq \theta < 360°$ are redundant because they merely duplicate the orientations of the plane defined by the angles $0° \leq \theta < 180°$.

(iii) The normal stress and shear stress components $(\sigma_n^{(P)}, \sigma_s^{(P)})$ acting on a plane P have a simple relationship on the Mohr diagram to the orientation in physical space of the normal $\mathbf{n}$ to the plane. In physical space, suppose that $\mathbf{n}$ (or $\sigma_n^{(P)}$) makes an angle θ from the axis of maximum principal stress $\hat{x}_1$ (or $\hat{\sigma}_1$) (Figure 8.7A). On the Mohr circle, the surface stress components on plane P, $(\sigma_n^{(P)}, \sigma_s^{(P)})$, plot at the end of the radius that lies at the angle 2θ from the radius to the maximum principal stress $(\hat{\sigma}_1, 0)$ (Figure 8.7B).

(iv) If there are two arbitrary planes in physical space P and P' whose normals are $\mathbf{n}$ and $\mathbf{n}'$ (Figure 8.8A), and if the angle from $\hat{x}_1$ to $\mathbf{n}$ is a counterclockwise angle θ and the angle from $\mathbf{n}$ to $\mathbf{n}'$ is a counterclockwise angle α, then on the Mohr diagram there are two points on the Mohr circle, $(\sigma_n^{(P)}, \sigma_s^{(P)})$ and $(\sigma_n^{(P')}, \sigma_s^{(P')})$, that define the normal stress and shear stress components on P and P', respectively (Figure 8.8B). The angle between radii to those points is 2α, measured counterclockwise from $(\sigma_n^{(P)}, \sigma_s^{(P)})$ to $(\sigma_n^{(P')}, \sigma_s^{(P')})$. If the angle in physical space is measured from $\mathbf{n}'$ to $\mathbf{n}$, it is clockwise and therefore a negative angle $-\alpha$, in which case a clockwise (negative) angle -2α is plotted on the Mohr circle from the radius at $(\sigma_n^{(P')}, \sigma_s^{(P')})$ to the radius at $(\sigma_n^{(P)}, \sigma_s^{(P)})$.

(v) The surface stress components that lie at opposite ends of any diameter of the Mohr circle ($2\alpha = 180°$) are the components acting on perpendicular planes in physical space ($\alpha = 90°$) (Figure 8.9A, B). Thus the principal stresses $\hat{\sigma}_1$ and $\hat{\sigma}_3$, which act on perpendicular planes, plot at opposite ends of a diameter of the circle, as do the two pairs of components $(\sigma_{xx}, \sigma_{xz})$ and $(\sigma_{zz}, \sigma_{zx})$ that specify the surface stresses acting on the perpendicular coordinate planes of an arbitrary coordinate system. Fundamentally, this statement is a corollary to the fact that angles measured in physical space are doubled when plotted on the Mohr diagram (item (ii) above).

4) Conjugate Planes of Maximum Shear Stress

(i) The stresses on the planes whose normals lie at $\theta = \pm 45°$ to the maximum principal stress $\hat{\sigma}_1$ in physical space (Figure 8.10A) occur on the Mohr circle

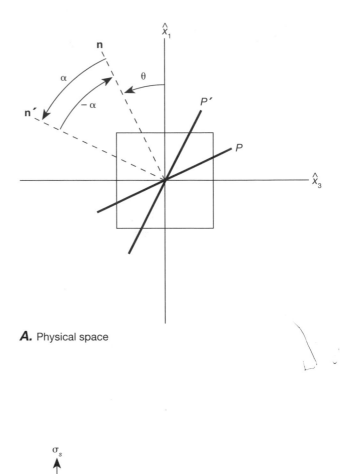

A. Physical space

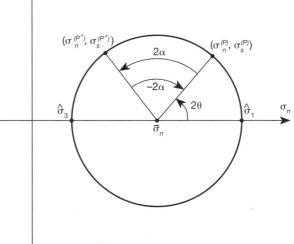

B. Mohr diagram

Figure 8.8 The geometric relationship between planes in physical space and the stress components on those planes. A. The orientations of the planes P and P' in physical space are determined by the orientations of their normals $\mathbf{n}$ and $\mathbf{n}'$, respectively. B. The geometry on the Mohr diagram of the stress components acting on the planes shown in part A. Note that the angles plotted are double the angles measured between normals to the planes in physical space but that the sense of rotation in measuring the angles is the same.

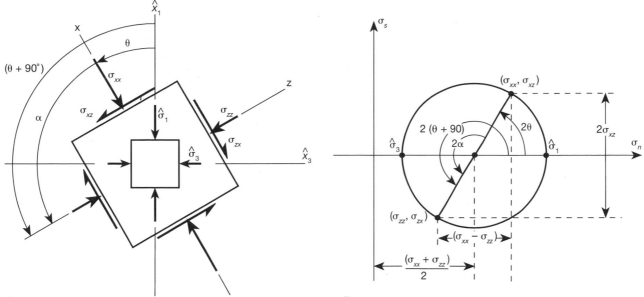

A. Physical space   **B.** Mohr diagram

Figure 8.9 Transferring stress components from a diagram of physical space to a Mohr diagram. A. Diagram of physical space showing stress components in two coordinate systems ($\hat{x}_1$, $\hat{x}_3$) and (x, z). The different sets of stress components represent the same stress and are shown on different sized coordinate squares for convenience. B. Representation on a Mohr diagram of the principal stresses as well as the stress components in the general coordinate system shown in part A. Components of surface stress acting on two planes that are perpendicular to each other in physical space plot on the Mohr circle at opposite ends of a diameter. The two scalar invariants of the stress are the center of the Mohr circle, defined by the mean of the normal stresses at opposite ends of a diameter, and by the length of the diameter, which is related by the Pythagorean theorem to the sum of the squares of $2\sigma_{xz}$ and ($\sigma_{xx} - \sigma_{zz}$).

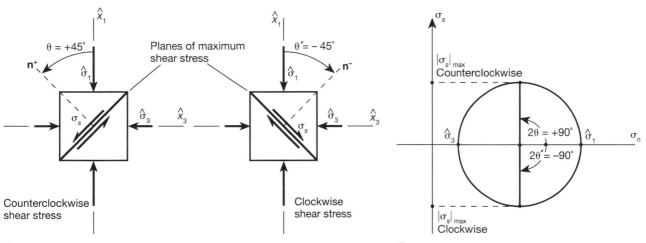

A. Physical space   **B.** Mohr diagram

Figure 8.10 Planes of maximum shear stress. A. Relationships in physical space between the planes of maximum absolute shear stress and the principal stresses. The two planes are said to be conjugate shear planes. B. Mohr diagram showing the plot of the maximum absolute shear stresses and their relationships to the principal stresses.

at $2\theta = \pm 90°$, measured from $(\hat{\sigma}_1, 0)$ (Figure 8.10B). On these planes, the absolute value of the shear stress $|\sigma_s|$ is a maximum. These planes are the **conjugate planes of maximum shear stress,** and in physical space, the planes themselves lie at $\pm 45°$ to the maximum compressive stress $\hat{\sigma}_1$. The stresses on these planes plot at opposite ends of a diameter of the Mohr circle, and therefore in physical space the normals to the planes are perpendicular, as are the planes themselves.

5) Scalar Invariants of the Stress

(i) The magnitude of the stress at a point is uniquely characterized by two **scalar invariants** of the stress, which are defined by the location of the center of the circle $\bar{\sigma}_n$, called the **mean normal stress,** and by the radius of the circle r, which is also the maximum possible absolute value of the shear stress, $|\sigma_s|_{(\text{max})}$ (Figure 8.7B). These two quantities are, respectively, half the sum and half the difference of the principal stresses.

$$\bar{\sigma}_n = \frac{\hat{\sigma}_1 + \hat{\sigma}_3}{2} \qquad r = |\sigma_s|_{(\text{max})} = \frac{\hat{\sigma}_1 - \hat{\sigma}_3}{2} \qquad (8.26)$$

$\bar{\sigma}_n$ and r are called scalar invariants because they are scalars whose values are the same for any set of components $(\sigma_{xx}, \sigma_{xz})$, $(\sigma_{zz}, \sigma_{zx})$ that define the same stress. In other words, if we know the end points of any diameter of the Mohr circle, we can construct the whole circle, because $\bar{\sigma}_n$ and r can always be determined. For the end points of an arbitrary diameter $(\sigma_{xx}, \sigma_{xz})$, $(\sigma_{zz}, \sigma_{zx})$ (Figure 8.9B),

$$\bar{\sigma}_n = \frac{\sigma_{xx} + \sigma_{zz}}{2}$$
$$r = 0.5[(\sigma_{xx} - \sigma_{zz})^2 + (2\sigma_{xz})^2]^{0.5} \qquad (8.27)$$

The second Equation (8.27) results from setting up a right triangle in Figure 8.9B with sides parallel to the axes and the diameter of the Mohr circle as the hypotenuse and then applying the Pythagorean theorem to calculate the diameter, which is twice the radius. In principal coordinates, the normal stresses become the principal stresses and the shear stresses are zero, so Equations (8.27) reduce to Equations (8.26). Our ability to construct the entire Mohr circle knowing only the surface stress components at the two end points of one diameter shows that the stress is completely defined by the surface stress components on any two perpendicular planes.

The scalar invariants of the stress describe fundamental geometric characteristics of the stress ellipse (Figure 8.5A). The mean normal stress $\bar{\sigma}_n$ is proportional to the mean radius of the ellipse, and the square of the radius of the Mohr circle, r^2, is proportional to the area of the ellipse.

6) Equations of the Mohr Circle

(i) The formulas for calculating the normal stress and shear stress components on any plane in physical space whose normal **n** is at an angle θ from the maximum principal stress $\hat{\sigma}_1$ are easily determined from the geometry of the Mohr circle (Figure 8.7B).

$$\sigma_n = \bar{\sigma}_n + r \cos 2\theta$$
$$= \left[\frac{\hat{\sigma}_1 + \hat{\sigma}_3}{2}\right] + \left[\frac{\hat{\sigma}_1 - \hat{\sigma}_3}{2}\right] \cos 2\theta \qquad (8.28)$$
$$\sigma_s = r \sin 2\theta = \left[\frac{\hat{\sigma}_1 - \hat{\sigma}_3}{2}\right] \sin 2\theta$$

Note that Equations (8.28) are written in terms of the scalar invariants of the stress.

The Mohr circle provides a very quick and convenient method for obtaining solutions to stress problems, and in Appendix 8.1, we give examples of some of the problems that can be solved by this means. Because it provides a simple way to visualize the stress at a point, we will use the Mohr circle repeatedly in our applications of stress to understanding brittle deformation in rocks.

8.4 The Stress Tensor

The stress at a point $\boldsymbol{\sigma}$ belongs to a group of mathematical quantities called second rank tensors, and it is therefore called the **stress tensor** (see Box 8.2). A second rank tensor can always be represented by a matrix of nine numbers in three dimensions, such as in Equation (8.23), or by a matrix of four numbers in two dimensions, such as in Equation (8.25).

Although the stress components $(\sigma_{xx}, \sigma_{xz})$, $(\sigma_{zz}, \sigma_{zx})$ that we defined in Section 8.2 are required for Mohr circle problems, the sign of the shear components is in fact not precisely defined. For example, a shear stress that is counterclockwise when viewed from one direction is clockwise when viewed from the opposite direction. Thus its sign in the Mohr circle convention depends on the direction from which it is viewed. This ambiguity does not cause us difficulty in solving two-dimensional Mohr circle problems, as long as we interpret the solution to a stress problem by using the same diagram of physical space in which the problem was defined. The ambiguity is intolerable, however, for more complex problems and for mathematical computation. In order to have an unambiguous definition of sign, we must use the set of stress components that represent the stress tensor (see Box 8.2).

To distinguish the components of the stress tensor from the Mohr circle stress components, we use a

Box 8.2 What Is a Tensor?

A tensor is a mathematical quantity that can be used to describe the state or the physical properties of a material. We represent a tensor by a set of scalar components referred to a particular coordinate system. Tensor components must change in a prescribed way if the coordinate axes are rotated (see Box 8.1 for this effect in vectors).

The **rank** of a tensor indicates how many scalar components are required to describe it completely. The number of components c equals the dimension d of the physical space raised to the power given by the rank r.

$$c = d^r$$

In three-dimensional space ($d = 3$), for example, a **scalar** is a tensor of zero rank ($r = 0$) and so has $3^0 = 1$ component. Common examples include temperature, mass, and volume. Scalars are defined simply by their magnitude and are invariant under a change of coordinates. We represent them mathematically by a single symbol, such as T for temperature and m for mass.

A **vector** is a first-rank tensor ($r = 1$) with $3^1 = 3$ components in three-dimensional space ($d = 3$). Force, velocity, and acceleration are all vector quantities. Vectors describe physical quantities that are characterized by magnitude and a single direction. The values of the vector components change under a rotation of coordinates as prescribed by Equations 8.1.6. The components are represented mathematically by a symbol with a single subscript, such as F_k. The subscript k is understood to take on the values 1, 2, and 3 in three-dimensional space and the subscripted symbol represents the three components (F_1, F_2, F_3), each of which is parallel to one of the coordinate axes. In two-dimensional space, k takes on just two values, such as 1 and 2.

A second-rank tensor ($r = 2$) in three-dimensional space ($d = 3$) has $3^2 = 9$ components; the most important examples of these in structural geology are stress, introduced in this chapter, and strain, introduced in Chapter 15. Second-rank tensors are used to describe physical quantities that have magnitudes and are associated with two directions. For the stress tensor, for example, the two directions associated with each component are the orientation of the normal to the plane on which the stress component acts and the orientation of the stress component acting on that plane. The transformation equations for two of the components of the second-rank stress tensor are given in essence by Equations (8.36) for transformation from principal coordinates and by Equations (8.3.3) and (8.3.4) for transformation from general

coordinates. Second-rank tensors, such as the stress $\sigma_{k\ell}$, are represented by a symbol with two subscripts. For three-dimensional space, both k and ℓ independently take on the values 1, 2, and 3. Thus for each value that k can have, ℓ can take on any of its values, thereby providing distinct symbols for each of the nine components. In two dimensions, k and ℓ take on only two values each, such as 1 and 2.

Note that for vectors, the terms in the transformation equations (Equations 8.1.6) involve the first power of the sine and cosine functions, whereas for second-rank tensors, the terms in the transformation equations (Equations 8.36, 8.3.3, and 8.3.4) involve the products of sine and cosine functions. The difference is due to the fact that transformation of vector components involves transformation of a single direction, whereas transformation of second-rank tensor components involves transformation of two directions. Thus tensors of different rank are characterized by different types of transformation equations for their components.

The magnitude of any physical quantity described by a tensor must be independent of the coordinate system in which we choose to describe it. Thus for each rank of tensor, there are a certain number of **scalar invariants** that define the magnitude of the quantity. For scalars, this fact is self-evident; the scalar is itself a magnitude and is invariant for any change of coordinate systems. Vectors have one scalar invariant, the magnitude, which we represent by the length of an arrow. This length is independent of the coordinate system in which we describe the vector. For second-rank tensors in three dimensions, three independent scalar invariants are needed to define the magnitude of the physical quantity; in two dimensions, a second-rank tensor has two scalar invariants. We discuss these invariants in Sections 8.3 and Box 8.4 as properties 5.i and 5.ii respectively.

Physical quantities that are described by tensors of higher rank also exist. For example, the piezoelectric material constants are represented by a third-rank tensor whose components can be symbolized by A_{ijk}, and the elastic constants of a material are defined, in general form, by a fourth-rank tensor symbolized by $A_{ijk\ell}$. These tensors are associated with three and four directions, respectively. In particular, the piezoelectric material constants describe the relationship between the stress on a material (two directions) and the associated electric field (one direction). The elastic constants describe the relationship between the stress on a material (two directions) and the associated strain (two directions). These higher-rank tensors, do not concern us in this book.

slightly different notation, which is explained at the end of Section 8.2. As a short hand notation for the stress tensor components, we refer to the three coordinate axes collectively as x_k, where the subscript k can take on the value 1, 2, or 3. The nine stress components are written collectively as $\sigma_{k\ell}$, where for each value of $k = 1, 2,$ or 3, ℓ can take on the value 1, 2, or 3 (compare Equation 8.23).

The surface stress on any of the coordinate surfaces consists of a pair of equal and opposite tractions or, equivalently, sets of equal and opposite traction components, which we represent as acting on opposite faces of an infinitesimal cube. We define the positive sides of the cube to be the ones facing in a positive coordinate direction, and we define the negative sides to be the ones facing in a negative coordinate direction (Figure 8.11A). The values of the stress tensor components are then defined to be equal either to the traction components acting on the negative cube faces, which gives the geologic sign convention, or to the traction components acting on the positive cube faces, which gives the engineering sign convention.[9] We use the same symbol for both the traction components and the stress tensor components, because they have the same absolute value and differ only in sign convention. As noted before, we will distinguish the traction components, where it is necessary, by using appropriate superscripts.

Using the geologic tensor sign convention, we see that the stress tensor component σ_{22} in Figure 8.11B is positive because the normal traction component acting on the negative side of the cube, σ_{22}^{-}, points in a positive coordinate direction. By the same token, σ_{22} in Figure 8.11C is negative because σ_{22}^{-} points in a negative coordinate direction. Thus compressive states of stress are positive, and tensile states of stress are negative. This convention gives the same sign as the Mohr diagram sign convention that we defined for σ_{xx}, σ_{yy}, and σ_{zz} (Section 8.2), so there is no ambiguity for the normal stress components. If the coordinate axes x, y, and z are parallel to x_1, x_2, and x_3, respectively, then $\sigma_{xx} = \sigma_{11}$, $\sigma_{yy} = \sigma_{22}$, and $\sigma_{zz} = \sigma_{33}$.

The same argument yields the signs for the shear stress components. Figure 8.11D shows that σ_{23} and σ_{32}

are *both positive* because the traction components acting on the negative side of the coordinate planes—σ_{23}^{-} and σ_{32}^{-}, respectively—both point in positive coordinate directions. By the same token, Figure 8.11E shows that σ_{23} and σ_{32} are *both negative* because σ_{23}^{-} and σ_{32}^{-} both point in negative coordinate directions. Here we see the difference from the Mohr circle sign convention (Section 8.2). In Figure 8.11D, for example, even though both shear couples are positive in the tensor sign convention, σ_{23} is clockwise and σ_{32} is counterclockwise, which in the Mohr convention means negative and positive signs respectively. In Figure 8.11E also, both shear couples are negative in the tensor sign convention, even though they are of opposite shear sense.

The tensor sign convention does not depend on the direction from which the diagram is viewed, and thus it is unambiguous.[10] In order to plot tensor components on a Mohr diagram, however, we must adopt special conventions to circumvent the ambiguity of sign for the shear stress components as defined for the Mohr diagram. We review these conventions in the next section.

Equations (8.24) show that the shear stress components in the stress tensor are not all independent. These equations are equivalent to Equations (8.22), as we can show if we equate the axes x, y, and z with x_1, x_2, and x_3, and then when we change to the Mohr circle notation and sign convention. Because of Equations (8.24), the stress tensor $\boldsymbol{\sigma}$ is called a **symmetric tensor** of second rank (compare Box 8.2), and it is necessary to specify only six of the nine numbers in the matrix in order to define completely the stress at a point.

We can gain some appreciation for the significance of a second rank tensor by comparing it with a vector. A force is a vector quantity that has a directional quality and is represented by a row array of three scalars. A stress is a second rank tensor quantity that has a bidirectional quality and is represented by a column array of three surface stresses, each of which is in turn represented by a row array of three scalars. The three

[9] The engineering sign convention gives tensor stress components with the opposite sign from those given by the geologic sign convention. This convention is used in engineering and physics and for most analytic applications of continuum mechanics, so it is also common in the geologic literature. Thus for the stress components, we have four different sign conventions: the geologic and the engineering conventions for the Mohr diagram, and the geologic and the engineering conventions for the stress tensor. Unfortunately, all are found in the geologic literature, and to avoid confusion, one must always be careful to determine which convention is employed. Often the convention is not stated explicitly.

[10] A mathematically more precise method of defining the geologic tensor sign convention is with reference to the **inward unit normal vectors**, which are vectors of unit length that are normal to the cube faces and point inward toward the center of the cube. If on any particular face, the traction component and the inward unit normal both point in positive coordinate directions or both point in negative coordinate directions, the stress tensor component is positive. If the component and the inward unit normal point in opposite coordinate senses, that is, one positive and the other negative, the stress tensor component is negative. This definition gives consistent results for any of the traction components on any face of the coordinate cube, so the sign need not be defined just in terms of the traction components on one side of the cube. For the engineering sign convention, the outward unit normal—that is, the unit normal vector to the cube faces that points away from the center of the cube—is used as a reference.

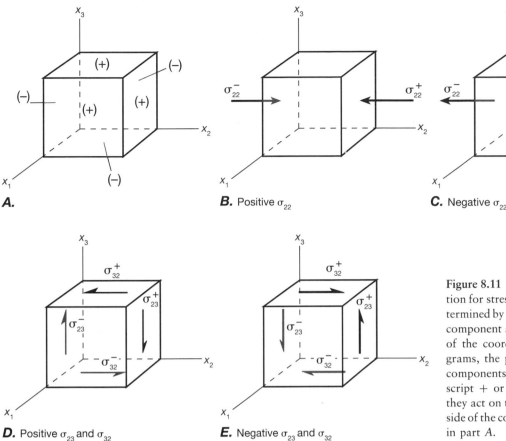

A.

B. Positive σ_{22}

C. Negative σ_{22}

D. Positive σ_{23} and σ_{32}

E. Negative σ_{23} and σ_{32}

Figure 8.11 The geologic sign convention for stress tensor components is determined by the direction of the traction component acting on the negative side of the coordinate surface. In all diagrams, the pairs of opposing traction components are labeled with a superscript + or −, depending on whether they act on the positive or the negative side of the coordinate surface as defined in part *A*.

surface stresses are those that act on the three coordinate surfaces, and the two directions involved in each stress component are the orientation of the surface on which each stress component acts and the orientation of the stress component itself (see Box 8.2).

For a given stress at a point, the stress tensor **σ** is a fixed physical quantity described, for example, by the stress ellipsoid (Figure 8.6A). The *representation* of that tensor in components such as in the matrix in Equation (8.24), however, requires the specification of a particular coordinate system. In two differently oriented coordinate systems, the three surface stresses on the three coordinate planes are in general different, as can be seen from the stress ellipsoid. Thus the stress components (Equation 8.24) also generally have different values. Because the stress ellipsoid is the same, however, the stress at the point is the same, and it is always possible to calculate one set of components from another, given the angular relationships between the coordinate frames (see Section 8.5 and Box 8.3). The situation is analogous to that for the components of a vector as described in Box 8.1, but the equations we use to calculate one set of components from another for vectors are different from those we use for second-rank tensors (compare Equations 8.1.6 with 8.3.3 and 8.3.4).

For any vector, it is always possible to define a coordinate system in which all components of the vector are zero except one. This is the case when one coordinate axis is parallel to the vector. The analogous situation for the stress tensor, as for any symmetric second rank tensor, is that there always exists a coordinate system of a particular orientation for which all the shear stresses on all three coordinate surfaces are simultaneously zero, and the normal stresses on these coordinate surfaces are extrema—that is, maximum, minimum, or minimax[11] (Box 8.3). This special coordinate system is the principal coordinate system, and in these coordinates, the stress tensor is completely represented by the three normal stress components that are the principal stresses. The axes of the principal coordinate system are parallel to the principal axes of the stress ellipsoid (Figure 8.6A), and the principal stresses are the surface stresses parallel to those axes.[12]

---

[11] A minimax is a quantity that is a minimum in the plane that contains the maximum and the minimax and is a maximum in the perpendicular plane that contains the minimum and the minimax.

[12] For those familiar with linear algebra, the principal stresses and principal directions are the eigenvalues and eigenvectors, respectively, for the matrix of stress components.

Box 8.3 **Derivation of Principal Stresses in Two Dimensions**

In order to show that principal stresses must exist for any stress tensor, we need to derive equations analogous to Equations (8.36) in terms of the stress components in a general coordinate system. The principles of the derivation are the same as those used to obtain Equations (8.36), and we merely outline the procedure here.

We limit ourselves to considering planes parallel to $\hat{x}_2$ so that our diagrams of physical space are only in the $\hat{x}_1$–$\hat{x}_3$ plane. We assume we know the orientation of the reference coordinate axes x_1 and x_3, both of which are normal to $\hat{x}_2$. Both normal and shear stresses act on the faces of the coordinate square (Fig-

ure 8.3.1A). The normal $\mathbf{n}$ to the plane P on which we determine the normal stress and shear stress components (σ_n, σ_s) makes an angle α with x_1. All stress components are drawn as positive components; α is drawn as a positive angle.

We isolate the shaded triangular element in Figure 8.3.1B and then construct a diagram of the forces acting on the triangular element (Figure 8.3.1C), where

$$F_{1n} = \sigma_{11}A_1 \qquad F_{3n} = \sigma_{33}A_3 \qquad F_n = \sigma_n A$$
$$F_{1s} = \sigma_{13}A_1 \qquad F_{3s} = \sigma_{31}A_3 \qquad F_s = \sigma_s A \qquad (8.3.1)$$

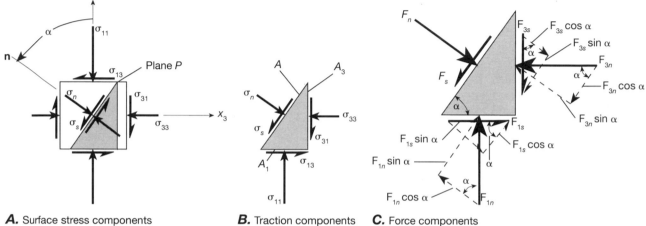

A. Surface stress components   **B.** Traction components   **C.** Force components

Figure 8.3.1 Geometric relationships used to deduce the transformation equations for components of two-dimensional stress. A. An infinitesimal coordinate square in an arbitrary coordinate system, showing the components of stress on the coordinate surfaces and on an arbitrary plane P. All quantities are shown as positive quantities. B. The traction components acting on the exterior surfaces of the shaded triangle in part A. Tractions are labeled with the associated stress components, because we want the transformation equations in terms of stress. Sign differences are accounted for in formulating the equations. Areas A, A_1, and A_3 can be thought of as the areas of the sides of a triangular prism of unit dimension normal to the diagram. C. Forces acting on the isolated triangular element, showing their components parallel and perpendicular to plane P.

8.5 A Closer Look at the Mohr Circle for Two-Dimensional Stress

In this section, we derive the equations for the Mohr circle. From this derivation, the relationship between the stress tensor components and the Mohr circle becomes evident. We restrict the following discussion to two-dimensional problems. Manipulation of the equation for two dimensions is significantly less complex than for three dimensions, and retaining the third dimension adds little to intuitive understanding. The analysis in three dimensions proceeds along similar lines, as summarized in Box 8.4 (readers should finish this section before reading the box).

In order for us to analyze a stress problem in two dimensions, the third dimension must be parallel to one of the principal stresses. Because the principal stresses are mutually perpendicular, two of the principal stresses must then lie in the plane of the analysis. In terms of the stress ellipsoid (Figure 8.6A), the surface stresses in a two-dimensional stress analysis must all lie in one of the principal planes, and the planes on which they act are all parallel to the third principal stress.

Figure 8.12A shows the most common geometry for a two-dimensional analysis of stress. If $x_2 = \hat{x}_2$ (the intermediate principal stress axis), then $\sigma_{22} = \hat{\sigma}_2$, and the x_2 plane must be a principal plane. Thus the shear

and where,

$$A_1 = A \cos \alpha \qquad A_3 = A \sin \alpha \qquad (8.3.2)$$

We resolve each force vector into two components parallel to F_n and F_s, respectively, and then require equilibrium by setting the sum of the forces in each of these two directions equal to zero. Then, expressing forces in terms of stresses with Equations (8.3.1), rearranging the equations to isolate σ_n and σ_s on the left, substituting Equations (8.3.2) for A_1 and A_3, dividing through by A to eliminate it from the equations, and using the symmetry condition of the stress tensor Equations (8.24), we find that

$$\sigma_n = \sigma_{11} \cos^2 \alpha - 2\sigma_{13} \sin \alpha \cos \alpha + \sigma_{33} \sin^2 \alpha \qquad (8.3.3)$$

$$\sigma_s = (\sigma_{11} - \sigma_{33})\sin \alpha \cos \alpha + \sigma_{13}(\cos^2 \alpha - \sin^2 \alpha) \qquad (8.3.4)$$

We now wish to determine the orientation α_0 of planes on which σ_n is a maximum or a minimum. To this end, we differentiate Equation (8.3.3) with respect to α and set the result equal to zero.

$$\frac{d\sigma_n}{d\alpha} = 0$$

$$= (\sigma_{11} - \sigma_{33}) \sin \alpha_0 \cos \alpha_0 + \sigma_{13}(\cos^2 \alpha_0 - \sin^2 \alpha_0) \qquad (8.3.5)$$

where we have used α_0 instead of α to indicate that the angle is no longer arbitrary. The right sides of Equations (8.3.5) and (8.3.4) are identical. This means that the condition for σ_n to be extreme is also the condition for σ_s to be zero.

We solve Equation (8.3.5) for α_0 by using the trigonometric identities:

$$\cos 2\alpha_0 = \cos^2 \alpha_0 - \sin^2 \alpha_0$$

$$\sin 2\alpha_0 = 2 \sin \alpha_0 \cos \alpha_0$$

$$\tan 2\alpha_0 = \sin 2\alpha_0/\cos 2\alpha_0 \qquad (8.3.6)$$

$$\tan 2\alpha_0 = \tan 2(\alpha_0 + 90°)$$

The result is

$$\tan 2\alpha_0 = \tan 2(\alpha_0 + 90°) = \frac{-2\sigma_{13}}{\sigma_{11} - \sigma_{33}} \qquad (8.3.7)$$

Thus Equations (8.3.5) shows that for planes on which σ_n is a maximum or a minimum, σ_s is zero. Equation (8.3.7) shows that there are two such planes. The normals to these planes make angles of α_0 and $(\alpha_0 + 90°)$ with x_1. The planes are therefore perpendicular to each other (Figure 8.3.2), and their normals are the orientations of the principal axes of stress.

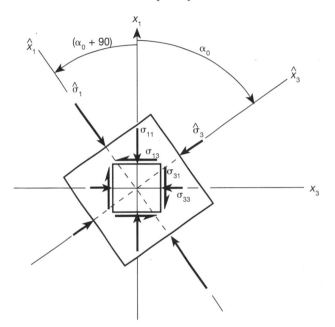

Figure 8.3.2 Stress components for a single stress shown on infinitesimal coordinate squares for a general coordinate system x_1–x_3 and for the principal coordinate system $\hat{x}_1$–$\hat{x}_3$. Both squares are infinitesimal, but they are drawn in different sizes for clarity. The angle α_0 is obtained from Equation (8.3.7).

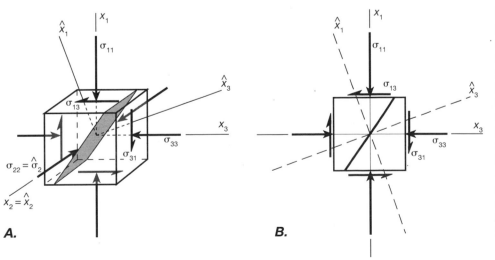

A.

B.

Figure 8.12 Geometry that permits a two-dimensional analysis of stress. A. If one coordinate axis, for example, x_2, is parallel to one of the principal coordinates, for example, $\hat{x}_2$, then the x_1–x_3 coordinate plane contains the principal axes $\hat{x}_1$ and $\hat{x}_3$, and the stresses can be analyzed in two dimensions in the x_1–x_3 plane. B. Appropriate two-dimensional diagram for analyzing the two-dimensional stress for the geometry shown in part A.

Box 8.4 **The Mohr Diagram for Three-Dimensional Stress**

In Section 8.5 we discuss determination of the surface stress acting on planes that are parallel to $\hat{x}_2$. The two-dimensional stress components are parallel to the $\hat{x}_1$–$\hat{x}_3$ plane. For the stresses in the other coordinate planes, exactly the same properties of the Mohr circle that are discussed in Section 8.5 apply. For planes parallel to any of the principal axes $\hat{x}_k$, a two-dimensional diagram of the $\hat{x}_i$–$\hat{x}_j$ plane is used, where $k \neq i < j \neq k$. Thus (i, j, k) can take on the values $(1, 3, 2)$ (Figure 8.4.1A), $(1, 2, 3)$ (Figure 8.4.1B), or $(2, 3, 1)$ (Figure 8.4.1C). The general forms of the equations analogous to Equations (8.38) and (8.41) are

$$\text{for}\quad (i, j, k) = (1, 3, 2), (1, 2, 3), \text{ or } (2, 3, 1) \quad (8.4.1)$$

$$\sigma_n = \frac{\hat{\sigma}_i + \hat{\sigma}_j}{2} + \frac{\hat{\sigma}_i - \hat{\sigma}_j}{2} \cos 2\theta_k$$

$$\sigma_s = \frac{\hat{\sigma}_i - \hat{\sigma}_j}{2} \sin 2\theta_k \qquad (8.4.2)$$

$$\left[\sigma_n - \left(\frac{\hat{\sigma}_i + \hat{\sigma}_j}{2} \right) \right]^2 + \sigma_s^2 = \left[\frac{\hat{\sigma}_i - \hat{\sigma}_j}{2} \right]^2 \qquad (8.4.3)$$

where here θ_k is positive, measured counterclockwise about the $\hat{x}_k$ axis from $\hat{x}_i$ in the $\hat{x}_i$–$\hat{x}_j$ plane (Figure 8.4.2A). When $(i, j, k) = (1, 3, 2)$, we recover Equations (8.38) and (8.41).

To the main properties of a single Mohr circle discussed in Section 8.3, we append the following properties that apply to a Mohr diagram of three-dimensional stress.

1) THE MOHR DIAGRAM

(iii) The three-dimensional stress plots on a Mohr diagram as a set of three Mohr circles each of which is a graph of the surface stress components on sets of planes parallel to one of the principal axes (Figure 8.4.2). The three circles are defined by Equations (8.4.2), with Equation (8.4.1), and each involves one pair of the principal stresses. All the properties 1–6 discussed in Section 8.3 apply to each of these circles.

2) PRINCIPAL STRESSES

(iii) All three principal stresses plot on the σ_n axis. Each principal stress plots at a point that is common to two of the Mohr circles. If all the principal stresses are unequal, there are no other common points among the circles. Each of the principal stresses is at the opposite end of a Mohr circle diameter from each of the other two principal stresses, which is consistent with the fact that the three principal stresses in physical space act on three mutually perpendicular surfaces (cf. property 3v in Section 8.3).

3) SURFACE STRESS AND THE ORIENTATION OF PLANES

(vi) Planes that are not parallel to one of the principal axes have normals that do not lie in any of the principal coordinate planes (Figure 8.4.3A). The components of the surface stress on all such planes must plot on the Mohr diagram within the largest Mohr circle and outside the two smaller circles in the area shaded in Figure 8.4.2B. The construction on the Mohr diagram for determining the stress components on such a plane is indicated in Figure 8.4.3B, for which the geometry in physical space is shown in Figure 8.4.3A. The complexity of such three-dimensional problems is beyond the scope of this book, and our interest is confined to problems involving planes that parallel one of the principal axes.

4) CONJUGATE PLANES OF MAXIMUM SHEAR STRESS

(ii) The maximum absolute values of the shear stress on any plane in three-dimensional space plot on the $\hat{\sigma}_1$–$\hat{\sigma}_3$ Mohr circle at $2\theta_2 = \pm 90°$ (Figure 8.4.4A). These stresses occur on a conjugate set of planes in physical space that are parallel to $\hat{x}_2$ and that have normals lying in the $\hat{x}_1$–$\hat{x}_3$ plane at $\theta_2 = \pm 45°$ from $\hat{x}_1$ (Figure 8.4.4B). Thus although each Mohr circle individually has maximum absolute values of the shear stress (Figure 8.4.2B), the maxima for the $\hat{\sigma}_1$–$\hat{\sigma}_2$ and the $\hat{\sigma}_2$–$\hat{\sigma}_3$ Mohr circles are maxima only for the particular set of planes that are parallel to $\hat{x}_3$ and $\hat{x}_1$, respectively. The true maxima for planes of all possible orientations occur only at the maxima for the $\hat{\sigma}_1$–$\hat{\sigma}_3$ Mohr circle (Figure 8.4.4A).

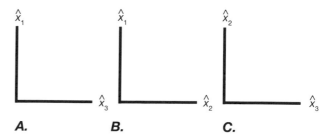

A.   **B.**   **C.**

Figure 8.4.1 The pairs of principal coordinate axes for the three Mohr circles in three-dimensional stress. The axes must be oriented such that there is a clockwise rotation from the positive axis parallel to the larger normal stress toward the positive axis parallel to the smaller normal stress. This convention standardizes the change between Mohr circle convention and tensor sign convention for the shear stress components.

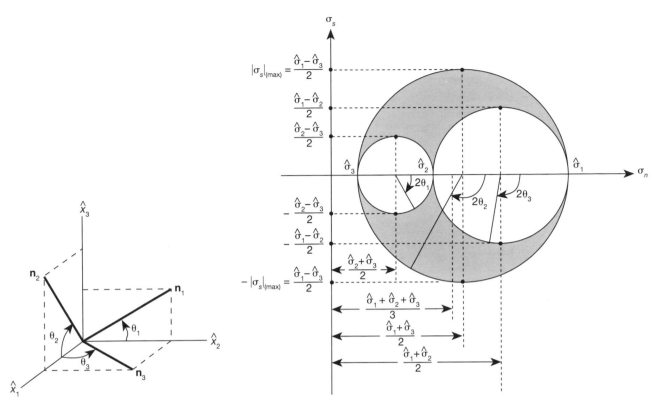

A. Physical space

B. Mohr diagram

Figure 8.4.2 Geometry of a three-dimensional stress on a Mohr diagram. A. Diagram of physical space. Here $\mathbf{n}_1$, $\mathbf{n}_2$, and $\mathbf{n}_3$ are the normals to three different planes (not shown), which are parallel respectively to $\hat{x}_1$, $\hat{x}_2$, and $\hat{x}_3$. θ_1, θ_2, and θ_3 are the angles between the appropriate principal axis and the normals $\mathbf{n}_k$, measured in one of the coordinate planes as shown. When viewed from the proper direction so that the axes are in the conventional orientation (property 2.i); Figure 8.4.1, all three angles are clockwise (negative) as shown. B. Mohr diagram. Mohr circles are shown for a three-dimensional stress. Each Mohr circle represents the two-dimensional stress in one of the principal coordinate planes. Angles plotted are those shown in part A. The surface stress components on the planes whose normals are shown in part A plot on the circumference of the appropriate Mohr circle.

5) SCALAR INVARIANTS OF THE STRESS

(ii) In three dimensions there are three scalar invariants of the stress tensor that characterize the stress at a point. In terms of the stress ellipsoid (Figure 8.6A), the three invariants are proportional to the mean of the three principal radii of the ellipsoid, the sum of the areas of the three principal planes of the ellipsoid, and the volume of the ellipsoid, respectively. Each of these invariants provides an independent measure of the size of the ellipsoid—and therefore of the magnitude of the stress.

With the added complexity of the third dimension, none of the invariants has the simple geometric interpretation, in terms of the Mohr circle, that we found for the two invariants of the two-dimensional stress tensor. We therefore discuss only the first invariant, which is still the mean normal stress defined by

$$\bar{\sigma}_n = \frac{\sigma_{11} + \sigma_{22} + \sigma_{33}}{3} = \frac{\hat{\sigma}_1 + \hat{\sigma}_2 + \hat{\sigma}_3}{3} \qquad (8.4.4)$$

where the first expression is for components in a general coordinate system, and the second is for the components in the principal coordinates. For this case, note that the mean normal stress is *not* the center of any of the three Mohr circles.

(*continued*)

Box 8.4 (continued)

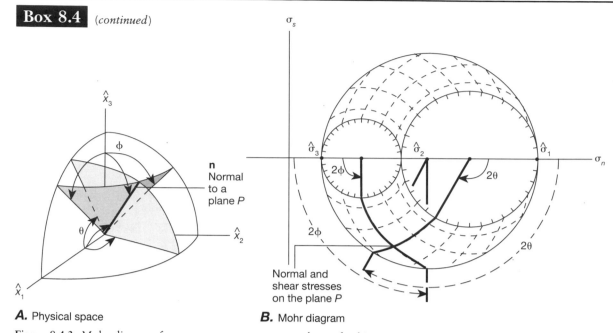

A. Physical space   **B.** Mohr diagram

Figure 8.4.3 Mohr diagram for stress components on a plane of arbitrary orientation in three dimensions. *A.* Physical space: The normal **n** to the plane *P* (not shown) is defined by the angles θ from $\hat{x}_1$ and ϕ from $\hat{x}_3$. Counterclockwise angles measured in the principal coordinate planes are positive when the coordinate axes are viewed according to convention 1 in Section 8.5 (Figure 8.4.1). The θ is negative (clockwise) in both the $\hat{x}_1 - \hat{x}_3$ plane and the $\hat{x}_1 - \hat{x}_2$ plane; ϕ is positive (counterclockwise) in both the $\hat{x}_1 - \hat{x}_3$ plane and the $\hat{x}_2 - \hat{x}_3$ plane. *B.* Mohr diagram: The angles in part *A* are transferred to the Mohr diagram to determine the normal stress and shear stress acting on the plane *P*. Families of dashed curves are arcs concentric with the two smaller Mohr circles.

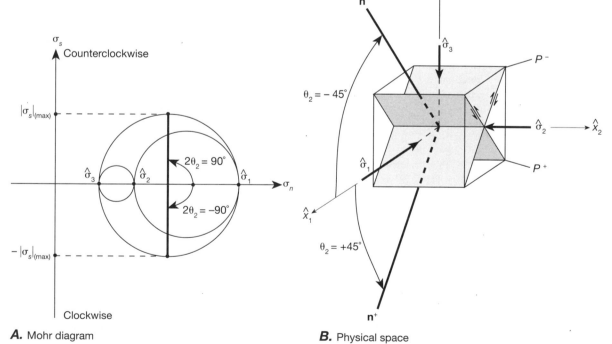

A. Mohr diagram   **B.** Physical space

Figure 8.4.4 Planes of maximum shear stress in three dimensions. *A.* Mohr diagram showing maximum absolute values of the shear stress. *B.* Diagram of physical space showing the conjugate planes of maximum shear stress and their relationship to the principal stresses.

stress components on that plane must be zero ($\sigma_{21} = \sigma_{23} = 0$). The matrix of stress components must therefore include at least one row and, by symmetry of the stress tensor (Equations 8.24), one column in which the shear-stress components are zero.

$$\boldsymbol{\sigma} = \sigma_{k\ell} = \begin{bmatrix} \sigma_{11} & 0 & \sigma_{13} \\ 0 & \hat{\sigma}_2 & 0 \\ \sigma_{31} & 0 & \sigma_{33} \end{bmatrix} \qquad (8.29)$$

All the nonzero stress components are shown in Figure 8.12A, and all except $\sigma_{22} = \hat{\sigma}_2$ lie in the x_1–x_3 plane. Under these conditions, the surface stress on any plane parallel to $x_2 = \hat{x}_2$ is completely determined by the components of the stress tensor that lie in the x_1–x_3 plane (Figure 8.12B). Thus none of the stress components with a 2 in the subscript affects the stress on the planes parallel to $\hat{x}_2$. This fact justifies our use of a two-dimensional analysis.

The matrix of components for the most common two-dimensional stress tensor is obtained simply by eliminating from the matrix in Equation (8.29) all components having a 2 as one of the subscripts, leaving

$$\boldsymbol{\sigma} = \sigma_{k\ell} = \begin{bmatrix} \sigma_{11} & \sigma_{13} \\ \sigma_{31} & \sigma_{33} \end{bmatrix} \qquad (8.30)$$

In order to derive the relationship among the normal stress and shear stress components and the orientation of the plane on which they act, we pose the following question: If we know the orientation of the principal axes, and we know the values of the principal stresses at a point, how can we determine the components of the surface stress that act on a plane of arbitrary orientation through that point? Consider the infinitesimal cube (Figure 8.13A) centered on the origin of the principal axes with faces parallel to the principal planes. The plane P is parallel to $\hat{x}_2$ but is otherwise of arbitrary orientation. With this geometry we can use a two dimensional analysis to determine the surface stress on P. The two dimensional stress tensor expressed in principal coordinates (the first Equation 8.25) is obtained from the first Equation (8.23) by deleting all stress components that have a 2 as a subscript.

$$\boldsymbol{\sigma} = \hat{\sigma}_{k\ell} = \begin{bmatrix} \hat{\sigma}_1 & 0 \\ 0 & \hat{\sigma}_3 \end{bmatrix} \qquad (8.31)$$

Figure 8.13B is the diagram of the stress components.

The following conventions, which we used in constructing Figure 8.13B, are crucial for establishing a consistent relationship between the stress tensor components and the values plotted on a Mohr diagram.

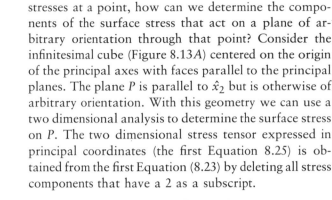

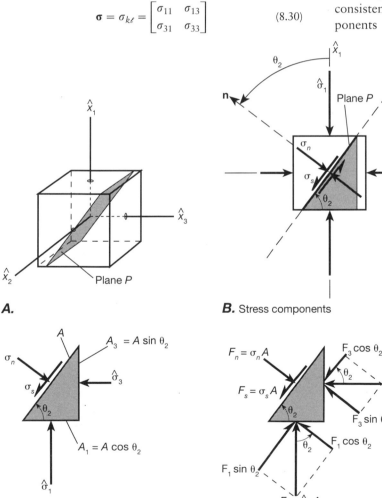

A.

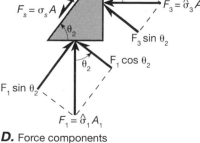

C. Traction components

D. Force components

Figure 8.13 Geometry for determining the normal stress and shear stress on a plane P of any given orientation through a point. *A.* The plane P through the principal coordinate cube is parallel to $\hat{x}_2$ but otherwise of arbitrary orientation. *B.* Two-dimensional view of the geometry in part *A*, showing the distribution of stress components. All stress components and angles are drawn as positive in this diagram. *C.* The triangular element shaded in part *B*, showing only those traction components that act on the exterior of the element. *D.* Diagram of the forces and force components derived from the traction components shown in part *C*.

Convention 1. The general convention for orienting any pair of coordinate axes requires that there is a clockwise sense of rotation from the positive coordinate axis paralleling the greatest normal stress component to the positive coordinate axis paralleling the least normal stress component, regardless of whether, for example, σ_{11} or σ_{33} is the largest. This convention fixes the direction from which we view the diagram and thereby eliminates the ambiguity about whether a shear couple is clockwise or counterclockwise. Thus the principal axes are drawn such that the 90° rotation from positive $\hat{x}_1$ to positive $\hat{x}_3$ is a clockwise rotation.

Convention 2. The orientation of the plane P is defined by the angle θ_2 between the positive $\hat{x}_1$ axis and **n**, where **n** is the vector of unit length that is *normal* to P. Positive angles are measured counterclockwise, and we construct the diagram with a positive angle θ_2. The subscript 2 on the angle θ_2 indicates that the angle measures a rotation about the $\hat{x}_2$ axis.

Convention 3. We draw the diagram with positive stress tensor components, according to the geologic sign convention. The normal stress and shear stress components on P are drawn as positive stress tensor components, considering that the vectors **n** and **p** (normal and parallel, respectively, to the plane P) are positive coordinate directions that coincide with the positive directions of $\hat{x}_1$ and $\hat{x}_3$ when $\theta_2 = 0$ (Figure 8.13B).

Note that with this convention, the positive shear stress component is automatically a counterclockwise shear couple, regardless of the value of θ_2. Thus on any two perpendicular planes for which θ_2 differs by 90°, counterclockwise shear couples are always positive. This result conflicts with the stress tensor sign convention (Figure 8.11D, E), which dictates that shear couples on perpendicular faces have equal values and opposite shear senses. Thus unavoidably there are different shear stress sign conventions for the Mohr circle and for the stress tensor. The need to shift from one convention to the other when plotting or determining stress tensor components on a Mohr circle is a common source of error.

We want to determine the normal and shear components (σ_n, σ_s) of the surface stress acting on P. To this end, we isolate in Figure 8.13C the shaded triangular element shown in Figure 8.13B, and we draw only those traction components that represent the action of the surrounding material on the triangle. Because the infinitesimal square in Figure 8.13B is in equilibrium, the triangular element in Figure 8.13C must also be in equilibrium, and we can determine the *surface stress components* on P by applying Newton's first law, which requires that the *forces* exerted on the triangular element be balanced.

We convert the traction components into force components by multiplying each traction by the area of the surface on which it acts (Figure 8.13D). Although we are dealing with tractions in this derivation, we will persist in using the components and sign convention for the surface stresses, taking care to account for the orientations of the tractions when we add or subtract the forces. In this way, our analysis will give the appropriate value for the surface stress on P. The areas of P and of the sides of the triangular element normal to $\hat{x}_1$ and $\hat{x}_3$ are A, A_1, and A_3, respectively, so the forces acting on the triangular element are

$$F_n = \sigma_n A \quad F_s = \sigma_s A \quad F_1 = \hat{\sigma}_1 A_1 \quad F_3 = \hat{\sigma}_3 A_3 \quad (8.32)$$

The force on A_1 can be resolved into a pair of components parallel to F_n and to F_s, which are the normal and tangential forces, respectively, on P (Figure 8.13D). The same is true of the force on A_3. Equilibrium of the triangular element is maintained if all forces perpendicular to P sum to zero and if all forces parallel to P sum to zero. From Figure 8.13D, these conditions imply that

$$F_n - F_1 \cos \theta_2 - F_3 \sin \theta_2 = 0 \qquad (8.33)$$
$$F_s - F_1 \sin \theta_2 + F_3 \cos \theta_2 = 0$$

where forces acting in the same direction as F_n or F_s in Figure 8.13D are added, and those acting in the opposite direction are subtracted. Rearranging Equation (8.33) to isolate F_n and F_s on the left side, and substituting for the force components from Equation (8.32), we get

$$\sigma_n A = \hat{\sigma}_1 A_1 \cos \theta_2 + \hat{\sigma}_3 A_3 \sin \theta_2 \qquad (8.34)$$
$$\sigma_s A = \hat{\sigma}_1 A_1 \sin \theta_2 - \hat{\sigma}_3 A_3 \cos \theta_2$$

We can eliminate the area terms from these equations by substituting the following relationships (Figure 8.13C)

$$A_1 = A \cos \theta_2 \quad A_3 = A \sin \theta_2 \qquad (8.35)$$

into Equation (8.34) and dividing through by A. By these manipulations, we express the force balance (Equations 8.33) strictly in terms of the stress components so that they give the results we seek.

$$\sigma_n = \hat{\sigma}_1 \cos^2 \theta_2 + \hat{\sigma}_3 \sin^2 \theta_2 \qquad (8.36)$$
$$\sigma_s = (\hat{\sigma}_1 - \hat{\sigma}_3) \sin \theta_2 \cos \theta_2$$

Note that all the terms with θ_2 involve products of sine and cosine functions. One of the trigonometric terms comes from resolving the force vectors (Equation 8.33) and the other from resolving the areas (Equation 8.35). The need to resolve both of these quantities to determine stress gives the stress the bi-directional quality that distinguishes it from the unidirectional quality of vectors such as force. Equations (8.12) and (8.13), derived in our numerical example, are the same as Equations (8.36) if in Equations (8.12) and (8.13) we replace σ_n' and σ_s' on the left sides with σ_n and σ_s, respectively, and on the right side we take $\sigma_n = \hat{\sigma}_1$ and realize that $\hat{\sigma}_3 = 0$.

Thus, given the orientation of any plane defined by θ_2, we can calculate the normal stress and shear stress components on that plane if we know only the principal stresses. These equations, then, justify our earlier assumption that the components of the stress tensor at a point are necessary and sufficient for determining the normal stress and shear stress components on a plane of any orientation through that point.

We can put the equations in a more easily interpreted form by using the standard trigonometric identities:

$$\cos^2 \theta_2 = 0.5(1 + \cos 2\theta_2) \quad \sin^2 \theta_2 = 0.5(1 - \cos 2\theta_2) \quad (8.37)$$

$$\sin \theta_2 \cos \theta_2 = 0.5 \sin 2\theta_2$$

Substituting Equations (8.37) into Equations (8.36) and rearranging gives

$$\sigma_n = \left[\frac{\hat{\sigma}_1 + \hat{\sigma}_3}{2}\right] + \left[\frac{\hat{\sigma}_1 - \hat{\sigma}_3}{2}\right] \cos 2\theta_2$$

$$\sigma_s = \left[\frac{\hat{\sigma}_1 - \hat{\sigma}_3}{2}\right] \sin 2\theta_2 \quad (8.38)$$

Here $(\hat{\sigma}_1 + \hat{\sigma}_3)/2$ is the mean normal stress, and $(\hat{\sigma}_1 - \hat{\sigma}_3)/2$ is the maximum possible shear stress, as can be seen from the fact that $\sin 2\theta_2$ in the second equation can be no greater than 1.

Equations (8.38) are identical to Equations (8.28), which we deduced from the geometry of the Mohr circle. Thus Equations (8.38) are the parametric equations for the Mohr circle, with σ_n and σ_s as the variables and θ_2 as the parameter.

We can obtain a more familiar form for the equation of a circle by eliminating θ_2. We rewrite the first Equation (8.38) as

$$\sigma_n - \left[\frac{\hat{\sigma}_1 + \hat{\sigma}_3}{2}\right] = \left[\frac{\hat{\sigma}_1 - \hat{\sigma}_3}{2}\right] \cos 2\theta_2 \quad (8.39)$$

then square both sides of the second Equation (8.38) and Equation (8.39), and add the resulting two equations together. Applying the trigonometric identity

$$\sin^2 2\theta_2 + \cos^2 2\theta_2 = 1 \quad (8.40)$$

yields the result

$$\left[\sigma_n - \left(\frac{\hat{\sigma}_1 + \hat{\sigma}_3}{2}\right)\right]^2 + \sigma_s^2 = \left[\frac{\hat{\sigma}_1 - \hat{\sigma}_3}{2}\right]^2 \quad (8.41)$$

This equation has the form

$$(x - a)^2 + y^2 = r^2 \quad (8.42)$$

which is the equation of a circle that has its center a distance a along the x axis and has a radius r.

We recommend the following procedure for plotting stress tensor components on the Mohr diagram: Draw a diagram of the coordinate square in physical space, with the coordinate axes oriented relative to each other according to convention (1) above and the stress components appropriately oriented according to the tensor sign convention. Make a table listing the values of the stress tensor components, and then, opposite each component, list its value according to the Mohr diagram sign convention. Normal stress components have the same sign as the tensor components. Determine the sign for the shear stress components by using the diagram of the coordinate square. A shear stress component is positive if it is a counterclockwise couple on the coordinate square, negative if it is a clockwise couple. Finally, plot the values of the components thus determined on the Mohr diagram (see the example given in Appendix 8A).

8.6 Terminology for States of Stress

A number of terms that refer to certain specific states of stress are common in the literature. They all have special characteristics, which are easy to describe in terms of the relevant stress tensor components and Mohr circle diagrams (Figure 8.14).

Hydrostatic pressure, $\hat{\sigma}_1 = \hat{\sigma}_2 = \hat{\sigma}_3 = p$ (Figure 8.14A). All principal stresses are compressive and equal. No shear stresses exist on any plane, so all orthogonal coordinate systems are principal coordinates. The Mohr circle reduces to a point on the normal stress axis.

Uniaxial stress. The Mohr diagram for the three dimensional stress is a single circle tangent to the ordinate at the origin. There are two possible cases:

1. *Uniaxial compression*, $\hat{\sigma}_1 > \hat{\sigma}_2 = \hat{\sigma}_3 = 0$ (Figure 8.14B). The only stress applied is a compressive stress in one direction. This geometry is commonly used in testing the strength of rock samples in the laboratory.
2. *Uniaxial tension*, $0 = \hat{\sigma}_1 = \hat{\sigma}_2 > \hat{\sigma}_3$ (Figure 8.14C). The only stress applied is a tension in one direction. Engineers often use this geometry to test the mechanical properties of metals.

Axial compression or confined compression, $\hat{\sigma}_1 > \hat{\sigma}_2 = \hat{\sigma}_3 > 0$ (Figure 8.14D). A uniaxial compression of magnitude $(\hat{\sigma}_1 - \hat{\sigma}_3)$ is superimposed upon a state of hydrostatic stress $(\hat{\sigma}_2 = \hat{\sigma}_3)$. This state is frequently used in laboratory experiments on the high-temperature, high-pressure properties of rock.

Axial extension, extensional stress, or extension, $\hat{\sigma}_1 = \hat{\sigma}_2 > \hat{\sigma}_3 > 0$ (Figure 8.14E). A uniaxial tension of magnitude $(\hat{\sigma}_1 - \hat{\sigma}_3)$ is superimposed on a hydrostatic stress $(\hat{\sigma}_1 = \hat{\sigma}_2)$. This state is also sometimes used in high-temperature, high-pressure laboratory deformation experiments. It is unfortunate that the term *extension* has a different meaning when applied to strain, and the distinction should always be made clear.

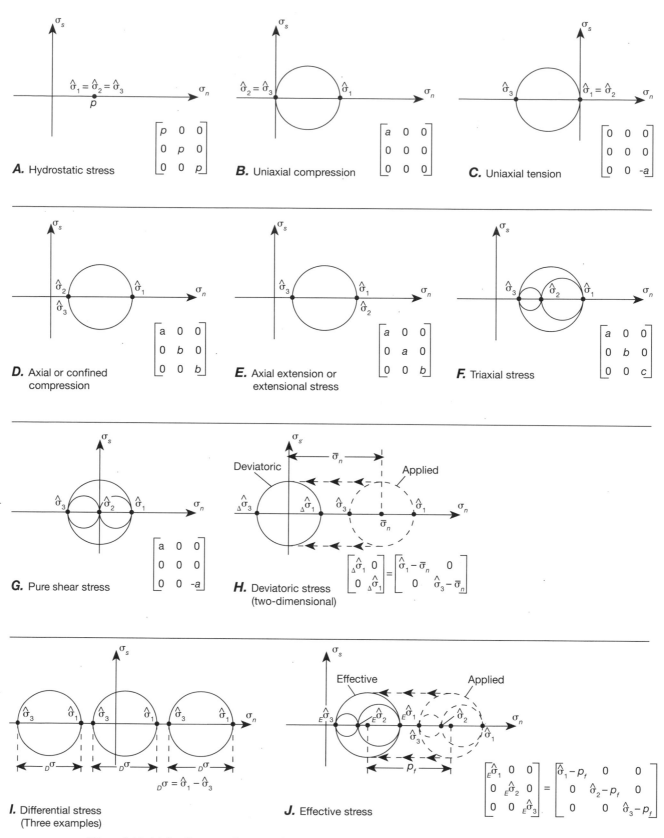

Figure 8.14 Mohr diagrams for special states of stress. The tensor components are shown for the principal coordinate system, with the principal stresses written in standard order from top left to bottom right along the principal diagonal. Here p, a, b, and c all take on positive values, and we assume $a > b > c$.

Triaxial stress, $\hat{\sigma}_1 > \hat{\sigma}_2 > \hat{\sigma}_3$ (Figure 8.14F). The principal stresses are all unequal and can be of either sign. The stress plots on the Mohr diagram as three distinct circles (see Box 8.4).

Pure shear stress or pure shear, $\hat{\sigma}_1 = -\hat{\sigma}_3$ and $\hat{\sigma}_2 = 0$ (Figure 8.14G). The maximum and minimum principal stresses are equal in magnitude and opposite in sign; the intermediate principal stress is zero. The normal stress on planes of maximum shear stress is zero—hence the name. The Mohr diagram is centered on the origin. The term *pure shear* has a different meaning when applied to strain, and the ambiguity can cause confusion.

Deviatoric stress (Figure 8.14H). The components of the deviatoric stress $\Delta\sigma_{k\ell}$ are defined by subtracting the mean normal stress $\bar{\sigma}_n$ from each of the normal stress components in the three or two-dimensional stress tensor. In three dimensions,

$$\Delta\sigma_{k\ell} = \begin{bmatrix} \sigma_{11} & \sigma_{12} & \sigma_{13} \\ \sigma_{21} & \sigma_{22} & \sigma_{23} \\ \sigma_{31} & \sigma_{32} & \sigma_{33} \end{bmatrix} - \begin{bmatrix} \bar{\sigma}_n & 0 & 0 \\ 0 & \bar{\sigma}_n & 0 \\ 0 & 0 & \bar{\sigma}_n \end{bmatrix}$$

$$\Delta\sigma_{k\ell} = \begin{bmatrix} \sigma_{11} - \bar{\sigma}_n & \sigma_{12} & \sigma_{13} \\ \sigma_{21} & \sigma_{22} - \bar{\sigma}_n & \sigma_{23} \\ \sigma_{31} & \sigma_{32} & \sigma_{33} - \bar{\sigma}_n \end{bmatrix} \quad (8.43)$$

where

$$\bar{\sigma}_n = \frac{\sigma_{11} + \sigma_{22} + \sigma_{33}}{3} \quad (8.44)$$

In two dimensions in the x_1–x_3 coordinate plane, the components of the deviatoric stress tensor are given by

$$\Delta\sigma_{k\ell} = \begin{bmatrix} \sigma_{11} - \bar{\sigma}_n & \sigma_{13} \\ \sigma_{31} & \sigma_{33} - \bar{\sigma}_n \end{bmatrix} \quad (8.45)$$

where

$$\bar{\sigma}_n = \frac{\sigma_{11} + \sigma_{33}}{2} \quad (8.46)$$

For the deviatoric stress in two dimensions, the center of the Mohr circle is shifted to the origin of the graph so that it appears to be a pure shear stress (Figure 8.13H). The deviatoric stress is useful in describing the behaviour of a material that depends only on the size of the Mohr circle, which is a measure of the maximum shear stress, and not on the location of the Mohr circle along the normal stress axis, which is a measure of the average pressure.

Differential stress (Figure 8.14I). The differential stress $_D\sigma$ is the difference between the maximum and minimum principal stresses:

$$_D\sigma = \hat{\sigma}_1 - \hat{\sigma}_3 \quad (8.47)$$

It is always a positive scalar quantity that is twice the radius of the largest Mohr circle and therefore twice the maximum shear stress. For a two dimensional stress, it is the diameter of the Mohr circle ($2r$; see the second Equation 8.26) and is therefore a scalar invariant of the stress tensor (Section 8.3, property 5). For a state of axial compression or axial extension (Figure 8.14D, E), it is the uniaxial stress that is applied in addition to the hydrostatic stress.

Effective stress (Figure 8.14J). The components of the effective stress tensor $_E\sigma_{kl}$ are defined in three dimensions by

$$_E\sigma_{k\ell} = \begin{bmatrix} \sigma_{11} & \sigma_{12} & \sigma_{13} \\ \sigma_{21} & \sigma_{22} & \sigma_{23} \\ \sigma_{31} & \sigma_{32} & \sigma_{33} \end{bmatrix} - \begin{bmatrix} p_f & 0 & 0 \\ 0 & p_f & 0 \\ 0 & 0 & p_f \end{bmatrix}$$

$$= \begin{bmatrix} \sigma_{11} - p_f & \sigma_{12} & \sigma_{13} \\ \sigma_{21} & \sigma_{22} - p_f & \sigma_{23} \\ \sigma_{31} & \sigma_{32} & \sigma_{33} - p_f \end{bmatrix} \quad (8.48)$$

where $\sigma_{k\ell}$ are the components of the applied stress, and p_f is the pressure of the pore fluid in the rock. As shown in the diagram, the effective stress is the result of a shift of the Mohr circle toward lower normal stresses by an amount equal to the pore fluid pressure p_f. We discuss the effective stress in greater detail in Section 9.5, where we show that the mechanical behavior of a brittle material depends on the effective stress, not on the applied stress.

Additional Readings

Eringen, A. C. 1967. *Mechanics of continua*. New York: Wiley.

Fung, Y. C. 1965. *Foundations of solid mechanics*. Englewood Cliffs, N. J.: Prentice-Hall.

Hubbert, M. K. 1972. *Structural geology*. Hafner Publishing Co.

Means, W. D. 1976. *Basic concepts of stress and strain for geologists*. New York: Springer-Verlag.

Appendix 8A: An Illustrative Problem

In this appendix, we discuss a specific numerical problem to illustrate the technique of using the Mohr circle and to illustrate the types of problems that one can solve using a Mohr circle. Students who have not read Section 8.4 should ignore, for each question, the discussions about the change from tensor to Mohr diagram sign convention and should simply start with the values of the stress components given for the Mohr diagram sign convention.

Consider a fault block that is 5 km thick and rests on a horizontal detachment associated with a listric

normal fault. The coordinate system is shown in Figure 8A.1A. Figure 8A.1B shows a free body diagram of a section of the detachment sheet. The action of the material that originally surrounded the free body is indicated by a distribution of traction components. These tractions arise from the force of gravity (the overburden), the applied tectonic stress (which we assume to be an east–west horizontal tensile stress), and the frictional resistance to sliding on the detachment. We determine the stress at the bottom left corner of the free body, where we know the tractions acting on both co-

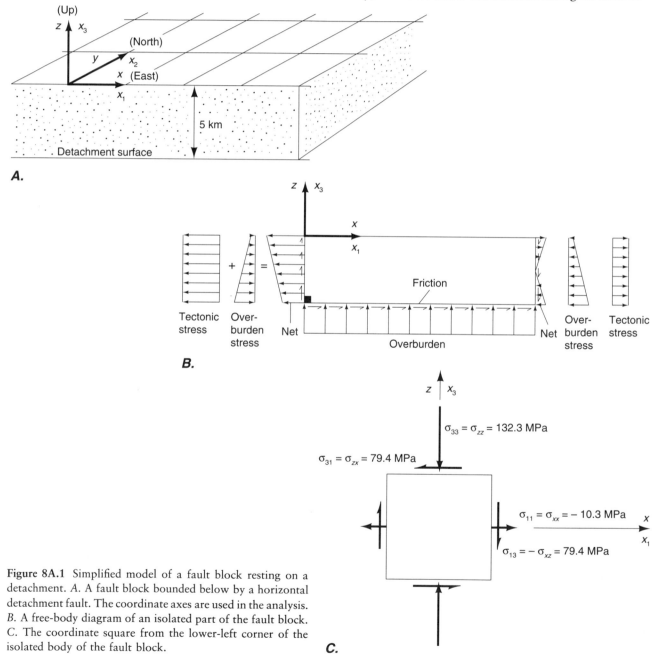

Figure 8A.1 Simplified model of a fault block resting on a detachment. A. A fault block bounded below by a horizontal detachment fault. The coordinate axes are used in the analysis. B. A free-body diagram of an isolated part of the fault block. C. The coordinate square from the lower-left corner of the isolated body of the fault block.

ordinate planes. The corner is shown enlarged in Figure 8A.1C.

The vertical normal stress σ_{33} is the **overburden pressure** due to gravity, and it equals the weight per unit area of the overlying rock. This is

$$\sigma_{33} = \rho g h \tag{8A.1}$$

where ρ is the mass density of the rock, g is the gravitational acceleration, and h is the distance to the top surface. The stress is compressive and therefore positive.

The horizontal normal stress σ_{11} is the sum of the horizontal compressive stress due to the overburden and the tectonically applied stress T. Because we assume the rock has some finite strength, the part due to the overburden is some fraction $\kappa < 1$ of the vertical normal stress (in a fluid, however, $\kappa = 1$). We assume, furthermore, that T is a tensile tectonic stress, constant with depth, that pulls the fault block to the west. Thus

$$\sigma_{11} = \kappa(\rho g h) - T \tag{8A.2}$$

where we subtract T because it is tensile. σ_{11} could be positive or negative, depending on the relative values of the overburden and of T.

The horizontal normal stress σ_{22} results only from the overburden, because there is no externally applied stress parallel to the x_2 axis.

$$\sigma_{22} = \kappa(\rho g h) \tag{8A.3}$$

Assuming that no shear stresses act on the plane of the diagram, we have

$$\sigma_{21} = \sigma_{23} = 0 \tag{8A.4}$$

The frictional shear stress along the detachment σ_{31} is given by the product of the coefficient of friction μ and the normal stress across the sliding surface. Applying the geologic tensor sign convention, we see that on the negative side of the coordinate surface, the frictional resistance acts in a positive coordinate direction (x_1). Thus this shear stress component must be positive.

$$\sigma_{31} = \mu(\rho g h) \tag{8A.5}$$

Because of the symmetry of the stress tensor (Equations 8.24), we have now determined all the independent components of the stress tensor in the given coordinate system.

In order to introduce definite numbers into the analysis, we adopt the following geologically reasonable values for the symbols in the equations:

$$\rho = 2700 \text{ kg/m}^3 \qquad h = 5000 \text{ m}$$
$$g = 9.8 \text{ m/s}^2 \qquad T = 50 \text{ MPa}$$
$$\kappa = 0.3$$
$$\mu = 0.6$$

Using these values with Equations (8A.1) to (8A.5), and using the symmetry condition for the stress tensor

(Equations 8.24), we obtain the following values in megapascals for the components of the stress tensor:

$$\sigma_{k\ell} = \begin{bmatrix} -10.3 & 0 & 79.4 \\ 0 & 39.7 & 0 \\ 79.4 & 0 & 132.3 \end{bmatrix} \tag{8A.6}$$

Because $\sigma_{21} = \sigma_{23} = 0$, the plane normal to x_2 must be a principal plane, x_2 must be a principal axis, and σ_{22} must be a principal stress. Without knowing the values of the other principal stresses, however, we do not know whether it is the maximum, the intermediate, or the minimum principal stress. We do know that the other two principal axes must lie in the x_1–x_3 plane. On the basis of the discussion at the end of Section 8.2 and that at the beginning of Section 8.5, we conclude that because x_2 is a principal axis, we can analyze the stresses on planes parallel to x_2 by using a two-dimensional analysis of stress components in the x_1–x_3 plane.

In the following discussion, we refer by number to the properties of the Mohr circle that we discussed in Sections 8.3 and Box 8.4 and to the conventions listed in Section 8.5, and we do not duplicate those discussions here.

Question 1

Construct the Mohr circle for the two-dimensional stress acting on planes normal to the x_1–x_3 plane—that is, parallel to x_2.

Procedure

To obtain the components of the two-dimensional stress tensor in the x_1–x_3 plane, we drop all components that have a 2 in the subscripts (see the discussion at the beginning of Section 8.5), which eliminates the second row and the second column of the matrix in Equation (8A.6), leaving, in megapascals,

$$\sigma_{k\ell} = \begin{bmatrix} \sigma_{11} & \sigma_{13} \\ \sigma_{31} & \sigma_{33} \end{bmatrix} = \begin{bmatrix} -10.3 & 79.4 \\ 79.4 & 132.3 \end{bmatrix} \tag{8A.7}$$

Next we must determine how these components plot on the Mohr diagram. The steps involved are as follows: (1) Draw a diagram of the stress components in physical space. (2) Use this diagram to change the stress components from the tensor sign convention (Section 8.4 and Figure 8.11) to the Mohr circle sign convention (Section 8.2 and Figure 8.3). (3) Plot the stress components on the Mohr diagram. (4) Construct the Mohr circle.

Discussion

1. Before constructing a diagram of the stress components, we must be sure the coordinates in Figure 8A.1C are drawn in accordance with convention 1 (see Section 8.5). If σ_{11} had been greater than σ_{33},

then we would have had to plot x_1 positive to the left and x_3 positive up, which is equivalent to viewing the diagram from the north instead of from the south as shown in Figure 8A.1A.

On the coordinate square (Figure 8A.1C), draw the pairs of arrows to represent the stress components, using the tensor sign convention and the values of the stress components (Equation 8A.7) to determine the correct orientations.

2. From Figure 8A.1C, we can determine the signs of the stress components appropriate for plotting on the Mohr diagram. We recommend constructing a table such as this:

Tensor value	Mode	Mohr diagram value
$\sigma_{11} = -10.3$	tensile	$\sigma_{xx} = -10.3$
$\sigma_{13} = 79.4$	clockwise	$\sigma_{xz} = -79.4$
$\sigma_{33} = 132.3$	compressive	$\sigma_{zz} = 132.3$
$\sigma_{31} = 79.4$	counterclockwise	$\sigma_{zx} = 79.4$

In the first column, list the symbols and values for the stress tensor components exactly as they are given in Equation (8A.7). Use Figure 8A.1C to check whether the normal stress components are tensile or compressive and whether the shear stress components are clockwise or counterclockwise; note this in the second column. This second column, with the sign conventions for the Mohr circle given in Figure 8.3, determines the sign of each stress component entered in the third column.

3. The pairs of values $(\sigma_{xx}, \sigma_{xz})$ and $(\sigma_{zz}, \sigma_{zx})$ from the last column plot as two points on the Mohr diagram (Figure 8A.2A).

4. A line connecting these two points on the Mohr diagram must be a diameter of the Mohr circle (property 3v), and the point where the diameter intersects the σ_n-axis is the center of the circle (property 5i). The circle can then be drafted with a drafting compass. Alternatively, we can use equations of the form of (8.27) to calculate the center and radius of the Mohr circle, from which the whole circle can be constructed.

$$\bar{\sigma}_n = \frac{\sigma_{xx} + \sigma_{zz}}{2} = \frac{-10.3 + 132.3}{2} = 61 \text{ MPa} \quad (8A.8)$$

$$r = 0.5[(\sigma_{xx} - \sigma_{zz})^2 + (2\sigma_{xz})^2]^{0.5}$$

$$= 0.5[(-10.3 - 132.3)^2 + 4(79.4)^2]^{0.5}$$

$$r = 106.7 \text{ MPa} \quad (8A.9)$$

Question 2

What are the values and orientations of the principal stresses in the x_1–x_3 (or x–z) plane? Draw a diagram

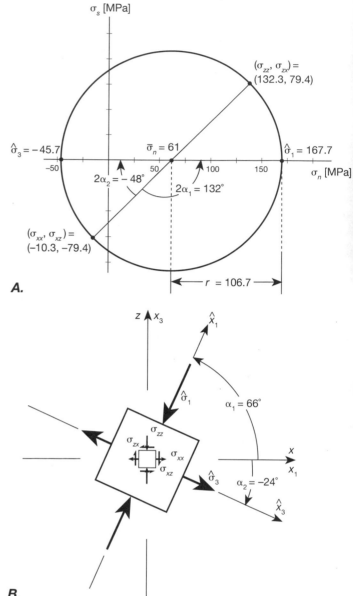

A.

B.

Figure 8A.2 Mohr circle construction for the illustrative problem. A. Mohr circle for the stress under consideration. B. Orientation of the principal axes in physical space and their relationship to the original coordinate system, as derived from the Mohr circle in part A. The two differently oriented coordinate squares both represent the infinitesimal point, and the components on each square represent the same stress. The squares are drawn different sizes for clarity; arrows are not to scale.

of physical space showing the relationship between the x_1 – x_3 (x – z) coordinates and the principal coordinates in that plane.

Procedure

The values of the principal stresses are read from Figure 8A.2A at the points where the Mohr circle intersects the σ_n-axis (property 2i). The values are 167.7 MPa and

−45.7 MPa. These values can also be obtained by adding and subtracting the magnitude of the radius of the Mohr circle (Equation 8A.9) to the value of $\bar{\sigma}_n$, the center of the circle (Equation 8A.8). Recalling that the third principal stress is 39.7 MPa (Equation 8A.6), we label the values in decreasing order such that the stress in principal coordinates is given by

$$\hat{\sigma}_{k\ell} = \begin{bmatrix} \hat{\sigma}_1 & 0 & 0 \\ 0 & \hat{\sigma}_2 & 0 \\ 0 & 0 & \hat{\sigma}_3 \end{bmatrix} = \begin{bmatrix} 167.7 & 0 & 0 \\ 0 & 39.7 & 0 \\ 0 & 0 & -45.7 \end{bmatrix} \quad (8A.10)$$

The orientations of the principal axes are also determined from Figure 8A.2A, using the properties 3i and ii and the plotting conventions used for Figure 8.13B. We recommend tabulating the measurements as shown below to avoid confusion and to ensure proper observance of the conventions. Measurements on the Mohr diagram are shown in Figure 8A.2A, and the corresponding measurements in physical space are shown in Figure 8A.2B.

On the Mohr Diagram

Angle	Sense of angle	Measured from	Measured to
$2\alpha_1 = 132°$	counter-clockwise	$(\sigma_{xx}, \sigma_{xz}) = (-10.3, -79.4)$	$(\hat{\sigma}_1, 0) = (167.7, 0)$
$2\alpha_2 = -48°$	clockwise	$(\sigma_{xx}, \sigma_{xz}) = (-10.3, -79.4)$	$(\hat{\sigma}_3, 0) = (-45.7, 0)$

In Physical Space

Angle	Sense of angle	Measured from	Measured to
$\alpha_1 = 66°$	counter-clockwise	$x (x_1)$	$\hat{x}_1$
$\alpha_2 = -24°$	clockwise	$x (x_1)$	$\hat{x}_3$

Discussion

On the Mohr diagram, the angle $2\alpha_1 = 132°$ is measured from the radius at $(\sigma_{xx}, \sigma_{xz})$ to the radius at $(\hat{\sigma}_1, 0)$. It is twice the angle $\alpha_1 = 66°$ in physical space measured from the coordinate axes x to $\hat{x}_1$, which are the respective normals to the planes on which the stress components act. We measure angles in physical space from x, because it is a coordinate axis whose orientation is known. The angles on the Mohr circle must therefore be measured from the point representing the stresses on the plane normal to x (property 3i). We could use the z-axis as a reference in the same way, in which case the corresponding angles on the Mohr circle would be measured from the radius at $(\sigma_{zz}, \sigma_{zx}) = (132.3, 79.4)$, and the corresponding angles in physical space would be measured from the z-axis.

In labeling the principal axes, we form a right-handed coordinate system (see Figure 8.1.2 in Box 8.1) with $\hat{x}_1$, $\hat{x}_2$, and $\hat{x}_3$ parallel respectively to the maximum, intermediate, and minimum compressive stresses and with the positive ends of the principal axes arranged such that they conform to convention 1, discussed in Section 8.5 (Figure 8A.2B). Here $\hat{x}_1$ and $\hat{x}_3$ are necessarily perpendicular, because they are the normals to planes whose stress components plot at opposite ends of a diameter of the Mohr circle (Figure 8A.2A) (properties 3v).

Question 3

What are the extreme absolute values of the shear stress acting on planes normal to the x_1–x_3 (x–z) plane—that is, parallel to x_2 (or y)—and what are the orientations of the planes on which these values occur?

Procedure

The maximum absolute values of the shear stress are read directly from the $\hat{\sigma}_1$–$\hat{\sigma}_3$ Mohr circle (property 4) (Figure 8A.3A).

$$|\sigma_s|_{(max)} = 106.7 \text{ MPa}$$

The orientations of the normals to the planes of maximum shear stress are determined from the angles on the Mohr circle (Figure 8A.3A) as tabulated below:

On the Mohr Diagram

Angle	Sense of angle	Measured from	Measured to
$2\theta = +90°$	counter-clockwise	$(\hat{\sigma}_1, 0) = (167.7, 0)$	$(61, 106.7)$
$2\theta = -90°$	clockwise	$(\hat{\sigma}_1, 0) = (167.7, 0)$	$(61, -106.7)$

In Physical Space

Angle	Sense of angle	Measured from	Measured to
$\theta = +45°$	counter-clockwise	$\hat{x}_1$	$\mathbf{n}^+$
$\theta' = -45°$	clockwise	$\hat{x}_1$	$\mathbf{n}^-$

where $\mathbf{n}^+$ and $\mathbf{n}^-$ are, respectively, the normals to the planes P^+ and P^- on which the maximum shear stress is respectively positive and negative in the Mohr circle sign convention (Figure 8A.3B, C).

Discussion

The value of the maximum shear stress is simply given by the length of the radius of the appropriate Mohr circle. In the $\hat{x}_1$–$\hat{x}_3$ plane, which is also the x–z plane, $|\sigma_s|$ is a maximum on the $\hat{\sigma}_1$–$\hat{\sigma}_3$ Mohr circle at those

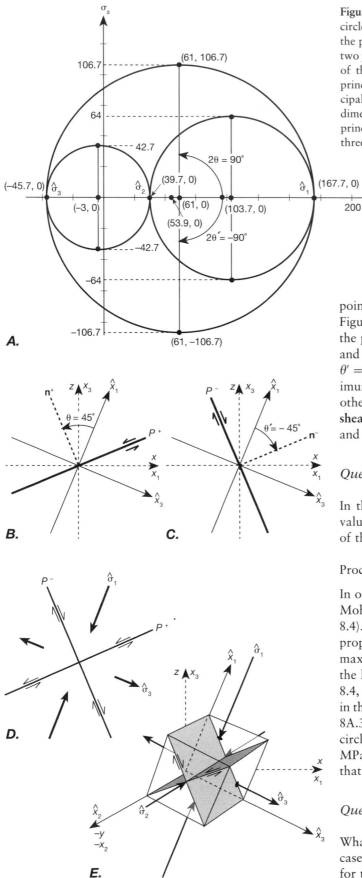

Figure 8A.3 (*Left*) Planes of maximum shear stress. *A.* Mohr circles for the three-dimensional stress, with radii drawn to the points of maximum shear stress on each circle. *B, C.* The two planes of maximum shear stress, showing the orientations of the normals to the planes relative to the reference and principal coordinate axes. *D.* Relative orientations of the principal axes and the planes of maximum shear stress in two dimensions. *E.* Relative orientation of the reference axes, the principal axes, and the planes of maximum shear stress in three dimensions.

points where the radius is normal to the σ_n-axis. From Figure 8A.3A, $2\theta = -2\theta' = 90°$. Thus in physical space, the planes of maximum shear stress are parallel to $\hat{x}_2$, and their normals are at $\theta = +45°$ (Figure 8A.3B) and $\theta' = -45°$ (Figure 8A.3C) from $\hat{x}_1$. The planes of maximum shear stress are therefore perpendicular to each other. They are called the **conjugate planes of maximum shear stress,** and the relationship between these planes and the principal stresses is shown in Figure 8A.3D.

Question 4

In the three-dimensional solid, what are the absolute values of the maximum shear stress and the orientations of the planes on which these values occur?

Procedure

In order to answer this question, we must consider the Mohr diagram for the three-dimensional stress (Box 8.4). When all three Mohr circles are plotted (Box 8.4, properties 1iii and 4ii; Figure 8A.3A), it is clear that the maximum absolute value of the shear stress occurs on the largest Mohr circle, which is the $\hat{\sigma}_1 - \hat{\sigma}_3$ circle (Box 8.4, property 4ii). The planes of maximum shear stress in three-dimensional physical space are shown in Figure 8A.3E. The maximum shear stress on the $\hat{\sigma}_1 - \hat{\sigma}_2$ Mohr circle (64 MPa) and on the $\hat{\sigma}_2 - \hat{\sigma}_3$ Mohr circle (42.7 MPa) are maxima only for the respective sets of planes that are parallel to $\hat{x}_3$ and $\hat{x}_1$.

Question 5

What is the mean normal stress for the two-dimensional case in the $\hat{x}_1 - \hat{x}_3$ plane? What is the mean normal stress for the three-dimensional case?

Procedure

The two-dimensional mean normal stress for the set of planes parallel to any of the principal axes can be determined from the Mohr diagram: Simply read off the value of the normal stress at the center of the appropriate Mohr circle (Figure 8A.3A). Alternatively, it can be calculated using an equation of the form of Equation (8.26). For the $\hat{\sigma}_1 - \hat{\sigma}_3$ Mohr circle, we have

$$\overline{\sigma}_n = \frac{\hat{\sigma}_1 + \hat{\sigma}_3}{2} = \frac{167.7 - 45.7}{2} = 61 \text{ MPa}$$

The three-dimensional mean normal stress cannot be read off the Mohr diagram in any simple way (Box 8.4, property 5ii). It must be calculated from Equation (8.4.4).

$$\overline{\sigma}_n = \frac{\hat{\sigma}_1 + \hat{\sigma}_2 + \hat{\sigma}_3}{3} = \frac{167.7 + 39.7 - 45.7}{3} = 53.9 \text{ MPa}$$

Discussion

In the two-dimensional case, $\overline{\sigma}_n$ is the center of the appropriate Mohr circle, and it is the value of the normal stress on the planes of maximum shear stress that parallel one of the principal axes. In the foregoing example, we considered the $\hat{\sigma}_1 - \hat{\sigma}_3$ Mohr circle, which shows the stresses on planes parallel to $\hat{x}_2$. In the three-dimensional case, the mean normal stress $\overline{\sigma}_n$ has neither of these properties (Figure 8A.3A).

Question 6

What are the values of the normal and shear components of the surface stress that acts on each of the following planes?

Plane A is parallel to $\hat{x}_2$, and its normal is at an angle $\alpha_A = 35°$ from x_1 (Figure 8A.4A, B).

Plane B is parallel to $\hat{x}_1$, and its normal is at an angle $\theta_B = -30°$ from $\hat{x}_3$ (Figure 8A.4C, D).

Procedure

Because plane A is parallel to $\hat{x}_2$, which is a principal axis, the normal $\mathbf{n}_A$ to the plane lies in the $\hat{x}_1 - \hat{x}_3$ plane, which is also the $x_1 - x_3$ (or $x-z$) plane (Figure 8A.4A, B). The stress components on this plane must therefore plot on the $\hat{\sigma}_1 - \hat{\sigma}_3$ Mohr circle (Figure 8A.4E). Similarly, plane B is parallel to the principal axis $\hat{x}_1$, so its normal $\mathbf{n}_B$ lies in the $\hat{x}_2 - \hat{x}_3$ plane (Figure 8A.4C, D). The stress components on plane B, therefore, must plot on the $\hat{\sigma}_2 - \hat{\sigma}_3$ Mohr circle (Figure 8A.4E). This geometry makes it possible to solve both problems by separate two-dimensional analyses (see the discussion at the beginning of Section 8.5). The geometric relationships in the appropriate two-dimensional planes are shown for planes A and B in Figures 8A.4B, D, respectively, where the conventions for plotting coordinate axes (convention (1), Section 8.5) have been used.

The relationships are summarized below in the table "In Physical Space" (Figure 8A.4B, D). The construction of the Mohr diagram that defines the stress components on the relevant planes is derived from properties 3i and ii and from the data in this table. It is summarized below in the table "On the Mohr Circle" (Figure 8A.4E).

In physical space (Figure 8A.4B), the normal $\mathbf{n}_A$ to plane A is defined by the angle $\alpha_A = 35°$ measured counterclockwise from x (x_1) in the $x-z$ (x_1-x_3) plane, which is the same as the $\hat{x}_1 - \hat{x}_3$ plane. Here x is the normal to the plane on which the stress components are $(\sigma_{xx}, \sigma_{xz}) = (-10.3, -79.4)$. Thus we can find the stress components on plane A on the $\hat{\sigma}_1 - \hat{\sigma}_3$ Mohr circle by measuring an angle $2\alpha_A = 70°$ counterclockwise from

In Physical Space

Angle	Coordinate plane containing the angle	Sense of angle	Measured from	Measured to
Plane A $\alpha_A = 35°$	$\hat{x}_1 - \hat{x}_3$ and $x_1 - x_3$ (or $x-z$)	counterclockwise	x_1 (or x)	$\mathbf{n}_A$
Plane B $\theta_B = 30°$	$\hat{x}_2 - \hat{x}_3$	clockwise	$\hat{x}_3$	$\mathbf{n}_B$

On The Mohr Diagram

Mohr circle	Angle	Sense of angle	Measured from	Measured to
Plane A $\hat{\sigma}_1 - \hat{\sigma}_3$	$2\alpha_A = 70°$	counterclockwise	$(-10.3, -79.4)$	(σ_n, σ_s) on plane A
Plane B $\hat{\sigma}_2 - \hat{\sigma}_3$	$2\theta_B = -60°$	clockwise	$(-45.7, 0)$	(σ_n, σ_s) on plane B

the radius at the point $(\sigma_{xx}, \sigma_{xz}) = (-10.3, -79.4)$ (Figure 8A.4E). A similar procedure is used to find the stress components on plane B (Figure 8A.4D), except that in this case we must use the $\hat{\sigma}_2$–$\hat{\sigma}_3$ Mohr circle, and angles are measured from $\hat{x}_3$ in physical space and from $(\hat{\sigma}_3, 0) = (-45.7, 0)$ on the Mohr diagram.

The normal stress and shear stress components on planes A and B are read from the appropriate Mohr circles in Figure 8A.5E. The results, using the Mohr circle sign conventions for the stress components, are

For plane A: $(\sigma_n, \sigma_s) = (111.1, -94.2)$

For plane B: $(\sigma_n, \sigma_s) = (-24.4, 37.0)$

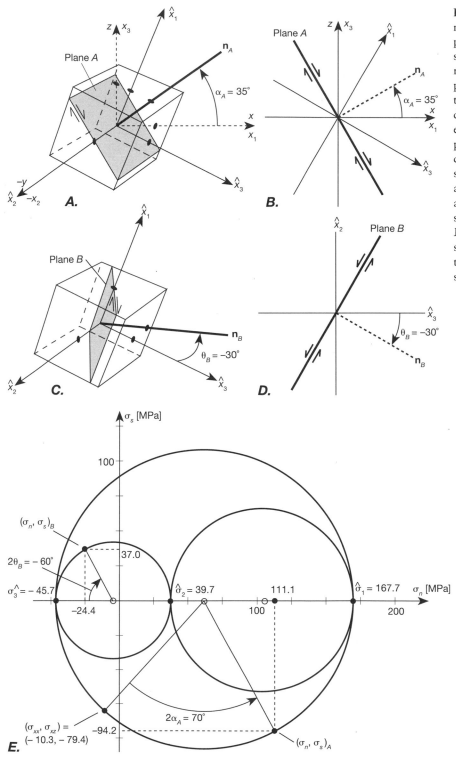

Figure 8A.4 Determining the components of surface stress on plane A and plane B. A. Three-dimensional diagram showing the orientation of plane A with respect to the principal coordinate planes. B. Two-dimensional diagram of the relationships in part A. C. Three-dimensional diagram showing the orientation of plane B with respect to the principal coordinate planes. D. Two-dimensional diagram of the relationships in part C. Note how the principal axes have been oriented relative to one another to conform to the convention shown in Figure 8.4.1C in Box 8.4. E. Mohr diagram of three-dimensional stress, showing the construction for determining the components of surface stress on planes A and B.

CHAPTER

Mechanics of Fracturing and Faulting

Theory and Experiment

In this chapter, we investigate the relationship between stress and the formation of rock fractures. We wish to understand the conditions under which fractures develop in Earth materials as a guide to understanding how and why natural fractures and faults form.

We begin with a discussion of **elastic deformation,** which characterizes brittle materials at stresses below those that cause fracture. In the laboratory, rock samples can be subjected to a variety of stress states that produce elastic deformation and that, if increased sufficiently, result in different types of fracturing. Data from such experiments make it possible to formulate **fracture criteria** that define both the stress state at fracture and the orientations of the fractures. Fracture criteria enable us to determine whether any given state of stress in the Earth will cause fracturing or will be stable. We next review the effects on fracture criteria of physical conditions such as confining pressure, pore pressure, rock anisotropy, the intermediate principal stress, and temperature. Finally, we consider the Griffith theory for brittle fracture, a unifying model that accounts, at the microscopic and submicroscopic level, for many of these observed effects.

9.1 Elastic Deformation and Experimental Fracturing of Rocks

Elastic Deformation

In a typical rock deformation experiment, a piece of rock[1] is cut into a cylinder that has a diameter ranging from less than 1 cm up to tens of centimeters, in some cases, and a length typically two to four times the diameter. The sample is placed between two pistons of hardened steel or similar material, which are forced together by a device such as a hydraulic ram. The applied stress changes the length, diameter, and volume of the sample. These changes are parts of the **strain,** which is measured by strain gauges attached to the sample. The

---

[1] For many experiments, samples representing a particular rock type are selected for their fine grain size and uniform characteristics. Examples of such rocks on which much experimental work has been done include the Solenhofen limestone, Yule marble, Hasmark dolomite, Berea sandstone, Martinsburg slate, and Westerly granite. The individual rocks are significant only insofar as they are representative of the general rock types.

primary information that such an experiment yields is the relationship between the axial force applied through the pistons and the associated change in dimensions of the sample.

When a material such as rock experiences a gradually increasing stress, the initial deformation is elastic, which means that changes in stress induce an instantaneous change in the sample dimensions, measured by the strain. When the stress is removed, the strain completely disappears. Thus the strain is **recoverable**. The **extensional strain** e_n, or simply the **extension**, is one measure of the strain. It is the change in length of the sample per unit of initial length—that is,

$$e_n = \frac{\ell - L}{L} = \frac{\Delta L}{L} \tag{9.1}$$

where L is the initial length, ℓ is the deformed length, and ΔL is the change in length. We also express the extension as a percent change in length by multiplying e_n by 100.

In a uniaxial state of stress, the magnitude of the elastic extension parallel to the applied stress is directly proportional to the magnitude of the stress (see zone II of the curves in Figure 9.21A):

$$\sigma_n = E e_n \qquad e_n = \frac{\sigma_n}{E} \tag{9.2}$$

where the constant of proportionality E is **Young's modulus,** which is one of two elastic constants we need to characterize the elastic behavior of an isotropic[2] material. For the geologic sign convention, a uniaxial compressive (positive) stress decreases the length of the specimen, thereby producing a negative extension. Conversely, a tensile (negative) stress produces a positive extension. Thus Young's modulus is a negative number.[3] For rocks, E characteristically has values in the range of -0.5×10^5 MPa to -1.5×10^5 MPa. The extreme value of the extension that most materials can reach before they fracture is generally quite small—a few percent at most and usually much less.

Like stress, strain is actually a symmetric second-rank tensor quantity, which we discuss further in Chapter 15. For present purposes, we note only that (1) the complete state of strain can be defined by three principal extensions, $\hat{e}_1 \geq \hat{e}_2 \geq \hat{e}_3$, where the circumflex above the symbols indicates that they represent principal values, consistent with our notation for stress, and (2) for the very small strains characteristic of elastic defor-

mation, the principal axes of strain are parallel to the principal axes of stress. This parallelism, however, does not hold in general for the large strains characteristic of ductile deformation.

In uniaxial compression, the sample shortens parallel to the applied stress. If the volume of the sample is conserved, the sample must also expand in a direction normal to the shortening (Figure 9.1A). Materials are never perfectly incompressible, however, so there is a net decrease in volume in any deformation caused by a compressive stress. **Poisson's ratio** ν is the absolute value of the ratio given by the extension $\hat{e}_\perp$ normal to an applied compressive stress, divided by the extension $\hat{e}_\parallel$ parallel to the applied compression.

$$\nu \equiv \left| \frac{\hat{e}_\perp}{\hat{e}_\parallel} \right| \tag{9.3}$$

Poisson's ratio is the second elastic constant that characterizes the behavior of an isotropic elastic material. If a material were perfectly incompressible, ν would equal 0.5. Poisson's ratio for most rocks, however, ranges from 0.25 to 0.33. The expansion normal to an applied compression is the **Poisson expansion**.

In a sample under confined compression, the magnitude of the axial extension depends not only on the

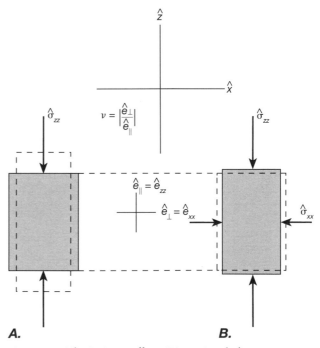

Figure 9.1 The Poisson effect. Dimensional changes are exaggerated for clarity. A. Comparing the unstressed shape (dashed rectangle) with the shape caused by uniaxial compression $\hat{\sigma}_{zz}$ (shaded rectangle) illustrates the Poisson expansion. B. If a radial pressure $p = \hat{\sigma}_{xx} = \hat{\sigma}_{yy}$ is added to the uniaxial stress in part A, the Poisson effect decreases the amount of shortening parallel to $\hat{\sigma}_{zz}$.

---

[2] An isotropic material is one whose mechanical properties are the same in all directions. A granite, for example, might well be mechanically isotropic, but a schist is definitely anisotropic.

[3] Young's modulus is generally reported as a positive number in tables of elastic constants, because for the engineering sign convention, the stress and the resulting extension both have the same sign.

axial stress but also, because of the Poisson expansion, on the radial pressure. The axial shortening produced by an axial compressive stress is reduced by the Poisson expansion associated with each of the compressive radial principal stresses (Figure 9.1B). Thus[4]

$$\hat{e}_{zz} = \frac{1}{E}\hat{\sigma}_{zz} - v\left(\frac{1}{E}\right)\hat{\sigma}_{xx} - v\left(\frac{1}{E}\right)\hat{\sigma}_{yy} \qquad (9.4)$$

$$\hat{e}_{zz} = \frac{1}{E}\hat{\sigma}_{zz} - \frac{v}{E}(\hat{\sigma}_{xx} + \hat{\sigma}_{yy}) \qquad (9.5)$$

Compare these relationships with Equation (9.2) for uniaxial stress. In Equation (9.4), the two terms containing v give the Poisson expansion that would develop in the $\hat{z}$ direction if each radial stress component parallel to $\hat{x}$ and to $\hat{y}$ were a uniaxial stress. We can write similar equations for the other two principal extensions simply by permuting the subscripts on the symbols for extension and stress.

The **volumetric extension** is the change in volume divided by the initial volume; for small strains, it is approximately related to the three principal extensions (see Sections 15.1 and 15.2) by

$$e_v = \frac{\Delta V}{V} \approx \hat{e}_1 + \hat{e}_2 + \hat{e}_3 \qquad (9.6)$$

Experimental Fracturing

Increasing the axial stress ultimately results in **failure** of a sample when the sample is unable to support a stress increase without permanent deformation. The stress at which failure occurs is a measure of the **strength** of the material. Because failure can occur in a number of different ways, there are a variety of different measures of material strength. **Brittle failure** occurs with the formation of a **brittle fracture,** which is a surface or zone across which the material loses cohesion. **Ductile failure** occurs when the material becomes permanently deformed without losing cohesion.

Experiments investigating the effects of pressure on failure employ a sample sealed in an impermeable jacket and surrounded in a pressure chamber by a fluid under pressure. The pressure of the surrounding fluid (the **confining pressure**) and that of the fluid in the pore

spaces in the rock (the **pore fluid pressure**) can be controlled independently to determine the influence of each on the behavior of the rock samples. In experiments on temperature effects, the temperature is controlled with a small wire-wound furnace that surrounds the sample in the pressure chamber.

Generally the applied forces are normal to the surfaces of the sample so that the principal axes of stress are either parallel to the cylinder axis (the **axial stress**) or perpendicular to the cylinder axis (the **radial stress** or confining pressure). Experiments usually are done in either uniaxial compression (Figure 8.14B) or confined compression (Figure 8.14D), the axial stress being the maximum compressive stress $\hat{\sigma}_1$. Axial extension experiments in which the axial stress is the minimum compressive stress $\hat{\sigma}_3$ (Figure 8.14E) are also common. Special modifications to the equipment permit experiments in uniaxial tension (Figure 8.14C), although these are relatively uncommon.

Experiments on brittle failure reveal two fundamentally different types of fracture: extension fractures (mode I) and shear fractures (modes II and III) (compare Figure 3.1). Each type exhibits a different orientation of the fracture plane relative to the principal stresses and a different direction of displacement relative to the fracture surface. These two fracture types mimic natural fractures in rocks (as described, for example, at the beginning of Chapter 3).

For **extension fractures,** the fracture plane is perpendicular to the minimum principal stress $\hat{\sigma}_3$ and parallel to the maximum principal stress $\hat{\sigma}_1$. Displacement is approximately normal to the fracture surface. Extension fractures are **tension fractures** (Figure 9.2A) if the minimum principal stress $\hat{\sigma}_3$ is tensile, as in uniaxial tension (Figure 8.14C). They form by **longitudinal splitting** (Figure 9.2B) if the minimum principal stress is equal or close to zero and the maximum compressive stress $\hat{\sigma}_1$ is the axial stress, as in uniaxial compression (Figure 8.14B). Fractures that form by longitudinal splitting tend to be more irregular in orientation and shape than other extension fractures. Extension fractures may also form under conditions of axial extension (Figure 9.2C), in which the axial stress is the minimum compressive stress $\hat{\sigma}_3$ (Figure 8.14E).

Shear fractures form in confined compression (Figure 8.14D) at angles of less than 45° to the maximum compressive stress $\hat{\sigma}_1$ (Figure 9.2D). Displacement is parallel to the fracture surface. If the state of stress is triaxial (Figure 8.14F), the shear fractures are parallel to the intermediate principal stress $\hat{\sigma}_2$ and form a conjugate pair of orientations at angles less than 45° on either side of the maximum compressive stress $\hat{\sigma}_1$. If $\hat{\sigma}_2 = \hat{\sigma}_3$ (Figure 8.14D), then the possible orientations of shear fractures are tangent to a cone of less than 45° about the $\hat{\sigma}_1$ axis (see Section 9.3).

---

[4] Because of the conventions $\hat{\sigma}_1 \geq \hat{\sigma}_2 \geq \hat{\sigma}_3$ and $\hat{e}_1 \geq \hat{e}_2 \geq \hat{e}_3$, and the sign conventions that require both a compressive stress and a lengthening extension to be positive, we encounter the inconvenient fact that the stress component $\hat{\sigma}_1$ produces the strain component $\hat{e}_3$. When relating stress to strain, we avoid confusion by using coordinate axes $\hat{x}, \hat{y}$, and $\hat{z}$ parallel to the principal axes and by relabeling the principal stresses and extensions such that $\hat{\sigma}_{xx}, \hat{\sigma}_{yy}$, and $\hat{\sigma}_{zz}$ are parallel to $\hat{e}_{xx}, \hat{e}_{yy}$, and $\hat{e}_{zz}$, respectively. Thus,

$$\begin{bmatrix} \hat{\sigma}_{xx} \\ \hat{\sigma}_{yy} \\ \hat{\sigma}_{zz} \end{bmatrix} = \begin{bmatrix} \hat{\sigma}_3 \\ \hat{\sigma}_2 \\ \hat{\sigma}_1 \end{bmatrix} \quad \text{and} \quad \begin{bmatrix} \hat{e}_{xx} \\ \hat{e}_{yy} \\ \hat{e}_{zz} \end{bmatrix} = \begin{bmatrix} \hat{e}_1 \\ \hat{e}_2 \\ \hat{e}_3 \end{bmatrix}$$

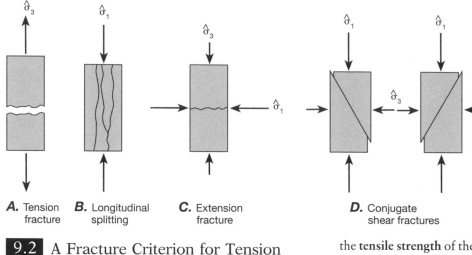

Figure 9.2 Types of fractures developed during experiments on rocks in the brittle field.

A. Tension fracture

B. Longitudinal splitting

C. Extension fracture

D. Conjugate shear fractures

9.2 A Fracture Criterion for Tension Fractures

Experiments on rocks under uniaxial tension show that there is, for each material or rock type, a characteristic value of tensile stress (T_0) at which tension fracturing occurs. The rock is stable at tensile stresses smaller than T_0, but it cannot support larger tensile stresses. T_0 is the **tensile strength** of the material. On a Mohr diagram, the boundary between stable and unstable states of tensile stress is called the **tension fracture envelope** (Figure 9.3A). It is a line perpendicular to the σ_n axis at T_0 and is represented by the equation

$$\sigma_n^* = T_0 \qquad (9.7)$$

where σ_n^* is the critical normal stress required to produce fracture. A Mohr circle that lies to the right of the line

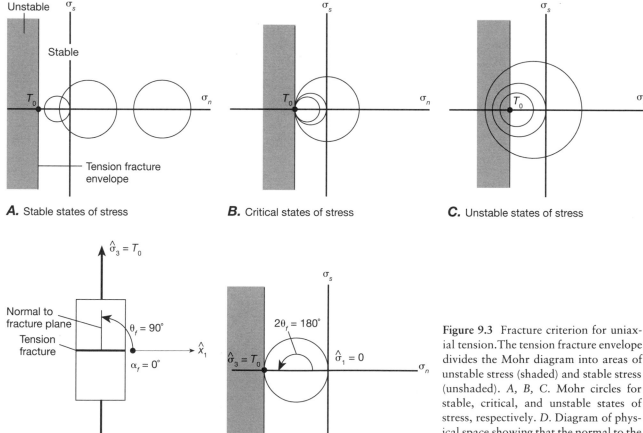

A. Stable states of stress

B. Critical states of stress

C. Unstable states of stress

D.

E. Critical uniaxial tension

Figure 9.3 Fracture criterion for uniaxial tension. The tension fracture envelope divides the Mohr diagram into areas of unstable stress (shaded) and stable stress (unshaded). *A, B, C.* Mohr circles for stable, critical, and unstable states of stress, respectively. *D.* Diagram of physical space showing that the normal to the tension fracture is parallel to $\hat{\sigma}_3$. *E.* Critical Mohr circle for tension fracture under a uniaxial tensile stress.

represents a stable stress state (Figure 9.3A). A Mohr circle tangent to the line (a **critical Mohr circle**) represents a state of stress that causes tension fracturing (Figure 9.3B). Mohr circles that cross the line represent states of stress that the material cannot support (Figure 9.3C).

We can describe the orientation of a fracture plane relative to the principal stresses by the **fracture plane angle** α_f, which is the angle between the maximum principal stress $\hat{\sigma}_1$ and *the fracture plane*, or by the **fracture angle** θ_f, which is the angle between the maximum principal stress $\hat{\sigma}_1$ and *the normal to the fracture plane*. For a given plane, $|\theta_f - \alpha_f| = 90°$, and if both angles are acute, they are opposite in sign (compare Figures 9.3D and 9.4D, E). In order to plot on a Mohr diagram the surface stresses that act on the fracture planes, we must define the orientation of the fracture plane by the orientation of its normal, so we use the fracture angle θ_f.

In experiments, the tension fracture plane is normal to the maximum tensile stress $\hat{\sigma}_3$. Thus the fracture plane angle α_f is $0°$ and the fracture angle θ_f is $90°$ (Figure 9.3D). On the Mohr diagram (Figure 9.3E), the stress on the fracture plane plots at an angle of $2\theta_f = 180°$ from $(\hat{\sigma}_1, 0)$. The normal stress and shear stress components on the fracture plane therefore plot exactly at the point of tangency between the critical Mohr circle and the tension fracture envelope.

Equation (9.7) thus provides a **fracture criterion,** because it defines both the stress required for fracturing and the orientation of the fracture: A tension fracture forms on any plane in the material on which the normal stress reaches the critical value T_0, and the fracture plane is perpendicular to the maximum tensile stress $\hat{\sigma}_3$.

This fracture criterion, however, applies only to tension fractures formed under conditions of tensile stress. It does not account for the occurrence of extension fractures that develop under conditions in which none of the principal stresses are tensile (Figure 9.2B, C), such as longitudinal splitting.

9.3 The Coulomb Fracture Criterion for Confined Compression

The relationship between the state of stress and the occurrence of shear fracturing for confined compression experiments (Figure 8.14D) is more complicated than for uniaxial tension. Fracture experiments on different samples of the same rock show that the initiation of fracturing depends on the differential stress ($_D\sigma = \hat{\sigma}_1 - \hat{\sigma}_3$) and that the magnitude of the differential stress necessary to cause shear fracture increases with confin-

ing pressure.[5] The fracture angle θ_f between $\hat{\sigma}_1$ and the normal to the fracture plane is generally around $\pm 60°$, so the fracture plane angle α_f between the fracture plane itself and the maximum compressive stress $\hat{\sigma}_1$ must be about $\pm 30°$ (Figure 9.2D).

Experimental data show that it is possible to construct on the Mohr diagram a **shear fracture envelope** that separates stable from unstable states of stress. This envelope is commonly approximated by a pair of straight lines that are symmetric across the $\hat{\sigma}_n$ axis (Figure 9.4A–C), although in fact the lines may be slightly concave toward that axis. Any Mohr circle contained between the two lines of the fracture envelope represents a stable stress state (Figure 9.4A). A Mohr circle tangent to the lines represents a critical state of stress that causes fracturing (Figure 9.4B). A Mohr circle that crosses the fracture envelope represents an unstable state of stress that cannot be supported (Figure 9.4C). Radii drawn to the points of tangency between each critical Mohr circle and the shear fracture envelope (Figure 9.4B) indicate the surface stress components on the actual fracture plane at the time of fracture.

The straight line approximation to the shear fracture envelope is known as the **Coulomb fracture criterion,** and it is described by the equation

$$|\sigma_s^*| = c + \mu\sigma_n \qquad (9.8)$$

where

$$\mu = \tan\phi \qquad (9.9)$$

and where σ_s^* is the critical shear stress, μ and c are the slope and intercept of the lines, respectively, and ϕ is the slope angle of the line, taken to be positive (Figure 9.4B). Because the equation is written in terms of the absolute value of the critical shear stress, it describes both lines of the fracture criterion.

The two constants in Equation (9.8), c and μ, characterize the failure properties of the material, and they vary from one material or rock type to another. The **cohesion** c is the resistance to shear fracture on a plane across which the normal stress is zero. We call μ the **coefficient of internal friction** and ϕ the **angle of internal friction** because of the similarity, when the cohesion is zero, between Equation (9.8) and the law of frictional resistance. The Coulomb fracture criterion can also be expressed in terms of the principal stress (see Box 9.1).

The Coulomb fracture criterion states that whenever the state of stress in a rock is such that on a plane of some orientation, the surface stress components (σ_n, σ_s) satisfy Equation (9.8), a shear fracture can develop on that plane. For any critical stress state, the criterion

---

[5]Remember that in confined compression, the confining pressure defines the magnitude of the radial stress that is exerted by the pressure medium, $p = \hat{\sigma}_2 = \hat{\sigma}_3$.

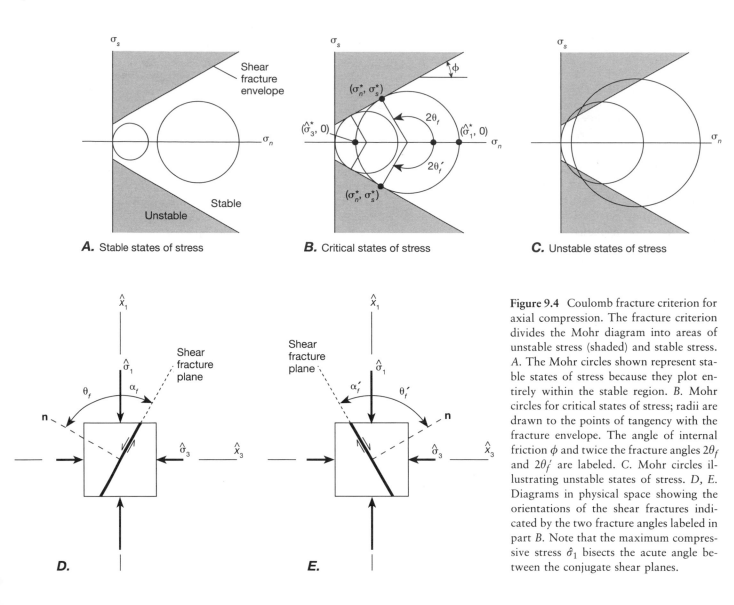

A. Stable states of stress

B. Critical states of stress

C. Unstable states of stress

D.

E.

Figure 9.4 Coulomb fracture criterion for axial compression. The fracture criterion divides the Mohr diagram into areas of unstable stress (shaded) and stable stress. A. The Mohr circles shown represent stable states of stress because they plot entirely within the stable region. B. Mohr circles for critical states of stress; radii are drawn to the points of tangency with the fracture envelope. The angle of internal friction ϕ and twice the fracture angles $2\theta_f$ and $2\theta_f'$ are labeled. C. Mohr circles illustrating unstable states of stress. D, E. Diagrams in physical space showing the orientations of the shear fractures indicated by the two fracture angles labeled in part B. Note that the maximum compressive stress $\hat{\sigma}_1$ bisects the acute angle between the conjugate shear planes.

Box 9.1 The Coulomb Fracture Criterion in Terms of Principal Stresses

The Coulomb fracture criterion is sometimes expressed in terms of the critical principal stresses at fracture. We can derive this relationship from Equation (9.8), using the positive value of σ_s^*. We substitute for σ_s and σ_n from Equations (8.28) or (8.38) with $\theta_2 = \theta_f$ and use the relationship from Equation (9.9) and the first Equation (9.10)

$$\mu = \tan\phi = -\frac{1}{\tan 2\theta_f} = -\cot 2\theta_f \qquad (9.1.1)$$

After some algebraic manipulation, using the standard trigonometric relations $\tan 2\theta_f = \sin 2\theta_f / \cos 2\theta_f$ and $\sin^2 2\theta_f + \cos^2 2\theta_f = 1$, we find that the maximum compressive stress required for fracture varies linearly with the minimum principal stress according to

$$\hat{\sigma}_1^* = S + K\hat{\sigma}_3 \qquad (9.1.2)$$

where

$$S = \frac{2c\sin 2\theta_f}{1 + \cos 2\theta_f} \qquad K = \frac{1 - \cos 2\theta_f}{1 + \cos 2\theta_f} \qquad (9.1.3)$$

and where the asterisk in $\hat{\sigma}_1^*$ indicates the critical maximum compressive stress for a given value of the minimum compressive stress. S is the fracture strength under uniaxial compression ($\hat{\sigma}_3 = 0$). Note, however, that setting $\hat{\sigma}_1 = 0$ does not give the tensile strength under uniaxial tension, because this equation is an expression of the Coulomb fracture criterion, which does not account for tensile fracture.

In this form, the fracture criterion is commonly plotted on a graph of maximum versus minimum principal stress. Although the Mohr circle does not plot on such a graph, the difference between $\hat{\sigma}_1^*$ and $\hat{\sigma}_3$ is the diameter of the critical Mohr circle.

is satisfied at the two points where the Mohr circle is tangent to the two lines given by Equation (9.8) (Figure 9.4B). These points define the stresses on two differently oriented planes called the **conjugate shear planes** (Figure 9.4D, E), and they indicate that there are two possible orientations of shear fractures that can develop. The fracture criterion does not predict which orientation should form.

On the Mohr circle (Figure 9.4B), the angle between the radii to $(\hat{\sigma}_1^*, 0)$ and to the stress components (σ_n^*, σ_s^*) on the critically stressed planes is $\pm 2\theta_f$. Thus in physical space, the normals to the two conjugate shear fractures must be at an angle of $\pm \theta_f$ to the $\hat{x}_1$ axis and to $\hat{\sigma}_1$, and the fracture planes themselves make an angle of $\pm \alpha_f$ with $\hat{\sigma}_1$ (Figure 9.4D, E). On the Mohr circle, the radius to the tangent point must be perpendicular to the fracture envelope, so the angles θ_f and α_f are related to the slope angle of the fracture envelope ϕ by

$$|2\theta_f| = (90 + \phi) \text{ degrees} = (\pi/2 + \phi) \text{ radians}$$
$$|2\alpha_f| = (90 - \phi) \text{ degrees} = (\pi/2 - \phi) \text{ radians}$$
(9.10)

Thus the fracture envelope defines both the critical stress required for fracture and the orientation of the shear fracture that develops.

If the three principal stresses are unequal, the line of intersection of the conjugate shear planes parallels the intermediate principal stress $\hat{\sigma}_2$. In uniaxial or confined compression where $\hat{\sigma}_2 = \hat{\sigma}_3$, there is no unique intermediate principal stress axis, and there are an infinite number of possible orientations for the shear fracture planes distributed as tangents to a cone whose axis is parallel to $\hat{\sigma}_1$.

The Coulomb fracture criterion that best fits the experimentally determined data for shear fracturing of Berea sandstone, for example, is, in units of megapascals (MPa),

$$|\sigma_s^*| = 24.1 + 0.49\sigma_n$$
(9.11)

By comparing Equations (9.8) and (9.11) and using Equations (9.9) and (9.10), we find that $\phi = 26°$, $|\theta_f| = 58°$, and $|\alpha_f| = 32°$.

The relationships expressed in Equations (9.9) and (9.10) indicate that μ, ϕ, θ_f, and α_f measure the same physical property. Figure 9.5A shows a histogram of experimentally determined values of α_f on a variety of rocks under widely differing experimental conditions. The mean of the fracture plane angles is about 29°, very close to the oft-quoted "constant" of 30°. These data suggest that the angle between the shear fracture plane and the maximum compressive stress $\hat{\sigma}_1$ is about 30°. Thus the fracture angle θ_f between the normal to the shear fracture plane and the maximum compressive stress $\hat{\sigma}_1$ is about 60°.

The large scatter in values of α_f in Figure 9.5A results partly from combining experiments performed under different conditions of stress, confining pressure, pore fluid pressure, temperature, and rock type. The data in the histogram indicated by dark shading are for jacketed dry samples of Berea sandstone. Here the mean fracture plane angle is 31° and the scatter is reduced, although the angles still range between 26° and 38°. Temperature apparently has no effect on these data, and although the fracture plane angle tends to increase with confining pressure, as shown by Figure 9.5B, indicating

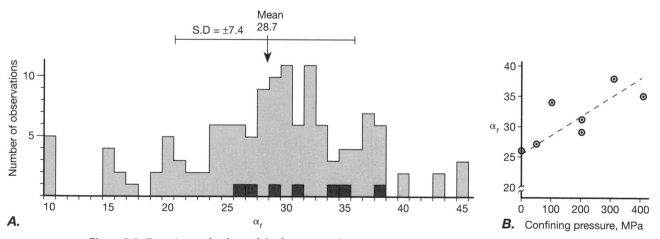

Figure 9.5 Experimental values of the fracture angle. A. Histogram of experimentally determined fracture plane angles α_f between the fracture plane and the maximum compressive stress $\hat{\sigma}_1$. The diagram includes data from jacketed and unjacketed plutonic, volcanic, sedimentary, and metamorphic rocks under confining pressures from about 0.5 to 500 MPa, temperatures from 20°C to 800°C, with and without pore fluid. Dark shading accents data angles for jacketed dry samples of Berea sandstone. B. Variation of the fracture plane angle with confining pressure for jacketed dry samples of Berea sandstone (dark shaded data in part A).

a fracture envelope concave toward the σ axis, the scatter is still considerable. Thus the experimental variation in the average value of 30° for α_f is considerable and is to be expected in nature as well.

It may seem strange that shear fractures form at an angle of approximately $\alpha_f = \pm 30°$ to the maximum compressive stress $\hat\sigma_1$ rather than parallel to the conjugate planes of maximum shear stress, which are oriented at $\pm 45°$. The lower angle results from competing effects of normal stress and shear stress on a given fracture orientation. The development of a shear fracture is promoted by both a minimum normal stress and a maximum shear stress on the fracture plane, but the normal stress on a plane is not a minimum at the same orientation for which the shear stress is a maximum. Thus the fracture angle is an optimization of these two effects.

Figure 9.6 illustrates this relationship for one of the fracture experiments on Berea sandstone. For this particular state of stress (inset, Figure 9.6A), the two solid curves in Figure 9.6A show how the normal stress σ_n (the first Equation (8.38)) and the shear stress σ_s (the second Equation (8.38)) vary with changing orientation of a plane (see also Equation 8.28). That orientation is defined by the angle θ_2 between the normal to the plane and the $\hat\sigma_1$ direction (Figure 9.6B). Note that it is impossible to minimize the normal stress and maximize the shear stress on the same plane. The curve for the

critical shear stress $|\sigma_s^*|$ required to cause fracture on any particular orientation of plane is calculated from the Coulomb fracture criterion for Berea sandstone (Equation 9.11) using values for σ_n from the curve in Figure 9.6A. Where the available shear stress equals the critical shear stress needed to cause fracture, the curves for σ_s and $|\sigma_s^*|$ touch, and that point defines the orientation of the critically stressed plane that becomes the fracture plane.

The predicted fracture angle where, within experimental error, the curves touch is $\theta_f = 58°$ ($\alpha_f = 32°$); the experimentally observed angle is $\theta_f = 55°$ ($\alpha_f = 35°$). The two curves, however, are almost parallel for angles of roughly $\pm 5°$ on either side of the ideal θ_f, so within this range of angles, all the planes are very nearly at the critical shear stress. Under such conditions, minor heterogeneities must play a significant part in determining which of these planes ultimately becomes the fracture plane, thereby accounting for some of the scatter in the observed data (dark shaded histogram bars in Figure 9.5A).

By far the most common stress state in natural environments is triaxial stress (Figure 8.14F). Because faults are shear fractures, the Coulomb fracture criterion leads us to expect faults to form parallel to the intermediate principal stress and at angles of $\alpha_f = \pm 30°$ to the maximum compressive stress $\hat\sigma_1$. In nature, one orientation of the conjugate pair tends to be dominant locally although both are present over large areas.

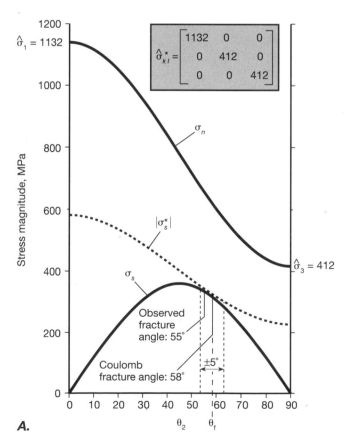

A.

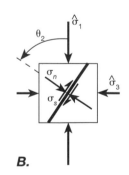

B.

Figure 9.6 A. The surface stress components σ_n and σ_s and the critical shear stress σ_s^* needed for fracture are plotted versus the orientation θ_2 of the surface for a particular critical state of stress (inset) on a sample of Berea sandstone. The fracture angle is the angle that minimizes the quantity $[|\sigma_s^*| - \sigma_s]$ as shown on the graph. B. Geometric relationships in physical space between the principal stresses, the orientation of the plane, and the surface stress components on the plane.

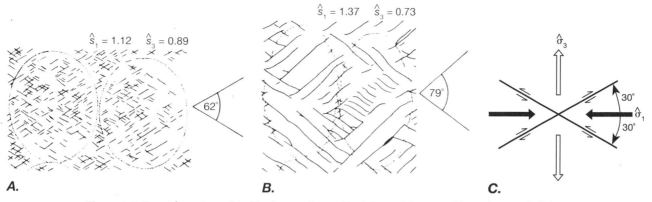

Figure 9.7 Pure shear in a clay block experimentally deformed by stretching of an underlying rubber sheet. Ellipses record the distortion of original circles *A*. Shear fractures form in two conjugate orientations separated by an angle of about 60°. *B*. Domains develop in which displacement is concentrated on one orientation of faults or the other. The angle between the conjugate fault sets increases. *C*. Interpretation of the faults in terms of the Coulomb fracture criterion.

We can get some insight into how natural deformation might proceed by examining scale models in the laboratory. The mechanical properties of dry sand and wet clay provide appropriate scaled analogues for the properties of the Earth's crust (see Sections 20.5 and 20.7). Figure 9.7 shows one experiment in which a layer of clay was shortened in one direction and stretched in a perpendicular direction by the deformation of an underlying sheet of rubber. The amount of deformation is indicated by the ellipses, which were circles. Faults initially develop in conjugate orientations that are approximately 60° apart (Figure 9.7A). Slip on one set of faults offsets faults of the other orientation, however, and therefore interferes with the deformation. Thus with continued deformation, domains tend to develop in which the faults of one or the other orientation dominate (Figure 9.7B). The angle between the conjugate sets of faults increases with increasing deformation. The conjugate shear fractures are consistent with the Coulomb fracture criterion, as indicated in Figure 9.7C.

A different geometry of faults develops if the layer of clay is subjected to a shear parallel to its boundaries (Figure 9.8; compare Figure 7.4). Again two conjugate sets of faults labeled *R* and *R′* appear (Figure 9.8A). Neither is parallel to the imposed direction of shearing, however, and they are referred to as **secondary shears** or **Riedel shears**. *R* shears are synthetic, having the same sense as the imposed shear (sinistral in Figure 9.8A) and are oriented about 15° from the direction of the imposed shear. *R′* shears are antithetic, having a sense of shear opposite to that imposed (dextral in Figure 9.8A), and are oriented about 75° to 80° from the direction of the imposed shear. With increasing deformation, *R* shears rotate to smaller angles, and another set of secondary synthetic shears (labeled *P*) develops, oriented at about −10° from the imposed shear direction (Figure 9.8B).

Only after this stage do shears parallel to the imposed direction form.

If the state of stress is assumed to be a pure shear stress (Figure 8.14G) and the plane of imposed shear is the plane of maximum shear stress at 45° to $\hat{\sigma}_1$, the Riedel shears *R* and *R′* can be accounted for as the conjugate shear fractures predicted by the Coulomb fracture criterion (Figure 9.8C; cf. Figure 7.4A). The secondary shear fractures labeled *P*, however, are not predicted by this simple analysis; rather, they probably result from Coulomb shear fracturing under a locally rotated orientation of the principal stresses.

Thus, although the Coulomb fracture criterion is a reasonable approximation to the data on fracturing, it applies only to a limited part of the Mohr diagram, and it does not account for all observed shear fracturing. Nevertheless, it is a useful predictive tool for brittle fracture in compression.

It is important to determine how changes in various physical conditions might affect both the mechanical behavior of rock and the range of conditions over which the failure criteria can be applied. In the following three sections, we discuss effects of confining pressure, pore fluid pressure, temperature, anisotropy, the intermediate principal stress, and the dimensions of the body of rock.

9.4 Effects of Confining Pressure on Fracturing and Frictional Sliding

Confining Pressure and Shear Failure

Experimental data show that the Coulomb fracture criterion does not apply in the tensile part of the Mohr diagram. In fact, the data indicate that rather than the

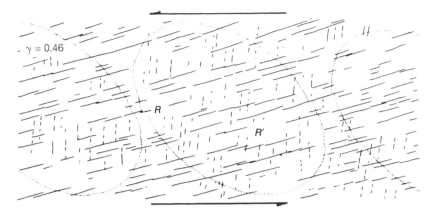

A.

B.

Figure 9.8 Sinistral simple shear in a clay block induced by shearing the substrate of the clay. *A.* Conjugate Riedel shears *R and R′* form at angles of about 15° and 75°, respectively, to the imposed direction of shear. *B. R′* shears rotate counterclockwise and *R* shears rotate clockwise. *R* shears become dominant and begin to become wavy. *P* shears begin to form at an angle of roughly −10° to −15° to the imposed shear direction. *C.* Interpretation of *R* and *R′* in terms of the Coulomb fracture criterion. The plane of imposed shear is the plane of maximum shear stress 45° from the maximum compressive stress. It is subjected to a pure shear stress. *P* shears and those parallel to the imposed direction of shear are not explained by this analysis.

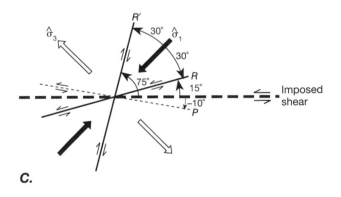

C.

straight-line fracture envelope shown in Figure 9.4*B,* a more comprehensive fracture criterion should actually approximate a parabolic curve on the negative side of the normal stress axis that connects with the Coulomb fracture criterion on the positive side (Figure 9.9). If the Mohr circle is tangent to the fracture envelope where the envelope crosses the σ_n axis, tension fractures form normal to the least principal stress ($2\theta_f = 180°$; Figure 9.9*A*). With increasing confining pressure, the Mohr circle shifts to the right, and its point of tangency with the fracture envelope shifts toward lower values of $2\theta_f$. As long as that point of tangency is in the tensile stress field, mixed-mode fractures develop that combine ex-

tensional and shear displacement across the fracture surface (Figure 9.9*B*). Higher confining pressures lead to shear fracture according to the Coulomb fracture criterion ($2\theta_f \approx 120°$; Figure 9.9*C*).

At still higher confining pressures, data indicate that the fracture envelope becomes concave toward the normal stress axis and therefore decreases in slope. As a consequence, $2\theta_f$ decreases with increasing confining pressure (Figure 9.9) and the fracture plane angle α_f increases (Figures 9.5*B* and 9.9*D*). This behavior is associated with a transition from brittle to ductile behavior. In the ductile region the Coulomb criterion no longer applies, and another failure criterion, the **von**

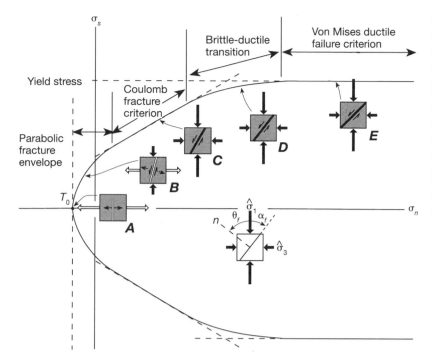

Figure 9.9 Schematic portrayal of failure envelopes and related fractures. The shaded boxes show the orientation of the failure plane at different points along the failure envelope. Double lines through the boxes indicate planes on which there is a component of extension fracturing; solid lines indicate shear failure planes. A. Tension fracture; B. mixed-mode fracture, including components of tension and shear; C. brittle shear fracture according to the Coulomb criterion; D. shear fracture in the brittle–ductile transition; E. ductile shear failure according to the von Mises criterion.

Mises criterion, becomes applicable. On a Mohr diagram, the von Mises criterion consists of a pair of lines of constant shear stress symmetric about the normal stress axis (Figure 9.9). This criterion implies that ductile deformation begins at a critical shear stress, the **yield stress,** which is independent of the confining pressure, and that planes of ductile failure are the planes of maximum shear stress ($2\theta_f = \pm 90°$; Figure 9.9E). The algebraic expression for the von Mises criterion is

$$|\sigma_s^*| = \text{constant} \qquad (9.12)$$

Figure 9.10, which records the results of a series of room temperature experiments on the Wombeyan marble from Australia, illustrates the transition from brittle fracture to ductile deformation with increasing confining pressure. At a confining pressure of about 0.1 MPa (atmospheric pressure), longitudinal splitting occurs (Figure 9.10A); at 3.5 MPa, standard shear fractures form (Figure 9.10B); at 35 MPa, the deformation is transitional and is characterized by more pervasive fracturing and by the development of conjugate shears with a large fracture angle (Figure 9.10C). Finally, at 100 MPa confining pressure, the marble cylinder deforms ductilely into a smooth barrel shape (Figure 9.10D).

Thus we can account for the failure of rocks over a broad range of pressure only by means of a composite failure criterion, as illustrated schematically on the Mohr diagram in Figure 9.9. In the Earth, of course, the situation is even more complex, because an increase

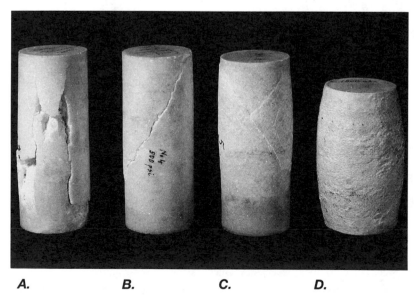

A.   B.   C.   D.

Figure 9.10 Effect of confining pressure p on the mode of deformation in Wombeyan marble at room temperature. All the samples had the same undeformed length. A. Longitudinal splitting at $p = 0.1$ MPa. B. Single shear fracture at $p = 3.5$ MPa. C. Brittle–ductile transition at $p = 35$ MPa. D. Ductile flow at $p = 100$ MPa.

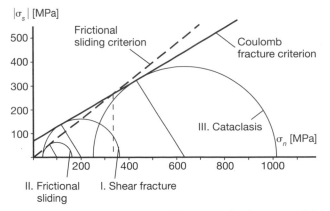

Frictional
sliding criterion

Coulomb
fracture criterion

III. Cataclasis

σ_n [MPa]

II. Frictional
sliding

I. Shear fracture

Figure 9.11 Example of Coulomb criterion for fracture (solid line: $|\sigma_s^*| = 70 + 0.6\ _E\sigma_n$) and for frictional sliding on an existing fault surface (dashed line: $|\sigma_s^*| = 0.81\ _E\sigma_n$). Data for Weber sandstone. Circle I: a critical stress for shear fracture. Circle II: a critical stress for frictional sliding on the fracture plane at constant confining pressure. Circle III: a critical stress for cataclastic flow during which fracturing requires a lower differential stress than frictional sliding at the same confining pressure $\hat{\sigma}_3$.

in pressure with depth goes hand in hand with an increase in temperature, which lowers the yield stress for ductile deformation.

Confining Pressure and Frictional Sliding

After a shear fracture develops in a rock at relatively low confining pressure, the fracture plane is a plane of weakness because the rock possesses no cohesion across it. Subsequent deformation occurs by frictional sliding on the fracture. The criterion for the onset of frictional sliding is given by an equation similar to the Coulomb fracture criterion, Equation (9.8):

$$|\sigma_s^*| = \overline{\mu}\sigma_n \qquad (9.13)$$

where $|\sigma_s^*|$ is the magnitude of the critical shear stress and $\overline{\mu}$ is the coefficient of sliding friction. Commonly, the coefficient of sliding friction is greater than the coefficient of internal friction ($\overline{\mu} > \mu$; compare Equation 9.8), so at low confining pressure, the differential stress $_D\sigma = \hat{\sigma}_1 - \hat{\sigma}_3$ required to produce sliding is less than that needed to form another fracture. Immediately upon fracture, therefore (Mohr circle I, Figure 9.11), the differential stress must drop to a level at which the frictional sliding criterion is not exceeded (Mohr circle II, Figure 9.11). Thus it is not surprising that faulting near the Earth's surface often results from sliding on preexisting faults.

At low confining pressure, frictional sliding occurs as a smooth, continuous motion called **stable sliding** (Figure 9.12A). As the compressive stress across the sliding surface increases with increasing confining pressure, the motion changes to **stick–slip** behavior (Figure 9.12B), which is characterized by "stick" intervals of no motion, during which the shear stress increases, alternating with "slip" intervals of rapid sliding that relieve the stress (Figure 9.12B). On a much larger scale than laboratory experiments, this same phenomenon may be responsible for the episodic nature of many earthquakes, as suggested by a detailed fault slip history deduced from young deformed sediments along the San Andreas fault in southern California (Figure 9.12C). The "stick" parts of the cycle represent the periods of qui-

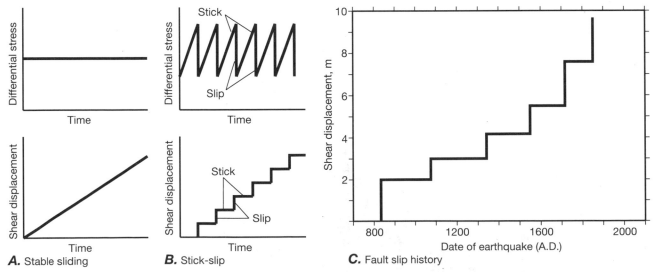

Figure 9.12 Variations of differential stress ($_D\sigma = \hat{\sigma}_1 - \hat{\sigma}_3$) and shear displacement with time *A.* for stable sliding and *B.* for stick-slip. *C.* A record of displacement versus time for the San Andreas fault in southern California.

escence between earthquakes, and the "slip" parts represent the earthquakes themselves.

As confining pressure increases still further, the frictional sliding criterion and the fracture criterion cross, and it requires less shear stress to form a new fracture in a rock than to slide along an existing one (Mohr circle III, Figure 9.11). The rock deforms by pervasive brittle fracturing and comminution of the grains, rather than by sliding on preexisting cracks. This process of **cataclasis,** or **cataclastic flow,** results in a **cataclasite** (see Figure 4.5B and Sections 18.3 and 19.1).

9.5 Effects of Pore Fluid Pressure on Fracturing and Frictional Sliding

The presence of pore fluid causes a rock to behave as though the confining pressure were lower by an amount equal to the pore fluid pressure. The mechanical behavior is described in terms of the **effective stress tensor,** whose components $_E\sigma_{k\ell}$ are defined in Equation (8.48) (see Figure 8.14J). Relative to the applied stress, each of the effective normal stress components ($_E\sigma_{k\ell}$, where $k = \ell$) is reduced by an amount equal to the pore fluid pressure p_f (Figure 9.13); shear stresses ($_E\sigma_{k\ell}$, where $k \neq \ell$) are unaffected. Thus the Mohr circle for the effective stress is the same size as for the applied stress, but it is shifted along the normal stress axis toward smaller compressive stresses by an amount equal to the pore fluid pressure (Figure 9.13).

The fracture criterion remains the same, except that the normal stress is replaced by the effective normal stress.

$$|\sigma_s^*| = c + \mu(_E\sigma_n) = c + \mu(\sigma_n - p_f) \quad (9.14)$$

where

$$_E\sigma_n = \sigma_n - p_f \quad (9.15)$$

The effect of shifting the Mohr circle to the left is that states of stress that are stable at zero pore fluid pressure may become unstable if the pore pressure is sufficiently high. If the differential stress $_D\sigma$ (the diameter of the Mohr circle) is small, as is commonly the case in the Earth, and if the pore fluid pressure p_f exceeds the minimum compressive stress $\hat{\sigma}_3$ by an amount equal to the tensile strength of the rock (that is, $\hat{\sigma}_3 - p_f = T_0$), then extension fracturing can occur, even at great depths (Figure 9.13A). If $_D\sigma$ is relatively large, on the other hand, shear fracturing can result if the pore fluid pressure is sufficiently high (Figure 9.13B).

Pore pressure has exactly the same effect on frictional sliding. An increase in the pore fluid pressure causes a decrease in the effective normal stress across the surface. Because frictional stress is proportional to the normal stress across the sliding surface, the critical shear stress necessary for sliding also decreases.

Thus pore fluid pressure is geologically important for several reasons. First, it lowers the differential stress necessary to cause failure and permits fracture at depths where the rock otherwise would be either stable or in the realm of ductile behavior. Second, it can shift the conditions of frictional sliding from those favoring

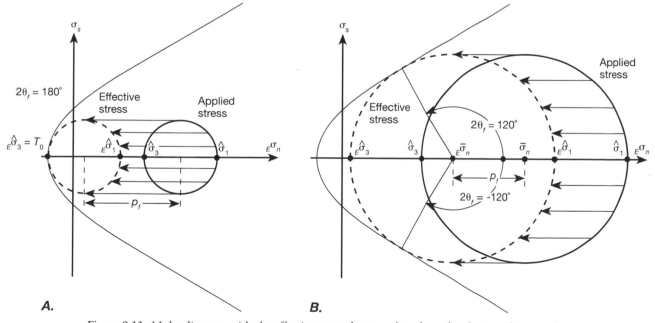

Figure 9.13 Mohr diagram with the effective normal stress plotted on the abcissa, showing the effect of pore fluid pressure on the fracture stability of rock. A. At small differential stress, an increase in pore pressure leads to tension fracture. B. At large differential stress, an increase in pore pressure leads to shear fracture.

stick–slip behavior to those favoring stable sliding. And third, it can shift deformation from cataclastic flow to frictional sliding. High pore fluid pressure is commonly an important factor in the development of joints and large-scale faults, as we discuss in Chapter 10.

The most obvious origins of pore fluid are water incorporated into a sediment during subaqueous deposition, and fluids released from minerals by dehydration reactions during metamorphism. If a rock at a given depth is permeable all the way to the surface, the pore spaces from the surface to that depth are interconnected, and the fluid pressure cannot exceed the weight of a column of water extending from the surface to that depth. The lithostatic pressure results from the weight of the overlying rock at a particular depth. Thus the hydrostatic pore fluid pressure (p_f) and the vertical lithostatic normal stress (σ_V) at a given depth are, respectively

$$p_f = \rho_w g h \quad \text{and} \quad \sigma_V = \rho_r g h \quad (9.16)$$

where ρ_w and ρ_r are the densities of water and rock, respectively, g is the acceleration due to gravity, and h is the given depth. If the density of water is 10^3 kg/m^3 and that of sediment is $(2.3)(10^3)$ kg/m^3, then the ratio λ of the hydrostatic pore pressure to the lithostatic pressure is

$$\lambda = \frac{p_f}{\sigma_V} = \frac{\rho_w g h}{\rho_r g h} = 0.4 \quad (9.17)$$

Despite the fact that this calculation cannot possibly be wrong by a factor as large as 1.5, values of λ approaching 1 are not uncommon in deep wells drilled in many sedimentary sequences, as well as in tectonic complexes along consuming plate margins. Thus in these areas, the assumptions underlying this calculation of λ must not apply. Such a fluid pressure buildup can occur only if impermeable barriers prevent free communication of the fluid with the surface.

9.6 Effects on Fracturing of Anisotropy, the Intermediate Principal Stress, Temperature, and Scale

Effect of Anisotropy

So far in this discussion, we have assumed that rocks have the same mechanical properties in all directions—that is, that they are **mechanically isotropic**. In this case the fracture criterion is the same regardless of the orientation of the principal stresses in the rock. Many rocks, however, are **mechanically anisotropic**; that is, their strength is different in different directions. A mechanical anisotropy may result, for example, from a

preferred planar alignment of platy minerals in a rock, called a cleavage, such as is characteristic of slates and schists. Such rocks break easily, or cleave, along these planes of weakness. A pervasive joint set has the same effect at a larger scale. These planes of weakness dominate the strength of the rock and the orientation of the fractures that develop for a wide range of orientations of the principal stresses relative to the anisotropy.

A series of fracture experiments performed on the Martinsburg slate illustrate these effects. Figure 9.14A shows examples of the fractured samples in copper jackets for different orientations (δ) of the slaty cleavage plane relative to the maximum compressive stress $\hat{\sigma}_1$. Figure 9.14B shows the relationship between the fracture plane angle α_f and the angle δ. If the cleavage plane and $\hat{\sigma}_1$ are either parallel ($\delta = 0°$) or perpendicular ($\delta = 90°$), there is no resolved shear stress on the cleavage plane because the cleavage is parallel to a principal plane of stress. In these cases, fracture strength is a maximum (Figure 9.14C), and shear fractures form at the usual angle of about 30° (Figure 9.14B). If the cleavage plane and $\hat{\sigma}_1$ are parallel ($\delta = 0°$), however, there is also a tendency to develop longitudinal splitting at low confining pressures. At values of δ between 15° and 60°, shear fractures tend to develop parallel to the cleavage, and even if δ is as high as 75°, the cleavage still has a substantial influence on the fracture plane orientation. Shear strength is lower for those orientations of slaty cleavage that affect the formation of shear fractures (Figure 9.14C), and it is a minimum when the cleavage is parallel to the usual shear fracture plane at an angle of approximately 30° to the maximum compressive stress $\hat{\sigma}_1$.

In simple terms, two different fracture criteria are necessary to account for the behavior of the rock. One criterion, plotted as the outer pair of solid lines in Figure 9.15, applies to fractures that develop across the plane of weakness. The Mohr circle cannot cross this criterion, because a fracture forms at the usual shear angle as soon as the surface stress components on any plane reach the critical values. The second criterion, plotted as the inner pair of dashed lines, applies only to the surface stress components acting on the cleavage plane.

The stress is stable as long as the Mohr circle is within the outer fracture envelope, and the surface stress components *on the cleavage plane* plot within the inner fracture envelope (Figure 9.15B, C). Note that in Figure 9.15C, the stable Mohr circle can cross the inner fracture envelope as long as the surface stress acting on the cleavage plane remains in the stable field. Unstable stresses occur either when the surface stress on the cleavage plane reaches the inner fracture envelope (Figures 9.15D, and 9.16B), in which case the fracture develops parallel to the cleavage, or when the surface stress on the cleavage is stable but the Mohr circle intersects the

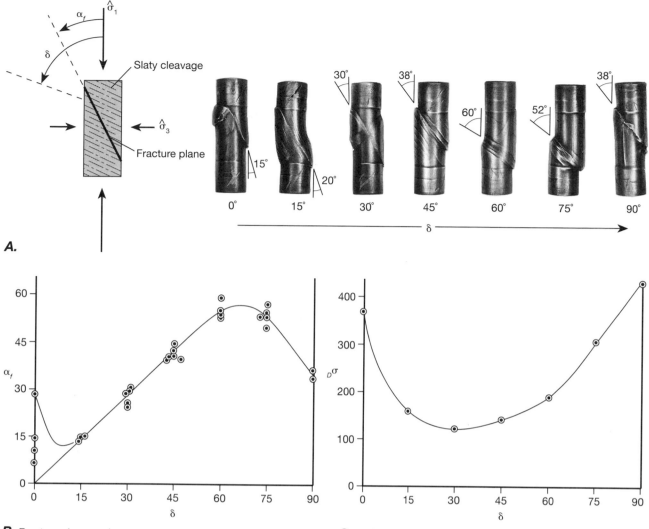

A.

B. Fracture plane angle

C. Shear strength (Differential stress at fracture)

Figure 9.14 The effect of anisotropy on fracture of Martinsburg slate at confining pressures of 50, 100, and 200 MPa. α_f is the fracture plane angle; δ defines the angle between the cleavage and the maximum compressive stress $\hat{\sigma}_1$. A. Copper-jacketed samples of Martinsburg slate showing the angle of shear fracture developed for the different values of δ. Samples with $\delta = 45°$ and $75°$ deformed at 100 MPa confining pressure; the other samples deformed at 200 MPa confining pressure. B. Variation of the fracture plane angle α_f with the orientation of the cleavage. C. Variation of the rock strength, measured in terms of the differential stress at fracture, with the orientation of the cleavage.

outer fracture envelope (Figure 9.17B), in which case the fracture develops across the cleavage.

The **shear strength** of a rock equals the differential stress at shear fracture (plotted in Figure 9.14C), which is the diameter of the critical Mohr circle. It is a minimum for those cleavage orientations in which the surface stress on the cleavage plane plots at the point of tangency between the Mohr circle and the inner fracture envelope (Figure 9.16B). It is a maximum when the shear stress on the cleavage plane is very small and the rock fractures across the cleavage (Figure 9.17B).

Effect of the Intermediate Principal Stress

So far in our discussion of fracture criteria, we have assumed that for isotropic rocks, shear fractures develop parallel to the intermediate principal stress $\hat{\sigma}_2$. In that orientation, $\hat{\sigma}_2$ contributes nothing to the normal stress or shear stress on the fracture plane, so it should have no effect on the fracture strength. This assumption is only approximately valid, however, because experiments indicate that $\hat{\sigma}_2$ does have a small effect on a rock's fracture strength. The strength is highest and

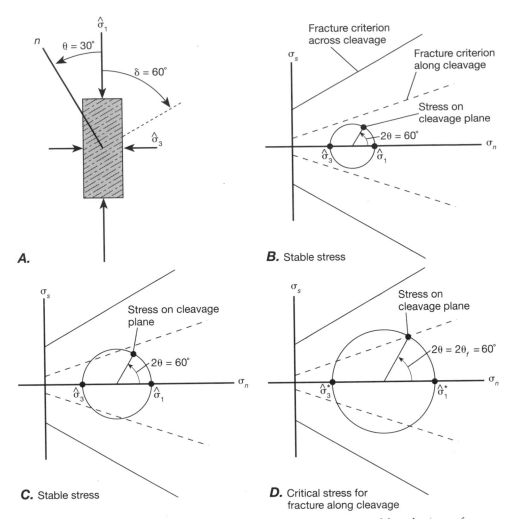

Figure 9.15 Model for the fracture behavior of an anisotropic material based on two fracture envelopes: one for fracture *across* the plane of weakness (outer pair of solid lines) and one for fracture *parallel* to the plane of weakness (inner pair of dashed lines). For these diagrams, the normal to the slaty cleavage is $\theta = 30°$ from the maximum compressive stress (the cleavage plane is $\delta = 60°$ from $\hat{\sigma}_1$). A. Diagram of physical space showing the relative orientations of the cleavage and the principal stresses. B. Stable state of stress. C. Stable state of stress. The inner envelope defines the critical stresses only for the surface stress on the cleavage plane, which in this case is in the stable zone. D. Critical state of stress. The surface stress on the cleavage plane reaches the critical condition defined by the inner envelope, and fracture occurs parallel to the cleavage.

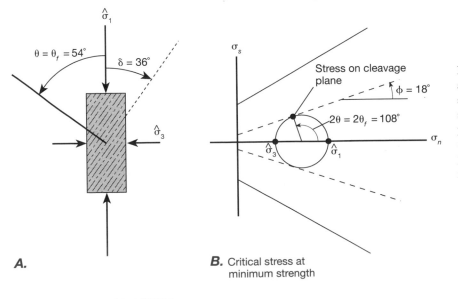

Figure 9.16 The minimum strength for a rock with a planar mechanical anisotropy. The fracture criteria are the same as in Figure 9.15A. Diagram of physical space: The plane of weakness is oriented parallel to the direction of preferred shear fracture. B. Mohr diagram: The plane of weakness is oriented so that the surface stress on the plane plots at the point of tangency between the Mohr circle and the inner (dashed) fracture criterion.

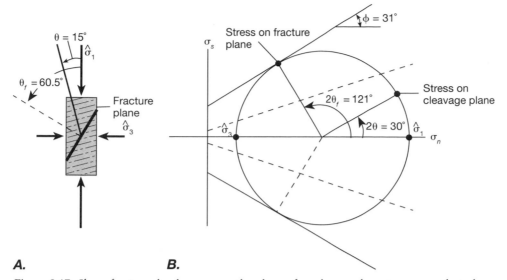

A.   **B.**

Figure 9.17 Shear fracture develops across the plane of weakness when stresses on that plane remain below the critical state. *A.* Stable orientation of the cleavage with respect to the principal stresses. Only one of the conjugate fracture planes is shown. *B.* Mohr diagram showing the same fracture criteria as in Figure 9.16. The orientation of the anisotropy is such that the surface stress on the plane of weakness is within the inner failure envelope—and is therefore stable—when the Mohr circle intersects the envelope for fracture across the anisotropy. Thus the rock fractures across the anisotropy. The dashed radius indicates the stress on the conjugate fracture planes.

the fracture plane angle α_f is lowest when the intermediate principal stress equals the maximum compressive stress $\hat{\sigma}_2 = \hat{\sigma}_1$ (extensional stress; see Figure 8.14E). Conversely, the strength is lowest and the fracture plane angle is highest when the intermediate principal stress equals the minimum compressive stress $\hat{\sigma}_2 = \hat{\sigma}_3$ (confined compression; see Figure 8.14D). These relationships imply that the angle of internal friction ϕ for the fracture envelope is highest for extensional stress and lowest for confined compression.

For anisotropic rocks, the plane of weakness need not be parallel to the intermediate principal stress. In that case, $\hat{\sigma}_2$ contributes to the normal and shear stresses on the plane of weakness and therefore affects the shear stress required for fracturing parallel to that plane.

Effect of Temperature

The effect of temperature on brittle fracturing is difficult to investigate experimentally, because above temperatures ranging from 200°C to 500°C (depending on the composition of the rock) ductile deformation mechanisms become important. The increase in temperature lowers the von Mises yield stress for ductile behavior (Figure 9.9), thereby lowering the pressure of the brittle–ductile transition and reducing the field of brittle behavior. Experimental data suggest, however, that there is also a small decrease in the brittle shear strength with increasing temperature.

Effect of Scale

Rock samples that are tested in the laboratory generally are homogeneous samples without flaws. In nature, however, such flaws as joints, faults, and compositional heterogeneities are a characteristic feature of large bodies of rock. Thus we should expect that the strengths determined from flawless samples in the laboratory might not describe the behavior of large bodies of rock. In pervasively jointed rock, for example, the strength may be determined more by the properties of the joints than by the rock material between them. In fact, experiments have demonstrated that as the scale of the sample tested increases, the measured strength decreases. Thus we must expect that the fracture strength of the Earth's brittle crust is less than that suggested by most measurements made on crustal rocks in the laboratory.

9.7   The Griffith Theory of Fracture

So far we have discussed empirical fracture criteria that relate the initiation of fracturing to stress and other physical conditions. These criteria have been reasonably successful in accounting for the macroscopic brittle behavior of most geologic materials, but they contribute little to our understanding of the physical mechanism of fracturing on a microscopic or molecular level.

One can calculate the theoretical tensile strength of a solid material on the basis of the strengths of the atomic bonds in the constituents of the solid. The strength derived in this manner, however, is generally about two orders of magnitude higher than the experimentally determined tensile strength of the material. In an attempt to account for this discrepancy, A. A. Griffith, in the early twentieth century, proposed that all solids contain a myriad of microscopic to submicroscopic, randomly oriented cracks, now called **Griffith cracks** (see stage I in Figure 9.21B), that reduce substantially the strength of the material.

A Griffith crack is a small, penny-shaped or slitlike crack that in cross section is much longer than it is thick and that has a very small radius of curvature at its tips. Griffith cracks may be imperfections within the crystal lattice of crystal grains in a rock, or they may be intragranular or grain boundary cracks. The cracks themselves are commonly modeled as extremely flattened ellipsoids, so the cross-sectional shape is elliptical with a large ratio of major axis to minor axis. (The diagrams of Griffith cracks given in Figures 9.18 through 9.20 show the minor axis $2c$ drawn very much larger in relation to the major axis $2a$ than is actually envisioned for a real Griffith crack.)

The ability of a Griffith crack to reduce substantially the strength of a material derives from the fact that an applied stress in general produces a local high concentration of tensile stress near a crack tip. Thus we distinguish the *applied stresses,* which are determined by the forces per unit area applied to the surfaces of the body, from the *local stresses,* which describe the state of stress immediately adjacent to a Griffith crack. The smaller the radius of curvature at the crack tip, and therefore the larger the ratio a/c of the ellipsoidal crack, the higher the local concentration of tensile stress near the crack tip.

To understand the mechanism, there are two factors that we must consider. The first factor is the way the local stresses are distributed around the surface of a Griffith crack. In general, the local tensile stress is a maximum near the crack tip at a point defined by the angle δ between the normal to the ellipsoidal surface of the crack and the major axis of the crack (Figure 9.18). The second factor is the crack orientation β relative to the applied principal stresses, which determines the magnitude and location of the local stress maximum. The orientation β^* of the most severely stressed Griffith crack and the location δ^* and orientation of the local maximum tensile stress on that crack surface govern how a fracture develops from a Griffith crack.

Formation of Tension Fractures

For a body under an applied tensile stress, the Griffith cracks are open. Thus any crack surface is a free surface, which cannot support a shear stress. A crack surface therefore must be a principal plane of the local stress. A free surface also cannot support a normal tensile stress. Thus the local maximum tensile stress σ_t^{max} must be parallel to the elliptical surface.

The orientation of the most critically stressed Griffith crack is perpendicular to the maximum applied tensile stress ($\beta^* = 90°$, Figure 9.19). In this orientation, the location of the maximum stress concentration is exactly at the crack tip ($\delta^* = 0°$), and the orientation of the local tensile stress σ_t^{max} is parallel to the applied tensile stress $\hat{\sigma}_3$. If the ellipticity of such a crack is, for example, $a/c \approx 100$, which is reasonable for a Griffith crack, then the magnitude of the local tensile stress at the crack tip is approximately 200 times that of an applied uniaxial tensile stress ($\sigma_t^{max} \approx 200\hat{\sigma}_3$). Thus the stress at the crack tip can be at the theoretical strength for the material when the applied stress is still roughly two orders of magnitude lower than the theoretical strength.

When the true strength of the material is exceeded at the crack tip, the crack propagates in a plane normal

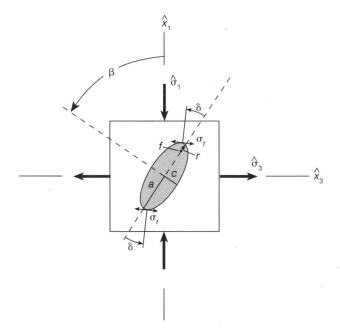

Figure 9.18 Schematic diagram of Griffith crack in two dimensions idealized as an ellipse with major semi-axis a, minor semi-axis c, and radius of curvature r at the crack tip measured from the focus f. $\hat{\sigma}_t$ is the *local* tensile normal stress that is parallel to the surface of the ellipse and that acts on a plane perpendicular to the surface at a point defined by the angle δ between the major axis of the ellipse and the normal to the ellipse. For the sake of clarity, the ratio of the minor axis to the major axis is much larger than expected for real Griffith cracks. β is the angle between the normal to the crack plane and $\hat{\sigma}_1$.

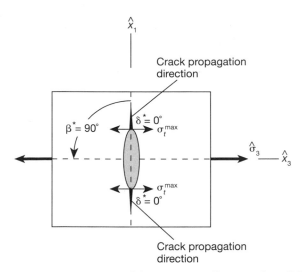

Figure 9.19 Orientation of the most critically stressed Griffith crack under applied uniaxial tensile stress ($\beta^* = 90°$). The maximum local tensile stress concentration is at the crack tips ($\delta^* = 0°$), and the orientation of the local maximum tensile stress σ_t^{max} is parallel to the applied tensile stress $\hat{\sigma}_3$. The crack grows perpendicular to the local maximum tensile stress σ_t^{max} and therefore perpendicular to $\hat{\sigma}_3$.

to the local tensile stress σ_t^{max}, and in this case the plane of propagation is parallel to the plane of the crack itself. As the crack propagates, the ellipticity of the crack a/c increases, which in turn increases the stress concentration. At constant applied stress, the growth of the crack therefore leads to an instability and the crack propagates rapidly, causing a tensile fracture to form (Figure 9.2A). Propagation of the crack ceases when the applied stress decreases to the point at which local stress concentrations are subcritical.

Longitudinal Splitting

Under conditions of uniaxial compression, any Griffith cracks that are not essentially parallel to the compressive stress are closed by the component of normal stress across their surfaces. Cracks parallel to the applied stress need not close, however, and even though a compressive stress is applied, the local stress concentration at the crack tip is tensile and oriented normal to the applied compressive stress. The situation is identical to that shown in Figure 9.19 except that the applied stress is a compressive stress parallel to $\hat{x}_1$ rather than a tensile stress parallel to $\hat{x}_3$ (compare Figures 9.2A and C). For the most critically stressed Griffith crack in uniaxial compression, we again have $\beta^* = 90°$ and $\delta^* = 0°$ (Figure 9.19). If the ellipticity $a/c \approx 100$, then the local maximum *tensile* stress on the crack is roughly 25 times the magnitude of the applied *compressive* stress

($\sigma_t^{max} \approx -25\ \hat{\sigma}_1$). When such cracks propagate, they grow roughly parallel to the applied compressive stress, leading to the development of longitudinal splitting (Figure 9.2B). This mechanism therefore accounts for the formation of extension fractures in compression at low to zero confining pressure. An increase in the confining pressure tends to close cracks of this orientation and to reduce the local stress concentrations that lead to this mode of fracturing.

Formation of Shear Fractures

The behavior of Griffith cracks in confined compression is more complicated. The cracks in general are closed, because the applied stress produces a compressive normal stress across the crack surfaces. Friction on the closed crack surfaces and the applied compressive stress cause the distribution of local stress around the crack to differ from that around Griffith cracks under applied tensile stress. Shear along the closed cracks results in tensile stress near the crack tips that reaches a maximum at a point not, in general, exactly at the tip. The crack grows by the opening of tensile cracks that accommodate the shear on the crack surface (Figure 9.20). The orientation of the most critically stressed Griffith crack

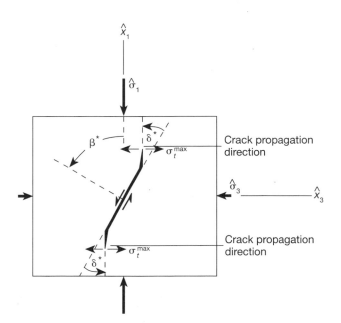

Figure 9.20 Orientation of the most critically stressed Griffith crack under applied confined compression. The crack is closed, and the orientation of the most critically stressed crack falls in the range $45° \leq \beta^* < 90°$. A local tensile stress concentration develops near, but not at, the crack tips and is a maximum at an angle $\delta^* > 0°$. The local tensile stress maximum σ_t^{max} is oriented such that the crack grows progressively toward parallelism with $\hat{\sigma}_1$. Crack growth must be accommodated by frictional sliding on the closed part of the crack surface.

in compression is at an angle β^* between $45°$ and $90°$, depending on the relative values of $\hat{\sigma}_1$ and $\hat{\sigma}_3$. Thus the most critically stressed crack plane is between $0°$ and $45°$ to the maximum compressive stress, which is the range of orientations within which shear fractures generally form.

The location of the point of maximum tensile stress concentration, however, is at a point where the surface of the crack is more nearly perpendicular to the maximum compressive stress (δ^* in Figure 9.20). Thus when the crack grows, an instability does not develop because the new part of the crack grows toward parallelism with the maximum compressive stress, which is a more stable orientation (Figure 9.20 and stage III of Figure 9.21B). The onset of Griffith crack growth in compression, therefore, does not lead immediately to shear failure. The differential stress must be increased considerably beyond that required for initial crack growth.

The mechanism of shear fracture formation is illustrated in Figure 9.21. The data for a typical fracture experiment are shown diagrammatically in Figure 9.21A, where the axial extension (Equation 9.1) and the volumetric extension (Equation 9.6) are plotted against the differential stress. The physical states of the material for the five different stages of the stress–strain curves (labeled I to V) are illustrated schematically in Figure 9.21B. In stage I, the initial undeformed material is filled with open Griffith cracks, and the low slope of the stress–strain curves results from the relatively large deformation associated with the closing up of the cracks. In stage II, the cracks are closed and stable; no crack growth occurs with increasing stress. Crack growth begins in stage III, where the volume of the opening cracks begins to offset the normal volumetric decrease associated with the increase in stress. As the applied stress increases, less critically oriented cracks begin to grow. Eventually, in stage IV, the volume decrease caused by increasing compression is completely offset by the volume increase caused by growing cracks, and the volume of the material actually begins to increase (a phenomenon called **dilation**). At this stage, the local stress fields around the cracks start to interact, the cracks begin to join together, and finally, in stage V, a through-going shear fracture develops. Thus shear fracturing under compressive stresses actually depends on the growth of tension cracks in the material.

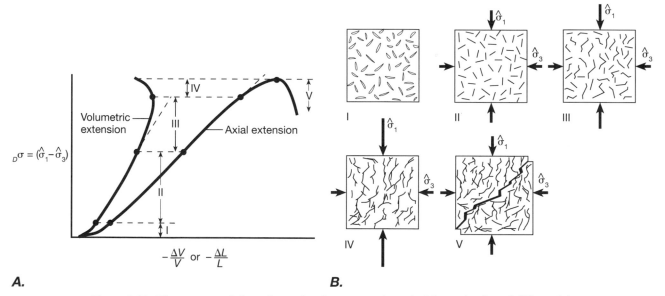

A.

B.

Figure 9.21 The process of shear fracturing in compression. A. Schematic plot of differential stress ($_D\sigma$) versus axial extension $\left(\frac{\Delta L}{L}\right)$ and volumetric extension $\left(\frac{\Delta V}{V}\right)$ for a characteristic shear fracture experiment in uniaxial compression. Stages I through V reflect changes in physical processes within the rock. B. Diagrammatic stages in the formation of a brittle shear fracture corresponding to the stages labeled on the stress–strain curves in part A. I. Griffith substance containing many open cracks. As stress increases, the cracks gradually close. II. The cracks are all closed and stable. Increase in stress does not cause crack growth. III. Under increasing load, the most critically oriented and the longest cracks begin to propagate toward the direction of maximum compressive stress. As the load increases, shorter and less favorably oriented cracks begin to grow. IV. The specimen is at this stage almost a granular solid. The local stress fields around cracks begin to interact. V. A through-going shear fracture forms by the coalescence of many small fractures.

Griffith Fracture Criteria

The Griffith crack models can be used to derive theoretical failure criteria for conditions of tensile and compressive stress. In the tensile stress regime, the predicted fracture criterion is a parabolic envelope, which corresponds reasonably well with experimental observation. The form of the fracture criterion for compressive stresses is similar to the Coulomb fracture criterion, but the predicted constants are different from the experimentally determined ones. The discrepancy exists because the Griffith theory provides a criterion for the first initiation of Griffith crack growth (Figure 9.21, beginning of stage III), whereas the development of a shear fracture occurs at considerably higher differential stresses (at the beginning of stage V). The actual process of shear fracture development is very complex, and the details of the mechanism are poorly understood.

The Griffith Theory and the Effects of Confining Pressure, Pore Fluid Pressure, and the Intermediate Principal Stress

The growth of tension cracks from Griffith cracks in a compressional stress regime is accompanied by frictional sliding on the Griffith crack surface (Figure 9.20). When confining pressure increases, the normal stress across the cracks increases, and the frictional resistance to sliding must also increase. To initiate crack growth, therefore, the shear stress on the crack must increase, and this requires an increase in the applied differential stress ($_D\sigma = \hat\sigma_1 - \hat\sigma_3$). Under higher confining pressures, then, higher differential stresses are required for shear fracturing, as specified by the Coulomb fracture criterion. Thus the angle of internal friction ϕ that defines the slope of the Coulomb fracture envelope is related to the friction on the surface of the Griffith cracks.

A confining pressure applied to a rock provides the same normal stress across Griffith cracks of all orientations. The presence of a pore fluid under pressure in the cracks directly counteracts the externally applied normal stress, so the net normal stress across a Griffith crack is exactly the effective normal stress defined in Section 8.6 and discussed in Section 9.5. In this manner, the Griffith crack theory accounts for the observed effect of pore pressure on brittle fracturing.

As the value of the intermediate principal stress $\hat\sigma_2$ varies from $\hat\sigma_1$ to $\hat\sigma_3$, the fracture strength of the rock decreases slightly (Section 9.6). The Griffith theory provides a model to account for this behavior. In a three-dimensional material, Griffith cracks are distributed in all orientations, and those that eventually coalesce into a shear fracture are not necessarily exactly parallel to $\hat\sigma_2$. On those cracks, $\hat\sigma_2$ contributes a small component of compressive stress. Thus the frictional resistance to sliding on the crack surface depends slightly on $\hat\sigma_2$, and it is higher for higher values of $\hat\sigma_2$. The increase in frictional resistance means that a larger differential stress is needed to cause fracturing.

Additional Readings

Johnson, A. 1970. *Physical processes in geology.* San Francisco: Freeman-Cooper.

Paterson, M. S. 1978. *Experimental rock deformation—The brittle field.* New York: Springer Verlag.

Scholz, C. H. 1990. *The mechanics of earthquakes and faulting.* New York: Cambridge University Press.

Mechanics of Natural Fractures and Faults

In this chapter we use the knowledge of stress and mechanics of brittle fracture that is presented in the preceding two chapters to gain a deeper understanding of brittle deformation. With that background we can reexamine fractures and faults described in Chapters 3 through 7 and draw some conclusions about the conditions under which they form.

It is worth mentioning at the outset that in cases of complex deformation history, structures may be difficult to interpret, either because the relative timing of the formation of different structures is obscure or because structures form under one set of conditions and are reactivated under another. Thus, for example, it may be difficult to determine whether a set of fractures were formed before or during folding; fractures may develop as extension fractures and subsequently be reactivated with shearing displacement along them; and faults initiated with thrust displacement may be reactivated as normal faults. Such complexities make the interpretation of structures challenging and sometimes controversial.

With that caution in mind, then, we examine first the magnitude and origin of stress within the Earth, next the formation of extension fractures, and finally the formation of faults.

10.1 Techniques for Determining Stress in the Earth

Techniques for determining the state of stress in the Earth were developed largely in the geological engineering, mining, and energy industries. In mining, knowledge of the state of stress is important for the design of safe tunnels and stable open pits. In dam construction, the stresses in the abutments are measured before, during, and after construction to ensure safe design and operation. In oil and geothermal energy production, artificially fracturing rocks at depth can increase their permeability, thereby enhancing the yield from wells. Control of artificial fracturing requires knowledge of the state of stress at depth.

We also need to know the state of stress within the Earth in order to understand how and why plates move; why, where, and when earthquakes occur; and why and how structures form. Because of these practical and fundamental needs, the determination of stress in the Earth has become a field of geologic investigation in its own right.

Techniques are available for determining the current state of stress in the Earth as well as for determining the "paleostress" that existed at some time in the geologic past. Some techniques provide a complete determination of the orientation and magnitude of the principal stresses; others provide only a partial determination of the state of stress. We discuss here techniques that rely on the effects of elastic or brittle deformation; those that rely on the effects of ductile deformation we discuss in Chapter 19 (see Box 19.2).

Stress Relief Measurements

Stress relief techniques of determining stress depend on the fact that the stress on an elastic material produces a proportional strain (see Section 9.1). When the stress is removed, the strain disappears, and measurement of the change in strain that accompanies unloading can be used to infer the original stresses, provided that the elastic constants of the rock are known independently.

Overcoring is a common technique that involves drilling a hole in the rock, attaching strain gauges to the surface of the hole, and then drilling an annulus around the hole to form a hollow cylinder of rock on which the stress from the surrounding rock has been released (Figure 10.1A). The release of stress causes elastic deformation of the cylinder, so its dimensions and its initial circular cross section change. Using the

theory of elastic deformation (see Equation 9.5, for example), we can calculate the magnitude and direction of the original stresses from measurements of this deformation. The calculation is not simple, however, because the presence of the first hole changes the stress from its value in the solid rock, and this effect must be accounted for. For technical reasons, the maximum practical depth of boreholes for overcoring is 30 m to 50 m.

The flat jack is an instrument used to measure the normal component of stress acting on a plane of a particular orientation. Reference pins are inserted into the rock to form a rectangular grid, and the distances d_i ($i = 1, \ldots, 6$) between pins are measured (Figure 10.1B). A slot is then cut into the rock which relieves the stress locally and changes the distances between the reference pins. A thin, hollow steel plate, the flat jack, is inserted into the slot, and the slot is filled with grout. When the grout has hardened, oil pressure in the hollow flat jack is increased until the reference pins return as closely as possible to their original relative positions. The measured oil pressure is then the normal component of stress acting in the rock across the plane of the flat jack. Several such measurements in different orientations can be used to determine the complete state of stress in the rock. This technique is commonly used in tunnels.

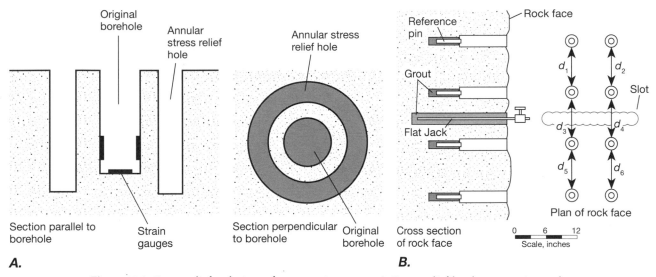

A.   **B.**

Figure 10.1 Stress relief techniques for measuring stress. *A.* Stress relief by the overcoring technique. A shallow borehole not exceeding 50 m in depth is drilled, and strain gauges are glued to the surfaces of the hole. An annular hole concentric with the first hole releases the stresses on the hollow cylinder of rock, and the resulting strain is measured by the strain gauges. Strain is converted to stress by means of the equations of elasticity and the elastic constants for the rock. *B.* The flat jack technique measures the component of stress normal to a plane of a particular orientation in the rock. An array of reference pins is inserted into the rock face, and the distances $d_1, d_2, \ldots, d_6$ between them are measured. A slot is cut, releasing the stress across the face of the slot. The flat jack is grouted into place in the slot. Pressurizing the flat jack with hydraulic fluid returns the reference pins to their original relative distances when the fluid pressure equals the original normal stress across the plane of the flat jack.

Hydraulic Fracturing (Hydrofrac) Measurements

Hydraulic fracturing, commonly abbreviated "hydro-frac," is a technique for fracturing the rock that, with some simplifying assumptions, makes it possible to determine both the magnitudes and orientations of the principal stresses. The technique was initially developed to increase the permeability of oil-bearing rocks penetrated by wells.

A section of a borehole is sealed off with two inflatable rubber packers (Figure 10.2A), and the fluid pressure between the pressure seals is pumped up until, at a critical pressure P_c, a tension fracture forms at the borehole. The fluid is sealed in immediately after fracturing occurs, and the pressure drops and stabilizes at a value called the instantaneous shut-in pressure P_s, which is the pressure that is just sufficient to keep the fracture open.

Because the surface of the Earth is a free surface on which the shear stress must be zero, the principal stresses there must be perpendicular and parallel to the surface, or approximately vertical and horizontal. We assume the stresses at depth have the same orientation. If we assume that the borehole is vertical and that the tension fracture is parallel to it, the instantaneous shut-in pressure equals the minimum horizontal compressive stress in the rock; that is, $P_s = \sigma_{H(min)}$. The critical pressure P_c is the sum of the minimum compressive stress tangent to the surface of the borehole and the tensile strength T_0 of the rock. The elastic theory of a hole in a stressed solid shows that P_c depends on T_0, $\sigma_{H(min)}$, and $\sigma_{H(max)}$. Knowing P_c, T_0, and $\sigma_{H(min)}$, we can calculate $\sigma_{H(max)}$ from that relationship. The vertical normal stress is assumed to be equal to P_c and to the stress caused by the weight of the overlying rock, the overburden stress $P_c = \sigma_V = \rho_r gh$. $\sigma_{H(min)}$ is perpendicular to the fracture, whose orientation is determined by using a downhole televiewer or by making an oriented impression of the surface of the borehole (Figure 10.2B). Thus all three principal stresses and their orientations are determined. Measurements at depths of up to 5 km have been achieved.

Stress Orientations from Earthquake First-Motion Studies

Earthquakes result from regional stresses in the Earth, and they occur at depths ranging from shallow up to

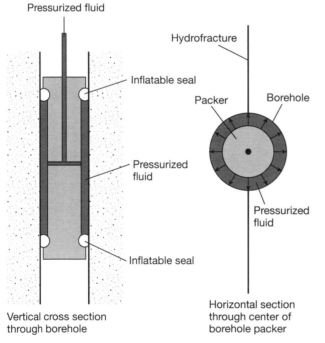

A.

Figure 10.2 Hydrofracting method of measuring stress. *A.* Vertical and horizontal sections of a borehole packer used to isolate a section of the borehole. Fluid pressure is increased between the seals to induce hydrofracture, which occurs at the critical pressure P_c. *B.* Impression of the inside of the borehole following hydraulic fracturing. The impression of the induced extension fracture is clearly visible roughly parallel to the borehole.

B.

several hundred kilometers—much greater than borehole techniques can reach. The radiation pattern of first motions of P waves and the location of earthquake aftershocks indicate, respectively, the sense of shear and the fault orientation at depth (see Box 2.4 and Figure 2.17). The maximum compressive stress $\hat{\sigma}_1$ lies in the rarefaction first-motion quadrant (*not* the compressive first-motion quadrant) perpendicular to the intersection of the fault plane and the nodal plane. The von Mises failure criterion (Figure 9.9) suggests that it bisects the angle between those planes. The orientation is approximate because the von Mises criterion is not necessarily the best one to use, and the mechanism for deep-focus earthquakes is not well understood. First-motion studies do not give reliable information about the stress magnitudes.

Field observation of faults and the slip direction on them can be used, under favorable circumstances, to infer the orientation of the principal paleostresses that caused the faulting. We discuss this very useful technique further in Section 10.10.

10.2 Stress in the Earth

Vertical Normal Stress

We commonly assume that the principal stresses are vertical and horizontal, because they must have that orientation at the horizontal surface of the Earth. A plot of principal stress orientations on a stereonet should therefore show a tight cluster of axes about the center of the net and a distribution of axes around the periphery. Figure 10.3A shows, for example, principal stress orientations determined in southern Africa. Although there is some clustering about the center of the net, there is a large amount of scatter, suggesting that the assumption is only a rough generalization.

Another common assumption is that the vertical normal stress should be equal to the overburden, which is determined by the density of the rocks. Figure 10.3B shows a set of measurements of the vertical normal stress compared with the overburden stress for a mean rock density of 2700 kg/m³. In fact, although the overburden stress is a good average of the vertical stresses, there is a great deal of variability, which again warns us that the common assumption is an oversimplification.

Nontectonic Horizontal Normal Stress

In a sedimentary basin that has not been subjected to tectonic deformation, we generally expect the state of stress to be dominated by the overburden. In such a case, the principal stresses should be vertical and horizontal, and the vertical normal stress should be the maximum compressive stress and should be equal to the overburden. The horizontal stress, however, is more difficult to estimate. We can suppose that the sediments in a basin behave as an elastic solid, and that the geometry of the Earth requires that the horizontal Poisson expansion be zero ($\hat{e}_{xx} = 0$). We can then calculate the magnitude of the horizontal stress σ_H that would exactly counteract the Poisson expansion due to the vertical stress σ_V as follows. Form the elasticity equation for $\hat{e}_{xx}$ from Equation (9.5) by changing subscripts z to x and x to z. Then set $\hat{\sigma}_{xx} = \hat{\sigma}_{yy} = \sigma_H$, $\hat{\sigma}_{zz} = \sigma_V$, and $\hat{e}_{xx} = 0$. Solving for the horizontal stress gives

$$\sigma_H = \frac{v}{1 - v} \sigma_V \tag{10.1}$$

For v between 0.25 and 0.33, which are common values of Poisson's ratio for rock, this equation implies that the horizontal stress should be only between about a third and half of the vertical stress. The constant $v/(1 - v)$ is one possible value for the constant κ in

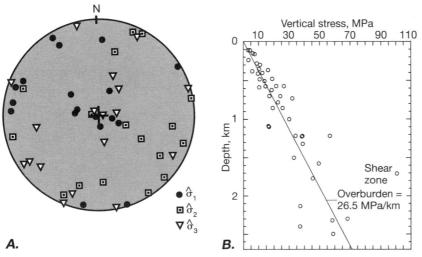

A.

B.

Figure 10.3 The orientation and magnitude of the vertical component of stress in the Earth. A. Plot of the orientations of the three principal stresses in southern Africa. Equal-area, lower-hemisphere projection. B. Plot of the magnitude of the vertical component of stress. The line is the lithostatic load for a rock density of 2700 kg/m³.

Equations (8A.2) and (8A.3) (Appendix to Chapter 8) and in Equation (10.3) (Section 10.9).

Figure 10.4 shows the values of the minimum horizontal compressive stress in sedimentary basins in the United States, determined by the hydrofrac technique. For comparison, the different lines indicate the overburden stress, the hydrostatic pressure, and the minimum compressive stress predicted from the Poisson effect for two values of v. Except for three measurements in granite, the stress calculated from the Poisson effect stress is too low, indicating that the assumptions we made for the calculation are not realistic.

If we had assumed that rocks were sufficiently ductile so that the flow would eliminate any differential stress, it would be equivalent to assuming that $v = 0.5$. In that case, the state of stress would be lithostatic and equal to the overburden ($\sigma_H = \sigma_V = \rho_r g h$). Figure 10.4 shows that this also is not a realistic assumption.

At best, the horizontal normal stress calculated from the extreme values of the Poisson ratio gives maximum and minimum bounds for a nontectonic stress.

Tectonic Horizontal Normal Stress

The only constraint we can put on horizontal stresses of tectonic origin is that the differential stress (the diameter of the Mohr circle) must not exceed the strength of the rock. We assume the strength is determined by the Coulomb fracture criterion, which we express as a relationship between the maximum and minimum principal stresses at fracture (Box 9.1 and Equation 9.1.2). For the sake of argument, we also assume that the principal stresses are horizontal and vertical and that the vertical stress is the overburden, although these are not necessarily accurate assumptions. We consider the cases for horizontal tectonic extension and horizontal tectonic compression with a fracture angle $\theta_f = 60°$ and a cohesion $c = 10$ MPa, which give $S = 34.6$ MPa and $K = 3$ in Equation (9.1.2).

For the condition of tectonic extension, the vertical normal stress is the maximum compressive stress, $\sigma_V = \hat{\sigma}_1$, and we can solve Equation (9.1.2) for the minimum possible value of $\hat{\sigma}_3$. The variation with depth for both the maximum and the minimum principal stresses for this case is shown in Figure 10.5A by the solid lines labeled overburden and min $\hat{\sigma}_3$, respectively. For the state of tectonic compression, the vertical normal stress is the minimum compressive stress, $\sigma_V = \hat{\sigma}_3$, and we can solve Equation (9.1.2) for the maximum possible value of $\hat{\sigma}_1$. The variation with depth for both the minimum and the maximum principal stresses is shown for this case in Figure 10.5A by the solid lines labeled overburden and max $\hat{\sigma}_1$, respectively.

The predicted strength of the rock is decreased if the effect of pore fluid pressure is taken into account (see Section 9.5). In Equation (9.1.2), the principal stresses $\hat{\sigma}_1$ and $\hat{\sigma}_3$ must be replaced by the effective principal stresses $_E\hat{\sigma}_1 = \hat{\sigma}_1 - p_f$ and $_E\hat{\sigma}_3 = \hat{\sigma}_3 - p_f$, respectively. We express the pore fluid pressure as a fraction λ of the overburden, $p_f = \lambda \sigma_V$ (Equation 9.17), and plot the results in Figure 10.5A as dashed lines for different values of λ in each of the fields for horizontal extension and horizontal compression.

Measurements of the minimum horizontal stress from an area of subsidence and normal faulting in southern Africa are shown in Figure 10.5B. The solid line is again the overburden, and the dashed line is the minimum possible stress for $\lambda = 0.4$, the value for permeable saturated rock (Equation 9.17). At shallow depths, several values of stress exceed the overburden. At greater depths, the predicted maximum and minimum stresses are better constraints to the data. Observed values falling below the $\lambda = 0.4$ line could be accounted for by a lower value of λ or by a larger value of the cohesion c, which has the effect of moving the stress axis intercept to more negative values.

Note that the plot of minimum values of $\hat{\sigma}_3$ in Figure 10.5A indicates that actual tensile stresses (neg-

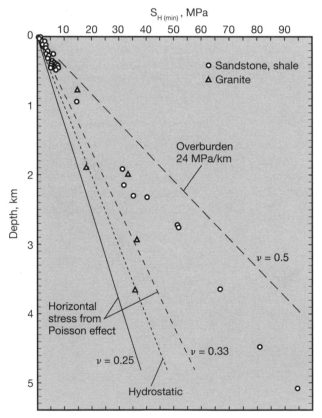

Figure 10.4 The minimum horizontal compressive stress measured by the hydrofrac technique in sedimentary basins in the United States (data points) compared with the overburden pressure, the hydrostatic pressure, and the minimum horizontal stress predicted by the Poisson effect for $v = 0.25$ and 0.33.

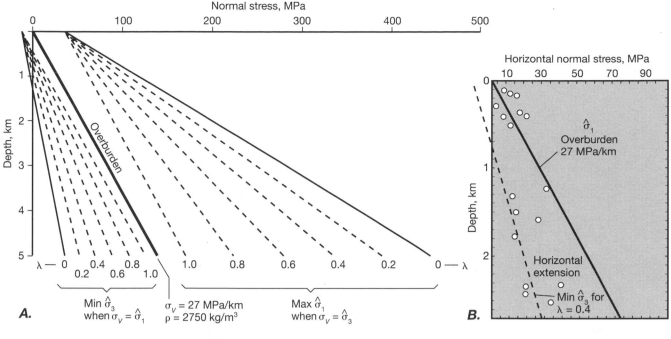

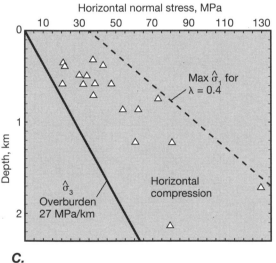

Figure 10.5 Constraints on maximum and minimum principal stresses based on the Coulomb fracture criterion and the pore fluid pressure effect. *A.* In horizontal extension, the overburden is the maximum compressive stress ($\sigma_V = \hat{\sigma}_1$), and values of the minimum compressive stress are indicated for various values of the pore fluid pressure ratio λ. Note that the minimum principal stress cannot be tensile below about 1 km. In horizontal compression, the overburden is the minimum principal stress ($\sigma_V = \hat{\sigma}_3$), and the maximum compressive stress is shown by the lines for the different values of λ. *B.* Data from an area of extension in southern Africa, showing the constraints provided by the overburden and the minimum principal stress for hydrostatic pore fluid pressure, $\lambda = 0.4$. *C.* Data from an area of tectonic compression in Canada with constraints on the stress provided by the overburden and the fracture criterion for hydrostatic pore fluid pressure, $\lambda = 0.4$.

ative values of the normal stress) cannot exist below a depth of about 1 km. In fact, tensile stresses have not been measured within the Earth at all.

Measurements of the maximum horizontal stress from a region of folding and thrust faulting in Canada are plotted in Figure 10.5C. The solid line is the over-burden stress, and the dashed line is the maximum possible compressive stress for $\lambda = 0.4$. All the measured stresses fall between the two lines.

This method of constraining the differential stress applies at best to the upper 15 km to 20 km of the crust, which is the range in which deformation is brittle. Below that, the increases in temperature and pressure induce a weakening of the rock because of the onset of ductile deformation processes (see Sections 18.4 to 18.6), and the Coulomb fracture criterion does not predict the strength of the rocks.

Regional Distributions of Stress

Stress orientations from all over the world measured with the techniques discussed above are summarized in Figure 10.6A. The figure shows the location of stress measurements, the orientation of principal stresses, and the type of stress determination. Over large regions, there is reasonable agreement among the different measurement techniques, but the orientations of the principal stresses change substantially even within a single continent. These stresses reflect major tectonic processes in the Earth and provide important constraints on models of the driving forces for plate tectonics, which must account for the observed stress distribution within the plates, at least to a first approximation.

Figure 10.6B presents a more detailed summary of stress orientation measurements in the United States.

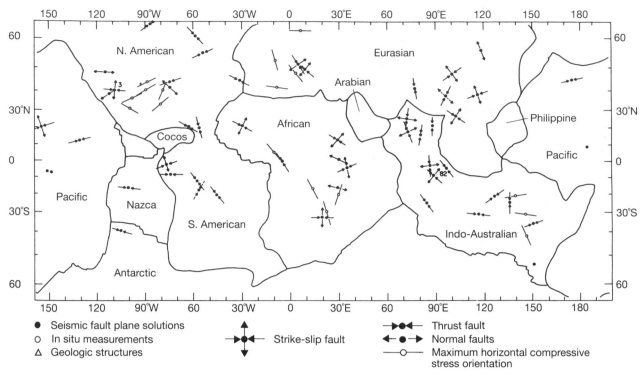

A.

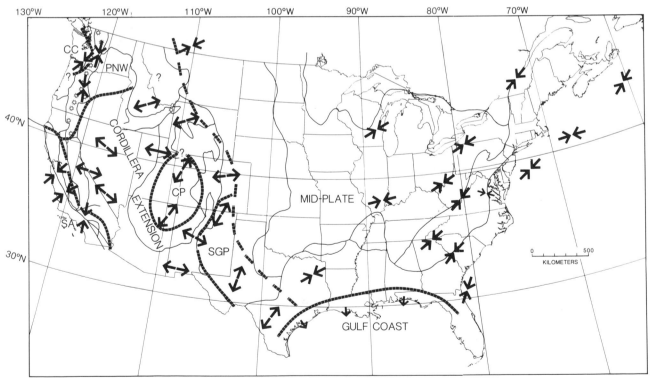

B.

Figure 10.6 Regional stress field determinations. *A.* Worldwide distribution of principal stress orientations. *B.* Stress orientations in the coterminus United States.

The boundaries separate regions of roughly similar states of stress, and to a significant extent, these regions correspond to geologic provinces in which the structures reflect the different tectonic regimes.

10.3 Mechanisms of Stressing the Earth's Crust

Stress arises in the Earth because of the overburden, the driving mechanisms of plate tectonic processes, horizontal and vertical motions, changes (over space and time) in temperature and pressure, the inhomogeneous mechanical properties of the crust, and pore fluid pressure. Once we understand the mechanisms by which stresses arise, we can begin to understand the possible origin of fractures in the Earth by making models of loading histories and their consequences. In this section, therefore, we examine the mechanisms by which the Earth's crust can be stressed.

The Overburden

Because the overburden stress results from the weight of the overlying column of rock, surface topography affects the stress distribution at depth. The greater the topographic relief, the greater the magnitude of the effect. The influence of topography on stress dies out with increasing depth and is generally negligible at depths greater than the horizontal length of the topographic feature, which may, however, be considerable.

The overburden may be increased by sedimentation or by tectonic thickening such as thrust faulting; conversely, it may be decreased by erosion or by tectonic thinning such as normal faulting. The resulting change in pressure should cause different amounts of deformation in different types of rocks, because each is characterized by its own elastic constants. If the rocks are constrained to deform the same amounts, however, stresses must be different in each rock type in order to satisfy these constraints.

Driving Processes of Tectonics

Stresses associated with plate motion are one of the major sources of regional stress in the lithosphere. Such stresses may arise from the pull of the down-going slab as it descends in a subduction zone, from the push of a midoceanic ridge associated with its relative topographic elevation above the adjacent sea floor, from the drag between the lithosphere and the underlying asthenosphere as the plates move relative to the underlying mantle, or from the interaction of adjacent plates. The stresses associated with subduction are particularly important to the formation of structures during collisions involving continents or island arcs with subduction zones.

Horizontal and Vertical Motions

Bending of the crust and of the lithosphere generates stresses whose extent is comparable to the wavelength of the bending. Plates bend at subduction zones where the plate enters the trench. Plates bend during isostatic response to surface loads, such as the huge volume of volcanic rocks of the Hawaiian Islands, the thick accumulations of ice in continental ice sheets, the thick sediments that accumulate in sedimentary basins, and the unloading caused by erosion. Deviatoric tensile stress should develop on the convex side of the bend, and deviatoric compressive stresses on the concave side (Figure 10.7). Bending of lithospheric plates also occurs as they drift from one latitude to another, because the Earth is roughly ellipsoidal in shape, and the surface has a greater curvature (a smaller radius of curvature) at the equator than at the poles. The plates must bend to accommodate this change in curvature.

Vertical motions, such as result from isostatic adjustment, can also induce stresses in rocks. As a segment of the crust is uplifted, for example, it should subtend a constant central angle β. As the radial distance from the Earth's center increases, however, the arc length increases, thereby stretching the rock in both horizontal directions.

Thermal and Pressure Effects

Thermal expansion or contraction of rocks in response to changes in temperature induces stresses in the rocks if they are not free to expand or contract. The stresses must be of sufficient magnitude to counteract the changes in dimension that would be caused by temperature changes in the unrestrained rock. Because different rocks have different coefficients of thermal expansion, a temperature change induces different stresses in two immediately adjacent but different rock types, such as a limestone and a sandstone. Stresses are also induced where different amounts of temperature change occur

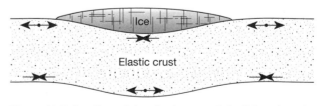

Figure 10.7 Bending of the elastic part of the lithosphere in response to the loading of a continental ice sheet causes deviatoric compression on the concave sides of the bends and deviatoric tension on the convex sides.

in adjacent rocks, such as where a magmatic intrusion cools off while the adjacent country rock warms up.

Because different types of rocks have different elastic coefficients, changes in pressure associated with the addition or removal of overburden induce different amounts of strain under unconstrained conditions. If the deformation is constrained, a differential stress must build up of sufficient magnitude to satisfy the constraints.

Pore Fluid Pressure

Finally, the existence of pore fluid pressure in rocks strongly affects their mechanical response and can cause extension fracturing even under conditions of purely compressive applied stresses (see Section 9.5 and Figure 9.13A). High pore fluid pressures can develop simply from the compaction of impermeable sediments, which decreases the pore volume. If this volume is filled with water, and if the water cannot escape from the sediment, then the pore fluid pressure must increase.

Water has a higher coefficient of thermal expansion than sediment, so if the pores are saturated with water that is trapped by impermeable layers, the pore fluid pressure must increase with temperature. This phenomenon is referred to as **aquathermal pressuring.**

Prograde metamorphic reactions, which occur under conditions of increasing temperature and pressure, are commonly dehydration or decarbonation reactions that release water or carbon dioxide, respectively, into the rock. Most crystalline rocks are highly impermeable, and if these fluids are produced faster than they can migrate away through the rock, the pore fluid pressure must increase. Hydrofractures may be a common feature of metamorphic terranes deep in the crust.

Partial melting during very high-grade metamorphism in deep crustal regions may also create high pore fluid pressure. In such a situation, the first melts to form are fluid-rich and generally of granitic composition. If the fluid cannot escape, the pressure of the melt can become very high. Some veins in the deepest core regions of mountain belts may originate as fractures induced by the fluid pressure of such melts.

10.4 Stress Histories and the Origin of Joints

Given the wide variety of mechanisms for inducing and changing stress conditions in the Earth's crust, it is not surprising that fractures in the crust have numerous possible origins. In this section, we look at possible loading histories that can lead to the formation of joints. Because joints are extension fractures, the tension fracture criterion is relevant to explaining their origin.

For sedimentary basins, we distinguish two principal sets of conditions: those that cause jointing during burial and those that cause jointing during uplift and erosion. The stress path associated with burial followed by uplift is not a reversible path, because the mechanical properties of the material change with time. During burial, the unconsolidated sediments gradually become compacted and lithified and may be affected by tectonic deformation. Thus when rocks are uplifted, they are very different materials from when they were buried, and they have different mechanical properties. This difference affects the way stresses accumulate (see Box 10.1).

All stresses that have been measured directly in the Earth are compressive; true tensile stresses are rare. For extension fractures to form, therefore, two conditions must be met. Pore fluid pressure must be large enough for the effective minimum principal stress to become tensile, and the differential stress must be small enough so that, at the critical pore fluid pressure, extension fractures form (Figure 9.13A) rather than shear fractures (Figure 9.13B).[1] Values of the tensile strength $|T_0|$ for small rock samples measured in the laboratory vary from a few megapascals for weak sedimentary rocks up to around 40 MPa for crystalline rocks. Widespread planes of weakness in crustal rocks, such as fractures and bedding planes, however, result in very low bulk tensile strengths. Measured differential stresses are generally small and tend to increase slightly with depth, being generally less than 20 MPa near the surface and, at 5 km depth, reaching values of no more than 50 MPa in sedimentary rocks and 70 MPa in crystalline rocks (Figure 10.5). Thus it is probable that hydrofracture in rocks should often result in extension fractures.

Joint Formation During Burial

In tectonically quiescent sedimentary basins, at depths less than about 3 km, measured fluid pressures are generally not greater than hydrostatic pressure, which suggests that flow of fluids through the rock is unrestricted above that level. With increasing depth of burial, flow becomes restricted, and compaction and aquathermal pressuring can increase the fluid pressure more rapidly than the minimum compressive stress increases. Eventually hydrofracture results.

In permeable rocks, the sudden local decrease of pore fluid pressure at the fracture causes a rapid flow of pore water into the fracture. If the sediment is unconsolidated, some of it may be carried into the fracture, producing clastic dikes.

---

[1] According to the Griffith theory of fracture, the differential stress (the diameter of the Mohr circle) that can cause extension fracturing is limited by $(\hat{\sigma}_1 - \hat{\sigma}_3) < 4|T_0|$, where T_0 is the tensile strength. This is the largest Mohr circle that can be tangent to the parabolic fracture criterion at the vertex of the parabola (see Figure 9.13).

The different mechanical properties of disparate rock types mean that in general they do not fracture at the same time. Consider an interlayered sandstone–shale sequence in which the overburden stress is the maximum compressive stress, $\hat{\sigma}_1 = \rho_r g h$. Because the sandstone can support a larger differential stress than the shales, the minimum principal stress $\hat{\sigma}_3$ is smaller in the sandstone than in the shale, and a smaller pore fluid pressure is required to cause hydrofracturing in the sandstone than in the shale (Figure 10.8; see Box 10.1). During burial, therefore, as the pore pressure gradually increases, hydrofractures develop first in the sandstone and do not extend into the shale. When pore pressure in the shale rises sufficiently to cause hydrofracture, the sandstone can also fracture, and fractures can cross the lithologic contacts. The extent of a set of joints can therefore be a significant factor in the interpretation of the history of joint development.

Joint Formation During Uplift and Erosion

During uplift and erosion, we assume that the principal stresses are horizontal and vertical and that the vertical stress equals the overburden. The vertical normal stress decreases as the overburden diminishes, and the temperature also decreases. The changes in horizontal stress components during uplift determine whether jointing occurs in this phase of the rock's history. The important factors determining horizontal stresses are the Poisson

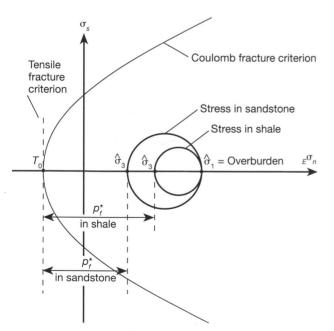

Figure 10.8 Stresses in shale and sand interbeds. The stronger sandstones support a larger differential stress than the weaker shales. p_f^* is the critical pore pressure required for hydrofracture, and it is smaller for the sandstones than for the shales.

and thermal effects, which tend to counteract each other, and bending stresses.

The uplift may be accompanied by doming of the crust, which creates stresses associated with bending. Without detailed knowledge of the geometry of the uplift, however, these stresses cannot be predicted.

If the rocks behave as an elastic material, the Poisson effect predicts that a decrease in the vertical load will cause expansion in the vertical direction and contraction in the horizontal direction (Figure 9.1A). The rocks are not free to change horizontal dimensions, however, so the horizontal components of stress decrease sufficiently to offset exactly the Poisson contraction (Equation 10.1). Starting, for example, from a lithostatic stress at the deepest point of burial, for which $\hat{\sigma}_1 = \hat{\sigma}_2 = \hat{\sigma}_3 = \rho_r g h$, uplift and erosion would result in a decrease in all components of compressive stress, but the horizontal stress would decrease less than the vertical component. Thus the horizontal stress would end up as the maximum compressive stress (see Box 10.1).

Thermal contraction of the rock associated with a decrease in temperature, however, competes with and commonly overwhelms the Poisson effect. Because the rocks cannot change horizontal dimension, the horizontal compressive stress decreases by an amount that exactly offsets the thermal contraction.

The net effect of most conditions of uplift and erosion is that the horizontal stress becomes the minimum compressive stress, and the vertical stress the maximum compressive stress (see Box 10.1). As long as one of the horizontal *effective* stresses is tensile, however, vertical joints that are normal to that stress component can form. Again, the effect of the pore fluid pressure plays a critical role in producing an effective tensile stress in a compressive stress regime.

The development of one set of vertical joints relieves the effective tensile stress normal to those joints. If the other horizontal principal effective stress is tensile, then it becomes the maximum effective tensile stress, and a second set of vertical joints may form orthogonal to the first set. Such systems of orthogonal vertical joints are a common feature, for example, of the flat-lying sediments in the midcontinent region of the United States. Determination of the relative timing of joints can be very difficult, however, and this interpretation of the origin of such orthogonal sets of joints is at present only an hypothesis.

Tectonic Joints

If tectonic stresses are imposed on a rock during burial, then compaction and restriction of the pore fluid circulation may occur at shallower depths than is possible under lithostatic loading. The resulting high pore fluid pressures can cause hydrofracturing at depths

Box 10.1 The Effect of Burial and Uplift on Stress

We consider a simple model of the evolution of stress in rocks during burial, lithification, and uplift. The model includes only the overburden, the Poisson effect, and the thermal effect, and we calculate the stress required to maintain a horizontal extension of zero. The change in the maximum and minimum horizontal normal stresses $\Delta\sigma_{H(\max)} = \Delta\sigma_{H(\min)}$, as a function of the changes in vertical stress $\Delta\sigma_V$ and temperature ΔT, is given by

$$\Delta\sigma_{H(\max)} = \Delta\sigma_{H(\min)} = \left(\frac{v}{1-v}\right)\Delta\sigma_V - \left(\frac{E}{1-v}\right)\alpha\,\Delta T$$

$$(10.1.1)$$

where α is the coefficient of thermal expansion, which gives the extension (Equation 9.1) per degree of temperature change. The first term gives the stress required to counteract the Poisson effect and comes from Equation (10.1). The second term gives the stress required to counteract the thermal effect and comes from the equations of elasticity (similar to Equation 9.5).[*]

The changes in stress and temperature indicated by the Δ in Equation (10.1.1), are the final minus the initial values, and we can express $\Delta\sigma_V$ and ΔT as a function of the change in depth.

$$\Delta\sigma_{H(\max)} = \Delta\sigma_{H(\min)} = \sigma_H^{(f)} - \sigma_H^{(i)} \qquad (10.1.2)$$

$$\Delta\sigma_V = \sigma_V^{(f)} - \sigma_V^{(i)} = \rho_r g(h^{(f)} - h^{(i)})$$

$$\Delta\sigma_V = (25\ \text{MPa/km})\,(h^{(f)} - h^{(i)}) \qquad (10.1.3)$$

$$\Delta T = (25°\text{C/km})\,(h^{(f)} - h^{(i)}) \qquad (10.1.4)$$

---

[*] In Equation (9.5) change subscripts z to x and x to z to obtain the equation for $\hat{e}_{xx}$. Assume that the two horizontal stress components are $\hat{\sigma}_{xx} = \hat{\sigma}_{yy}$ and that $\hat{\sigma}_{zz} = 0$. Solving for $\hat{\sigma}_{xx}$ gives $\hat{\sigma}_{xx} = [E/(1-v)]\hat{e}_{xx}$. Then set $\hat{e}_{xx} = -\alpha\,\Delta T$ so that it is the contraction required to cancel out the thermal expansion, that is, it is the negative of the thermal extension.

where the superscripts (f) and (i) indicate "final" and "initial" respectively, and where h is the depth in kilometers. We then substitute Equations (10.1.2) through (10.1.4) into Equation (10.1.1).

With this model, we determine the history of stress for both a sandstone and a shale that are buried as unconsolidated sediments to a depth of 1 km, lithified at the maximum depth of burial, and uplifted back to the surface. The elastic and the thermal expansion constants listed in Table 10.1.1 show that the clay and shale have different mechanical properties, as do the sand and sandstone. This fact ensures that the stress history during burial is different from that during uplift, as shown in Figure 10.1.1.

For the burial, we take $\sigma_H^{(i)} = 0$ MPa, $h^{(f)} = 1$ km, and we solve for $\sigma_H^{(f)}$ by using the constants for sand and clay listed in Table 10.1.1. At a depth of 1 km, the final horizontal stresses on the sand and clay, respectively, are 7 MPa and 25 MPa, and the vertical stress for both is 25 MPa (Figure 10.1.1). For uplift, we use the final horizontal stress from burial as the initial horizontal stress for uplift, $\sigma_H^{(i)} = 7$ MPa or 25 MPa, we use $h^{(i)} = 1$ km and $h^{(f)} = 0$ km, and we use the constants for sandstone and shale from Table 10.1.1. The final horizontal stresses at the surface are -12 MPa and 9 MPa for the sandstone and shale, respectively (Figure 10.1.1).

For the sand in this simple model, the horizontal stress is compressive during burial and is the minimum principal stress; the vertical stress is the overburden and is the maximum principal stress (Figure 10.1.1A). Because lithification from sand to sandstone changes the elastic properties of the material (Table 10.1.1), uplift of the sandstone carries it along a stress-depth path of shallower slope than for burial. Thus the horizontal stress decreases more rapidly with decreasing depth, and it actually becomes tensile during

---

much shallower than 3 km, and the orientation of the joints should reflect the orientation of the principal tectonic stresses. Tectonic stresses may be applied either before or after the formation of burial joints. Because tectonic deformation is commonly accompanied by the formation of a foliation in the rocks (see Chapter 14), the cross-cutting relationships of joints with foliations can be an important element in reconstructing the sequence of deformational events.

Tectonic stresses can, of course, affect rocks during uplift as well as during burial. Such stresses can govern the orientation of new joints by changing the value of one of the horizontal components of stress. If a horizontal stress became the minimum compressive stress

as proposed above, and to that were added a horizontal tensile tectonic stress, then vertical joints would form normal to the tectonic stress. If the horizontal tectonic stress were the maximum compressive stress, vertical joints would form parallel to it.

The Origin of Sheet Joints

As we noted in Chapter 3, sheet joints are subparallel to the topographic surface. We mentioned two mechanisms for their formation: Either the topography controls the orientation of the sheet joints, or the orientation

Table 10.1.1 Mechanical Properties of Sediment During Burial and Uplift[a]

	Burial		Uplift	
	Sand	Clay	Sandstone	Shale
E, in MPa	-1.0×10^3	small	-16.5×10^3	-4.9×10^3
v	0.21	0.5	0.33	0.36
α, in $°C^{-1}$	10.0×10^{-6}	—	10.8×10^{-6}	10.0×10^{-6}

[a] Data assembled from various sources by Engelder (1985)

uplift. The tensile strength T_0 is exceeded after only about half the overburden has been removed (Figure 10.1.1A), at which point joints could form.

The assumption for clay that $v = 0.5$ results in all stress components being lithostatic along the stress-depth path for burial (Equation 10.1 and Figure 10.1.1B). After lithification, uplift causes a decrease in the horizontal stress, but this decrease is less than that of the overburden. Thus the horizontal stress remains compressive throughout the history and is the maximum principal stress during uplift (Figure 10.1.1B).

Different lithologies therefore can have very different stress histories in response to the same externally applied conditions, and the same fractures do not necessarily develop in all rock types, even in the same location.

Tensile stresses have not been measured in rocks, lithification is not likely to occur only at the greatest depth of burial, and we have neglected the effects of pore fluid pressure, so this model is oversimplified. It does, however, illustrate some of the variability that is inherent in the evolution of stress at depth in different rocks.

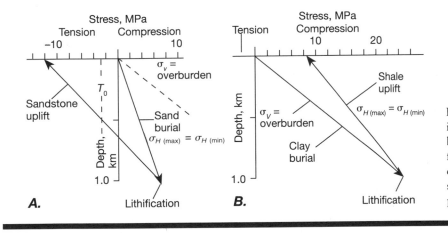

Figure 10.1.1 Stress histories during burial as a sediment, followed by lithification and uplift as a rock for (A) sand/sandstone and (B) clay/shale. The graphs are constructed by using the physical properties from Table 10.1.1.

of the sheet joints is controlled by preexisting stresses in the rock, and the joints affect the evolution of the topography.

Under some circumstances, the maximum compressive stress can remain horizontal and the minimum compressive stress vertical during uplift, as might result if a tectonic compression were applied. As the vertical stress approaches zero, horizontal joints could propagate in a manner similar to longitudinal splitting. The topography would affect the local orientation of the stress field, because the topographic surface is a free surface that must be a principal surface of stress. Thus the principal stresses locally must be perpendicular or parallel to the topography. This model predicts that the joints should tend to parallel topography.

The alternative hypothesis posits that topography is controlled by the orientation of the joints, which are in turn the result of residual stresses in the rock. For example, in a plutonic igneous body, cooling at depth concentric with the boundary of the pluton could produce residual thermal stresses within the body with the maximum compressive stress ($\hat{\sigma}_1$) subparallel to the boundary. As the minimum compressive stress decreases toward zero during uplift, sheet jointing could develop by longitudinal splitting; the orientations of the joints would reflect the shape of the boundary or cooling surfaces in the pluton. Subsequent erosion is controlled by the orientations of the joints. The interpretation of sheet joints is not clear-cut, and both hypotheses could be correct in different cases.

The Origin of Columnar Joints

The polygonal fracture patterns, or columnar joints, that are common features of many igneous extrusions and shallow intrusions probably result from thermal stresses set up by unequal cooling and thermal contraction between the igneous body and the country rock. After solidification, the higher temperature of the igneous rock means that its thermal contraction would be considerably greater than that of the adjacent country rock if the contact were free to slip. A welded contact makes any relative displacement between the two rock masses impossible. In this case, as the two rocks cool, stresses build up on both sides of the contact sufficient to prevent displacement along the contact. Normal stress components that are parallel to the contact are tensile in the igneous rock, preventing it from contracting as much as thermal contraction would require; these stress components are balanced by a compressive stress in the country rock, which forces it to contract more than thermal contraction would require. In general, the tensile stresses in the igneous rock become oriented parallel to the isothermal surfaces during cooling. Because rocks are weaker in tension than in compression, the igneous rocks tend to form tensile fractures perpendicular to the surfaces of equal temperature.

The origin of the hexagonal shape of the columns is not well understood. More than one set of fractures is required to relieve the tensile stress in two orthogonal directions. Such a system of fractures can fill a volume with close-packed fracture-bounded prisms if the prism cross section is triangular, rectangular, or hexagonal. Of these, the hexagonal prisms have the smallest fracture surface area per unit volume of prism. Thus fracture-bounded prisms with a hexagonal cross section require less energy to produce than other prism shapes, and this form of columnar joint is dominant. In principle, however, two sets of fractures should suffice to relieve tensile stresses in two orthogonal directions. Consequently, we do not at present understand the mechanism of development of the three sets of tensile fractures that define the hexagonal prisms.

A similar process must account for the development of hexagonal mud cracks, which form during the desiccation and associated contraction of the surface layers of mud.

10.5 The Spacing of Extension Fractures

The regular spacing of joints and the dependence of that spacing on layer thickness (Figure 3.11) are characteristics that any proposed mechanism of formation must account for. Several explanations have been suggested, but there is no definite proof which, if any, is correct.

One hypothesis that has been proposed to explain the characteristic fracture spacing involves the pore fluid pressure. When a fracture forms, the pore fluid pressure in the neighborhood of the fracture decreases as pore fluid flows into the open fracture. As the pore fluid pressure declines, the effective Mohr circle moves away from the failure criterion, so further fracture in the vicinity of the initial fracture is impossible. A second fracture can form in the rock only beyond the zone of reduced pore pressure, thereby defining the minimum spacing for the formation of hydrofractures. This distance must depend on the permeability of the rock, so highly permeable rocks should have a larger fracture spacing than less permeable rocks.

Layers can also be fractured by the contact forces imposed by adjacent layers. To illustrate this process, consider three layers with welded contacts (Figure 10.9). Suppose that upon uplift, the two outside layers tend to extend more than the central layer. The normal component of stress parallel to the layers is compressive in the outside layers and tensile in the central layer. The force F_t resulting from the tensile stress across the thickness of the layer must balance the forces F_s exerted by the shear stresses along the surfaces of the layer: $F_t = 2F_s$. F_s increases with the length ℓ of the layer. Thus the spacing of the extension fractures that can form within the layer is determined by the length of layer necessary to build up a tensile stress equal to the fracture strength, $F_t = T_0$. For a thicker layer, the fracture spacing should be larger because the force required to fracture the layer is larger. In principle, this also must be the type of process involved in the formation of columnar joints.

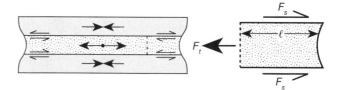

Figure 10.9 Changes in pressure, stress, or temperature can induce stresses in a layer imbedded in a rock of a different type if the coefficients of thermal expansion or elastic constants differ between layer and matrix. In a three-layer sequence, if the central layer tends to expand less than the layers on either side, the central layer will be in a state of tension. The force created by the shear stress on the boundaries of the central layer increases with length of the boundary ℓ; ℓ must be long enough so that the tensile force in the layer divided by its cross-sectional area equals the tensile strength. That distance represents the smallest possible spacing of the fractures.

If a tensile fracture develops, the tensile stress normal to the fracture surface is relieved in the neighborhood of the fracture. In an isolated homogeneous elastic body, the stress relief is negligible at a distance away from the crack equal to about five to ten times the crack depth. Beyond that distance, then another crack may develop. If fracture depth is limited by the thickness of a layer, this relationship suggests that fracture spacing should vary with layer thickness, as is indeed observed. Fracture spacing, however, is usually much less than that predicted by this relationship (Figure 3.11), indicating that this mechanism is not the dominant one.

10.6 Distinguishing Extension Fractures from Shear Fractures

It is often difficult to tell the difference between fractures that have formed as extension fractures and those that have formed as shear fractures, unless some distinguishing characteristic of the mode of formation is present.

The presence of plumose structure on the fracture surface is clear evidence of formation by extension fracturing. Lack of any offset, even down to the microscopic scale, is also clear evidence of extension fracturing.

The presence of pinnate fractures along a fracture is good evidence that the fracture originated as a shear fracture. Pinnate fractures may be extensional cracks that tend to form approximately parallel to the maximum compressive stress. They may also be secondary shear fractures that possibly form at Coulomb fracture angles under locally rotated orientations of the principal stresses relative to the main shear fracture. The orientation of such fractures is not necessarily a reliable indication of their origin. Fractures that display ridge-and-groove lineations (see Figure 4.8A, B and Section 14.6, Figure 14.6A, B) also must have formed as shear fractures. Such features, however, commonly are not present or are not easily observed on all shear fracture surfaces.

The ambiguity in the interpretation of fracture origin is particularly troublesome for those fractures along which there is shear displacement. Such fractures may originate as shear fractures, or they may be extension fractures that are subsequently reactivated as shear planes. Reactivated fractures could even have mineral fiber slickenside lineations (see Chapter 14). On shear fractures that have a very small displacement, however, slickenside lineations might not develop.

The angular relationship between sets of fractures in rocks is not diagnostic of the origin of the fractures, although many interpretations in the literature assume otherwise. Two sets of fractures intersecting in an acute angle often are interpreted to be conjugate shear fractures, and sets of three fractures in which one set bisects

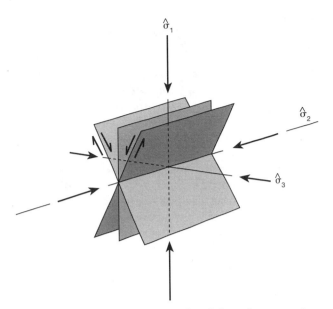

Figure 10.10 Principal stresses inferred from fracture orientations. The maximum compressive stress $\hat{\sigma}_1$ bisects the acute angle between conjugate shear planes; intermediate compressive stress $\hat{\sigma}_2$ is parallel to the intersection line of the conjugate shear planes; the minimum compressive stress $\hat{\sigma}_3$ bisects the obtuse angle between the conjugate planes. The acute angle between conjugate shear planes is bisected by an extension fracture.

the acute angle between the other two are interpreted as sets of conjugate shear fractures bisected by an extension fracture. On the basis of the Coulomb fracture criterion and the tensile fracture criterion, the principal stresses are inferred to have had the following orientation with respect to the fractures (Figure 10.10): The maximum compressive stress $\hat{\sigma}_1$ bisects the acute angle between the conjugate shear fractures and parallels the extension fracture; the intermediate principal stress $\hat{\sigma}_2$ parallels both the line of intersection of the conjugate shear fractures and the extension fracture itself; and the minimum compressive stress $\hat{\sigma}_3$ bisects the obtuse angle between the conjugate fractures and is perpendicular to the extension fracture.

Interpretation of the origin of fractures on the basis of their relative orientations, however, is unjustified without independent evidence of the nature of the fractures and their relative times of formation. In several well-documented examples, careful investigation of the relative timing of joints has revealed that all the fractures forming a pattern similar to that expected for conjugate shear fractures are in fact extension fractures that developed at different times and under the influence of different orientations of stress (Figure 3.12). Thus all interpretations in which fracture angle is cited as the only evidence for a shear fracture origin should be regarded with suspicion.

10.7 Fractures Associated with Faults

The fractures that are parallel and conjugate to faults (Figure 3.16) may represent conjugate shear fractures corresponding to the two fracture orientations predicted by the Coulomb fracture criterion (Figure 9.4*B*, *D*, *E*). In such a case, the approximate stress orientations are as shown for the conjugate shear planes in Figure 10.10.

Many pinnate fractures that are arrayed *en echelon* along a shear fracture (Figures 3.7 and 4.16*A*) form as extension fractures during shearing, and they are oriented approximately perpendicular to the minimum compressive stress when they form. Some pinnate fractures may also originate as secondary shear fractures such as the *R* Riedel shears (Figures 9.8 and 4.16*C*, *D*) or the *P* secondary shears (Figures 9.8 and 4.16*E*, *F*). The acute angles between the fault and both the extension fractures and the *R* Riedel shears point in the direction of relative motion of the fault block containing the secondary fractures. This fact accounts for the sense-of-shear criteria discussed in Section 4.3 (Figure 4.16).

The relationship of gash fractures to the associated shear zone is comparable to that of feather fractures and can be accounted for by assuming the gash fractures form as extension fractures perpendicular to the minimum compressive stress $\hat{\sigma}_3$ (Figure 10.11*A*). The gash fractures, however, may be rotated by ductile deformation during or after formation (Figure 3.8). Gash fractures that initiate at different times during the ductile shear should show different amounts of rotation (Figures 10.11*B* and 3.8). Because the minimum compressive stress $\hat{\sigma}_3$ is normal to any unrotated part of the gash fracture, either the tips of the sigmoidal fractures or the latest-formed fractures provide the best estimate of the stress orientations.

In some cases, *en echelon* gash fractures occur parallel to a conjugate shear zone, an orientation that is not accounted for by this analysis (Figure 3.8). Such orientations may be consistent with the fracture criteria we have discussed and may record locally rotated stress axes. Their geometry may be better accounted for, however, by assuming the fractures form perpendicular to the direction of greatest incremental extension, a possibility we discuss further in Section 17.4. In other cases, each gash fracture may have been completely rotated by ductile deformation, or the set of fractures may have formed as hybrid shears, which have components of both extension and shear across their surfaces (Figure 9.9*B*).

A word of caution is in order concerning the use of fracture orientations to infer the orientations of the principal stresses. Not all fractures near faults form at the same time as the faults. Fractures that predate a fault may actually influence its orientation, because they

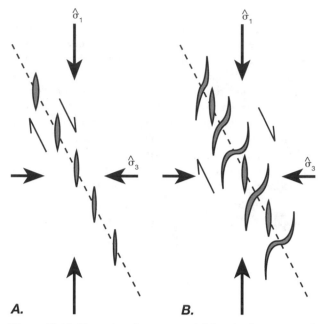

Figure 10.11 Extension fracture model for the formation of gash fractures. *A*. Gash fractures form an *en echelon* array along a shear zone with each fracture perpendicular to the minimum compressive stress. *B*. Ductile shearing along the shear zone rotates the central portions of the fractures, leaving a sigmoidal fracture with the tips of the fractures perpendicular to the minimum compressive stress $\hat{\sigma}_3$. Fractures formed at different times during the ductile shearing show different amounts of rotation, and the smallest, youngest fractures may not be rotated at all.

are preexisting planes of weakness that give the rock a mechanical anisotropy. Such anisotropies are common and can lead to fractures at orientations different from those predicted by the Coulomb fracture criterion. On the other hand, fractures can postdate an adjacent fault and can have an orientation totally unrelated to it.

10.8 Fractures Associated with Folds

The fracture orientations associated with folds (Figure 3.17) have been interpreted as sets of conjugate shear fractures with or without a set of extension fractures (Figure 10.10). This interpretation seems to be based largely on the relative orientations of the different fracture planes, which in the last section we argued is not a reliable criterion. Some studies of fractures associated with folds, in fact, have found that at least some of the fractures existed in the rocks before the folding.

On the other hand, because the orientation and magnitude of stresses in layers undergoing folding vary radically, both from one place to another in the fold and through time as the fold develops, it is possible to account, at least qualitatively, for most of the observed fracture orientations.

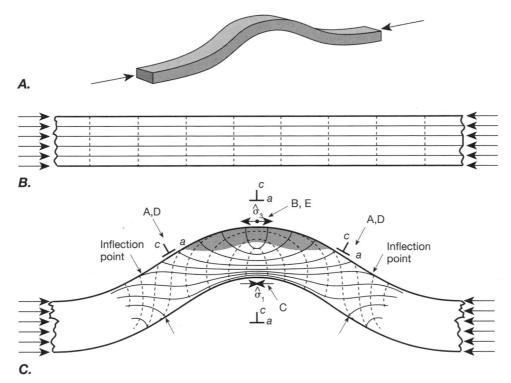

A.

B.

C.

Figure 10.12 Stress distribution in a bar of gelatin undergoing buckling by layer-parallel compression. *A.* Perspective view of the folding experiment. *B.* Stress trajectories in the bar before folding. Solid lines parallel the maximum compressive stress $\hat{\sigma}_1$; dashed lines parallel the minimum compressive stress $\hat{\sigma}_3$. *C.* Stress trajectories in the bar after folding. The shaded area shows where the layer-parallel normal stress component is tensile. The letters A through E show locations where different fracture orientations can develop; they correspond to the fracture patterns shown in Figure 3.17 and listed in Table 10.1. Fracture patterns shown in Figure 10.10 oriented with respect to the reference axes *a, b,* and *c* as indicated in Table 10.1 can account for the observed fracture patterns.

Figure 10.12 shows the evolution of stress in an elastic layer folded by layer-parallel compression (Figure 10.12*A*).[2] The lines in Figure 10.12*B* and *C* are **stress trajectories,** which are everywhere parallel to the principal stresses. Solid lines are trajectories for the maximum principal stress $\hat{\sigma}_1$, and the dashed lines are the trajectories for the minimum principal stress $\hat{\sigma}_3$. The closer the trajectories are to one another, the greater the magnitude of the stress. Before the bar buckles (Figure 10.12*B*), the maximum principal stress is everywhere parallel to the length of the bar, and the minimum principal stress is everywhere perpendicular to the top and bottom of the bar. After buckling (Figure 10.12*C*) the stress orientations are more complex. The maximum principal stress on the concave side of the fold is roughly parallel to the bar and considerably larger than the applied stress, but on the convex side it is at a high angle to the bar. The minimum principal stress on the convex side is parallel to the bar and is actually a tensile stress in the shaded area.

The important points to emphasize in this example are that the orientations of the principal stresses change through time during the buckling process and that the magnitudes—and even the signs—of the principal stresses also change. The existence at the same place of different stresses at different times, and the existence at the same time of different stresses in different places in the fold, can account for the variety of fractures that are observed.

Table 10.1 summarizes the interpretation of the different observed fracture sets (Figure 3.17) in terms of the states of stress that develop during folding (Figure 10.12). The labels of the different fracture sets A through E correspond to the same labels in Figure 3.17 and Table 10.1, and they are used in Figure 10.12*C* to indicate the locations on the fold where the different fracture sets are commonly found. The reference axes *a, b,* and *c* are defined in Table 10.1 and Figure 3.17 and shown in

---

[2] The experiment was performed on a bar of gelatin illuminated from behind by plane polarized light. Because gelatin is a "photoelastic" material, it rotates the plane of polarization by an amount proportional to the elastic strain. By observing the bar through a polarizer set perpendicular to the original plane of polarization, one can determine the amount of rotation of the polarized light and interpret it in terms of the strain magnitude, which, by the equations of elasticity, is proportional to the magnitude of the stress.

Table 10.1. Stress Interpretation of Fractures in Folds

Fracture set[a]	Principal stress parallel to reference axes[b]			Time of formation[c]	Place of Formation
	a	b	c		
A	$\hat{\sigma}_1$	$\hat{\sigma}_3$	$\hat{\sigma}_2$	before folding	throughout fold
B	$\hat{\sigma}_3$	$\hat{\sigma}_1$	$\hat{\sigma}_2$	during folding	convex areas of max curvature
C	$\hat{\sigma}_1$	$\hat{\sigma}_2$	$\hat{\sigma}_3$	during folding	concave areas of max curvature
D (conjugate pair)	$\hat{\sigma}_1$	$\hat{\sigma}_3$	$\hat{\sigma}_2$	before folding	throughout fold
bc fractures	$\hat{\sigma}_3$	$\hat{\sigma}_1$	$\hat{\sigma}_2$	during folding	convex side
E	$\hat{\sigma}_3$	$\hat{\sigma}_2$	$\hat{\sigma}_1$	during folding	convex areas of max curvature

[a] The letters correspond to the fracture sets shown in Figure 3.17 and to the locations around the fold shown in Figure 10.12C.
[b] Here c is normal to bedding; a and b are in the plane of the bedding; b is parallel to the fold axis; a is normal to b and c (see Figures 3.17, 10.12C).
[c] In general, "before folding" corresponds to the stress state in Figure 10.12B, and "during folding" corresponds to the stress state in Figure 10.12C.

Figure 10.12C, where b is everywhere perpendicular to the plane of the diagram.

By orienting the principal stresses and the fracture planes in Figure 10.10 parallel to the reference axes as indicated in Table 10.1, we can account for all the fracture orientations commonly associated with folds in terms of Coulomb shear fractures or extension fractures (Figure 3.17). Note that some fractures (sets A and D) are formed in the stress field that exists before folding (Figure 10.12B), and others (sets B, C, E and bc fractures) are formed in the stress field that exists during folding (Figure 10.12C); that some fracture sets that formed at different times occur in the same places; and that the different stresses that exist in different places at the same time during folding account for different fracture sets (for example, sets C and E). The difference between sets B and E and between sets A and C is that the stresses parallel to b and c exchange positions, presumably depending on local deformation in the direction parallel to b.

The interpretation of fractures associated with folds is currently in need of considerable study. We discuss the stress field associated with folding in more detail in Chapter 20.

10.9 Stress Distributions and Faulting

Anderson's Theory of Faulting

The Coulomb fracture criterion provides a useful theoretical explanation for the threefold classification of faults into normal, thrust, and strike-slip faults. This explanation, called Anderson's theory of faulting after the British geologist, E. M. Anderson, who proposed it,

depends on the fact that the surface of the Earth is a free surface which can support no shear stress. It must therefore be a principal plane of stress, and at the surface the principal stresses must be normal and parallel to the surface. The Coulomb criterion requires that shear fracture planes contain the intermediate principal stress $\hat{\sigma}_2$ and that the fracture plane angle α_f between the fracture plane and the maximum compressive stress $\hat{\sigma}_1$ be less than 45° (Figures 9.4B, D, E, 9.5, and 9.6). The type of fault that develops in a given situation depends on which of the three principal stresses is vertical.

The various possibilities are illustrated diagrammatically in Figure 10.13, where we assume a fracture plane angle of $\alpha_f = 30°$. If the maximum compressive stress $\hat{\sigma}_1$ is vertical, the faults that form should have dips of 60°, and the sense of shear should be hanging-wall-down (Figure 10.13A); these are characteristics of normal faults. If the minimum compressive stress $\hat{\sigma}_3$ is vertical, the faults should dip at 30°, and the shear sense should be hanging-wall-up (Figure 10.13B); these are characteristics of thrust faults. If the intermediate principal stress $\hat{\sigma}_2$ is vertical, faults should be vertical with horizontal shear directions (Figure 10.13C); these are characteristics of strike-slip faults.

The stress orientations measured in the Earth in regions of active tectonics are generally consistent with this interpretation. For example, the Basin and Range province in Nevada is characterized by roughly north-south–oriented normal faults (Figure 5.10). The minimum horizontal stress is oriented approximately east-west (Figure 10.6B), which is consistent with the maximum compressive stress being vertical and thus with the requirements of Anderson's theory for normal faults. The tectonics of the Himalayas are characterized by north-south–directed thrusting. Near the northern

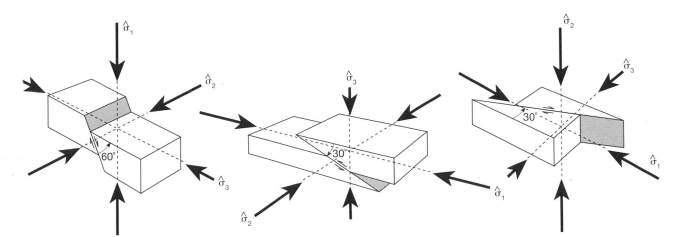

Figure 10.13 Anderson's theory of faulting, showing the relationship between the orientation of the principal stresses and the different ideal fault types. *A.* Normal fault with maximum compressive stress $\hat{\sigma}_1$ vertical. *B.* Thrust fault with minimum compressive stress $\hat{\sigma}_3$ vertical. *C.* Strike-slip fault with intermediate compressive stress $\hat{\sigma}_2$ vertical.

boundary between the Indian and Asian plates, the maximum compressive stress is oriented approximately north-south (Figure 10.6A), which is consistent with the minimum compressive stress being vertical and thus with Anderson's model for thrust faults. Right-lateral strike-slip motion occurs along the northwest-southeast–oriented San Andreas fault in California (Figure 7.2A). The maximum compressive stress measured in this area is oriented roughly north-south (Figure 10.6B), consistent with the intermediate compressive stress being vertical as required by Anderson's theory for strike-slip faults. In this case, however, the fault is at a much higher angle to the maximum compressive stress than Anderson's theory would predict. Thus the shear stress on the fault is much lower than expected, which may result from high pore fluid pressure along the fault.

Faulting and the Distribution of Stress with Depth

The Coulomb fracture criterion provides a concise explanation for the existence of the three major types of faults observed at the Earth's surface. Strictly speaking, however, it applies only near the surface of the Earth, and it assumes strictly planar faults in isotropic material. Most faults are curved and are not just confined to the shallow parts of the crust. In addition, the principal stresses may not be parallel to the horizontal and vertical directions.

Thus it is of interest to examine the possible orientation of the stress field with depth. To investigate this question, we isolate a block of the Earth's crust and consider the distribution of stresses along the boundaries of the block and within it.

We begin by considering the stresses on rock that arise only from the overburden. The vertical normal

stress σ_{33} at any depth x_3 in the block is simply the overburden,

$$\sigma_{33} = \rho_r g x_3 \qquad (10.2)$$

where ρ_r is the average density of the rock, g is the acceleration due to gravity, and x_3 is the depth (positive values for depth below the surface). The corresponding horizontal stress is equal to a fraction κ of the vertical stress,

$$\sigma_{11} = \kappa \rho_r g x_3 \qquad (10.3)$$

where κ is a factor less than 1 that depends on the effective Poisson ratio of the rock (see Section 10.3 and Equation 10.1).

Because no shear stresses exist at the surface and none are applied to any other surface of the block, the principal planes of stress must be parallel and perpendicular to the sides of the block. In other words, the maximum compressive stress $\hat{\sigma}_1 = \sigma_{33}$ is everywhere vertical, and the minimum compressive stress $\hat{\sigma}_3 = \sigma_{11}$ is everywhere horizontal.[3] The stress trajectories, which are lines everywhere parallel to the orientations of these principal stresses, are horizontal and vertical throughout the block. The state of stress arising only from the overburden is often called the **standard state.** We discuss various faulting situations by superimposing additional stresses on the standard state.

Consider, first of all, superposition of a tectonic horizontal compressive stress, adequate to cause faulting, on the standard state (Figure 10.14). In the figure, a supplementary horizontal compressive stress K is added to the standard state. If the added stress is suf-

---

[3] Remember that the principal axes are always numbered such that the principal stresses are $\hat{\sigma}_1 \geq \hat{\sigma}_2 \geq \hat{\sigma}_3$. In cases such as this, therefore, the numbering of the principal coordinate axes may be different from that of the general coordinate axes.

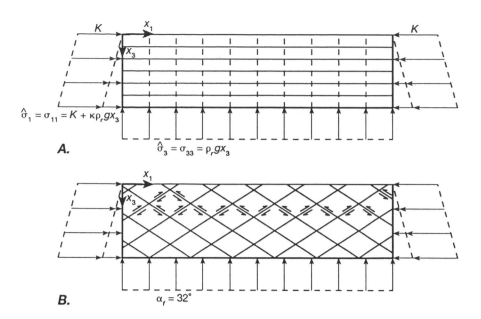

Figure 10.14 Free-body diagram for a horizontal compressive tectonic stress (K) constant with depth added to the standard state stress, which consists of a horizontal stress that increases with depth, and the vertical overburden. *A.* Tractions and stress trajectories. Solid lines are trajectories of $\hat{\sigma}_3$. *B.* Attitudes of potential shear fractures, assuming a fracture plane angle $\alpha_f = 32°$.

ficiently large, the horizontal stress becomes the maximum compressive stress $\hat{\sigma}_1 = \sigma_{11}$, as shown in Figure 10.14*A*. The potential faults, shown in Figure 10.14*B*, are a conjugate set of thrust faults, and the geometry corresponds to that assumed in Anderson's theory.

This model of the stress distribution is unrealistically simple. In particular, we have assumed that no shear stresses exist on the boundaries of the crustal block. As a result, all the stress trajectories are straight lines. If a crustal block were extending or shortening, for example, we would expect shear stresses opposing the motion to be present along the base of the block.

Consider the effect, therefore, of adding a horizontal shear stress that increases with depth and has a constant magnitude along the base of the block (Figure 10.15*A*).

This assumption also is overly simple, but the results are interesting. The symmetry of the stress tensor requires that vertical shear stresses balance the horizontal ones. Shear stresses must exist on the vertical boundaries of the block. Moreover, the requirement that all horizontal forces must sum to zero means that the horizontal normal force on the right side of the block must be less than that on the left, the difference being made up by the force contributed by the shear stress on the base.

The fact that shear stresses exist on the sides and bottom of the block means that these boundaries are no longer principal planes of stress and that the principal axes in general are no longer horizontal and vertical. The top of the block, however, still supports no shear stress, so it must be a principal plane, and the vertical

$$\sigma_{31} = kx_3 \qquad\qquad (10.4)$$

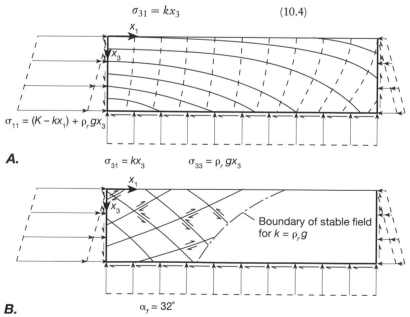

Figure 10.15 Free-body diagram for a compressive tectonic stress added to the standard state stress and including shear stresses on the boundaries of the block. *A.* Tractions and stress trajectories. Solid lines are trajectories of $\hat{\sigma}_1$; dashed lines are trajectories of $\hat{\sigma}_3$. *B.* Potential fault surfaces. The blank area indicates the region of stability where stresses are subcritical, as determined by the fracture criterion given in terms of the principal stresses by $\hat{\sigma}_1 = 4\hat{\sigma}_3 + 100$[MPa], where $\rho_r g = 25$ MPa/km and the value of $k = \rho_r g$ is chosen as an example.

boundaries just at the free surface must also be principal planes. Thus the shear stress on both horizontal and vertical planes must diminish to zero at the surface, as indicated by Equation (10.4). The principal stress trajectories, therefore, are horizontal and vertical at the surface and curve with depth to provide the shear stress on the vertical and horizontal surfaces that increases with depth (Figure 10.15A). With this stress distribution, the potential fault surfaces (Figure 10.15B) show a curvature comparable to that found on natural faults (compare Figure 6.12). Two possible directions of faulting are shown; one concave upward, reminiscent of many listric thrust faults, and one concave downward, reminiscent of faults along some basement uplifts (compare Figure 6.9).

Other possible boundary conditions, consisting of different stress distributions applied to the boundaries of the block, can of course be considered, and they lead to models of different tectonic environments. The diagram in Figure 10.16, for example, shows a stress distribution along the base of the block consisting of a sinusoidally varying vertical normal stress as well as a cosinusoidally varying horizontal shear stress. The standard state of stress is not shown in the diagram but is assumed to be added to the stress shown. The imposed stresses cause a bending of the block, and the stress trajectories are comparable to those in the folded layer in Figure 10.12C. This stress distribution is a possible model for a midoceanic spreading center where upwelling and laterally spreading material provide a vertical tectonic stress that decreases laterally from the spreading axis. The potential fault surfaces form a conjugate set of normal faults symmetrically oriented about the center of the block. Listric normal faults dip toward the center on both sides, and the conjugate faults dip away from the center and steepen with depth. The listric faults are comparable to faults observed on either side of the spreading axis of midoceanic ridges (compare also with the distribution of fractures on folds: Figures 3.17E and 10.12C and Table 10.1).

10.10 Determination of the Stress Field from Faults

The relationship between faults and principal stress directions implied by the Coulomb fracture criterion (Figures 9.4B, D, E, and 9.5A) suggests that we can use faults to estimate the orientation of the principal stresses. The reliability of such estimates depends on the nature of the faulting and the preservation of features indicating movement along them. For two sets of faults to be firmly identified as conjugate faults, the angle between them should be between about 40° and 90°, they must have opposite senses of shear, and there must be good evidence—such as mutual cross-cutting—that the two fault orientations were active at the same time. In such cases, we assume that the line of intersection of the conjugate faults is the intermediate principal stress direction $(\hat{\sigma}_2)$, that the maximum compressive stress direction $(\hat{\sigma}_1)$ bisects the acute angle between the fault planes, and that the minimum compressive stress $(\hat{\sigma}_3)$ bisects the obtuse angle (Figure 10.10).

Shear fractures can develop, and shearing can occur, however, on preexisting fractures, faults, bedding planes, or other planes of weakness that are not in the orientation predicted by Coulomb theory. Moreover, the shear fractures predicted by Coulomb theory can accommodate extension or shortening only in the $\hat{\sigma}_1$–$\hat{\sigma}_3$ plane. If there is a component of extension or shortening parallel to $\hat{\sigma}_2$, the Coulomb criterion cannot predict the orientation of the fractures that form.

Fractures associated with a major fault commonly have a wide variety of orientations on which slickenside lineations, or slickenlines, are developed. Such lineations are parallel to the direction of slip on the different shear planes. If we assume the slip directions are parallel to the direction of maximum resolved shear stress on each plane, we can calculate the orientation of slickenline that should develop on any given plane whose orientation relative to the principal stresses, is known. We

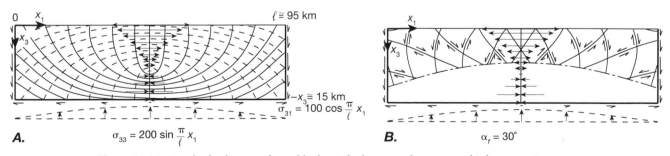

Figure 10.16 Free-body diagram for a block, with the normal stress on the base varying as a sine function, and shear stress on the base varying as a cosine function. The standard state is not shown but is assumed to be part of the stress state. A. Tractions and stress trajectories. Solid lines are trajectories of $\hat{\sigma}_1$; dashed lines are trajectories of $\hat{\sigma}_3$. B. Potential fault surfaces. The blank area shows the field of stability if the maximum value of σ_{33} is 200 MPa.

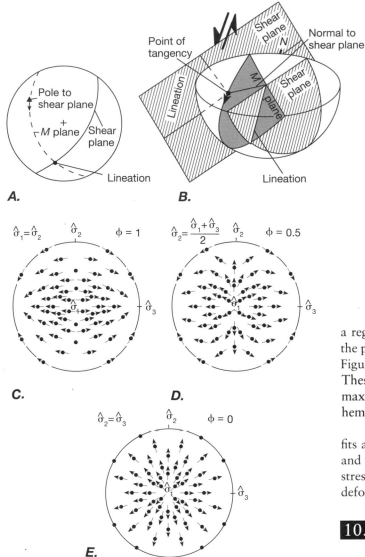

Figure 10.17 Construction and interpretation of tangent-lineation diagrams for slickenside lineations. *A.* The tangent lineations are plotted as arrows through the pole to the shear plane, tangent to the great circle that contains the shear plane pole and the lineation orientation in the shear plane. The arrow points in the direction of shear of the footwall. *B.* Interpretation of the tangent-lineation diagram. The shear plane is tangent to the outside of the plotting hemisphere at the shear plane pole. The arrow is parallel to the slickenline in the shear plane and indicates the shear sense of the footwall block on the shear plane. Note that both equal-area and equal-angle projections distort the true trend of the slickenline. *C, D, E.* Directions of maximum resolved shear stress on a set of planes having a uniform distribution of orientations over the plotting hemisphere for different values of the stress difference ratio ϕ. $\hat{\sigma}_1$ is vertical, $\hat{\sigma}_2$ is horizontal at the top and bottom of the stereogram, and $\hat{\sigma}_3$ is horizontal at the right and left sides of the stereogram. Lower-hemisphere, equal-angle projection.

a regular pattern whose details depend on the ratio of the principal stress differences $\phi = (\hat{\sigma}_2 - \hat{\sigma}_3)/(\hat{\sigma}_1 - \hat{\sigma}_3)$. Figure 10.17C–E shows examples of such patterns. These diagrams therefore show the orientations of the maximum shear traction over the outside of the plotting hemisphere.

Using a computer, we can find the pattern that best fits a given set of field data on slickenline orientations, and we can thereby infer the orientation of the principal stresses and the stress difference ratio ϕ that caused the deformation.[4]

10.11 The Mechanics of Large Overthrusts

In Chapter 6 we discussed the existence of large overthrust sheets that extend for distances of up to hundreds of kilometers along strike and over 100 km across strike. Such large thrusts have been known since the end of the nineteenth century. Soon after they were recognized, however, it became clear that to push such a large mass would seem to require forces that the rocks would be unable to withstand.

M. S. Smoluchowski first formulated the problem in elemetary form in 1909. Consider a rectangular block of height H (parallel to coordinate axis x_3), width W parallel to the thrusting direction (and to coordinate

---

[4] A more general theory of slickenline orientations is obtained by assuming they are parallel to the direction of maximum rate of shear on any particular surface and by accounting specifically for the rotation of the shear planes. The patterns shown in Figure 10.17C–E emerge for a special case.

plot these directions on a tangent-lineation diagram, which combines information about the shear plane orientation, the orientation of the slickenline in that plane, and the sense of shear on the plane.

To plot slickenline data on a tangent-lineation diagram (Figure 10.17A), we construct on a lower hemisphere projection the great circle that contains both the pole (the normal) to the shear plane and the orientation of the lineation in the shear plane, and we draw an arrow tangent to this great circle at the shear plane pole. The arrow points in the direction of relative shear of the footwall block. The pole to the shear plane is the point on the plotting hemisphere where the shear plane would be *tangent* to the *outside* of the hemisphere (Figure 10.17B). The arrow is then parallel to the slickenline in the tangent plane and points in the direction of relative shear of the footwall block for that plane. The directions of maximum resolved shear stress on a set of planes uniformly distributed on the plotting hemisphere form

x_1), and length L perpendicular to the direction of motion (and parallel to coordinate x_2). Its weight per unit volume is $\rho_r g$, and the coefficient of sliding friction on the block's base is $\bar{\mu}$. The frictional force that resists the motion of the block (F_f) equals the normal force across the base (F_n) times the coefficient of friction ($\bar{\mu}$). That is,

$$F_f = \bar{\mu} F_n \tag{10.5}$$

$$= \bar{\mu} \times (\text{normal force per unit area}) \times (\text{area})$$

$$F_f = \bar{\mu}(\rho_r g H)(WL) \tag{10.6}$$

The driving force required to move the block must be greater than or equal to the frictional resistance. If the driving force is applied across the back vertical face of the block, the stress on that face is the driving force per unit area,

$$\sigma_{11} = \frac{F_f}{LH} = \bar{\mu}\rho_r g W \tag{10.7}$$

where we introduced Equation (10.6) for F_f. This stress cannot exceed the fracture strength of the block. Choosing average values for the coefficients of friction, density, and strength ($\bar{\mu} = 0.6$, $\rho_r = 2500 \text{ kg/m}^3$, and $\sigma_{11}^* = 250$ MPa), we can solve Equation (10.7) for W.

$$W = \frac{\sigma_{11}^*}{\bar{\mu}\rho_r g} = 17{,}007 \text{ m} = 17 \text{ km} \tag{10.8}$$

Thus this model predicts that the maximum possible dimension of an overthrust sheet in the direction of thrusting is $W = 17$ km. For larger dimensions, the fracture strength of the rock is exceeded at the rear face of the sheet before the frictional resistance can be overcome. Large overthrusts, however, are known to have widths W of over 100 km, so something must be wrong with this model. A more sophisticated analysis yielding more general but comparable results is given in Box 10.2.

There are several assumptions in this simple model that may be inappropriate for explaining the mechanics of emplacement of large thrust sheets: (1) The force of friction on the base of the thrust could be lower than we assumed. (2) The very assumption that resistance to motion is frictional in origin may be incorrect, and the shear along the décollement in some cases may be accommodated by ductile flow of weak rocks. (3) The thrust sheet may be driven not by a push from the rear but by gravitational forces. (4) Thrust sheets in general are not rectangular blocks, as assumed in our model, but instead taper to smaller thicknesses toward the foreland. (5) Thrust sheets do not move en masse as a single sheet, but rather caterpillar style, by the propagation of localized domains of slip along the fault. All of these factors may be important in explaining aspects of the mechanics of thrust sheets.

Basal Friction

The force of frictional resistance can be reduced in two possible ways: The coefficient of friction $\bar{\mu}$ on the base can be significantly smaller than we assumed, either intrinsically or because of lubrication, or the effective normal stress across the décollement can be less than we assumed.

Laboratory measurements of the coefficient of friction of rock on rock consistently give values near $\bar{\mu} = 0.85$ and do not leave much possibility for significant reduction. The presence of water on a rock interface actually seems to increase the coefficient of friction; it does not act as a lubricant.

A high pore fluid pressure along the décollement, with λ approaching 1, would reduce the effective normal stress across the surface and thereby lower the frictional resistance (Equation 10.5; see Section 9.5). If the frictional resistance decreases, then a horizontal normal stress that is equal to the critical fracture stress can move a greater width of thrust sheet. For zero resistance, the possible width of the thrust sheet is unlimited. Sedimentary basins in active tectonic regions are prime locations for the formation of high pore fluid pressure (Section 10.3), and large overthrust sheets are common in such environments. This explanation has been accepted as a fundamental mechanism associated with the emplacement of large thrust sheets, but it is not a complete explanation.

Ductile Flow

Thrust faults commonly follow layers of weak rock in the stratigraphic section. Evaporites, and especially halite (common rock salt), are among the weakest rocks known. For conditions characteristic of geologic deformation, halite has a yield stress in the range of 0.1 to 1 MPa. Even at shallow depths and low temperatures, the differential stress that makes halite flow is one to two orders of magnitude less than frictional stresses and the yield stresses of other rocks.

Large accumulations of evaporites (such as halite, gypsum, and anhydrite) underlie many sedimentary basins, including the Gulf Coast of the United States, southwestern Iran, and the Appalachian plateau in western Pennsylvania and adjacent states. The resistance to the motion of thrust sheets can therefore be determined by the yield stress of halite rather than the considerably higher frictional stresses. The yield stress, moreover, is relatively insensitive to pressure, unlike friction. Thus where thrust faults can occupy salt beds, the resistance to motion is significantly less than where the salt is absent, and our model would suggest that in those areas, the thrust sheets can extend much farther out toward the foreland. This explanation accounts for the major

Box 10.2 Simple Model of a Thrust Sheet

We adopt a model of a thrust sheet composed of cohesionless material underlain by a horizontal décollement on which motion occurs by frictional sliding. In general, the height of the sheet as a function of the horizontal distance in the direction of displacement x_1 is $h(x_1)$, and for the maximum dimensions of the sheet, when $x_1 = W$, then $h(W) = H$ (Figure 10.2.1). The tractions acting on the external surfaces of the thrust sheet are as shown in Figure 10.2.1, and they include a horizontal tectonic traction σ_t^+ applied to the rear vertical face of the sheet, a vertical traction σ_V^- applied along the bottom of the thrust sheet, and a frictional shear traction σ_f^- also applied along the bottom of the thrust sheet. The superscript $+$ and $-$ indicate that we are considering the traction components acting on the positive and negative sides of the coordinate planes, respectively.

The sum of the horizontal tectonic force F_T and the total force of frictional resistance on the base F_F must be zero if the thrust wedge is moving as a block but not accelerating.

$$F_T + F_F = 0 \qquad (10.2.1)$$

where

$$F_T = \int_0^H \sigma_t^+ \, dx_3, \quad \text{at } x_1 = W \qquad (10.2.2)$$

$$F_F = \int_0^W \sigma_F^- \, dx_1, \quad \text{at } x_3 = 0 \qquad (10.2.3)$$

The frictional traction σ_f^- is related to the effective vertical traction $(\sigma_V^- - p_f)$ by the coefficient of friction on the base $\bar{\mu}_b$.

$$\sigma_F^- = \bar{\mu}_b[\sigma_V^- - p_f^{(b)}], \quad \text{at } x_3 = 0 \qquad (10.2.4)$$

where $p_f^{(b)}$ is the pore fluid pressure on the base of the sheet.

We now make the following assumptions: (1) The horizontal stress is approximately the maximum compressive stress. (2) The horizontal stress is as large as possible and thus is given by the Coulomb fracture criterion. (3) The vertical stress is approximately the

minimum compressive stress and equals the overburden pressure. We express assumption 2, the Coulomb fracture criterion, in terms of the principal stresses by using Equation (9.1.2) with the effective principal stresses $_E\hat{\sigma}_k = \hat{\sigma}_k - p_f^{(i)}$ substituted for the principal stresses. Thus

$$\hat{\sigma}_1^* - p_f^{(i)} = S + K(\hat{\sigma}_3 - p_f^{(i)})$$

where $p_f^{(i)}$ represents the internal pore fluid pressure of the wedge. Setting $S = 0$ to represent the fracture criterion for a cohesionless material, and rearranging the equation, we find that

$$-\sigma_t^+ = \hat{\sigma}_1^* = K\hat{\sigma}_3 - (K-1)p_f^{(i)}, \quad \text{at } x_1 = W \quad (10.2.5)$$

where the minus sign is introduced because the traction σ_t^+ has the opposite sign from the stress component. (It acts on the positive side of the coordinate surface, and we use the geologic sign convention, which assigns the stress component the sign of the traction acting on the negative side of the surface.) We express assumption 3 as

$$\sigma_V^- = \hat{\sigma}_3 = \rho_r g(h - x_3) \qquad (10.2.6)$$

These assumptions imply that the principal stresses are everywhere horizontal and vertical, which cannot actually be true because there is a shear stress on the horizontal base of the sheet. Thus the stress trajectories should be inclined, but our simplifying assumption should be a reasonable first approximation if the shear stress is relatively small and therefore the inclination of the maximum principal stress is small.

In order for sliding to occur on the base, rather than faulting to occur within the wedge, the coefficient of sliding friction on the base must be less than the coefficient of internal friction for the Coulomb fracture criterion; that is, $\bar{\mu}_b < \mu_i$. The pore fluid pressures internal to the wedge and along the base can be expressed as a fraction λ of the vertical stress:

$$p_f^{(i)} = \lambda_i \rho_r g(h - x_3) \qquad p_f^{(b)} = \lambda_b \rho_r g h \quad (10.2.7)$$

where i and b as superscripts or subscripts indicate the variable *internal* to the wedge or along the *base*, respectively, and where in the second equation x_3 does not appear because it is zero along the horizontal décollement. We assume that λ_b is constant along the décollement and that λ_i is constant within the thrust wedge.

We now combine Equations (10.2.1) through (10.2.7) to obtain

$$[K - (K-1)\lambda_i] \int_0^H (H - x_3) \, dx_3 = \bar{\mu}_b(1 - \lambda_b) \int_0^W h \, dx_1$$

$$(10.2.8)$$

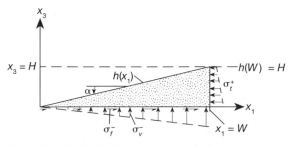

Figure 10.2.1 Model for a thrust sheet, showing the geometry and the tractions acting on the surfaces.

Integrating the left side of the equation, which is the tectonic traction across the height of the rear face of the thrust wedge (where $x_1 = W$ and $h = H$) and collecting constants on the left side of the equation, we find that Equation (10.2.8) becomes

$$0.5 \, CH^2 = \int_0^W h \, dx_1 \qquad (10.2.9)$$

where

$$C = \frac{K - (K-1)\lambda_i}{\bar{\mu}_b(1 - \lambda_b)} = \frac{(K-1)(1-\lambda_i) + 1}{\bar{\mu}_b(1 - \lambda_b)} \qquad (10.2.10)$$

Note that if there is no pore fluid pressure, then λ_i and λ_b are both zero, and C is the ratio of the fracture strength constant of the thrust sheet, K, to the frictional resistance on the base, $\bar{\mu}_b$. In general, then, C is just this ratio modified by the effects of pore fluid pressure. Equation (10.2.9) can be interpreted in two different ways, which we discuss in turn below.

We can assume that the thrust sheet must be everywhere below the critical Coulomb fracture stress, and we assume a particular shape $h(x_1)$ for the thrust sheet. With H a given constant, we can interpret Equation (10.2.9) as determining the limiting cross-sectional width of the thrust sheet W for which the stress in the thrust sheet remains below the critical value. Supposing the thrust sheet to be a rectangular block, we choose $h(x_1)$ to be a constant H for the whole thrust sheet. We considered this problem in a simplified way at the beginning of Section 10.11. Upon integrating the right side of Equation (10.2.9) and rearranging, we find that

$$W = 0.5CH \qquad (10.2.11)$$

Thus for a given thickness H of a block-shaped thrust sheet, and for given values of K, $\bar{\mu}_b$, λ_i, and λ_b, which determine C, this relationship gives the maximum width W for a thrust sheet that can be moved over the décollement. We assume for simplicity that $\bar{\mu}_b = \mu_i$ and that $\lambda_b = \lambda_i$, and we use Equation 10.2.10 and the values from Table 10.2.1 to graph the dependence of $C/2(=W/H)$ on λ (Figure 10.2.2A). For zero pore fluid pressure, $C/2$ is between 2.75 (for $\theta_f = 65°$) and 3.85 (for $\theta_f = 50°$). Thus for a thrust

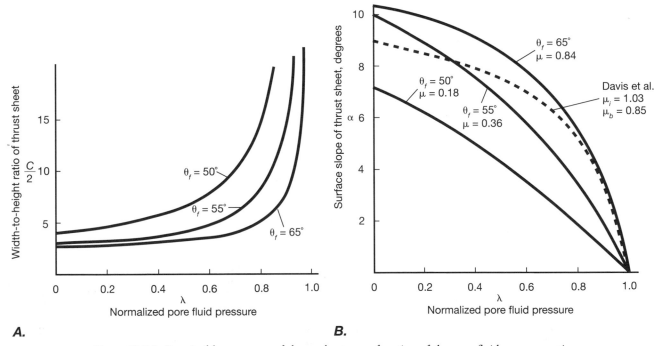

A.

B.

Figure 10.2.2 Permissible geometry of thrust sheets as a function of the pore fluid pressure ratio λ. A. Maximum possible ratio of width to height ($C = 2W/H$) for a thrust sheet shaped like a rectangular block, plotted as a function of λ. B. Equilibrium surface slope α for a wedge-shaped thrust sheet with a horizontal décollement, plotted as a function of λ. Solid lines result from the present analysis; the dashed line results from the more detailed model of Davis et al., 1983.

Box 10.2 (*continued*)

sheet of thickness $H = 5$ km, the width W must be between about 13.7 km and 19.7 km, for the different values of the fracture angle θ_f. This result is of the same order as the more approximate solution in Equation (10.8). Note that the possible length of the thrust sheet increases without limit as λ approaches 1.

For an alternative interpretation of Equation (10.2.9), we can assume that the entire thrust sheet must be just at the critical Coulomb fracture stress, and we take H and W to be variables. Equation (10.2.9) then defines the shape of the thrust sheet, because it prescribes how the height H must vary with the cross-sectional length W across the thrust sheet in order that the thrust sheet be everywhere at the critical Coulomb stress. Equation (10.2.9) can be satisfied only if $h(x_1)$ is a linear function of x_1, because the integral must have dimensions of [length]^2.

$$h(x_1) = Ax_1 + B \qquad (10.2.12)$$

We require that

$$h(W) = H \qquad A = \tan \alpha \qquad (10.2.13)$$

where the first Equation (10.2.13) is implicit in the way the quantities are defined for the problem, and where α is defined as the surface slope of the wedge. Substituting this equation into Equation (10.2.12) shows that

$$H = AW + B \qquad (10.2.14)$$

$$\int_0^W h \, dx_1 = \int_0^W (Ax_1 + B) \, dx_1 = 0.5AW^2 + BW \qquad (10.2.15)$$

Substituting Equations (10.2.14) and (10.2.15) into Equation (10.2.9) and simplifying, we get

$$0.5[A^2C - A]W^2 + [ABC - B]W + 0.5[B^2C] = 0 \qquad (10.2.16)$$

This relationship must hold for a thrust wedge of any width W, and for this to be true, each of the coefficients in brackets must independently be zero:

$$A^2C - A = 0 \qquad ABC - B = 0 \qquad B^2C = 0 \qquad (10.2.17)$$

To satisfy the third Equation (10.2.17), either $B = 0$ or $C = 0$. Taking $C = 0$ implies from Equation (10.2.10) that K is a function of λ_i or, through the second Equation (9.1.3) that the fracture angle θ_f is a function of the pore pressure ratio. Experimental work shows that this is a physically unacceptable solution, so we must choose $B = 0$. This result implies, from Equation (10.2.12), that the thrust wedge tapers to a point, which is physically reasonable for a material with no cohesion (Figure 10.2.1).

Both the first and second Equations (10.2.17) give exactly the same condition,

$$A = \frac{1}{C} \qquad (10.2.18)$$

Introducing Equation (10.2.10) and the second Equation (10.2.13) into Equation (10.2.18), we find that

$$\tan \alpha = \frac{\bar{\mu}_b(1 - \lambda_b)}{(K - 1)(1 - \lambda_i) + 1} = \frac{1}{C} \qquad (10.2.19)$$

where α is the topographic slope angle of the thrust wedge.

If $\lambda_b = \lambda_i = 0$—that is, if there is no pore fluid pressure—the surface slope of the thrust wedge is determined by $\bar{\mu}_b/K$, the ratio of frictional resistance on the base of the sheet to the fracture strength constant of the thrust wedge. As the pore fluid pressure internal to the wedge increases, the effect is to decrease the strength of the wedge. Thus for higher values of λ_i, the denominator of Equation (10.2.19) decreases, the ratio increases, and the surface slope of the thrust wedge increases. As the pore fluid pressure along the base increases, there is less resistance to frictional sliding. Thus for higher values of λ_b, the numerator in Equation (10.2.19) decreases, the ratio decreases, and the surface slope of the thrust wedge decreases.

For purposes of simplification, we assume $\mu_i = \bar{\mu}_b = \mu$ and $\lambda_i = \lambda_b = \lambda$. Figure 10.2.2B shows the relationships then predicted by Equation (10.2.19) between the topographic slope α of the thrust sheet and the magnitude of the pore fluid pressure ratio λ for values of the constants in Table 10.2.1. The predicted slopes are all less than about 10°. The slopes approach 0°, and the frictional resistance to sliding on the décollement decreases toward zero as λ approaches 1. These results are comparable to the angles calculated from more sophisticated analyses for the same angle of the décollement (Figure 10.2.2). More thorough analyses include the dip of the décollement as a variable.

Table 10.2.1 Relationships Among Fracture Angle, Coefficient of Internal Friction, and K^a

θ_f	μ_i	K
65	0.84	4.60
60	0.58	3.00
55	0.36	2.04
50	0.18	1.42
45	0.00	1.00

^{a} Given θ_f, the tabulated values of μ_i and K are calculated from Equations (9.1.1) and (9.1.3).

salient in the northwestern Appalachians (Figure 6.11A). Here the large belt of very gentle folding in the Appalachian plateau northwest of the Valley and Ridge province is almost coincident with the extent of Silurian salt beds at depth.

Anhydrite and gypsum also have relatively low yield stresses, and strata rich in these minerals also commonly act as décollement zones. Where evaporites are not present, shales are generally the weakest rocks, and at greater depths and higher temperatures, limestone (marble) and even quartzite may be sufficiently weak to localize major zones of ductile shear in a décollement.

Gravitational Driving Forces

One problem with our simple model arises from the need to drive the thrust sheet forward by means of a stress transmitted through the thrust sheet from the rear. If the force of gravity were the driving force, however, this restriction would not arise, because gravitational forces act independently on every point in a body.

Gravitational sliding occurs if the shear force provided by the force of gravity (F_s) is at least equal to the frictional resistance on the décollement ($F_s = F_f$; Figure 10.18A). If we know the resistance, we can determine the slope necessary to cause such a thrust sheet to slide. From Equation (10.5), therefore, we have

$$\bar{\mu} = F_s/F_n \qquad (10.9)$$

If gravity is the only force driving the sheet, then the normal force across the décollement (F_n) and the shear force parallel to it (F_s) are related to the dip δ of the thrust surface by

$$\tan \delta = F_s/F_n$$

Using Equation (10.9), and assuming that the coefficient of friction $\bar{\mu} = 0.6$, which is actually a low value compared with most experimental data, we find that

$$\tan \delta = \bar{\mu} \approx 0.6$$

$$\delta = 31°$$

Thus a slope of at least 31° is required to move the thrust block gravitationally against a conservative value of the frictional resistance. A 100-km thrust sheet would need to slide off a topographic high of at least 51.5 km altitude for this mechanism to explain some of the larger thrust sheets (Figure 10.18B). Given that Mt. Everest is less than 9 km above sea level, this solution does not appear to be satisfactory. Moreover, evidence for steep dips over significant lengths of large thrust sheets is utterly lacking. This mechanism could account for the observations only if it were effective on slopes on the order of a few degrees at most. Such slopes imply a very small resistance along the décollement, and we must therefore include in the model either high pore fluid pressure or ductile flow to make it acceptable.

If tectonic processes thicken the crust and create a topographic high, gravitational collapse of the thickened part of the crust could result in the formation of thrust sheets. This mechanism requires ductile flow throughout much of the thickened part of the crust, which spreads outward under its own weight rather like a mound of silicon putty spreads out into a puddle, or, to draw an even more apt analogy, like a continental ice sheet spreads out from its center (Figure 10.18C). The driving force is provided by the topographic slope of the thickened region of crust, and the slope of the décollement is not restricted; it could even slope upward in the direction of thrusting, as is a common feature of thrust sheets.

Intuitively, gravitational forces may not seem strong enough to cause rocks to deform significantly. We must not forget, however, that ultimately, gravitational forces drive the whole plate tectonic machine

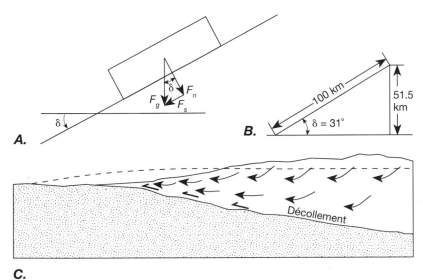

A.

B.

C.

Figure 10.18 Models of gravitationally driven thrust sheets. *A.* Resolution of the gravitational force on a thrust sheet to determine the driving force available (F_s) and the normal force across the décollement (F_n). *B.* Normal rock friction would require too steep a slope to account for the size of thrust sheets and dips of décollement observed. *C.* Gravitational collapse of a tectonically produced topographic high by ductile flow within the thrust sheet. The solid lines indicate the tectonically uplifted topography; the dashed line indicates the topography after gravitational collapse of the uplift. Arrows indicate the general pattern of flow within the collapsing sheet.

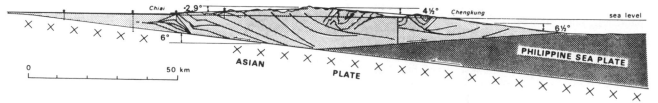

through mantle convection. Given the great lengths of time available, and the ability of rocks to creep slowly in response to relatively small differential stresses, emplacement of thrust sheets by gravitational collapse cannot be discounted.

Tapered Thrust Sheets

Active thrust sheets, such as occur in western Taiwan and in the Himalayas, and active submarine accretionary prisms over subduction zones are wedge-shaped, rather than rectangular, in cross section, with thickness increasing with increasing distance from the front of the thrust sheet (Figure 10.19). We can account for this tapered shape by means of a simple model. We assume that the rocks in the thrust sheet are everywhere just at the critical stress for failure. Furthermore, we require that the driving force on a vertical face through the sheet just balances the frictional resistance to sliding on that part of the décollement that lies ahead of the vertical face. The force resisting sliding on the décollement must increase with increasing distance from the front of the thrust sheet. Thus the driving force on a vertical face must also increase with increasing distance from the front. Because the driving *stress* is limited by the strength of the rock, the driving *force* can increase only if the area of the vertical face increases, and this means the thickness of the thrust sheet must increase. Thus the thickness of the thrust sheet at any point depends on the length of the sheet ahead of that point that must be moved, resulting in a thrust sheet that has a wedge shape (see Box 10.2).

The determination of a proper mechanical model for such a thrust wedge is complex (see Box 10.2). The surface slope of the wedge actually is affected not only by the resistance to motion of the décollement but also by the slope of the décollement. Moreover, resistance to sliding can be affected by the pore fluid pressure or by the presence of weak ductile rock along the décollement. This simple analysis of the driving force also ignores the small horizontal pressure gradient created by the surface slope of the thrust sheet. The mechanics of such wedges, however, is similar to the mechanics of dirt and snow wedges that would form in front of bulldozer and snowplow blades if the blades were flat

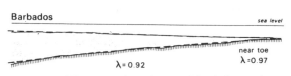

Figure 10.19 The tapered wedge model of thrust sheets is supported by observations of the geometry of active continental fold-and-thrust belts such as in western Taiwan and by submarine accretionary prisms that overlie active subduction zones such as east of Barbados in the eastern Carribbean Sea.

and vertical.[5] Such a wedge is the thinnest body of a given width parallel to the direction of thrusting that can be slid over the décollement.

If material is added to the front, or toe, of the wedge, the whole wedge deforms to maintain the critical taper. The deformation takes the form of thrust faults, folds, and fault ramp folds internal to the wedge, all of which result in a net shortening and thickening of the wedge (Figure 6.12). If the taper of the wedge becomes too large, then the thrust fault propagates out in front of the wedge to lengthen the thrust sheet and decrease the taper; internal faulting and folding provide the adjustments to the taper throughout the rest of the sheet.

A comparison of the tapered wedge model with observations is indicated in Figure 10.20, which plots the dip of the décollement β against the dip of the surface slope α. The lines are the theoretically predicted relationship for a variety of values for the pore fluid pressure ratio λ. The boxes indicate the approximate geometries of active wedges as labeled. It is clear from this figure that most thrust wedges require a value of λ considerably above the hydrostatic value of about 0.4, implying significant overpressure of the pore fluid. Such values of λ are consistent with measurements made in wells that penetrate into some of these wedges, which lends credence to the theory. The presence of salt along the

---

[5] In fact, however, such plow blades are vertically curved, a design that forces the snow or dirt to slide up the blade and fall forward, creating a pile whose taper is the angle of repose of the material rather than the critical taper under discussion here. The angle of repose is the angle of steepest slope that loose material can support, and it is generally about 30°, whereas the steepest slopes predicted by the tapered thrust sheet model are about 10°. Thus this analogy can be somewhat misleading.

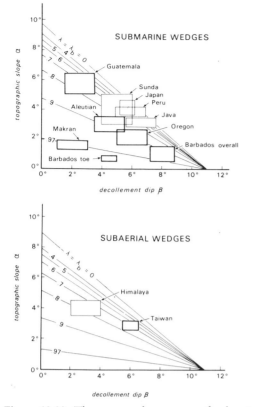

Figure 10.20 The measured geometry of subaerial and submarine thrust wedges compared with the theoretically predicted relationships among the topographic slope angle α, the slope of the décollement β, and the pore fluid pressure ratio λ.

décollement, however, could result in surface slopes as low as 1° and thus could also account for some of the very low slopes observed.

The Propagation of Slip Domains

The foregoing discussion of the mechanics of a thrust sheet assumes the entire mass is at the critical stress for fracture that is predicted by the Coulomb fracture criterion. This is a simplifying assumption, however, because the entire thrust sheet does not move as a rigid block or undergo pervasive deformation at one time. Rather, the deformation is accommodated by the propagation of discontinuous slip events over finite areas of faults within and at the base of the sheet. Such slip events, which are localized in time and space, commonly cause earthquakes that we can observe. Only by averaging these events over a long period of time — perhaps tens to hundreds of thousands of years — would we see the pattern of pervasive deformation and the slip of the entire thrust sheet on the décollement that we assume for the model. The applied stress needed to make slip events propagate across the fault is lower than that required to make the entire thrust fault slip at once, an effect that should be accounted for in mechanical models of faulting. The effect is similar to the propagation of dislocations in a crystal lattice, which we discuss in Chapter 19.

10.12 Cause and Effect: A Word of Caution

Our interpretations in this chapter of the origin of brittle deformation structures, including application of the fracture criteria that we discussed in Chapter 9, implicitly assume that stress is the cause of the deformation. Although this is often a very useful assumption, it is not necessarily appropriate in all situations.

The cause of a mechanical process essentially is determined by the **boundary conditions,** which are the conditions that are externally imposed both on the boundaries of a body and throughout it as distributed sources, such as the force of gravity. If stresses are imposed and maintained on the boundaries of the body, then stress is the cause of the process, and deformation develops in response to the imposed stress. If, however, the boundaries of the body are required to move a prescribed amount or at a prescribed rate in a prescribed direction—that is, the deformation is prescribed on the boundaries of the body—then the deformation is the cause of the process, and the stresses develop in response to the imposed deformation. Under these circumstances, the origin of different structures is better understood with reference to the deformation.

We discuss strain, a measure of deformation, and its application to the interpretation of structures in Chapters 15 through 17; the relationships between stress and deformation are the topic of Chapter 18; and the role of boundary conditions are discussed further in Section 20.1.

Additional Readings

Stress in the Plates

Brace, W. F., and D. L. Kohlstedt. Limits of lithospheric stress imposed by laboratory experiments. 1980. *J. Geophys. Research* 85: 6248–6252.

McGarr, A. 1980. Some constraints on the levels of shear stress in the crust from observations and theory. *J. Geophys. Research* 85: 6231–6238.

McGarr, A., and N. C. Gay. 1978. State of stress in the earth's crust. *Ann. Rev. Earth and Planet. Sci.* 6: 405–436.

Richardson, R. M., S. C. Solomon, and N. H. Sleep. 1979.

Tectonic stress in the plates. *Rev. Geophys. and Sp. Phys.* 17: 981–1019.

Zoback, M. D., H. Tsukahara, and S. Hickman. 1980. Stress measurement at depth in the vicinity of the San Andreas fault: Implications for the magnitude of shear stress at depth. *J. Geophys. Research* 85: 6157–6173.

Zoback, M. L., and M. D. Zoback. 1980. State of stress in the coterminus United States. *J. Geophys. Research* 85: 6113–6156.

Zoback, M. L., et al. 1989. Global patterns of tectonic stress. *Nature* 341: 291–298.

Formation and Interpretation of Joints

Bahat, D. 1979. Theoretical considerations on mechanical parameters of joint surfaces based on studies on ceramics. *Geol. Mag.* 116: 81–92.

Bahat, D. 1991. Plane stress and plane strain fracture in Eocene chalks around Beer Sheva. *Tectonophysics* 196: 61–67.

Bahat, D., and T. Engelder. 1984. Surface morphology on joints of the Appalachian plateau, New York and Pennsylvania. *Tectonophysics* 104: 299–313.

Brown, E. T., and E. Hoek. 1978. Trends in relationships between measured *in-situ* stresses and depth. *Int. J. Rock Mech. Mining Sci.. and Geomech. Abstr.* 15: 211–215.

Degraff, J. M., and A. Aydin. 1987. Surface morphology of columnar joints and its significance to mechanics and direction of joint growth. *Geol. Soc. Am. Bull.* 99: 605–617.

Engelder, T. 1985. Loading paths to joint propagation during a tectonic cycle: An example from the Appalachian plateau, U.S.A. *J. Struct. Geol.* 7: 459–476.

Engelder, T., and P. Geiser. 1980. On the use of regional joint sets as trajectories of paleostress fields during the development of the Appalachian plateau, New York. *J. Geophys. Research* 85: 6319–6341.

Haxby W. F., and D. L. Turcotte. 1976. Stresses induced by the addition or removal of overburden and associated thermal effects. *Geology* 4(3): 181–184.

Ladeira, F. L., and N. J. Price. 1981. Relationship between fracture spacing and bed thickness. *J. Struct. Geol.* 3(2): 179–184.

Kulander, B. R., and S. L. Dean. 1985. Hackle plume geometry and joint dynamics. In *Fundamentals of rock joints.* Proceedings of the International Symposium of Rock Joints, Bjorkliden, Sweden, 85–94.

Pollard, D. D., P. Seagall, and P. T. Delaney. 1982. Formation and interpretation of dilatant *en echelon* cracks. *Geol. Soc. Am. Bull.* 93: 1291–1303.

Fractures Associated with Faults

Beach, A. 1975. The geometry of *en-echelon* vein arrays. *Tectonophysics* 28: 245–263.

Conrad, R. E., II, and M. Friedman. 1976. Microscopic feather fractures in the faulting process, *Tectonophysics* 33: 187–198.

Engelder, T. 1974. Cataclasis and the generation of fault gouge. *Geol. Soc. Am. Bull.* 85: 1515–1522.

Friedman, M., and J. M. Logan. 1970. Microscopic feather fractures. *Geol. Soc. Am. Bull.* 81: 3417–3420.

Fractures Associated with Folds

Currie, J. B., H. W. Patnode, and R. P. Trump. 1962. Development of folds in sedimentary strata. *Geol. Soc. Am. Bull.* 73: 655–674.

Friedman, M., R. H. H. Hugman, III, and J. Handin. 1980. Experimental folding of rocks under confining pressure: Part VIII. Forced folding of unconsolidated sand and of lubricated layers of limestone and sandstone. *Geol. Soc. Am. Bull. Part I,* 91(5): 307–312.

Handin, J., M. Friedman, K. D. Min, and L. J. Pattison. 1976. Experimental folding of rocks under confining pressure: Part II. Buckling of multilayered rock beams. *Geol. Soc. Am. Bull.* 87: 1035–1048.

Norris, D. K.. 1967. Structural analysis of the Queensway folds, Ottawa, Canada. *Canad. J. Earth Sci.* 4: 299–321.

Spang, J. H., and R. H. Groshong, Jr. 1981. Deformation mechanism and strain history of a minor fold from the Appalachian Valley and Ridge. *Tectonophysics* 72: 323–342.

Stearns, D. W. 1968. Certain aspects of fractures in naturally deformed rocks. In *NSF advanced science seminar in rock mechanics for college teachers of structural geology,* ed. R.E. Riecker, pp. 97–118. Bedford, Mass.: Terrestrial Sciences Laboratory, Air Force Cambridge Research Laboratories.

Mechanics of Faulting and Thrust Sheets

Angelier, J. 1984. Tectonic analysis of fault slip data sets. *J. Geophys. Res.* 89(B7): 5835–5848.

Hafner, W. 1951. Stress distributions and faulting. *Geol. Soc. Am. Bull.* 62: 373–398.

Hubbert, M. K., and W. W. Rubey. 1959. Role of fluid pressure in mechanics of overthrust faulting. *Geol. Soc. Am. Bull.* 70: 115–206.

Chapple, W. M. 1978. Mechanics of thin-skinned fold-and-thrust belts. *Geol. Soc. Am. Bull.* 25: 1189–1198.

Dahlen, F. A., J. Suppe, and D. Davis. 1984. Mechanics of fold-and-thrust belts and accretionary wedges: Cohesive Coulomb theory. *J. Geophys. Research* 89: 10,087–10,101.

Dahlen, F. A. 1984. Noncohesive critical coulomb wedges: An exact solution. *J. Geophys. Research* 89: 10,125–10,133.

Davis, D., J. Suppe, and F. A. Dahlen. 1983. Mechanics of fold-and-thrust belts and accretionary wedges. *J. Geophys. Research* 88: 1153–1172.

Davis, D., and T. Engelder. 1985. The role of rock salt in fold-and-thrust belts. *Tectonophysics* 119: 67–88.

Elliot, D. 1976. The motion of thrust sheets. *J. Geophys. Research* 81: 949–963.

Emerman, S., and D. Turcotte. 1983. A fluid model for the shape of accretionary wedges. *Earth and Planet. Sci. Letts.* 63: 379–384.

Scholz, C. H. 1990. *The mechanics of earthquakes and faulting.* New York: Cambridge University Press.

Stockmal, G. S. 1983. Modeling of large-scale accretionary wedge formation. *J. Geophys. Research* 88: 8271–8287.

Twiss, R. J., G. M. Protzman, and S. D. Hurst. 1991. Theory of slickenline patterns based on the velocity gradient tensor and microrotation. *Tectonophysics* 186: 215–239.

PART
III Ductile Deformation

WE TURN in this section to structures in rocks that form as a result of ductile deformation. The word *ductile* is used in the literature in several different ways, which creates considerable confusion and misunderstanding. Much of the problem can be traced to the fact that there are at least three different criteria by which ductile deformation can be recognized. It can be recognized by (1) the characteristic structures that are preserved in rocks; (2) the rheology of the deformation—that is, the form of the relationship among stress, strain rate, pressure, and temperature; or (3) the microscopic mechanisms that operate to produce the deformation.

Our use of the term is based on the first set of criteria, which is consistent with our emphasis on describing rocks using nongenetic terminology. We use the term **ductile deformation** to refer to a permanent, coherent, solid-state deformation in which there is no loss of cohesion on the scale of crystal grains or larger and no evidence of brittle fracturing. Thus there is evidence for distributed smoothly varying deformation with no evidence for discontinuities such as open cracks or pores along grain boundaries or within grains, discrete shear planes on the scale of crystal grains or larger, or angular grain fragments that indicate brittle fracturing. Our definition specifically excludes cataclastic flow, which many would consider a ductile deformation but which we consider to be characteristic of the brittle–ductile transition. It also in principle excludes soft-sediment deformation, which is not a coherent deformation at the grain scale.

Many writers refer instead to plastic or crystal plastic deformation. These terms imply a mechanism of deformation that may not be appropriate. For example, it is not clear that they would appropriately describe deformation accomplished by solution–diffusion phenomena. The term *plastic* carries implications of a specific type of rheological behavior that does not include, for example, dependence of the strain rate on the first power of the stress (see Part IV).

The value of a descriptive and nongenetic term is that the criteria for using the term can be agreed upon on the basis of observable structures and characteristics so differences in interpretation do not affect the use of the word. Nevertheless, we expect that the identification of appropriate descriptive features should be useful for inferring the rheology and mechanism of the deformation. Thus in Part IV, we discuss those characteristics of rheology and mechanism that are associated with the structures of ductile deformation. In particular, we find that ductile deformation, in the sense in which we use the term, is associated with the dependence of strain rate on stress raised to a power generally between 1 and 5, that it is a thermally activated process that occurs at elevated temperatures roughly above half the absolute melting point of the material, and that the rheology of ductile deformation is only weakly dependent on the confining pressure. In terms of mechanism, ductile deformation is accomplished by the motion of defects called dislocations through crystal lattices and/or by diffusion. Thus in practice, observational evidence for any of these conditions or phenomena may provide additional justification for applying the term *ductile deformation*.

In the end, we must admit that there is no completely satisfactory and unambiguous term to use. Moreover, the processes by which rocks deform range from brittle to ductile, and imposing arbitrary boundaries on such a gradation is always to some extent unsatisfactory.

Terminology aside, the most important concept of this part of the book, is that many structures occur in rocks that could form only by flow of the rocks in the solid state. Solid-state flow at first may seem to be an oxymoron: Liquids flow, but do solids? In fact they do. Much of our modern use of metals, for example, depends on solid-state flow. Bars, rods, and sheets of steel, as well as copper and aluminum wire, are all produced by rolling out blocks of metal or drawing it through dies, essentially forcing it to flow in the solid state into the desired configuration. Glaciers also flow slowly yet inexorably downhill by processes that include solid-state flow. Metals and glacier ice, like rocks, are polycrystalline materials; they are made up of an aggregate of crystals. By analogy, it may not seem so surprising, then that rocks also can undergo large amounts of solid-state flow when subjected to the appropriate conditions.

Our aim in Part III is to document the evidence for the solid-state flow of rocks; provide a means for objectively describing the characteristics of the resulting structures; and introduce the concept of strain, by which we can measure ductile deformation and begin to understand how the different types of structures form.

First we describe folds in rocks (Chapter 11) and various kinematic models that can account for their formation (Chapter 12). We then describe foliations and lineations (Chapter 13) and models for their formation (Chapter 14). With these fundamental structures providing evidence for pervasive ductile deformation in rocks, we next investigate how we can describe that deformation quantitatively through the concept of strain (Chapter 15) and how that concept helps us evaluate and interpret different models for the formation of folds, foliations, and lineations (Chapter 16). To conclude Part III, we describe methods for actually measuring strain in rocks and discuss examples of its application to the study of natural structures (Chapter 17).

CHAPTER

11 The Description of Folds

Folds are wavelike undulations that develop during deformation of rock layers, such as sedimentary strata. They are the most obvious and common structures that demonstrate the existence of ductile deformation in the Earth (see Section 1.3). In fact, as long ago as 1669, the Danish naturalist Nicholas Steno described folds and attributed them to Earth movements. Folds occur on all scales, ranging from huge features that dominate the regional structure of orogenic core zones (Figure 11.1A) and form entire mountain sides (Figure 11.1B), through mesoscopic folds on the scale of an outcrop (Figure 11.1C), to folds visible only under a microscope.

Orogenic belts are all characterized by a number of fold systems. The flanks of orogenic belts are generally marked by large fold and thrust belts in unmetamorphosed to lightly metamorphosed sedimentary rocks, which are underlain by major décollements (see Chapter 6). These belts, exemplified by the Appalachian Valley and Ridge province (Figure 11.2A; see Figures 6.12A and 6.13A), the Canadian Rockies (Figures 6.12B and 6.13B), the Himalaya front, and the Jura mountains north of the Alps (Figure 11.2B; see Figure 7.11 and 6.21) commonly contain folds that are continuous for tens of kilometers and that in cross section are characterized by layers of relatively constant thickness.

In the central regions, or core zones, of orogenic belts, the exposed rocks were generally deformed at

greater depth, where the temperature is higher than in the outer fold and thrust belts. The deformation there is associated with pervasive metamorphism and recrystallization of the rocks, and the folding is more intense, resulting in folds with a different appearance from those in the fold and thrust belts (Figure 11.1B, C). Cross sections of the core zones of the Alps (Figure 11.1A) and of the New England Appalachians (see Figure 12.35) indicate the large-scale character of such fold systems. Shapes of folds similar to those in metamorphic rocks (Figure 11.1B, C) also typify deformed salt deposits (see Figure 12.34B) and glaciers, which deform at much lower temperatures than the high-grade metamorphic silicate rocks. Such structures apparently imply a high degree of mobility of the rocks and their component minerals during deformation.

Although most folds we observe are in bedding or former bedding surfaces, folds also affect other types of layers, including dikes, veins, metamorphic or igneous compositional layering, and foliations, which are planar structures defined, for example, by the preferred orientation of platy minerals in the rock (see Chapter 13).

Folds are usually studied strictly to reveal their geometry. The shape, orientation, and extent of folds can be of critical importance in finding economically valuable deposits and in predicting continuations of known deposits. Oil and gas are commonly trapped in

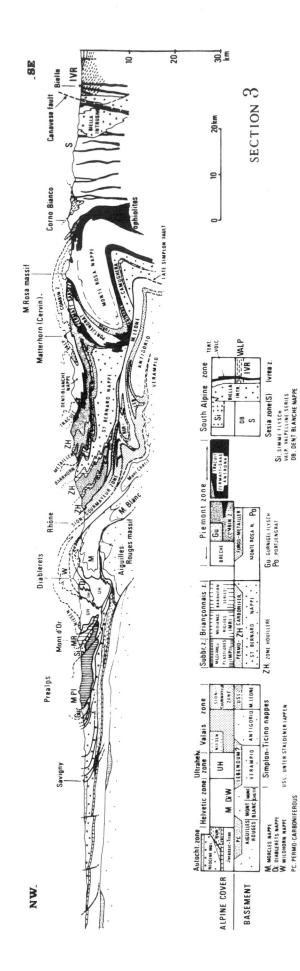

Figure 11.1 Scales of folding in ductile metamorphic rocks. *A.* Cross section of metamorphosed rocks in central region of western Alps. *B.* This fold in metamorphic gneiss (Grandjeans-Fjord, east Greenland) is typical of folding in the metamorphic cores of orogenic belts. Height of the cliff is 800 m. *C.* Fold in banded marbles in the Snake Range detachment, western Nevada.

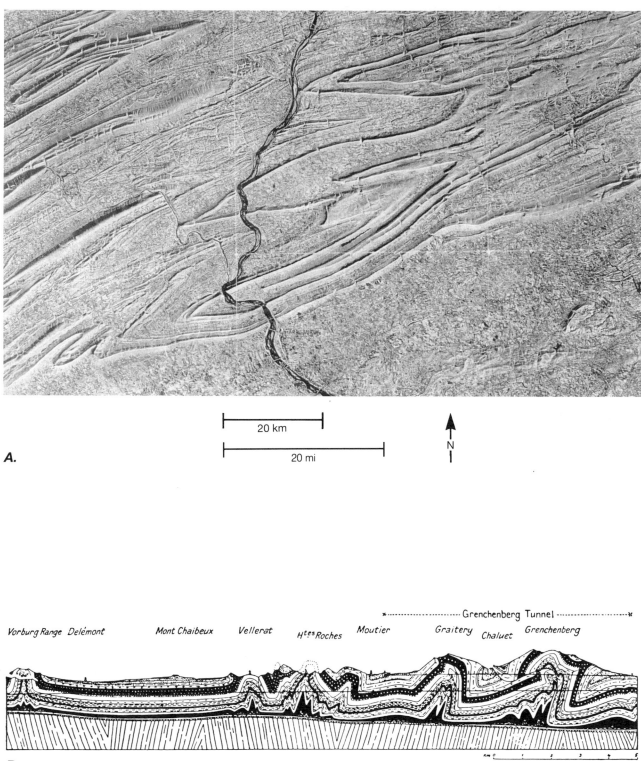

Figure 11.2 Folds in sedimentary rocks of fold and thrust belts. *A.* Folds in sedimentary rocks of Appalachians. Ridges are formed by erosion-resistant sandstones and conglomerates. Note the fold train of doubly plunging anticlines and synclines in the northwest. *B.* Cross section of the Jura Mountains north of the Alps, showing the folds in the sedimentary layers (largely limestones) and the décollement, or sole fault, below the fold. Folds are class 1B to class 1C.

the up-bowed parts of folds. Ore deposits may be concentrated in certain parts of folds, such as hinge zones, which are the most sharply curved areas, or they may be located in particular layers that have been folded.

Beyond their economic importance, however, folds provide a record of tectonic processes in the Earth. The great variety of fold shapes in rocks must reflect both the physical conditions (such as stress, temperature, and pressure) and the mechanical properties of the rock that existed when the folds developed. If we could understand the significance of fold geometry, then we would have a valuable key to understanding conditions of deformation in the Earth.

The description of folds should be free of genetic implications, because genetic terms require interpretation of the origin, which may not be well understood. Ultimately, however, we wish to associate fold geometry with the mechanism of formation so that accurate description can lead to useful interpretation. Thus the geometric description in this chapter serves as the basis for the discussion, in subsequent chapters, of the kinematics (Chapter 12, Sections 16.1 and 17.3) and the mechanics (Sections 20.2 and 20.3) of fold formation. The terminology for describing folds has evolved and accumulated over the past century or so of geologic investigation, and it is extensive and not always consistent. We introduce the most useful terms in this chapter and present an objective system of describing fold geometry in terms of elements of fold style.

11.1 Geometric Parts of Folds

The simplest part of a fold that displays the characteristic fold geometry is a single folded surface such as a bedding surface, which is the interface between two layers of rock. A folded layer can be viewed as the volume contained between two such surfaces. Most folds consist of a stack of layers folded together, and they can be described as a nested set of folded surfaces. We discuss the parts of folds by looking first at folded surfaces, then at folded layers and multilayers.

Parts of a Single Folded Surface

Figure 11.3 shows several features of folds in a single surface. A single fold is bounded on each side by an **inflection line** where the surface changes its sense of curvature—for example, from convex up to concave up. (Fold I in the figure is bounded by inflection lines i_1 and i_2.) If the fold surface is planar in the region of the inflection, then by definition we take the inflection line to be the midline of the planar segment. A **fold**

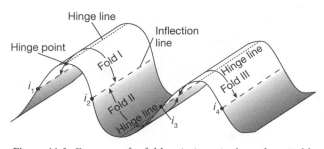

Figure 11.3 Features of a fold train in a single surface. Folds I and III are convex up; fold II is concave up. Folds I and III are unshaded; fold II and the incomplete parts of folds at either end are shaded. Inflection lines (dashed) delimit individual folds; points i_1, i_2, i_3, and i_4 are the inflection points. Dotted lines are the hinge lines of each fold.

train is a series of folds characterized by alternating senses of curvature. Folds that are convex upward (folds I and III) are **antiforms,** and folds that are concave upward (fold II) are **synforms.** A **fold system** is a set of folds of regional extent characterized by a comparable geometry and presumably a common origin.

The **curvature** of any surface is a measure of the change of orientation per unit distance along that surface. A circular arc has constant curvature, and a flat plane has no curvature. In general, the curvature measured along the folded surface from one inflection line to the next is not constant, and the **hinge line,** or more simply the **hinge,** is the line in the folded surface along which the curvature is a maximum (Figures 11.3 and 11.4A, B). A single fold may have more than one hinge (Figure 11.4B). If the maximum curvature is constant along an arc of finite length, then we take the midpoint of the arc to be the location of the hinge point (Figure 11.4C). The curvature may also vary in magnitude *along* any given hinge, and the hinge need not be a straight line (see, for instance, Figure 11.5A).

A fold with a single hinge **closes** where the limbs converge at the hinge zone (Figure 11.4A). On an outcrop pattern of such a fold, the **closure** is also sometimes called the **nose** of the fold. For a double-hinge fold, the closure is in the region of minimum curvature between the two hinges (Figure 11.4B).

The **hinge zone** is the most highly curved portion of a fold near the hinge line (Figure 11.4A); the **limbs** (sometimes called the **flanks**) are regions with lowest curvature and include the inflection lines. Technically, the hinge zone can be defined as that portion of the folded surface having a greater curvature than the reference circle that is tangent to both limbs at the inflection points of the fold (Figure 11.4A). In the unusual case of a fold with constant curvature, the areas near the hinge and those near the inflection lines are still referred to loosely as the hinge zone and limbs, respectively.

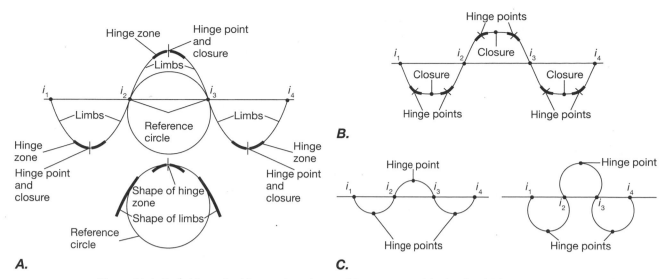

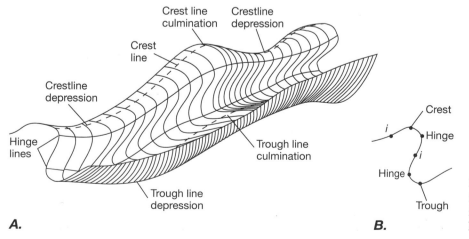

Figure 11.4 Definition of a hinge point, closure, hinge zone, and limb of a fold. *A.* The hinge points are points of maximum curvature. The closure point is the hinge point on a single-hinge fold. The *hinge zone* and *limb* are defined with reference to a circle that is tangent to both sides of the fold at two adjacent inflection points. The part of the fold that has a curvature greater than that of the reference circle is the hinge zone; the parts between the hinge zone and the inflection points that have a curvature less than that of the reference circle are the limbs. *B.* Individual folds may have two hinges. The closure point is the point of minimum curvature between the two hinges. *C.* Fold trains in which each fold has constant curvature and thus is the arc of a circle (perfect circular folds). Hinge points are the midpoints of each of the arcs.

The **crest line** and **trough line** on a fold are the lines of highest and lowest elevation, respectively, on the folded surface (Figure 11.5). These lines may, but do not necessarily, coincide with the hinge (Figure 11.5B), and they need not be straight lines (Figure 11.5A). **Culminations** and **depressions** are areas where crest or trough lines go through maximum and minimum elevations, respectively.

We generally portray the form of a fold by its **profile**, which is the trace of the folded surface on a plane normal to the hinge line (Figure 11.6A). The profile is the form of the fold seen when it is viewed looking parallel to its hinge. The curvature of most folds is greatest along their profiles. The hinge, inflection lines, and crest and trough lines appear on the profile, of course, as points.

A **cylindrical fold** is one for which a line of constant orientation, called the **fold axis**, can be moved along the folded surface without losing contact with it at any point (Figure 11.6A). Thus it is a line of fixed orientation that makes an angle of 0° with every orientation of the folded surface. Folds that do not possess this property are called **noncylindrical** folds. A **conical fold** is one whose surface is everywhere at a constant nonzero angle to a line of fixed orientation, which is also called the fold axis (see Figure 11.7A). A fold axis is thus an

Figure 11.5 Crest and trough of a fold. *A.* Three-dimensional view of fold. *B.* Cross section normal to the hinge.

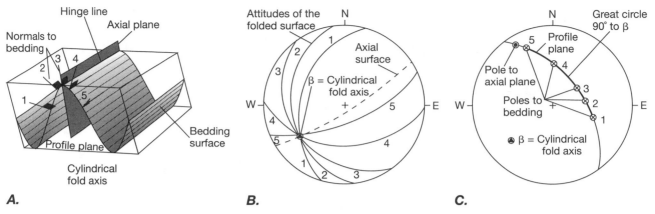

Figure 11.6 Geometry of a cylindrical fold in three dimensions and on a spherical projection. *A.* Diagram of a cylindrically folded surface, showing the fold axis, the profile plane, and the perpendiculars to the folded surface, which are parallel to the profile plane. *B.* Schmidt net (equal area) plot of several orientations of the cylindrically folded surface in part *A*, all of which intersect at the fold axis. *C.* Schmidt net (equal area) plot of the poles to the surface orientations plotted in part *B*. All the poles must lie along a great circle perpendicular to the fold axis. The great circle also defines the orientation of the profile plane.

imaginary geometric property of certain kinds of folds. Folds that are neither cylindrical nor conical in geometry do not, strictly speaking, possess a fold axis.

Although natural folds are never geometrically perfect, many have approximately cylindrical geometry, at least locally, and so can be described by the orientation of an approximate fold axis. Even irregular folds generally can be divided into local segments each of which is approximately cylindrical so the fold axis can be defined locally. The irregularity of the fold can be described by the variation, from place to place, in the orientation of the fold axis.

At the hinge of a cylindrical fold, the fold axis coincides with the hinge line, and for this reason the two terms are used interchangeably. It is useful to maintain the distinction, however, because the term *hinge line* refers to a linear feature having a specific orientation at a *specific location* on the folded surface, whereas the term *fold axis* refers to a line having only *a specific orientation* that characterizes the fold geometry, at least locally. Moreover, noncylindrical folds have a hinge but (with the exception of conical folds) no fold axis.

The geometry of a cylindrical fold and its fold axis has a particularly simple representation on a stereographic projection, which can be extremely useful in the analysis of folding in a region. A line of constant orientation plots as a point on a stereographic projection. A plane plots as a great circle. If the line lies in the plane, then on the projection, the point must lie somewhere on the great circle. The fold axis is by definition a common orientation to all attitudes of a cylindrically folded surface. Thus on a stereographic projection the point representing the attitude of the fold axis must lie on each of the great circles representing

attitudes of the folded surface. Thus these great circles must all intersect at the fold axis orientation (Figure 11.6*A, B*). Moreover, any line perpendicular to the folded surface must also be perpendicular to the fold axis. On a stereographic projection, all lines perpendicular to a reference line must lie along the great circle normal to the reference line. Thus the locus of lines plotted normal to the folded surface (called the poles to the surface) must be along the great circle normal to the fold axis (Figure 11.6*C*). This great circle also defines the orientation of the profile plane (Figure 11.6*A*).

These geometric relationships, and the fact that many folds are at least locally almost cylindrical, enable a field geologist to deduce the orientation of a fold axis from measurements of two or more different attitudes of a folded surface. Plotting these attitudes as either great circles (Figure 11.6*B*) or their poles (Figure 11.6*C*) makes it possible to determine the orientation of the fold axis; this technique is especially useful in areas where, owing to the scale of the folds or to limited exposure, the fold axis is not directly observable.

For a conical fold (Figure 11.7*A*), the great circles representing orientations of the folded surface do not intersect in a point, and they do not exhibit an easily recognized relationship to the fold axis (Figure 11.7*B*). The poles to the folded surface however must lie along a small circle at a constant angle from the fold axis (Figure 11.7*C*).

Parts of Folded Layers and Multilayers

The geometry of a folded layer or a stack of folded layers is equivalent to that of a nested set of two or more folded surfaces. A single multilayer fold is delim-

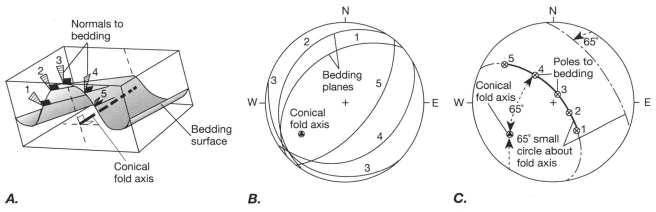

Figure 11.7 Geometry of a conical fold in three dimensions and on a spherical projection. *A.* Diagram of a conically folded surface. The fold axis is a line of constant orientation that is at a constant nonzero angle to the folded surface—in this case, 25°. The front face of the block is perpendicular to the fold axis, not the hinge line, and therefore is not the profile plane. The perpendiculars to the bedding surface (1 through 5) are in this case all 25° from the plane normal to the fold axis and 65° from the fold axis itself. *B.* Schmidt net (equal area) plot of various attitudes of a conically folded surface. *C.* Schmidt net plot of the poles to the folded surface and the fold axis shown in part A. The poles lie on a small circle around the fold axis.

ited by two **inflection surfaces** that join the inflection lines on adjacent folded surfaces in the nested stack (Figure 11.8).

The surface joining all hinge lines in a particular nested set of folds is variously called the **hinge surface,** the **axial surface,** and (if the surface is planar) the **axial plane.** In field studies, we usually recognize folds by their outcrop pattern on a topographic surface. The intersection of the axial surface with a surface of exposure is a linear feature called the **axial surface trace,** which in general is very different from both the hinge

line and the fold axis and must not be confused with either (Figure 11.9). The axial surface trace is never parallel to the hinge line unless the hinge is parallel to the surface of exposure.

If the folded layers are sedimentary beds, and if we can determine their relative ages, then we can distinguish anticlines from synclines. **Anticlines** (derived from the Greek *anti,* which means "against," and *klinein,* which means "to slope") are folds in which the older layers are on the concave side of a bedding surface and the younger layers are on the convex side. **Synclines**

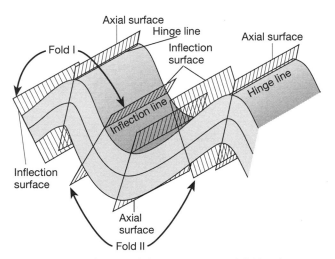

Figure 11.8 Folds in multilayers. A train of folds, showing the inflection surfaces, each of which contains the inflection lines of all the folded surfaces on one limb of a nested set of folds, and the hinge or axial surfaces, each of which contains all the hinge lines in a single nested set of folds.

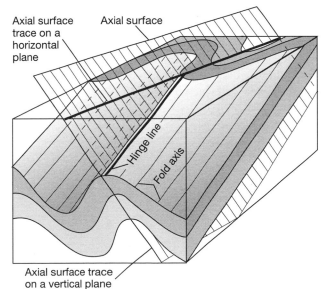

Figure 11.9 Block diagram of folds, showing the distinction among the axial surface trace on a vertical and a horizontal surface, the hinge line, and the fold axis.

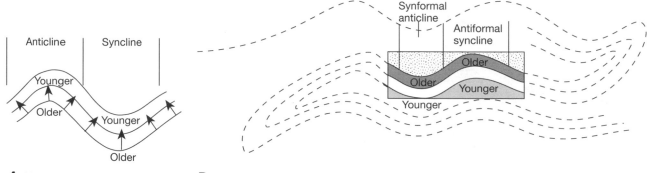

A. Upright beds   **B.** Overturned beds

Figure 11.10 Distinction between an anticline and a syncline in a fold cross section. Arrows point from oldest to youngest beds—that is, in the stratigraphic up direction. The dashed structure in part *B* suggests one way in which large sections of strata could be overturned.

(the Greek *syn*, means "with, to") are folds in which the younger layers are on the concave side of a bedding surface and the older layers are on the convex side.

Most anticlines are convex up (antiforms) and most synclines are concave up (synforms) (Figure 11.10*A*), although this geometry is not universal. In areas of complex deformation where the entire stratigraphy has been overturned, anticlines may be synformal and synclines antiformal (Figure 11.10*B*).

11.2 Fold Scale and Attitude

The Scale of Folds

Scale is a measure of the size of a fold in a layer or stack of layers. There are two components of the scale: the amplitude *A* and the wavelength λ (Figure 11.11). We define them with reference to the enveloping surfaces and the median surface. The **enveloping surfaces** are the two surfaces that bound the fold train developed in a single folded surface. The **median surface** includes all

the inflection lines of a fold train in a single surface. The **amplitude** of any fold is the distance from the median surface to either of the enveloping surfaces, measured parallel to the axial surface. The **wavelength** is the distance, measured parallel to the median surface, between one point on a fold and the geometrically similar point on a neighboring fold—from one antiformal hinge to the next, for example, or from one synformal hinge to the next.

The Attitude of Folds

The orientation in three-dimensional space of a fold or train of folds is an important factor in any geologic study of folded rocks. Accordingly, an extensive nomenclature has been based on the attitude of folds. We express the attitude of a fold by the trend and plunge of the hinge line or fold axis and the strike and dip of the axial surface. A fold is **upright** if the dip of the axial surface is close to vertical; it is **steeply, moderately,** or **gently inclined** as the dip angle progressively decreases; and it is **recumbent** if the axial surface is close to horizontal. Depending on the plunge of the hinge, a fold

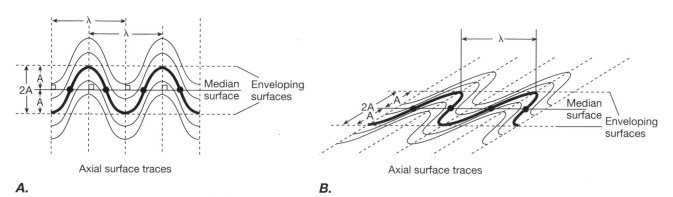

Figure 11.11 The scale of folding is defined by the wavelength λ and the amplitude *A* for (*A*) symmetric folds and (*B*) asymmetric folds.

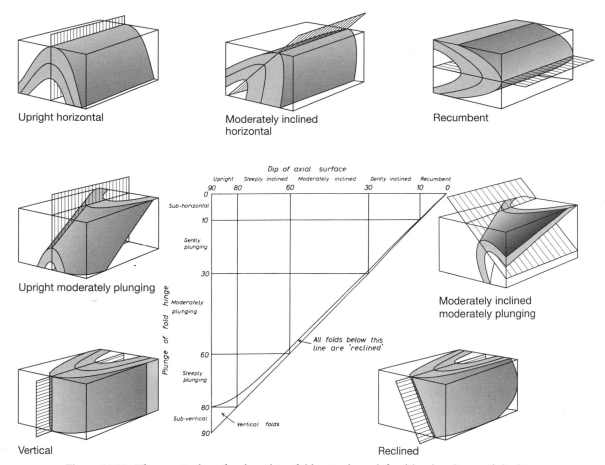

Figure 11.12 The terminology for describing fold attitude as defined by the plunge of the hinge (vertical axis) and the dip of the axial surface (horizontal axis). The center graph showing the ranges of angles associated with each term, and the surrounding diagrams of folds in varying attitudes, corresponding to the categories in the graph.

is **horizontal; subhorizontal; gently, moderately,** or **steeply plunging; subvertical;** or **vertical.** A **reclined** fold is one whose hinge plunges down the dip of the axial surface. Thus a fold could be upright horizontal; moderately inclined; moderately plunging; recumbent; and so forth. Figure 11.12 (the central triangular diagram) displays graphically the conventional definitions of these terms, and the surrounding diagrams give examples of the various categories.

No folds are indefinite in length; all eventually die out along the hinge by decreasing in amplitude or terminating against a fault. Where upright or inclined horizontal folds die out, the hinge line must plunge. If a fold hinge plunges at both ends and the hinge line is at least a few times as long as the half-wavelength, it is a **doubly plunging** fold (see folds in the northwest part of Figure 11.2*A*). As the length of the hinge becomes comparable to the half-wavelength of the fold, the fold is called a **dome** or a **basin,** depending on whether it is antiformal or synformal.

Several other common terms specify relative orientations of the limbs of folds. A **homocline** (derived from the Greek *homo,* which means "same," and *klinein,* which means "to slope") is characterized by a surface such as bedding that has a nonhorizontal attitude, uniform over a regional scale with no major fold hinges (Figure 11.13*A*). A **monocline** (the Greek *mono* means "single, only") is a fold pair characterized by two long horizontal limbs connected by a relatively short inclined limb (Figure 11.13*B*). A **structural terrace** is a fold pair with two long planar inclined limbs connected by a relatively short horizontal limb (Figure 11.13*C*). An inclined or recumbent fold in which one limb is overturned—that is, rotated more than 90° from its original horizontal position (Figure 11.13*D*)—is sometimes called an **overturned fold.** Note that the term *overturned* refers to only one limb of the fold, not to the whole fold. Thus an overturned anticline (Figure 11.13*D*) is not the same as an upside down, or synformal, anticline (Figure 11.10*C*).

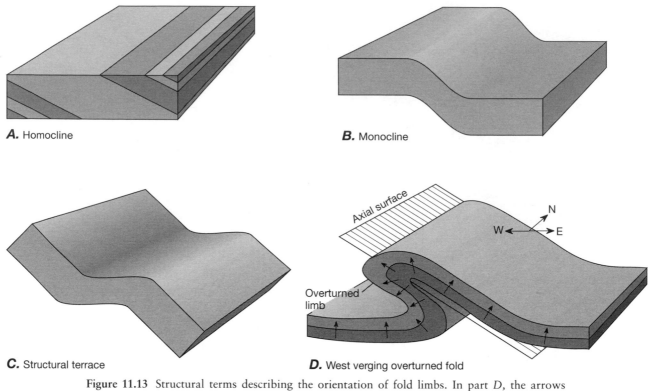

A. Homocline

B. Monocline

C. Structural terrace

D. West verging overturned fold

Figure 11.13 Structural terms describing the orientation of fold limbs. In part *D*, the arrows show the stratigraphic up direction on the sedimentary beds.

11.3 The Elements of Fold Style

The **style** of a fold is the set of characteristics that describe its form. It is analogous, for example, to the architectural style of a building. Over years of working with folds, geologists have identified certain features as particularly useful in describing folds and understanding how they develop. We refer to these features, which are summarized in Table 11.1, as the **elements of fold style**. In this section, we briefly define and discuss these elements. In Section 11.5, we apply these definitions to describe the most common fold styles that appear in deformed rocks.

Table 11.1 Elements of Fold Style

1. Cylindricity
2. Symmetry
3. Style of a folded surface
    Aspect ratio
    Tightness
    Bluntness
4. Style of a folded layer (Ramsay's classification)
    Relative curvature: dip isogon pattern
    Orthogonal thickness
    Axial trace thickness
5. Style of a folded multilayer
    Harmony
    Axial surface geometry

We must first define two angles that describe the amount that a surface has been folded (Figure 11.14). The **folding angle** ϕ is the angle between the normals to the folded surface constructed at the two inflection points of a fold. It is the angle through which one limb has been rotated relative to the other by the folding. The more commonly used **interlimb angle** ι is the angle between the tangents to the two fold limbs constructed at the inflection points. It measures the dihedral angle between the two limbs and is the supplement of the folding angle (that is, $\iota = 180 - \phi$).

Cylindricity

The degree to which a fold approximates the geometry of a cylindrical fold (Section 11.1) is a feature that characterizes different styles of folding. The cylindricity is represented qualitatively on a stereonet by how closely the poles to planes around a fold fit a great circle distribution (Figure 11.6C). The distance along the hinge for which the cylindrical geometry is maintained, measured as a proportion of the half-wavelength, is also a significant characteristic of the cylindricity. A multilayer fold can be described as cylindrical if the attitudes from all surfaces in the multilayer conform to the geometry of a cylindrical fold (Figure 11.6C). The term **cylindroidal** is sometimes used to describe a fold that closely approximates an ideal cylindrical geometry.

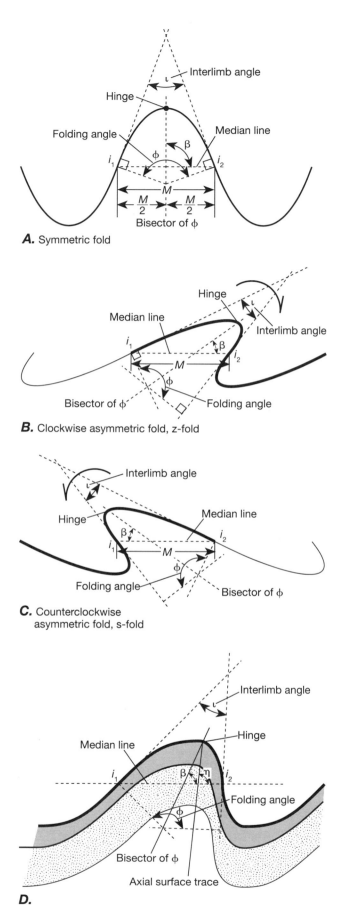

A. Symmetric fold

B. Clockwise asymmetric fold, z-fold

C. Counterclockwise asymmetric fold, s-fold

D.

Figure 11.14 (*Left*) The folding angle, the interlimb angle, and the symmetry of folds. In part *A*, the fold profile from the hinge to i_1 is the mirror image of the profile from the hinge to i_2. The bisector of the folding angle and the interlimb angle is the mirror plane, and it is the perpendicular bisector of the median line i_1i_2. β equals 90° in part *A*, but it is not equal to 90° in parts *B* and *C*. In part *D*, an asymmetric multilayer fold is characterized by the inclination β of the folding angle bisector as well as by the inclination η of the axial surface with respect to the median surface.

Symmetry

A folded surface forms a **symmetric fold** if in profile, the shape on one side of the hinge is a mirror image of the shape on the other side, and if adjacent limbs are identical in length (Figure 11.14*A* and 11.11*A*). For folded layers and multilayers, the axial plane is the mirror plane of symmetry. It is the perpendicular bisector of the median surface between the inflection points, and it bisects both the folding angle ϕ and the interlimb angle ι.

Asymmetric folds in profile have no mirror plane of symmetry, and the limbs are of unequal length (Figures 11.11*B* and 11.14*B, C*). The degree of asymmetry is determined for a folded surface by the angle of inclination β between the bisector of the folding angle ϕ (or the interlimb angle ι) and the median surface (Figure 11.14*B, C*). For a multilayer fold, the axial plane is not generally parallel to the bisector of the interlimb angle, and the angle of inclination η between the axial surface and the median surface is another independent characteristic of the asymmetry (Figure 11.14*D*).

The sense of asymmetry of a fold changes depending on whether we view the fold from one direction along the hinge or from the other. By convention, we specify the sense of asymmetry on a plunging fold when looking *down the plunge* of the hinge line. An asymmetric fold is a **clockwise fold**, or **z-fold**, if the short limb has rotated clockwise with respect to the long limbs, and the short limb with its two adjacent long limbs therefore defines a z-shape (Figure 11.14*B*). An asymmetric fold is a **counterclockwise fold**, or **s-fold**, if the short limb has rotated counterclockwise with respect to the long limbs, and the short limb with its two adjacent long limbs therefore defines an s-shape (Figure 11.14*C*).

If the fold hinge is horizontal, the geographic direction of viewing must be part of the description of the asymmetry; for example, the fold is counterclockwise (or an s-fold) looking north. For an inclined fold with a horizontal to gently plunging hinge, however, the sense of asymmetry is more conveniently specified by the **vergence** (from the German word *Vergenz,* which means "overturn"). The vergence is the direction of "leaning" of the axial surface or the up-dip direction on the axial surface of an asymmetric fold. In Figure 11.13D, for example, the vergence of the fold is to the west.

Small symmetric folds, especially if they are within the core of a larger fold, are sometimes called **m-folds**.

The Style of a Folded Surface

We describe the geometry of a folded surface by specifying three style elements: **aspect ratio, tightness,** and **bluntness.** To define these characteristics, it is first convenient to construct a quadrilateral around the fold in question such that the sides are tangent to the limbs of the fold at the inflection points, the base is the line M between the inflection points, and the top is tangent to the fold and normal to the bisector of the folding angle ϕ. For symmetric folds, the quadrilateral is a trapezoid, as shown in Figure 11.15A and B for folds with a folding angle $\phi = 130°$ and 230°, respectively.

The **aspect ratio** P is the ratio of the amplitude A of a fold, measured along the axial surface, to the distance M, measured between the adjacent inflection points that bound the fold (Figure 11.15). In other words, P is the ratio of the height of the quadrilateral to its base. For a periodic fold train, in which successive folds have the same wavelength λ (Figure 11.11), M is the half-wavelength ($\lambda/2$). Folds of increasing aspect ratio have a wide, broad, equant, short, or tall aspect, as defined in Table 11.2.

The **tightness** of folding is defined by the folding angle ϕ or the interlimb angle ι (Figure 11.15). As the degree of folding increases, the folding angle increases and the interlimb angle decreases. Folds are gentle, open, close, tight, isoclinal, fan, or involute folds, as defined in Table 11.3. Isoclinal folds, which have essentially parallel limbs, fall on the boundary between acute folds ($\phi/2 < 90°$) and obtuse folds ($\phi/2 > 90°$).

The **bluntness** b measures the relative curvature of the fold at its closure (Figure 11.15). It is defined by

$$b = \begin{cases} r_c/r_0 & \text{for } r_c \leq r_0 \\ 2 - r_0/r_c & \text{for } r_c \geq r_0 \end{cases}$$

where r_c is the radius of curvature at the fold closure, and r_0 is the radius of the circle that is tangent to the limbs at the inflection points. Folds are sharp, angular, subangular, subrounded, rounded, or blunt (Table 11.4 and Figure 11.16). A bluntness of $b = 0$ describes folds that have perfectly sharp hinges ($r_c = 0$); $b = 1$ describes perfectly circular folds, which, for both acute and obtuse folds, consist of a single circular arc; $b = 2$ describes a double-hinged fold with a flat closure ($r_c = \infty$). One can picture the folds having $b = 2$ by looking at the

right-hand end of the series of folds in the shaded trapezoids in Figure 11.16A and B and imagining the radius of closure curvature to increase indefinitely. Thus all folds must have a bluntness between 0 and 2. For double-hinged folds, a complete description must include the bluntness of the hinges in addition to the bluntness of the closure.

We show the range of fold styles defined by the aspect ratio and the bluntness for two constant folding angles, $\phi = 130°$ (Figure 11.16A) and 230° (Figure 11.16B). The folds along the horizontal line of shaded trapezoids in each diagram show the styles that can occur within a single shape of trapezoid. They are distinguished only by different values of the bluntness. The folds along any vertical line show how fold style changes for different aspect ratios at constant bluntness. The folds along the inclined line labeled **perfect folds** are

Table 11.2 Aspect Ratio

| | Aspect Ratio P | |
Descriptive Term	$P = A/M$	Log P
Wide	$0.1 \leq P < 0.25$	$-1 \leq \log P < -0.6$
Broad	$0.25 \leq P < 0.63$	$-0.6 \leq \log P < -0.2$
Equant	$0.5 \leq P \leq 2$	$-0.2 \leq \log P < 0.2$
Short	$1.58 \leq P < 4$	$0.2 \leq \log P < 0.6$
Tall	$4 \leq P < 10$	$0.6 \leq \log P < 1$

Source: After Twiss (1988).

Table 11.3 Tightness of Folding

Descriptive Term	Folding Angle ϕ, deg	Interlimb Angle ι, deg
Acute		
Gentle	$0 < \phi < 60$	$180 > \iota > 120$
Open	$60 \leq \phi < 110$	$120 \geq \iota > 70$
Close	$110 \leq \phi < 150$	$70 \geq \iota > 30$
Tight	$150 \leq \phi < 180$	$30 \geq \iota > 0$
Isoclinal	$\phi = 180$	$\iota = 0$
Obtuse		
Fan	$180 < \phi < 250$	$0 > \iota > -70$
Involute	$250 \leq \phi < 360$	$-70 \geq \iota \geq -180$

Source: Modified after Fleuty (1964).

Table 11.4 Bluntness of Folds

Descriptive Term	Bluntness
Sharp	$0.0 \leq b < 0.1$
Angular	$0.1 \leq b < 0.2$
Subangular	$0.2 \leq b < 0.4$
Subrounded	$0.4 \leq b < 0.8$
Rounded	$0.8 \leq b \leq 1$
Blunt	$1 < b \leq 2$

Source: After Twiss (1988).

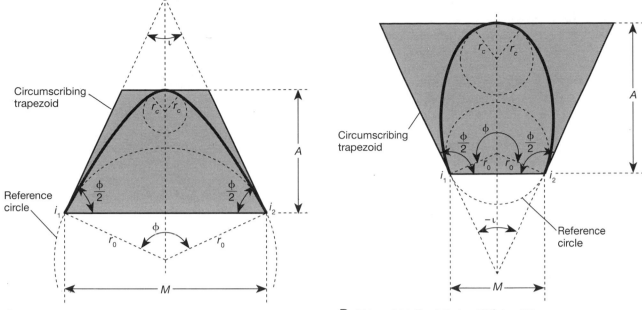

A. Acute fold: $P = 0.6$; $\phi = 130°$; $b = 0.18$

B. Obtuse fold: $P = 1.7$; $\phi = 230°$; $b = 0.7$

Figure 11.15 The style of a folded surface is characterized by the quadrilateral formed by the tangents to the limbs at the inflection points, by the line $i_1 i_2$, and by the normal to the folding-angle bisector that is tangent to the fold near the closure. The curvature of the folded surface within the quadrilateral further defines the style. The aspect ratio $P = A/M$. The bluntness b is defined in terms of the relative values of the closure radius r_c and the reference radius r_0.

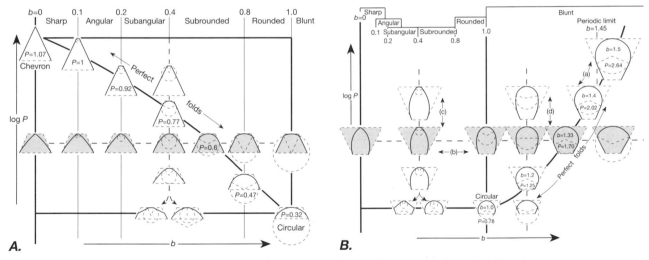
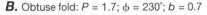

Figure 11.16 Variation of possible fold styles on planes of constant ϕ through fold style space. Axes are not to scale. Bluntness categories are indicated across the top of each diagram. The reference radius for all folds in each diagram is the radius of the perfect circular fold at the lower end of the line of perfect folds. Shaded trapezoids show bluntness categories for the shaded trapezoids in Figure 11.17. Perfect folds have perfectly straight limbs tangent to hinge zones that are perfectly circular arcs. A. Acute folds for $\phi = 130°$. Perfect folds plot along the diagonal from the perfect chevron fold in the upper left to the perfect circular fold in the lower right. All single-hinged folds plot within the shaded area. Outside this area, folds are multiple-hinged. B. Obtuse folds for $\phi = 230°$. Perfect folds plot along the diagonal curve from the perfect circular fold in the lower center through the limit for periodic folds at the vertical dashed line in the upper right. All single-hinged folds plot within the shaded area. Outside that area, folds are multiple-hinged.

idealized folds having perfectly planar limbs that are tangent to perfectly circular hinge zones. Note that increasing either the bluntness or the aspect ratio beyond the value for perfect folds (that is, outside the shaded area of the diagram) results in a fold with two hinges.

Figure 11.17, a plot of aspect ratio versus tightness, shows the various possible quadrilateral shapes that define fold style. The folds shown in the quadrilaterals are all perfect folds. The area between the lines for $b = 0$ and $b = 1$ indicates the possible range for all single-hinged folds. Perfect chevron folds, for which $b = 0$, are an upper bound limiting the geometry of all folds. Perfect circular folds, for which $b = 1$, are a lower bound for all possible single-hinged folds. Any obtuse, single-hinged, perfect fold that is part of a periodic fold train must have a closure radius less than or equal to the half-wavelength of the fold ($r_c \leq M$). These folds therefore provide an upper bound for single-hinged periodic folds.

The heavy dashed lines and the solid trapezoids show where Figure 11.16A and B, and the shaded trapezoids in those figures, projects onto this plot. The shaded trapezoids in Figure 11.17 expand in the third dimension into the row of shaded trapezoids shown in Figure 11.16A at a value of $P = 0.6$, and in Figure 11.16B at a value of $P = 1.7$.

This brief outline of the three-parameter method for classifying the style of a folded surface gives some idea of the wide variety of fold shapes that can be simply described. With a fuller investigation of the three-dimensional geometry of "fold style space," we can show, for example, that all perfect folds plot on a single surface in the space that defines the boundary between all possible single- and double-hinged folds. Note that we have restricted our discussion to symmetric folds. Asymmetric folds can also be included in this scheme, although they are considerably more complex.

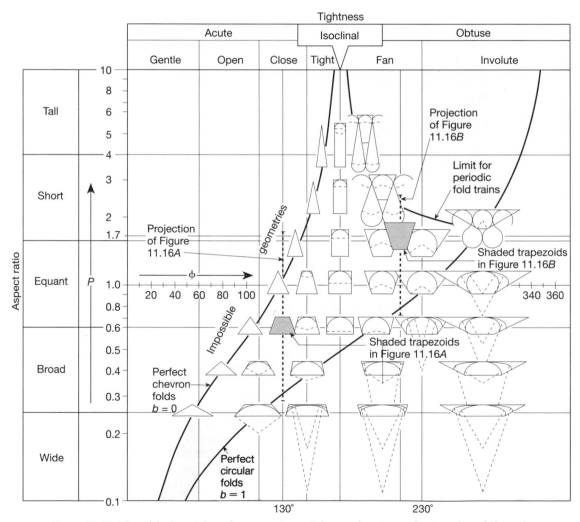

Figure 11.17 Plot of the logarithm of aspect ratio vs. tightness showing a selection of quadrilaterals and a few representative perfect and double-hinged folds. Also shown are areas of impossible geometries, areas of double-hinged folds, and lines for circular folds and periodic obtuse fold trains. Figure 11.16A, B projects onto this diagram along the heavy vertical lines.

The Style of a Folded Layer: Ramsay's Classification

The style of a folded layer is determined by comparing the fold styles of the two surfaces of the layer. The comparison is conveniently made by using three geometric parameters that are defined relative to a given pair of parallel lines tangent, respectively, to the inner (concave) and outer (convex) surfaces of the layer on the fold profile (Figure 11.18). The inclination of the fold surface at the point of tangency is given by α, the angle between the tangent line and the line normal to the axial surface trace. The three geometric parameters are as follows: (1) the **dip isogon,** which is the line across the layer connecting two points of equal dip on opposite surfaces of the layer; (2) the **orthogonal thickness** t_α, which is the perpendicular distance between the two parallel tangents; and (3) the **axial trace thickness** T_α, which is the distance between the two tangents, measured parallel to the axial surface trace. The two measures of layer thickness t_α and T_α are related by $t_\alpha = T_\alpha \cos \alpha$.

The elements of style for a folded layer are defined according to how these geometric parameters vary across the fold from the hinge to the limbs, or with increasing values of the surface inclination α.

1. The **relative curvature** or the variation in dip isogon. The relative curvature of the convex and concave surfaces is revealed by constructing a set of dip isogons at regular intervals from the hinge to the limb, each of which connects points of identical inclination α on the inner (concave) and outer (convex) surfaces (Figure 11.19). If the dip isogons converge toward the inner side of the fold, the curvature of the inner surface is greater than that of the outer surface; if they diverge toward the inner surface, the opposite is true.

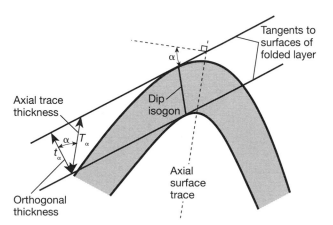

Figure 11.18 Definition of the layer inclination α, the dip isogon, the orthogonal thickness t_α, and the axial trace thickness T_α used to define the style of folded layers.

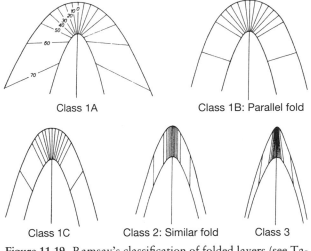

Figure 11.19 Ramsay's classification of folded layers (see Table 11.5).

For most folded layers, the relative curvature is consistent and defines three styles of folds. Dip isogons that converge toward the inner side of the fold characterize **class 1 folds** (Figure 11.19A, B, C). Parallel dip isogons, which also are parallel to the axial surface, characterize **class 2 folds** (Figure 11.19D). Folds of this style are referred to as **similar folds,** because adjacent fold surfaces ideally are identical (similar) in form. Dip isogons that diverge toward the concave side of the fold characterize **class 3 folds** (Figure 11.19E). The relative curvature is most obvious in the hinge zone, which generally makes it possible to classify folds by visual inspection.

2. Variation in the **orthogonal thickness.** The variation in the orthogonal thickness from hinge to limb is characteristic of different styles of folds and is the basis for the subdivision of class 1 folds (Figures 11.19A, B, C and 11.20A).

For **class 1A folds,** the orthogonal thickness increases from hinge to limb (Figure 11.19A). For **class 1B folds,** the orthogonal thickness is constant from hinge to limb (Figure 11.19B); These folds are referred to as **parallel folds** because t_α is constant all around the fold. **Concentric folds** are parallel folds whose inner and outer surfaces both have a bluntness of $b = 1$. Thus, they are folds defined by two circular arcs having a common center. For **class 1C folds,** the orthogonal thickness decreases from hinge to limb (Figure 11.19C). The orthogonal thickness also decreases from hinge to limb for class 2 and class 3 folds (Figure 11.20A).

3. Variation in the **axial trace thickness.** The three classes of fold style also differ in the way the axial trace thickness varies. From hinge to limb—that is, with increasing α—the axial trace thickness increases in class 1 folds, is constant in class 2 folds, and decreases in class 3 folds (Figure 11.20B).

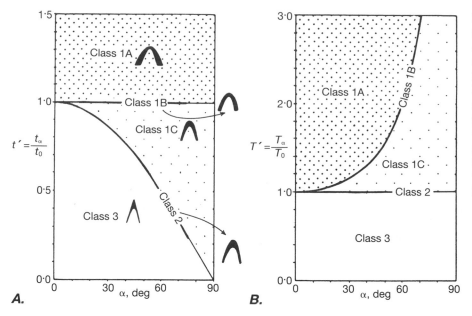

Figure 11.20 The classification of folded layers according to the thickness variation of the layer with increasing α (that is, from hinge to limb). A. Fold classes distinguished by the normalized orthogonal thickness $t' = t_\alpha/t_0$, where t_0 is the orthogonal thickness at the hinge where $\alpha = 0$. B. Fold classes distinguished by the normalized axial trace thickness $T' = T_\alpha/T_0$, where T_0 is the axial trace thickness at the hinge where $\alpha = 0$.

Table 11.5 summarizes the characteristics of the different fold classes in terms of the dip isogon geometry and the variation in orthogonal thickness and axial trace thickness. The characteristics of thickness shown in Figure 11.20 demonstrate that fold classes 1B and 2 are idealized geometries that form the boundaries between the other classes of folds. Thus some combinations of the geometries that fall in classes 1A and 1C closely approach class 1B style. Similarly, some combinations of fold geometries in classes 1C and 3 closely approach class 2 style. Not all possible folds are included in this classification, but most of the commonly observed geometries are included, and the characteristics of other styles can be presented on graphs such as in Figure 11.20.

The Style of a Folded Multilayer

A multilayer fold is composed of a stack of layers folded together. Its fold style can be defined in terms of the harmony of the folding and the axial surface geometry.

1. Harmony of Folding. In profile, all multilayer folds must die out in both directions along the axial surface trace (Figure 11.21) unless the folded sequence includes a free surface such as the Earth's surface. The depth of folding D is the distance along the axial surface trace over which the folding persists. The harmony H is a scale-independent measure of the rate at which the fold dies out along the axial surface trace and is equal to the ratio of the depth of folding D to the half-wavelength $\lambda/2$.

$$H = 2D/\lambda$$

A **harmonic** fold is continuous along its axial trace for many multiples of the half-wavelength (Figure 11.21A). A **disharmonic** fold dies out within a couple of half-wavelengths or less (Figure 11.21B).

In general, because multilayer folds die out along the axial surface trace, dip isogons must form closed contours between two adjacent hinges (Figure 11.22A). As the fold amplitude increases, reaches a maximum, and then decreases along the axial surface

Table 11.5 **Style of a Folded Layer**

Class	Dip Isogon Geometry (from convex to concave surface)	Orthogonal Thickness (from hinge to limb)	Axial Trace Thickness (from hinge to limb)
1	Convergent		Increases
1A	Convergent	Increases	Increases
1B	Convergent	Constant	Increases
1C	Convergent	Decreases	Increases
2	Parallel	Decreases	Constant
3	Divergent	Decreases	Decreases

Source: After Ramsay (1967).

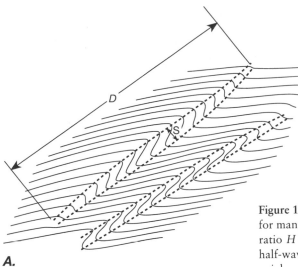

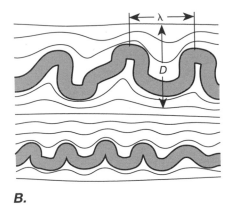

B.

A.

Figure 11.21 Harmony of folded multilayers. *A.* Harmonic folds affect layers for many times the half-wavelength along the axial surface. They have a large ratio H of the depth of folding D to half the wavelength λ: $H = 2D/\lambda$. The half-wavelength may conveniently be approximated by the spacing of adjacent axial surfaces S. This fold style is approximately a multilayer class 2 fold. *B.* Disharmonic folds die out within a distance on the order of a wavelength along the axial surface and thus have a small ratio of depth to half-wavelength. Nearby layers fold independently of one another.

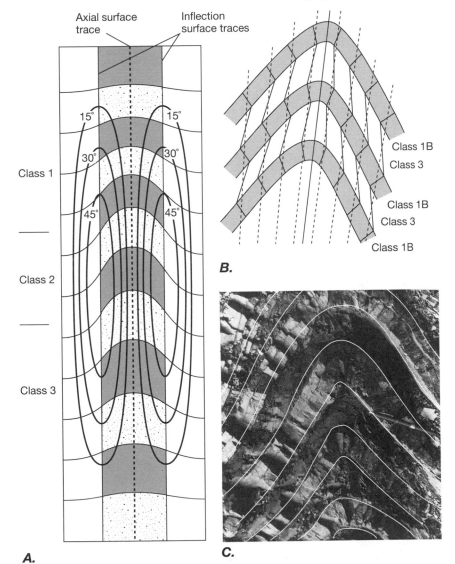

A.

B.

C.

Figure 11.22 Profiles of multilayer folds, showing dip isogon patterns in successive layers. *A.* The dip isogon pattern in a multilayer fold that dies out in both directions along the axial surface trace. In the shaded fold, isogons show regions of convergence (class 1), parallelism (class 2), and divergence (class 3). *B.* Diagram of a fold in which the dip isogons are alternately converging (class 1b folds) and diverging (class 3 folds) in successive layers. The average isogon pattern is approximately parallel to the axial surface, giving an approximately class 2 geometry. *C.* Folds in an interlayered chert–shale sequence that approximates the geometry shown in part *B* (lines drafted on photo emphasize bed contacts).

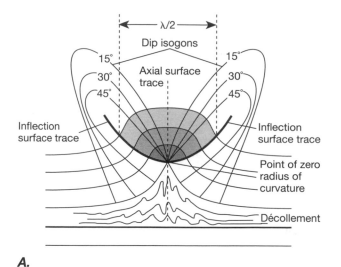

A.

B.

Figure 11.23 Disharmonic nature of class 1B folding resulting in a surface of disharmony or décollement. *A.* Diagram of a class 1B fold showing the surface of disharmony, the décollement. The half-wavelength is measured on the surface that has the maximum amplitude. Dip isogons converge strongly on the point of zero radius of curvature. *B.* A class 1B fold formed in a banded gneiss during deformation well after peak metamorphism.

trace, the dip isogons converge, are roughly parallel, and then diverge. Thus all three of Ramsay's fold styles must occur in every multilayer fold. Although the strict definitions of the fold class nomenclature are more difficult to apply to multilayer fold classification, the dip isogon pattern still reveals important characteristics of the folds.

For harmonic folds, the average convergence or divergence of the dip isogons is very small, so the folds approximate a class 2 (similar) geometry. Dip isogons constructed for each layer, however, may vary smoothly from layer to layer (Figure 11.22A) or change radically, in some cases alternating from convergent to divergent in succeeding layers (Figure 11.22B, C). In the latter cases, the harmony is determined by the trend of the dip isogons averaged over several adjacent layers.

For disharmonic folds (Figure 11.21B), the dip isogons converge or diverge very strongly along the axial surface trace. For multilayer folds that approximate class 1B (parallel) folds, for example, the radius of curvature decreases toward the concave side of the fold, and the dip isogons converge strongly (Figure 11.23A). Where the radius of curvature approaches zero, the fold must die out rapidly along the axial surface at a décollement or a sole fault (Figure 11.23). In fold and thrust belts this décollement commonly corresponds to the basal thrust fault into which thrusts converge (Figure 11.2B; see also Chapter 6).

2. Axial Surface Geometry. Throughout our discussion so far, we have assumed that the axial surface is

planar. Many folds, in fact, display parallel or subparallel axial surfaces that are planar or only slightly curved. It is not unusual, however, for folds to have a nonplanar axial surface; such folds are called **convolute folds.** In some cases the axial surface itself describes a cylindrical fold (see Figures 12.31 and 12.32), whereas in others it is more irregular. The convolution generally is the result of deformation of earlier folds by one or more subsequent generations of folding. We discuss the geometry of such superposed folding in Section 12.7.

Some folds of a single generation develop with axial surfaces that have widely disparate orientations or that split into two or more surfaces. Such folds are usually called **polyclinal folds** (derived from the Greek *poly*, which means "many" and *klinein*, which means "slope").

11.4 The Order of Folds

Folds characteristically develop simultaneously at different scales, so large folds include smaller-scale folds in their limbs and hinge zones. We generally distinguish among these different scales of related folds in terms of their **order,** the largest-scale folds being first-order folds, and successively smaller scale folds being of higher order (Figure 11.24). First-order folds are generally regional-scale features. Folds observed on the outcrop scale are commonly second- or higher-order folds. Higher-order

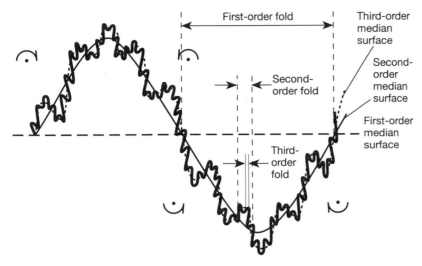

First-order fold

Second-order fold

Third-order fold

Third-order median surface

Second-order median surface

First-order median surface

Figure 11.24 Illustration of different orders of folding. A fold train showing three orders of folds. The median surface of third-order folds defines the second-order folds, and the median surface of the second-order folds defines the first-order folds. The asymmetry of the second-order folds changes across the axial surfaces of the first-order folds, and the asymmetry of the third-order folds changes across the axial surfaces of the second-order folds. Because of the different limb lengths of the second-order folds, the predominant asymmetry of the third-order folds is different on opposite sides of first-order axial surfaces.

folds are sometimes called **parasitic folds**. The median surface of a set of high-order folds defines the folds of the next lower order. Thus the median surface of a train of third-order folds defines the second-order fold train, and the median surface of second-order folds describes the first-order fold shape (Figure 11.24; see Figure 12.18).

The asymmetry of higher-order folds changes across the axial surface of the next lower-order fold, as seen in Figure 11.24, and this feature is a very convenient field mapping tool for identifying the presence and location of low-order folds. The style and attitude of higher-order folds are generally very close to those of lower-order. This correspondence, known as **Pumpelly's rule,** is also a valuable aid in deducing the geometry of large structures.[1]

11.5 Common Styles and Structural Associations of Folding

Some combinations of style elements occur together so often in deformed rocks that these fold styles have been given names. Moreover, certain styles of folds are characteristic of particular tectonic settings. In this section we describe some of the more common of these associations.

Parallel Folds

This style of fold is strictly defined as class 1B for either single or multilayer folds. In standard usage, however, the term applies to class 1A and class 1C folds whose

geometry is very close to that of class 1B (Figure 11.20). Folds of this style characterize the geometry of fold and thrust belts, which lie on the margins of orogenic belts (Figure 11.2).

Rocks of these deformed belts are mostly unmetamorphosed to lightly metamorphosed layered sediments. Generally, the folds are approximately cylindrical over distances along the hinge that are large compared with the wavelength. Hinges are horizontal to gently plunging, and in the outer regions near the foreland they tend to have upright axial surfaces, wide aspect, and gentle to open limbs. In the inner part of the belt closer to the hinterland, the aspect ratio tends to increase, limbs are tight or isoclinal (Figure 11.17), and the axial surfaces become inclined or recumbent (Figure 11.12), with vergence toward the foreland. Hinges are rounded in some cases and angular in others.

At a depth comparable with their dominant wavelength, the parallel folds of these belts die out at a sole fault, or décollement, as required for the geometry of class 1B multilayer folds (Figures 11.2*B* and 11.23). This décollement tends to rise to progressively higher stratigraphic levels toward the foreland in a series of steps or ramps that alternately parallel and cross-cut the bedding. Some of these folds, called fault-ramp folds, develop as the thrust sheet slides up these ramps (Figures 6.6 and 6.11).

The structure of fold and thrust belts in map view is exemplified by that of the Appalachian Valley and Ridge province (Figures 11.2*A* and 6.12*A*). The folds are continuous for up to tens or hundreds of kilometers. They typically die out as plunging structures, and the shortening accommodated by a fold that dies out is taken up either by adjacent folds or by thrust faults. The higher-amplitude folds are toward the interior of the range, and both the amplitude and the abundance of folds decreases toward the foreland. In the Appalachian Plateau, for example, the folding angle of the dominant folds is typically only a few degrees.

---

[1] The rule is named for Raphael Pumpelly, the geologist for the U.S. Geological Survey who first proposed this relationship, which he recognized from mapping in the metamorphic rocks of the Green Mountains, western Massachusetts, in 1894.

Similar Folds

As strictly defined, similar folds have the geometry of class 2 single and multilayer folds. In common usage, however, the term is applied to fold styles that are very close to the class 2 style but that range from class 1C to class 3 (Figure 11.20). These folds are typical of the regionally metamorphosed central core zones of orogenic belts (Figure 11.1). They vary in attitude, and many are recumbent, although upright and reclined folds are not unusual. The folds are approximately cylindrical, although the distance along the hinge for which the cylindrical geometry is consistent is highly variable. Asymmetric folds are typical. The folds tend to have large aspect ratios, close to isoclinal limbs (see Figure 11.17), and angular to subangular hinges (see Figure 11.16A). Fold axial surfaces commonly are convolute and themselves describe fold systems. An axial surface foliation is often associated with the folds.

Folds of this style that are large-scale, recumbent, and isoclinal are called **fold nappes** (Figure 11.1A and B).[2] In some cases, the overturned limbs of these folds

---

[2] The term *nappe* is a French word meaning "cover sheet" or "tablecloth" and refers to any allochthonous sheetlike body of rock that has moved on a shallowly dipping surface. A nappe may originate as a recumbent isoclinal fold or as a thrust fault.

become sheared out so that the fold is further displaced by faulting, thus becoming a **thrust nappe** (an example is the Morcles nappe shown in Figure 11.1A).

Folds in salt domes and glaciers tend also to be similar folds. In both settings, the folds are generally harmonic and tight to isoclinal, with subangular to angular hinges. Folds in salt domes are steeply reclined with their axes parallel to the margins of the structure, whereas in glaciers the folds tend to be gently plunging, recumbent features.

Other Styles of Folds

Chevron and **kink folds** are cylindrical, harmonic, multilayer class 2 folds that have angular to sharp hinges, equant aspect, and gentle to close limbs. Chevron folds are symmetric (Figure 11.25A, C) and kink folds are asymmetric (Figure 11.25B). Both fold styles commonly develop in rocks that have a strong planar mechanical anisotropy such as phyllites and schists, which are characterized by a strong preferred orientation of abundant platy minerals, and finely laminated rocks such as interbedded sandstones or cherts with shales. In the latter case, the multilayer class 2 geometry is provided by alternations between class 1 and class 3 folds in the

A.

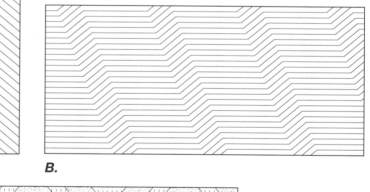

B.

C.

Sandstone
Shale

Figure 11.25 *A.* Chevron folds, *B.* kink folds, *C.* chevron folds in a sequence of alternating layers such as sandstone and shale.

Figure 11.26 Ptygmatically folded layers in a banded marble, Bishop Creek roof pendent, Sierra Nevada, California.

sandstones and shales, respectively (Figures 11.25C and 11.22C).

Ptygmatic folds (the Greek word *ptygma* means "fold") are disharmonic folds that develop in individual layers. The folds tend to be equant in aspect with close to fanning limbs, rounded to subrounded hinges, and class 1B or 1C layer geometry. They typically develop in layers, dikes, or veins in metamorphic rocks (Figure 11.26) and in sandstone layers or dikes in some sedimentary sequences.

Additional Readings

Fleuty, M. J. 1964. The description of folds. *Proc. Geol. Ass. Lond.* 75: 461–492.

Hudleston, P. J. 1973. Fold morphology and some geometrical implications of theories of fold development. *Tectonophysics* 16: 1–46.

Ramsay, J. G. 1967. *Folding and fracturing of rocks.* New York: McGraw-Hill.

Twiss, R. J. 1988. Description and classification of folds in single surfaces. *J. Struct. Geol.* 10(6): 607–623.

Whitten, E. H. Timothy. 1966. *Folded rocks.* Chicago: Rand McNally.

CHAPTER

12

Kinematic Models of Folding

In Chapter 11 we described the geometric characteristics displayed by folds in naturally deformed rocks. Geologists would like to understand the significance of these geometric features in terms of the mechanism of folding. To this end, they propose models to account for how folds might develop and then compare the characteristics of the model folds with natural folds. In this chapter, therefore, we discuss kinematic models of folding. We first consider various two-dimensional models for folding single layers (Sections 12.1 through 12.4), by which we can account for much of the geometric variation included in Ramsay's classification of folded layers. We then discuss two-dimensional models of multilayer folding (Sections 12.5 through 12.7), including kink and chevron folding, and fault-bend and fault-propagation folding. Finally, we discuss models for some three-dimensional aspects of folding, including the relationship between "drag folds" and the slip direction on associated faults, the geometry of superposed folds, and diapiric flow (Sections 12.8 through 12.10). Models such as these specify the motion of the deforming body but not, in general, the cause of the motion. For complete mechanical models of folding we must understand how stress and deformation are related, and we discuss these models in Chapter 20.

In order to discuss kinematic models of folding, we must introduce some basic concepts of deformation,

and we restrict ourselves here to deformation that occurs in only two dimensions. After a **homogeneous deformation,** straight and parallel lines remain straight and parallel (Figure 12.1A, B; 12.2A, B), whereas after an **inhomogeneous deformation,** straight and parallel lines become curved and nonparallel (Figures 12.1A, C; 12.2A, C).

Simple shear is a two-dimensional constant-volume (in two dimensions, constant cross-sectional area) deformation that resembles the sliding of cards in a deck (Figure 12.1). If the deformation is homogeneous, the rectangular shape of the deck changes into a parallelogram (Figure 12.1B); if the deformation is inhomogeneous, the two sides of the deck normal to the cards become curved (Figure 12.1C). If a layer of rock is parallel to the shear planes, it is sheared, but it is not rotated by the deformation, and its length and thickness remain unchanged. Layers that are cross-cut by the shear planes are rotated, and they may be shortened and thickened, or lengthened and thinned, depending on their initial orientation.

Flattening is a deformation that can be represented by taking a square and shortening it parallel to one side while lengthening it parallel to the perpendicular side (Figure 12.2). If, in two dimensions, the deformation is homogeneous and the area stays constant, homogeneous flattening is called **pure shear.** After homogeneous flat-

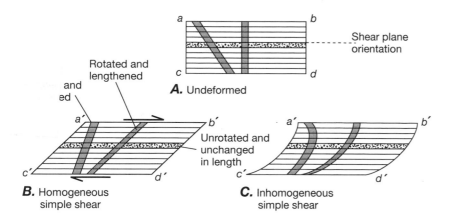

Figure 12.1 Geometry of simple shear. Rectangle *abcd* becomes deformed into *a'b'c'd'*. All displacements are parallel to the shear plane. Shaded bands represent different possible orientations of a layer relative to the shear planes.

tening, layers parallel to sides *ab* and *cd* are all unrotated and are shortened and thickened; layers parallel to *ac* and *bd* are also unrotated but are lengthened and thinned. Layers of any other orientation are rotated and may be shortened and thickened, or lengthened and thinned, depending on their initial orientation. In Chapter 15 we examine the geometry of deformation and strain in more detail.

The mechanical properties of the rocks involved in folding have a profound effect on the style of fold

that develops. Qualitatively, we describe the relative rate at which a ductile material is able to flow at a particular differential stress in terms of its **competence**.[1] Under the same differential stress, a **competent** material deforms ductilely at a relatively low rate compared with an **incompetent** material. If similar-sized layers of competent and incompetent rock are forced to deform at a given rate, the differential stress is higher in the competent material than in the incompetent material. We discuss these mechanical properties more thoroughly in Chapters 18 through 20.

12.1 Flexural Folding of a Layer

Class 1B folds are a common feature of many fold belts (Section 11.5). The geometry of this class of folds may be explained by **orthogonal flexure, flexural shear,** and **volume-loss flexure.** Collectively, these models are called **flexural folding.** In all three models, the orthogonal thickness of the layer remains constant during folding, thereby producing class 1B folds. The class of the fold, therefore, cannot be used to distinguish the different mechanisms. The fold mechanisms differ, however, in whether the convex side of a fold is lengthened or remains constant and in whether its concave side is shortened or remains constant. Because the volume-loss mechanism can produce several geometries of fold, we consider it in a separate section (Section 12.3). Here we discuss orthogonal flexure and flexural shear.

Flexural folding of layers of rock can result from bending or buckling, which are two different ways of applying forces to the layers. **Bending** of a layer results from the application of pairs of forces that produce

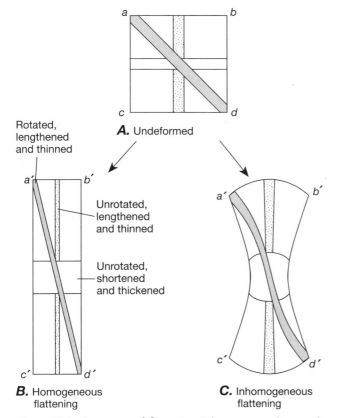

Figure 12.2 Geometry of flattening. The points *a, b, c,* and *d* become *a', b', c',* and *d'* after deformation. Shaded bands parallel to *ac, ab,* and *ad* represent layers in different orientations with respect to the direction of flattening.

---

[1] Some authors have used the term *ductility* in this sense. We eschew this usage because of the common engineering definition, also used by some experimental geologists, in which the ductility is the amount of ductile strain a material can accumulate before it fractures.

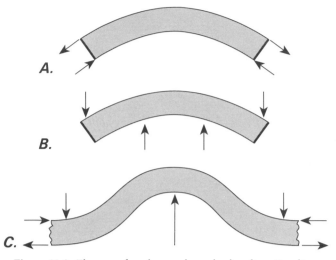

Figure 12.3 Flexure of a plate or layer by bending. Bending is caused by application of pairs of torques. A torque is a force that is applied normal to a lever arm and tends to make a body rotate. *A.* Bending moments created by pairs of equal and opposite forces applied parallel to the plate. *B.* Four-point loading: torques are created by forces applied perpendicular to the plate. *C.* One possible distribution of forces required to bend an infinite layer into a single localized fold.

equal and opposite torques that bend the layer into a fold. In pure bending, there is no net tension or compression, averaged over the layer, either parallel or perpendicular to it. Three possible systems of applied forces that provide such torques are shown in Figure 12.3.

Flexural folds can form by bending where a vertical force acts from below a layer to lift it into a fold. For example, the beds above a lenslike magmatic intrusion (called a laccolith) may fold in this manner. The fluid pressure of the magma provides a uniformly distributed upward pressure along part of the base of the layer. Monoclines or drape folds may also develop by bending (Figure 12.4) where faulting in the basement rocks provides the vertical force that bends the overlying strata into a monoclinal fold.

Buckling results from the application of compressive stresses parallel to the layer (Figure 12.5*A, B*). If the compressive stress is sufficiently large, the layer becomes unstable and buckles into a fold, either under compressive stresses alone (Figure 12.5*C*), or in association with additional torques (Figure 12.5*D*).

Buckling may be important in fold and thrust belts in which the compressive stress that drives the thrusting causes the layers to buckle, thereby shortening and thickening the thrust sheet. Folds in such belts, however, can also result from the sliding of thrust sheets up thrust ramps (Sections 6.2 through 6.4) and thus may form by a combination of bending and buckling (Section 12.7).

Buckling also is of prime importance in the formation of ptygmatic folds.

A layer may respond to either bending or buckling loads by **orthogonal flexure** (Figure 12.6). In this kinematic process, all lines that were perpendicular to the layer before folding remain perpendicular to the layer after folding. In the profile plane, the surface of the layer on the convex side of a fold is stretched, and the surface on the concave side is shortened. The surface within the layer that does not change length during the folding is called the **neutral surface.** The orthogonal thickness of the layer remains constant all around the fold.

Orthogonal flexure should be characteristic of folds with low curvature developed in competent layers that are resistant to ductile deformation. As the curvature increases to high values, the orthogonality condition cannot be maintained.

A layer can also respond to bending or buckling by **flexural shear,** which is also called **flexural flow.** Folding is accommodated by simple shear parallel to the layer, and there is no stretching and shortening,

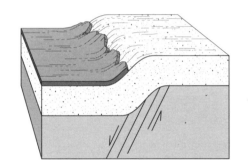

A.

B.

Figure 12.4 Monoclinal folds developed by bending. *A.* Diagram showing the formation of a monocline in sediments overlying a normal fault in basement rocks. *B.* Photograph of the Rattlesnake Mountain monocline, Wyoming.

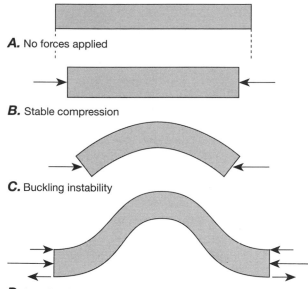

A. No forces applied

B. Stable compression

C. Buckling instability

D. Localized buckling instability

Figure 12.5 Flexure of a plate or layer by buckling. Equal and opposite forces are applied to opposite ends of the plate causing a compression of the plate.

respectively, of the convex and concave sides of the fold, as there is in orthogonal flexure.

Flexural-shear folding is analogous to the bending of a deck of cards (Figure 12.7) in that all the motion is parallel to the shear planes (represented by the cards) and the material on the convex side of a shear plane shears toward the fold hinge relative to that on the concave side. The sense of shear on the limbs of a fold therefore changes across the fold axial surface, and the magnitude of the shear decreases toward the hinge. The thickness of the body measured perpendicular to the shear planes is constant. Lines that were perpendicular

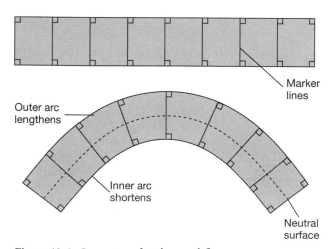

Figure 12.6 Geometry of orthogonal flexure.

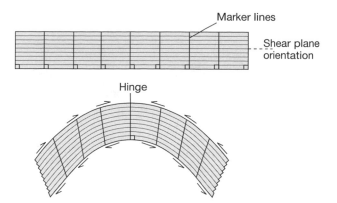

Figure 12.7 Geometry of flexural-shear folding.

to the surface of the layer before folding, however, do not remain perpendicular, except exactly at the hinge. During folding of a deck of cards, the length of individual cards is constant. Similarly, in flexural-shear folding, any length measured in the profile plane parallel to the shear planes is constant, so neither the convex nor the concave surface of the layer changes length.

Flexural-shear folding may occur instead of orthogonal flexure if the layer is less competent and therefore able to undergo ductile deformation more readily, or if the layer has a strong planar mechanical anisotropy,[2] such as fine interbedding of chert and shale or a strongly developed schistosity parallel to the layer.

12.2 Passive-Shear Folding of a Layer

In **passive-shear folding,** which is also called **passive-flow folding** or simply **flow folding,** the layer is highly incompetent and exerts no influence on the process of folding; it simply acts as a marker that records the deformation. Deformation takes place by inhomogeneous simple shear on shear planes that cross-cut the layer, and the amount and sense of shear vary systematically across the shear planes to produce the folded geometry. This process results in class 2, or similar, folds.

To illustrate the kinematics of the folding process, we can again refer to the model of the shearing of a deck of cards (Figure 12.1). In this case, however, the shear planes represented by the cards are not parallel to the layer being folded, as they are in flexural-shear folding, but instead cross-cut the layer (Figure 12.8). Along a given axial surface, the folded shape—and therefore the curvature—of the convex side of the folded

---

[2] The mechanical properties of a mechanically anisotropic material are different in different directions in the material.

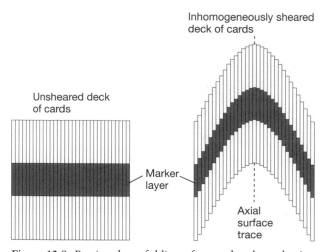

Figure 12.8 Passive-shear folding of a marker layer by inhomogeneous simple shear is approximated by "deck-of-cards" shear. The axial surface of the fold is parallel to the shear planes. The thickness of the layer parallel to the shear planes is constant. The shape of the fold is exactly the same on the convex and the concave side of the layer.

layer is exactly the same as that of the concave side. Thus the hinge lines of folded surfaces along the same axial surface must also lie on the same shear plane, and the shear planes are therefore parallel to the axial surface. Because there is no deformation within any given shear plane (none of the cards in the deck changes size or shape), the fold is cylindrical, and the axial trace thickness of the layer, which is measured parallel to the shear planes, is constant around the fold. These geometric characteristics are exactly those of class 2, or similar, folds.

In passive-shear folds, the fold hinge and the fold axis must parallel the intersection of the shear planes

with the original layer orientation (Figure 12.9). The shear planes can be oriented at any nonzero angle to the layer, and the shear direction within the shear planes can be in any orientation except parallel to the layer being folded. As long as there is a component of shear across the layer, a fold can form. Thus the fold axis or hinge is not related to the direction of shear.

Natural fold geometries that come close to the geometry of class 2 folds are characteristic of deformation in high-grade metamorphic rocks, in salt domes, and in glaciers (Figure 11.1; see Sections 11.5 and 12.10) which suggests that this class of folds characterizes the deformation of incompetent materials. The model of passive-shear folding certainly requires incompetent behavior, but as we show in the next two sections, it is not the only mechanism that produces folds having a geometry very close to that of class 2.

12.3 Volume-Loss Folding of a Layer

Volume-loss folding is a mechanism by which folds can form or be amplified by the gradual removal of material from particular zones in a folded layer. The loss generally results from solution, so the folding process is also called **solution folding**. The volume-loss mechanism, however, does not result in a unique class of folds, because the fold geometry depends on the orientation of the zones of volume loss relative to the layer. Folds may form with class 1B, class 1C, or class 2 geometry. Volume loss from discrete zones may result in the offset of beds, giving an appearance of shearing along the zones although in fact no shearing at all is required.

Three ideal fold geometries can result from volume-loss folding (Figure 12.10). Removing wedges of ma-

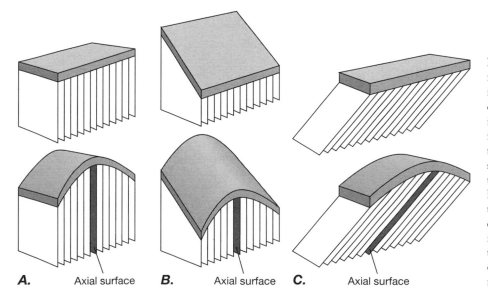

A. Axial surface **B.** Axial surface **C.** Axial surface

Figure 12.9 The orientation of the fold hinge for a passive shear fold is determined by the intersection of the shear planes with the original orientation of the layer to be folded. In parts A through C, the top diagram shows the relationship between the shear planes and the original orientation of the layer. The bottom diagram shows the layer after folding. The shear directions could be any orientation in the shear plane except parallel to the surface being folded. The orientation of the fold hinge does not indicate the direction of shear.

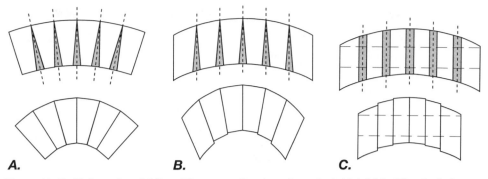

Figure 12.10 Volume-loss folding. The upper diagrams show the initial folds. The shaded areas are removed during intensification of the folds. *A.* Wedge-shaped areas of volume loss are symmetric about lines (dashed) that are normal to the surfaces of the layer. Both the convex and concave surfaces of the resulting fold are continuous and smooth. *B.* Wedge-shaped areas of volume loss are symmetric about lines (dashed) that are not normal to the surfaces of the layer. The convex surface of the resulting fold is smooth, but the concave surface has offsets that suggest shearing of the layer along the surface of volume loss. *C.* Lath-shaped areas of volume loss are parallel to one another and in general oblique to the layer. Both the convex and the concave surface of the fold show offsets that suggest shearing comparable to passive-shear folding.

terial symmetric about a line normal to the layer surface results in a class 1B fold in which both the concave and the convex surfaces are smooth (Figure 12.10*A*). Removing wedges of material symmetric about a line oblique to the layer produces a fold with the approximate geometry of a class 1C fold, but the concave surface of the fold is not smooth, and the discontinuous offsets along zones of volume loss could be misinterpreted as evidence of shearing (Figure 12.10*B*). For both models, the length of the convex side of the fold is unchanged by the loss of material, but the concave side of the fold is shortened.

Volume loss from parallel zones of constant thickness oriented oblique to an initial irregularity or gentle fold in the bedding can amplify a preexisting fold or irregularity, although it cannot produce a fold from a flat layer (Figure 12.10*C*). To this extent, it is geometrically comparable to deformation by homogeneous flattening, a process we discuss further in the next section. The result of this geometry of volume loss is a fold that approaches a class 2 style. The discontinuous offsets in both the convex and the concave surfaces suggest a fold formed by shearing on discrete shear surfaces, but no shearing is required.

Figure 12.11 provides an example of a fold that has been amplified by solution of material with a geometry comparable to that shown in Figure 12.10*B*. In the two photographs that make up Figure 12.11*A*, the bedding and the solution surfaces at a high angle to the bedding are visible, especially near the hinge zone of the fold. Figure 12.11*B* shows the fold restored to a more open configuration: The deformation caused by solution is undone, and the geometry of the concave

side of the fold is returned to a smooth surface. The empty wedge-shaped gaps in the photo illustrate the volume of material removed along major solution surfaces.

12.4 Homogeneous Flattening of Folds in a Layer

With the models of folding considered so far, we have succeeded in producing class 1B, class 1C, and class 2 folds. Other kinematic models of deformation can also account for some of these classes, as well as for the other classes of folds in Ramsay's system. Flexural folding can accommodate only a limited amount of shortening before the folds are so tight that they cannot take up any further shortening. The model for passive-shear folding, moreover, does not permit any shortening whatsoever normal to the shear planes. We consider here the effects of homogeneous flattening (Figure 12.2*B*) superimposed on folds formed by the mechanisms discussed above.

It is impossible to create a fold in a perfectly flat layer by a homogeneous flattening, because in any homogeneous deformation, such a layer remains planar with parallel surfaces. Homogeneous flattening can, however, amplify an initial irregularity and change the geometry of a fold. An initial class 1B fold in a layer is contained within the square *abcd* (Figure 12.12). During homogeneous flattening normal to the axial surface, any part of the layer not exactly parallel to the direction of shortening is rotated away from that direction. In the

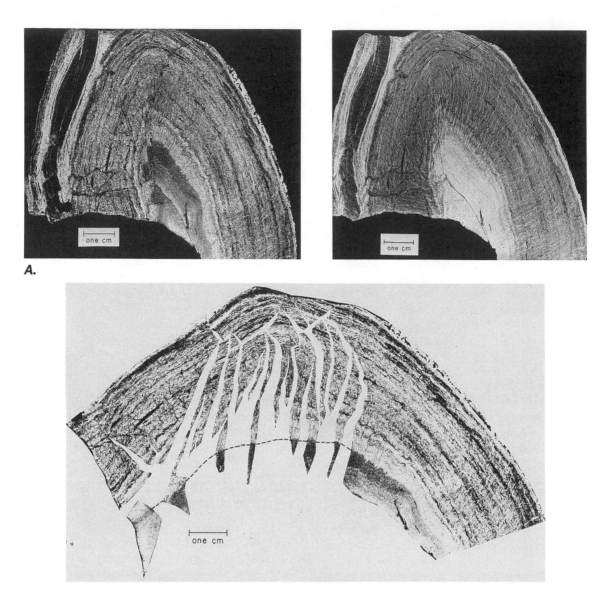

A.

Figure 12.11 Partial unfolding of a fold tightened by volume-loss (solution) folding. *A.* Negative photos of acetate peels emphasizing the layering (left) and the foliation (right). *B.* The fold shown in part *A* restored to the condition of having a smooth surface on the concave side by opening the fold along major solution seams. The volume of material lost is indicated by the blank areas in the photo, along which the fold has been opened.

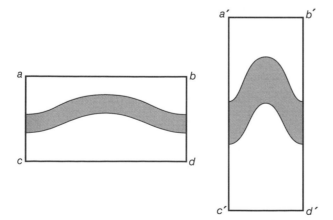

Figure 12.12 Intensification of folding by homogeneous flattening normal to the axial surface. The initial fold is a gentle class 1B fold. Progressive homogeneous flattening increases the amplitude and changes the fold into a class 1C fold.

fold hinges, where the layer is parallel to the shortening direction, the layer thickens in proportion to the change in length of the vertical sides ac and bd (Figure 12.12; see also Figure 12.2B). In the limbs of the fold, where the layer rotates toward a high angle from the shortening direction, its thickness decreases in a manner analogous to the decrease in length of the horizontal sides ab and cd. Thus after deformation, the layer is no longer of constant orthogonal thickness but is thicker in the hinge zone than in the limbs. The dip isogons still converge, so the curvature on the concave side of the fold is still greater than on the convex side. The resulting fold has a class 1C geometry.

Consider now an initial fold of class 2 (Figure 12.13). If the axial surface trace is initially parallel to the vertical sides ac and bd, it remains parallel to these sides throughout the deformation. Although the axial trace thickness T changes during the deformation, the change is the same everywhere and is proportional to the change in length of sides ac and bd. Thus the initial class 2 fold remains a class 2 fold under homogeneous flattening.

If a fold is subjected to a homogeneous flattening in a direction parallel to its axial surface and perpendicular to its hinge (Figure 12.14A), the layer thickness decreases in the hinge area and increases on the limbs (Figure 12.14B). Dip isogons still converge, but the orthogonal thickness increases from hinge to limb. The resulting geometry is that of a class 1A fold.

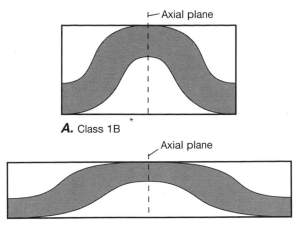

A. Class 1B

B. Class 1A

Figure 12.14 Homogeneous flattening of a class 1B fold parallel to the axial surface transforms the fold into a class 1A fold.

12.5 Flexural-Shear and Passive-Shear Folding of Multilayers

Most natural folding involves multilayered sequences of rocks that develop a more complex folding geometry than single layers. An important cause of this complexity is the difference in mechanical properties that can exist between adjacent rock layers. We take account of this factor in our models of fold formation by considering the mean competence for the whole multilayer and the contrast in competence among individual layers (Figure 12.15). First we consider a simple fold model that involves many layers of essentially the same high competence (high mean competence, low competence contrast). Then we consider the effect of alternating thin incompetent and thick competent layers (high mean competence, high competence contrast). And finally, we

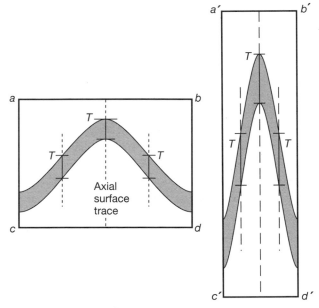

Figure 12.13 Intensification of a class 2 fold by homogeneous flattening normal to the axial surface. The initial square $abcd$ is deformed into the rectangle $a'b'c'd'$. The axial trace thickness T is changed by the deformation to T' but remains equal all around the fold. The fold remains class 2.

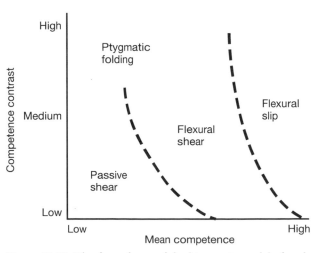

Figure 12.15 The dependence of the kinematic model of multilayer folding on the mean competence of the multilayer and on the contrast in competence between adjacent layers.

consider the effect of increasing the ratio of incompetent to competent material in the multilayer (decreasing mean competence, high competence contrast).

If the competence contrast is zero, then the multilayer behaves as a single layer according to the models discussed in Sections 12.1 through 12.4. Even for sequences of layers of different competence, however, the package of layers may behave like a single unit with an effective thickness greater than any one of the individual layers. In that case, if a neutral surface develops within the package, layers on the convex side of the neutral surface may be stretched and thinned at the hinges, giving them a class 1A geometry.

A stack of layers can respond to either bending or buckling by **flexural-slip folding** if the layers have essentially the same high competence (high mean competence) and if the friction between the layers is relatively low, allowing them to slide freely (this creates what is in effect a high competence contrast between the layer surfaces and the layer interiors) (Figures 12.15 and 12.16). If each layer folds by orthogonal flexure, the concave side of each layer is shortened and the convex side is stretched. Thus, across a bedding surface, the layer on the convex side must slip toward the fold hinge relative to the layer on the concave side (Figure 12.16). This relative slip between layers is greatest on the limbs and decreases to zero at the hinge line, where it changes shear sense. The geometry of deformation is similar to flexural-shear folding (Figure 12.7), except that in flexural-shear folding, the shear is distributed uniformly across the folding layers, whereas in flexure-slip folding it is concentrated along the interfaces between layers. This type of folding produces a class 1B multilayer fold.

Sliding of the layers past one another commonly results in the development of linear striations or mineral fibers (slickenside lineations or slickenlines) perpendicular to the fold axis on the bedding surfaces. The lineations are best developed on the limbs where the slip is a maximum, and they do not develop at all at the hinge (Figure 12.16A). The lineations (such as ℓ', ℓ'', and ℓ''' labeled on the fold in Figure 12.16A) plot on a stereonet along a great circle perpendicular to the fold axis f (Figure 12.16B).

If some degree of flexural shear occurs during folding (moderate mean competence, moderate competence contrast), then some of the potential slip between layers can be taken up by shear within the layer, and the amount of interlayer slip decreases. If all the slip is distributed within the layers (moderate mean competence, zero competence contrast), the result is simply a multilayer flexural shear fold with class 1B geometry.

Many folds, however, consist of interlayered competent and incompetent lithologies of comparable thicknesses (moderate mean competence, high competence

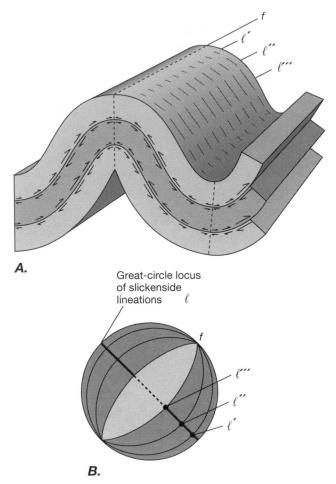

A.

Great-circle locus
of slickenside
lineations ℓ

B.

Figure 12.16 Flexural-slip folding in a multilayer. *A*. Fold formed from an originally planar multilayer, showing relative displacement on layer surfaces. Layers on the convex side of a surface slip toward the hinge line relative to those on the concave side. The shear sense reverses across the hinge line. The lines on the surface of the layer indicate the orientation of slickenside lineations, and their lengths indicate relative amounts of slip. *B*. A stereonet diagram showing the range of orientations of the folded surfaces (shaded region) and the orientations of the lineations in those surfaces. The lineations lie on a great circle normal to the fold hinge f. The labeled lineations correspond to those shown in part *A*.

contrast). The competent layers as a group deform by flexural-slip folding, and the interlayer slip is taken up by deformation in the incompetent layers (Figure 12.17A). A multilayer class 1C fold develops, as indicated by the fact that on the average, the dip isogons converge toward the concave side of the fold, but not as strongly as for a class 1B fold. On the limbs of the fold, the incompetent layers are strongly sheared, whereas in the hinge zone they are simply flattened.

As the thickness of the incompetent layers increases relative to that of the competent layers (decreasing mean competence), the requirement that the adjacent competent layers nest tightly against one another becomes

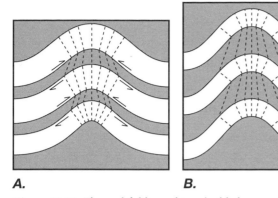

A.   **B.**

Figure 12.17 Flexural folding of interbedded competent and incompetent layers. *A.* A multilayer comprising three competent layers (unshaded) separated by thin incompetent layers (shaded). The dashed lines are dip isogons. Flexural folding of the competent layers is accommodated in the incompetent layer by shearing on the limbs and by flattening in the hinge of the fold. *B.* Flexural folding of a multilayer in which the incompetent layer is comparable in thickness to the competent layers. The multilayer class 2 fold comprises class 1B folds in the competent layers alternating with class 3 folds in the incompetent layers. Dashed lines are dip isogons.

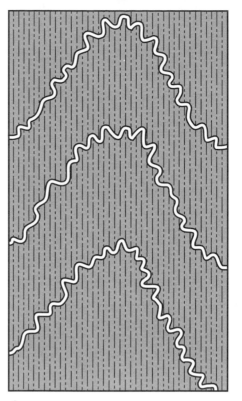

A.

less stringent. Competent layers still fold by the flexural folding mechanisms discussed for single layers, and they still dominate the development of the fold. The curvatures of the adjacent surfaces of two competent layers need not be the same, however, because the incompetent layer between the two competent layers flows in whatever manner is required to accommodate the difference in geometry; this generally involves layer-parallel shearing on the limbs and flattening in the hinge zone. Thus on the average, the dip isogons are not strongly convergent and could be parallel or even divergent. In an incompetent layer, the fold has a smaller radius of curvature on its convex side than on its concave side, the dip isogons diverge, and the axial trace thickness T_α decreases from hinge to limb (Figure 12.17B). These features characterize single-layer class 3 folds.

Thus a layered sequence can form multilayer class 1C, class 2, or class 3 folds by alternate development of class 1B folds in the competent layers and class 3 folds in the incompetent layers. The pattern of the dip isogons averaged over a number of layers may be convergent, parallel, divergent, or irregular, and this pattern defines the actual style of the multilayer fold.

If the incompetent layers are much thicker than the competent layers (low mean competence, high competence contrast), they dominate the large-scale deformation. The spacing between the competent layers is so large that flexural folding of one competent layer does not affect the next one, and disharmonic ptygmatic folds develop (Figures 12.15 and 12.18). Although high-

B.

Figure 12.18 Folding of a multilayer in which the incompetent layers are much thicker than the competent layers. Fold geometry is dominated by flow in the incompetent layers. *A.* A diagram of ptygmatic folds in thin competent layers in a fold whose geometry is dominated by flow of the incompetent material. *B.* Photograph of black amphibolitic layers ptygmatically folded in a metasedimentary rock from the Matterhorn Peak roof pendant, Sierra Nevada. The geometry is the same as in part *A.*

order folds in the individual competent layers are classes 1B and 1C, the geometry of the lower-order multilayer folds is close to class 2 and is dominated by ductile flow of the incompetent layers.

If the entire multilayer is made up of incompetent material with negligible difference in competence from layer to layer (low mean competence, low competence contrast), and if the layers do not slip past one another on their interfaces, then the multilayer is mechanically homogeneous and the layers simply act as passive markers of the deformation. Under these circumstances, the material should deform by passive shear with homogeneous flattening, rather than by bending or buckling, thereby forming folds that approximate class 2 style.

Thus flexural folding in multilayers requires competent layers and a planar mechanical anisotropy such as is provided by low-friction interfaces or thin incompetent interlayers (high mean competence, high competence contrast). Passive-shear folding in multilayers requires that an incompetent material dominate the mechanical behavior, and the effect of any competent layers is negligible (low mean competence, high to low competence contrast).

12.6 Formation of Kink and Chevron Folds

Folds with straight limbs and sharp hinges are **chevron folds** if they are symmetric and **kink folds** if they are asymmetric (see Section 11.5). They develop in strongly layered or laminated sequences that have a strong planar mechanical anisotropy, and they accommodate a component of shortening parallel to the layering or laminations.

Kink Folds

Kink folds occur in pairs with one short limb connecting two longer limbs (Figure 12.19). A **kink band** is the short limb between the two axial surfaces, which are the **kink band boundaries.** In the kink band, laminations are deformed and are rotated with respect to the undeformed material by an angle κ called the kink angle. We describe four different kinematic models of kink band formation, each of which involves a component of shearing parallel to the laminations as well as preservation of continuity of the laminations across the kink band boundaries. The models differ from one another in the way the kink grows and in the geometry of the deformation.

In two models (Figure 12.20), the kink develops by migration of the kink band boundary into the undeformed material. Folding by the migration of axial sur-

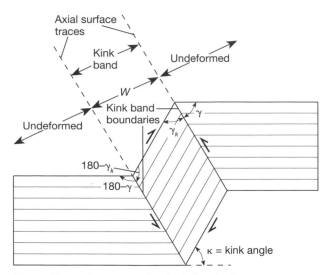

Figure 12.19 Geometry of a kink band, illustrating terminology. κ is the kink angle, w is the width of the kink band, γ and γ_k are the angles between the kink band boundary (the axial surface) and the undeformed and deformed material, respectively.

faces is different from any of the kinematic models of folding we have considered, although passage of the kink band boundary is accompanied by shearing of the material parallel to the laminations. Laminations in the undeformed and the kinked parts of the material maintain equal angles with the kink band boundary ($\gamma = \gamma_k$, Figures 12.19 and 12.20), and both the line lengths parallel to the laminations and the cross-sectional area remain constant. In Figure 12.20A, the kink nucleates along a line (AB) normal to the laminations and grows by rotation of the right boundary (Ab) counterclockwise about A and rotation of the left boundary (aB) counterclockwise around B, while A and B remain fixed points in the material. The kink angle κ increases continuously with kink growth. In Figure 12.20B, the kink nucleates along a line (AB) oblique to the laminations, and the two margins migrate in opposite directions while maintaining the same orientation. The kink angle κ is fixed by the angle between the laminations and AB, and it does not change with growth of the kink band. In both these cases, the deformation is of constant volume.

In the two other models (Figure 12.21), the kink band boundaries do not migrate but mark the fixed boundaries of a shear zone. As the kink develops, the kink angle κ increases, but the angles γ and γ_k are not equal: γ remains constant whereas γ_k decreases. In Figure 12.21A, kinking produces a deformation equivalent to homogeneous simple shear parallel to the kink band boundaries—and therefore is essentially like the passive-shear model. The width w of the kink band is constant, and the laminations are deformable. The lam-

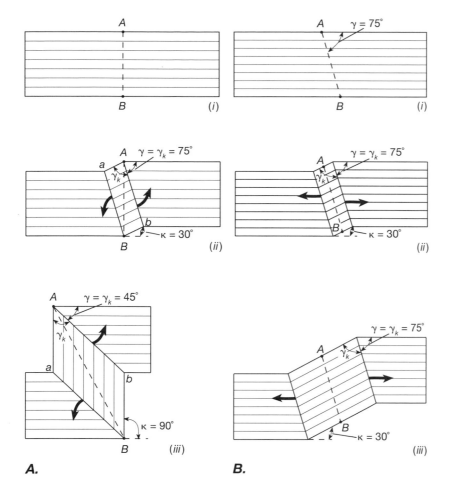

A. **B.**

Figure 12.20 Kinematic models for the growth of a kink band by migration of the kink band boundary through the material. A. The kink band nucleates along the dashed line AB (i). It grows by rotation of the kink band boundary Ab counterclockwise about the fixed material point A and by rotation of the opposite boundary Ba counterclockwise about the fixed material point B (ii to iii). As the kink band grows, the kink angle κ increases. The angles γ and $γ_k$ both decrease during kink band growth, but they remain equal. B. The kink band nucleates along the dashed line AB (i). The kink band grows by migration of the kink band boundaries in opposite directions into the undeformed material (ii to iii). As the kink band grows, the kink angle κ remains constant. The angles γ and $γ_k$ remain equal and constant during kink band growth.

inations first rotate toward an orientation perpendicular to the kink band boundary, becoming shorter and thicker. With further rotation, the laminations lengthen and thin. The cross-sectional area remains constant throughout.

In Figure 12.21B, folding essentially involves a flexural-shear mechanism with shearing parallel to the laminations, which maintain constant length and width. As the kink develops, the laminations rotate toward an orientation perpendicular to the kink band boundary (Figure 12.21B, ii). The kink band becomes wider, and gaps open up between the lamination. With further rotation, the kink band becomes thinner again, and the gaps between laminations close. When $γ_k$ decreases to the value of γ, no further kinking is possible (Figure 12.21B, iii). Because of the opening and closing of the gaps between the laminations, the cross-sectional area is not constant during the kinking.

Experiments on kink band formation indicate that kink bands do not develop along planes of high shear stress. Because the third and fourth models assume the kink band to be a zone of shear, the experiments suggest that these models may not be appropriate for describing natural deformation. Evidence points most strongly to the operation of the first and second models, either singly

or together, in kink band formation. Some natural kink bands, however, show evidence of an increase in volume during deformation, such as accumulation of later minerals between separated layers. This indicates that in some cases, at least, model B in Figure 12.21 represents a component of the kinking mechanism. Model A in Figure 12.21 may account for some kink formation in high-grade metamorphic rocks.

Chevron Folds

Two kinematic models exist to account for the formation of chevron folds. In the first model, chevron folds develop where kink bands of conjugate orientation intersect (Figure 12.22). Transformation of the entire undeformed body into one completely filled with chevron folds requires a shortening of 50 percent. Although this mechanism has been observed to operate during the experimental deformation of phyllites, in naturally deformed rocks the observed shortening that results from kink folding rarely exceeds 25 percent, which is insufficient to form chevron folds by this mechanism.

Chevron folds can also develop by a process that is similar to flexural-shear folding (Section 12.1). In this

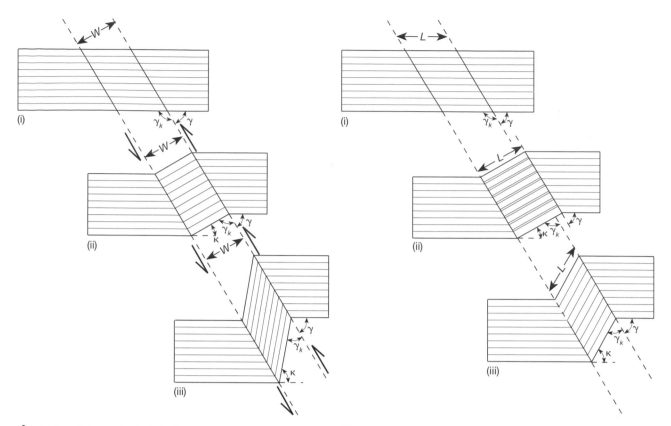

A. Kink band of constant width *W*   **B.** Laminations of constant length *L*

Figure 12.21 Models for formation of a kink band in which the kink band boundary remains fixed in the material. *A.* Kink band growth by simple shear parallel to the kink band boundaries. The length of the laminations first decreases (i to ii), and their thickness increases until the laminations are perpendicular to the kink band boundary. With further rotation, their length increases and their thickness decreases (ii to iii). The width w of the kink band remains constant. As the kink band develops, the kink angle κ increases, γ remains constant, and γ_k decreases. Thus γ and γ_k do not remain equal. *B.* Formation of a kink band by rigid rotation of the laminations. The length of the laminations L in the kink band remains constant during kink band formation. Thus from (i) to (ii), where the width of the kink band increases as the laminations become perpendicular to the kink band boundary, and the volume increases as spaces open up between the laminations. From (ii) to (iii) the width decreases and the spaces between laminations close. Thus this is not a constant-volume deformation. As the kink angle κ increases, γ remains constant, and γ_k decreases until $\gamma_k = \gamma$, at which point the spaces between laminations are completely closed and further kinking is impossible.

case, however, because the idealized laminations of the model are infinitesimally thin, the radius of curvature of the hinge does not have to change along the axial surface, as it does for flexural-shear folding of beds of finite thickness. The result is a class 2 chevron fold formed by a flexural-shear mechanism.

Kink or Chevron Folding of Layered Sequences

Our idealized models have assumed that the kinked material is made of infinitesimally thin laminations and that shearing on the laminations results in a homoge-

neously distributed deformation. Such a condition is most closely approached in nature by foliated rocks such as slates, phyllites, and schists. Kink and chevron folds, however, also occur in thinly bedded rocks such as interbedded chert and slate. Whether a symmetric chevron fold or an asymmetric kink fold forms depends on whether the direction of shortening is parallel or oblique to the layering. The geometry of a chevron fold formed in a layered sequence that has low-friction bedding planes is illustrated in Figure 12.23. Formation of a class 2 fold by class 1B folding of the individual layers requires the opening of voids between the layers at the hinge. (When such voids are filled with secondary mineral de-

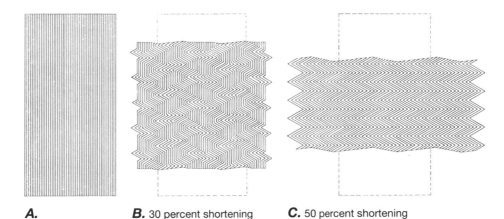

A. **B.** 30 percent shortening **C.** 50 percent shortening

Figure 12.22 Development of chevron folds by kinking. *A.* An undeformed block with a strong planar mechanical amisotropy. *B.* Shortening of the block parallel to the plane of weakness results in the formation of two sets of kink bands that have conjugate orientations. Chevron folds develop at the zones of interference between conjugate kink bands. *C.* As the widths of the conjugate kink bands increase, the area of interference, where the chevron folds develop, also increases until the entire block is filled with chevron folds.

posits, they are called *saddle reefs.*) If the competent layers are separated by incompetent material instead of low-friction surfaces, the incompetent material may flow from the limbs to the hinge zone to accommodate the mismatch in the fold form of the competent layers. This process once again produces multilayer class 2 fold geometry by alternate class 1B and class 3 folding in the competent and incompetent layers, respectively, as described in Section 12.5.

Our kinematic models do not explain why and under what conditions different folding mechanisms should operate. For example, they cannot resolve the question of why rounded folds form in some cases, and chevron folds form in others. In fact, we have not even explained why folds form at all, instead of the layers simply shortening and becoming thicker. To approach these questions, we must consider the mechanics of fold formation, which involves the mechanical properties of

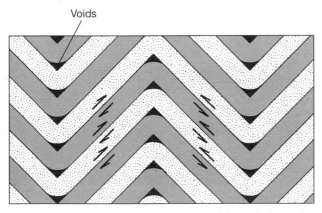

Voids

Figure 12.23 Chevron folding of layers of finite thickness by the flexural-slip mechanism introduces voids in the hinge zone (black areas).

the material and the relationship between the stress and the deformation. We discuss the mechanics in Chapter 18 and Section 20.1 and address the application to folding in Sections 20.2 and 20.3.

12.7 Fault-Bend and Fault-Propagation Folding of a Multilayer

We describe in Chapters 5 and 6 on normal faults and thrust faults how bends in the fault surface (changes in dip or fault ramps) result in folding of the hanging wall block where it rides over the bend. Such fault-bend folds include the rollover anticlines on normal faults and the complex anticlinal stacks above thrust duplexes. We can explain the fold geometry in many natural examples of fault-bend folding by using some fairly simple geometrical constraints and assumptions about the kinematics of layers being displaced over fault bends. Such constraints have proved extremely useful in interpreting the structure of several fold and thrust belts, and they are applicable to normal faulted terranes as well.

The kinematic analysis requires that we make the following assumptions:

1. No gaps are introduced as a result of slip along the fault plane.
2. Fault bends are sharp.
3. The orthogonal thicknesses of layers in the deformed block are preserved.
4. The lengths of layers in the deformed block are preserved.
5. Layers that have not been transported across a fault bend are undeformed.

These assumptions imply a bending model of folding that is accommodated by layer-parallel shear (that is, by the flexural-shear mechanism), that the folds have straight limbs and sharp hinges with a bluntness of zero, and that the folds are multilayer class 2 folds. Note that this is the one geometry for which flexural folding can produce a class 2 fold, and it is a basic property of models for kink and chevron folds (Section 12.6). In nature, the flexural shear actually may be approximated by flexural slip.

The geometry of the deformation in a fault-bend fold is illustrated for a fold concave toward the fault (an anticline) in Figure 12.24. For simplicity, the footwall block does not show the stratigraphy. The layers in the hanging wall block are shown folded across the fault bend. Dashed extensions of the layers indicate the layer geometry that would prevail if no deformation occurred in the hanging wall block. The angle through which the fault bends is β. The initial cutoff angle between the fault and the layers is θ, and ψ is the final cutoff angle. The interlimb angle ι is bisected by the axial plane, which is the geometry required to preserve constant bed thickness, and the folding angle is ϕ. From Figure 12.24, we can see that

$$\phi = 180 - \iota \qquad \phi = \psi + (\theta - \beta) \qquad (12.1)$$

This geometry leads to the following equation, relating the fault-bend angle β to the interlimb angle ι of the associated fold if the initial cutoff angle is θ.

$$\tan \beta = \frac{-\sin (0.5\iota - \theta) \, [\sin (\iota - \theta) - \sin \theta]}{\cos (0.5\iota - \theta) \, [\sin (\iota - \theta) - \sin \theta] - \sin 0.5\iota} \qquad (12.2)$$

For a simple ramp in a décollement for which $\beta = \theta$, this relationship reduces to

$$\tan \beta = \tan \theta = \frac{\sin \iota}{2 + \cos \iota} \qquad (12.3)$$

These relationships determine the geometric evolution of a fault-bend fold. Figure 12.25 illustrates the two phases in the development of a fault-bend fold at a simple ramp. At the initial increment of displacement on the fault, two kink bands form, with kink band boundaries A and A' for one and B and B' for the other. Axial planes A' and B' are fixed in the hanging wall block at X' and Y', respectively, and they migrate with the block as displacement accumulates on the fault. Axial planes A and B are fixed in the footwall block at X and Y, respectively. Thus as displacement continues, material in the hanging wall block migrates through the axial surfaces A and B, and the kink folds grow. The first phase of development continues until the point Y' in the hanging wall block, to which axial plane B' is attached, reaches the point X at the top of the ramp. At this instant, the fold reaches its maximum amplitude, the axial surface B' becomes fixed at the point X in the

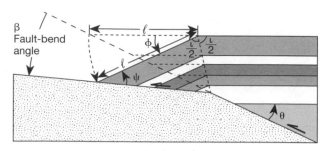

Figure 12.24 Geometry of deformation in a fault-bend fold. The angles are drawn in the positive sense for fault-bend folds that are concave toward the fault (anticlines). For fault-bend folds convex toward the fault (synclines), the angles ψ and θ, measured as shown, are considered to be negative. $\theta =$ the initial cutoff angle of the beds against the fault ($90° \geq \theta \geq -90°$); $\psi =$ the final cutoff angle ($180° \geq \psi \geq -90°$); $\beta =$ the angle of the bend in the fault plane ($90° \geq \beta \geq 0°$); $\iota =$ the interlimb angle ($180° \geq \iota \geq 0°$); $\phi =$ the folding angle ($180° \geq \phi \geq 0°$).

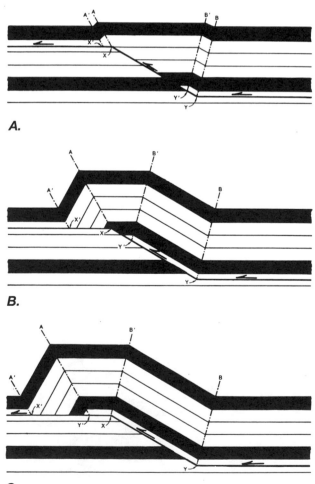

Figure 12.25 Development of a fault-bend fold at a simple fault ramp.

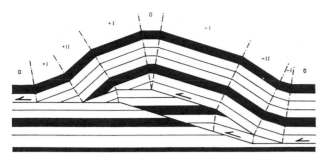

Figure 12.26 A fault-bend fold resulting from imbrication of a thrust fault at a simple ramp and initiating a duplex structure at depth. In this case, the number of increments of dip at the front and back of the fold indicates the number of imbrications on the fault.

footwall block, and the axial surface A becomes fixed to the point Y' in the hanging wall block. The second phase of development begins with further displacement on the fault. Axial planes A' and A are now fixed in the hanging wall block and migrate with it, and axial planes B and B' are now fixed with respect to the footwall block at the bottom and top of the ramp, respectively. Material in the hanging wall block migrates through these axial planes, becoming sheared as it passes through B and unsheared as it passes through B'.

More complex models can also be treated, such as the development of fault-bend folds above imbricated thrust faults and duplexes (Figure 12.26). Note that the dips of the layers change in a stepwise manner at the axial planes and that, in this case, the number of stepwise increases in dip at the front and back of the fold is an indication of the number of fault imbrications at depth. Thus under favorable circumstances the analysis of dip domains on fault bend folds at the surface can help constrain the geometry of complex fault structures at depth.

This same model applied to normal faults (Figure 5.5) predicts the existence of fault bend anticlines and synclines that reflect the geometry of fault surface at depth. A simple model of a listric normal fault with a roll-over anticline is provided by the left half of Figure 5.5B including only the main fault that cuts the surface and the connecting flat. The deformation associated with folding is accommodated by shearing on a set of synthetic faults parallel to both the main fault and the axial surfaces of the kink fold above the flat (cf. Figure 5.3).

Comparable folds also form in association with the propagation of a fault across a layered sequence, as illustrated in Figure 12.27. Where the fault turns upward to cut across the layering, a pair of kink folds form with kink band boundaries A and A' for one kink and B and B' for the other. The axial plane A' terminates at the tip line of the fault but is not parallel to the fault ramp.

Thus it migrates through the material as the fault tip propagates. The kink band between A' and A accommodates the slip ahead of the fault. Axial plane B is fixed relative to the footwall block at the bend in the fault, and displacement on the fault causes material in the hanging wall block to migrate through B. Axial plane B' intersects axial plane A at the same stratigraphic level where the fault tip is located at any given time. Below this stratigraphic level, folding is complete because further displacement is taken up by slip on the fault, not by folding. Axial planes A and B' also migrate through the material as the fault tip propagates, but the axial plane formed from the merging of A and B' remains fixed in the hanging wall block and is displaced with it.

For the formation of a simple ramp in a thrust fault, under assumptions 1 through 4 above for fault-bend folds, a unique relationship can be obtained between the cutoff angle θ and the interlimb angle of the resulting fold ι.

$$2 \sec \theta - \cot \theta = -\cot \iota \qquad (12.4)$$

If folding becomes impossible at some point in this process (because, for example, of the resistance of a particular layer), the fault may propagate between the axial planes A' and A. If it cuts through above A', it leaves a tight syncline in the footwall block, a feature commonly observed in nature and ascribed to "fault drag" rather than to fault-propagation folding.

Comparing Figures 12.25 and 12.27 reveals similarities in the folds formed by fault-bend folding and fault-propagation folding. The relationship between the interlimb angle ι and the initial cutoff angle θ, however, is different, as shown by Equations (12.3) and (12.4).

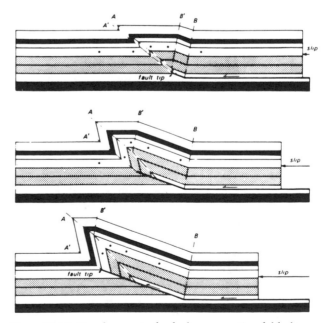

Figure 12.27 Development of a fault-propagation fold above the tip of a propagating thrust fault.

For fault-bend folding, the maximum possible cutoff angle is $\theta = 30°$. For higher angles, the necessary deformation is impossible within the assumptions of the model. For fault-propagation folds, there is a unique relationship between interlimb angle and cutoff angle, and cutoff angles as high as 60° are permitted. In general, for a given cutoff angle, the interlimb angle for fault-propagation folds is smaller than is possible for common fault-bend folding, so the origin of a fault-related fold can in principle be determined. For most cutoff angles, which are less than 30°, tight folds result from fault propagation, and open folds from fault-bend folding.

These kinematic and geometric models for fault-related folding have proved very useful in the interpretation of the deep structure in a number of fold and thrust belts. Such interpretations must be based on surface mapping, well data, seismic data, and regional stratigraphic data, and they are not unique. The geometric requirements of the fold models, however, constrain how these data can be fitted into a viable model of the structure at depth.

If any of the assumptions for the model are violated in natural deformation, of course, the model does not provide reliable constraints on the reconstruction, and the distinction between the two fold origins may become blurred. Beds may deform by nonlayer-parallel shear, they may thicken or thin by homogeneous deformation, or the volume of part of the section may be changed by solution of material. Some aspects of nonlayer-parallel shear can be included in the model, but most other types of deformation do not yield unique geometric constraints. In such cases, inconsistencies in the reconstructions can point to situations in which the assumptions of the model do not apply.

12.8 "Drag Folds" and Hansen's Method for Determining the Slip Line

When rocks are subjected to shear, layers in the rock commonly form asymmetric folds whose sense of asymmetry reflects the sense of shear of the deformation. Such folds are commonly called **drag folds**, the implication being that the velocity gradient in the shear zone has dragged the layer into a fold. Characteristically they are noncylindrical, asymmetric, and disharmonic. Because hinge orientations depend on the original orientation of the layer relative to the shear plane and on local inhomogeneities in the flow, they can vary widely and need not be linear (see, for example, the folds in the salt bed in Figure 12.34A). Thus the hinge orientations do not indicate the slip direction. The hinges may form parallel to the shear plane; if they do not,

subsequent simple shearing tends to rotate them toward parallelism with both the shear plane and the slip direction. More complex geometries of flow than simple shearing, however, can rotate fold hinges toward being either parallel or perpendicular to the direction of flow.

The sense of asymmetry of any "drag fold," whether its hinge is curved or straight, must be consistent with the sense of shear in the zone. This relationship of fold asymmetry to shear sense is the basis of the Hansen method of determining the slip direction.[3] If all hinge orientations are plotted on a stereonet with their appropriate shear senses, then they should lie approximately along the shear plane (Figure 12.28A, B). The separation angle, across which the shear sense changes, contains the slip direction, and the asymmetry of the folds defines the sense of shear of the deformation (Figure 12.28B). The hinges closest to being parallel to the slip direction (hinges 1 and 3, Figure 12.28) constrain the possible slip orientation, because fold hinges on opposite sides of the slip line must have opposite shear senses.

---

[3] A complete discussion of the application and pitfalls of this method can be found in Hansen (1971).

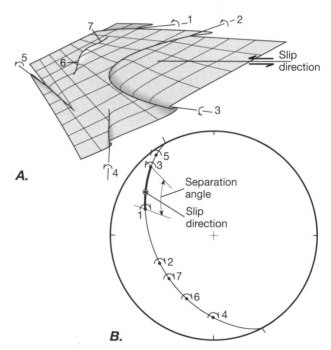

Figure 12.28 Hansen's method for slip line determination in folded layers subparallel to the shear plane. *A.* Fold hinges numbered 1 through 7 have a variety of orientations. The asymmetry of any part of a fold hinge, however, is consistent with the sense of shear. Because fold hinges 2 and 3 lie on opposite sides of the slip direction, they have opposite shear senses. *B.* On a stereonet, the hinge orientations plot parallel or subparallel to the shear plane, and the asymmetry of the folds changes sense across the separation angle, which must contain the slip direction. The sense of fold asymmetry defines the shear sense of the deformation.

12.9 The Geometry of Superposed Folding

In complexly deformed areas such as the central core regions of orogenic belts, folded layers of rock commonly display a geometry indicating that earlier folds have been folded by one or more sets of later folds. Such multiple foldings are referred to as **superposed folding,** and the different sets of folds are called **generations of folds.** A first generation of folds is refolded by a second generation and by all subsequent generations.

In discussing superposed folds, we need a notation to describe the successive surfaces and hinges formed. The terminology is illustrated in Figure 12.29. Surfaces are labeled S. Bedding is S_0, and the axial surfaces of first, second, and higher generations of folds, which are assumed to form as planar surfaces, are designated S_1, S_2, and so on. Fold hinges are labeled f. The fold hinges of successive generations are f_1, f_2, and so on. It is also useful to include with the fold hinge symbol a designation of the surface being folded. Thus the fold hinge $f_1^{S_0}$ means a first-generation fold hinge in the S_0 surface (bedding). Second-generation folds develop in the already folded bedding S_0 and in the first-generation axial surfaces S_1. Thus second-generation fold hinges in these surfaces are labeled $f_2^{S_0}$ and $f_2^{S_1}$, respectively (Figure 12.29).

The basic patterns of orientations of fold hinges and axial surfaces that result from the superposition of two generations of folding can be analyzed according to fairly simple geometric rules. In general, the youngest generation of folding has planar axial surfaces. The axial surfaces of older generations are folded by all

younger generations. After two generations of folding, for example, second-generation folds are developed in both the bedding S_0 and the earlier generation axial surface S_1, and both have the same planar second-generation axial surfaces S_2 (Figure 12.29). Earlier generation fold axes commonly behave like passive linear features and are rotated by later generations of folding. The rotated axes develop predictable patterns that depend on the initial fold axis orientation and the geometry of the later deformation. We discuss some of these patterns in Section 16.1 (Figures 16.1–16.3). Although the youngest generation axial surfaces are commonly planar, the associated fold axes develop in a range of orientations depending on the initial orientation of the surface being folded (compare fold axes in Figure 12.9 A, B). They are related to one another only in that all the youngest generation fold axes must lie within the youngest generation axial surface. Although real folds can be considerably more complicated than these idealized relationships suggest, analyses of the geometrical relations have allowed complex sequences of superposed structures to be unraveled. The details of the procedures are beyond the scope of this book.

If second-generation folding is flexural folding of approximately the same scale as first-generation folds, which would be expected for competent layers, the geometry resulting from the superposition does not conform neatly to the principles we describe above. Figure 12.30A, B shows the results of the experimental superposition of two generations of flexural folds in a stiff, puttylike material, and they illustrate the complexity of the superposed geometry.[4] Although in Figure 12.34B many of the second-generation fold hinges ($f_2^{S_0}$) lie close to the second-generation axial surface (S_2) (Figure 12.30C), the first-generation folds are widely scattered and show no simple geometric pattern.

If the second generation of folding occurs by passive shear and is similar in scale to the first generation, which is an approximate model for folding of incompetent rocks, the resulting outcrop patterns of the superposed folds, called **interference patterns,** have characteristic styles that depend on two angles ψ and θ. ψ is the angle between $\mathbf{a}_2$, the second-generation slip direction, and S_1, the first-generation axial surface ($\psi = \mathbf{a}_2 \wedge S_1$). θ is the angle between f_1, the first-generation fold axis, and the second-generation axial surface ($\theta = f_1 \wedge S_2$) (see Figure 12.31).

The interference patterns are shown in the right-hand diagrams in Figure 12.31A, B, and C. The first diagram in each part shows the geometry of first generation (f_1) folding. The second diagram shows the

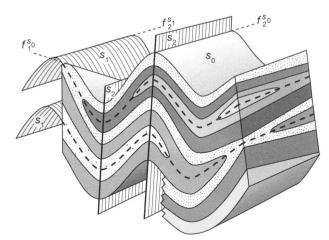

Figure 12.29 Geometric elements of a refolded fold. The first-generation folds are folds in S_0 with fold axis $f_1^{S_0}$ and axial surface S_1. The first-generation folds are refolded by second-generation folds that have axial surface S_2. Second-generation fold axes develop in S_0 ($f_2^{S_0}$) and in S_1 ($f_2^{S_1}$).

---

[4] For an everyday analogue, imagine how difficult it would be to fold a sheet of corrugated sheet metal or plastic in a direction not parallel to the corrugations. The corrugations are put into the material precisely to give it flexural rigidity.

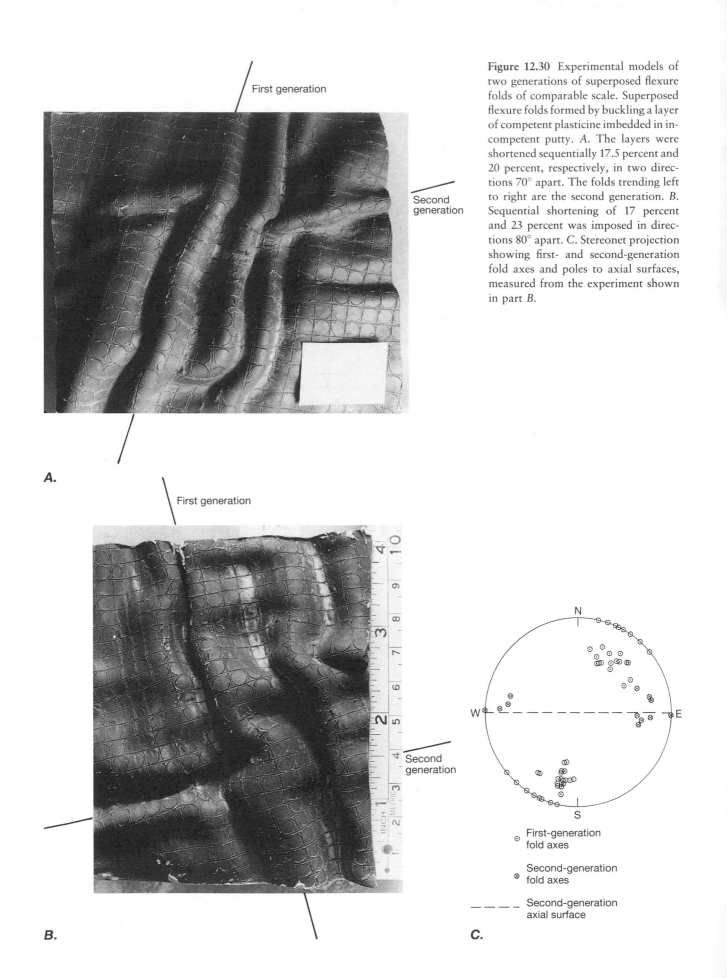

Figure 12.30 Experimental models of two generations of superposed flexure folds of comparable scale. Superposed flexure folds formed by buckling a layer of competent plasticine imbedded in incompetent putty. *A.* The layers were shortened sequentially 17.5 percent and 20 percent, respectively, in two directions 70° apart. The folds trending left to right are the second generation. *B.* Sequential shortening of 17 percent and 23 percent was imposed in directions 80° apart. *C.* Stereonet projection showing first- and second-generation fold axes and poles to axial surfaces, measured from the experiment shown in part *B.*

First generation

Second generation

A.

First generation

Second generation

B.

⊙ First-generation fold axes

⊗ Second-generation fold axes

— — — Second-generation axial surface

C.

First-generation folding   Second-generation folding   Interference patterns

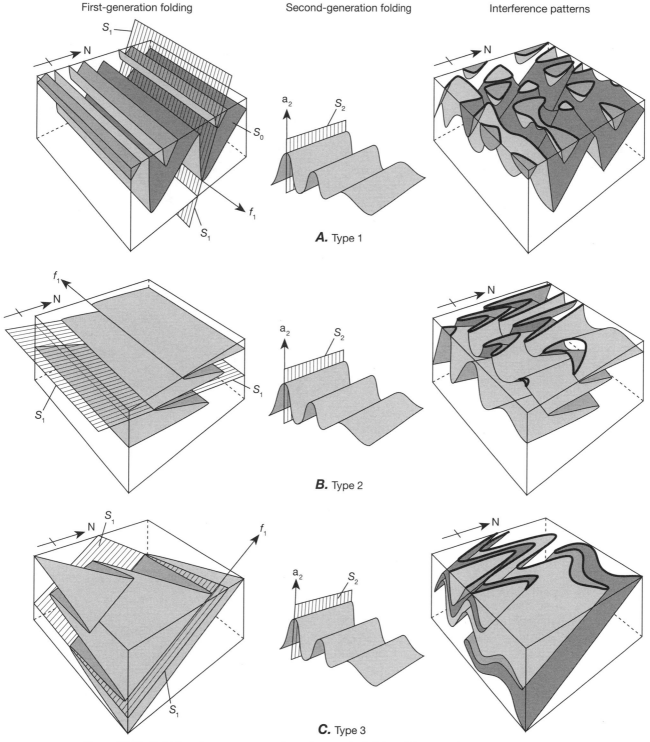

Figure 12.31 Fold interference patterns. In each case, the left-hand diagram shows first-generation folds in S_0 and the axial surface S_1; the middle diagram shows the geometry of second-generation folding as it would appear in an initially horizontal surface; the right-hand diagram shows the superposition of the second-generation folding on the folds in the left-hand diagram, with the surface eroded down to a flat plane to reveal the characteristic outcrop patterns, which are shown in heavy lines. A. Type 1 interference folds, showing dome-and-basin interference patterns. Here a_2 is at a small angle to S_1, and f_1 is at a high angle to S_2. B. Type 2 interference folds, showing arrowhead- or mushroom-shaped patterns. Here a_2 is at a high angle to S_1, and f_1 is at a high angle to S_2. C. Type 3 interference folds, showing the wavy outcrop pattern of the S_1 axial surface. Here a_2 is at a high angle to S_1, and f_1 is at a small angle to S_2.

geometry of the second-generation folding as it would appear in an initially horizontal surface. It is the same cylindrical fold train for A, B, and C. When this folding geometry is superposed on the first-generation folds, the result is the refolded folds that appear as interference patterns. The intersection of these superposed fold styles with a horizontal surface of erosion produces characteristic interference patterns, which are emphasized with heavy lines in the interference pattern diagrams.

The different types of interference patterns shown in Figure 12.31 depend on two angles: the angle between the first-generation axial surface S_1 and the second-generation slip direction $\mathbf{a}_2$, and the angle between the first-generation fold axis f_1 and the second-generation axial surface S_2. These angles may vary from near $0°$ (small) to near $90°$ (large).

The type 1 interference folds (Figure 12.31A) are characterized by complete closures of the outcrop pattern of individual S_0 layers (Figure 12.32A). This pattern reflects the presence of domes, basins, and intervening saddles in the folded surface. It develops when the slip direction $\mathbf{a}_2$ is contained between the limbs of the first-generation folds and thus is generally at a small angle to S_1, and when the angle between f_1 and S_2 typically is large.

The type 2 interference folds (Figure 12.31B) are characterized by arrowhead-, crescent-, and mushroom-shaped outcrop patterns of the folded surfaces (Figure 12.32B). These patterns develop when the $\mathbf{a}_2$ direction is not contained between the limbs of the first-generation folds and is therefore generally at a high angle to S_1, and when the angle between f_1 and S_2 typically is large, as for the type 1 pattern.

The type 3 interference folds (Figure 12.31C) are characterized by an undulating axial surface trace of first-generation folds (Figure 12.32C). This pattern develops if the slip direction $\mathbf{a}_2$ is not contained between the limbs of the first-generation folds and is therefore typically at a high angle to S_1, as for type 2 interference patterns, and when the angle between f_1 and S_2 is small. These types of interference patterns are end members of a continuous gradation of patterns.

Interpretations

It is important to remember that to describe a second generation of folds as being superposed on a first generation implies only a sequence of deformational events. It says nothing about the interval of time between those events, and it does not necessarily imply that all folds of a particular generation developed at the same time everywhere. Moreover, the same number of fold generations do not necessarily appear everywhere, so any possible correlation is probably more reliable if it is

A.

B.

C.

Figure 12.32 Natural examples of interference fold patterns A. Type 1 style, showing domes and basins in a gneiss. B. Type 2 style developed in banded marble. C. Type 3 style developed in interlayered silicates and marble.

determined by fold style rather than generation number. These are very important restrictions on the interpretation of superposed folding. The deformations associated with two generations could be associated with two distinct orogenic events separated by tens or hundreds of millions of years, or they could be the result of two separate phases of a single orogenic event. In the latter case, different generations could represent separate chronological phases of a single orogeny, or they could represent changes in the geometry of deformation from place to place along the flow line for the rocks.

12.10 Diapiric Flow

Diapirs are generally circular to elliptical structures on a horizontal section that form when relatively low density rock at depth rises through overlying rock of higher density, driven by buoyant forces (the word "diapir" comes from the Greek *diapero* meaning "I pierce, I penetrate"). As the low-density material rises, there is a complementary sinking of the overlying higher-density material. The net effect is a lowering of the potential energy of the system, which makes it more stable. This process is an extremely important one in geology. It is associated with the formation of salt domes, metamorphic gneiss domes, and igneous plutons and with solid-state convective flow in the mantle.

Salt diapirs were the first such structures to be recognized and are the best understood, in part because of their economic importance as oil traps and sources for salt and sulfur. They are widespread in areas such as the north German Plain, western Iran, the Gulf Coast region of the United States and Mexico (Figure 5.14), the southwestern Soviet Union, west central Africa, and the Canadian Arctic. Salt is deposited in oceanic basins in which circulation is very restricted, and evaporation concentrates salts in solution until they precipitate. In a rifted margin tectonic setting, salt deposits accumulate after ocean water first enters the rift but before the rift widens into an open ocean. Thus salt commonly lies at the base of a section of denser marine sediments. It can also be deposited in a restricted closing ocean basin, such as developed in late Miocene time in the Mediterranean.

Diapirs begin as anticlinal or domal uplifts and evolve into walls, columns, bulbs, or mushroom shapes (Figure 12.33). The diapirs may become detached from the original low-density layer. As the salt moves upward it pierces through the overlying sediments, which become bent upward along the margins of the diapir. These upturned sediments, truncated against the impermeable salt, provide the excellent traps for hydrocarbons that make salt domes so economically important. The tops of salt domes usually have been dissolved away by ground water (see the top of Figure 12.34C) and are characterized by broad, subhorizontal, insoluble residues from the salt and by brecciated fragments of overlying rock called the caprock. A basin commonly forms above the diapir because of the solution. The sinking of the surrounding sediments to compensate for the rise of the salt often produces a rim syncline surrounding the diapir.

When we study the structures in rocks, our only clue to the geometry of the original deformation is usu-

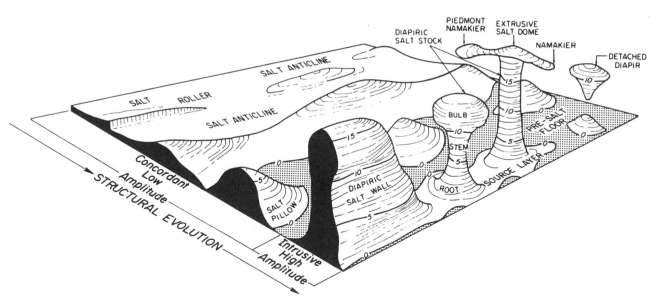

Figure 12.33 Common forms of salt intrusions. Diapirs originate from a layer of salt and then rise as salt pillows, salt stocks, or a salt wall, depending upon the extent of intrusion.

ally the geometry of features such as folds. In salt diapirs, however, we have an unusual opportunity to examine the folding that results from a fairly well understood pattern of flow. Such examples provide us with models that we can use to understand other deformational environments in which a comparable style of folding is produced.

Field relationships and models (see Section 20.7; Figures 20.19–20.21) suggest that horizontal radial flow converging toward the rising salt column initially forms a set of circumferential folds whose hinges become radially oriented (Figure 12.34A). Minor shifts in the flow geometry produce refolding of earlier generations of folds (lower inset, Figure 12.34A). As the salt moves up into the stock, the folds rotate into a vertical plunge, parallel to the main axis of the salt dome. A map of layers in the Grand Saline salt dome shows a complex geometry of class 2 refolded folds and sheath folds that

have subvertical to vertical hinges (Figure 12.34B; compare upper inset in Figure 12.34A). In the bulbs and mushroom caps of the domes, lateral spreading of the salt and drag along the margins causes complex refolding of the salt layers and the possible entrainment of adjacent sediments into the folds (Figure 12.34C).

This example shows how different generations of folds can form from a change in the flow regime along the flow path. It also illustrates that different generations of folds can form in various places at the same time. In mountain belts, therefore, one must be cautious about the significance ascribed to generations of folds.

Shale diapirs are present in some areas where folding of unconsolidated sediments has taken place or where rapid sedimentation and compaction have generated high fluid pressures in unlithified shales and caused them to move upward through the overlying rocks. The general form of these structures resembles

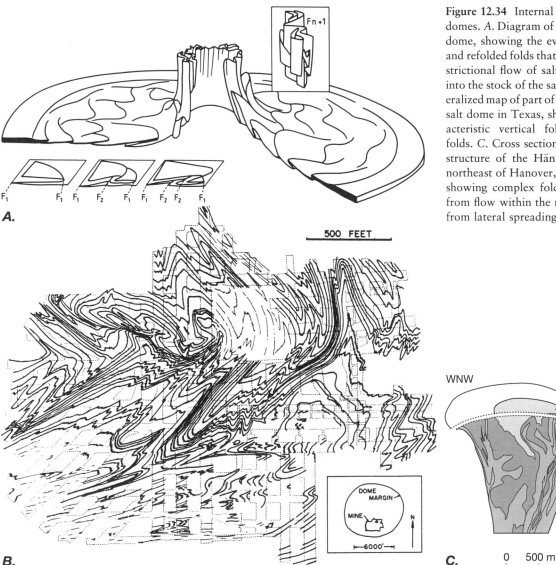

Figure 12.34 Internal structure of salt domes. *A.* Diagram of the base of a salt dome, showing the evolution of folds and refolded folds that result from constrictional flow of salt from the layer into the stock of the salt dome. *B.* Generalized map of part of the Grand Saline salt dome in Texas, showing the characteristic vertical folds and sheath folds. *C.* Cross sections of the internal structure of the Hänigsen salt dome northeast of Hanover, West Germany, showing complex folding that results from flow within the rising diapir and from lateral spreading of the bulb.

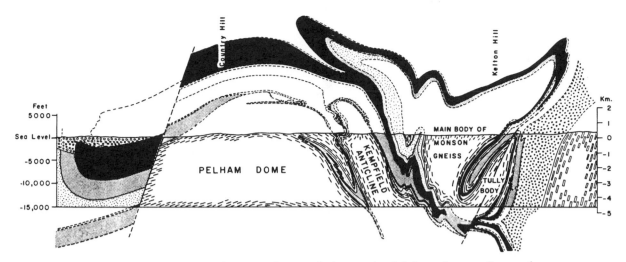

Figure 12.35 Cross section of a gneiss dome with the mantle of deformed metasediments from the Bronson Hill Anticlinorium of west-central New England. The metasedimentary rocks were strongly deformed into recumbent nappes before deformation associated with the emplacement of the gneiss domes.

that of salt domes. In some cases the shale diapirs even reach the surface, where they form "mud volcanoes."

Mantled gneiss domes are domical bodies of gneissic rock found in highly metamorphosed core zones of orogenic belts (Figure 12.35). They commonly display foliation parallel to the walls of the body, and they are surrounded, or "mantled," by a sheath of metamorphosed sedimentary rocks. These bodies may be diapirs of gneiss intruded into overlying rocks during intense regional metamorphism.

Additional Readings

Donath, F. A., and R. B. Parker. 1964. Folds and folding. *Geol. Soc. Am. Bull.* 75: 45–62.

Groshong, R. H. 1975. Strain, fractures and pressure solution in natural single-layer folds. *Geol. Soc. Am. Bull.* 86: 1363–1376.

Hansen, E. 1971. *Strain facies.* New York. Springer-Verlag.

Jackson, M. P. A., and C. J. Talbot. 1989. Anatomy of mushroom-shaped diapirs. *J. Struct. Geol.* 11: 211–230.

Patterson, M. S., and L. Weiss. 1966. Experimental deformation and folding in phyllite. *Geol. Soc. Am. Bull.* 77: 343–374.

Ramsay, J. G. 1967. *Folding and fracturing of rocks.* New York: McGraw-Hill.

Skjernaa, L. 1975. Experiments on superimposed buckle folding. *Tectonophysics* 27: 235–270.

Suppe, J. 1983. Geometry and kinematics of fault bend folding. *Am. Jour. Sci.* 283: 684–721.

Suppe, J. 1985. *Principles of structural geology.* Englewood Cliffs, N. J.: Prentice-Hall.

Talbot, C. J., and M. P. A. Jackson. 1987. Internal kinematics of salt diapirs. *Am. Assoc. Petrol. Geol. Bull.* 71 (9): 1068–1093.

Thiessen, R. L., and W. D. Means. 1980. Classification of fold interference patterns: A reexamination. *J. Struct. Geol.* 2: 311–316.

Weiss, L. E. 1980. Nucleation and growth of kink bands. *Tectonophysics* 65: 1–38.

CHAPTER

13 Foliations and Lineations in Deformed Rocks

A **foliation**[1] is a homogeneously distributed planar structure in a rock.[2] Examples of foliations include sedimentary bedding; the planar alignment of sedimentary clasts; the planar structure defined by the parallel alignment of platy minerals in a schist, a slate, a shale, or a volcanic rock; the parallel alignment of flattened mineral grains and conglomerate pebbles; and compositional banding defined by the concentration of particular minerals into layers, which is common in gneisses, ultramafic rocks, and some volcanic rocks.

A **lineation** is a homogeneously distributed linear structure. Lineations are **surficial** if they are present along discrete surfaces, **penetrative** if they occur throughout the volume of a rock. Examples of surficial lineations include sedimentary groove casts in a bedding surface and the parallel alignments of mineral fibers that develop along some fault surfaces. Examples of penetrative lineations include the hinges of pervasive small crenulations in a foliation, the preferred alignment of elongate mineral grains such as amphiboles or quartz, and the linear alignment of elongate clusters of grains of a particular mineral such as quartz or mica.

Foliations and lineations are primary if they originate by primary sedimentary or igneous processes. Primary sedimentary processes such as sediment transport and deposition produce, for example, linear tool marks, the preferred orientation of sedimentary clasts, and bedding. Primary igneous processes such as flow and crystallization result in the preferred orientation of bubbles and pumice fragments or in compositional streaks and bands. Foliations and lineations are secondary if they originate by secondary processes such as tectonic deformation or metamorphism. We sometimes use terms such as *sedimentary foliation, igneous lineation,* and *tectonite foliation* to specify the inferred origin of a foliation. Because the origin of a structure is an interpretation, however, it is an inappropriate basis for classification, and we define *foliation* and *lineation* in strictly descriptive terms.

In this chapter we discuss foliations and lineations that are characteristic of **tectonites,** which are rocks whose structure is a product of deformation and which are commonly, but not necessarily, metamorphosed. Most such foliations and lineations are secondary in origin, although some may be inherited primary features. **S-tectonites** and **L-tectonites** are rocks dominated by planar and linear preferred orientation, respectively, of mineral grains.

Generally we consider a structure to be homogeneously distributed, or **penetrative,** if the spacing or the

---

[1] Derived from the Latin *folium,* which means "leaf." The definition of the term *foliation* is not universally agreed on. Some authors use it in a more specific sense than we have adopted.

[2] A feature that is homogeneously distributed in a body has the same characteristics in any arbitrary volume of the body.

scale of the structure in a rock is very small compared to the size of the rock volume under consideration. To qualify as a foliation or lineation, a structure must be penetrative within a volume that has a dimension on the order of tens of centimeters. Planar features that have an average spacing on the order of meters are not foliations but structures such as fractures, faults, or shear zones. The spacing of many foliations, for example in slates, is so small as to be unresolvable in hand sample. If a lineation occurs on a penetrative planar structure such as a foliation, then the lineation also is penetrative.

Several other terms are used to describe penetrative planar features in rocks. The term **S-surface** is generally synonymous with *foliation* (the S comes from the German *schiefer* which means "schist"). It refers to any penetrative planar feature of a rock and therefore includes sedimentary bedding, schistosity, and axial surfaces of folds, which may be simply geometric constructs rather than actual physical features in the rock. Bedding is commonly designated S_0, and other penetrative planar features such as foliations and axial surfaces are labeled $S_1, S_2, \ldots$, where the subscripts generally indicate the sequence in which the different features developed. We have, in fact, already used this notation in our discussion of superposed folding (Section 12.8).

Rock cleavage, or simply **cleavage,** is the tendency of a rock to break or cleave along surfaces of a specific orientation. All cleavages are foliations, and the two terms are often used to describe the same structure. *Foliation* is a more general term than *cleavage*, however, because it includes planar geometric features that might not result in a cleavage. Planar alignment of slightly flattened grains (for example, of quartz in a quartzite, olivine in a peridotite, or compositional banding in a gneiss) would define a foliation but could provide so small a mechanical anisotropy that it would not result in a cleavage.

The terms **layer** and **banding** describe planar tabular features in rocks that are characterized by differences in composition, or possibly in texture, from adjacent rock. The terms are commonly used in descriptions of plutonic igneous rocks and high-grade metamorphic gneisses.

We outline a morphological classification for foliations in tectonites in Figure 13.1 and Sections 13.1 to 13.4. The classification is based on the shape and/or arrangement of components of the rock. This approach is preferable to the use of numerous older terms that are poorly defined, are imprecise, or have a genetic connotation (see Section 13.6).

Many foliations have a structure characterized by laminar to lenticular **domains,** which are restricted volumes that are uniform with regard to a particular structure and that differ in that regard from adjacent volumes.

Foliation and cleavage	Spaced	Compositional	Diffuse
			Banded
		Disjunctive	Stylolitic
			Anastomosing
			Rough
			Smooth
		Crenulation	Zonal
			Discrete
	Continuous	Fine	Microcrenulation
			Microdisjunctive
			Microcontinuous
		Coarse	

Figure 13.1 Morphological classification scheme for foliations.

Foliation domains are distinguished by differences in preferred orientation of mineral grains, in structure, or in composition. Foliations defined by domains that have a spacing of 10 μm or more are **spaced foliations** (Figure 13.1; see also Figures 13.2 to 13.10). Foliations that exhibit a finer domainal structure, or no domainal structure at all, are **continuous foliations** (Figure 13.1; see also Figures 13.11 and 13.12).

Spaced foliations are categorized on the basis of four features: (1) domain shape, (2) domain spacing, (3) distinguishing characteristics of individual domains, such as mineral composition or the preferred orientation of mineral grains, and (4) the proportion of the rock occupied by the different types of domains. We recognize three categories of spaced foliation: compositional, disjunctive, and crenulation foliations (Figure 13.1).

A morphological classification of lineations in tectonites is outlined in Figure 13.18 and in Sections 13.7 and 13.8. We divide tectonite lineations into two major categories: Structural lineations are defined by geometric structures, mineral lineations by mineral grains or aggregates (Figure 13.18). Both types of lineations may be either surficial or penetrative.

13.1 Compositional Foliations

Compositional foliations (Figure 13.2) are marked by layers, or laminae, of different mineralogical composition. A planar alignment of platy or needle-shaped crystals may also be present, but the rock has at most a weak tendency to cleave parallel to the foliation. We subdivide these structures on the basis of mineralogical variation and the spacing and relative thicknesses of the compositional layers. **Diffuse foliations** are characterized by widely spaced weak concentrations of a mineral

A.

B.

Figure 13.2 Compositional foliations. *A.* Sparse compositional foliation in a dunite, Klamath Mts., Oregon. The rock is composed mainly of olivine. Concentrations of pyroxene crystals in sparse layers define the foliation. Scale bar is 6 in. *B.* Banded compositional foliation in a high-grade metamorphic gneiss, Wopmay Orogen, NW Canada.

in a rock of predominantly one lithology. They are common in ultramafic rocks such as dunites in which diffuse layer concentrations of pyroxene crystals define a weak compositional layering (Figure 13.2*A*). Diffuse foliations also occur in deformed granites in which concentrations of mafic minerals define the foliation. **Banded foliations** are composed of relatively closely spaced compositional layers that are mineralogically distinct and of comparable abundance. They are common in high-grade metamorphic gneisses (Figure 13.2*B*).

13.2 Disjunctive Foliations

Disjunctive foliations (Figures 13.3 through 13.8; the Latin word *disjunctus* means "disjoined or detached") are characterized by thin domains, called **cleavage domains or seams,** marked by concentrations of oxides and/or strongly aligned platy minerals. The cleavage domains are separated by tabular to lenticular domains called **microlithons** in which platy minerals may be less abundant or more randomly oriented (Figures 13.3 and

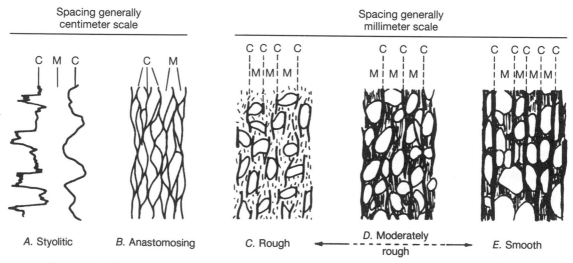

Figure 13.3 Sketches showing characteristics of the various types of disjunctive foliation. C marks cleavage domains; M marks microlithons. Note the change from the centimeter to millimeter scales.

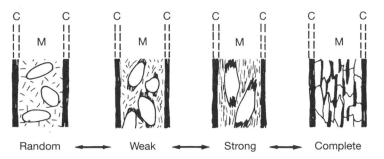

Random ⟷ Weak ⟷ Strong ⟷ Complete

Figure 13.4 Preferred orientation within a microlithon bounded on either side by cleavage domains. C marks cleavage domains; M marks microlithons. In random fabric, the large grains have no preferred orientation, and the fine platy minerals in the matrix are also not oriented. Weak fabric is characterized by a slight elongation of coarse mineral grains and weak preferred orientation of their long axes, weak development of mica "beards" at the ends of the coarse mineral grains, and a weak preferred orientation of the platy minerals in the matrix. Strong fabric is characterized by distinct elongation of the coarse mineral grains and strong alignment of their long axes, well-developed and oriented mica "beards," and strong alignment of the platy minerals in the matrix. In completely oriented fabric, detrital grain shapes are not preserved; mineral grains are elongated and show a strong preferred orientation. The fabric is transitional to a continuous foliation.

13.4). Disjunctive foliations commonly form in previously unfoliated rocks such as limestones or mudstones, although in some foliated rocks they may also develop cross-cutting the earlier foliation.

We divide disjunctive foliations into four groups—**stylolitic, anastomosing, rough,** and **smooth**—on the basis of the smoothness or regularity of the cleavage domains (Figure 13.3). This order corresponds to a general increase in smoothness of cleavage domains and to a decrease in spacing, as well as to a tendency toward stronger preferred mineral orientations within the microlithons. The fabric of the microlithons ranges from random to complete, which provides the basis for further subdivision (Figure 13.4).

Stylolitic foliation (in Greek, *stylos* means "stalk" and *lithos* means "rock") exhibits long, continuous, but very irregular cleavage domains that commonly have a distinct toothlike geometry in cross section (Figures 13.3A and 13.5A). This type of foliation is typical in limestones in which the cleavage domains characteristically are thin, dark, clay-rich seams. In some limestones, fossil fragments are truncated by a stylolite (Figure 13.5B). The spacing of the cleavage domains ranges from 1 to 5 cm or more. There is generally no preferred orientation visible in the microlithons.

Anastomosing foliation is distinguished by long, continuous, wavy cleavage domains that form an irreg-

A.

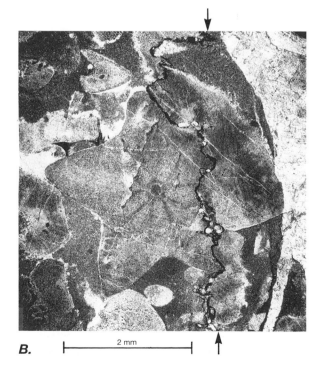

B.

Figure 13.5 Stylolitic foliation. *A.* Stylolitic foliation in limestone layers. Stylolites are the dark irregular lines in the rocks. *B.* Stylolite (see arrows) truncating a pentacrinoid fossil in a limestone. The stylolite is the roughly vertical, irregular black seam. The fossil is shaped like a five-pointed star, but two of the points on the right side are truncated by the stylolite and have been largely removed by solution.

Figure 13.6 Anastomosing foliation in a limestone. Bedding is parallel to the ruler.

ular network outlining lenticular microlithons (Figures 13.3*B* and 13.6). Such foliations are common in limestones and in phyllites and schists. The spacing of the cleavage domains tends to be smaller than for stylolitic foliation, averaging perhaps 0.5 to 1 cm. The cleavage domains contain concentrations of platy minerals with a strong preferred orientation parallel to domain boundaries. The fabric within the microlithons is usually random to weak.

Rough foliation typically develops in rocks containing abundant sand-size mineral grains. The cleavage domains are short, discontinuous concentrations of highly oriented platy minerals that bound or envelope the coarse grains (Figures 13.3*C* and 13.7). The spacing of the cleavage domains is generally 1 mm or less. Microlithons exhibit a wide range of preferred orientations from random to strongly oriented (Figure 13.4).

Smooth foliation represents the planar end of the spectrum from irregular to planar cleavage domains (Figures 13.3*D, E*). Cleavage domains are long, continuous, and smooth and have concentrations of highly oriented platy minerals. The spacing of the cleavage domain is generally less than a millimeter. Fabric development within the microlithons commonly ranges from random to completely oriented (Figure 13.4). With decreasing domain spacing, this type of foliation is transitional with microdomainal, fine, continuous foliations characteristic of some slates (Figure 13.8), as described in Section 13.4.

13.3 Crenulation Foliations

Crenulation foliations are formed by harmonic wrinkles or chevron folds that develop in a preexisting foliation. The new foliation cuts across the old foliation and is

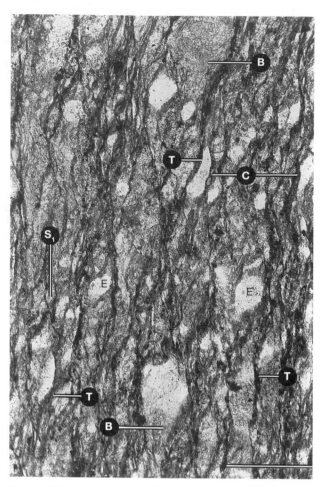

Figure 13.7 Rough foliation (S$_1$) in a deformed wacke. Dark seams are the cleavage domains composed of insoluble residues. At C, remnant detrital sand grains are truncated against cleavage domains. At T, thin plate-like quartz grains result from solution of the grains along cleavage domains (see Section 14.3). B marks "mica beard" overgrowths on detrital grains. Scale bar is 1 mm.

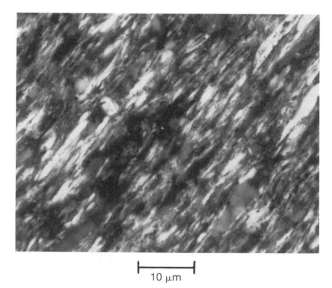

10 µm

Figure 13.8 Smooth foliation in a slate.

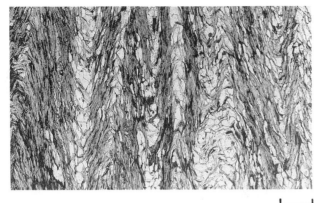

A.   0.63 mm

B.

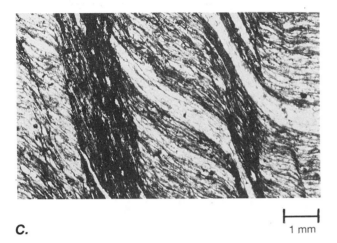

C.   1 mm

Figure 13.9 Zonal crenulation foliations. Note that the laminations and the preferred orientation of the platy minerals varies continuously from microlithon to cleavage domain and that within the cleavage domain, the laminations and platy minerals are not strictly parallel to the new cleavage domain. *A.* If the crenulations are symmetric, both limbs define crenulation cleavage domains and the hinge zone is preserved in the microlithons. Note the compositional differentiation. Limbs of crenulations (dark bands) are rich in mica and poor in quartz. Hinge zones (light bands) are rich in quartz and poor in mica. *B.* Asymmetrical crenulation foliation in schistose metagreywacke. Coin diameter is about 2.5 cm. Rotmell, Grampian Highlands, Scotland. *C.* Asymmetric crenulations in a quartz–rich schist. A loss of quartz from the cleavage domain results in a compositional differentiation of the domains.

defined by both limbs of symmetric crenulations (Figure 13.9*A*) or by one of the limbs of asymmetric crenulations (Figure 13.9*B, C*). The old foliation is preserved in the microlithons either as the hinges of symmetric crenulations (Figure 13.9*A*) or as one of the limbs of asymmetric crenulations (Figure 13.9*B, C*). The microlithon width is comparable to the half-wavelength or wavelength of the crenulations.

The orientation pattern of platy minerals in the cleavage domain is the basis for the further subdivision of crenulations. In a **zonal crenulation foliation,** the platy minerals in the new cleavage domain are oriented at a small angle to the domain and form a continuous variation of orientations from the platy minerals in the microlithons (Figure 13.9). The microlithon boundaries are gradational. In many cases, there is a compositional difference between cleavage domains and microlithons characterized by a higher proportion of platy minerals in the cleavage domains then in the microlithons.

In a **discrete crenulation foliation,** the orientation of platy minerals in the new cleavage domains is parallel to the domains and sharply discordant with the orientations of platy minerals in the microlithons (Figure 13.10). The crenulations are preserved in the microli-

1cm

Figure 13.10 Discrete crenulation cleavage in a slate. Note the sharp discontinuity in orientation that marks the boundary between cleavage domain and microlithon. The orientation of the platy minerals in the cleavage domain is parallel to the domain boundary.

thons. The cleavage domains are generally narrow and may or may not correspond to limbs of crenulations in the microlithons. Differences in mineralogy between the two domains are similar to those of zonal foliations.

All variations between these two "extremes" of crenulation foliation can be observed. In fact, it is not uncommon to find both morphologies in the same sample.

Continuous Foliations

Continuous foliations are defined either by domains with a spacing less than 10 μm (Figure 13.11A, B) or by a nondomainal structure (Figure 13.12). They are divisible by grain size into **fine** and **coarse continuous**

foliations (Figure 13.1), as exemplified by slates and schists, respectively. Fine continuous foliations may be either microdomainal or microcontinuous. The **microdomainal** fine foliations may be microcrenulation (Figure 13.11A) or microdisjunctive (Figure 13.11B), and they have the same characteristics as their macroscopic counterparts except that the microdomain spacing is less than 10 μm. A **microcontinuous** fine foliation is characterized by the parallel alignment of all platy or inequant grains in a rock, and it lacks any domainal structure. The terms *microdomainal* and *microcontinuous* are impractical to use as field classification terms, because in fine-grained rocks only an electron microscope can reveal the distinction between the structures.

Coarse continuous foliations are characterized by the complete orientation of homogeneously distributed platy minerals (Fig. 13.12A) or by the alignment of

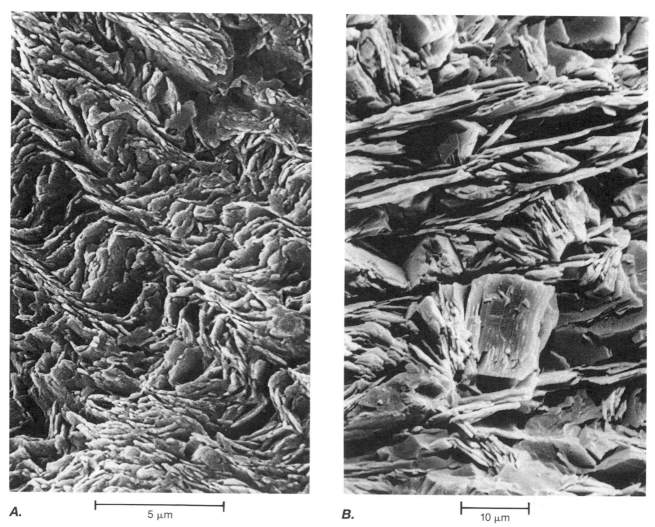

A. |⊢——— 5 μm ———⊣|

B. |⊢——— 10 μm ———⊣|

Figure 13.11 Scanning electron micrographs of continuous fine foliations. *A*. Microdomainal continuous fine foliation in a slate with a microcrenulation structure. *B*. A microdisjunctive continuous fine foliation in a slate.

A.

B.

Figure 13.12 Coarse continuous foliation showing a strictly continuous structure. *A.* A schist with the foliation defined by mica. *B.* A grain-shape foliation parallel to the pencil in a very coarse-grained marble layer.

flattened mineral grains (Fig. 13.12*B*). They have no domainal structure, which would be easily revealed by the coarse grain size.

13.5 The Relationship of Foliations to Other Structures

Secondary foliations so commonly occur parallel or subparallel to the axial surfaces of folds that the association is almost axiomatic. Such foliations are called **axial surface foliations** or **axial plane cleavages.** The orientation of such foliations characteristically changes progressively from one side of the fold to the other, or **fans** across the fold, and is actually parallel to the axial surface only at the hinge surface. Foliation fans are **convergent** or **divergent,** depending on whether the foliation orientations converge toward one another (Figure 13.13*A*, layers I and III) or diverge from one another (Figure 13.13*A*, layer II) in passing from the convex to the concave side of a fold.

It is important to distinguish foliation fans from fans of dip isogons on folds. The terminology and the geometry are the same for both, and diagrams of the two features look similar. The foliation fan, however, is an actual physical structure that can be observed in the rock (Figures 13.13*B*) whereas the dip isogon fan is a geometric construction (Figures 11.18 and 11.19). In general, the two features are not parallel.

The extent of fanning of an axial surface foliation is typically associated with the composition of the rock in which the foliation is developed. Foliations tend to be most strongly convergent across folds in rocks that contain only small proportions of platy minerals, such as sandstones, and they are least convergent or divergent in rocks rich in platy minerals, such as schists and slates. The orientation of the foliation commonly changes significantly at a lithologic contact (Figure 13.13*A, B*). We call this feature a **refracted foliation** or **refracted cleavage** by analogy with the bending, or refraction, of a light ray as it passes obliquely across an interface between two media. The analogy has no significance, however, beyond the similarities of geometry.

The relationship of foliations subparallel to the axial plane of folds is so consistent that it can be used in field mapping to help determine the geometry of the folding. In an area that has been subjected to only one generation of folding, it can also be a valuable indicator of whether sedimentary beds are overturned or right side up, because a given surface of axial foliation can cut a particular folded surface only once (Figure 13.14*A*). Thus the relative orientations of bedding and foliation (Figure 13.14*B*) permit us to determine the general location and direction of the fold closures, as shown in Figure 13.14*A* (see also 13.13*A, B*). An interpretation of Figure 13.14*B* that has the fold closing to the left at the top of the photo and to the right at the bottom would be incorrect, because it would require the foliation plane to cut a single bedding surface more than once.

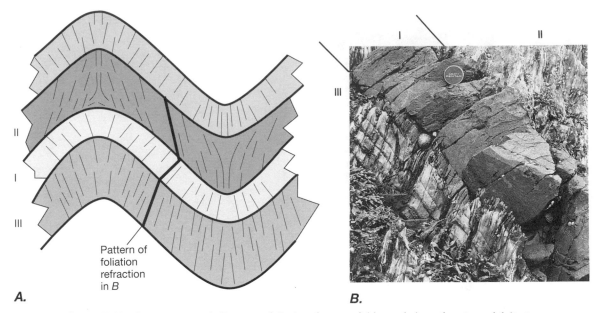

Figure 13.13 Convergent and divergent foliation fans on folds, and the refraction of foliation across lithologic contacts. *A.* The typical pattern of foliations in a folded sequence of sandstone, shale or slate, and siltstone. The foliation pattern is convergent in the sandy (I) and silty (III) layers and divergent in the slate (II). Foliation orientation is "refracted" at the contacts between the layers. The heavy line across the middle limb emphasizes the foliation orientations shown in part *B.* *B.* A sandstone (I), shale (II), siltstone (III) sequence showing the "refraction" of the foliation at the contacts. The photo illustrates the orientations of the beds shown in the middle limb of the diagram in part *A.* Silurian beds, Port Allen, Southern Uplands, Scotland.

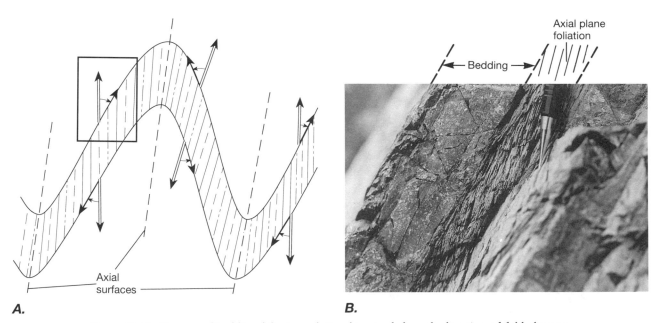

Figure 13.14 The use of bedding-foliation relationships to deduce the location of fold closures and axial surfaces. A foliation plane cannot cut a given bedding surface more than once. *A.* Folded bedding with an axial foliation. Hollow arrows point along the foliation. Solid arrows point along bedding planes. The sense of rotation of the hollow arrow through the acute angle from the foliation toward the bedding (solid arrow) changes across an axial surface. The box outlines the foliation–bedding relationship shown in part *B,* and the fold indicates the correct inference for the direction of fold closure. *B.* Bedding–foliation relationship in interbedded sandstone and shale. The foliation is obvious in the shale layer. The bedding–foliation relationship in the photograph indicates that the fold closes upward to the right and downward to the left, as shown in part *A.* Marathon thrust belt, Texas.

A useful rule of thumb by which to remember this relationship is to imagine an arrow drawn along the foliation surface (Figure 13.14A) and then rotated through the acute angle from the foliation to the bedding. The sense of rotation is the same as the sense of asymmetry that a higher-order fold would have at the same location (see Figure 11.24): clockwise (z) on the left side of an antiformal fold, and counterclockwise (s) on the right side (Figure 13.14A). The sense of rotation changes across an axial surface. Thus it can be used to map the locations of axial surfaces and to infer the direction of closure of the lower-order (larger) fold, even when the exposure does not permit direct observation of these features.

Inferring the location of the fold closures in this manner enables us to deduce whether the bed is overturned or not, provided we know that only one generation of folds has affected the rocks (Figure 13.15A–

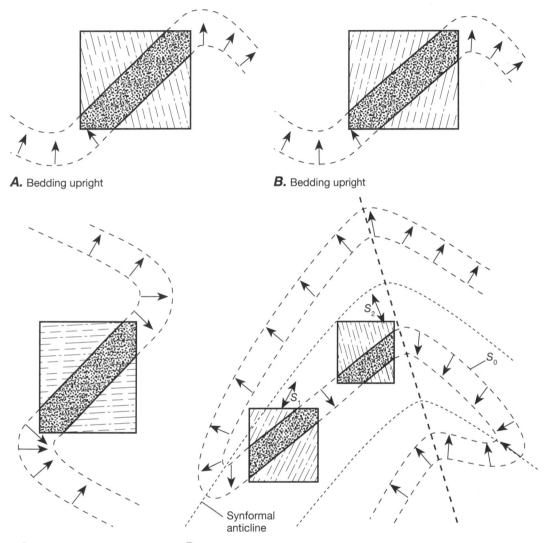

A. Bedding upright   **B.** Bedding upright

C. Bedding overturned   **D.** Complex folding with overturned beds

Figure 13.15 Use of bedding-foliation relationships to deduce the stratigraphic up direction in simply folded layers. The boxes outline the part of the structure that is assumed to be observable in the field. The dashed continuation of the structure shows the unobservable portion. The arrows in the folded layer indicate the stratigraphic up direction. A. Upright bedding is indicated if bedding and foliation dip in opposite directions. B. Upright bedding is indicated if bedding and foliation dip in the same direction and bedding has the shallower dip (see also Figure 13.14B). C. Overturned bedding is indicated if bedding and foliation dip in the same direction and bedding has the steeper dip. D. For complex folding, the relationships between bedding and foliation do not give reliable results for the stratigraphic up direction. In this example, note that the foliation could be S_1, parallel to the first-generation axial surface (lower box), or S_2, parallel to the second-generation axial surface (upper box), and in both cases the bedding is overturned. Comparison of the upper box with part A and of the lower box with part B shows that the technique does not work in this case.

C). Another rule of thumb is helpful: If bedding and foliation dip in opposite directions, the bedding must be upright (Figure 13.15A). If bedding and foliation dip in the same direction, the bedding is upright if it has a shallower dip than the foliation (Figure 13.15B) and is overturned if it has a steeper dip than the foliation (Figure 13.15C).

This method for determining the "stratigraphic up" direction does not work if multiple generations of folding have affected the rocks. In this case the folding is complex, and different foliations may develop in association with the different fold generations. In places on a synformal anticline (for example, the boxes in Figure 13.15D), neither the S_1 nor the S_2 foliation provides the correct indication that the bedding is overturned (see Figure 13.15A, B). Thus the method must be applied with caution. When deformation is complex, we must rely on geopetal structures such as those described in Section 2.2.

Rocks that display multiple generations of deformation commonly have two foliations, which may be the same type or different types, and which may or may not be equally well developed. The earlier foliation commonly becomes folded with a second foliation developed subparallel to the axial surface of the second-generation folds. This relationship is most obvious in the hinge zone of the second-generation folds. On the limbs, the two foliations may be parallel and completely indistinguishable such that the rock appears to contain only one foliation. If multiple foliations are present, it is important to recognize and distinguish them while mapping, because they provide information about different parts of the deformation history and must be separated in the analysis of the structure of the area.

In ductile shear zones the rocks may contain two foliations (labeled S and C in Figure 13.16), both of which develop during a single deformation. Such rocks are called **S-C tectonites.** The **S-foliation** is a continuous coarse foliation defined by the preferred orientation of mica grains and commonly by elongate quartz grains; its dominant orientation is oblique to the ductile shear zone. The **C-foliation** (the C comes from the French *cisaillement,* which means "shear") is a set of shear bands in the rock that develop subparallel to the boundaries of the shear zone. The C-foliation surfaces may have fibrous crystals (slickenfibers, see Section 13.11) lying

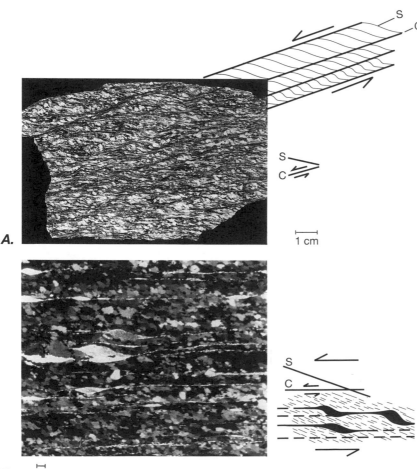

A.

1 cm

B. 100 μm

Figure 13.16 Foliations in an S-C tectonite. *A.* Type I S-C tectonite in a ductile shear zone in a granodiorite. The S-foliation is continuous coarse mica foliation that curves toward an orientation parallel to the C-foliation, which in turn is parallel to the shear zone boundaries. This sense of curvature is the same as the general shear sense and indicates a sinistral shear. The idealized geometry is shown in the adjacent diagram. *B.* Type II S-C tectonite in a quartz-rich mylonite. The S-foliation is defined by the grain-shape foliation of the quartz and by the preferred orientation of large mica porphyroclasts ("mica fish"). The C-foliation is the shear plane defined by the trails of fine micas commonly connected to the tips of the porphyroclasts. The sense of curvature from the mica porphyroclasts to the mica trails is the same as the sense of shear on C, and it indicates sinistral shear. The idealized relationships are illustrated in the adjacent diagram, in which the micas are shown in black.

on and subparallel to them, indicating that they were shear surfaces during the deformation. If platy minerals are relatively abundant, a type I S-C tectonite develops in which the C-foliation cross-cuts an S-foliation that has a sigmoidal shape between adjacent C-surfaces (Figure 13.16A). If platy minerals are relatively sparse, as in some micaceous quartzites, a type II S-C tectonite develops (Figure 13.16B). The S-foliation is defined by the preferred orientation of the large mica grains (called mica porphyroclasts, or "fish") and by a grain-shape foliation in the quartz. The C-foliation is defined by thin seams of very fine-grained mica connected to the ends of the mica "fish." In both cases, micas in the S-foliation curve toward parallelism with C, and the sense of curvature defines the shear sense on the shear zone: Counterclockwise indicates sinistral shear, clockwise indicates dextral shear (see Figure 4.17C). With large amounts of shearing, the S- and C-foliations may become essentially parallel and indistinguishable, and a new foliation, labeled C′, may develop that has characteristics similar to those of the C-foliation but is oriented at a low angle to the shear zone boundaries.

Although the S-C morphology is similar to some examples of crenulation foliation (Figure 13.9C, for example), it is important to recognize that the S and C surfaces form during the same deformation, whereas crenulation foliations result from the superposition of two separate deformations.

A **transposition foliation** results from the superposition of a tectonite foliation on an earlier compositional layering, such as bedding or a compositional foliation. With progressive deformation, the compositional layering becomes isoclinally folded and dismembered (Figure 13.17). The earlier layering is transformed into a discontinuous banding parallel to the new foliation and the folds are no longer recognizable except possibly for scattered **rootless folds,** which are isolated isoclinal fold hinges that have axial surfaces parallel to the foliation and are not connected to any other hinges.

13.6 Special Types of Foliations and Nomenclature

Many terms for various types of foliations exist in the geologic literature. Some terms are strictly descriptive and are therefore useful in referring to specific morphologic features. Others have genetic connotations that may be misleading. We strongly recommend abandoning the use of genetic terms for descriptive purposes because many are not well defined, and their use may lead to incorrect assumptions about the origin of structures. Interpretation, of course, has a valid place in any scientific investigation, but the descriptive use of inter-

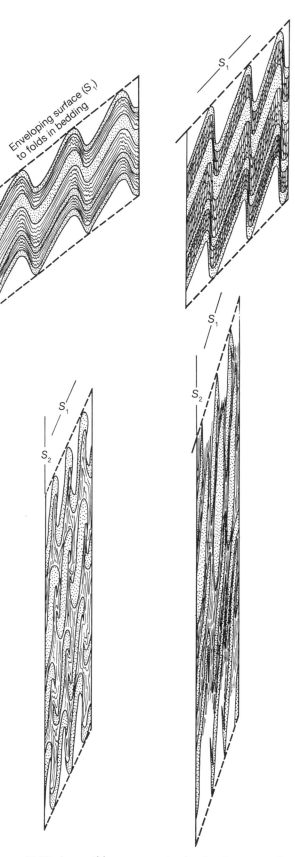

Figure 13.17 A possible sequence in the development of a transposition foliation. S_1 is the enveloping surface to the folds in bedding. S_2 is the transposition foliation.

pretive terms inevitably leads to confusion. In this section we review some common terms and indicate what we believe to be the equivalents in the morphological classification presented in Sections 13.1 through 13.5.

Four terms—*slaty cleavage, phyllitic cleavage, schistosity,* and *gneissic foliation*—are not so specific as categories in the morphological classification given earlier, but they remain useful terms for general and field description. The first three of these describe a continuum in grain size for foliations in rocks containing abundant platy minerals. The last pertains to rocks in which platy minerals are not abundant.

Slaty cleavage refers to fine continuous foliations characteristic of slates. Slates are very fine-grained, low grade metamorphic rocks that contain abundant sheet silicates (generally clays, chlorites, and micas). They may also contain subordinate amounts of silty and carbonaceous material. The foliation may be either continuous or micro-spaced, but in the latter case, the micro-domain spacing certainly cannot be recognized in the field. The foliation provides a very strong cleavage to the rock, along which the rock breaks easily and tends to weather preferentially. Rocks with slaty cleavage traditionally have been a valuable source of materials such as roofing slates and blackboards.

Phyllitic cleavage resembles slaty cleavage except that the grain size of the rock is slightly coarser. It characterizes phyllites, which are low (greenschist) grade, fine-grained, metamorphic rocks that contain abundant micas and/or chlorite. In hand sample, the surface of the foliation has a sheen, and individual sheet silicate flakes may be just resolvable with a good hand lens. The foliation is generally intermediate between a fine and a coarse continuous foliation, although some phyllitic cleavages may be smooth disjunctive foliations. The foliation strongly affects the rock's weathering pattern.

Schistosity refers to the foliation found in coarse-grained, mica-rich, medium-to-high-grade metamorphic rocks. Chlorite, biotite, or muscovite defines the foliation, and the mineral grains are coarse enough to be visible with the unaided eye. This foliation may appear as an anastomosing to smooth disjunctive foliation or as a coarse continuous foliation. It provides a strong cleavage to the rock.

Gneissic foliations are foliations that develop in gneisses, which are coarse-grained, high-grade metamorphic rocks in which platy minerals are sparse or absent. The term includes compositional foliations as well as coarse continuous foliations defined by the alignment of sparse platy minerals, by flattened mineral grains, or by needle-shaped mineral grains. The foliation generally provides at best a weak cleavage.

The following terms all have a genetic connotation and will not be used in this book. We include them for the sake of completeness and, we hope, for strictly historical interest and reference in understanding the older literature.

Flow cleavage is a loosely defined term that seems to have been applied to continuous axial surface foliations interpreted to have been the result of a large amount of ductile deformation in the rocks. It was commonly, and erroneously, interpreted to represent the orientation of flow (shear) planes in the rock during ductile deformation.

Fracture cleavage refers to a variety of disjunctive foliations or discrete crenulation foliations. The term has most often been applied to disjunctive foliations in which the microlithon has little or no fabric and the cleavage domains are thin and can have the superficial aspect of a penetrative set of fractures, especially on weathered surfaces. The term is misleading because the fractures that are observed are in general secondary structures that form along previously developed foliation planes.

Shear cleavage, solution cleavage, and **strain–slip cleavage** are terms that have been used to describe a variety of spaced foliations. Solution cleavage refers to disjunctive foliations, especially at the more irregular end of the scale. Shear cleavage and strain–slip cleavage both refer to crenulation foliations. None of these terms is well defined, and none should be used descriptively. If solution or shearing has been independently demonstrated, the use of *solution cleavage* or *shear cleavage* as an interpretive term is acceptable.

13.7 Structural Lineations

Figure 13.18 shows a morphological classification of lineations in deformed rocks. We discuss the two main subdivisions, structural lineations and mineral lineations, in this and the following section.

Structural lineations are defined by the preferred orientation of a linear structure contained within a rock. They include **discrete lineations,** which are formed by the deformation of discrete objects such as ooids, pebbles, fossils, and alteration spots, and **constructed lineations,** which are formed from planar features constructed or deformed during the deformations and include the intersection of two foliations, crenulation hinge lines, boudin lines, structural slickenlines, and mullions.

Discrete Lineations

Ductile deformation of the rock may distort discrete objects in the rock into well-aligned elongate shapes. Discrete lineations of this nature include stretched peb-

Lineations in tectonites (surficial or penetrative)	Structural	Discrete	Pebbles Ooids Fossils Alteration spots
		Constructed	Hinge lines Intersections Boudin lines Mullions Structural slickenlines
	Mineral	Polycrystalline	Rods Mineral clusters Mineral slickenlines Nonfibrous overgrowths

		Mineral grain	Acicular habit grains
			Elongated grains
			Mineral fibers
			    Fibrous vein filling
			    Slickenfibers
			    Fibrous
			        overgrowths

Figure 13.18 Morphological classification scheme for lineations.

ble conglomerates (Figure 13.19*A*), deformed oolitic limestone, and slates with alteration spots (Figure 13.19*B*). In these cases, objects that were roughly spherical before deformation are deformed into ellipsoidal shapes whose long and intermediate axes (*a* and *b*, respectively, in Figure 13.19*C*) may define a foliation, and whose long axes define a lineation. The true orientation of the lineation is apparent only on planes that contain the *a* axis of the ellipsoid. Although other sections through an ellipsoid are generally elliptical in shape, the long axis of such an ellipse is not the true lineation.

Alteration spots, also called reduction spots, are volumes in rock distinguished mainly by color differences caused by chemical alteration of some of the rock's components (Figure 13.19*B*). They may be initially spherical features that develop in the sediment shortly after deposition; they are most common in slates.

Constructed Lineations

A variety of lineations fall into the category of constructed lineations, and they have in common the characteristic that the structures originated during deformation of the rock.

The intersection of two planar elements such as two foliations, one of which may be bedding, defines an **intersection lineation.** If one foliation is defined by platy minerals or flattened mineral grains, the lineation appears on the intersecting foliation as the parallel alignment of the edges of platy mineral grains or of the long axes of the flattened mineral grains (for example, the

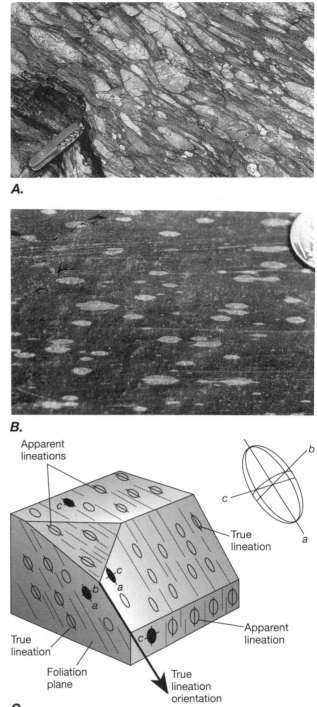

A.

B.

C.

Figure 13.19 Discrete lineations. *A.* A stretched pebble conglomerate showing quartzite pebbles flattened parallel to the foliation and elongated to define a lineation. *B.* Alteration spots in a slate. The foliation is perpendicular to the plane of the photo, and the maximum and minimum axes of the ellipsoids are exposed (*a* and *c*; see also part C). *C.* True and apparent lineations associated with ellipsoidal structures. The true lineation orientation is shown on any plane containing the *a* axis (longest axis) of the ellipsoid. Planes of other orientations show elliptical sections through the ellipsoidal structures that do not define the true orientation of the lineation.

intersection of S_1 with the S_0 surface in Figure 13.20A). If one foliation is bedding or a spaced foliation in which the cleavage domains and the microlithons differ in mineralogy, the intersection lineation may appear on the intersecting foliation as streaks of different composition (for example, the intersection of S_0 with the S_1 surface in Figure 13.20A). Two intersecting foliations can produce a lineation called a **pencil cleavage** (Figure 13.20B), so named because the tendency of the rock to cleave along both foliation surfaces produces elongate rhombic prisms or "pencils."

Fold hinge lineations are defined by the preferred orientation of microfold or crenulation hinges devel-

oped in foliations. The crenulations may, but need not, be associated with a crenulation foliation. On a regional scale, the orientation of hinges of outcrop-sized folds may be treated as a regionally penetrative lineation.

Boudins (the French word *boudin* means "blood sausage") are linear segments of a layer that has been pulled apart along periodically spaced lines of separation called **boudin lines** (Figure 13.21A). The boudins may be completely separated from one another, or they may be connected by an attenuated **neck** in the layer, in which case the boudin line may also be called a **neck line**. Boudins are most easy to recognize in an exposure that is at a high angle to the boudin line (Figure 13.21B, C). The process of forming boudins is called **boudinage**.

Boudins display a wide variety of shapes. **Pinch-and-swell structures** are rather gentle oscillations in the thickness of a bed. The necks between boudins may be smoothly curved, or they may look more like fractures. In some cases the necks appear broken, and the space between the broken ends may be filled in with a sec-

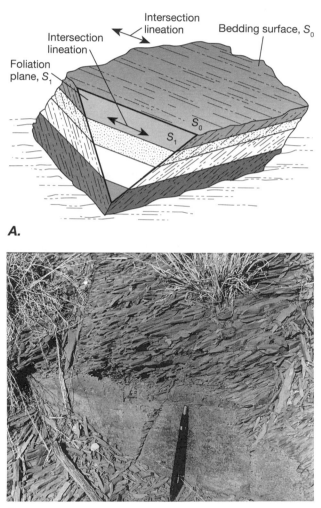

A.

B.

Figure 13.20 Intersection lineations. *A.* The intersection of a foliation and a bedding surface. The trace of the secondary foliation S_1 on the bedding S_0 and the trace of S_0 on S_1 are essentially the same lineation. *B.* Pencil cleavage in argillite, an intersection lineation defined by the intersection of two foliations, one of which may be bedding. Cleavage of the rock along both foliations produces elongate prisms, or pencils, of rock.

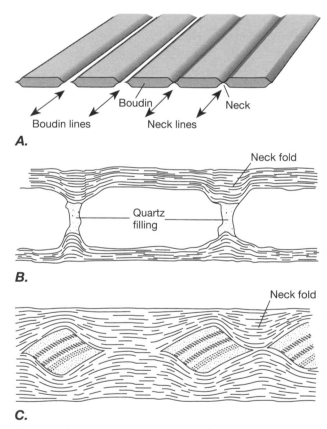

A.

B.

C.

Figure 13.21 Boudins. *A.* Diagram of a boudinaged layer, showing the relationship between the boudins, the necks, and the boudin and neck lines. When viewed in a two-dimensional section normal to the boudin line, boudins often look like a chain of link sausages. *B.* Boudins separated by fractured necks with concentrations of quartz filling the gap between. *C.* Boudins offset along their boudin lines and rotated.

ondary mineral, commonly quartz or calcite (Figure 13.21*B*). In some boudins there is little or no thinning of the layer. Instead it may shear along a surface oblique to the layering. Subsequent separation and rotation of the segments form a string of boudins shaped, in cross section, like rhombs or parallelograms (Figure 13.21*C*).

Where finely laminated layers are present on either side of the boudinaged layer, laminations near the necks commonly describe disharmonic folds, called **neck folds,** that conform to the interface between the layers (Figure 13.21*B, C*) and die out a short distance away from the interface. Any fine laminations in the boudinaged layer itself also bend and thin into the neck area.

Most boudin lines define a pronounced lineation. Some, however, display much scatter in orientation, and others may occur in two intersecting sets, with tablet-shaped boudins called **tablet boudinage,** or **chocolate tablet structure,** because of their resemblance to the tablets of a chocolate bar.

Structural slickenlines (Figure 4.8*A, B*) are grooves and ridges that appear on **slickensides,** the polished surfaces that develop along faults.[3]

Mullions are linear fluted structures developed within a rock or at lithologic interfaces. The name derives from the resemblance of the geologic structure to the vertical fluted architectural structures, called mullions, that separate windows in Gothic cathedrals. They are characterized in cross section by convex surfaces with intervening cusps (Figure 13.22*A*) or by alternating convex and concave surfaces (Figure 13.22*B*). Characteristically, they have a cross-sectional dimension of a few to several tens of centimeters and an indefinite linear dimension.

The surfaces of **fold mullions** are defined by parting along the cylindrically folded surfaces of layers of foliations. At a boundary between two thick layers of very different competence, such as a sandstone and a shale,

A.

---

[3] Slickensides are surfaces, although a number of authors use the term to refer to the lineations in the surface. Three types of lineations develop on slickensides: structural slickenlines, mineral slickenlines, and slickenfibers. The first is described here; the other two are described in the next section. The mechanisms by which these lineations are produced are discussed in Chapter 14.

B.

Figure 13.22 Mullions. *A.* Fold mullions in a sandstone at the contact with a shale (now eroded away). The mullion surface is restricted to the bedding surface; it does not form a closed cylindrical surface. *B.* Irregular mullions showing the irregular cross section and the strongly cylindrical structure of the lineation. The mullion surfaces may be coated with a thin film of mica.

the mullions appear in the more competent member as cylindrical surfaces convex toward the incompetent rock, joined by cusps that point into the competent rock (Figure 13.22A). This type of mullion is restricted to the bedding surface.

Irregular mullions are long fluted structures showing an irregular cross section that conforms in general neither with bedding nor with foliation (Figure 13.22B). The surfaces of the mullions may be covered with a thin film of mica, and the surface of one mullion fits exactly against that of its neighbors. Some structures of this nature are fault mullions and result from irregularities in the fracture surface (Figure 4.8B). Other irregular mullions, such as in the one shown in Figure 13.22B, are not well understood.

13.8 Mineral Lineations

Mineral lineations consist of a preferred orientation of either individual elongate mineral grains or elongate polycrystalline aggregates (Figure 13.18). **Mineral grain lineations** are formed by the parallel alignment of individual acicular (from the Latin word *aciculus,* which means "needle-like") mineral grains such as amphibole, by grains of minerals that have been stretched into an elongate shape, or by mineral fibers that have grown in a preferred orientation. **Polycrystalline mineral lineations** are formed by the preferred orientation of elongate clusters of grains of a particular mineral measuring at least a few grains in diameter. A preferred orientation of a crystallographic axis of the mineral is commonly associated with both types of lineation, although it need not be parallel to the orientation of the lineation. Mineral lineations may occur as surficial lineations on lithologic contacts, foliation surfaces, or fault surfaces and as penetrative lineations in the rock.

Polycrystalline Mineral Lineations

A variety of structures fall into the general category of polycrystalline mineral lineations.

Rods are polycrystalline mineral lineations formed by rod-shaped concentrations of a particular mineral, commonly quartz (Figure 13.23). The rods may appear in cross section to be isolated cylindrical masses, rootless fold hinges, or boudinaged layers or fold limbs. Thus they can in some cases also be classified as constructed structural lineations. They vary from approximately one to several tens of centimeters in diameter and typically occur parallel to a foliation plane and to the local orientation of fold hinges.

In many metamorphic rocks, **mineral cluster lineations** form small elongate concentrations or clusters

Figure 13.23 Quartz rod lineations. Rods are generally parallel to local fold hinges, and they may be isolated fold hinges or boudinaged fold limbs. In some cases, therefore, they could be classified as structural lineations as well.

of individual minerals on the scale of a millimeter to a several centimeters (Figure 13.24). The texture of the minerals in the clusters is no different from that in any other part of the rock. The lineations may be quite subtle, as when they are defined by small polycrystalline trains of muscovite and of quartz in a quartz–feldspar–muscovite schist, or they may be strikingly obvious, as when they are defined by elongate clusters of quartz, of feldspar, and of biotite in a gneiss (Figure 13.24). The lineations generally lie in a foliation plane, but in the case of a so-called **pencil lineation,** the rock fabric is dominated by a strong mineral cluster lineation, and no foliation is evident.

Mineral slickenlines appear as streaks developed on slickensides in fault zones (Figure 13.25). The streaks are probably the remnants of mineral grains or aggregates sheared out in the slickenside material, but the grain size is so small that individual mineral grains usually cannot be identified, even with a hand lens.

Figure 13.24 Mineral cluster lineation in a quartz–feldspar–biotite schist defined by elongate concentrations of quartz and feldspar and of biotite.

Figure 13.25 Mineral slickenlines on the slickenside of a fault surface.

These lineations may not always be distinguishable from structural slickenlines such as those shown in Figure 4.8*A, B*, and the two types of lineations commonly occur together.

Nonfibrous overgrowths are concentrations of one mineral—commonly quartz—around inclusions or grains of another mineral such as garnet or pyrite. Both nonfibrous and fibrous overgrowths are often referred to collectively by the genetic term **pressure shadows.** Such overgrowths may define a polycrystalline mineral lineation if the overgrowths are elongate and have a preferred orientation. Mineral grains in the overgrowth do not necessarily have a dimensional or crystallographic preferred orientation, so the lineation is defined strictly by the dimensional preferred orientation of the overgrowth (see Figure 13.26*B*).

Mineral Grain Lineations

Three types of mineral grain lineations commonly occur in rocks. They are formed by acicular (needle-shaped) minerals, by elongate mineral grains, and by mineral fibers, respectively. The grain shape may, but need not, be simply related to the crystallography.

Some mineral grains, such as amphiboles and sillimanite, naturally grow with a prismatic or acicular habit. If their long axes have a preferred orientation, such minerals define an **acicular habit lineation.** If one crystallographic axis (the *c* axis in amphiboles and sillimanite, for instance) parallels the long axis of each

mineral grain, the lineation is parallel to a crystallographic preferred orientation.

Under some conditions, **elongated grain lineations** may form in a rock by deformation of preexisting equant mineral grains into aligned elongate forms. Such mineral grains approximate triaxial ellipsoids in shape, and the lineation is parallel to the longest axis of the ellipsoids (see Figure 13.19*C*). These lineations are similar to the discrete lineations described in Section 13.7. Crystallographic axes usually are aligned as well, but that alignment need not be parallel to the morphologic alignment of the mineral grains.

Mineral fiber lineations are formed by very elongate crystal grains of a particular mineral—commonly quartz, calcite, chlorite, or serpentine. The structure and composition of the mineral fibers are so distinct from those of the rock in which they occur that it is clear the fibers have grown in the rock during deformation. They

A.

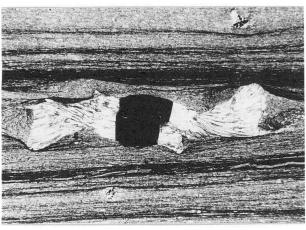

B.

Figure 13.26 Mineral fiber lineations. *A.* Curvilinear serpentine slickenfibers on a fault surface. *B.* Quartz fiber overgrowths on a pyrite grain in phyllite.

occur packed densely together in fibrous sheets or bunches in which all the fibers are strongly aligned in either a linear or a curvilinear arrangement (Figure 13.26).

Mineral fiber lineations are commonly found as surficial lineations both in **fibrous vein fillings** in veins and as **slickenfiber** lineations along fault planes (Figure 4.8C, and 13.26A). In both cases, the mineral fibers have a very strong preferred orientation, which is generally at a high angle to the vein wall and at a low angle to the fault surface. Mineral fiber lineations may also form a penetrative lineation, where they occur in strongly oriented **fibrous overgrowths** on crystals or particles throughout the rock (Figure 13.26B). These lineations are common in low-grade metamorphic rocks.

If the fibers in any of these mineral fiber lineations are strongly curved (Figure 13.26A), or if they occur on planes that have a wide diversity of orientation (Figure 10.17C–E) it may be difficult to define a unique lineation for the rock. Nevertheless, the study of these mineral fibers can yield significant information about the deformation and its history during the fiber growth.

The very strong linear preferred orientation of the fibers need not reflect a comparable preferred orientation of their crystallographic axes. Many mineral fiber lineations display nearly random distribution of crystallographic axes, although most quartz and calcite fiber lineations have a strong crystallographic preferred orientation.

13.9 Associations of Lineations with Other Structures

Lineations rarely are the only structure in an area, and the way they are related to other structures can help us understand the structural history. The fabric of some rocks is completely dominated, at least locally, by a lineation. Pencil gneisses, are characterized by a strong pencil lineation. Many lineations, however, are parallel to and lie within foliations or other planar features, and many are geometrically related to fold axes. A given area commonly contains different types of lineations, which may all have the same orientation, although that is not necessary.

Lineations and Foliations

Some lineations are defined at least in part of foliations, and of course these types must be parallel to that foliation. Intersection lineations (Figure 13.20A), including pencil cleavage (Figure 13.20B), must be parallel to the surfaces that defines them.

Other lineations are defined by features that characteristically lie in a foliation. Acicular mineral grains may be oriented parallel to a plane, defining a foliation, and they may also have a preferred orientation within the plane, defining an acicular habit lineation. Fold hinge lineations, fold mullions (Figure 13.22A), and in many cases rods (Figure 13.23) depend on folding for their linear character. If the folds are associated with an axial surface foliation, then these lineations must be parallel to that foliation. Discrete lineations and mineral cluster, acicular habit, and elongated grain lineations also commonly lie in a foliation defined by platy minerals.

Some lineations, such as boudin lines, mineral fiber lineations, and structural and mineral slickenlines, are not defined by a foliation and do not contribute to the definition of one. Whether such lineations parallel a foliation depends on the geometry of the deformation.

Lineations, of course, may develop on surfaces other than foliations. Slickenlines and slickenfibers are often found on fault surfaces, and slickenfibers may be found on bedding surfaces in some circumstances, especially associated with flexural-slip folds. Fold hinge lineations, boudin lines, and fold mullions must be parallel to the lithologic layers in which they develop. Intersection lineations involving lithologic layering, of course, must lie in the plane of the layering.

Lineations and Folds

The relationship between folds and lineations can be of major importance in our efforts to decipher the structural geometry of an area and interpret the conditions under which the structures formed. Some lineations, such as fold hinge lineations, fold mullions (Figure 13.22A), and rods (Figure 13.23) are generally parallel to the regional distribution of fold hinges. An intersection lineation defined by a folded surface and by the axial foliation to the folds also parallels the fold axis if the folding is close to cylindrical. Mineral lineations also are commonly parallel to fold hinges.

Because lineations are generally smaller-scale structures than folds, and because small-scale structures commonly reflect the geometry of large-scale structures, it may be easier to map the geometry of fold hinges by mapping the orientation of the appropriate lineations. The parallelism of a particular lineation with the hinges of a particular generation of folds, however, must be established independently.

Lineations such as boudin lines, acicular mineral grains, elongate mineral grains, and mineral cluster lineations, as well as discrete lineations and overgrowth lineations, may be found either parallel or perpendicular to fold axes. Some lineations, such as acicular habit lineations, have also been observed to be parallel to fold axes in hinge zones but perpendicular to them on the limbs. Slickenfiber lineations on folded bedding surfaces are usually perpendicular to the associated fold hinge. They are most strongly developed on the limbs and fade to nonexistent in the hinge zone.

Less often, lineations are found at arbitrary angles to fold axes. Such a geometry is usually the result of the deformation of earlier lineations, as discussed in more detail in Chapter 16. Other possibilities cannot be dismissed, however, and each situation requires individual investigation.

Lineations often can be used to infer the distribution and geometric characteristics of the deformation in an area. We postpone discussion of these topics until after Chapter 15, where we introduce the concept of strain as a measure of deformation (see Section 16.7).

Additional Readings

Alvarez, W., T. Engelder, and P. A. Geiser. 1978. Classification of solution cleavage in pelagic limestones. *Geology* 6: 263–266.

Cloos, E. 1957. Lineation, a critical review and annotated bibliography. *Geol. Soc. Am. Mem.* 18.

Engelder, T., and S. Marshak. 1985. Disjunctive cleavage formation at shallow depths in sedimentary rocks. *J. Struct. Geol.* 7(3/4): 327–343.

Fleuty, M. J. 1975. Slickensides and slickenlines. *Geol. Mag.* 112: 319–322.

Glen, R. A. 1982. Component migration patterns during the formation of a metamorphic layering, Mount Franks area, Willyama Complex, N.S.W., Australia. *J. Struct. Geol.* 4: 457–468.

Gray, D. R. 1977. Morphologic classification of crenulation cleavage. *Jour. Geol.* 85: 229–235.

Gray, D. R. 1978. Cleavages in deformed psammitic rocks from southeastern Australia: Their nature and origin. *Geol. Soc. Am. Bull.* 89: 577–590.

Lister, G. S., and A. W. Snoke. 1984. S-C mylonites. *J. Struct. Geol.* 6: 617–638.

Powell, C. Mc A. 1979. A morphological classification of rock cleavage. *Tectonophysics* 58: 21–34.

Turner, F., and L. Weiss. 1963. *Structural analysis of metamorphic tectonites*. New York: McGraw-Hill.

Weber, K. 1981. Kinematic and metamorphic aspects of cleavage formation in very low grade metamorphic slates. *Tectonophysics* 78: 291–306.

CHAPTER 14

Formation of Foliations and Lineations

The diverse foliations and lineations described in the previous chapter form in many different ways, and their significance is not always obvious. The principal causes of foliations and lineations are ductile flattening and elongation of the rock itself; mechanical rotation; solution and precipitation; and recrystallization. The latter two mechanisms commonly involve chemical reaction. The composition of the rock also influences the type of foliation developed. Stylolitic foliations, for example, are largely restricted to limestones and marbles, although they also form in other rocks, principally calcareous or argillaceous sandstones. A quartz-rich sandstone characteristically develops a rough to smooth disjunctive foliation, but never a fine continous foliation. Similarly, a crenulation foliation generally forms in rocks that contain a high proportion of platy minerals, although it may also develop in finely laminated rocks.

In this chapter we discuss the principal mechanisms of formation of foliations and lineations. To do so, we need first to introduce a few concepts of the geometry of deformation, which is treated in more detail in Chapter 15.

14.1 Shortening and Lengthening

At the beginning of Chapter 12, we introduce the concepts of flattening and of shear in two dimensions by considering the deformation of a square of material.

When the square is homogeneously flattened, it changes into an elongate rectangle (see Figure 12.2*A, B*). When the square is homogeneously sheared, it changes into a parallelogram (see Figure 12.1*A, B*).

Generalized to three dimensions, a **flattening** deformation changes a cube of material into a rectangular prism (Figure 14.1*A*), of which one dimension has been shortened and the other two have both lengthened. If two dimensions of the cube are shortened and only one is lengthened the deformation is a **constriction** (Figure 14.1*A*). Shear of a cube parallel to one face and one edge produces a rhombohedron (Figure 14.1*A*). In each case the original volume of the cube can be, but is not necessarily, conserved. The shaded faces of the blocks show the two-dimensional deformation in the plane that contains the directions of maximum shortening and lengthening.

Instead of using a cube, it is more convenient to represent the effect of deformation by looking at a circle or a sphere, because all lines in all directions start out the same length. The left sides of the blocks in Figure 14.1*A* illustrate the relationship between the deformation of a square and of a circle inscribed in the square. Figure 14.1*B* shows the deformation of a sphere equivalent to the deformation shown in Figure 14.1*A*. The circle is deformed into an ellipse called the **strain ellipse,** and the sphere is deformed into an ellipsoid called the **strain ellipsoid.** We can define the geometry of an ellipse (or an ellipsoid) by the lengths of the two (or three) principal axes. These axes are the **principal axes of strain.** In three

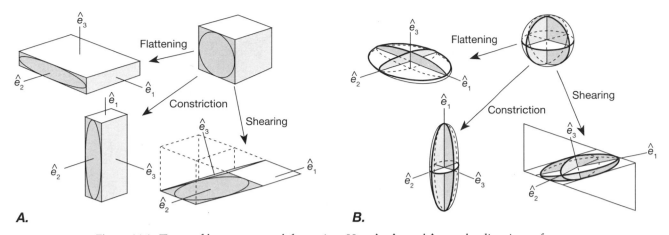

Figure 14.1 Types of homogeneous deformation. Here $\hat{e}_1$, $\hat{e}_2$, and $\hat{e}_3$ are the directions of maximum, intermediate, and minimum extension. A. A cube undergoes flattening if two of its dimensions are lengthened and one is shortened; it undergoes constriction if two of its dimensions are shortened and one is lengthened; and it undergoes shearing if the cube is deformed so that its cross section is a rhomboid. The shaded faces of the blocks show the deformation in the $\hat{e}_1$–$\hat{e}_3$ plane. A circle is deformed into the strain ellipse whose maximum and minimum axes are parallel to $\hat{e}_1$ and $\hat{e}_3$. B. The same geometries of deformation as in part A imposed on a sphere produce ellipsoids called strain ellipsoids whose principal axes are the principal directions of extension $\hat{e}_1 \geq \hat{e}_2 \geq \hat{e}_3$. If subjected to flattening, a sphere is deformed into a pancake-shaped (oblate) ellipsoid; if subjected to constriction, it becomes a cigar-shaped (prolate) ellipsoid; and if subjected to shearing, it becomes an ellipsoid with axes inclined relative to the shear plane and with no deformation parallel to the $\hat{e}_2$ direction. The strain ellipses in part A are sections through the strain ellipsoids.

dimensions, each of the three principal extensions[1] $\hat{e}_1 \geq \hat{e}_2 \geq \hat{e}_3$ (see Equation 9.1) is the change in length of the principal ellipsoid radius divided by the length of the original radius of the sphere. In two dimensions, we generally use only $\hat{e}_1$ and $\hat{e}_3$. The **plane of flattening** is the $\hat{e}_1$–$\hat{e}_2$ plane, and the direction of maximum principal extension $\hat{e}_1$ is also called the direction of maximum **stretch**. We discuss these concepts in more detail in Chapter 15.

A rock is made up of mineral grains[2] and commonly contains fossils or other deformable objects. Objects that initially are spherical including ooids, radiolarian tests, and alteration spots, are changed into ellipsoids by a deformation such as a flattening. These deformed objects are aligned parallel to the plane of flattening ($\hat{e}_1$–$\hat{e}_2$) and to the axis of maximum stretching ($\hat{e}_1$) (Figure 14.2A), providing the most straightforward mechanism for formation of a foliation or lineation, respectively. A lineation that is parallel to $\hat{e}_1$ is called a **stretching lineation.**

Other features, such as equant mineral grains, clusters of mineral grains, and clasts in a conglomerate, may

not have an initially spherical shape or random distribution of orientations. A deformation such as a flattening, however, also changes these features to produce foliations and lineations (Figure 14.2B, C), although the effect of an initial preferred orientation can never be completely eliminated (Figure 14.2C).

Boudins form during deformation if there is a component of lengthening parallel to a competent layer in an incompetent matrix (Figure 14.3A). Ductile extension of the competent layer cannot keep pace with that of the incompetent matrix, and the layer tends to pull apart into boudins. As the difference in competence increases from small to large, the form of the boudins changes from a pinch-and-swell structure, through separated boudins with pronounced necks, to boudins with sharp ends that may actually be fractures (as illustrated in the progression from top to bottom in Figure 14.3A). If the matrix is too competent to flow around the separating boudins, the region of low stress between the boudins becomes a favorable site for precipitation of a mineral such as quartz or calcite.

In simplified terms (Figure 14.3B), the incompetent material exerts a shear stress σ_s on each side of the layer, which provides a total force $2F_s$ parallel to the layer and proportional to the length L on which it acts ($F_s = \sigma_s L$). A deviatoric tensile stress $_\Delta\sigma_n$ acting across the thickness T of the competent layer provides a force

---

[1] As in the notation for stress, we use a circumflex to indicate a principal value.
[2] Excepting, of course, coal and volcanic rocks composed of glass.

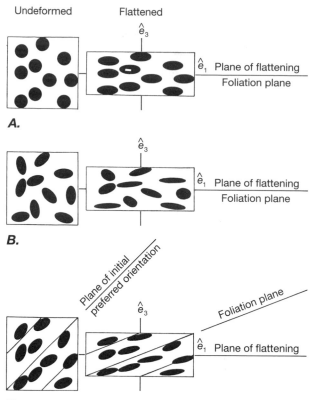

A.

B.

C.

Figure 14.2 Foliations formed by the deformation of discrete objects in the rock. *A.* Flattening of initially spherical objects produces a foliation parallel to the plane of flattening (the $\hat{e}_1$–$\hat{e}_2$ plane). *B.* Flattening of randomly oriented elliptical objects produces a foliation statistically parallel to the plane of flattening. *C.* Flattening of initially elliptical objects with an initial preferred orientation produces a foliation in an orientation different from the plane of flattening.

$F_n = {}_\Delta\sigma_n T$ that balances the force $2F_s$. The length L^* at which ${}_\Delta\sigma_n$ reaches the yield stress ${}_\Delta\sigma_n^*$ of the competent material determines the location of failure of the layer and the characteristic length of the boudin in cross section.

$$F_n = 2F_s$$

$$_\Delta\sigma_n^* T = 2\sigma_s L^*$$

$$L^* = \frac{{}_\Delta\sigma_n^*}{2\sigma_s} T$$

14.2 Mechanical Rotation

The occurrence of mechanical rotation of mineral grains during deformation is demonstrated by "snowball" garnets (Figure 4.17E) and curvilinear fibrous overgrowths. Several lines of evidence indicate that rotation of mineral grains into a preferred orientation is an important mechanism of formation of foliations and lineations. For example, originally detrital mica grains, which are commonly parallel to bedding in undeformed sediments, are parallel to a tectonite foliation in the deformed equivalents. Mica grains in some crenulation foliations are rotated or bent into parallelism with the new foliation. The experimental deformation of rocks that contain randomly oriented micas has produced a preferred orientation under conditions for which rotation is the only possible mechanism of reorientation.

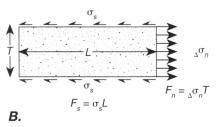

B.

Figure 14.3 Formation of boudins. *A.* Competent layers imbedded in an incompetent matrix and oriented parallel to the axis of maximum stretch accommodates the lengthening by segmenting into boudins. The contrast in competence between the layer and the matrix increases from zero (top) to high (bottom), resulting in a progression from (1) uniform stretching and thinning through (2) pinch and swell and (3) necked boudins to (4) fractured boudins. *B.* The length of each boudin is determined by the yield or tensile strength of the layer and the balance of forces created by the shear stresses σ_s on the layer surface and the deviatoric tensile stress within and parallel to the layer ${}_\Delta\sigma_n$.

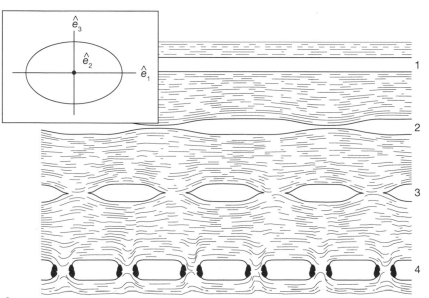

A.

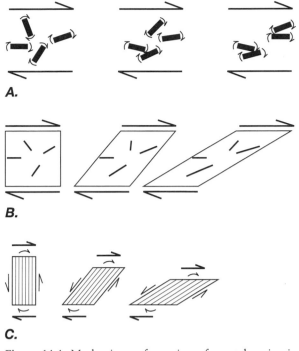

A.

B.

C.

Figure 14.4 Mechanisms of rotation of crystal grains in a deforming rock. *A*. The Jeffrey model: rotation of rigid elongate mineral grains in a ductile matrix. *B*. The March model: rotation of mineral grains that act as passive markers during the deformation. *C*. The Taylor-Bishop-Hill model: rotation of a mineral grain accompanying ductile shear on a single set of slip planes within the grain.

Such a rotation can occur in three ways: A mineral grain can rotate as a rigid particle surrounded by a ductile matrix (the **Jeffrey model**,[3] Figure 14.4*A*); it can act as a strictly passive marker in the deforming rock (the **March model**,[4] Figure 14.4*B*); or it can shear on crystallographic slip planes and rotate such that its deformation is compatible with that of the surrounding matrix (the **Taylor-Bishop-Hill model**,[5] Figure 14.4*C*).

The consequences of the Jeffrey model for a simple shearing deformation are apparent when a rigid rectangular plate is suspended in a shearing fluid. The plate rotates continually about an axis parallel to the shear plane and perpendicular to the shear direction. Although it never comes to rest, the plate rotates most slowly when it is parallel to the shear plane because in this orientation, the torque applied to the plate by the shearing fluid is a minimum. The plate rotates most rapidly

---

[3] After the British physicist G. B. Jeffrey, who in 1923 investigated theoretically the motion of rigid grains suspended in a deforming viscous fluid.
[4] After the German physicist A. March, who in 1932 analyzed theoretically the development of preferred orientation in deformable rods and plates.
[5] Named for the metallurgists G. I. Taylor, J. F. W. Bishop, and R. Hill who developed the theory to explain the formation of crystallographic preferred orientations in ductilely deformed metals.

when it is perpendicular to the shear plane because in this orientation the applied torque is a maximum. The shearing of a matrix could thus produce a concentration of the orientations of suspended rigid particles parallel to the shear plane. Such a preferred orientation defines a foliation.

For the March model, the rotation of passive particles in a continuum is different from that of rigid particles in a viscous matrix (Figure 14.4*B*). For example, during simple shearing of the medium, the passive markers do not rotate continually but instead approach a limiting orientation parallel to the shear plane (Figure 14.4*B*). They cannot rotate past the shear plane, and the rotation rate approaches zero as the markers approach parallelism with the shear plane. The resulting concentration of passive markers subparallel to the shear plane produces a preferred orientation and thus a foliation.

The Taylor-Bishop-Hill model shows how ductile shear on internal crystallographic slip planes of a mineral grain can result in rotation of the grain. For example, Figure 14.4*C* shows a hypothetical crystal grain that can deform by shear parallel to only one slip plane in one slip direction. Slip on this slip system, however, does not conform to the externally imposed geometry of shear, so the crystal must combine shear on its slip system with a rotation to make the internal and external deformations compatible. When the internal slip system coincides in orientation with the externally imposed shear, the crystal does not rotate further. This mechanism therefore tends to rotate crystals into a preferred orientation that can define a foliation.

During homogeneous flattening (Figure 14.1), any of the three rotation mechanisms results in the rotation of platy or elongate particles, such as mica plates or amphibole needles, toward parallelism with the plane of flattening (the $\hat{e}_1$–$\hat{e}_2$ plane) to produce a foliation. Figure 14.5 shows this effect for randomly oriented grains that behave passively during the deformation. During homogeneous constriction (Figure 14.1), platy or elongate grains rotate toward parallelism with the extension direction $\hat{e}_1$ to produce a lineation. Grains parallel to any of the principal planes do not rotate, and those initially subparallel to $\hat{e}_3$ remain at high angles to the foliation.

14.3 Solution, Diffusion, and Precipitation

Rock deformation that produces foliations and lineations depends in part on the mobility of mineral species through the rock. The mechanisms involved include the breakdown of minerals by solution and chemical reaction, the migration of the chemical components

Figure 14.5 Rotation of passive particles in a ductile matrix. If the deformation is homogeneous flattening, the particles rotate toward the plane of flattening and statistically define a foliation parallel to that plane

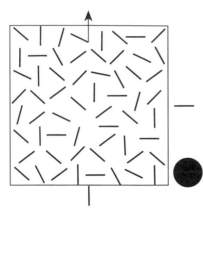

often reprecipitates locally, possibly in microlithons where it accommodates a local dilation (a volume increase); as overgrowths on preexisting minerals or particles in the rock (B in Figures 13.7 and 13.26B); as slickenfibers on shear surfaces (Figure 13.26A); or as fibrous or massive deposits in veins. In some cases, however, the bulk composition of the rock may be permanently changed by the removal or introduction of one or more chemical components.

Two major factors affect the dissolution of minerals. First, any deformed mineral is more soluble than an undeformed one because of its higher locked-in strain energy (see Chapter 19). Second, a crystal subjected to a differential stress tends to dissolve more readily at surfaces on which the normal-stress component is a maximum. This mechanism is commonly called **Riecke's principle**,[6] and the process is called **pressure solution** or **solution transfer**. A corollary of Riecke's principle is that minerals tend to precipitate at surfaces on which the normal-stress component is a minimum.

Volume-loss folding is accomplished by solution and removal of material from the folding layers (Section 12.3 and Figure 12.10). In the resulting folds (see, for example, Figure 12.11), the foliation develops as an axial foliation composed of solution seams.

Crenulation foliations commonly exhibit a compositional banding associated with the cleavage domains and microlithons. Cleavage domains tend to be enriched in platy minerals and depleted in quartz, compared both to the microlithons and to the uncrenulated rock (Figure 13.9A, C). In some cases the microlithons are enriched in quartz, suggesting solution of quartz in the cleavage domains and precipitation in the microlithons.

The banding shown in Figure 13.9A, C, for example, could result either from preferential solution of more highly deformed minerals in the cleavage domains or from pressure solution, particularly if quartz dissolved preferentially at a quartz–mica interface. If the maximum compressive stress $\hat{\sigma}_1$ were at a high angle to the limbs of the crenulations, and therefore to the quartz–mica interfaces, quartz would readily dissolve in the crenulation limbs. In the microlithons, however, the quartz–mica interfaces would be subparallel to $\hat{\sigma}_1$ and at a high angle to the minimum compressive stress $\hat{\sigma}_3$, and on those surfaces the quartz would tend to precipitate. This mechanism could therefore account for the migration of quartz from cleavage domain to microlithon.

Oriented overgrowths, or pressure shadows, on mineral grains and particles may originate in the same way. If the particles in question behaved relatively rigidly during ductile deformation of the surrounding ma-

through the rock, and the formation of new mineral grains by precipitation and recrystallization. Many spaced foliations result in part from such processes.

Stylolitic foliations, commonly found in deformed limestones and marbles, are perhaps the most familiar example. The irregular stylolite cleavage domains may truncate fossils, which indicates that part of the fossil has been dissolved (Figure 13.5B), and that solution has accommodated shortening across the stylolite. The material that fills these stylolites—largely clay minerals, iron oxides, and carbonaceous matter—is the insoluble residue from limestone solution and may include some secondary minerals as well.

In some deformed sandy argillites that have a rough disjunctive foliation, truncation of detrital sand grains against cleavage domains (C in Figure 13.7) results from solution rather than from shear displacement along the cleavage. Originally equant detrital grains, such as quartz, may ultimately be almost completely dissolved away into thin, platelike grains parallel to and partly defining the foliation (T in Figure 13.7). An insoluble residue of platy minerals and oxides forms the cleavage domains.

The dissolved material migrates through the rock, probably by grain boundary diffusion over short distances or by transport in a fluid flowing through pores or fractures over large distances. The dissolved material

---

[6] After the nineteenth-century German physicist E. Riecke.

terial, a zone of abnormally low stress would develop where the particle boundary is at a high angle to the minimum compressive stress $\hat{\sigma}_3$. Minerals in solution diffuse to the low-stress area and precipitate as either nonfibrous or fibrous overgrowths (B in Figures 13.7 and 13.26B).

14.4 Recrystallization

Recrystallization is the creation of new crystal grains out of old ones. During deformation, recrystallization can result in a preferred orientation of mineral grains. Two kinds of recrystallization are important in structural geology. In **coherent recrystallization** either old deformed grains are progressively transformed into new undeformed grains as a grain boundary migrates through the old crystal lattice, or old grains are subdivided into many new grains by the rotation of small internal domains called subgrains. The crystal structure and the composition of old and new grains are the same, although new grains have different lattice orientations from the old. In **reconstructive recrystallization**, the old crystal structure breaks down—for example, during a chemical reaction—and a new structure forms that generally has a different composition. The distinction between the solution/precipitation process and reconstructive recrystallization is not always well defined.

Both types of recrystallization can change the shape and arrangement of grains. A foliation or lineation can develop by solution or chemical reaction, for example, either by a selective destruction of old grains that leaves only grains with a particular orientation, or by the production of new grains that grow in a preferred orientation.

The evolution of slaty cleavage shown in Figure 13.11 is an example of the effect of reconstructive recrystallization. The initial foliation is a zonal crenulation foliation (Figure 13.11A). With increasing amounts of recrystallization of the platy clay minerals, the structure of the foliation changes to a discrete crenulation foliation, and then gradually to a disjunctive foliation with a very strong fabric in the microlithons but no remnant of the initial crenulations (Figure 13.11B). These slates are still low-grade metamorphic rocks.

Some phyllites contain relict sedimentary mica grains as well as newly crystallized micas. The relict grains are characterized by large, irregular grain shapes; the new grains are smaller with very regular grain boundaries and a strong preferred orientation that defines the new foliation. If recrystallization is extensive, it can obliterate all clues to the nature and preferred orientation of the original grains.

Mimetic growth (the Greek word *mimetikos* means "imitation") is the growth of new crystals that nucleate on older crystals of similar structure in an orientation governed by the orientation of the older crystal. In this way, the growth of new crystals can enhance any preexisting preferred orientation of the old crystals.

An existing foliation in a rock can also control the orientation of new mineral grains, because growth may occur parallel to the foliation more easily than across it. Micas, for example, grow most rapidly parallel to their cleavage plane, and new mica grains that nucleate with cleavage planes parallel to a foliation grow more rapidly than micas in other orientations, thereby enhancing the preexisting foliation.

Without some external controlling factor, reconstructive recrystallization generally does not produce a preferred orientation, and it can even destroy a preexisting one. Recrystallization in association with deformation, however, can produce a preferred orientation or enhance an existing one, and because these processes are very commonly associated, the interaction between them strongly influences the fabrics that result.

14.5 Steady-State Foliations

Different types of foliation are not independent of one another and in fact may develop from one another, as is evident from Figures 13.9, 13.11, and 13.17. Thus crenulation foliations develop from the deformation of an earlier foliation, and with progressive deformation and recrystallization, a crenulation foliation can evolve into a continuous foliation. Continuous foliations in turn can become crenulated (Figure 13.9) and evolve into a crenulation foliation.

Such sequences of development raise the question of whether a foliation ever reaches a "final" state of evolution, and if so, under what circumstances that could occur. In some regions, two or more cycles of foliation evolution have been decifered, each of which includes the crenulation of an initial foliation, whether of sedimentary or tectonic origin, followed by the formation of a crenulation foliation, commonly with compositional differentiation accentuating the difference between cleavage domains and microlithons, and followed in turn by increasing recrystallization and the development of a new continuous foliation.

Such circumstances are probably common, and make the realistic identification of different generations of foliation, and the correlation of particular generations from one area to another, highly suspect. The designation of different generations of foliation by numerical subscripts such as S_1, S_2, and so on thus may only have very local significance. If applied over a large area, such a system of notation indicates a possibly erroneous conclusion concerning the correlation of different foliations and the simplicity of the deformational history.

14.6 Slickenside and Mineral Fiber Lineations

Slickensides on faults generally contain lineations that parallel the direction of slip on the fault surface (Section 4.2). The various types include structural slickenlines, mineral slickenlines, and slickenfibers. Mineral fiber lineations occur not only as slickenfibers but also as fibrous vein fillings and fibrous overgrowths. These lineations originate by a variety of different mechanisms, which we discuss below.

Structural Slickenlines

Fault surfaces are never perfectly flat but contain minor irregularities or protrusions called **asperities** (the Latin word *asper* means "rough") (see Figure 19.2). If asperities are particularly strong and resistant to abrasion and fracture, they can scratch and gouge the opposite surface of the fault, giving rise to one type of structural slickenline (Figure 14.6A). Scratches and gouge marks end where an asperity breaks off the opposite surface of the fault. The length of these lineations is a lower bound for the displacement on the fault surface.

Small ridges can develop where the fracture plane is deflected behind a hard asperity. A corresponding groove, of course, must form on the opposite surface. Similarly, **ridge-in-groove lineations,** or **fault mullions,** form if the fault surface is an irregular surface rather than planar, and if the irregularities are linear and par-

allel to the slip direction (Figure 14.6B; see also Figure 4.8B). In both cases, the length of the lineation is not necessarily related to the amount of displacement on the fault, because the ridge and the matching groove form as part of the fracture propagation process, not as a result of the displacement on the fault. Thus they cannot be used to constrain the magnitude of the displacement.

The displacement on some fault surfaces is accommodated by solution where there is a component of shortening across the fault. The mechanism is similar to the production of stylolites, except that the solution surface, called a **slickolite surface,** is subparallel to the displacement direction rather than approximately normal to it, as for stylolites. The counterpart of the tooth structure on stylolites (Figure 13.5) is a **spike** on a slickolite surface (Figure 14.6C).

Mineral Slickenlines

Slickenlines defined by streaks on a slickenside result from the smearing out of mineral grains and soft asperities (Figure 14.6D). They may also accumulate behind hard asperities and may form in combination with scratches and gouge lineations.

Mineral Fiber and Slickenfiber Lineations

Mineral fiber lineations occur as slickenfibers on fault surfaces (Figure 14.7A, B), as fibrous vein fillings (Figures 14.7C, D and 14.9), and as fibrous overgrowths on grains or particles (Figure 14.8). The continuity and morphology of the mineral fibers across the vein or shear surface imply that fiber growth accompanied and kept pace with gradual displacement and crack opening.

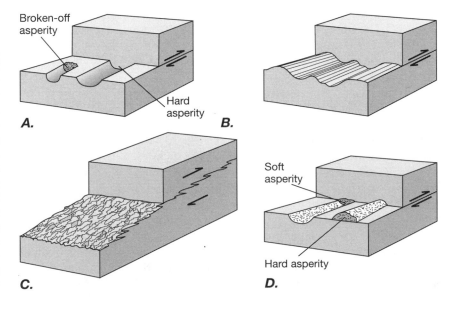

Figure 14.6 Structural and mineral slickenlines. *A.* Structural slickenlines formed by scratching and gouging of one side of the fault by hard asperities in the other side. *B.* Ridge-in-groove structural slickenlines formed by linear irregularities in the fault surface that parallel the slip direction. *C.* Spikes on a slickolite. The slickolite is a solution surface subparallel to the direction of displacement. The spikes are irregularities in the solution surface that parallel the slip direction and are comparable in origin to the teeth on stylolites. *D.* Mineral streak lineations form from the wearing down and smearing out of mineral grains and soft asperities or from the collection of gouge behind a hard asperity.

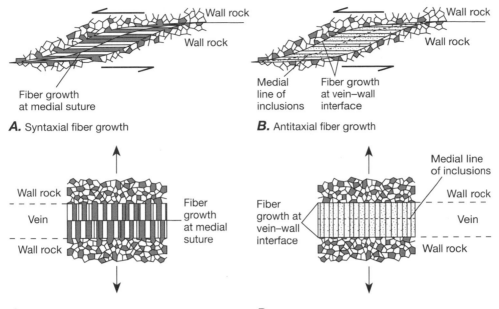

A. Syntaxial fiber growth

B. Antitaxial fiber growth

C. Syntaxial fiber growth

D. Antitaxial fiber growth

Figure 14.7 Comparison of syntaxial and antitaxial mineral fiber lineations in faults and veins. Syntaxial growth occurs if the mineral making up the fibers is also a common mineral in the host rock. If the mineral fibers are different from minerals in the host rock, antitaxial growth occurs. The arrows indicate the direction of displacement. *A.* Syntaxial growth of slickenfiber lineations on a fault surface. Growth occurs along the medial suture. *B.* Antitaxial growth of slickenfiber lineations on a fault surface. Growth occurs along the interface between the fibers and the wall of the fault. *C.* Syntaxial fiber growth in a vein occurs at the medial suture in the vein. *D.* Antitaxial fiber growth occurs at the interface between fibers and wall rock.

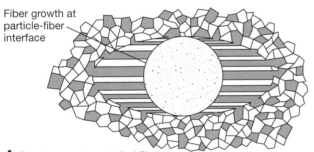

A. Displacement-controlled fibers

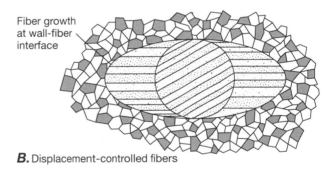

B. Displacement-controlled fibers

Figure 14.8 Fibrous overgrowths or pressure shadows. *A.* If the fiber mineral is similar to host rock minerals and different from the particle mineral, fibers grow in optical continuity with similar mineral grains in the wall rock. Growth occurs at the fiber–particle interface. *B.* If the fiber mineral is the same as the particle but different from the minerals in the host rock, fibers grow in optical continuity with the particle, here illustrated by a twinned calcite grain. Growth occurs at the fiber–wall rock interface. *C.* Face-controlled fiber orientation in a fibrous overgrowth on a pyrite cube. The mineral fibers grow perpendicular to the crystallographic faces of the pyrite. Growth occurs at the fiber–pyrite interface, and the suture line between differently oriented groups of fibers indicates the displacement of the corner of the grain.

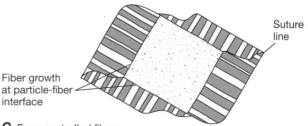

C. Face-controlled fibers

Transport of the mineral constituents to the ends of the fibers occurs either by diffusion along grain boundaries or, more probably, through a fluid phase in which the constituents are dissolved. Slickenfibers on fault surfaces therefore imply slow aseismic creep rather than large, rapid displacements that would produce earthquakes.

Whether the fibers are at low angles to fault surfaces (Figure 14.7A, B), at high angles to vein surfaces (Figure 14.7C, D; see also Figure 14.9), or attached to mineral grains or particles as overgrowths (Figure 14.8), they generally grow in an orientation parallel to the direction of displacement at the time of fiber growth (see Sections 16.3 and 17.5). In some cases, however, the fibers grow normal to the crystal faces of a mineral grain such as pyrite (Figure 14.8C). In this case, the suture line between the differently oriented fibers records the displacement history of the corner of the grain. The mechanism is not well understood, but the different fiber orientations at different crystal faces make this case easy to identify. Four types of displacement-controlled fiber growth are recognized that reflect the process by which growth has taken place.

In **syntaxial growth,** fibers tend to grow in optical continuity with mineral grains of the same composition. (In Greek, *syn* means "with, together" and *tassein* means "to arrange.") Thus for this structure to develop, the fiber mineral must be a mineral that is present in the host rock. For slickenfibers (Figure 14.7A) and vein fillings (Figure 14.7C), fibers extending from mineral grains in opposite walls meet at a medial suture where there is both a structural and an optical discontinuity. The suture is the site of latest growth of the fibers.

In fibrous overgrowths around a particle, the fibers can grow syntaxially on mineral grains in the host rock, with fiber growth occurring at the fiber–particle interface (Figure 14.8A), or they can grow syntaxially on the particle itself, as illustrated for a particle of twinned calcite in Figure 14.8B. In this case, fiber growth occurs at the interface between fiber and wall rock.

Antitaxial growth occurs when the fiber mineral is absent or uncommon in the host rock. In slickenfibers (Figure 14.7B) and vein fillings (Figure 14.7D), a medial suture may contain inclusions of host rock composition, but the fibers are optically and structurally continuous across the suture. Fiber growth occurs along the margins of the vein or fault, where there is a discontinuity in mineral composition.

Composite growth occurs when fibers of two different minerals grow, one of which is common, and one rare or absent, in the host rock. The fiber structure in veins may then show a central antitaxial band of the mineral that is rare in the host, flanked on both sides by syntaxial bands of the mineral that is common in the host. The fibers in both types of bands grow at the interface between the different bands.

Although we have discussed only straight fibers here, curved fibers on fault planes (Figure 13.26A) and in veins and overgrowths (Figure 13.26B) are relatively common. The curvature of the fibers in most cases records a component of rotation in the deformation during fiber growth.

Fibers may also grow by the **crack–seal mechanism,** which involves repeated microfracturing of the fiber along its length, followed by the deposition of optically continuous overgrowths that heal the fracture. Evidence for this mechanism includes abundant subplanar arrays of microscopic fluid inclusions crossing the fibers at the healed cracks (Figure 14.9). The result is sometimes called **stretched crystals.** They can occur if the fiber mineral is the same as the dominant mineral in the host rock. Crystal fibers are structurally and optically continuous across the whole vein, and they often connect, and are optically continuous with, two fragments of crystal grain on opposite sides of the vein that were originally a single grain. Because an increment of growth may occur at any place along the fiber, the orientation and shape of the fiber do not necessarily record the history of the displacement—only the net result.

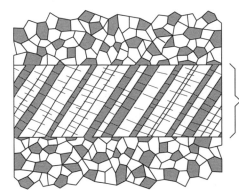

Fiber growth at randomly located cracks

Figure 14.9 Crack–seal growth forming a mineral fiber lineation of "stretched" crystals in a vein. Growth does not occur at any particular surface but by repeated cracking followed by deposition of fiber mineral to close the crack.

Additional Readings

Durney, D. W. 1972. Solution transfer, an important geological deformation mechanism. *Nature* 235: 315.

Durney, D. W., and J. G. Ramsay. 1973. Incremental strains measured by syntectonic crystal growth. In *Gravity and tectonics,* ed. K.A. DeJong, and R. Scholten. New York: John Wiley and Sons, pp. 67–96.

Grey, D. R. 1978. Cleavages in deformed psammitic rocks. *Geol. Soc. Am. Bull.* 89A: 5677–5690.

Mancktelow, N. S. 1979. The development of slaty cleavage, Fleurieu peninsula, South Australia. *Tectonophysics* 58: 1–20.

Marlow, D. C., and Etheridge, M. A. 1977. Development of a layered crenulation cleavage in mica schists of the Kanmantoo group near Macclesfield, South Australia. *Geol. Soc. Am. Bull.* 88: 873–882.

Means, W. D. 1981. The concept of steady-state foliation. *Tectonophysics* 78: 179–199.

Means, W. D. 1987. A newly recognized type of slickenside striation. *J. Struct. Geol.* 9: 585–590.

Ramsay, J. G. 1976. Displacement and strain. *Phil. Trans. Roy. Soc. Lond.* A283: 3–25.

Ramsay, J. G., and M. I. Huber. 1983. *The techniques of modern structural analysis.* Vol. 1, Strain Analysis. New York: Academic Press.

Ramsay, J. G., and M. I. Huber. 1987. *The techniques of modern structural analysis.* Vol. 2, Folds and Fractures. New York: Academic Press.

Tobisch, O. T., and S. C. Paterson. 1988. Analysis and interpretation of composite foliations in areas of progressive deformation. *J. Struct. Geol.* 10(7): 745–754.

Tullis, T. E. 1976. Experiments on the origin of slaty cleavage and schistosity. *Geol. Soc. Am. Bull.* 87: 745–753.

Weber, K. 1981. Kinematic and metamorphic aspects of cleavage formation in very low grade metamorphic slates. *Tectonophysics* 78: 291–306.

Willis, D. G. 1977. Kinematic model of preferred orientation. *Geol. Soc. Am. Bull.* 88: 883–894.

CHAPTER 15

Geometry of Homogeneous Strain

To further our understanding of the origin and significance of the folds, foliations, and lineations discussed in the last four chapters, we need to become more familiar with the nature of strain, as manifested in rocks. We introduced some concepts of strain in Chapters 7, 9, 12, and 14, but we need a more thorough and systematic understanding in order to evaluate theoretically the models proposed for formation of ductile structures, as well as to test these models against observations of natural deformation.

Our approach is largely geometric and qualitative, because our intent is to provide intuition into the physical characteristics of deformation, and strain lends itself easily to geometric description. The quantitative analysis of the ideas discussed in this chapter requires a rigorous mathematical treatment of strain, which we introduce in Box 15.1, and which is developed in depth in more advanced books on continuum mechanics and its geologic applications (see the list of readings at the end of this chapter). Readers interested in this approach should read through Section 15.2 before reading Box 15.1.

The **strain** of a body is simply the change in *size* and *shape* that the body has experienced during deformation. The strain is **homogeneous** if the changes in size and shape are proportionately identical for each small part of the body and for the body as a whole (Figure 15.1*A, B*). A consequence of these conditions is

that for any homogeneous strain, planar surfaces remain planar, straight lines remain straight and parallel planes and lines remain parallel. The strain is **inhomogeneous** (Figure 15.1*A, C*) if the changes in size and shape of small parts of the body are proportionately different from place to place and different from that of the body as a whole. Straight lines become curved, planes become curved surfaces, and parallel planes and lines generally do not remain parallel after deformation.

The strain must be inhomogeneous during folding, because in such a deformation, planes and lines do not generally remain planar, straight, or parallel. Within very small volume-elements, however, the strain is statistically homogeneous, and we describe an inhomogeneous strain as a variation of homogeneous strain from place to place in the structure. We discuss how big such a "small" volume-element must be in Section 15.7.

The **progressive deformation** of a body refers to the motion that carries the body from its initial undeformed state to its final deformed state. The strain states through which the body passes during a progressive deformation define the **strain path**. The **state of strain** of a body is the net result of all the deformations the body has undergone. Although all states of strain are the result of progressive deformation, the final state of strain provides no information about the particular strain path that the body experienced.

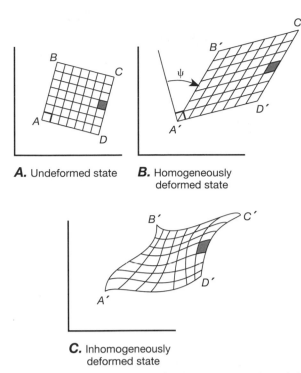

A. Undeformed state   **B.** Homogeneously deformed state

C. Inhomogeneously deformed state

Figure 15.1 Homogeneous and inhomogeneous plane deformation of a material square. A and B. Homogeneous strain. The small black square is strained in exactly the same way as the whole square and as all the other squares. ψ is the angle of shear. A and C. Inhomogeneous strain. The small black square is sufficiently small that its strain is essentially homogeneous, but it is not identical to the strain of the whole square or to that of any of the other small squares.

Strain in general must be described in three dimensions, because the size and shape of a body are three-dimensional characteristics. In much of our discussion, however, we consider only a two-dimensional deformation called **plane strain**, in which the strain is completely described by changes in size and shape in a single orientation of plane through the body, and no deformation occurs normal to that plane. Although plane strain is commonly used to analyze deformation, its application to many situations in natural rock deformation is, strictly speaking, unjustified. Nevertheless, the geometry of two-dimensional deformation is intuitively easier to understand, and the generalization to three dimensions adds considerable complexity but little insight into the geometric characteristics of deformation. For these reasons we concentrate on the properties of two-dimensional strain.

In discussing the geometry of strain, we refer to geometric objects such as lines, planes, circles, and ellipses. Such geometric objects are called **material objects** if they are always defined by the same set of material particles. A bedding plane, for example, is a material plane because no matter how it moves and deforms, it is always defined by the same set of material particles. A coordinate plane defined by two reference axes, on the other hand, is a nonmaterial plane because as a body deforms, its material particles can move through the coordinate plane and, consequently, different sets of material particles occupy the coordinate plane at different times. This distinction is important in the subsequent discussion.

15.1 Measures of Strain

Linear Strain

The size of a body is measured by its volume, which in turn is proportional to the product of three characteristic lengths of the body. For example, the volume V of a rectangular block that has edges of lengths ℓ_1, ℓ_2, and ℓ_3 is $V = \ell_1\ell_2\ell_3$, and the volume of an ellipsoid that has semiaxes of lengths r_1, r_2, and r_3 is $V = [4/3]\,\pi\,r_1r_2r_3$. In Cartesian coordinates, the description of the change in size requires specification of the change in length of line segments in the three coordinate directions.

The change in absolute length is an inadequate measure of the deformational state of a line segment, because for a given change in length, the intensity of the change is much greater for a short line segment than for a long one. Thus the lengthening is expressed as a proportion of the original line length. Two measures in common use are the stretch s_n and the extension e_n, which was introduced at the beginning of Chapter 9. The subscript n indicates that the stretch or extension is measured in a direction parallel to a specified unit vector **n**. The **stretch** s_n is the ratio of the deformed length ℓ of a material line segment to its undeformed length L.

$$s_n \equiv \frac{\ell}{L} \tag{15.1}$$

(We often use upper-case letters when referring to the undeformed state and lower-case letters when referring to the deformed state.) The **extension** e_n of a material line segment is the ratio of its change in length, ΔL, to its initial length L, where the change in length is the final length minus the initial length.[1]

$$e_n \equiv \frac{\ell - L}{L} = \frac{\Delta L}{L} \tag{15.2}$$

---

[1] Note that with the sign conventions we have adopted, a positive value for extension measures a lengthening, whereas a positive value for stress measures a compression. We thus end up with a positive stress causing a negative extension. This incompatibility does not arise with the engineering sign convention for stress, which is why it is generally used in analytic applications of continuum mechanics.

A homogeneous transformation of any material point from the undeformed state to the deformed state is represented mathematically by a linear relationship between the coordinates of any point in the undeformed state (X_1, X_3) and its coordinates in the deformed state (x_1, x_3), where we use upper-case letters to describe the undeformed state and lower-case letters to describe the deformed state. If we restrict our analysis to plane deformation, the general form of such a transformation is

$$x_1 = AX_1 + BX_3 + C \qquad x_3 = DX_1 + EX_3 + F \quad (15.1.1)$$

where A, B, C, D, E, and F are constants. The parts of the transformation defined by C and F are the same for all particles, and therefore these constants describe a rigid-body translation. If any or all of these constants vary with time, then these equations describe the motion of the material particles.

The equations say that given the original location of any material particle in the undeformed state (X_1, X_3), we can calculate its final location in the deformed state (x_1, x_3). The equations may be solved for X_1 and X_3 so that given the deformed location of a material particle (x_1, x_3), we can also calculate its original location (X_1, X_3). These equations define the inverse transformation.

$$X_1 = ax_1 + bx_3 + c \qquad X_3 = dx_1 + ex_3 + f \quad (15.1.2)$$

where

$$a \equiv \frac{E}{AE - BD} \qquad b \equiv \frac{-B}{AE - BD} \qquad c \equiv \frac{BF - CE}{AE - BD}$$

$$d \equiv \frac{-D}{AE - BD} \qquad e \equiv \frac{A}{AE - BD} \qquad f \equiv \frac{DC - AF}{AE - BD}$$

$$(15.1.3)$$

and where, again, c and f describe a rigid body translation.

As examples of such a transformation and its inverse, the following equations describe a pure shear, which transforms a square with sides parallel to the principal coordinates into a rectangle (Figure 15.9B)

$$\begin{aligned} x_1 &= AX_1 & x_3 &= (1/A)X_3 \\ X_1 &= (1/A)x_1 & X_3 &= Ax_3 \end{aligned} \quad (15.1.4)$$

A simple shear, which transforms a square into a parallelogram (Figure 15.11B), and its inverse are described by

$$\begin{aligned} x_1 &= X_1 + BX_3 & x_3 &= X_3 \\ X_1 &= x_1 - Bx_3 & X_3 &= x_3 \end{aligned} \quad (15.1.5)$$

When the constants A in Equation (15.1.4) and B in Equation (15.1.5) are linear functions of time, the motions are steady and these equations describe

progressive pure shear and progressive simple shear, respectively (see Section 15.4).

With Equations (15.1.2), it is easy to show that a homogeneous deformation transforms a circle into an ellipse. A circle of unit radius in the undeformed state is represented by the equation

$$(X_1)^2 + (X_3)^2 = 1 \quad (15.1.6)$$

If we substitute for X_1 and X_3 from Equations (15.1.2), we find the locus in the deformed state of all material particles that lie on the circle in the undeformed state. Because a rigid-body translation does not contribute to the strain, we assume $c = f = 0$. Then, making the substitution, we find

$$(a^2 + d^2)(x_1)^2 + 2(ab + de)x_1 x_3 + (b^2 + e^2)(x_3)^2 = 1$$

$$(15.1.7)$$

Equation (15.1.7) is the equation of an ellipse with its principal axes tilted with respect to the coordinate axes, and it is, in fact, the *strain ellipse*.

The components of the strain tensor are related to the displacement vectors for the material particles. A displacement vector connects the position of a particle in the undeformed state to its position in the deformed state. The vector and its components (U_1, U_3) parallel to the X_1 and X_3 coordinate axes are (Figure 15.1.1A)

$$\mathbf{U} \equiv \mathbf{x} - \mathbf{X} \quad (15.1.8)$$

$$U_1 = x_1 - X_1 \qquad U_3 = x_3 - X_3 \quad (15.1.9)$$

When a material deforms, the displacement vectors for two neighboring material points are different. If they were the same, the "deformation" would be a rigid body motion. The difference in these displacement vectors therefore describes the deformation. Thus we consider two neighboring points A and B that are displaced by the deformation to a and b, respectively. The displacement vectors for the two points are $\mathbf{U}^{(A)}$ and $\mathbf{U}^{(B)}$, and the difference between them is $d\mathbf{U}$ (Figure 15.1.1B). The material line segment $d\mathbf{X}$ connecting A to B is deformed into $d\mathbf{x}$ connecting a to b. The change in that line segment due to the deformation $\Delta d\mathbf{X}$ is also described by the vector $d\mathbf{U}$ (Figure 15.1.1B). Thus,

$$d\mathbf{U} \equiv \mathbf{U}^{(B)} - \mathbf{U}^{(A)} = \Delta d\mathbf{X} \equiv d\mathbf{x} - d\mathbf{X} \quad (15.1.10)$$

The relationship between the first and last terms in this equation is just the differential of Equation (15.1.8).

We can consider the components dX_1 and dX_3 of the line segment $d\mathbf{X}$ to be two material line segments that are initially perpendicular to each other and parallel to the coordinate axes X_1 and X_3 respectively. If we restrict our analysis to infinitesimal strain,

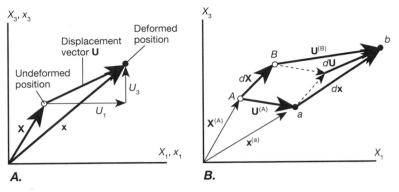

A.

B.

Figure 15.1.1 The displacement vector. *A.* The displacement vector connects the position of a material particle in the undeformed state to its position in the deformed state. *B.* If a material is deformed, the displacement vectors for two neighboring points are different. Point *A* is deformed to the position *a*; *B* is deformed to the position *b*. The difference in the displacement vectors *d*U describes the deformation of the material.

characterized by the conditions $dU_1 \ll 1$ and $dU_3 \ll 1$, the displacement associated with each of these line segments due to the deformation can be expressed using Equation (15.1.10) and the chain rule of differentiation for $d\mathbf{U}$

$$\Delta d\mathbf{X} = \Delta dX_1 + \Delta dX_3 = d\mathbf{U} = \frac{\partial \mathbf{U}}{\partial X_1} dX_1 + \frac{\partial \mathbf{U}}{\partial X_3} dX_3$$

$$(15.1.11)$$

Thus the changes ΔdX_1 and ΔdX_3 in each of the line segments due to the deformation is given in terms of the components of the displacement vector $\mathbf{U}$ by (Figure 15.1.2).

$$\Delta dX_1 = \frac{\partial \mathbf{U}}{\partial X_1} dX_1 = \frac{\partial U_1}{\partial X_1} dX_1 + \frac{\partial U_3}{\partial X_1} dX_1$$

$$\Delta dX_3 = \frac{\partial \mathbf{U}}{\partial X_3} dX_3 = \frac{\partial U_1}{\partial X_3} dX_3 + \frac{\partial U_3}{\partial X_3} dX_3$$

$$(15.1.12)$$

For each of the material line segments dX_1 and dX_3, the extensional strains are labeled e_{11} and e_{33} respectively, and each one is the change in length divided by the initial length, as defined in Equation (15.2). For dX_1, for example, the change in length is $(\partial U_1/\partial X_1)\, dX_1$, and the initial length is dX_1 (Figure 15.1.2). Similar relations hold for dX_3. Thus

$$e_{11} \equiv \frac{1}{dX_1}\left[\frac{\partial U_1}{\partial X_1} dX_1\right] = \frac{\partial U_1}{\partial X_1}$$

$$e_{33} \equiv \frac{1}{dX_3}\left[\frac{\partial U_3}{\partial X_3} dX_3\right] = \frac{\partial U_3}{\partial X_3}$$

$$(15.1.13)$$

The shear strain of dX_1 relative to dX_3 and vice versa are labeled e_{13} and e_{31}, respectively, and are defined in Equation (15.7) to be half the tangent of the shear angle $\psi_{13} = \psi_{31} = \psi = \alpha + \beta$. For very small strains, $\alpha \ll 1$ and $\beta \ll 1$, and the standard trigonometric identity for the tangent of the sum of two angles gives

$$\tan\psi \equiv \tan(\alpha + \beta) = \frac{\tan\alpha + \tan\beta}{1 - \tan\alpha\,\tan\beta} \approx \tan\alpha + \tan\beta$$

$$(15.1.14)$$

because the product $\tan\alpha\,\tan\beta$ is negligibly small. The tangent of an angle is the length of the side opposite the angle divided by the length of the adjacent side. For infinitesimal strains, the side opposite the angle α is approximately $(\partial U_1/\partial X_3)dX_3$, and the adjacent side is dX_3 (Figure 15.1.2). Similar relationships hold for the angle β. Thus we have

$$\tan\alpha \approx \frac{1}{dX_3}\left[\frac{\partial U_1}{\partial X_3} dX_3\right] = \frac{\partial U_1}{\partial X_3}$$

$$\tan\beta \approx \frac{1}{dX_1}\left[\frac{\partial U_3}{\partial X_1} dX_1\right] = \frac{\partial U_3}{\partial X_1}$$

$$(15.1.15)$$

(Continued)

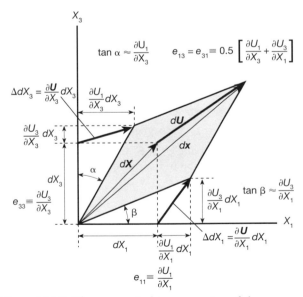

Figure 15.1.2 The geometrical interpretation of the components of infinitesimal strain for two-dimensional strain. For clarity, the strain is greatly exaggerated in the diagram. The vectors $d\mathbf{X}$, $d\mathbf{x}$, and $d\mathbf{U}$ are the same as the vectors having the same labels that appear in Figure 15.1.1B. The strain components thus are defined by the change in the displacement vector $d\mathbf{U}$ for two neighboring points.

Box 15.1 *(Continued)*

Then using the definition of the shear strain (Equation 15.7) with Equations (15.1.14) and (15.1.15) gives

$$e_{13} = e_{31} = 0.5 \tan \psi \approx 0.5\left(\frac{\partial U_1}{\partial X_3} + \frac{\partial U_3}{\partial X_1}\right)$$

$$= 0.5\left(\frac{\partial U_3}{\partial X_1} + \frac{\partial U_1}{\partial X_3}\right) \qquad (15.1.16)$$

These relations for the extensions and shear strains associated with the material line segments dX_1 and dX_3 are the components of the infinitesimal strain tensor. In shorthand component notation, we summarize Equations (15.1.13) and (15.1.16) by

$$e_{k\ell} \equiv 0.5\left(\frac{\partial U_k}{\partial X_\ell} + \frac{\partial U_\ell}{\partial X_k}\right), \qquad k, \ell = 1, 2, 3 \quad (15.1.17)$$

This expression for $e_{k\ell}$ remains exactly the same if k and ℓ are interchanged, which shows that $e_{k\ell} = e_{\ell k}$ and that the strain tensor is a symmetric tensor (compare Equation 15.12). Thus $e_{k\ell}$ is the symmetric part of the displacement gradient tensor $\partial U_k / \partial X_\ell$.

The antisymmetric part of the displacement gradient tensor can be shown to be the infinitesimal rotation tensor, defined by

$$r_{k\ell} \equiv 0.5\left(\frac{\partial U_k}{\partial X_\ell} - \frac{\partial U_\ell}{\partial X_k}\right), \qquad k, \ell = 1, 2, 3 \quad (15.1.18)$$

The antisymmetric character of $r_{k\ell}$ is evident from this equation, because interchanging the subscripts k and ℓ gives the relation

$$r_{k\ell} = -r_{\ell k} \qquad (15.1.19)$$

Components on the principal diagonal of the matrix $r_{k\ell}$ must therefore be zero. In two-dimensional strain, there is only one independent off-diagonal component $r_{13} = -r_{31}$. Thus from Equations (15.1.15) and (15.1.18) we can see that

$$r_{13} \approx 0.5(\tan \alpha - \tan \beta) \qquad (15.1.20)$$

For very small angles, the tangent of the angle is approximately equal to the angle measured in radians, so we can write

$$r_{13} \approx 0.5(\alpha - \beta) \qquad (15.1.21)$$

Thus r_{13} is half the difference in the components of the shear angle, and $r_{k\ell}$ is thus a measure of the net rotation of the material line segment $d\mathbf{X}$.

The displacement components (U_1, U_3) can be expressed solely in terms of the coordinates of the material point in the undeformed state by substituting Equations (15.1.1) into (15.1.9), assuming the rigid translations are zero ($C = F = 0$)

$$U_1 = (A - 1)X_1 + BX_3 \qquad U_3 = DX_1 + (E - 1)X_3$$

$$(15.1.22)$$

Using Equations (15.1.22) in (15.1.17), we find the values of the strain components in terms of the constants that define the motion of the material particles:

$$\begin{bmatrix} e_{11} & e_{13} \\ e_{31} & e_{33} \end{bmatrix} = \begin{bmatrix} (A-1) & 0.5(B+D) \\ 0.5(D+B) & (E-1) \end{bmatrix} \quad (15.1.23)$$

As indicated above, the relationships given here are correct only for very small strains. The analysis of large strains is considerably more complex, although this geometric interpretation of the strain components remains intuitively useful.

For a line segment of arbitrary orientation in the undeformed state, given by the angle θ with respect to the principal coordinate axis $\hat{X}_1$, it can be shown that the extension and the shear strain for infinitesimal plane strain are given in terms of the principal extensions by

$$e_n = \hat{e}_1 \cos^2 \theta + \hat{e}_3 \sin^2 \theta$$

$$(15.1.15)$$

$$e_s = (\hat{e}_1 - \hat{e}_3) \sin \theta \cos \theta$$

These equations are identical in form to Equations (8.36), which we found for the stress components, and the mathematical characteristics of the stress and the infinitesimal strain tensors are identical, including the possibility of deriving a Mohr circle for infinitesimal strain.

The relationships for large deformations are somewhat more complex, but a Mohr circle that is useful in solving strain problems can nevertheless be defined for large strains. We refer the reader to books containing more quantitative analyses (see the works by Means, Ramsay and Huber, and Eringen in the list of additional readings at the end of this chapter).

Comparing Equations (15.1) and (15.2) shows that these two measures of extensional strain are related:

$$e_n = \frac{\ell}{L} - \frac{L}{L} = s_n - 1 \qquad (15.3)$$

Values of $s_n > 1$ and of $e_n > 0$ represent increases in the length of material lines, and values where $0 < s_n < 1$ and $e_n < 0$ represent decreases in length (Table 15.1).

Other measures are also used, including the quadratic elongation and the natural strain. The **quadratic elongation** is simply the square of the stretch, and it is often given the symbol λ, although some authors use this symbol to designate the stretch. The **natural strain** $\bar{e}_n$, also called the **logarithmic strain,** is the integral of all the infinitesimal increments of extension required to make up the deformation, where the reference length

Table 15.1 Extensional Strain of a Material Line

		Length Change ΔL	Stretch $s_n \equiv \ell/L$	Extension $e_n \equiv (\ell - L)/L$
Undeformed		$\Delta L = 0$	$s_n = 1$	$e_n = 0$
Shortened		$\Delta L = \ell - L < 0$	$0 < s_n < 1$	$e_n < 0$
Lengthened		$\Delta L = \ell - L > 0$	$s_n > 1$	$e_n > 0$

for each increment in length $d\ell$ is taken to be the instantaneous deformed length ℓ.

$$\bar{\varepsilon}_n \equiv \int_L^{\ell_f} \frac{d\ell}{\ell} = \ln\left(\frac{\ell_f}{L}\right) = \ln s_n \qquad (15.4)$$

where L is the initial length, ℓ_f is the final length, and ln indicates the natural logarithm. Notice that the natural strain is the natural logarithm of the stretch. The natural strain is sometimes convenient for discussion of strain history (see Figure 15.20). It also provides a symmetric measure of shortening and lengthening.[2] The time derivative of the natural strain is also often used as a measure of the strain rate (see Box 18.1).

Volumetric Strain

We can now consider measures of the volumetric strain, which we refer to as the volumetric stretch (s_v) and the volumetric extension[3] (e_v). If the undeformed volume is V and the deformed volume is v,

$$s_v \equiv \frac{v}{V} \qquad e_v \equiv \frac{v - V}{V} = \frac{\Delta V}{V} = s_v - 1 \qquad (15.5)$$

A rectangular block that undergoes only volumetric strain has undeformed sides (L_1, L_2, and L_3) and deformed sides (ℓ_1, ℓ_2, and ℓ_3). The volumetric stretch is

$$s_v = \frac{\ell_1 \ell_2 \ell_3}{L_1 L_2 L_3}$$

$$s_v = s_1 s_2 s_3 = (e_1 + 1)(e_2 + 1)(e_3 + 1) \qquad (15.6)$$

We consider further aspects of volumetric strain in the next section.

Shear Strain

A body can also change shape without changing volume. For example, a cube can deform into a rhombohedron, or a sphere into an ellipsoid. Changes in shape are described by the changes in the angle between pairs of lines that are intially perpendicular (Figure 15.2). The change in angle is called the **shear angle** ψ, and the **shear strain** e_s is defined by

$$e_s \equiv 0.5 \tan \psi \qquad (15.7)$$

As defined here, e_s is the tensor shear strain. It differs from another common measure of the shear strain, the engineering shear strain γ, by a factor of 2 ($\gamma \equiv \tan \psi = 2e_s$). For two material line segments originally oriented along the positive coordinate directions (Figure 15.2A), a decrease in angle between the two lines is considered a positive shear strain (Figure 15.2B, C) and an increase in angle is a negative shear strain (Figure 15.2D, E). Both γ and e_s increase from 0 in the unstrained state to ∞, where $\psi = 90°$ (Figure 15.2F).

15.2 The State of Strain: The Strain Ellipsoid and the Strain Tensor

The Strain Ellipsoid

We know the **state of strain at a point** if, for a material line of any orientation, we can determine its extension, as well as its shear strain with respect to any other line initially perpendicular to it. Any homogeneous strain always deforms a material sphere into an ellipsoid called the **strain ellipsoid** (Figures 14.1 and 14.2A) or, in plane strain, a material circle into the **strain ellipse** (see Box 15-1).

The stretch, extension, and shear strain all have a simple geometric interpretation related to the strain ellipsoid. We describe these relationships here for two dimensions, but they are essentially the same when extended to three dimensions.

Assume that a material circle in the undeformed state has a radius $R = 1$ (Figure 15.3A). After the deformation, any radius of the circle is transformed into a radius r of the strain ellipse whose length varies with orientation. Although R and r are lines made up of the same material points, they differ in length and orientation because of the deformation. If we superimpose the original unit circle on the strain ellipse (Figure 15.3A), we can see how much any radius of the strain

---

[2] For example, for a line segment stretched to twice its initial length and one shortened to half its initial length, $s_n = 2$ and 0.5, and $e_n = 1$ and 0.5, but $\bar{\varepsilon}_n = 0.693$ and -0.693, respectively.

[3] The volumetric extension is commonly given the symbol Δ and called the **dilation**, or even the **dilatation**. We reserve Δ to indicate the change in a variable.

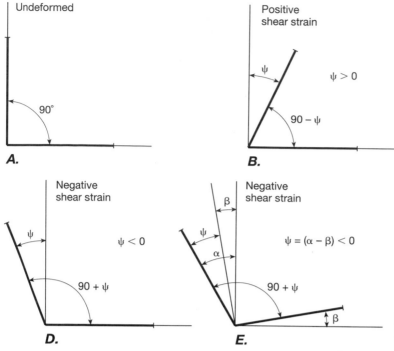

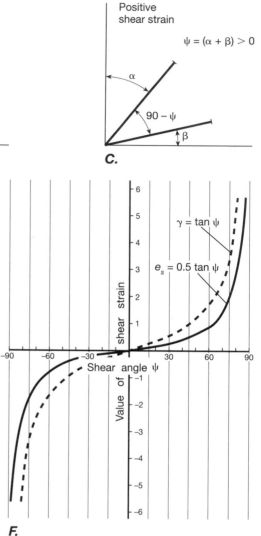

Figure 15.2 The tensor shear strain $e_s = 0.5 \tan \psi$ and the engineering shear strain $\gamma = \tan \psi$ of a material line, where ψ is the shear angle. *A.* The undeformed state. Shear of a material line is defined with reference to another material line initially normal to the first. *B* and *C.* Definition of a positive shear strain: $(90 - \psi) < 90$. *D* and *E.* Definition of a negative shear strain: $(90 - \psi) > 90$. *F.* Tensor and engineering shear strains as a function of shear angle ψ.

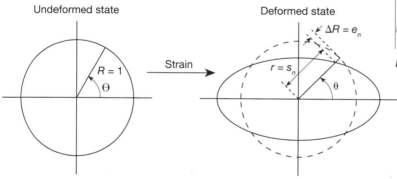

A. Extension and stretch

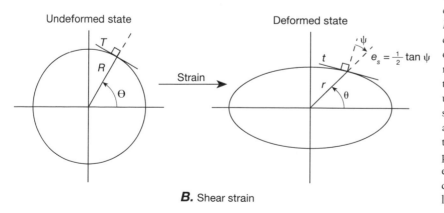

B. Shear strain

Figure 15.3 The relationship of the stretch, extension, and shear strain to the geometry of the strain ellipse. *A.* A homogeneous strain transforms the unit circle into an ellipse. An undeformed radius $R = 1$ is transformed into a deformed radius r, which has a different length and orientation. The stretch is the length of the radius of the ellipse, and the extension is the difference in radius between the initial unit circle and the ellipse. *B.* The shear strain is determined from the change in angle between a radius and a tangent at the end of the radius. The two lines are perpendicular on the circle but not, in general, on the ellipse. The change in angle ψ defines the shear strain for that pair of lines.

ellipse has been shortened or lengthened. Using the definitions of the stretch (Equation 15.1) and the extension (Equation 15.2) and the fact that $R = 1$, we find that

$$s_n = \frac{r}{R} = r \qquad e_n = \frac{r - R}{R} = \frac{\Delta R}{R} = \Delta R \qquad (15.8)$$

Thus for the deformation of the unit circle, the radius of the strain ellipse is the stretch, and the difference between the radius of the ellipse and that of the unit circle is the extension.

The shear strain of a line is determined with reference to another line initially normal to it. On a circle, the line T drawn perpendicular to any radius R at its end point is tangent to the circle (Figure 15.3B). After deformation, the lines T and R are transformed into the lines t and r, respectively. Although r and t are no longer perpendicular in the deformed state, t is still tangent to the ellipse at the end point of the radius. Accordingly, any radius and the associated tangent to the strain ellipse define the angle between two material lines that were perpendicular in the undeformed state. The change in angle ψ is thus easily constructed (Figure 15.3B), and it is a measure of the shear strain for that pair of lines.

The Strain Tensor

The strain ellipsoid is a complete representation of the state of strain at a point. We can describe that state if we know the extension and the two shear strains for each of only three material line segments that were mutually orthogonal in the undeformed state. We consider the volumetric and the shear components of the strain separately.

For an orthogonal coordinate system (X_1, X_2, X_3) in the undeformed state, the extension of a material line segment of length L_1 initially parallel to X_1, for example, is (Figure 15.4A).

$$e_{11} = \frac{\Delta L_1}{L_1} \qquad (15.9)$$

where the first subscript on e_{11} indicates that the line is initially parallel to X_1, and the second subscript indicates that the change in length is also parallel to X_1. Similar relations define the extensions e_{22} and e_{33} for material lines initially parallel to X_2 and X_3 respectively (Figure 15.4A).

For the shear component of the strain, material lines initially parallel to X_1, X_2, and X_3 are, after deformation, parallel to x_1, x_2, and x_3 respectively (Figure 15.4B). The two shear strain components for the material line parallel to x_1 are e_{12} and e_{13},

$$e_{12} = 0.5 \tan \psi_{12} \qquad e_{13} = 0.5 \tan \psi_{13} \qquad (15.10)$$

In each case the first subscript indicates that the shear strain is for the line initially parallel to X_1, and the second subscript indicates that the shear strain is determined relative to a line initially parallel to X_2 and to X_3, respectively (Figure 15.4B). Each angle ψ_{12} and ψ_{13} is the difference between $90°$ and the deformed angle

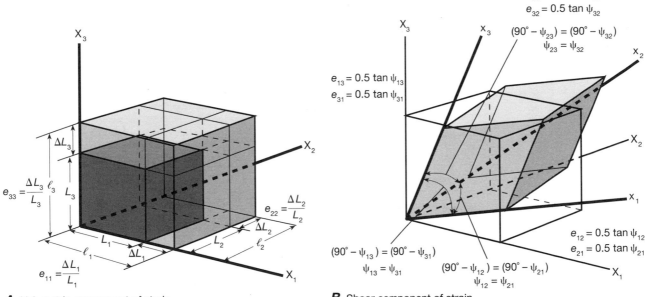

A. Volumetric component of strain

B. Shear component of strain

Figure 15.4 Geometric significance of the strain tensor components in three dimensions. *A.* Volumetric part of the strain. The small cube increases in volume to the larger cube by the equal lengthening of all sides of the cube. *B.* Shear part of the strain. The shear strain describes the change in shape from a cube into a rhombohedron (shaded). x_1, x_2, and x_3 are parallel to the deformed edges of the rhombohedron. All the tensor shear strain components are defined by three independent angles $\psi_{12} = \psi_{21}$, $\psi_{13} = \psi_{31}$, $\psi_{23} = \psi_{32}$.

$x_1 \wedge x_2$ and $x_1 \wedge x_3$, respectively. The comparable strain components for the material line segment initially parallel to X_2 are e_{21} and e_{23}; for the material line segment initially parallel to X_3, they are e_{31} and e_{32}.

Thus there are a total of nine strain components. The strain components for each material line are written in a separate row, forming an ordered array.

$$e_{k\ell} = \begin{bmatrix} e_{11} & e_{12} & e_{13} \\ e_{21} & e_{22} & e_{23} \\ e_{31} & e_{32} & e_{33} \end{bmatrix} \qquad (15.11)$$

PRINCIPAL DIAGONAL

The components on the principal diagonal of the array, which have both subscripts the same, are the extensions (Figure 15.4A). The off-diagonal components, which have two different subscripts, are the shear strains (Figure 15.4B). This array of strain components represents the **strain tensor**, which provides enough information for us to calculate the extension and shear strain for a line segment of any specified orientation (see Box 15.1).[4]

The strain tensor is symmetric about the principal diagonal, because for a given pair of material lines initially parallel to X_1 and X_2, for example, the shear angle (ψ_{12}) of X_1 with respect to X_2 is the same as the shear angle (ψ_{21}) of X_2 with respect to X_1 (Figure 15.4B). Thus

$$e_{12} = e_{21} \qquad e_{23} = e_{32} \qquad e_{31} = e_{13} \qquad (15.12)$$

and there are only six independent strain components in three-dimensional strain. Thus the strain, like the stress, is a second-rank symmetric tensor.

For plane strain, we have $e_{21} = e_{22} = e_{23} = 0$, and by Equations (15.12), $e_{12} = e_{32} = 0$. Thus if we drop from Equation (15.11) all terms that necessarily become zero for plane strain, the plane strain tensor is represented by only four strain components, three of which are independent.

$$e_{k\ell} = \begin{bmatrix} e_{11} & e_{13} \\ e_{31} & e_{33} \end{bmatrix} \qquad (15.13)$$

Therefore, in order to describe the state of plane strain, we need only the extension and one shear strain for each of the two material lines that originally are parallel to X_1 and X_3, respectively.

Principal Strains and Stretches

Parallel to the principal axes of the strain ellipsoid, the extensions and stretches are a maximum, minimax,[5] and minimum, which we designate[6]

$$\hat{e}_1 \geq \hat{e}_2 \geq \hat{e}_3 \quad \text{and} \quad \hat{s}_1 \geq \hat{s}_2 \geq \hat{s}_3 \qquad (15.14)$$

Tangents to the ellipsoid at the ends of the principal radii are perpendicular to the radii (Figure 15.5), and these are the only points on the ellipsoid where this is true. Because these radii and tangents must have been perpendicular before deformation, the shear strains for those radii and tangents all must be zero. Thus if we define a set of **principal coordinates** parallel to the principal axes of the strain ellipsoid, the representation of the strain tensor reduces to a particularly simple form in which the extensions are the principal values, and the shear strains are zero. For three- and two-dimensional strains, respectively

$$e_{k\ell} = \begin{bmatrix} \hat{e}_1 & 0 & 0 \\ 0 & \hat{e}_2 & 0 \\ 0 & 0 & \hat{e}_3 \end{bmatrix} \qquad e_{k\ell} = \begin{bmatrix} \hat{e}_1 & 0 \\ 0 & \hat{e}_3 \end{bmatrix} \qquad (15.15)$$

It is very important to remember that in general the principal axes of finite strain are *not* parallel to the principal axes of stress. We discuss this further in Section 15.4 and in Chapter 18.

We now see that, for any general deformation, the volumetric stretch s_v (Equation 15.6) can be expressed in terms of the principal stretches and extensions as follows:

$$s_v = \hat{s}_1 \hat{s}_2 \hat{s}_3 = (\hat{e}_1 + 1)(\hat{e}_2 + 1)(\hat{e}_3 + 1) \qquad (15.16)$$

---

[5] $\hat{e}_2$ and $\hat{s}_2$ are each a minimax because each is a minimum in the $\hat{e}_1 - \hat{e}_2$ (or $\hat{s}_1 - \hat{s}_2$) plane and a maximum in the $\hat{e}_2 - \hat{e}_3$ (or $\hat{s}_2 - \hat{s}_3$) plane, which is perpendicular to the first.
[6] Consistent with our notation for stress, we use the circumflexes and a single subscript to indicate principal values. The subscript indicates the principal axis to which the extension or stretch is parallel.

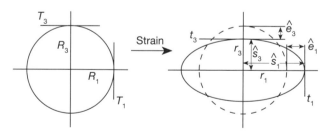

Figure 15.5 Representation of the principal stretches and the principal extensions on the strain ellipse formed from the unit circle. The shear strains are zero for the material lines parallel to the principal axes of strain, because the tangents at the ends of the principal radii are perpendicular to those radii both before and after deformation.

---

[4] Our definitions of the tensor strain components are correct only for small strains. For large strains, additional nonlinear terms must be added to our definitions, which makes the theory more complex. Nevertheless, the results discussed hereafter are true for both small and large strains.

Although derived for the example of a deformed cube, Equations (15.16) are completely general.[7] In plane strain, $\hat{s}_2 = 1$ and $\hat{e}_2 = 0$, so Equation (15.6) reduces to

$$s_\nu = \hat{s}_1 \hat{s}_3 = (\hat{e}_1 + 1)(\hat{e}_3 + 1) \qquad (15.17)$$

Thus the condition for constant-volume deformation is given for three-dimensional and plane strains, respectively, by

$$s_\nu = \hat{s}_1 \hat{s}_2 \hat{s}_3 = 1 \quad \text{and} \quad s_\nu = \hat{s}_1 \hat{s}_3 = 1 \qquad (15.18)$$

The last equation implies

$$\hat{s}_1 = \frac{1}{\hat{s}_3} \qquad (15.19)$$

The Inverse Strain Ellipse

In analyzing large strains such as are common in ductilely deformed rocks, it may be more convenient to measure the stretches and shear strains of three material lines that are mutually perpendicular in the *strained* state, rather than in the unstrained state as we described above. This analysis requires a different strain ellipse called the **inverse strain ellipse,** which is the ellipse in the undeformed state that is transformed into a circle in the deformed state (Figure 15.6). The lengths of its principal axes are the inverse of the principal axes of the strain ellipse, and the material lines parallel to the principal axes of inverse strain in the undeformed state become parallel to the principal axes of strain in the deformed state. For the purposes of our descriptive discussion, however, we deal mostly with the strain ellipse.

Why Study Strain?

All this discussion of circles and ellipses may seem academic and far removed from the study of real rocks. It is not, however, because structures that are initially approximately circular or spherical are relatively common in some rock types. Where these rocks have been deformed, those structures provide a fascinating record of the distribution of strain throughout the rock. Ooids, for example, are small, almost spherical, pelletlike bodies common in limestones (Figure 15.7A), and they deform with the rock to record the shape and orientation of the strain ellipsoid (Figure 15.7B). Radiolaria and

---

[7] We derive Equation (9.6) from the equation for e_ν in Equations (15.5) by substituting for s_ν from the second Equation (15.16), multiplying out the indicated product, and ignoring second- and third-order terms. The result is the sum of the components on the principal diagonal of the strain tensor matrix (Equation 15.11), which is a scalar invariant of the strain tensor (see the definition of the scalar invariants of the stress tensor in Equations 8.4.4 and 8.26) and hence is the same for the representation of strain in any coordinate system. That is, $e_\nu = \hat{e}_1 + \hat{e}_2 + \hat{e}_3 = e_{11} + e_{22} + e_{33}$.

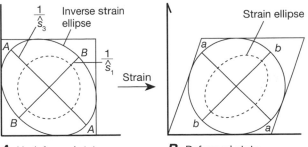

A. Undeformed state   **B.** Deformed state

Figure 15.6 Definition of the inverse strain ellipse and its relationship to the strain ellipse. Solid lines show how the inverse strain ellipse in the undeformed state (part A) is transformed into a circle in the deformed state (part B). The dashed lines show how a circle in the undeformed state (part A) is transformed into the strain ellipse in the deformed state (part B). Material lines A and B, which are parallel to the principal axes of inverse strain ellipse in part A, are transformed by the deformation to lines a and b, which are parallel to the principal axes of the strain ellipse in part B. In general, A and B are not parallel to a and b, respectively.

foraminifera, which are tiny spherical or disk-shaped fossils found in cherts or limestones, and alteration spots in slates (Figure 13.19B) may also serve as strain indicators. Other fossils, such as cephalopods and brachiopods, as well as pebbles and cobbles in conglomerates, (Figure 13.19A) can provide information about the strain, even though they are not originally spherical and may have an original preferred dimensional orientation in the undeformed rock (see Figure 14.2C). We discuss the significance of strain for interpreting the origin of structures in Chapter 16, and the measurement and observation of strain in deformed rocks in Chapter 17.

Some structures, such as folds and boudins, also record components of the strain. Consider, for example, a competent layer imbedded in an incompetent matrix. A variety of structures can develop (Figure 15.8). A set of folds develops if the layer is parallel to a principal axis of shortening and normal to an axis of lengthening (Figure 15.8A–D). Boudins develop if the layer is parallel to a principal axis of lengthening (Figure 15.8C–F). Two interfering sets of folds form if the layer is parallel to two principal directions of shortening and normal to an axis of lengthening (Figure 15.8A). Folds develop that are boudinaged parallel to the fold axis if the layer is perpendicular to a principal axis of lengthening, and the two principal axes parallel to the layer are axes of lengthening and shortening respectively (Figure 15.8C, D). Finally, tablet boudinage develops if the layer is parallel to two principal axes of lengthening and perpendicular to one of shortening (Figure 15.8F). Thus the orientation of the layer relative to the principal stretches is a major factor in determining what structures can develop.

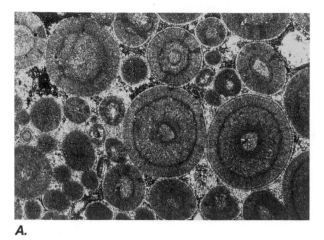

A.

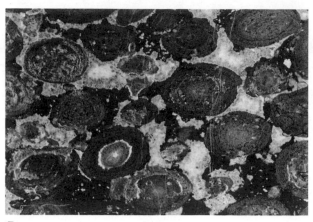

B.

Figure 15.7 Ooids serve as strain markers in deformed limestone. *A.* An undeformed oolitic limestone. *B.* A deformed oolitic limestone. The ratio of the principal stretches is $(\hat{s}_1/\hat{s}_3) \approx 1.4$. The larger ooids are approximately 4 mm in diameter.

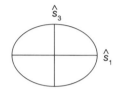

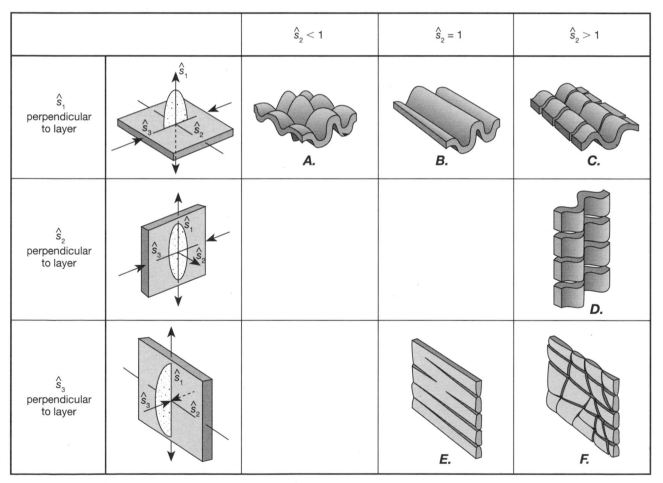

		$\hat{s}_2 < 1$	$\hat{s}_2 = 1$	$\hat{s}_2 > 1$
$\hat{s}_1$ perpendicular to layer		A.	B.	C.
$\hat{s}_2$ perpendicular to layer				D.
$\hat{s}_3$ perpendicular to layer			E.	F.

Figure 15.8 Structures that could develop in a competent layer imbedded in an incompetent matrix depend on the orientation of the layer relative to the principal stretches, and on the value of $\hat{s}_2$. In this diagram, we assume that lengthening the layer causes boudinage, shortening causes folding, and deformation is at constant volume, so that $\hat{s}_1 > 1$, $\hat{s}_3 < 1$, and $\hat{s}_2$ can take on any value.

15.3 Examples of Homogeneous Strains

Various simple geometries of homogeneous strain are given specific names.

Pure strain is any strain for which the principal axes of strain are constant in orientation relative to the reference coordinate system. Thus the principal axes of strain and the principal axes of inverse strain are parallel. Strain geometries belonging to this class (and described below) include uniform dilation, pure shear, simple extension, simple flattening, and uniaxial strain.

Uniform dilation is a pure volumetric strain with no change in shape of the deforming body. A cube or a square is transformed into a body that is of the same shape but has either a larger dimension (uniform expansion) or a smaller dimension (uniform contraction). The same statement, of course, applies to both a sphere or a circle. The stretch has the same value in all directions, as does the extension, and the shear strains are zero in all directions; that is, $\psi = 0$ for all orientations of line. All material lines change length, but none changes orientation.

Pure shear is a constant-volume ($s_v = 1$) plane strain ($\hat{s}_2 = 1$) that changes the shape of the deforming body (Figure 15.9). Material lines parallel to the principal axes of strain do not rotate and experience no shear strain. Material lines of all other orientations in the plane of strain (the $\hat{s}_1$–$\hat{s}_3$ plane) are rotated toward $\hat{s}_1$. Two orientations of line in the plane of strain have the same length as their initial length; these are the **lines of no finite extension.** They divide the ellipse into sectors within which all radial lines are either shortened (sectors S in Figure 15.9C) or lengthened (sectors L), depending on their orientation.

Simple extension involves lengthening parallel to one principal axis of strain and axially symmetric short-

ening in all directions perpendicular to that axis. **Simple flattening** involves shortening parallel to one principal strain axis and axially symmetric lengthening in all directions perpendicular to that axis. The volume of the body in either case is not necessarily constant.

Uniaxial strain is characterized by having two of the principal stretches equal to 1. The third principal stretch may be either greater than 1 (uniaxial extension; Figure 15.10A) or less than 1 (uniaxial shortening, Figure 15.10B). Volume is not conserved. Lines perpendicular to the unique axis of stretch are unchanged in length. Lines in all other orientations are lengthened in uniaxial extension and shortened in uniaxial shortening.

Simple shear is a type of strain we discussed briefly at the beginning of Chapter 12 (Figure 12.1). It is a constant-volume ($s_v = 1$) plane strain ($\hat{s}_2 = 1$) whose characteristics resemble the shearing of a deck of cards; thus, for a homogeneous deformation, the side of the deck changes from a rectangle to a parallelogram (Figure 15.11A). It is not a pure strain, because the orientations of principal strain axes change with the magnitude of shear, and the principal axes of strain and of inverse strain are not parallel (Figure 15.11B). Displacement of all material particles is parallel to the shear plane (the x_1–x_2 plane in Figure 15.11) and all material lines are rotated except those parallel to the shear plane. There are two orientations of no finite extension in the plane of strain (the x_1–x_3 and $\hat{s}_1$–$\hat{s}_3$ plane), one of which is always parallel to the shear plane. These lines divide the strain ellipse into sectors of shortened radii (S in Figure 15.11C) and lengthened radii (L).

These states of strain are all special cases of the infinite variety of possible states. They have no special qualities that make them uniquely applicable to the interpretation of rock deformation, but they are used because the geometry of each is simple and well defined. An arbitrary deformation, however, can *always* be ex-

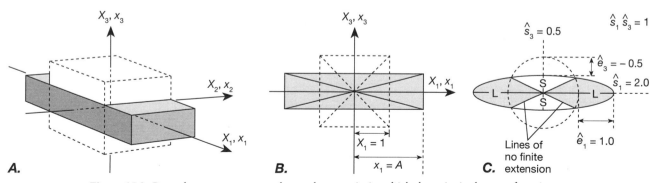

Figure 15.9 Pure shear: a constant-volume plane strain in which the principal axes of strain are not rotated by the deformation. A. Pure shear of a cube into a rectangular prism (shaded). B. Pure shear of a two-dimensional square to form a rectangle (shaded). The diagonals of the square are material lines that are rotated and stretched to become the diagonals of the rectangle; they are *not* the same as the lines of no finite elongation. C. Pure shear of a unit circle to form an ellipse. The lines of no finite extension divide the strain ellipse into sectors in which all radii are shortened (sectors S) and those in which all radii are lengthened (sectors L).

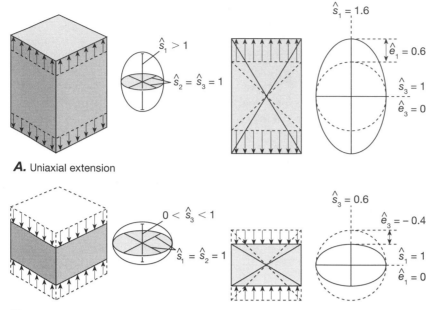

Figure 15.10 Uniaxial strain: two principal stretches are both equal to 1. Dashed lines indicate the undeformed state, solid lines the deformed state.

A. Uniaxial extension

B. Uniaxial shortening

pressed as the sum of a pure strain that has stretches parallel to the axes of inverse strain (Figure 15.12*A*, *B*), a rigid rotation of the body that brings the principal axes of strain into the proper orientation (Figure 15.12*C*), and a rigid translation of the body that brings it into the proper location (Figure 15.12*D*). These components of the deformation can in principal be applied in any order. The net result of a simple shear strain (Figure 15.11), for example, can be reproduced by the sum of a pure shear (Figure 15.9) parallel to the axes of inverse strain, a rotation of the principal axes, and a translation (Figure 15.12). Other geometrically more complex deformations can be similarly reproduced.

15.4 Progressive Deformation

So far in our discussion, we have simply related the deformed state to the undeformed state, without implying anything about the intermediate strain states that develop during the deformation. In rocks, we generally can observe only the final strained state and must infer the initial undeformed state. The history of the deformation is also of great interest, and in some cases it is recorded by features in deformed rocks. Understanding the consequences of different strain paths can provide insight that is useful in interpreting strain in rocks.

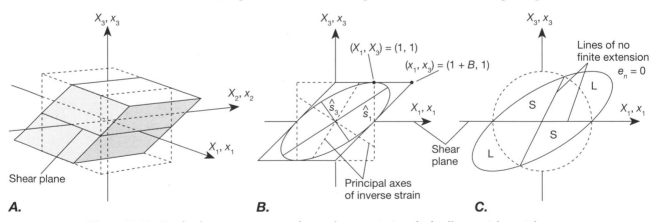

Figure 15.11 Simple shear: a constant-volume plane strain in which all material particles are displaced strictly parallel to the shear plane. Dashed lines indicate the undeformed state, solid lines the deformed state. *A.* Simple shear of a cube. *B.* Simple shear in two dimensions of a square. The principal axes of inverse strain in the undeformed state are dashed; the principal axes of strain in the deformed state are solid. Material lines parallel to the axes of inverse strain are rotated by the deformation into parallelism with the principal axes of strain. *C.* Lines of no finite extension in the strain ellipse divide the ellipse into sectors of shortened (S) and lengthened (L) radii of the ellipse.

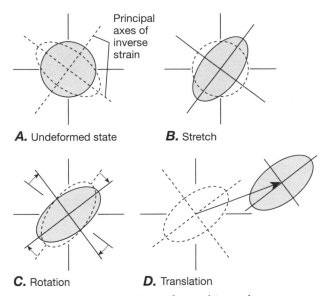

A. Undeformed state   **B.** Stretch

C. Rotation   **D.** Translation

Figure 15.12 Decomposition of an arbitrary homogeneous strain into a pure strain, a rigid rotation, and a rigid translation. These components may be applied in any sequence. *A. The undeformed state, showing the unit circle, the inverse strain ellipse (dashed), and the principal axes of inverse strain. B. Stretches are imposed parallel to the principal axes of inverse strain to reproduce the final shape of the strain ellipse. The inverse strain ellipse becomes a circle. C. Rigid-body rotation brings the principal axes into the correct final orientation. D. Rigid-body translation brings the body into the correct final location.*

We refer to the nonrigid motion of a body as a **progressive strain** or **progressive deformation,** and we can describe the motions of all material particles in the body by describing the deformed position of the particles as a function of their original position and of time (see Box 15.1).

Structures such as folds, boudins, foliations, and lineations develop in rock in response to progressive deformations. Folds and boudins develop in material layers in the rock, such as sedimentary layers, cross-

cutting veins, or dikes. Most spaced foliations are also defined by material surfaces. Therefore, in order to understand the relationship between such structures and the principal axes of strain, we investigate what happens to material lines of various orientations during different progressive plane deformations.

We can conceptualize the geometry of the progressive deformation by stopping it, marking a material circle on the body, and allowing the deformation to continue for a unit increment of time. The ellipse formed from the circle represents the increment of strain for that increment of time and is therefore called the **incremental strain ellipse.** Thus the incremental extension ε_n, the incremental shear strain ε_s, and the incremental stretch ζ_n (the Greek letter zeta) are defined in terms of the instantaneous length of a material line ℓ, its incremental change $d\ell$, and the incremental shear angle $d\psi$ of two instantaneously perpendicular lines.

$$\varepsilon_n \equiv \frac{d\ell}{\ell} \qquad \varepsilon_s \equiv 0.5 \tan d\psi \qquad \zeta_n \equiv \frac{\ell + d\ell}{\ell} \quad (15.20)$$

The incremental strain ellipse is represented by the incremental strain tensor $\varepsilon_{k\ell}$, which has the same properties as the infinitesimal strain tensor.[8] The half-lengths of the principal axes are the principal incremental stretches, $\hat{\zeta}_1 \geq \hat{\zeta}_2 \geq \hat{\zeta}_3$. If the incremental strain ellipse is constant for every unit increment in time, the motion of the material particles is called a **steady motion.**

To illustrate the effects of different motions on material lines, consider two special steady motions: progressive pure shear and progressive simple shear. The particle paths during these progressive deformations are shown in Figure 15.13A and B, respectively. (See Equations (15.1.4) and (15.1.5) in Box 15.1, for the quanti-

---

[8] Because the incremental strain ellipse represents the strain in a unit increment of time, it is similar to the strain rate tensor (see Box 18.1). Note that the natural strain $\bar{\varepsilon}_n$ is the integral of the incremental strain over time (see Equation 15.4).

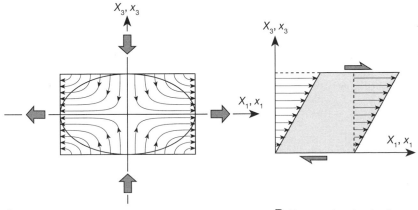

A. Progressive pure shear   **B.** Progressive simple shear

Figure 15.13 Particle motions during two progressive deformations. *A. Particle motions during progressive pure shear. The lines with the arrowheads are parallel to the velocity vectors of the particles in the body. B. Particle motions during progressive simple shear are all strictly parallel to the shear plane (X_1 direction). The velocity varies linearly with distance normal to the shear plane (X_3 direction).*

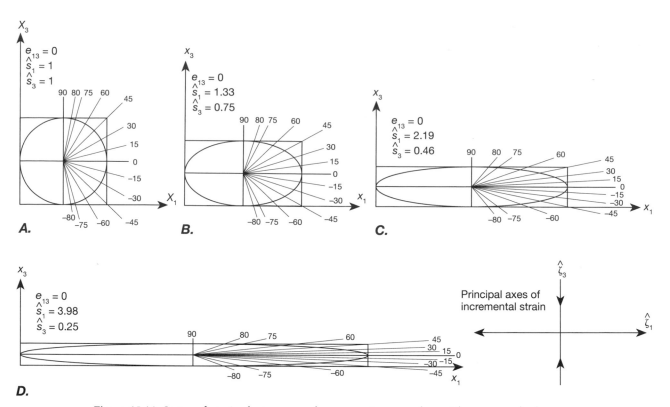

Figure 15.14 States of strain during a steady progressive pure shear. The axes at the bottom right of the figure indicate the constant orientation of the principal axes of incremental stretch. Material lines are labeled by the angle they make with X_1 in the undeformed state. The lines 0 and 90 are the only ones that do not rotate during the deformation, and they are always parallel to the principal axes of strain. The magnitudes of the principal stretches in each diagram are the same as for the corresponding diagram in Figure 15.15.

tative description of these motions.) For these examples, the incremental strain ellipse has the geometric properties of either pure shear (Figure 15.9) or simple shear (Figure 15.11) for each increment of strain through time.

Figures 15.14 and 15.15 illustrate the consequences of progressive pure shear and progressive simple shear, respectively. Part *A* in each figure is the undeformed state, showing a sheaf of material lines. In Figure 15.14*A*, the material lines are oriented at regular angular intervals, and each line is labeled with the angle it originally makes with the X_1 axis. In Figure 15.15*A*, the material lines are parallel to the axes of inverse strain for the state of strain in the diagram labeled with the corresponding letter. For example, the material lines *C* and *C'* in part *A*, are parallel to the principal axes of inverse strain for the strain state shown in part *C*. These lines are rotated by the deformation into the orientations shown by *c* and *c'*, which become parallel to the principal axes of strain in part *C*. Parts *B* through *D* in both figures show the evolution of both the strain ellipse and the orientations of the same material lines as appear in part *A*. The corresponding diagrams in the two figures show the same states of strain, although the orientations of the principal axes are different (see Figure 15.12).

A comparison of Figures 15.14 and 15.15 shows the following significant differences in behavior:

1. With respect to the coordinate axes, the principal axes of strain do not rotate in progressive pure shear, but in progressive simple shear they do. Thus the former is an **irrotational,** and the latter a **rotational** progressive deformation. The difference in behavior of the principal strain axes is described by the **vorticity** of the deformation,[9] which is a measure of the average rate of rotation of material lines of all orientations about each coordinate axis.

   The vorticity is zero for irrotational deformations and nonzero for rotational deformations. In Figure 15.14, for example, the material lines in the upper-right quadrant rotate in the opposite sense to those in the lower right.

---

[9] Technically, the vorticity vector ω is the curl of the velocity ($\omega = \nabla \times V$), which has the three components
$$[\omega_1, \omega_2, \omega_3] \equiv [(\partial v_3/\partial x_2 - \partial v_2/\partial x_3),$$
$$(\partial v_1/\partial x_3 - \partial v_3/\partial x_1), (\partial v_2/\partial x_1 - \partial v_1/\partial x_2)]$$
It is related to the spin tensor, which is the antisymmetric part of the velocity gradient tensor (see footnote in Section 19.7).

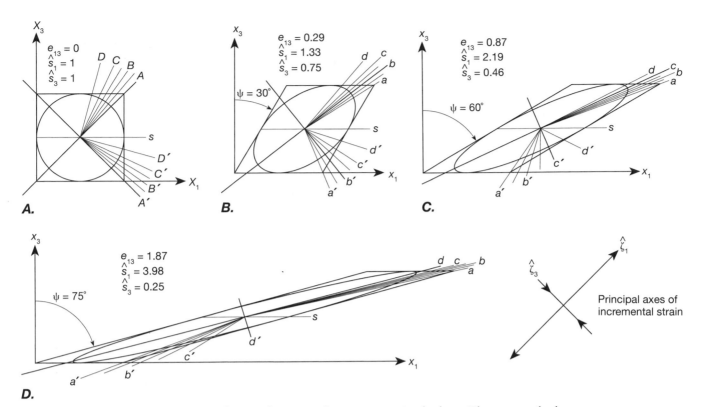

Figure 15.15 States of strain during steady progressive simple shear. The axes at the bottom right of the figure indicate the constant orientation of the principal axes of incremental stretch. The pairs of material lines in the undeformed state labeled (*B* and *B'*), (*C* and *C'*), and (*D* and *D'*) are parallel to the principal axes of inverse strain for the strain states shown in parts *B*, *C*, and *D*, respectively. These pairs of material lines take on the orientations in the deformed states indicated by the lines labeled in the equivalent lower-case letters, and each pair becomes parallel to the principal axes of strain in the diagram labeled with the same letter as the line pair. Thus the material lines rotate past the principal axes of strain, which themselves are not material lines. *S* and *s* indicate a material line parallel to the shear plane. This is the only orientation of line for which the orientation and length are constant throughout the deformation.

Because material lines oriented symmetrically relative to the x_1 axis have exactly opposite rates of rotation, the average over all orientations must be zero. In contrast, all the material lines in Figure 15.15 rotate in the same sense, so the average rate of rotation is nonzero.

2. In progressive pure shear, the principal axes of finite strain are always parallel to, or coaxial with, the principal axes of incremental strain. The deformation is therefore a **coaxial** progressive deformation. In progressive simple shear, the principal axes of finite strain rotate with respect to those of incremental strain, and this characteristic defines a **noncoaxial** progressive deformation. Note that for progressive simple shear the principal axes of incremental strain are always at a 45° angle to the shear plane.

   The terms *irrotational* and *coaxial* are not synonymous, nor are *rotational* and *noncoaxial*. The difference is in the reference frame from which the rotation is determined. A deformation is rotational or irrotational depending on how the principal axes of finite strain behave with respect to the coordinate system, which is always somewhat arbitrarily defined by the observer. A deformation is coaxial or noncoaxial depending on how the principal axes of finite strain behave with respect to the principal axes of incremental strain. This reference frame is intrinsic to the geometry of the deformation itself and is therefore not arbitrary. Thus the description of a progressive deformation as coaxial or noncoaxial is somewhat more fundamental than the description as rotational or irrotational, especially in geologic situations in which the best choice of an external coordinate system is not obvious.

3. In progressive pure shear, all material lines rotate during the deformation except those parallel to the principal axes of strain. The lines rotate toward parallelism with the $\hat{s}_1$ direction.

Note that the term *irrotational* refers only to the behavior of the principal axes of strain and to the average motion of all material lines, not to the motion of a specific material line. In progressive simple shear, all lines except those parallel to the shear plane rotate during the deformation, and the rotation rate of any line decreases with decreasing angle between the line and the shear plane.

4. The lines that rotate most rapidly in progressive pure shear are those at an angle of 45° to the principal axes of the incremental strain ellipse. In progressive simple shear, the lines that rotate most rapidly are normal to the shear plane, and these lines are also at a 45° angle from the principal axes of incremental strain. Lines parallel to the shear plane, however, do not rotate at all, and they too are 45° from the principal axes of incremental strain.

5. In progressive pure shear, the same pair of material lines remains parallel to the principal axes of strain throughout the deformation. In progressive simple shear, material lines rotate through the principal axes of strain. This characteristic shows that the principal axes of strain are not in general material lines. During progressive simple shear material lines that are parallel to the principal axes at any time were originally orthogonal in the undeformed state. During the deformation, however, any such pair of lines is sheared out of orthogonality, then back into orthogonality when they are parallel to the principal axes, and finally out of orthogonality again (lines C and C′ in Figure 15.15).

6. In both progressive pure and progressive simple shear, the stretch of material lines depends on their orientation. Some lines experience a history only of shortening, others experience only lengthening, and still others experience initial shortening followed by lengthening and can end up being either shorter or longer than they were originally. The pattern of variation determines what types of structures can develop. We discuss this further in the next section.

If the deformation stops at any time, the final state of strain can always be related to the initial state in Figure 15.14 by a pure shear strain or in Figure 15.15 by a simple shear strain. The converse of this statement, however, is not true: If a final state of strain can be related to the initial state either by a pure shear strain or by a simple shear strain, it does not follow that the final state of strain was the result of a progressive pure shear or a progressive simple shear, respectively. There are an infinite number of strain paths that lead from an undeformed state to a deformed state, and the final state

of strain does not by itself provide sufficient information for any of the paths to be distinguished. It is very important to remember this when interpreting the strain in rocks.

From the foregoing discussion, it is evident that if a progressive deformation is noncoaxial, the principal axes of finite strain rotate relative to those of incremental strain, and that the principal axes of incremental strain are constant in orientation only if the deformation is steady. It should not be surprising, therefore, that the principal axes of finite strain are not in general parallel to the principal axes of stress. In fact, we see in Chapter 18, where we discuss the relationships between stress and strain, that for steady motions of homogeneous isotropic materials, the principal stress axes are parallel to the principal axes of *incremental* strain or of strain rate. Because most natural deformations are probably not steady, even this relationship may not be accurate for interpreting deformation that we observe in rocks. Thus as a general rule, structures should always be interpreted in terms of the principal axes of strain. Only under very special circumstances can useful inferences be made about the orientations of the principal stress axes.

15.5 Progressive Stretch of Material Lines

If the unit circle is superposed on the finite strain ellipse, the radii to the intersection points define lines of no finite extension ($e_n = 0$), which are lines that are the same length as they were in the undeformed state ($s_n = 1$). These lines divide the ellipse (Figure 15.16A) into sectors in which radii are longer than they were originally ($s_n > 1$, labeled L) and sectors in which the radii are shorter ($0 < s_n < 1$; labeled S).

We can also examine a similar superposition of the unit circle on the incremental strain ellipse. For generality, we show in Figure 15.16A and *B* the finite and incremental principal strains in a relative orientation that can occur in nature only if the incremental principal axes have changed orientation during the deformation. The intersection of the circle with the incremental ellipse defines a pair of lines that instantaneously are not changing length. These lines divide the incremental strain ellipse (Figure 15.16B) into sectors in which lines are becoming longer ($[ds_n/dt] > 0$; labeled $\dot{\text{L}}$) and sectors in which the lines are becoming shorter ($[ds_n/dt] < 0$; labeled $\dot{\text{S}}$).

The sector boundaries on the incremental strain ellipse (Figure 15.16B) are not in the same orientation as those on the finite strain ellipse (Figure 15.16A), and

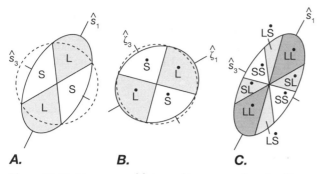

A.   **B.**   **C.**

Figure 15.16 Geometry of finite and incremental strain ellipses for a deformation in which the incremental strain is superposed on a preexisting homogeneous strain. For generality, we have chosen an orientation of the finite strain ellipse that can be formed only from an unsteady deformation, characterized by an incremental strain ellipse whose principal axes change orientation during the deformation. *A.* The strain ellipse, showing lines of no finite extension that define sectors in which radial material lines have been lengthened (L) or shortened (S) by the deformation. The unit circle is shown dashed. *B.* The incremental strain ellipse, showing lines of no rate of extension that divide the ellipse into sectors in which radial material lines are being lengthened ($\dot{L}$) (positive rate of change of stretch) and sectors in which radial material lines are being shortened ($\dot{S}$) (negative rate of change of stretch). *C.* The combination of the two sets of sectors from parts *A* and *B* on the strain ellipse defines sectors in which radial material lines have different combinations of stretch and rate of stretch.

because material lines in general rotate during a deformation, they can pass from one sector into another. Thus the finite strain ellipse can be divided into sectors in each of which the material lines have a different history of stretching (Figure 15.16C). The different possible histories are illustrated in Figure 15.17, where shortening of material lines is represented as folding or imbrication, and lengthening of material lines is represented as boudinage. In sectors labeled $S\dot{S}$, lines are shorter than the original length and have a history of continuous shortening (Figure 15.17A). In sectors labeled $L\dot{S}$, lines are longer than the original length indicating an initial history of lengthening, but they are now shortening (Figure 15.17B); with continued deformation they may end up shorter than their initial length and therefore positioned in the $S\dot{S}$ sector (see Figure 15.17E). In sectors $L\dot{L}$, lines are longer and have a history of continuous lengthening (Figure 15.17C); and in sectors $S\dot{L}$, lines are shorter, indicating an initial history of shortening, but they are now lengthening (Figure 15.17D). With continued deformation they may end up longer than their initial length and therefore in the ($L\dot{L}$) sector. Thus ($S\dot{S}$) sectors may be subdivided according to whether or not the lines had an initial history of lengthening (compare Figure 15.17A, B). Similarly, ($L\dot{L}$) sectors may be subdivided according to whether or not

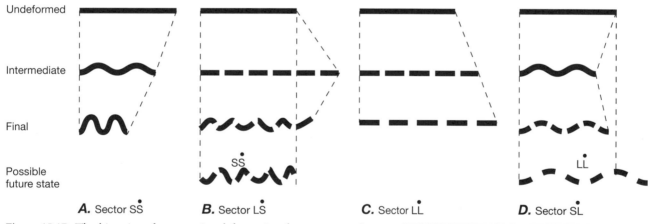

Undeformed

Intermediate

Final

Possible future state

A. Sector $S\dot{S}$ **B.** Sector $L\dot{S}$ **C.** Sector $L\dot{L}$ **D.** Sector $S\dot{L}$

Figure 15.17 The histories of progressive deformation for competent layers oriented within the different sectors shown in Figure 15.16C. The undeformed, intermediate, and final states are points along the deformation path. *A.* Sectors $S\dot{S}$: shorter and being shortened. The layer is continuously folded. *B.* Sectors $L\dot{S}$: longer and being shortened. The layer was initially boudined and subsequently shortened, which caused folding and imbrication of the boudins. The "final" overall length is greater than the initial length, but continued shortening could make it less, thereby transferring the line into the $S\dot{S}$ sector. *C.* Sectors $L\dot{L}$: longer and being lengthened. The layer is continuously boudinaged. *D.* Sectors $S\dot{L}$: shorter and being lengthened. The layer is initially folded and subsequently boudinaged. The "final" overall length is smaller than the original length, but continued lengthening could make it longer, thereby transferring the line into the $L\dot{L}$ sector. *E.* Boudins that have been shortened after formation, illustrating the deformational history in part *B.*

E.

the lines had an initial history of shortening (compare Figures 15.17C, D: see lighter grey portions of LL̇ sectors in Figure 15.18A, B).

Thus, depending on the orientation of the material line with respect to the strain axes, the same deformation can produce folds, boudinage, boudinaged folds, or folded and imbricated boudins. The distribution of such sectors for progressive pure shear and for progressive simple shear is shown in Figure 15.18A, B, respectively. The main difference in the distribution of sectors about the principal axes of strain is the absence of an (SL̇) sector for progressive simple shear subparallel to the shear plane. Thus the sectors of the strain ellipse for progressive pure shear have an overall orthorhombic symmetry, whereas the sectors for progressive simple shear have an overall monoclinic symmetry. When these aspects of the deformation are taken into account, an arbitrary deformation cannot be reproduced by the sequence of operations indicated in Figure 15.12 because, for example, the rotation of the sectors with the strain ellipse produced by progressive pure shear (Figure 15.18A) does not reproduce the sectors in the strain ellipse formed by progressive simple shear (Figure 15.18B), even though the strain ellipses themselves are the same shape.

In principle, then, it should be possible to distinguish some features of the strain history, such as coaxial and noncoaxial progressive deformations, by examining the relationship between the deformational structures in the rock and their orientations. For example, if veins are intruded into a rock in a variety of orientations,

subsequent deformation could cause veins to form folds and/or boudins depending on their orientation relative to the principal stretches. The observed distribution of these structures defines the sectors of the finite strain ellipse (Figure 15.18C). In practice, however, the sector patterns are difficult to establish. The distribution of orientations of deformed layers is usually not ideal (Figure 15.18D), and layers can shorten and thicken without folding or can lengthen and thin without boudinage. Despite its limited practical application, this analysis demonstrates the important fact that no single type of structure is uniquely indicative of a particular geometry of deformation.

15.6 The Representation of Strain States and Strain Histories

It is often useful to compare various states of strain in order to show, for example, how they are related to one another in heterogeneously deformed rocks or to illustrate the sequence of strain states that represents a particular progressive deformation. Such a comparison is easily made by plotting the information on a **Flinn diagram**, on which the ordinate and abcissa are the ratios a and b of the principal stretches, defined by

$$a = \frac{\hat{s}_1}{\hat{s}_2} \qquad b = \frac{\hat{s}_2}{\hat{s}_3} \qquad (15.21)$$

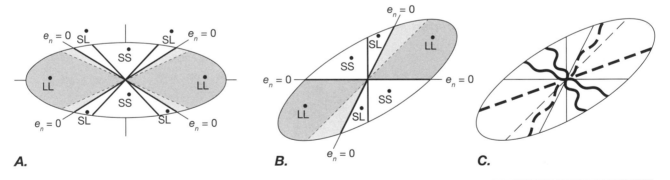

A.

B.

C.

D.

Figure 15.18 Distribution of sectors of stretch and stretching. Material lines in the lighter grey parts of the LL̇ sectors have an initial history of shortening followed by lengthening (see bottom of Figure 15.17D). A. Progressive pure shear. B. Progressive simple shear. This case differs from progressive pure shear mainly in the lack of symmetry of the (SL̇) sectors about the principal axes of strain. C. Structures developed in competent layers in an incompetent matrix consistent with the sectors for progressive simple shear. D. Folding of a layer (left) and simultaneous boudinage of a perpendicular layer (horizontal above pencil) during deformation of a marble.

The study of geologic strains rarely includes the volumetric strain, because it is very uncommon to know the original *size* of a strained object such as a fossil, even though its original *shape* may be known. Thus we can frequently determine the relative lengths of the principal axes of the strain ellipsoid but not the absolute lengths. Because the Flinn diagram is a plot of the ratios of the principal stretches, it can be used to show the shape of a strain ellipsoid, but not the size.

The origin of the coordinate axes for the Flinn diagram is generally taken to be (1, 1) because a and b cannot be less than 1, as can be seen from the second Equation (15.14) and Equation (15.21). Any strain ellipsoid plots at a particular point on the Flinn diagram, and the slope k of the line from the origin (1, 1) to that point is

$$k = \frac{a-1}{b-1} = \frac{\hat{s}_1 \hat{s}_3 - \hat{s}_2 \hat{s}_3}{(\hat{s}_2)^2 - \hat{s}_2 \hat{s}_3} \qquad (15.22)$$

The value of k provides a useful way of classifying the types of constant-volume ellipsoids (Figure 15.19). Three lines, for $k = 0$, $k = 1$, and $k = \infty$, divide the graph into two fields, with ellipsoids of different characteristics plotting along each line and within each field. The field of flattening strain comprises the region for which $0 \leq k < 1$. The line $k = 0$ characterizes oblate uniaxial ellipsoids (pancake-shaped; $\hat{s}_1 = \hat{s}_2 > 1 > \hat{s}_3$), and the range $0 < k < 1$ characterizes oblate triaxial ellipsoids ($\hat{s}_1 > \hat{s}_2 > \hat{s}_3$). The line $k = 1$ characterizes all plane strain ellipsoids ($\hat{s}_1 > \hat{s}_2 = 1 > \hat{s}_3$). The field of constrictional strain includes the values $1 < k \leq \infty$. The

range $1 < k < \infty$ describes prolate triaxial ellipsoids ($\hat{s}_1 > 1 > \hat{s}_2 > \hat{s}_3$), and the line $k = \infty$ describes prolate uniaxial ellipsoids (cigar-shaped, $\hat{s}_1 > 1 > \hat{s}_2 = \hat{s}_3$). The values of the stretches given here apply only to constant-volume strains (Equations 15.18).

The Flinn diagram lends itself well to the representation of strain paths, which define the sequence of strain states through which a body passes in a progressive deformation. Steady motions produce strain paths that plot as straight lines. In geologic deformation, however, steady motions over long periods of time are probably the exception, and curved paths, which may even cross from the constrictional field into the flattening field, or vice versa, are probably common. The diagram makes no distinction, however, between coaxial and noncoaxial progressive deformations. Progressive pure shear and progressive simple shear, for example, are both constant-volume progressive plane deformations that plot along the line $k = 1$. This fact shows that the rotational component of any deformation, which distinguishes pure shear from simple shear, for example, is not represented on a Flinn diagram.

Volumetric deformation is easy to represent on the Flinn diagram. Because plane strain geometry ($\hat{s}_2 = 1$) must always separate the field of constriction ($\hat{s}_2 < 1$) from the field of flattening ($\hat{s}_2 > 1$), the location of this boundary separates constrictive from flattening strains even when the volume is not constant. In order to determine the equation for the line of plane strain when the volume is not constant, we take $\hat{s}_2 = 1$ in Equation (15.21), which gives

$$a = \hat{s}_1 \qquad b = \frac{1}{\hat{s}_3} \qquad (15.23)$$

We can then express the equation for volumetric stretch in plane strain (the first Equation 15.17) in terms of a and b by using Equation (15.23).

$$s_v = a/b \qquad \text{or} \qquad a = s_v b \qquad (15.24)$$

The second Equation (15.24) is the equation for the plane strain line on the Flinn diagram in terms of the volumetric stretch. Taking the natural logarithm of both sides gives an alternative form:

$$\ln a = \ln s_v + \ln b \qquad (15.25)$$

The base-10 logarithm could also be used. The second Equation (15.24) shows that the volumetric stretch s_v determines the slope of the line through the point $(a, b) = (0,0)$ on the Flinn diagram that separates constrictional strain from flattening strain. Note that in general, these lines do not pass through the origin of the Flinn diagram $(a, b) = (1, 1)$. Only when the volume is constant ($s_v = 1$) is the slope of the plane strain line equal to 1, in which case $k = 1$ also, and the line passes through the origin of the Flinn diagram $(a, b) = (1, 1)$.

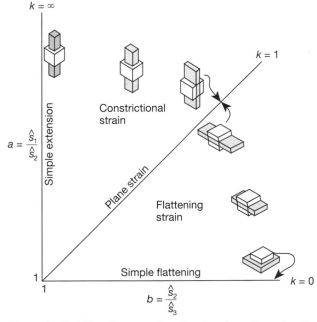

Figure 15.19 Flinn diagram showing the three lines ($k = 0$, $k = 1$, and $k = \infty$) and the two fields ($0 < k < 1$ and $1 < k < \infty$) of finite strain ellipsoids for constant-volume deformation.

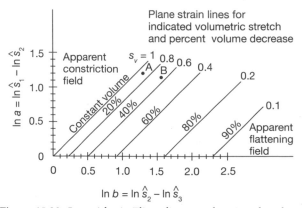

Figure 15.20 Logarithmic Flinn diagram showing the plane strain boundary lines (Equation 15.25) between the fields of flattening and constrictional strain for various amounts of volumetric stretch s_v (Equation 15.16).

The logarithmic Flinn diagram, on which the axes are $\ln a$ and $\ln b$, is more convenient for showing the effects of volumetric deformation (Figure 15.20). Equation (15.25) shows that on this form of the diagram, the plane strain line maintains a constant slope of 1, and the volumetric stretch determines the intercept. Each line on Figure 15.20 represents the plane strain line for a different volumetric stretch, as labeled, and each line therefore separates the field of constrictive strain above from the field of flattening strain below.

The danger of interpreting strain measurements without knowing the volumetric stretch is evident from Figure 15.20. A strain ellipsoid that plots at point A, for example, would be in the flattening field for $s_v = 1$ but in the constrictive field for $s_v \leq 0.8$. Similarly, a strain ellipsoid that plots at point B would be in the flattening field for $1 \geq s_v \geq 0.8$ but in the constrictive field for $s_v \leq 0.6$. Thus plotting strain ellipses on the Flinn diagram without knowing the volumetric stretch can be misleading, and the common assumption of constant-volume deformation for rocks can lead to incorrect interpretations.

15.7 Homogeneous and Inhomogeneous Deformation

So far in this chapter we have restricted our discussion to homogeneous strains. As we noted at the beginning of this chapter, if we are interested in the inhomogeneous distribution of strain, such as in the formation of a fold, we assume the deformed body can be divided into vol-

umes that are sufficiently small for the deformation to be described as locally homogeneous. The variation of these local strains across the body describes the inhomogeneous strain distribution. For any real material, we must realize that the description of a deformation as homogeneous at any particular scale is the result of averaging the deformation over volumes that are large compared with the scale of inhomogeneities that are of no immediate interest, but small compared with the scale at which the inhomogeneous distribution of strain is of interest.

Figure 15.21, for example, shows the so-called deck-of-cards model for forming a passive shear fold (see also Figure 12.8). As discussed in Section 12.2, the deformation is accomplished by a discontinuity in the shear displacement at the card surfaces, with no deformation at all of the individual cards. On the scale of a fold limb, however, the deformation in this example can be regarded as homogeneous simple shear, and it produces the average strain ellipse shown on each fold limb in the figure. Thus the description of the strain as homogeneous results from averaging the strain over a region that is large compared with the thickness of the cards, but small compared with the wavelength of the fold. In other words, the homogeneity depends on scale.

The variety of scales on which we could consider a deformation to be homogeneous is illustrated in Figure 15.22. In Figure 15.22A, the body of folded rock measures about 1 km in length. The scale of the whole block is large compared with the wavelength of the folds, but small compared with the dimension of a mountain belt. At this scale, the average deformation is homogeneous and is represented by the strain ellipse shown beside the block.

When we look at a scale comparable to the fold wavelength, however, the strain is no longer homogeneous (Figure 15.22B). We then describe the deformation in terms of the variation in local strain, which is considered homogeneous on a scale, for example, of about

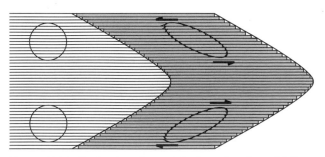

Figure 15.21 Deck-of-cards model of passive-shear folding. On each card, the arcs of the undeformed circle are displaced so as to approximate the shape of the strain ellipse.

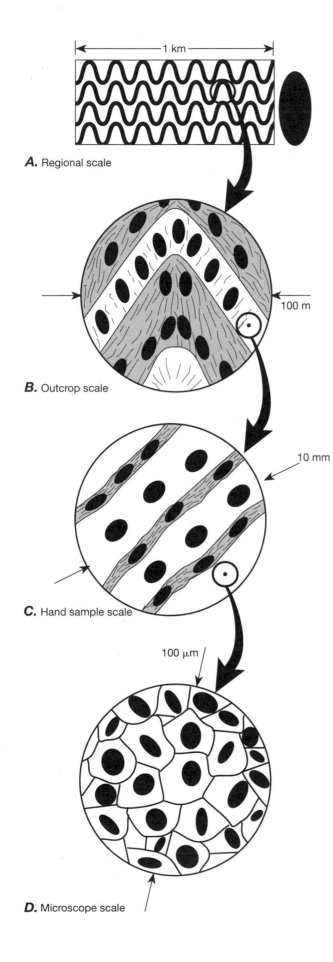

A. Regional scale

B. Outcrop scale

C. Hand sample scale

D. Microscope scale

Figure 15.22 Scales of homogeneous and inhomogeneous strain. In each diagram, the volume over which the strain is averaged to form a locally homogeneous strain ellipse can be viewed at a smaller scale at which the strain distribution is inhomogeneous.

a meter. That scale is small compared with the wavelength of the fold, but large compared with the inhomogeneities in strain that might be present, for example, if the layer were a sandstone containing a spaced foliation.

When we shift scales again, down to the level of the spaced foliation (Figure 15.22C), we again find an inhomogeneous distribution of local strain. In this case, the local strain is averaged over a volume small relative to the spacing of the foliation domains, but large relative to the grain size.

Another shift in scale brings us down to the scale of the grains (Figure 15.22D), where the local strain is again inhomogeneous and the strain in each grain is averaged over a volume that is large compared with the scale of crystal lattice imperfections.

Thus we can consider the strain to be "homogeneous" on a scale that is small compared with the particular structure within which we want to determine the strain distribution, but large compared with the scale of inhomogeneities in which we are not interested and over which we want to average the deformation.

Additional Readings

Eringen, A. C. 1967. *Mechanics of continua.* New York: Wiley.

Flinn, D. 1962. On folding during three dimensional progressive deformation. *Q. J. Geol. Soc. Lond.* 118: 385–428.

Means, W. D. 1976. *Stress and strain: Basic concepts of continuum mechanics for geologists.* New York: Springer-Verlag.

Passchier, C. W. 1990. Reconstruction of deformation and flow parameters from deformed vein sets. *Tectonophysics* 180(2–4): 185–199.

Ramsay, J. G. 1967. *Folding and fracturing of rocks.* New York: McGraw-Hill.

Ramsay, J. G. 1976. Displacement and strain. *Phil. Trans. Roy. Soc. Lond.* A283: 3–25.

Ramsay, J. G., and M. I. Huber. 1983. *The techniques of modern structural geology.* Vol. 1: Strain analysis. New York: Academic Press.

Wood, D. S., and P. E. Holm. 1980. Quantitative analysis of strain heterogeneity as a function of temperature and strain rate. *Tectonophysics* 66: 1–14.

CHAPTER

16 Strain and Models for the Formation of Ductile Structures

We now return to folds, foliations, and lineations and consider the relationship between the strain distribution and the various mechanisms of formation of ductile structures outlined in Chapters 12 and 14. We consider first folds and then foliations and lineations.

In Chapter 12, we found that several different mechanisms of folding could lead to the same or a comparable style of fold and that, therefore, fold style does not uniquely indicate folding mechanism. The distribution of strain in a fold, however, can help us distinguish the different folding mechanisms. We focus on strain in the profile plane of a fold and in the surface of the folded layer. The profile plane contains the maximum and minimum principal strains for all the models of folding discussed in Chapter 12 except passive shear. The components of strain in the surface of a folded layer affect the patterns of deformation of preexisting lineations on these surfaces and offer insight into the folding mechanism.

16.1 Patterns of Strain and Lineations in Folds

We consider the folding of a layer (Figure 16.1A(i)) into an anticline–syncline pair. The layer is characterized by a lineation that makes a constant angle

$\beta_A = \beta_I = \beta_S = \beta$ (Figure 16.1A) with the incipient fold axis f. The subscripts A, I, and S refer to the lines that become the anticlinal hinge, the inflection line, and the synclinal hinge, respectively. For all the fold types, we show the fold profile on a three-dimensional sketch of the fold. The top surface of the fold is also shown as it would look flattened out but with all strain and angular relationships preserved, and a stereonet plot gives the folded lineation orientations in the top surface. Strain ellipses show the total strain accumulated from the initial state shown in Figure 16.1A.

Layer Bending by Orthogonal Flexure

Bending by orthogonal flexure results in all lines that are initially normal to the surface of the layer remaining so after folding (Figure 16.1B(i)). On the fold profile, the convex side of the layer lengthens and the concave side shortens. Because of the resulting strain gradient, there is a surface within the layer called the **neutral surface** at which there is no strain.

We assume plane strain, and we assume that the intermediate principal stretch ($\hat{s}_2 = 1$) is everywhere parallel to the hinge. Thus all the strain appears in the profile plane of the fold, which contains the two principal stretches $\hat{s}_1$ and $\hat{s}_3$. Along the convex side of the fold profile, the $\hat{s}_1$ axes ($\hat{s}_1 > 1$) are parallel to the surface and normal to the hinge line (Figure 16.1B(i)). On the concave side of the profile, the $\hat{s}_1$ axes form a convergent fan normal to the folded surface. Thus the profile plane and the surface of the layer are both principal planes of strain.

The geometry of a lineation on a folded surface depends on the strain in the surface. On a neutral surface, the folding rotates the lineation into different orientations, but the angle between the lineation and the fold axis f is everywhere constant and unchanged from the original angle, because there is no strain in the plane of the neutral surface (Figure 16.1B(ii)). Thus on a stereonet, the lineation plots along the arc of a small circle of angle β about the fold axis (Figure 16.1B(iii)).

On the top surface of the layer, however, the stretch normal to the fold axis consists of a lengthening where it is on the convex side of a fold and a shortening where it is on the concave side (Figure 16.1B(i), (iv)). Thus the strain in the surface is inhomogeneous. Lineations lying in the surface rotate toward the axis of maximum stretch in the surface and away from the axis of minimum stretch. The top surface of the layer is lengthened most at the hinge on the convex side of a fold (A in Figure 16.1B(i), (iv)) and is shortened most at the hinge on the concave side of a fold (S in Figure 16.1B(i), (iv)). It is not stretched at all at the inflection line (I in Figure 16.1B(i), (iv)). Thus the angle between the fold axis f and the lineation (Figure 16.1B(iv)) is a maximum along the line A ($\beta'_A > \beta$), is unchanged at I ($\beta'_I = \beta$), and is a minimum along the line S ($\beta'_S < \beta$).

On a stereonet, the lineations therefore fall between the small circles for the maximum and minimum angles (Figure 16.1B(v)). In general, then, we expect measurements of folded lineations from folds formed during bending by orthogonal flexure to plot within a small-circle band concentric about the fold axis (Figure 16.1B(v)).

Layer Bending by Flexural Shear

Bending by flexural-shear folding involves internal shearing of the layer parallel to the layer surface (Figure 12.7). The distribution of strain in the fold profile is illustrated in Figure 16.1C(i). The profile plane is a principal plane of strain, but the maximum and minimum principal stretches, $\hat{s}_1$ and $\hat{s}_3$, are neither parallel nor perpendicular to the layer. The $\hat{s}_1$ axis makes an angle of 45° or less with the layer, and the maximum angle is near the hinge surface at A and S, where the shear strain across the entire layer is zero. The sense of shear reverses from one side of the hinge surface to the other. The amount of shear increases along a limb from zero at the hinge to a maximum at the inflection surface. With increasing shear, the $\hat{s}_1$ axis rotates to lower angles

with the layer surface, which parallels the shear plane (Figure 16.1C(i)). As a result, $\hat{s}_1$ axes tend to form a curved pattern across the fold profile that can vary from a divergent fan in the hinge zone to a convergent fan along the limbs.

Surfaces parallel to the layer are planes of simple shear. They are therefore surfaces of no strain, which are parallel to a circular section through the strain ellipsoid (Figure 16.1C(ii)) and are not parallel to a principal plane. On any of these surfaces, the original angle β between a lineation and the fold axis f remains unchanged by folding. On a stereonet, the orientations of folded lineations plot along a single small circle centered on the fold axis (Figure 16.1C(iii)), just as for the neutral surface in the orthogonal flexure fold (Figure 16.1B(iii)).

Layer Bending by Volume-Loss Flexure

Layer bending by volume-loss flexure results from solution along seams concentrated on the concave side of the fold (Figure 12.10A, B). Thus the layer is shortened in a direction normal to the hinge on the concave side of the fold, whereas the convex side undergoes no strain and forms the neutral surface. The profile plane is a principal plane of strain, with the maximum shortening $\hat{s}_3 < 1$ normal to the seams, and $\hat{s}_1 = 1$ parallel to the seams (Figure 16.1D(i)). If the seams are initially at a high angle to the layer surface, then $\hat{s}_1$ axes in the fold form a convergent fan around the fold. Only if the seams are normal to the layer, as in Figure 16.1D(i)), is the layer surface a principal plane.

On the convex surface of a folded layer, the original angle between lineation and fold axis f is preserved because the surface is not strained ($\beta'_A = \beta'_I = \beta$; Figure 16.1$D$(ii)). On the concave surface, the shortening normal to the fold axis rotates the lineation to smaller angles with the fold axis f, reaching a minimum angle ($\beta'_S < \beta$) at the hinge where the shortening is a maximum. On a stereonet, lineations from the convex side of the fold plot along a small circle of angle β about the fold axis f, and those from the concave side plot at smaller angles (Figure 16.1D iii). Thus in general, the lineations plot within a small-circle band between the angles β and β'_S from the fold axis.

Layer Buckling by Flexural Folding

Buckling occurs after initial homogeneous layer-parallel shortening, as shown by the strain distribution across the layer in Figure 16.2A(i). Thus, depending on the kinematic mechanism, the final strain in a buckle fold is the sum of the initial shortening (Figure 16.2A(i)) and one of the strain distributions shown in Figures 16.1B, C, and D, and it depends on the relative amounts of strain contributed by each.

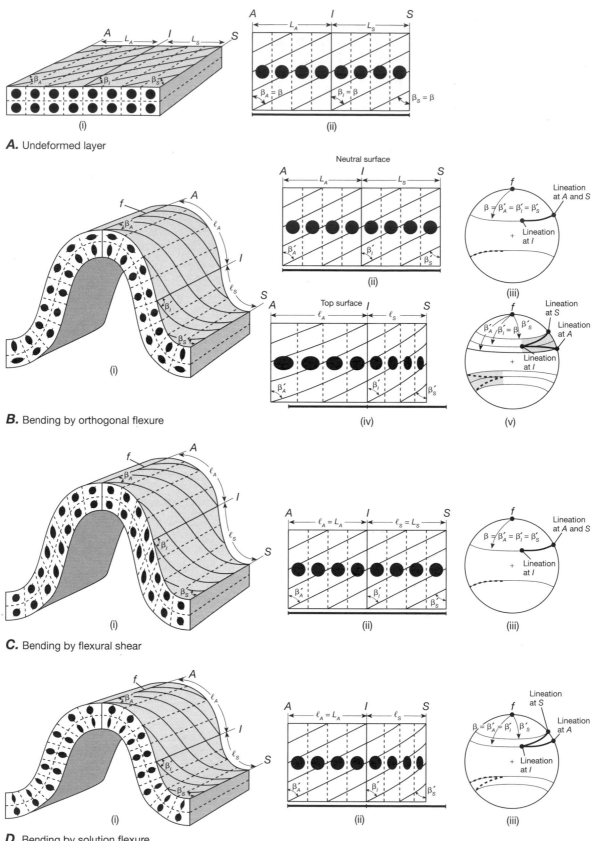

A. Undeformed layer

B. Bending by orthogonal flexure

C. Bending by flexural shear

D. Bending by solution flexure

Figure 16.1 Models of bending by flexural folding and the strain distribution in the profile planes and on the folded surfaces. In each of the fold models, the first column shows the strain in the profile plane. The second column shows the folded top surface, from the anticlinal hinge to the synclinal hinge, flattened out so as to preserve the strain and lineation orientations in the surface. The bar beneath each diagram shows the original undeformed length of the surface. The third column shows the geometry of the lineation rotation on a lower-hemisphere equal-area projection. *A* indicates the antiformal hinge, *I* the inflection line, and *S* the synformal hinge. *A.* The undeformed layer that becomes folded into the half of the fold extending from the anticlinal hinge to the synclinal hinge. *B.–D.* The results of different modes of bending.

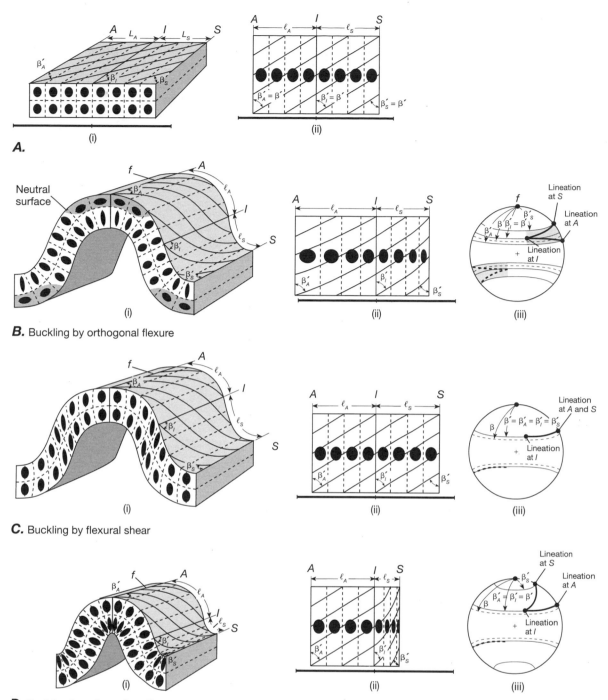

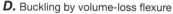

Figure 16.2 Models of buckling by flexural folding and the strain distribution in the profile planes and on the folded surfaces. See the legend to Figure 16.1 for details. *A.* The layer to be folded into half the antiform–synform pair after shortening but just before the onset of buckling. The initial dimensions before shortening are the same as in Figure 16.1*A. B.–D.* The results of different modes of buckling.

If buckling occurs by orthogonal flexure, the initial layer-parallel shortening counteracts the lengthening that develops on the convex side of the fold and increases the shortening that accumulates on the concave side (Figure 16.2*B*(i)). If the initial shortening is greater than the subsequent lengthening, the $\hat{s}_1$ axis is normal to the

layer surface around the entire fold profile, forming a convergent fan, and no neutral surface exists within the layer at all. If the lengthening due to the folding exceeds the initial shortening (Figure 16.2*B*(i)), then a neutral surface develops through part of the layer (edge of the shaded zone in Figure 16.2*B*(i); compare Figure 10.12*C*).

If buckling occurs by flexural shear, the resulting strains added to the layer-parallel shortening tend to form a convergent fan of $\hat{s}_1$ axes around the fold (Figure 16.2C(i)). Compared with pure flexural shear (Figure 16.1C), the axes of maximum stretch are rotated toward higher angles with the layer.

For volume-loss flexure, the zero strains on the convex side of the layer (Figure 16.1D(i)) are replaced by layer-parallel shortening normal to the hinge; as a result, $\hat{s}_1$ is normal to the layer (Figure 16.2D(i)). On the concave side of the fold, the initial shortening strains are increased by the folding, and $\hat{s}_1$ forms a convergent fan across the fold.

The pattern of lineations folded by buckling, however, shows little difference from those folded by bend-ing. Before the initiation of buckling, the layer-parallel shortening homogenously rotates a lineation on the layer surface to a smaller angle with the incipient fold axis ($\beta_A = \beta_I = \beta_S = \beta' < \beta$; Figure 16.2A(ii)). Thus the initial angle between lineation and hinge line is smaller if folding results from buckling rather than bending. Otherwise, the patterns from the different models of flexure remain the same (Figures 16.2B(iii), C(iii), and D(iii)).

Folding by Passive Shear

The distribution of strain across the profile of a passive-shear fold reflects the inhomogeneous simple shear that produces the fold (Figure 16.3; see also Figure 12.8).

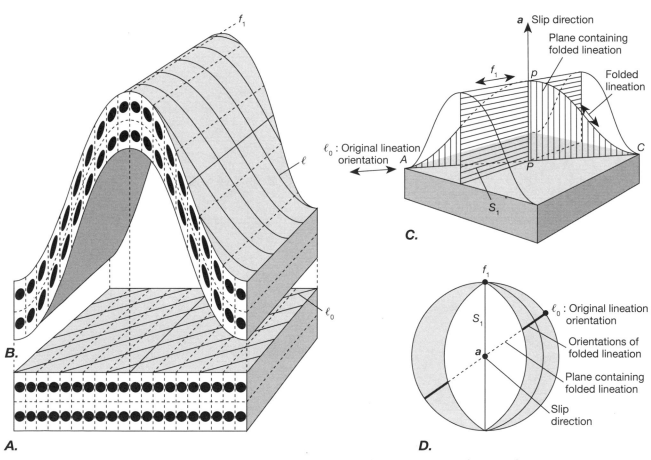

Figure 16.3 Passive-shear folding. *A.* Initial layer to be folded into the antiform–synform pair. *B.* Inhomogeneous simple shear on a set of parallel shear planes intersecting a layer produces a class 2 fold. An original lineation ℓ_0 is folded into the form shown by the lines ℓ. The $\hat{s}_1$ directions of the strain ellipses on the profile plane of the fold diverge across the fold. Sections of the strain ellipsoids on the shear planes are circular sections. *C.* The original lineation *AC* is folded by inhomogeneous passive shear on planes parallel to S_1 in the direction parallel to *Pp*. The folded lineation lies on the folded surface in the plane *ApC*, which contains the original orientation of the lineation *AC* and the shear direction *Pp* = **a**. *D.* A stereonet diagram showing the orientation of the fold axis f_1, the slip direction **a**, the shear plane and axial surface of the fold S_1, the orientation of the undeformed lineation ℓ_0, and the range of orientations of the folded lineation, which lie along a great circle representing the plane that contains ℓ_0 and **a**.

The shear planes are parallel to the hinge surface, and the shear sense changes across that surface. Near the hinge surface, the angle between the $\hat{s}_1$ axis and the surface is 45°. The acute angle between $\hat{s}_1$ and the shear plane decreases with increasing shear strain (Figure 16.3A), so the $\hat{s}_1$ axis rotates toward parallelism with the shear plane and hence with the axial surface of the fold, especially in the limbs where the shear is highest. In principle, however, the $\hat{s}_1$ direction can never actually become parallel with the shear planes, and the long axes of the strain ellipses form a divergent fan across the profile of a passive-shear fold (Figure 16.3B).

In simple shear, the $\hat{s}_2$ direction lies in the shear plane perpendicular to the shear direction, but it is not in general parallel to the fold axis unless it is initially parallel to the layer being folded (see Figure 12.9). Thus the $\hat{s}_1$ and $\hat{s}_3$ axes do not in general lie in the profile plane of the fold. The shear planes always contain a circular section of the strain ellipsoid.

Passive shear deforms a linear feature into a range of orientations all of which lie in a plane (Figure 16.3B, C). The orientations of lineations in a folded surface all plot on a stereonet along the great circle that contains the original lineation orientation ℓ_0 and the shear direction **a** (Figure 16.3C, D). The lineation is folded from its original orientation in both directions toward the slip direction. The slip direction also must lie in the shear plane, which is the axial surface. Thus on a stereonet, the intersection of the plane containing the lin-

eation orientations with the plane of the axial surface defines the slip direction if the fold formed by passive shear. Both of these planes can be determined by field measurements.

The Effects of a Superposed Homogeneous Flattening

Superposition of homogeneous flattening perpendicular to the axial surface changes class 1B folds into class 1C, but class 2 folds do not change class (Section 12.4). All $\hat{s}_1$ axes rotate toward the superposed plane of flattening, which in this case is the axial surface of the fold. We consider the effects on folds that result from two mechanisms: bending by orthogonal flexure, and passive shear.

Figure 16.4 illustrates the effect of superposed flattening on the strain distribution in a simplified orthogonal flexure fold of constant curvature formed by bending. Note that a true concentric fold having constant curvature is possible in nature only for the flexural shear mechanism, because for other mechanisms, a discontinuity in strain would occur at the inflection surfaces (compare Figure 16.4A with Figure 16.1B(i)). Flattening of the initial fold (Figure 16.4A) by 20 percent (Figure 16.4B) produces orientations of $\hat{s}_1$ axes comparable to those in a buckled flexural-shear fold (compare Figure 16.2C(i)).

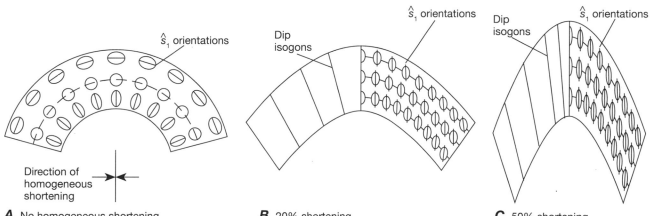

A. No homogeneous shortening   **B.** 20% shortening   **C.** 50% shortening

Figure 16.4 Homogeneous flattening normal to the axial surface of an orthogonal flexure fold formed in a layer by bending. Dip isogons are plotted on the left limbs, strain distributions on the right, with the line through the ellipses parallel to $\hat{s}_1$. A. Bending of a plate by orthogonal flexure. The fold is class 1B. This simple fold form, which is characterized by constant curvature, would give rise to strain discontinuities at the inflection surface if it were part of a fold train, so it is an oversimplified model. A more realistic model, in which the strain diminishes to zero at the inflection surface, is shown in Figure 16.1Bi. B. Twenty percent homogeneous shortening normal to the axial surface of the fold in part A. The fold is class 1C. C. Fifty percent homogeneous shortening normal to the axial surface of the fold in part A. The fold is class 1C.

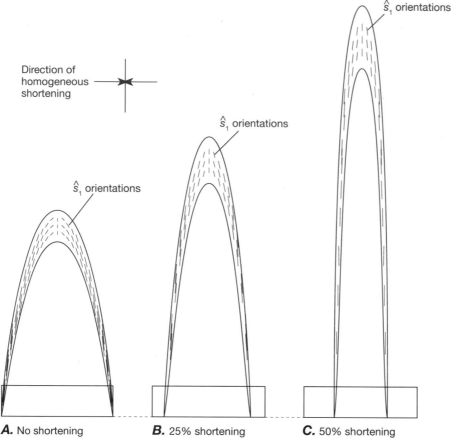

$\hat{s}_1$ orientations

$\hat{s}_1$ orientations

$\hat{s}_1$ orientations

Figure 16.5 Homogeneous flattening normal to the axial surface of a class 2 passive-shear fold. The lines in the folded layer show the orientation and relative magnitude of $\hat{s}_1$; the undeformed length of the line is shown at the hinge of the fold in part A. The initial dimensions of the undeformed plate are shown as the blank rectangle. *A.* A class 2 fold formed by passive shear of the undeformed plate. *B.* A class 2 fold formed by 25 percent shortening of the fold in part A. *C.* A class 2 fold formed by 50 percent shortening of the fold in part A.

A. No shortening

B. 25% shortening

C. 50% shortening

The flattened fold is class 1C, however, and the flexural-shear fold is class 1B. Flattening by 50 percent (Figure 16.4C) rotates the $\hat{s}_1$ axes subparallel to the axial surface, and the fold approaches class 2 geometry (compare Figure 16.5). Note that in general, the dip isogons do not parallel the $\hat{s}_1$ axes of the strain ellipse (Figure 16.4B), although the difference in orientation may not be obvious if the amount of flattening is large (Figure 16.4C). Other types of flexural folding show similar effects.

Figure 16.5 illustrates the effects of superposed flattening on a class 2 fold formed by passive shear. Although the fold remains class 2, the $\hat{s}_1$ axes rotate toward the axial surface. The divergent fanning of $\hat{s}_1$ across the fold becomes less pronounced, and at 50 percent shortening it is difficult to detect in practice.

Superposed homogeneous flattening destroys the simple relationship between a flexure-folded lineation and a fold axis. With pronounced flattening, lineations that initially lie along a small circle are reoriented such that their distribution approaches a great circle—that is, a planar distribution.

Thus the geometry, strain distribution, and lineation orientations of a strongly flattened class 1B fold approach those of a class 2 fold. With incomplete or inexact data, the lineation distribution on a flattened class 1B fold could be mistaken for a class 2 distribu-

tion, and erroneous conclusions about the mechanism of folding could be drawn.

Superposed homogeneous flattening of a lineation rotated by passive-shear folding preserves its distribution in a plane, although the plane is rotated toward the superposed plane of flattening. The construction for determining the orientation of the slip direction of the passive-shear folding (Figure 16.3) is still valid. The slip direction, however, is also rotated by the superposed flattening, unless it is parallel to one of the principal axes of the flattening strain.

Strain in Multilayer Folds

Folding of multilayers of similar competence by any of the mechanisms we discuss above produces the strain distributions already described. When a multilayer consists of alternating layers of comparable thickness and of very different competence, however, a different mechanism of folding in the incompetent layers leads to the development of class 3 folds (see Section 12.5 and Figures 12.17A, B, 11.22B, C) and a different distribution of strain.

If competent layers form flexure folds by buckling, a general pattern of convergent $\hat{s}_1$ axes develops across them (Figure 16.6). These competent layers control the

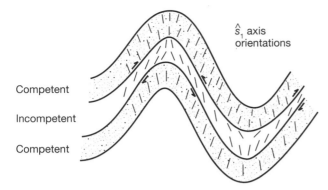

Figure 16.6 Buckling of interlayered competent and incompetent layers of comparable thickness. The short lines indicate the orientation of the $\hat{s}_1$ axes in the multilayer fold.

geometry of the incompetent layers. Thus the fold limbs in the incompetent layer undergo strong shearing parallel to the layer and flattening approximately normal to it. In the hinge area, the incompetent layer is strongly flattened where it is compressed between the closing limbs of the adjacent competent layer, and there is some shear along the contact on the concave side of the fold. The resulting distribution of strain shows a divergent fan geometry of $\hat{s}_1$ axes in the incompetent layer (Figure 16.6).

If kink bands form by either of the mechanisms shown in Figure 12.20 (see Section 12.6) material lines parallel to the laminations, as well as to the kink band boundary, all are the same length after kinking as they were before. Thus these two directions must be lines of no finite extension, and the principal axes of the strain ellipse must bisect the angle between them. For the models in Figure 12.21, the orientations of the principal axes are determined by the geometry of simple shear and are not uniquely related to the laminations or the kink band boundary.

16.2 Strain, Foliations, and Lineations

Field data suggest that many foliations are parallel to the plane of flattening (the $\hat{s}_1$–$\hat{s}_2$ plane) of the finite strain ellipsoid. Some lineations are stretching lineations, which parallel the $\hat{s}_1$ direction, whereas others may be normal to $\hat{s}_1$ and possibly parallel to $\hat{s}_2$. Although in principle the parallelism for many cases cannot be exact, in practice it is usually close enough to provide a useful indicator of the orientation of the principal strain axes. It is, however, important to understand the conditions under which this relationship does or does not hold. We distinguish here between material and nonmaterial foliations and lineations, because the relationships to strain are different in each case.

Material foliations and **material lineations** are defined by planes or lines that are composed of sets of specific material points. The orientation and motion of material foliations and lineations are therefore determined by the motions of the material particles in the deforming continuum (see the beginning of Chapter 15). Material foliations include compositional layers such as mica films, compositionally differentiated cleavages, and compositional foliations, and may also include pervasive sets of microfractures. Spaced foliations and microspaced continuous foliations generally belong in this category. Material lineations include most constructed lineations (Figure 13.18), such as hinge lines, intersection lineations formed by material surfaces, boudin lines, and mullions.

Nonmaterial foliations and **nonmaterial lineations** are defined by the preferred orientation of planar or linear constituents of the material. Such a preferred orientation is not itself a material plane or line, and therefore nonmaterial foliations and lineations behave differently from material planes and lines. The preferred orientation of deformed ooids, for example, defines a foliation or lineation that is parallel, respectively, to the $\hat{s}_1$–$\hat{s}_2$ plane or the $\hat{s}_1$ axis of the finite strain ellipsoid. The principal axes of the strain ellipsoid are not material lines, because material lines rotate through the principal strain axes during progressive noncoaxial deformations (see Section 15.4 and Figure 15.15). Thus the foliations or lineations defined by these principal axes cannot be material planes or lines. Foliations and lineations that fall in this category include those defined by the preferred orientation of passively strained markers and by the preferred orientation of rigidly rotated particles.

In the following three sections, we discuss how strain is related to the orientation of foliations and lineations, considering the motion of foliations and lineations during deformation, the specific mechanisms discussed in Chapter 14 by which foliations form, and the characteristics of particular types of foliation described in Chapter 13.

16.3 Relation of Strain to Material Foliations and Lineations

Passive Reorientation During Deformation

The relationship between strain and material foliations and lineations can be discussed with reference to the geometric properties of the deformation. There are three conceivable associations of material foliations and lineations with strain (refer to Figures 15.14 and 15.15):

1. If a material foliation forms parallel to the plane of flattening, or if a material lineation forms parallel to

the maximum principal stretch, it can maintain that orientation throughout the progressive deformation only if the deformation is coaxial.

2. If a material foliation or lineation forms in the same orientation as described in argument 1, a progressive simple shear (a noncoaxial deformation) rotates those structures out of their initial orientation and toward the shear plane or the shear direction, respectively. At the same time, however, the plane of flattening ($\hat{s}_1$–$\hat{s}_2$) rotates toward the shear plane, and the $\hat{s}_1$ axis rotates toward the shear direction, although at a different rate from the material foliation or lineation. After a large deformation, the angle between the $\hat{s}_1$–$\hat{s}_2$ plane or the $\hat{s}_1$ axis and the material foliation or lineation becomes negligible.

3. If a material foliation or lineation forms in an orientation not parallel to the principal axes of the finite strain ellipsoid, in general it is rotated during progressive deformation. After a large deformation, whether coaxial or noncoaxial, it may end up at negligibly small angles with the $\hat{s}_1$–$\hat{s}_2$ plane or the $\hat{s}_1$ axis, respectively.

It is always tempting to infer that boudin lines form normal to the direction of maximum principal stretch $\hat{s}_1$ and that fold hinges form normal to the direction of minimum principal stretch $\hat{s}_3$. This relationship can occur, however, only if the layers in which these structures form are parallel to the appropriate strain axis (see Figure 15.8). Because boudin lines and fold hinges must be parallel to the layers in which they develop, they need not be parallel to any particular axis of strain. After formation, both boudins and fold axes undergo the same passive rotation as any other material line.

Coaxial deformation is a very special geometry of deformation, and the circumstances in which it can occur are therefore much more restricted than those for noncoaxial deformation. Thus we expect noncoaxial deformation to be the most common condition in nature. Because material foliations and lineations can be strictly parallel to a principal plane or axis of strain only for coaxial deformation (see argument 1 in the foregoing list), we expect from arguments 2 and 3 that, in general, they will only approximate such an orientation. For large strains, however, the deviation from parallelism is undetectable if it is less than the precision of measurement.

Solution and Precipitation

The processes of solution and precipitation commonly produce material foliations and lineations, such as solution seams filled with an insoluble residue, and mineral fibers on shear planes and in veins. In some cases, such features have not been subjected to large homogeneous deformations. Thus they may not have rotated into orientations that are simply related to the finite strain ellipse, as happens to material foliations and lineations subjected to large deformations (see arguments 1–3 above).

Stylolites and slickolites result from solution during low-temperature deformation, and the teeth on stylolites (Figures 13.3 and 13.5) and the spikes on slickolites (Figure 14.6C) are interpreted to be parallel to the direction of relative displacement across the surface. For stylolites, the teeth are generally at a high angle to the surface and probably indicate the direction of maximum shortening (minimum stretch $\hat{s}_3$). If the stylolite forms by pressure solution, however, it should be oriented normal to the maximum compressive stress $\hat{\sigma}_1$. The two interpretations could be equivalent only if the strain is small. In that case, if the principal axes of stress are parallel to those of incremental strain, which is a simple but reasonable assumption, the principal axes of finite strain $\hat{s}_k$ and incremental strain $\hat{\zeta}_k$ are essentially parallel. For slickolites, however, the spikes are generally at a low angle to the surface and indicate the direction of relative shear. Thus there appears to be no general strain interpretation for the orientation of stylolite and slickolite planes.

Mineral fibers grow during progressive deformations as a result of mineral precipitation, and the fibers are parallel to the direction of relative displacement across the surface of growth. In the case of slickenfibers on fault surfaces, the fibers are parallel to the direction of relative shear across the fault surface. In the case of fibers in veins and in most overgrowths, they are parallel to the axis of maximum incremental stretch $\hat{\zeta}_1$. If an overgrowth lineation forms by precipitation in pressure shadows, it should be parallel to the minimum compressive stress $\hat{\sigma}_3$. These interpretations are compatible assuming the principal axes of incremental strain and stress are parallel.

Because mineral fibers grow during a progressive deformation, they preserve information about the deformation history and in some cases can be used to distinguish coaxial from noncoaxial deformations. In veins, straight fibers generally indicate a coaxial deformation because they imply that the same material lines have remained parallel to the principal axes of incremental strain throughout the deformation (Figure 14.7C, D). Conversely, curved mineral fibers indicate noncoaxial deformation, because they imply that the earliest fibers rotated progressively away from the direction of growth, which remained parallel to the axis of maximum incremental stretch $\hat{\zeta}_1$. Detailed interpretation must be based on a knowledge of whether the fibers form by syntaxial or antitaxial growth (Figure 14.7), composite growth, or by crack-seal growth (Figure 14.9) because the growth mechanism affects the fiber pattern that develops.

16.4 Relation of Strain to Nonmaterial Foliations and Lineations

For nonmaterial foliations and lineations, the orientation relative to the principal axes of strain depends on the mechanism of development. We must therefore consider individually the effects of flattening and elongation of passive markers, and the effects of the rotation of rigid particles and ductile crystal grains.

The Preferred Orientation of Flattened and Elongated Markers

Many passive markers record flattening and elongation, including pebbles, ooids, spherical fossils such as radiolarian tests, alteration spots, mineral grains, and mineral grain clusters. Such passively deformed features are flattened parallel to the $\hat{s}_1$–$\hat{s}_2$ plane and elongated parallel to the $\hat{s}_1$ direction, regardless of whether the deformation is coaxial or noncoaxial. Although these features are material objects, the principal axes of strain that they record are nonmaterial, and the preferred orientation of those axes therefore defines a foliation or lineation that is nonmaterial.

Three limitations prevent such foliations and lineations from being ideal indicators of finite strain. First, most such features in rocks are spheroidal (that is, not exactly spherical), so the deformed shapes are not true strain ellipsoids. Second, many features have an initial preferred orientation, the effect of which is never completely eliminated by deformation; thus the principal axes of the deformed features are not the true principal axes of strain (see, for example, Figure 14.2C). Third, many "passive" markers are of different composition and competence from the surrounding matrix and therefore might not behave as strictly passive markers. Thus the strain they record might not be the total strain in the rock, and the foliation or lineation defined by their preferred dimensional orientation might not be parallel to the total strain axes. Moreover, if the markers are not strictly passive, their preferred orientation could be affected by a component of rigid rotation, which we discuss below.

Despite these problems, at sufficiently large deformations the inaccuracies become minor or even undetectable, and foliations and lineations defined by deformed spheroidal features indicate the approximate orientation of the total finite strain ellipsoid.

Preferred Orientation Formed by Rotation

Each of the different rotation mechanisms discussed in Section 14.2 can produce preferred orientations of plate-like or needlelike particles in the deforming material. The foliations and lineations so defined have different relationships to the principal axes of strain. Particles can rotate according to the March model (Figures 14.4B and 14.5); competent or rigid particles can rotate according to the Jeffrey model (Figure 14.4A); and crystal grains that deform internally rotate according to the Taylor-Bishop-Hill model (Figure 14.4C).

If an initially random-orientation distribution of deformable or rigid platy or elongate grains is carried passively parallel to material planes in a deforming matrix, the March model predicts that regardless of the geometry of the deformation, the grains adapt a preferred orientation defining either a foliation or a lineation nearly or exactly parallel to the plane of flattening (Figures 14.4B, 14.5, 15.14, and 15.15) or to $\hat{s}_1$, respectively. The preferred orientations increase in strength with increasing strain. If, for example, mica flakes or amphibole needles behave rigidly but rotate with the grain boundaries of adjacent ductile mineral grains such as quartz, then their preferred orientation is predicted by the March model.

If particles in an initially random distribution of orientations are free to rotate independently like rigid particles in a viscous fluid, the Jeffrey model predicts that only in a coaxial deformation could a stable preferred orientation develop. That preferred orientation would define a nonmaterial foliation parallel to the plane of flattening or a nonmaterial lineation parallel to the maximum principal stretch.

During progressive simple shear (a noncoaxial deformation), the Jeffrey model predicts that rigid elongate grains should develop a weak preferred orientation subparallel to $\hat{s}_2$. Rigid platy grains should develop a preferred orientation subparallel to the shear plane, but this preferred orientation is not stable, and it continually oscillates in both orientation and intensity because the particles never stop rotating (see Figure 14.4A). Because the plane of flattening never becomes exactly parallel to the shear plane, such foliations and lineations are not exactly parallel to the principal axes of strain, although for large strains they can be close.

Deformation of crystals by slip on their internal slip planes produces preferred orientations of the crystallographic axes that depend on which slip planes are active, the Taylor-Bishop-Hill model (see Figure 19.26). If the crystallographic axes are associated with a dominant platy or elongate crystal habit, this process can produce a preferred dimensional orientation of the crystal grains that defines a foliation or lineation. For example, muscovite generally forms thin platy crystal grains in which the dominant crystallographic slip plane is parallel to the plates. We show in Section 19.7 that if such crystal grains deform in a coaxial deformation, their crystallographic slip planes tend to rotate toward parallelism with the plane of flattening; in a noncoaxial simple shear, they rotate toward parallelism with the plane of shear. Thus a preferred dimensional orientation would result because of the geometry of the internal slip mechanism of the crystal, thereby defining a nonmaterial foliation that, depending on the geometry of the deformation, could be parallel to the plane of flattening or to the plane of shear.

Unfortunately, it is difficult or almost impossible to determine whether the rotation that produced a preferred orientation resulted from the March mechanism, the Jeffrey mechanism, the Taylor-Bishop-Hill mechanism, or some combination of these. The only factor that simplifies the interpretation of such foliations and lineations is that for large deformations, all these preferred orientations tend to approach parallelism with the plane of flattening, although they might not be strictly parallel.

16.5 Strain and Kinematic Models of Foliation Development

Disjunctive Foliations

The thin films of clay or mica and/or insoluble oxides and carbonaceous material that characterize disjunctive foliations are largely the residue from the solution and removal of components such as quartz or calcite. In principle, the amount of shortening could be determined by measuring the amount of insoluble material in the cleavage domain and comparing that to its concentration in the original rock. Mineralogical and chemical variation and alteration, however, can make such an estimate uncertain. Evidence is clear, however, that much of the strain associated with disjunctive foliations is accommodated by solution. We discuss two models for the formation of disjunctive foliations that relate the amount of strain in the rock to the characteristics of the foliation. One applies to limestones and the other to argillites.

The first model proposes that the variation in disjunctive foliation in limestones from stylolitic foliation through anastomosing and rough foliations to smooth foliation (see Figure 13.3) is associated with a progressive increase in the amount of shortening normal or at a high angle to the foliation. The foliation is therefore approximately parallel to the plane of flattening.

Figure 16.7 shows a particularly vivid example of the relationship between disjunctive foliations in limestone and the amount of shortening in the rock. The shortening is recorded by the amount of imbrication of the insoluble chert layers (Figure 16.7A). An increase in the amount of shortening recorded in the chert layers correlates with a decrease in the spacing of the cleavage domains and with an increase in the smoothness of those domains (top to bottom in Figure 16.7B). In weakly deformed rocks, the foliation is defined by a widely spaced, irregular stylolitic foliation. Moderately deformed rocks show a more closely spaced, anastomosing foliation. The most strongly deformed rocks show a closely spaced, subplanar foliation.

A similar example is shown in Figure 16.7C, where the shortening in the limestone is recorded by the imbrication of relatively insoluble chert nodules. In the undeformed limestone, these nodules are present in distinct layers and are nowhere imbricated. Thus measurement of the amount of overlapping of the imbricated nodules gives a minimum amount of shortening parallel to the bedding, which neglects any original spacing between nodules. Again, there is a strong correlation between the degree of development of the foliation and the amount of bedding-parallel shortening.

The foliation diverges around individual chert nodules and converges toward those areas where the imbrication of the nodules is a maximum (Figure 16.7C). Because the nodules did not deform, there is no strain in the limestone immediately adjacent to them, and the shortening of the limestone becomes concentrated in the intervening spaces, which causes a deflection of the $\hat{s}_1$–$\hat{s}_2$ plane around the nodules. Thus the foliation pattern is qualitatively consistent with the pattern that would be expected if foliation developed parallel to the plane of flattening of the finite strain ellipsoid.

The second model is for the formation of disjunctive foliations in argillite. It is illustrated in Figure 16.8, which assumes plane coaxial deformation, with the stretch parallel to x_2 always equal to 1. An undeformed argillite may have a foliation parallel to bedding, a **bedding plane fissility,** caused by the orientation of platy minerals during deposition and subsequent compaction parallel to x_1. This process results in an oblate (pancake-shaped) strain ellipsoid with the $\hat{s}_1$–$\hat{s}_2$ plane parallel to the bedding (x_2–x_3) (Figure 16.8A).

Initial tectonic shortening parallel to x_3 occurs by further compaction and by solution, transforming the oblate total strain ellipsoid into a prolate (cigar-shaped) one (Figure 16.8B). This deformation gradually destroys the primary foliation and develops a new disjunctive foliation that appears first as an anastomosing foliation and then as a pencil cleavage defined by the intersection of the new disjunctive foliation with the primary foliation and bedding (see Figure 13.20B). Although the lineation is technically a stretching lineation parallel to the $\hat{s}_1$ axis and to x_2, the deformation takes place largely by compaction and solution, and little if any real stretching is required (that is, $\hat{s}_1 \approx 1$).

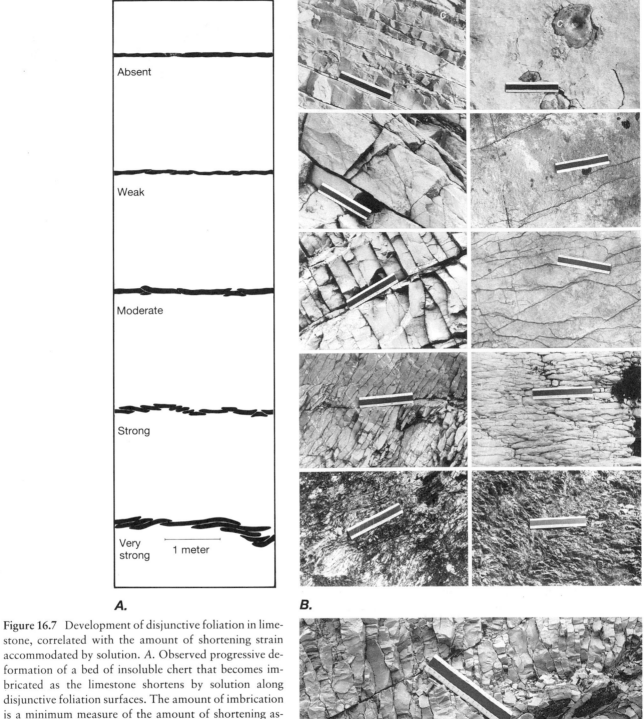

A.

B.

C.

Figure 16.7 Development of disjunctive foliation in limestone, correlated with the amount of shortening strain accommodated by solution. *A.* Observed progressive deformation of a bed of insoluble chert that becomes imbricated as the limestone shortens by solution along disjunctive foliation surfaces. The amount of imbrication is a minimum measure of the amount of shortening associated with solution. *B.* Foliation morphologies on surfaces perpendicular to bedding (left column; rulers parallel bedding trace) and parallel to bedding (right column; rulers parallel to foliation trace) in a sequence of increasingly deformed limestones. The shortening indicated by the diagrams of chert beds in part *A* is associated with the adjacent foliation morphology shown in part *B.* Scale bar in each photo is 6 in. long. *C.* Spaced disjunctive foliation in a limestone. The foliation diverges around insoluble chert nodules and converges toward regions of maximum imbrication of the nodules.

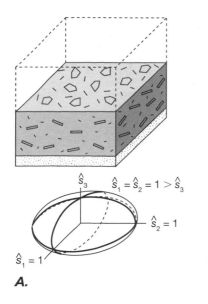

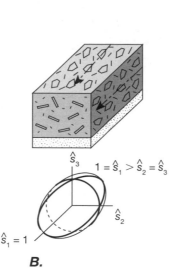

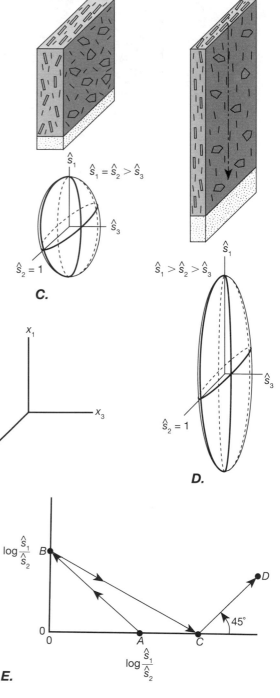

Figure 16.8 Model for development of disjunctive foliation in an argillaceous sediment with increasing strain. *A*. Flattening parallel to x_1 is caused by sedimentary compaction. A shaly fissility develops parallel to x_2–x_3. Strain ellipsoid is oblate (pancake-shaped). *B*. Tectonic shortening parallel to x_3 is accommodated initially by volume loss through continued compaction and solution. Stretching parallel to x_1 and x_2 is minimal. A spaced disjunctive foliation develops parallel to x_1–x_2 with a lineation parallel to x_2. The strain ellipsoid is prolate (cigar-shaped). *C*. Continued tectonic shortening parallel to x_3 is accommodated by stretching parallel to x_1. The disjunctive foliation is well developed but contains no lineation when the strain ellipsoid is essentially oblate, as it has become again in this diagram. *D*. Tectonic shortening and stretching continue. The foliation continues to strengthen. A stretching lineation develops in the foliation parallel to x_1. The strain ellipsoid becomes triaxial. *E*. Logarithmic Flinn diagram illustrating the finite strain path shown in parts *A* through *D*.

Further tectonic shortening parallel to x_3 occurs by constant-volume deformation, which is accommodated by extension normal to bedding (Figure 16.8*C*). As this process continues, the foliation becomes increasingly planar and regular, the stretching normal to bedding eventually becomes the maximum principal stretch $\hat{s}_1$ parallel to x_1, and a new stretching lineation develops parallel to the new $\hat{s}_1$, leaving the pencil lineation preserved as an intersection lineation parallel to what has become the intermediate principal stretch $\hat{s}_2$ parallel to x_2 (Figure 16.8*D*), and commonly parallel to the fold

axes. The progression of the strain states is indicated on a logarithmic Flinn diagram in Figure 16.8*E*.

The onset of noncoaxial deformation, such as fold formation, would rotate material foliation surfaces and would alter, at least somewhat, the relationship between the foliation and the finite strain ellipsoid.

Crenulation Foliations

Crenulation foliations are common features that may form by a variety of kinematic mechanisms. Different

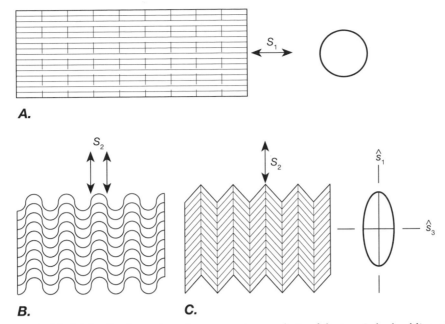

Figure 16.9 Model for the formation of a symmetric crenulation foliation S_2 by buckling of S_1. Chevron-shaped crenulations may also develop. Strain ellipses reflect the macroscopic strain. A. Initial foliation S_1 is parallel to the direction of maximum shortening. B. Progressive shortening produces symmetric crenulations by buckling. The limbs become the cleavage domains for the new crenulation foliation S_2. C. Chevron style of crenulation. The axial surfaces define the new foliation S_2.

mechanisms lead to different relationships between the orientation of the foliation and that of the principal strain axes. Thus the interpretation of strain orientations from crenulation foliations is not straight forward. We distinguish zonal from discrete crenulation foliations (see Figure 13.1) and symmetric from asymmetric crenulations (see Figure 13.9). Symmetric zonal crenulations probably form in response to shortening parallel to a preexisting foliation S_1 (Figure 16.9). The old foliation is rotated toward low angles to the axial surfaces of the new crenulations, which form the new foliation S_2. Solution of material from the limbs may strengthen the S_2 foliation. Such symmetric crenulations commonly occur in the core of a lower-order fold, where the crenulation foliation is subparallel to the axial surface of the larger fold. In this model, the direction of minimum principal stretch $\hat{s}_3$ is normal to the crenulation foliation.

Models for the formation of asymmetric crenulations and the resulting foliations include shortening oblique to a preexisting foliation, shear across an earlier foliation, and rotation and asymmetric deformation of symmetric crenulations. Solution of material from cleavage domains may be a common adjunct to the mechanism. We discuss these models in turn.

Asymmetric crenulations can form by shortening at a low angle to the initial foliation (Figure 16.10). The axial surfaces of the kinks form at a high angle to the

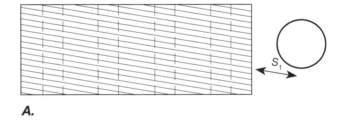

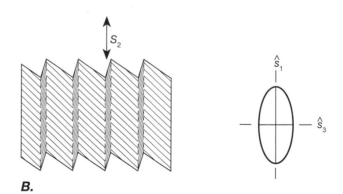

Figure 16.10 Formation of an asymmetric crenulation foliation S_2 by buckling of S_1. Strain ellipses reflect the macroscopic strain. A. The initial foliation S_1 is at a low angle to the direction of maximum shortening. B. Asymmetric crenulations form by buckling of S_1. The short limbs of the crenulations, in which the original foliation has rotated to a low angle with the crenulation axial surface, define the new crenulation foliation S_2. Hinges may be sharp or rounded.

original foliation and are parallel to a new crenulation foliation S_2. S_2 is defined by the short limbs of the crenulations in which the old foliation S_1 is rotated into subparallelism with the new foliation. There is no net shearing of the body parallel to the new foliation, and the axis of maximum shortening (minimum principal stretch $\hat{s}_3$) is normal to the new foliation.

Asymmetric crenulation foliations can also form as **shear bands,** which are bands of concentrated shear that cut across a preexisting foliation and define the new cleavage domains. A crenulation foliation that is inferred to have formed by this mechanism is sometimes called a **shear band foliation,** a genetic term. If the maximum incremental shortening direction $\hat{\zeta}_3$ is initially at a low angle to a preexisting foliation S_1, shear bands may form in which the old foliation undergoes a large rotation to produce close to tight crenulations (Figure 16.11A, B). If the maximum incremental stretching direction $\hat{\zeta}_1$ is at a low angle to S_1, shear bands form in which S_1 rotates only a small amount, forming open to gentle crenulations (Figure 16.12A, B). This geometry of shear band foliation is sometimes called an **extensional crenulation foliation,** which implies that the maximum extension is subparallel to the original foliation. In both cases, the preexisting foliation S_1 rotates toward parallelism with the shear band, thereby defining the new crenulation foliation S_2 (Figures 16.11A, B and 16.12A, B). The rotation may be enhanced by volume loss in the shear band, which accommodates shortening normal to its boundaries (Figures 16.11C and 16.12C).

The shear bands form in orientations of high shear strain relative to the principal axes of incremental strain, but there is no general and unique relationship between the principal axes of finite strain and the foliation. With progressive deformation, the shear bands behave as a material foliation, so at very high strains they may rotate into low angles with the plane of flattening of the finite strain ellipsoid (see the total strain ellipses in Figures 16.11 and 16.12). If this occurs, a second generation of shear bands could develop in an orientation of higher incremental shear strain.

An extensional crenulation foliation resulting from sequential superposed deformations should not be confused with an S-C fabric in which both S and C develop in a single deformational event (see Figure 13.16 and Section 13.5). In some cases, however, the distinction is difficult to make. For example, during the formation of an S-C fabric in a shear zone, S rotates progressively toward parallelism with C with increasing strain. When the two become essentially parallel, shear bands (labeled C′) may develop at a low angle to the main shear zone. The morphology of the shear-band foliation may be difficult to distinguish from the original S-C morphology. Fortunately, it is possible to infer the sense of shear on the shear zone from the shear-band foliation by using

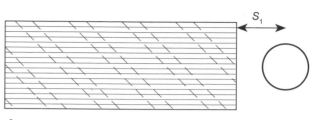

A.

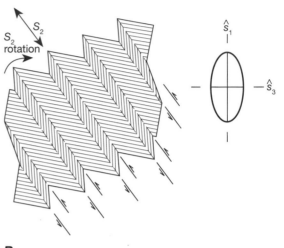

B.

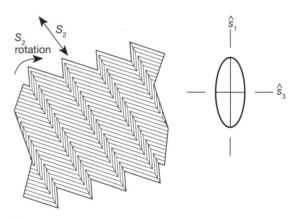

C.

Figure 16.11 Production of an asymmetric crenulation foliation S_2 by the formation of shear bands crossing a preexisting foliation which is oriented at a low angle to the maximum shortening axis of the incremental strain ellipse. Strain ellipses reflect the total strain. A. Undeformed state with foliation initially parallel to the incremental shortening direction. B. Simple shear within shear bands produces asymmetric close to tight crenulations. Shear bands rotate toward higher angles with the axis of maximum finite shortening $\hat{s}_3$. The shear bands become the S_2 cleavage domains. Rounded hinges are also common. C. Volume loss in the cleavage domains strengthens the crenulation foliation S_2.

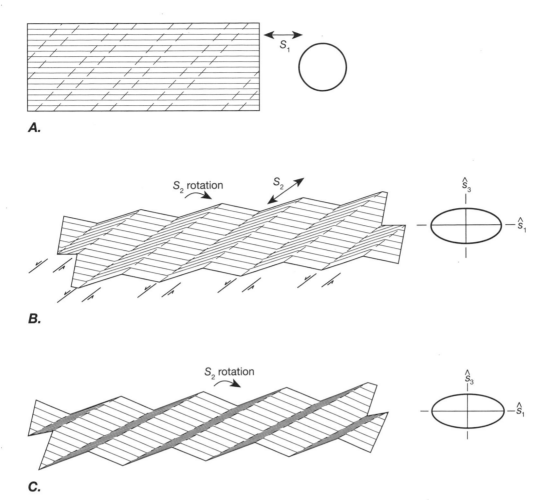

Figure 16.12 Production of an asymmetric extensional crenulation foliation S_2 by formation of shear bands. Strain ellipses indicate the total strain. *A.* Undeformed state with S_1 foliation initially parallel to the direction of maximum incremental stretch. *B.* Extension is accommodated by the formation of shear bands producing open to gentle asymmetric crenulation foliation S_2. Rounded hinges are also common. *C.* The crenulation foliation is enhanced by volume loss from the S_2 cleavage domains.

the same criterion that is applied to the original S-C relationship (see Figure 13.16), because even though the S-C and C′ foliations form sequentially, they are related to the same deformational event.

An asymmetric crenulation also could develop from a symmetric one by solution of components preferentially from one set of limbs. This mechanism might occur, for example, if a layer containing an initial symmetric crenulation (Figure 16.13A) were buckled into a lower-order fold by the flexural-shear mechanism. Subsequent rotation of the crenulations in the limbs of a lower order fold orients one set of crenulation limbs at a higher angle to the overall maximum shortening direction than the other set. Solution from these limbs and deposition in the adjacent limbs could accommodate both shortening of the rock normal to the axial plane of the low-order fold and shear parallel to the

low-order fold limbs. (Figure 16.13B, C). The result could be interpreted as a consequence of apparent shear along the new crenulation foliation S_2, which would not be an accurate interpretation of the actual deformation. No unique relationship between foliation and strain arises from this mechanism, although the foliation would probably end up at a low angle to the large-scale plane of flattening.

In general, then, asymmetric crenulation foliations cannot easily be interpreted in terms of the orientation of the principal axes of strain, although if the strains are sufficiently large, the foliations tend to rotate toward parallelism with the plane of flattening ($\hat{s}_1$–$\hat{s}_2$).

Discrete crenulation foliations develop from zonal crenulations by recrystallization of the platy minerals in the cleavage domains. New platy mineral grains replace the old grains and grow parallel to the cleavage

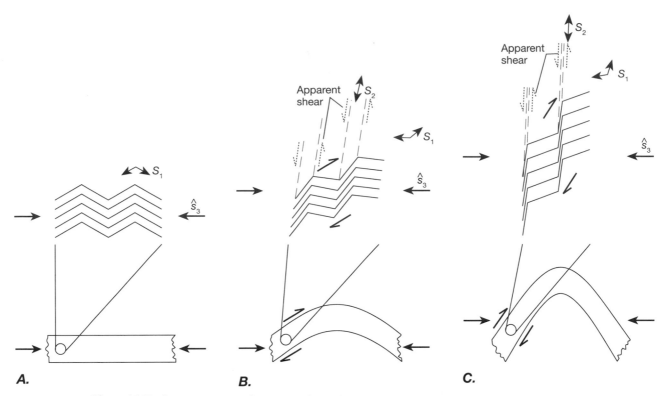

Figure 16.13 Asymmetric crenulations produced from symmetric crenulations by rotation during lower order folding accompanied by solution in those crenulation limbs at a high angle to the shortening direction and by precipitation in the adjacent limbs. *A*. Initial symmetric crenulations. *B* and *C*. Solution of material in the crenulation limbs that are at a high angle to $\hat{s}_3$ and precipitation in the adjacent limbs accommodate overall shortening and relative shear of the crenulated S_1 laminations. The set of crenulation limbs at a high angle to $\hat{s}_3$ become the cleavage domains of the S_2 foliation, and the other set of limbs becomes the microlithons. The apparent shear on S_2, indicated by dashed arrows, is an artifact.

domain, producing a discontinuity between it and the microlithon.

Continuous Foliations

Although slaty cleavage gives the impression of continuity down to the finest scales, microscopic and scanning electron microscopic examination of many slates reveals a domainal structure that has the characteristics of crenulation or disjunctive foliations (see Figure 13.11). Progressive recrystallization of the platy minerals in slates has been correlated with progressive changes from a microzonal crenulation foliation, through a microdiscrete foliation, and finally to a microdisjunctive foliation exhibiting a strong preferred orientation in the microlithons.

Such studies suggest that microcontinuous slates, as well as coarse continuous foliations (see Figure 13.12A), develop as the final products of recrystallization of earlier crenulation or disjunctive foliations. The processes of deformation, rotation, solution, and re-

crystallization can ultimately wipe out evidence of the initial structure. In rare circumstances, the origin of a coarse continuous foliation is preserved within porphyroblastic minerals such as garnets that overgrew and included the earlier crenulation foliation before it was transformed by recrystallization.

Although the complicated nature of the process does not lead to a simple prediction of the relationship between strain and foliation, field studies in regions where the strain can be determined independently indicate consistently that slaty foliation is parallel to the plane of flattening of the strain ellipsoid, within the error of measurement (see Sections 17.2, 17.3).

16.6 Foliations and Shear Planes

The relationship between foliations and shear planes has long been a source of confusion and misinterpretation in structural geology. This situation stems in part from the classical passive-shear model for producing similar folds, which requires inhomogeneous simple shear on planes parallel to the axial surface (Section 12.2 and Figure 12.8). Because a foliation is commonly subparallel to the axial surface of similar style folds, it was interpreted as the plane of simple shear. There is considerable evidence, however, that foliations are usually parallel or subparallel to the plane of flattening of the strain ellipsoid, an interpretation generally incompatible with foliations as shear planes. Under what circumstances, then, can shear strain accumulate across foliation planes? Under what circumstances can foliations be planes of simple shearing? And how reliable is the interpretation that shear has occurred on a foliation plane?

Any material foliation is a plane across which shear strain accumulates during a noncoaxial deformation. This fact is indicated by the initially perpendicular pairs of material lines such as a and a' through d and d' in Figure 15.15. We can take each pair of lines to represent a material plane and a material line initially perpendicular to it. In general, any two such lines are sheared into a nonperpendicular orientation, and they return to being perpendicular only for the instant that they are parallel to the principal axes of strain. At that instant, the total shear strain for that pair of lines is zero. The incremental shear strain, however, is not zero because the principal axes of the incremental strain are not parallel to those of the total strain. Thus shear strain can accumulate across any material foliation plane throughout the deformation. This result is simply a consequence of the geometry of deformation of material lines and planes. It does not imply that such foliations are planes of simple shearing.

Many foliations, both material and nonmaterial, are defined by strong planar mechanical anisotropies in the rock. As in the case of brittle fracture (see Figures 9.14–9.17), such anisotropies can have a significant effect on the geometry of ductile deformation. As an example, let us consider a state of pure shear stress (see Figure 8.14G) causing a progressive simple shear in which the shear plane is the plane of maximum shear stress. The principal axes of incremental strain ($\hat{\zeta}_1$ and $\hat{\zeta}_3$) are parallel to the principal axes of stress ($\hat{\sigma}_3$ and $\hat{\sigma}_1$, respectively; Figure 16.14A). A foliation that either is a material plane or is parallel to the plane of flattening of the finite strain ellipse rotates toward the shear plane as strain increases. If the yield stress for shear on the foliation is lower than the yield stress for shear across the foliation, then the foliation can become the plane of simple shearing when it rotates sufficiently close to the plane of maximum shear stress.

Figure 16.14 illustrates this situation. On the Mohr diagrams of deviatoric stress, the higher yield criterion (solid lines) is for ductile shear across the foliation plane, and the smaller yield criterion (dashed lines) is for duc-

tile shear along the foliation plane. The surface stress on any given plane cannot exceed the yield stress for shear along that plane, which limits the possible diameter of the Mohr circle. With the foliation at a high angle δ to the maximum shear stress plane (δ is technically the angle between the normals to the planes; Figure 16.14A), stress on the foliation is subcritical, and ductile shear occurs along the plane of maximum shear stress at a yield stress given by the solid line in Figure 16.14B. With increasing strain, the angle of the foliation decreases to δ' (Figure 16.14C). At this point, the stress on the foliation and the stress on the plane of maximum shear stress are both at the critical value. With further decrease in the foliation angle to δ'' (Figure 16.14D), the differential stress must decrease so that the surface stress on the foliation does not exceed the yield stress. Thus the stress must relax to a value below the yield point for ductile flow on the original shear plane. In this situation, the foliation becomes the active shear plane even though it did not form as one.

If the foliation was initially parallel to the plane of flattening (Figure 16.15A), then as soon as shearing on the foliation begins, the strain ellipse is rotated to a different orientation that is no longer parallel to the foliation (Figure 16.15B). Note that the sense of rotation is initially in an opposite sense to that normally expected for the finite strain ellipse in simple shear.

The inference that shear has occurred on foliation surfaces commonly relies on the observation of displaced lithologic boundaries. The same problem of distinguishing separation from displacement that confronts us when we are interpreting fault displacement (Figures 4.21, 4.22) also presents itself in this case. Beyond that, however, there exist several mechanisms that result in apparent shear of lithologic boundaries where none has actually occurred.

If a lithologic layer is cut obliquely by a disjunctive foliation along which there has been a significant amount of solution, the effect can be an apparent shearing of the layer on the foliation, whereas in fact the displacement may have been strictly normal to the foliation (Figure 16.16). If a thin layer is initially ptygmatically folded in association with deformation of the surrounding rock, subsequent preferential solution of one set of limbs of these folds along a disjunctive foliation can produce an apparent shear of the layer along the foliation (Figure 16.17). Again, however, the displacement may have been strictly normal to the foliation. Note that in Figure 16.16C, one might infer opposite senses of shear on the foliation to explain the apparent displacement, even for layers such as 2, 3, and 4 that had the same initial orientation. Figure 16.17 shows an example in which the selective solution of ptygmatic fold limbs can be seen in varying stages of advancement.

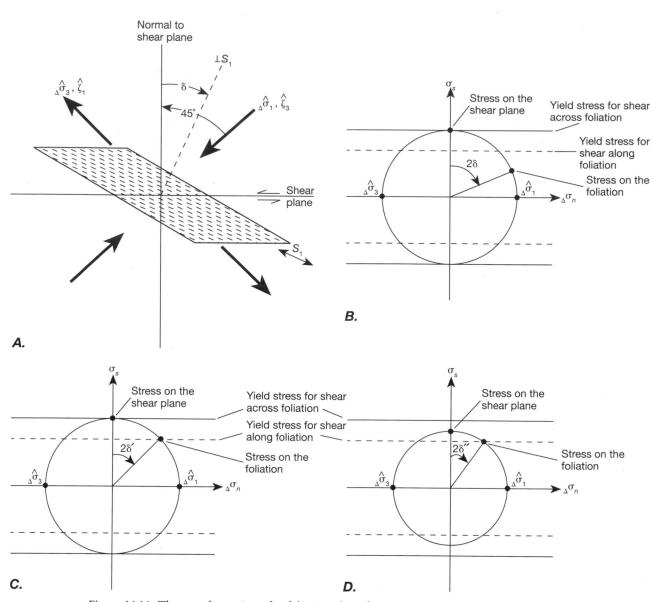

B.

A.

C.

D.

Figure 16.14 The transformation of a foliation plane from a passive plane to an active shear plane. In the Mohr diagrams, the solid lines are the von Mises criterion for ductile flow across the foliation; the dashed lines are the von Mises criterion for ductile flow on the foliation. *A.* During progressive simple shear, a material foliation S_1 rotates toward the shear plane: δ decreases progressively. *B.* Mohr circle for deviatoric stress at large foliation angles. Shearing does not occur on the foliation because the stress on the foliation plane is below the yield stress. *C.* The foliation angle has decreased to the value δ' at which stress on the foliation plane is at the yield stress. *D.* Further decrease in the foliation angle to δ'' requires the differential stress to decrease so that the stress on the foliation does not exceed the yield stress. Stress on the original shear plane drops below the yield stress, and shearing on that plane ceases.

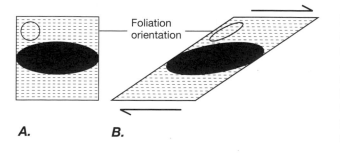

A.　　**B.**

Figure 16.15 Shear parallel to the plane of flattening of the finite strain ellipse causes the ellipse to change shape so that the shear plane is no longer parallel to the plane of flattening. *A.* Finite strain ellipse (solid black) parallel to a foliation. The circle in the upper left records the subsequent deformation. *B.* Shear parallel to the plane of flattening in part *A* initially causes the principal axes of finite strain to rotate away from the shear plane. The small ellipse in the upper left records the amount of shear.

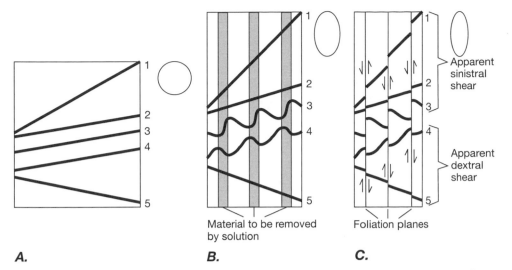

A.   **B.**   **C.**

Figure 16.16 Apparent shear along a disjunctive foliation resulting from solution along the surfaces of the foliation. Displacement is strictly normal to the foliation. The strain ellipses show the bulk strain. *A.* The undeformed state. Layers 1, 2, and 5 are incompetent and behave passively during the deformation. Layers 3 and 4 are competent. *B.* After homogeneous constant-volume flattening (pure shear), the competent layers 3 and 4 have buckled. The shaded areas indicate where solution will remove material, producing a disjunctive foliation. *C.* After deformation by solution along the foliation planes, layers 1, 2, and 3 suggest sinistral shear on the foliation, and layers 4 and 5 suggest dextral shear. The apparent shear direction is determined by the orientation of the layer with respect to the foliation (compare layers 1 and 2 with layer 5) or by which limb of the fold train is systematically removed by solution (compare layers 3 and 4).

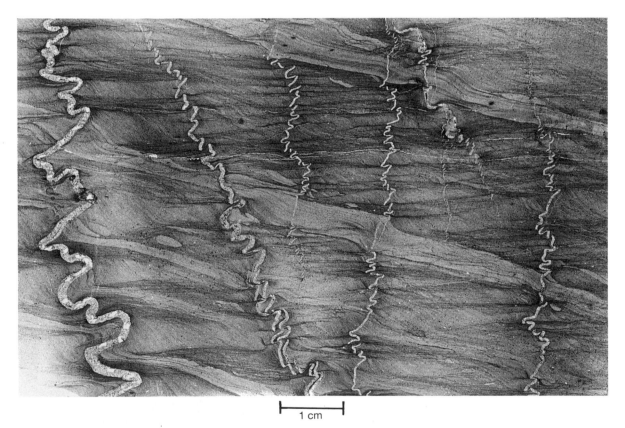

1 cm

Figure 16.17 Photograph showing various stages in the progressive solution of limbs of ptygmatic folds. Compare with layer 3 in Figure 16.16C.

Although some foliation planes are demonstrable shear planes, such as C planes in shear zones (see Figure 13.16) on which slickenfibers may develop, the association of a foliation plane with the dominant shear plane for the deformation cannot be assumed, and even evidence of apparent shear displacement on the foliation planes must be examined critically.

16.7 Strain and Lineations

Relatively few studies of the relationship between strain and lineations have been published. Discrete lineations composed of the long axes of initially spherical structures, of course, are generally parallel to the $\hat{s}_1$ axis of the finite strain ellipsoid (stretching lineations). Any deviation from an initial spherical shape and any associated initial preferred orientation can alter this relationship, but at large strains, the difference is not great.

Constructed lineations such as hinge lines, intersections of foliation and bedding, and boudin lines must lie within the plane of the bedding in which they form. Thus unless the geometry of the strain is related to the orientation of the bedding (as, for example, in compaction of sediment normal to the bedding surface or tectonic shortening parallel to bedding), there is no necessity for such lineations to be parallel to one of the principal axes of strain. The structural slickenlines that form on fault surfaces parallel to the slip direction cannot ever form parallel to a principal direction of strain, because they form parallel to a surface and a direction of shear.

Acicular and elongate mineral grain lineations may form parallel to one of the principal axes, and it is not uncommon for them to be parallel to the $\hat{s}_1$ direction. This relationship is not universal, however, and the parallelism of such lineations with the strain axes should be demonstrated rather than assumed.

Lineations are commonly associated with folds—they are especially likely to be oriented parallel to the fold hinges—but the significance of such lineations in terms of the strain is not always clear. Some such lineations may be stretching lineations that are sub-parallel to the $\hat{s}_1$ axis. This could occur if the fold hinges formed or were rotated toward $\hat{s}_1$, as could happen, for example, as a result of large shear strain or if the rock were constricted along the lines of flow (Figure 16.18; cf. Figure 15.8D).

A mineral lineation parallel to a fold axis and to $\hat{s}_1$ may be the result of dissolution of mineral grains. In this case it would be possible to have $\hat{s}_1 = 1 \geq \hat{s}_2 > \hat{s}_3$, for which no lengthening would have occurred parallel to the $\hat{s}_1$ direction (see Figure 16.8A, B).

In other cases, the lineation may be constrained to develop parallel to a bedding plane. If so, it may form parallel to the maximum stretch in the bedding plane. It need have no direct relationship to the principal axes of strain and may change its geometric relationship to the fold axis from hinge to limb.

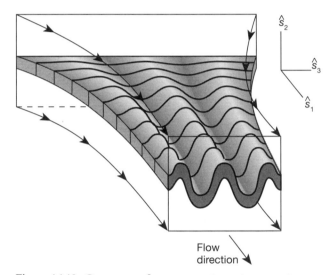

Figure 16.18 Convergent flow can result in the axis of maximum principal stretch being parallel to the flow direction. Stretching lineation and fold hinges form parallel to the maximum stretching direction.

Additional Readings

Alvarez, W., T. Engelder, and P. A. Geiser. 1978. Classification of solution cleavage in pelagic limestones. *Geology* 6: 263–266.

Alvarez, W., T. Engelder, a᠎d W. Lowrie. 1976. Formation of spaced cleavage and folds in brittle limestone by dissolution. *Geology* 4: 698–701.

Chapple, W. M., and J. H. Spang. 1974. Significance of layer parallel slip during folding of layered sedimentary rocks. *Geol. Soc. Am. Bull.* 85: 1523–1534.

Gray, D. R. 1979. Geometry of crenulation folds and their relationship to crenulation cleavage. *J. Struct. Geol.* 1(3): 187–205.

Gray, D. R., and D. W. Durney. 1979. Investigations on the mechanical significance of crenulation cleavage. *Tectonophysics* 58(1–2): 35–80.

Gratier, J. P. 1983. Estimation of volume changes by comparative chemical analyses in heterogeneously deformed rocks (folds with mass transfer). *J. Struct. Geol.* 5 (3/4): 329–339.

Hobbs, B. E. 1971. The analysis of strain in folded layers. *Tectonophysics* 11: 329–375.

Marlow, P. C., and M. A. Etheridge. 1977. Development of a layered crenulation cleavage in mica schists of the Kanmantoo Group near Macclesfield, South Australia. *Geol. Soc. Am. Bull.* 88: 873–882.

Platt, J. P., and R. L. M. Vissers. 1980. Extensional structures in anisotropic rocks. *J. Struct. Geol.* 2(4): 397–410.

Ramsay, J. G.. 1967. *Folding and fracturing of rocks.* New York: McGraw-Hill.

Ramsay, J. G., and M. I. Huber. 1983. *The techniques of modern structural geology.* Vol. 1: Strain analysis. London: Academic Press.

Reed, L. J., and E. Tryggvason. 1974. Preferred orientations of rigid particles in a viscous matrix deformed by pure shear and simple shear. *Tectonophysics* 24 (1/2): 85–98.

Rees, A. I. 1979. The orientation of grains in a sheared dispersion. *Tectonophysics* 55 (3/4): 275–288

Williams, P. F. 1976. Relationships between axial plane foliations and strain. *Tectonophysics* 39: 305–328.

Williams, P. F., and Chr. Schoneveld. 1981. Garnet rotation and the development of axial plane crenulation cleavage. *Tectonophysics* 78: 307–334.

Observations of Strain in Deformed Rocks

17.1 Techniques of Measuring Strain

Our description of the geometry of strain during progressive deformation would be of limited application if it were not possible to measure in deformed rocks at least some of the parameters that characterize the strain. An extensive literature exists on various techniques for measuring strain in rocks and for determining the progressive deformation that the rocks have experienced.[1] We describe a few of the more common and straightforward methods below, both to illustrate the types of data that can be used and to impart some intuitive understanding of the concepts involved.

We limit our discussion of techniques for determining strain to two dimensions, because the geometry and the principles are most easily understood in two dimensions. Extension to three dimensions adds little to conceptual understanding but increases the practical difficulty considerably.

Techniques for measuring two-dimensional strain must be applied with caution. The intersection of a plane of any orientation with an ellipsoid is always an ellipse.

Thus there is no way to determine, from a single plane, whether the observed maximum and minimum axes of that ellipse are parallel to the maximum and minimum principal stretches, or even whether they lie within one of the principal planes. Careful measurements on more than one arbitrary orientation of plane can be used to determine the ratios of the three principal stretches and their orientations. The measurements and the required calculations, however, are difficult and time-consuming. In some cases, the existence of particular structures enables us to infer the orientation of one or more of the principal axes or the principal planes. In such cases a two-dimensional analysis can be used with more confidence, and the problem of determining the ratios of the principal stretches in three dimensions also is simplified.

One of the frustrating problems we encounter in trying to determine the strain in rocks is that for many strain markers available, the original shape of the marker is known but the original dimension is unknown. In particular, we can often determine the lengths of material lines parallel to the principal axes of strain, $\hat{\ell}_1$ and $\hat{\ell}_3$, and know that before the deformation, those lines were both the same length L (Figure 17.1). We do not know, however, what that initial length L was, so we cannot determine the absolute value of the principal stretches $\hat{\ell}_i/L$. Thus the volumetric strain is a quantity

---

[1] Ramsay and Huber (1983) provide the best summary, as well as an extensive bibliography.

that can be determined only rarely and under very special circumstances. We can describe the shape of the strain ellipse, however, by determining the ratio of the principal axes, which is the ellipticity R. As shown in Equation (17.1), R does not involve the original lengths L.

$$R \equiv \frac{\hat{s}_1}{\hat{s}_3} = \frac{\hat{e}_1 + 1}{\hat{e}_3 + 1} = \frac{\dfrac{\hat{\ell}_1}{L}}{\dfrac{\hat{\ell}_3}{L}} = \frac{\hat{\ell}_1}{\hat{\ell}_3} \qquad (17.1)$$

The two-dimensional state of strain is characterized by the components of a two-by-two symmetric matrix. In general coordinates,

$$e_{k\ell} = \begin{bmatrix} e_{11} & e_{13} \\ e_{31} & e_{33} \end{bmatrix} \qquad e_{13} = e_{31} \qquad (17.2)$$

where the second equation expresses the symmetry of the matrix. Thus there are only three independent components in this strain tensor matrix, so only three independent measurements in the $\hat{s}_1$–$\hat{s}_3$ principal plane (which is also the $\hat{e}_1$–$\hat{e}_3$ plane) are required to define the shape, size, and orientation of the strain ellipse.

The three simplest independent measurements to deal with are the lengths $\hat{\ell}_1$ and $\hat{\ell}_3$ parallel to the principal coordinates $\hat{x}_k$, and the orientation α of the maximum principal stretch with respect to a reference axis x_1 in the plane (Figure 17.1). If we assume that the deformation was at constant volume, the radius L of the undeformed circle that has the same area as the deformed elliptical object is

$$L = \sqrt{\hat{\ell}_1 \hat{\ell}_3} \quad \text{for constant area (volume)} \qquad (17.3)$$

The ratios of the deformed to the undeformed lengths define the two principal stretches $\hat{s}_1$ and $\hat{s}_3$, which define $\hat{e}_1$ and $\hat{e}_3$ (Figure 17.1). With these values, we require only a coordinate transformation to determine the three independent components of strain $e_{k\ell}$ in any general coordinate system (Equation 17.2). Generally we cannot be sure the deformation was at constant volume, however, so the original radius L of the feature is unknown. Thus we can determine only the *ratio* of the maximum and minimum stretches (Equation 17.1), along with the orientation α, which means we can determine only two of the three quantities necessary to define the strain. For that reason our description is incomplete, and the volumetric strain is usually indeterminate.

Deformed Objects of Initially Circular Cross Section

The simplest way to determine the strain in a plane is to observe directly the deformed shape of an object that is known to have been spherical before deformation.

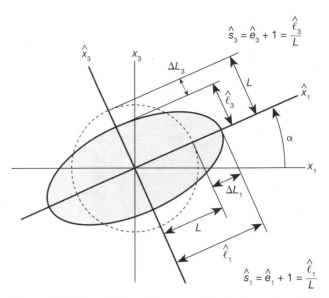

Figure 17.1 Three independent quantities $\hat{s}_1$, $\hat{s}_3$, and α with which we can define the shape and orientation of the strain ellipse. If L is unknown, we cannot determine the principal stretches independently but only their ratio $R = \hat{s}_1/\hat{s}_3 = \hat{\ell}_1/\hat{\ell}_3$.

Common examples of such objects include ooids and spherulites (Figure 15.7), alteration spots (Figure 13.19B), and radiolaria and foraminifera shells. For some structures only a cross section is initially circular so the strain in the plane of the cross section can be determined easily, although this plane does not necessarily coincide with a principal plane of strain. Such features include circular disc-shaped segments of crinoid stems, commonly deposited flat on a bedding plane, and scolithus tubes, sediment-filled cylindrical worm holes that are initially oriented normal to bedding with a circular cross section in the bedding plane.

The initial shape of natural objects is never perfectly round, deformation is not always perfectly homogeneous, and our measurements are not so precise as we would like. Thus it is always necessary to measure a large number of deformed objects and to take some average ellipticity R and some average orientation of $s_{\max}$ as the best description of the state of two-dimensional strain.

Deformed Linear Objects

In some cases, the stretch parallel to a deformed linear object may be evident either as folding ($s_n < 1$) or as boudinage ($s_n > 1$) of the object, and in these cases, measurement of the stretch in different directions enables us to determine the strain. This method has been applied, for example, to deformed acicular crystals such

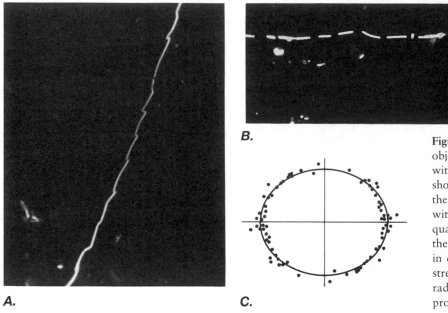

Figure 17.2 Examples of stretched linear objects. *A.* Needles of rutile included within a grain of quartz and folded by shortening during ductile deformation of the quartz. *B.* Needle of rutile stretched within a ductilely deformed grain of quartz. *C.* Strain ellipse determined from the stretching of rutile needles imbedded in quartz crystals. The magnitude of the stretch of each needle is plotted along a radius parallel to the needle orientation to produce each data point in the plot.

as tourmaline, amphibole, and rutile (Figure 17.2*A*, *B*) and to deformed linear fossils such as belemnites. Ptygmatic folding and boudinage of layers, such as quartz–feldspar veins in a schist, can also be used to estimate two-dimensional strain if the deformation is measured in a principal plane of strain.

To determine the stretch of the object, we take the original length to be the arc length of folds, or the sum of the segment lengths of boudins, and we take the final length to be the length of the fold wave train measured along the median surface or, for boudins, the total of the segment lengths plus the spacings between them. Such measurements probably provide a minimum estimate of the stretch, however, because homogeneous ductile deformation of the linear object may have lengthened and thinned it or shortened and thickened it, and because the object may differ in competence from the matrix and may not record all the stretching that occurred in the rock.

If the linear objects have a wide variety of orientations, then determination of the stretches in the different orientations can define the strain ellipse. In principle, as noted above, we need to measure the stretches in only three independent directions in order to calculate a unique strain ellipse. In practice, however, random errors and uncertainties require that a much larger number of measurements be made.

We can determine the shape and orientation of the strain ellipse by plotting the values of the stretches as radii in the directions that correspond to the orientations of the linear objects measured. Figure 17.2*C*, for example, shows the strain in a quartz grain, determined

from the deformation of imbedded rutile needles (Figure 17.2*A*, *B*). Each point is the stretch of a particular needle, plotted in the orientation of that needle, relative to a prescribed reference axis.

Some fossils have a well-defined characteristic dimension, such as the spacing between segments of certain species of graptolite (see Figure 17.9). Because an original dimension is well known, these fossils provide one of the few structures from which absolute magnitudes of the strain can be determined (see Section 17.2).

The Nearest-Neighbor Center-to-Center Technique

If the particles in an undeformed rock were initially distributed such that the distances between the centers of nearest neighbors were statistically constant and isotropic, then the distances between these centers in a deformed rock provides a measure of the accumulated strain. This technique can be applied, for example, to sand grains in a sandstone, to conglomerate pebbles in a conglomerate, and to ooids in an oolitic limestone. It can also be used to compare the bulk strain with the strain recorded by the shapes of individual sand grains, conglomerate pebbles, and ooids.

Two particles are nearest neighbors, if the center-to-center line does not cross any other particles. During deformation, the center-to-center lines behave as material lines and are rotated and stretched according to the shape and orientation of the strain ellipse (Figure 14.5). Thus this technique is a variation of the one described in the previous section for determining strain

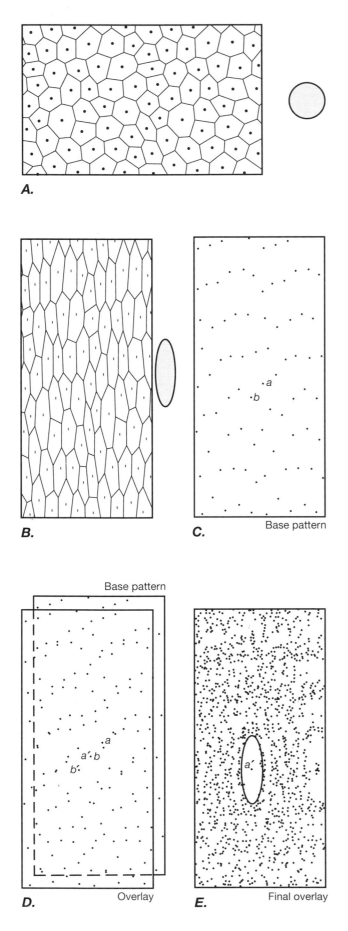

A.

B.

C. Base pattern

Base pattern

D. Overlay

E. Final overlay

Figure 17.3 (*Left*) Strain determination using the nearest-neighbor center-to-center technique of Fry (1979). *A.* Undeformed polycrystalline aggregate. The dots are the centers of the grains. *B.* Deformed polycrystalline aggregate with strain ellipse. *C.* Base pattern of grain centers. Point *a* is chosen as the reference point. *D.* Transparent overlay, with the base pattern copied on it, is placed over the base pattern such that the reference point *a′* on the overlay is superimposed on point *b* in the base pattern. The base pattern is then copied onto the overlay. *E.* Placing the reference point *a′* on the overlay over a large number of points in the base pattern and copying the base pattern onto the overlay each time defines the strain ellipse about the reference point.

from linear objects. Measurements of the orientation and length of each nearest-neighbor center-to-center line about a given particle make it possible to determine the strain ellipse. In order to average out irregularities in the initial fabric and in the deformation, we must make such sets of measurements about many particles.

A graphical technique called the Fry method is useful for determining the strain ellipse from a large number of points. In essence, it involves plotting the length and orientation of a large number of center-to-center lines relative to a single reference point. The surface on which the strain is determined is commonly a photographic enlargement of an oriented thin section. Figure 17.3*A*, *B* shows the deformation of a polycrystalline aggregate with the centers of each grain marked. The strain is indicated by the adjacent strain ellipse. Analysis of the strain in Figure 17.3*B* by the Fry method proceeds as follows: On a sheet of tracing paper, make a base pattern of points by marking the centers for a large number of particles (Figure 17.3*C*). On a second transparent overlay, copy the base pattern and mark a central reference point *a′*. Place the reference point *a′* of the overlay over each of the points on the base pattern in succession, maintaining a constant relative orientation of the two sheets of paper (Figure 17.3*D*), and repeatedly trace onto the overlay the locations of the points on the base pattern. The final plot (Figure 17.3*E*) will show an elliptical central area, empty or nearly empty of points, or an elliptical concentration of points that defines the shape and orientation of the strain ellipse. The technique works because there is a minimum possible distance in any direction between two nearest-neighbor particles, and the variation with azimuth of this minimum distance shows up as the empty elliptical area in the plot.

Sheared, Initially Orthogonal Pairs of Lines

Many fossils have features that are perpendicular to one another. Examples include the hinge line and symmetry plane of brachiopod shells (Figure 17.4) and the body segments and symmetry plane of trilobites. When such

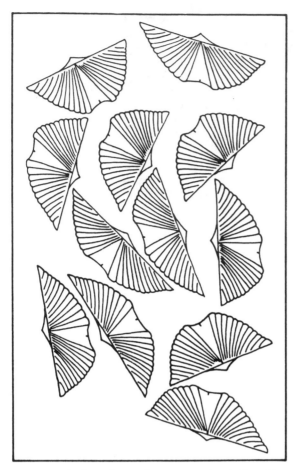

Figure 17.4 The hinge line and symmetry plane of brachiopod shells are initially perpendicular, and they therefore permit determination of the shear strain in the deformed state.

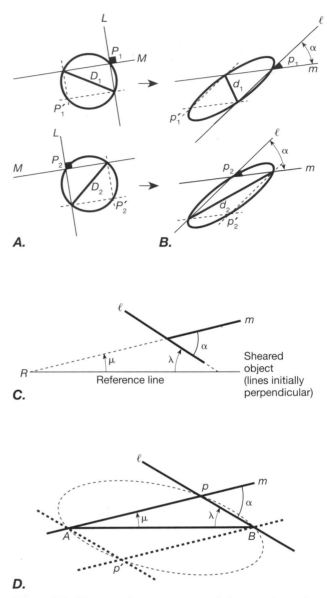

Figure 17.5 The use of measurements of shear strain to determine the ellipticity and orientation of the strain ellipse. A. If the orientations of two perpendicular lines L and M are constructed from opposite endpoints of *any* diameter of a circle, such as D_1 or D_2, their points of intersection P_1, P'_1, P_2, and P'_2 *must lie on the circle.* B. Deformation changes the circle into an ellipse and shears the perpendicular lines. However, if the deformed orientations of two lines ℓ and m that were initially perpendicular are constructed from opposite end points of a line representing an arbitrary diameter of the strain ellipse, the points of their intersections p_1, p'_1, p_2, and p'_2 *must lie on the ellipse.* C. Sheared material lines ℓ and m were initially perpendicular. Angles λ and μ define the orientations of these lines relative to an arbitrarily defined reference line R. D. The line orientations from part C are constructed at opposite ends of a reference line AB parallel to R, which represents a diameter of the strain ellipse. Two intersections (p and p') are defined, both of which must lie on the strain ellipse. The size of the ellipse is not constrained by data of this type.

fossils are deformed, the angle between the initially orthogonal lines makes it possible to determine the shear strain. If only shear strains can be measured, however, it is impossible, even in principle, to determine the actual magnitudes of the principal stretches, because the shear strain depends only on the difference between the principal stretches.

Theoretically, measurement of two shear strains in different directions is enough to determine the ratio of the principal stretches and their orientations. In practice, of course, we rely on multiple determinations to estimate the average strain.

The graphical construction technique known as the Wellman method is intuitively the clearest method for determining the strain. It is based on the fact that angles inscribed in a semicircle are always right angles (Figure 17.5A). Thus two perpendicular lines (L and M) of any given orientation, constructed from the opposite ends of any diameter of a circle, intersect on the circumference of the circle. After a homogeneous strain, the semicircle becomes a semi-ellipse, but the deformed lines (ℓ and m) must still intersect on the circumference of the ellipse, *regardless of the diameter on which they are constructed* (Figure 17.5B). All the perpendicular pairs of chords are sheared except one, and the lines of this pair are parallel to the principal stretches. Thus any possible shear strain is represented by some pair of chords inscribed within the semi-ellipse.

Reversing the argument, we see that if two deformed lines (ℓ and m, Figure 17.5C) were initially perpendicular, then lines constructed parallel to ℓ and m (Figure 17.6D) from both end points of a line AB representing a diameter of the strain ellipse intersect at p_1 and p_2 on the circumference of the ellipse.

This geometry enables us to construct the shape of the strain ellipse if we are given a sufficiently large number of deformed objects. For each object (Figure 17.5C), the orientation of the initially orthogonal pair of lines is determined relative to a convenient reference line. To represent a diameter of the ellipse, choose a line AB (Figure 17.5D) of convenient length parallel to the reference line. Because shear strain data constrain only the ratio of the principal strains and not their magnitudes, the length of the line, and thus the size of the ellipse to be constructed, is arbitrary. From both end points of this diameter, construct the orientations of the pair of lines to determine their intersections (Figure 17.5D). Each pair of orientations generates two intersections that must lie on the strain ellipse. When points are constructed for a large number of strained objects of different orientations, the shape and orientation of the ellipse become apparent. The ellipticity, given by the ratio of maximum to minimum principal axes, is then the ratio of the maximum to minimum stretch, and the principal axes of the constructed ellipse define the orientation of the principal stretches.

The R_f–ϕ Method

Some rocks, such as deformed conglomerates, contain deformed objects that were not initially spherical. If the undeformed objects were approximately ellipsoidal, it is possible to separate the initial ellipticity from that imposed by the strain and thereby to estimate the strain. The technique is based on calculating the theoretical distribution of final ellipticities and orientations that result from imposing different strains on objects that have a known initial ellipticity and orientation (Figures 17.6, 17.7; see also Figure 14.2).

The final ellipticity R_f and the orientation ϕ of a deformed object depend on the initial ellipticity R_i, on the initial orientation θ of the undeformed object, and

on the ellipticity R_s of the imposed strain ellipse (see, for example, ellipses, *a, b,* and *c* in Figure 17.6). We choose the direction of the maximum principal axis of strain as the reference direction for our examples. Generally we do not know this direction a priori, but any other direction can be used as a reference, and the analysis determines the orientation of $\hat{s}_1$.

As a simple example, we first consider the case in which the undeformed objects all have the same ellipticity $R_i = 2$ and are randomly oriented (Figure 17.6A), and we investigate the effect of imposing strains for which the ellipticity of the strain ellipse is $R_s = 1.5$ (Figure 17.6B) and $R_s = 3$ (Figure 17.6C). If the long axis of an object is parallel to the direction of maximum lengthening $\hat{s}_1$ of the imposed strain (ellipse *b* Figure 17.6A), then the final ellipticity R_f of the deformed object is a maximum (ellipse *b* Figure 17.6B, C), given by[2]

$$R_{f(max)} = R_i R_s \tag{17.4}$$

If, on the other hand, the long axis of an object is parallel to the maximum shortening direction $\hat{s}_3$ of the imposed strain (ellipse *c* in Figure 17.6A), the final ellipticity R_f of the deformed object is a minimum (ellipse *c* in Figure 17.7B, C) given by

$$R_{f(min)} = \begin{cases} R_i/R_s & \text{if } R_s < R_i \\ R_s/R_i & \text{if } R_s > R_i \end{cases} \tag{17.5}$$

For any other initial orientation of the object, the final deformed ellipticity is intermediate between these two values.

The graphs on the right in Figure 17.6 are the R_f–ϕ graphs which show how the final ellipticity R_f varies with the final orientation ϕ of the deformed ellipses for the different ellipticities of strain. The heavy line in each graph is the curve for the initial ellipticity of $R_i = 2.0$ that characterizes the ellipses in the diagrams on the left. It shows the variation of the final ellipticity R_f as a function of orientation ϕ of the deformed objects. The R_i-curve varies with increasing strain from a straight line for the constant initial ellipticity R_i at zero strain (Figure 17.6A), to an open curve (Figure 17.6B), and to a closed curve (Figure 17.6C). The transition from an open to a closed curve occurs when the ellipticity of the strain equals the initial ellipticity of the deforming objects, $R_s = R_i$. The open curve means deformed ellipses of any orientation can occur (Figure 17.6B), whereas the closed curve indicates that orientations of the deformed ellipses are restricted to orientations of $|\phi| < 45°$ (Figure 17.6C). The other curves in the R_i family are for other initial ellipticities. The entire family of R_i-

---

[2] These equations were first obtained by Ramsay (1967), Sections 5.5–5.6. See Lisle (1985) for a detailed description of the whole method.

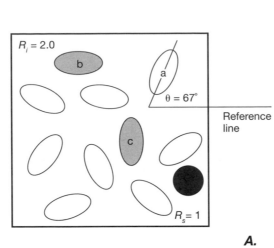

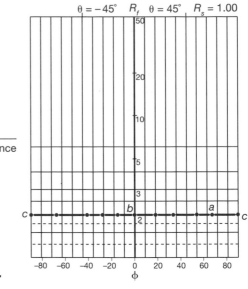

A.

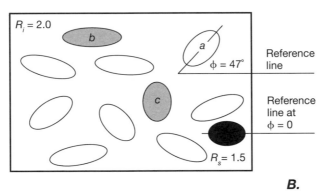

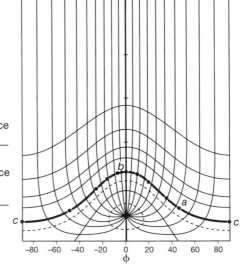

B.

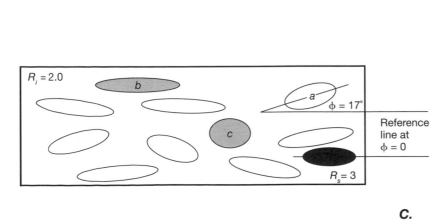

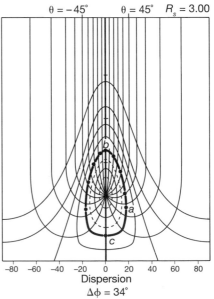

C.

Figure 17.6 (*Opposite page*) The effects of initial orientation θ of elliptical objects in the undeformed state on the final ellipticity R_f and final orientation ϕ of the objects in the deformed state. A. The undeformed state in which elliptical objects that have $R_i = 2$ are randomly oriented. The orientation θ is shown for ellipse a relative to a reference line, which for this example we take to be parallel to the maximum principal stretch in the deformed state. Ellipses b and c are in special orientations; their major and minor axes are parallel to the principal stretches in the deformed state. The R_f–ϕ plot shows that all these ellipses plot along one line, which is the $R_i = 2$ curve from the family of R_i curves for the undeformed state ($R_s = 1$). B. A deformed state for which the strain ellipticity $R_s = 1.5 < R_i$. The final ellipticity R_f is a maximum for ellipse b and a minimum for ellipse c. For $R_s < R_i$, the major axes of the deformed objects are dispersed within $\phi = \pm 90°$ of the maximum principal stretch. The graph shows the R_f–ϕ plot for $R_s = 1.5$. All the deformed ellipses plot along a single curve for which $R_i = 2$. This curve extends from $\phi = -90°$ to $\phi = +90°$, as do all R_i curves for $R_i > R_s$, reflecting the dispersion of the major axes about the direction of the maximum principal stretch. C. A deformed state for which $R_s = 3 > R_i$. The final ellipticity R_f is a maximum for ellipse b and a minimum for ellipse c. For $R_s > R_i$, the major axes of the deformed objects are dispersed within $\phi = \pm 45°$ of the maximum principal stretch, and in this case the dispersion is $\pm 17°$. The graph shows the R_f–ϕ plot for $R_s = 3.0$. All the deformed ellipses plot along a single curve for which $R_i = 2$. This curve is a closed curve (as are all R_i curves for $R_i < R_s$) that includes a range of ϕ values from $-17°$ to $+17°$, reflecting the dispersion of the major axes about the direction of the maximum principal stretch.

curves takes on shapes that vary with the ellipticity of the different values of imposed strain.

In nature it is unlikely of course that any set of objects would have identical initial ellipticities. Thus we consider as a second simple example the deformation of a group of objects having a constant initial orientation θ but variable initial ellipticities R_i (Figure 17.7). The heavy line in each of these R_f–ϕ graphs is one of a family of θ-curves. It shows for the objects in the diagrams on the left how the final ellipticities vary as a function of their final orientations for the particular initial orientation of $\theta = 63°$. In the undeformed state (Figure 17.7A), the line is a straight line at constant ϕ (63° in this example) indicating that all ellipses have the same initial orientation. The shape of the θ-curve changes with increasing strain. For any given strain ellipticity, there is a family of θ-curves that characterize different initial orientations. The pattern of the family of θ-curves changes with increasing strain (compare Figures 17.7A, B, C). For an initially random distribution of orientations, the θ-curves for $\theta = +45°$ and $\theta = -45°$ must enclose half the data between them.

Because natural objects do not have the simple distribution of shapes or orientations considered in our two examples, we can expect natural data to be scattered on an R_f–ϕ plot, (as shown, for example, in Figure 17.8A). For a random initial distribution of ellipticities and orientations, the final data should be symmetrically distributed about some line of constant ϕ that divides the data into two equal groups. That value of ϕ defines the orientation of the maximum principal axis of strain. Moreover, because all the orientations initially are randomly distributed within a 180° angle, 50 percent of the

data should lie within the 90° angle that is bisected by the maximum principal strain axis. On the R_f–ϕ diagrams, therefore, 50 percent of the data should lie between the lines for $\theta = 45°$ and $\theta = -45°$ (Figure 17.6).

R_f–ϕ data measured from a large number of deformed ooids in a limestone are plotted in Figure 17.8A, where ϕ is measured relative to a convenient reference orientation. The data cluster is roughly symmetric about the line $\phi \approx 30°$, implying an initial random orientation of the markers and an orientation of the maximum principal stretch axis $\hat{s}_1$ at 30° from the reference line. The value of the strain ellipticity R_s is defined by finding the family of θ-curves for which the $\theta = \pm 45°$ lines divide the data into two equal groups. The R_i curve that forms a tight envelope about most of the data defines the maximum initial ellipticity that is common among the deformed ooids. Because of the scatter in the data, the fitting of the curves provides only an approximate solution. A curve that shows a reasonable fit to the shape of the data cluster is for $R_s \approx 1.7$, $R_i \approx 1.6$ (Figure 17.8B). These values therefore provide an estimate of the strain and the maximum initial ellipticity of the ooids.

We can also determine the ellipticity of the strain ellipse R_s and the maximum initial ellipticity R_i analytically by combining Equations (17.4) and (17.5) to find

$$R_s^2 = \begin{cases} R_{f(max)}R_{f(min)} & \text{if } R_s > R_i \\ R_{f(max)}/R_{f(min)} & \text{if } R_s < R_i \end{cases}$$
$$R_i^2 = \begin{cases} R_{f(max)}/R_{f(min)} & \text{if } R_s > R_i \\ R_{f(max)}R_{f(min)} & \text{if } R_s < R_i \end{cases} \quad (17.6)$$

We take $R_{f(max)}$ and $R_{f(min)}$ to be the maximum and

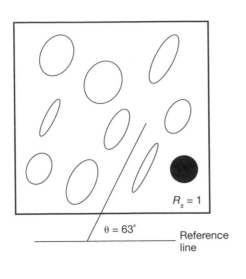

$R_s = 1$

$\theta = 63°$

Reference line

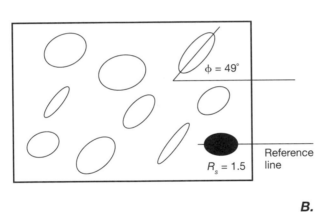

$\phi = 49°$

$R_s = 1.5$

Reference line

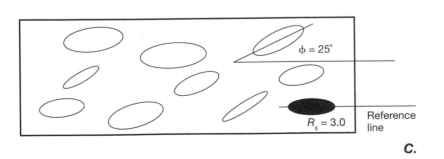

$\phi = 25°$

$R_s = 3.0$

Reference line

$\theta = -45°$ R_f $\theta = 45°$ $R_s = 1.00$

$\theta = 63°$

ϕ

A.

$\theta = -45°$ R_f $\theta = 45°$ $R_s = 1.50$

$\theta = 63°$

ϕ

B.

$\theta = -45°$ R_f $\theta = 45°$ $R_s = 3.00$

$\theta = 63°$

ϕ

C.

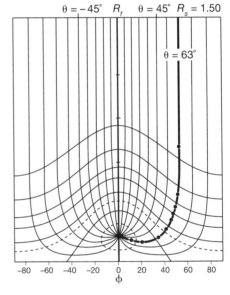

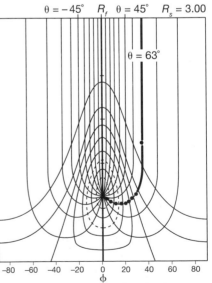

Figure 17.7 (*Opposite page*) The effect of varying initial ellipticity R_i of undeformed objects that all have the same initial orientation θ. A. The undeformed state ($R_s = 1$) with $\theta = 63°$ and R_i varying between 1.1 and 7.3. The R_f–ϕ plots shows the initial distribution of the data. B and C. Two deformed states with $R_s = 1.5$ and $R_s = 3.0$, respectively. The principal axes of the deformed objects do not remain parallel, and those with the smallest initial ellipticity change orientation the most. The R_f–ϕ plots show how the data plot asymmetrically relative to the direction of maximum principal stretch ($\phi = 0°$) along a curve of constant θ, and they suggest that the θ curves are different for different values of strain ellipticity. The $\theta = \pm45°$ curves divide data from a random initial distribution of orientations into two equal portions (compare Figure 17.6).

minimum values of R_f, measured on a set of deformed objects, ignoring isolated and widely scattered data point such as are found in Figure 17.8A.

If there was an initial preferred orientation to the objects being measured, the analysis is more difficult. The R_f–ϕ data then are not symmetric about any line of constant ϕ. Nevertheless, if the initial orientations are normally distributed about a preferred orientation, then we can find a θ-curve for some value of strain ellipticity R_s that still splits the data symmetrically into two equal groups. In principle, then, we can find a value of R_s, a value of ϕ, and an associated curve for constant θ that best fits a given data set, as well as an R_i curve

that most closely bounds the data. These values define the ellipticity R_s of the imposed strain ellipse, the orientation ϕ of the maximum imposed stretch axis $\hat{s}_1$, the initial preferred orientation θ relative to $\hat{s}_1$, and the maximum initial ellipticity R_i of the deformed objects.

The R_f–ϕ technique has been applied to deformed objects such as ooids, conglomerate pebbles, breccia fragments, quartz grains, porphyroblasts, mineral aggregates, amygdules, pumice fragments in tuff, the cross sections of worm burrows, and alteration spots in slates. Because of the abundance of rocks containing such objects, it is one of the most useful and reliable methods for determining tectonic strain.

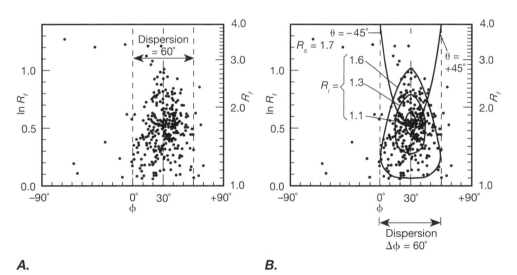

A.　　　　　　　　　　　　**B.**

Figure 17.8 Interpretation of an R_f–ϕ plot for ooids in a deformed oolite. A. The approximate symmetry of points about the value of ϕ at which R_f is a maximum means that initially, the orientations of the elliptical ooids were nearly randomly distributed. Thus the value of ϕ at the symmetry line defines the orientation of $\hat{s}_1$ relative to the chosen reference line and divides data for an originally random distribution into two equal halves. B. Inference of the orientation of $\hat{s}_1$ ($\phi = 30°$), the strain ellipticity ($R_s = 1.7$), and the initial ellipticities ($R_i < 1.6$) of the deformed objects for the data in part A. For the best value of imposed strain R_s, the lines for $\theta = 45°$ and $\theta = -45°$ should separate data from an initially random distribution into two equal halves, and an R_i curve should form a tight envelope around the data. The dispersion of ϕ for the data is an indication of the largest initial ellipticities R_i of the undeformed objects. For the solution shown, the $\theta = \pm45°$ line actually lies too high to divide the data into two equal halves. This indicates that the estimate for R_s is slightly high.

17.2 Relationship of Strain to Foliations and Lineations

Determination of the relationship of strain to foliations or lineations requires that there be a good strain marker in the same rock that contains the foliation or lineation. This restriction makes it possible to study the relationship only in special circumstances. Understanding the relationship is important, however, because it helps determine how a foliation forms during deformation and because, if the relationship is simple, foliations and lineations may be useful as strain indicators where other more obvious methods are not available.

Strain in Slates

The most extensively studied rocks, from this point of view, have been slates. These rocks have a very well-developed foliation, but they generally are not so highly metamorphosed that fossils and other strain markers are obliterated. If the slate is micro-domainal in structure, the spacing of the foliation domains is very small compared with the size of most strain markers. Thus the strain markers effectively average the strain over many domains. Studies of alteration spots (see Figure 13.19B) show that the plane of flattening of the strain ellipsoid is parallel to the slaty foliation within a few degrees. Similar results have been reported from essentially every study of slaty foliation in which the independent determination of strain has been possible. The result is an empirically established parallelism between slaty foliation and the plane of flattening of the finite strain ellipsoid.

Determination of the state of strain associated with disjunctive foliations is more difficult than with slates. For stylolitic foliations, the spacing between domains is usually larger than most common strain markers, so the averaging inherent in strain determinations with slate is rarely possible. We have already discussed the association between the amount of bedding-parallel shortening and the morphology of a disjunctive foliation in a limestone (Section 16.5; Figure 16.7). Such data, while suggestive, are insufficient for a complete determination of the orientation and ellipticity of the strain ellipsoid.

The similarity between the microstructure of slaty foliations and that of other foliations (especially crenulation foliations), and the evidence that these foliations commonly transform into continuous foliations through recrystallization, suggest that the relationship between slaty foliations and strain could be applied to other crenulation and continuous foliations. Our understanding of the mechanisms of foliation formation, however, reminds us to be cautious in doing so. Some mechanisms

do not lead necessarily to parallelism between foliation and the plane of flattening, and foliations that have acted as a plane of simple shearing *cannot* have exactly this orientation (see Section 16.6; Figure 16.15).

Volumetric Strain in the Martinsburg Shale

Figure 17.9 presents results from a unique study of the formation of foliation in the Martinsburg shale in the central Appalachian Valley and Ridge province. The strain indicators used were fossil graptolites, which were originally composed of chitinous material and are preserved as carbon films on bedding surfaces. It is unlikely that this material ever had much strength, so the graptolites probably record accurately the total strain in the rock. For certain graptolite species, the original spacing between segments (thecae) is constant and well known, so measurement of this spacing on deformed fossils can be used to determine *absolute* values for the stretches. In most studies of strain in rocks, only the ratio of the stretches can be determined. The graptolites lie parallel to the bedding plane (Figure 17.9A) and therefore record only the part of the strain that is represented by a bedding-parallel section through the strain ellipsoid. However, because of folding, the bedding planes have a wide range of orientations with respect to the foliation, and they thereby provide information about the three-dimensional geometry of the strain relative to the foliation.

The trace of the foliation on the bedding plane is an intersection lineation, generally parallel to the fold axis, that provides a convenient reference orientation (Figure 17.9A). The open circles in Figure 17.9B show that the extension parallel to the foliation trace is always very small regardless of the angle θ between bedding and foliation. The solid circles show that the extension normal to the foliation trace approaches zero in a direction parallel to the foliation plane and decreases to a minimum normal to the foliation plane. These data therefore demonstrate that there is almost no extension in any direction parallel to the foliation and that, normal to the foliation, the shortening is a maximum (stretch is a minimum). This result can be achieved only by a loss of volume. The trend defined by the solid circles can be fit with a theoretical curve indicating a volume loss for the rock of approximately 50 percent, and the volume loss was accomplished almost entirely by shortening perpendicular to the foliation.

This astounding result means that half of the original volume of the rock has been lost during deformation and the development of foliation! Other evidence, such as the partial solution of shells, indicates that most of this volume loss occurred by solution of material, although other processes, such as decrease in porosity,

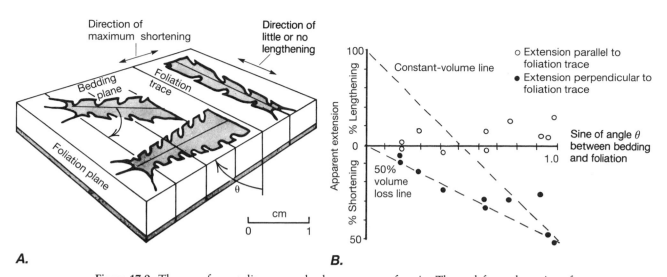

Figure 17.9 The use of graptolites as an absolute measure of strain. The undeformed spacing of thecae, or segments, on the graptolites is known and relatively constant for a given species. *A.* Graptolites lying in the bedding plane are shortest and thickest where the stipe (the long dimension of the graptolite) is normal to the trace of the foliation, and they are longest and thinnest where the stipe is parallel to the trace of the foliation. *B.* Percent extension in the bedding plane measured parallel to the foliation trace (open circles) and perpendicular to the foliation trace (solid circles) plotted against the sine of the angle θ between bedding and foliation. The dashed lines represent the theoretical relationships for plane strain with constant volume and with 50 percent volume loss.

loss of surface water on clay particles, and dehydration reactions of clay to micas, may have contributed.

This study proves conclusively that solution is a mechanism of major importance for low-temperature deformation and for the formation of foliations. It also indicates that solution can produce foliations oriented parallel to the plane of flattening of the finite strain ellipsoid. Moreover, this study demonstrates the potential for error in assuming that deformation is constant-volume simply because the volumetric strain cannot be determined.

17.3 Measurement of Strain in Folds

Studies of strain distribution throughout folds indicate that folding is a more complex process than is assumed in simple models. Although the strain distributions do not have a unique interpretation, they constrain the possible folding mechanisms.

A Study of Orthogonal Flexure

Figure 17.10 shows an example of strain analysis in a close fold that approximates class 1B in geometry and that developed in a layer of limestone pebble conglom-erate. Evidence of deformation includes deformed pebbles, pressure solution seams, twinned calcite in veins, extensional veins filled with calcite, stylolites parallel to bedding associated with compaction, and tectonic stylolites at high angles to bedding. Extensional veins are present on the limbs of the fold, and pressure solution films and twinned calcite crystals are most prominent along the concave side of the fold. Other extensional veins of calcite are found normal to the fold axis, indicating extension parallel to the fold axis.

The strain distribution around the fold is shown in Figure 17.10A, where measurements of deformed pebbles in each of the sectors are combined to define the average strain ellipse in that sector. In each sector in the hinge zone, between 10 and 50 clasts were measured; a few hundred were measured in each limb.

Figure 17.10B–E shows the strain distribution around the hinge area predicted by four different models of folding. The limb strains are not shown. The model for orthogonal flexure by bending (Figure 17.10B; see also Figure 12.6 and 16.1B) shows a distribution of $\hat{s}_1$ orientations similar to that observed in the natural fold, but the magnitudes of the principal stretches are markedly different. Models for flexural shear (Figures 17.10C; see also Figures 12.7 and 16.1C) and for volume-loss flexure from pressure solution (Figure 17.10D; see also Figures 12.10A, B and 16.1D) do not at all reproduce

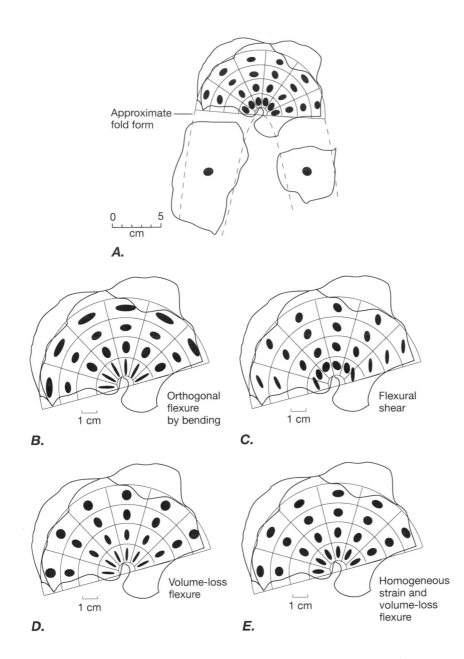

Approximate
fold form

0 5
cm

A.

Orthogonal
flexure
by bending
1 cm

B.

Flexural
shear
1 cm

C.

Volume-loss
flexure
1 cm

D.

Homogeneous
strain and
volume-loss
flexure
1 cm

E.

Figure 17.10 The strain distribution determined on a natural fold in a limestone pebble conglomerate, compared with the theoretical distribution of strain for different idealized models of folding. A. Distribution of the strain ellipses in the profile plane of the natural fold. A large number of measurements of deformed clasts within each sector of the fold were combined for each strain determination. B. Strain distribution for orthogonal flexure by bending. C. Strain distribution for flexural-shear model of folding. D. Strain distribution for volume-loss flexural folding. The outer arc of the fold is assumed to remain constant in length. E. Strain distribution resulting from a homogeneous flattening strain normal to the bedding, as would result from compaction, followed by volume-loss flexural folding.

the essential features of stain in the natural fold and therefore are inappropriate models. Figure 17.10E shows a more complex model in which a homogeneous flattening normal to the bedding is followed by folding through volume loss. This model most nearly reproduces the orientation and magnitude of the strains in the natural fold.

In the limbs of the natural fold, the strain is small and homogeneous over a relatively large area. The models for orthogonal flexure by bending (see Figure 16.1B) or by volume loss (see Figure 16.1D) are consistent with very small limb strains, but the models for flexural-shear folding (see Figure 16.1C) or for an initial component of flattening normal to bedding are inconsistent.

Thus the best model for the hinge area is homogeneous flattening normal to bedding followed by volume-loss flexure, but the best models for the limbs are either orthogonal or volume-loss flexure. The discrepancy suggests either that the mechanism of folding was more complex than assumed by the models or that the strained clasts do not record the total strain. If measured values for the strain are too low, the model for orthogonal flexure by bending is probably the most appropriate, although the distribution of solution seams suggests some folding by volume loss. Such ambiguities are not uncommon in the interpretation of rocks, and the uncertainty inherent in studying the uncontrolled experiments performed by nature is a problem with which the geologist must always contend.

The South Mountain Fold

A classic study of the relation of strain to folds and foliation is the investigation of deformed oolitic limestones in the South Mountain fold at the western edge of the Blue Ridge Mountains in Maryland (see Figure 15.7). Figure 17.11*A* is a map and cross section showing the regional setting of the fold, and Figure 17.11*B* is a generalized profile of the fold showing the relationships among the fold geometry, the strain distribution, and the foliation geometry. The $\hat{s}_1$–$\hat{s}_2$ plane of the strain ellipsoid inferred from the ooids is parallel to the foliation and forms a convergent fan around the fold. A stretching lineation is parallel to the $\hat{s}_1$ axes.

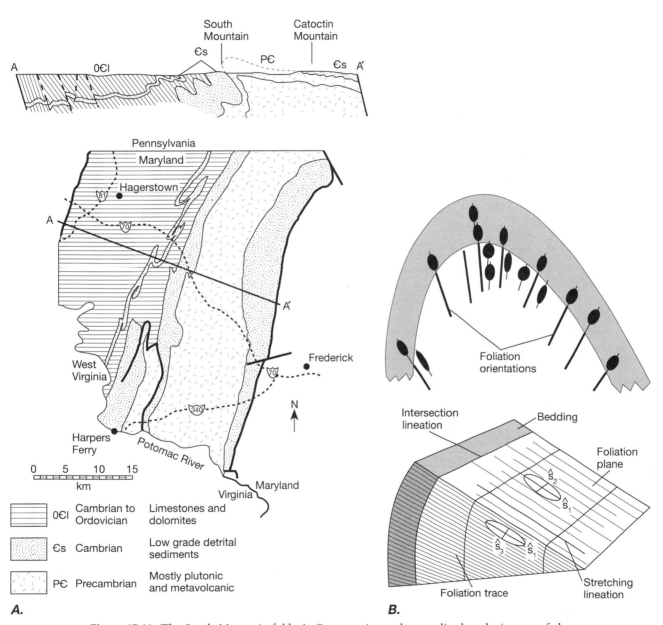

Figure 17.11 The South Mountain fold. *A.* Cross section and generalized geologic map of the Appalachian Mountains in western Maryland in the region of the South Mountain fold. *B.* Variation of strain across the South Mountain fold and its relationship to the foliation orientation. Fold profile is a composite section produced by down-structure projection.

The convergent fan of $\hat{s}_1$ axes is consistent with a flexural-folding mechanism by volume loss and/or layer-parallel shortening and folding by buckling. It is not consistent with passive-shear folding. The maximum extensions indicated on the cross section by the shapes of the strain ellipses vary irregularly around the fold. The deformation of the ooids may not reflect the total deformation in the rock, and additional deformation might have been contributed—for example, by solution of the matrix. Thus studying the ooids alone might result in underestimation of the total strain associated with folding.

Foliation Patterns in Folds

The patterns of foliation orientation in folded layers generally show considerable diversity, much of which may reflect the strain distributions. In the absence of

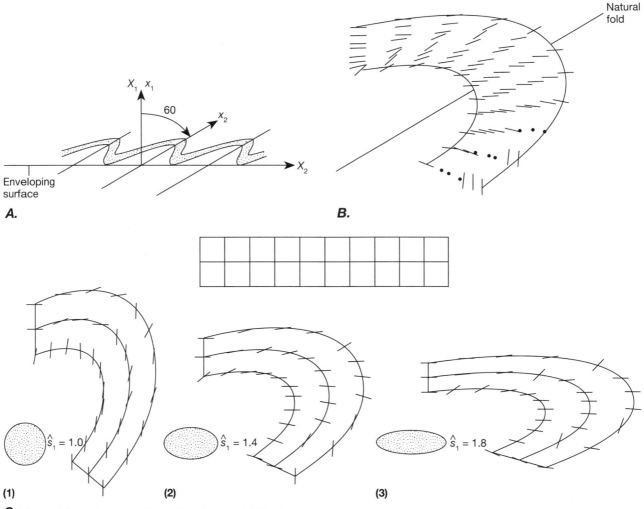

C. Model of flexural-shear folding with subsequent flattening

Figure 17.12 Correspondence between mica orientations in a class 1C fold and orientations of $\hat{s}_1$ in geometrically comparable theoretical folds. A. The fold that is analyzed is one of the antiformal folds in an asymmetric fold train developed in a layer of quartz schist. B. Observed orientations of mica in the fold. C. Orientations of $\hat{s}_1$ for a fold model combining flexural shear folding with different amounts of subsequent homogeneous flattening normal to the enveloping surface. The rectangular bar shows the undeformed layer. Strain determinations for the different models are made at the corners of the squares. The homogeneous flattening component is indicated by the ellipses and the values of the maximum stretch. The orientations of $\hat{s}_1$ axes in model 2, for which the homogeneous flattening has a maximum stretch of $\hat{s}_1 = 1.4$, most successfully reproduce the distinctive orientations of the micas observed in the natural fold, although the model is not unique.

independent strain measurements, the comparison of theoretically predicted strain distributions with such foliation patterns suggests a parallelism between foliation and the flattening plane of finite strain. We describe below several examples of such inconclusive but suggestive studies of folding.

The foliation in folded, interlayered competent and incompetent beds generally is refracted from one orientation to another across the bedding planes, the angle between bedding and foliation being higher in competent beds than in incompetent beds (Section 13.5 and Figure 13.13). In places where the composition—and presumably the competence—varies gradually, the orientation of the foliation also changes gradually.

In Figure 16.6, the strain distribution shows exactly the same pattern. In this case, the competent layers buckle predominantly by flexural shear, and the relative slip between the competent layers is accommodated by layer-parallel shear within the incompetent layers. The difference between competent and incompetent layers in the geometry and amount of strain leads to the strong refraction of the $\hat{s}_1$ orientations across the bedding, especially in the limbs of the folds. In the incompetent layers, the strong divergent foliation fan near the hinges, and the low intensity of foliation development at the concave side of the hinge zone (Figure 13.13), are also mirrored by the orientation of $\hat{s}_1$ axes and by the fact

that the ellipticity of the strain ellipse is very small in this region.

The orientation of mica flakes in two folds of substantially different geometry is shown in Figures 17.12 and 17.13. The first fold is an asymmetric fold of class 1C geometry in a relatively competent layer imbedded in a less competent matrix (Figure 17.12A). The orientations of mica flakes (Figure 17.12B) are unusual in the lower limb, where the dots represent areas of low preferred orientation, and in a zone through the upper limb, where some micas seem discordant with the dominant preferred orientation. Figure 17.12C shows three fold models consisting of initial flexural-shear folding followed by varying amounts of homogeneous flattening perpendicular to the enveloping surface. The orientations of the $\hat{s}_1$ axes are plotted in each model. The distribution of mica flake orientations is most closely mimicked by the orientations of $\hat{s}_1$ in model 2 (Figure 17.12C). The model is not unique, however; for example, it does not consider the possibility of any homogeneous shortening parallel to the layer. Thus the similarity of patterns does not prove that the micas are oriented parallel to the plane of flattening of the strain ellipsoid, although it is consistent with such an interpretation.

The second fold is one with class 2 geometry formed in a shear zone in a layered granulite (Figure

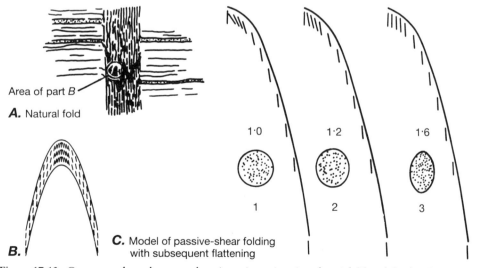

Figure 17.13 Correspondence between the mica orientations in a class 2 fold and the $\hat{s}_1$ orientations in theoretical folds of comparable geometry. A. The setting of the similar fold in the shear zone. The folded layer in a quartzo-feldspathic schist. B. The distribution of mica flakes around the fold profile. C. Orientations of $\hat{s}_1$ for three fold models combining passive shear (inhomogeneous simple shear) with various superposed components of homogeneous flattening normal to the axial plane. The homogeneous flattening component is indicated by the ellipses and the values of the maximum stretch. Short lines are parallel to the orientations of $\hat{s}_1$ around the fold. Model 2, for which the homogeneous flattening has a maximum stretch of $\hat{s}_1 = 1.2$, most nearly reproduces the pattern defined by the mica orientations in part B.

17.13A). A stretching lineation is oriented oblique to the hinge line. On a section through the fold normal to the lineation, the micas in the hinge zone form a divergent fan. In the limbs they approach parallelism with the folded layer (Figure 17.13B). Figure 17.13C shows orientations of $\hat{s}_1$ for several fold models consisting of passive shear followed by varying amounts of homogeneous flattening normal to the axial surface. The pattern of $\hat{s}_1$ axes that most closely reproduces the pattern of the mica orientations is shown as model 2 (Figure 17.13C). Although the similarity in patterns is striking, the model is again not unique. Thus the correspondence is suggestive—but not conclusive—evidence that the mica foliation is parallel to the plane of flattening of the finite strain ellipsoid.

Although the models presented are consistent with the data, they are based on untested assumption, and other models might work equally well. Thus in trying to interpret these structures, ideally we must try to test the assumptions, not simply accept them as being appropriate.

Folds in the Cambrian Slate Belt of Wales

We often tend to think that folds reflect an overall shortening of the crust, and it is easy to assume that the fold axes form parallel to the $\hat{s}_2$ direction, with $\hat{s}_1$ in the axial plane normal to the fold axis and $\hat{s}_3$ normal to the axial plane. This interpretation may be reasonable for symmetric parallel folds in foreland fold and thrust belts, for example, but in other situations the validity of such an interpretation cannot necessarily be presumed.

Strain studies of alteration spots in the Cambrian slate belt in Wales (Figure 17.14A) illustrate the point. Figure 17.14B shows a longitudinal section along the hinge surface of one of the folds. The upper heavy line shows the orientation of the hinge line in the axial surface, and the distance between the two lines represents the layer thickness in the fold hinge. Note that the diagram covers a section over 20 kilometers in length. The hinge goes through several culminations and depressions of varying amplitude along its length. The strain ellipses illustrated were measured in a vertical plane normal to the axial surface. Wherever there is a culmination in the hinge line, layer thickness increases and the strain ellipse shows an increase in ellipticity; maximum ellipticities correspond to maximum culminations. Conversely, hinge depressions correspond to the minimum layer thicknesses and the least elliptical strain ellipses. The $\hat{s}_1$ axes of the strain ellipsoids are vertical, and the orientation of the fold hinge is only locally parallel to the $\hat{s}_2$ direction, deviating in some places as much as 25°. In higher grade metamorphic terranes, such deviation may become even more extreme.

Figure 17.14C shows a kinematic model that accounts for the associated variations in strain and in layer thickness and hinge line undulations. In the model, folding reflects an overall shortening of the body of rock perpendicular to the axial surface. The shortening is accommodated by vertical extension. The strain is inhomogeneous across the axial planes, leading to formation of the folds, and inhomogeneous along the axial surface, leading to the formation of culminations and depressions in the fold hinge. Where the shortening normal to the axial surface is greatest, the vertical extension is also the greatest, resulting in a thickening of the layers and a culmination in the fold hinge. Where shortening is least, the vertical extension is also small, resulting in a minimum in the layer thickness and a depression in the fold hinge.

These observations reinforce the conclusion that the orientations of many foliations and lineations are related to the orientation of the strain ellipsoid in the rocks. Our theoretical understanding of the mechanisms involved in the formation of foliations and lineations can help us make sense of the observed relationships, but in many cases it is not adequate to enable us to predict them. We still need more definitive studies of the relationships among strain, folds, and the various types of foliations and lineations in rocks, and we need a more detailed understanding of the mechanisms by which these structures form.

17.4 Strain in Shear Zones

Shear zones show a wide variety of characteristics ranging from brittle through ductile features. Brittle shear zones, commonly associated with faulting near the Earth's surface (see Part II on brittle deformation), are characterized by pervasive brittle fractures. Brittle–ductile shear zones show features that have characteristics of both brittle and ductile deformation, such as extensional gash fractures that are rotated by ductile deformation (see Figures 3.8, 10.11, and 17.19). Ductile shear zones show features such as sigmoidally shaped foliation traces that indicate coherent deformation and a smooth variation of strain across the zone. In these ductile shear zones, the total amount of shear that has occurred is recorded by the strain distribution across the shear zone, and if appropriate strain markers are deformed in the shear zone, a reconstruction of the deformation may be possible.

The simplest model for shear zone deformation is inhomogeneous progressive simple shear (see Figure 15.15). This model of a ductile shear zone requires that the magnitude of the shear strain be inhomogeneously distributed varying from zero at both shear zone boundaries to a maximum in the central part of the zone. The

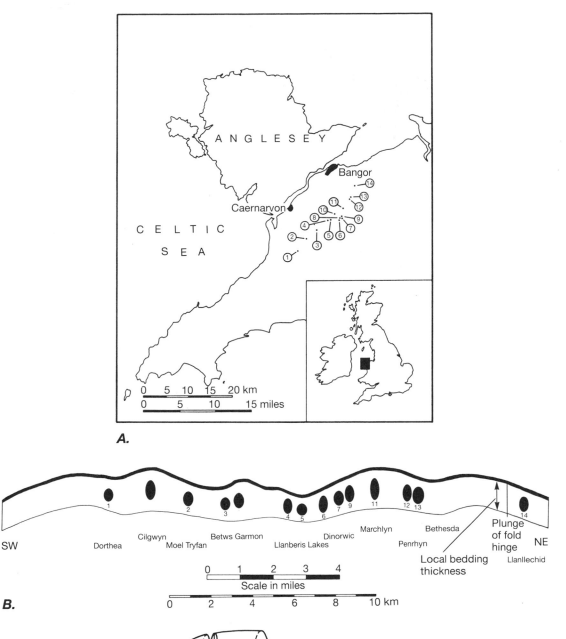

A.

B.

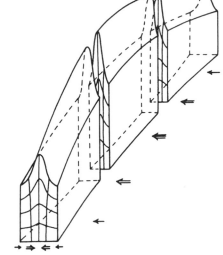

C.

Figure 17.14 Deformation in the Cambrian slate belt in Wales. *A.* Location map for the slate belt in Wales. *B.* Longitudinal profile of the hinge surface of a fold, showing the form of the hinge line (top line) and the variation in layer thickness (distance between top and bottom lines) along the surface. The strain ellipses show the principal axes of strain in a plane *perpendicular* to the hinge surface. Where the ellipticity is a maximum, the fold hinge goes through a culmination; where the ellipticity is a minimum, the fold hinge goes through a depression. *C.* Model to account for the variation in strain and the associated changes in the fold hinge orientation. For a constant-volume plane strain deformation, the most extreme horizontal shortening must be associated with a maximum in vertical extension.

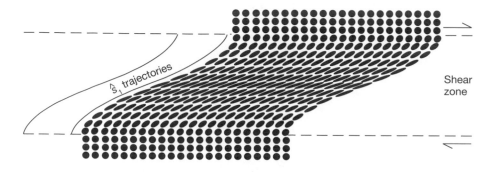

orientations and shapes of the strain ellipses that result from such a deformation, and in particular the characteristic sigmoidal trajectories of the $\hat{s}_1$ axes, are illustrated in Figure 17.15. At the boundaries of the shear zone where the strain approaches zero, the orientation of the $\hat{s}_1$ axes is at an angle of 45° to the boundary (compare the maximum principal axis line A in Figure 15.15A). As the central part of the zone is approached, the magnitude of the shear strain increases, and the angle between the shear zone boundary and the $\hat{s}_1$ axes decreases in a predictable relationship (compare the maximum principal axes in Figure 15.15B–D). The greater the magnitude of the shear strain, the smaller the angle between the shear plane and the $\hat{s}_1$ axes. Thus the exact trajectory of the $\hat{s}_1$ axes depends on the amount of shear that has accumulated across the shear zone.

Investigations of naturally occurring shear zones in rocks are often based on the assumption that inhomogeneous progressive simple shear approximates the natural deformation, although in some cases this assumption is not justified. In massive crystalline rocks

that have been deformed subsequent to their formation, the mode of deformation is commonly one of concentrated shear along well-defined ductile shear zones. The development of a schistosity is associated with these shear zones, and the massive rock characteristically becomes increasingly foliated from the boundary into the center of the shear zone. Figure 17.16A shows an example of such a shear zone developed in a massive gneiss. At the boundaries the schistosity is very weakly developed and is oriented at an angle of 45° to the boundary. Progressing in toward the central part of the shear zone, the schistosity becomes increasingly strongly developed, and the angle it makes with the shear zone boundary decreases.

These features can be explained readily by using the model of inhomogeneous progressive simple shear for the shear zone and assuming the schistosity is parallel to the $\hat{s}_1$–$\hat{s}_2$ plane of the finite strain ellipsoid, the plane of flattening. The degree of development of schistosity correlates with the magnitude of strain, and the orientation pattern of the schistosity corresponds with the trajectories of the $\hat{s}_1$ axes of finite strain (Figure 17.15).

This interpretation of shear zones and their associated schistosity is reinforced by observations of a deformed aplite dike and deformed xenoliths in a granite

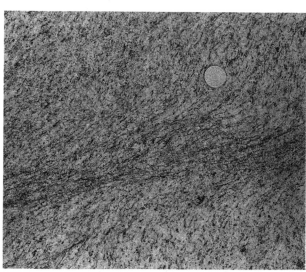

A.

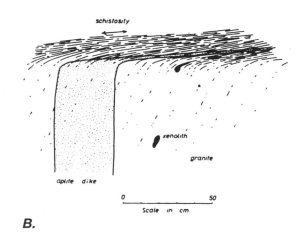

B.

Figure 17.16 Deformation features of ductile shear zones. A. Development of schistosity in a small ductile shear zone in a metagabbro. B. Deformation of an aplite dike and mafic xenoliths in a granite cut by a ductile shear zone normal to the dike.

that is cut by a ductile shear zone (Figure 17.16B). In the undeformed part of the granite, the thickness of the dike is 32 cm, there is no schistosity, and the xenoliths are roughly equidimensional. Within the boundary zone of the shear zone, the dike is only slightly deflected, a weak schistosity is present oriented at an angle of approximately 45° to the boundary, and the xenoliths are slightly elongate. Within the central part of the shear zone, the dike is very strongly rotated and its thickness is reduced to 2 cm, the schistosity is strongly developed and oriented at a low angle to the shear zone boundary, and the xenoliths are stretched out into long, thin shapes.

The xenoliths can be use to obtain a rough measure of the magnitude of the strain in the different parts of the shear zone. This strain corresponds well with that deduced from the angle between the schistosity and the shear zone boundary, assuming that the schistosity parallels the plane of flattening of the finite strain ellipse and that the orientation can be predicted by the model of progressive simple shear (see Figure 15.15).

The consequences of deforming a tabular body, such as the dike, in a ductile shear zone partly depend on the relative orientations of the shear zone and the dike, on the sense of shear, and on the relative competence of the tabular body and the matrix. Figure 15.17A, C, and D and Figure 15.18B, C illustrate the possible histories of lengthening, shortening, or shortening followed by lengthening during simple shear. Note, however, that lengthening cannot be followed by shortening. If the competence of the layer is comparable to that of the matrix, the layer would lengthen and thin homogeneously (Figure 17.16B) or shorten and thicken homogeneously. If the layer is significantly more competent than the matrix, it would lengthen by forming boudins and shorten by forming ptygmatic folds.

Although progressive simple shear is commonly a successful model for ductile shear zones, it is by no means unique, and other geometries of deformation must be considered as well. Heterogeneous volume loss alone can produce structures similar to those produced by simple shear (see Figure 16.16). The superposition of homogeneous and/or heterogeneous strains on a ductile shear zone can also produce geometries that are similar to those of simple shear but that have different strain distributions.

Figure 16.16 illustrates how a volume loss concentrated along individual planar features can produce a geometry similar to that of a brittle shear zone. Similarly, a distributed inhomogeneous volume loss can produce a geometry similar to that of a ductile shear zone (compare the deformation of the line in Figure 17.17B, C). The addition of a component of inhomogeneous volume loss to the inhomogeneous simple shear would provide a geometrically more complex model of a shear zone, for which the resulting deformation is illustrated in Figure 17.17D. The superposition of a homogeneous deformation on any of these models would further modify the strain distribution across the ductile shear zone.

Such considerations suggest a model to explain why in some cases, gash fracture planes do not bisect the angle between conjugate shear zones as required by the stress interpretation (Figure 10.11). In this case we assume that gash fractures develop normal to the direction of maximum incremental extension $\hat{\zeta}_1$, rather than normal to the minimum compressive stress $\hat{\sigma}_3$ as we assumed before (Figure 10.11). If we then add to the simple shearing a component of inhomogeneous volume increase or decrease normal to the shear zone (Figure 17.18), different orientations of $\hat{\zeta}_1$ relative to the shear zone result, thereby accounting for the different orientations of gash fractures that are observed.

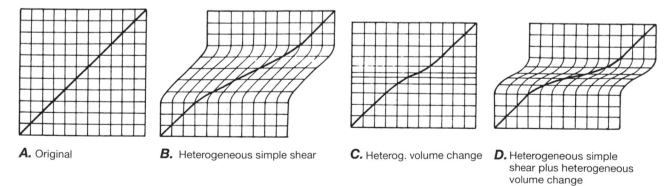

A. Original   **B.** Heterogeneous simple shear   **C.** Heterog. volume change   **D.** Heterogeneous simple shear plus heterogeneous volume change

Figure 17.17 Displacement fields suggesting ductile shear zones. *A.* Undeformed state. *B.* Shear zone formed by a progressive inhomogeneous simple shear. *C.* Apparent ductile shear zone formed by a distributed inhomogeneous volume loss. *D.* Shear zone formed by the superposition of the displacement fields in parts *B* and *C*.

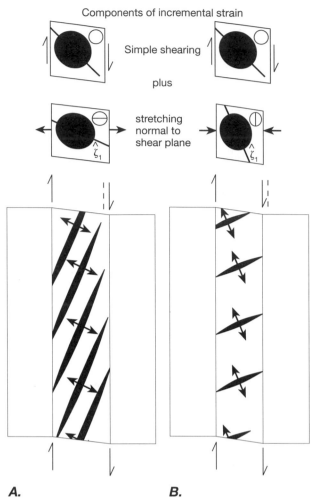

Components of incremental strain

Simple shearing

plus

stretching normal to shear plane

A.   **B.**

Figure 17.18 Assuming that gash fractures form normal to the orientation of the maximum incremental stretch $\hat{\zeta}_1$, the addition to the incremental shear of an incremental component of inhomogeneous volume increase (*A*) or decrease (*B*) normal to the shear zone can account for the various observed orientations of gash fracture relative to the shear zone.

17.5 Deformation History

In order for us to infer the geometry of progressive deformation (that is, the deformation path), we require information from the rocks about the deformation history (see the beginning of Chapter 15 and Section 15.5). Determination of the finite strain ellipsoid is insufficient, because it is the net result of the whole preceding history of the deformation, and it does not define the deformation path.

Distinguishing between coaxial and noncoaxial deformation also requires some knowledge of the deformation history. To infer a coaxial geometry for the deformation, for example, we must be able to show that the principal axes of the finite strain ellipsoid have re-mained parallel to the same material lines throughout the deformation or, equivalently, that the principal axes of the finite strain ellipsoid and the incremental strain ellipsoid are parallel (see Sections 15.4 and 15.5).

Features that record different portions of the strain history can help us distinguish between coaxial and noncoaxial deformation. For example, features that are introduced into the rock at different times during its deformation (such as dikes, veins, and metamorphic segregations) record only the strain that accumulates after their introduction. An example of such a record can be found in brittle–ductile shear zones in which *en echelon* gash fractures are developed. We assume that the fractures open initially in an orientation either perpendicular to $\hat{\sigma}_3$, the axis of minimum compressive stress, or perpendicular to $\hat{\zeta}_1$, the maximum extension axis of the incremental strain ellipsoid (Figure 17.19*A*). The two directions may be the same, but that is not necessary, especially if the material is not mechanically isotropic. As shear strain accumulates by a combination of ductile flow and fracturing, the older parts of the gash fractures rotate, while the fractures continue to grow at the tips in a direction perpendicular to $\hat{\sigma}_3$ or $\hat{\zeta}_1$ (Figure 17.19*B*). The result is a sigmoidal-shaped gash fracture. With sufficient rotation, extension fracturing on the original set of fractures ceases, and a second generation of fractures develops in the same original orientation as the first generation with respect to the stress or incremental strain axes. Cross-cutting relationships between the material filling the first- and the second-generation fractures can clearly indicate the sequence of fracture development. After growth at the tips of the old fractures ceases, the tips may be rotated, along with the rest of the fracture, out of the unique relationship with the stress or incremental strain axes. Further shearing may result in the development of still other generations of fractures along the shear zone (Figure 17.19*C*).

This case is an example of successively forming structures recording different amounts of the ductile strain along a shear zone. The clear evidence of the sequential development of the fractures, the rotation of the older fractures relative to the newer ones, and the inferred relationship between the direction of fracture growth and the local stress or incremental strain axes indicate that material lines are rotated past the orientations of the principal axes of incremental strain. Thus the geometry of strain accumulation must have been noncoaxial. The curvature and the accumulated widths of the gash fractures reflect a minimum amount of shear strain that has occurred on the shear zone.

A different technique for constraining the deformation history depends on the observation of mineral fibers that grow in the rock during the deformation (Section 14.6; Figures 14.7 through 14.9). Where fibers

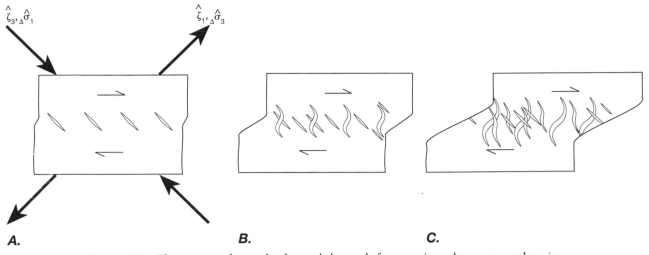

$\hat{\zeta}_3, _\Delta\hat{\sigma}_1$ $\hat{\zeta}_1, _\Delta\hat{\sigma}_3$

A.   **B.**   **C.**

Figure 17.19 The pattern of growth of *en echelon* gash fractures in a shear zone undergoing both fracturing and ductile shearing. The arrows indicate the direction of principal incremental stretches and deviatoric stress. *A.* Initial development of gash fractures normal to $\hat{\zeta}_1$ and to $\hat{\sigma}_3$. *B.* Growth of rotated gash fractures ceases, and a second generation of fractures develops in the original orientation, in places cross-cutting the older fractures. *C.* Growth of the rotated second generation of fractures ceases, and a third set begins to develop.

form the filling of extension fractures or veins, or form overgrowths on particles in the rock, and fiber growth is displacement-controlled, the orientations of the fibers record the orientation history of the maximum incremental stretch axis $\hat{\zeta}_1$. Consider again a shear zone in which gash fractures form and are filled by mineral fibers. Because the deformation in this case is noncoaxial, the material planes defined by the fracture wall rotate with respect to the shear zone boundaries and the principal axes of incremental strain (see Figure 15.15). The fibers consistently grow parallel to the maximum incremental stretch axis $\hat{\zeta}_1$, however, which is constant in orientation with respect to the shear zone boundaries. Thus the fibers progressively change orientation with respect to the fracture wall and end up being curved. The pattern of curvature for a given non-

coaxial deformation depends on whether fiber growth is syntaxial or antitaxial (Figure 14.7). The first orientation of fibers in later-generation gash fractures is perpendicular to the fracture wall and parallel to the last orientation of fibers in the earlier-generation fractures. This relationship confirms the interpretation that the fibers continually grow parallel to the maximum incremental stretch axis $\hat{\zeta}_1$.

Thus whether in extension fractures, shear fractures, or oriented overgrowths, fibers preserve a record of the history of displacement, and the length of the slickenfibers records the total displacement across the fracture or overgrowth during the time of fiber growth. Curved fibers indicate the existence of component of rotation during deformation relative to the incremental strain axes and thereby imply a noncoaxial deformation.

Additional Readings

Alvarez, W., T. Engelder, and P. A. Geiser. 1978. Classification of solution cleavage in pelagic limestones. *Geology* 6: 263–266.

Alvarez, W., T. Engelder, and W. Lowrie. 1976. Formation of spaced cleavage and folds in brittle limestone by dissolution. *Geology* 4: 698–701.

Cloos, E. 1947. Oolite deformation in the South Mountain fold, Maryland. *Geol. Soc. Am. Bull.* 58: 843–918.

Cloos, E. 1971. *Microtectonics along the Western Edge of the Blue Ridge. Maryland and Virginia.* Baltimore: Johns Hopkins Press.

Durney, D. W., and J. G. Ramsay. 1973. Incremental strain measured by syntectonic crystal growths. In K. A. DeJong and R. Scholten, eds., *Gravity and tectonics*. New York: Wiley, pp. 67–96.

Fry, N. 1979. Random point distributions and strain measurement in rocks. *Tectonophysics* 60: 89–105.

Grey, D. R., and D. W. Durney. 1979. Investigations on the mechanical significance of crenulation cleavage. *Tectonophysics* 58: 35–79.

Hobbs, B. E. 1971. The analysis of strain in folded layers; *Tectonophysics* 11: 329–375.

Hudleston, P. J., and T. B. Holst. 1984. Strain analysis and fold shape in a limestone layer and implications for layer rheology. *Tectonophysics* 106(3/4): 321–347.

Lisle, R. J. 1985. *Geological strain analysis: A manual for the R$_f$/ϕ technique*. Oxford: Pergamon.

Mitra, G. 1978. Microscopic deformation and flow laws in quartz within the South Mountain anticline. *Jour. Geol.* 86: 129–152.

Ramsay, J. G. 1967. *Fracturing and folding of rocks*. New York: McGraw-Hill.

Ramsay, J. G. 1980. Shear zone geometry, a review. *J. Struct. Geol.* 2: 83–89.

Ramsay, J. G., and Graham, R. H. 1970. Strain variation in shear belts. *Can. Jour. Earth Sci.* 7: 786–813.

Ramsay, J. G., and M. I. Huber. 1983. *The techniques of modern structural geology*. Vol. 2: Folds and fractures. London: Academic Press.

Wellman, H. W. 1962. A graphical method for analyzing fossil distortion caused by tectonic deformation. *Geol. Mag.* 99: 348–352.

Wood, D. S. 1973. Patterns and magnitudes of natural strain in rocks. *Phil. Trans. Roy. Soc. Lond* A274: 373–382.

Wright, T. O., and L. B. Platt. 1982. Pressure dissolution and cleavage in the Martinsburg shale. *Am. J. Sci.* 282: 122–135.

PART

IV Rheology

SO FAR in our investigation of deformation in the Earth, we have discussed the relationship between stress and the fracture of rocks, and we have discussed the application of strain to the understanding of ductile deformation in rocks. Except for a brief introduction during the discussion of brittle fracturing, however (for example in Section 9.1 and applications in Chapter 10), we have not discussed the relationships between stress and strain. Broadly speaking, these relationships are the subject of Part IV of this book, which comprises the next three chapters.

Ultimately, if we are to have a deeper understanding of the formation of deformational structures in rocks, we must be able to use the relationships between stress and strain to make mathematical models of the deformation. With such models, we can calculate how structures develop. Our models, of course, must be based on the real behavior of rocks, and rock deformation experiments are essential in obtaining this information. Therefore, in the next chapter (Chapter 18), we describe these models and the experimental evidence for how rocks behave.

In Chapter 19 we take a closer look at the microscopic and submicroscopic mechanisms that give rise to the ductile deformation of rocks and at the structures that preserve the evidence for the operation of those mechanisms. By this means we gain an understanding of the underlying physical principles of ductile deformation. This knowledge enables us to evaluate extrapolations from laboratory conditions to conditions in the Earth, as well as to understand and interpret the microscopic structures associated with ductile flow.

Models of deformation in the Earth may be both mathematical and physical. The equations that describe the behavior of different types of materials can be combined with the fundamental conservation, or balance, laws of physics to enable us to calculate mathematically the motion and deformation of a body of rock, given the conditions to which it is subjected. In this way, we can determine how material properties affect the formation of structures. In Chapter 20, we give some examples of how such mathematical models can improve our understanding of the formation of folds. Scale models provide physical analogues to deformation in the Earth, so in Chapter 20 we also introduce the theory of scale models and illustrate their application to investigations of geometrically complex structures.

Models, of course, must always be simplifications and idealizations of reality. The accepted approach (known as Occam's razor) is to employ models of minimum complexity, incorporating only those features that are necessary to account for the observations of interest. To the extent that the characteristics of various models correspond with our observations of the Earth, we accept them as reasonable representations of the conditions and processes that occur in the Earth. Such models, of course, can never be more than approximations to the real world.

CHAPTER

 Macroscopic Aspects of
Ductile Deformation

Rheology and Experiment

In this chapter we consider the relationships between stress and strain that are useful in describing the behavior of rocks on a macroscopic scale. At this scale, we assume rocks can be described as continua (see the beginning of Chapter 1) so that the inhomogeneities and anisotropies associated with their polycrystalline nature are averaged out. Our goal is to determine a set of mathematical relationships that are of practical use in describing the mechanical behavior of different rock types under the various conditions of temperature, stress, and pressure that are common within the Earth.

Different materials behave differently under the same state of stress. For example, under a uniaxial compression, one type of material might shorten slightly and then stabilize, whereas another might flow continuously like putty. We must express such differences mathematically if we are to be able to calculate the response of a material to stress. We first discuss, therefore, a few simple mathematical idealizations of material behavior that we can use to describe the relationship between stress and either strain or strain rate for different types of materials. These equations are called **constitutive equations,** because their nature depends on the constitution of the material.

Experimental deformation of rocks provides us with much of the information on which we base our choice of constitutive equations. From experiments we can observe the different types of mechanical behavior

in rock and how each depends on stress, temperature, pressure, grain size, composition, and chemical environment. Primarily, we emphasize one-dimensional constitutive relationships, because these contain the essence of the mechanical behavior without the complexity that the complete three-dimensional equations require. (The three-dimensional relationships are discussed briefly in Box 18.3.)

18.1 Continuum Models of Material Behavior

Elastic Materials

One of the simplest and perhaps most familiar stress–strain equations is for a linear elastic solid. Such a material deforms by an amount proportional to the applied stress. When the stress is released, the material returns to its original undeformed state, so the deformation is said to be **recoverable.** The relationship is represented by a linear equation, which states either that a normal stress σ_n is proportional to the amount of extension e_n or that a shear stress σ_s is proportional to the amount of shear strain e_s (see Equation 9.2).

$$\sigma_n = E e_n \qquad \sigma_s = 2\mu e_s \qquad (18.1)$$

The constant E is **Young's modulus**; μ is the **shear modulus** or the **modulus of rigidity**. The graphical form of the equations is shown in Figure 18.1A.

These equations are identical in form to Hooke's law, which describes the behavior of a spring (Figure 18.1B): The applied force F is equal to the product of a spring constant k_1 and the displacement of the spring Δx. For this reason, a linear elastic solid is sometimes called a Hookean solid.

A variation of stress with time (Figure 18.1C) consisting of the application of stress at t_1, a linear increase to t_2, a constant value to t_3, and a linear decrease to zero at t_4 results in a similar-shaped strain–time history (Figure 18.1D).

Equations (18.1) may be rewritten to express the strain in terms of the stress (Equation 9.2; see also Equation 9.5 in three dimensions). Thus we may consider that the stresses cause the strains or that the strains cause the stresses (see Section 10.12). We discuss in Chapters 9 and 10 the importance of stress and elastic deformation in the formation of fractures and faults in the Earth's crust.

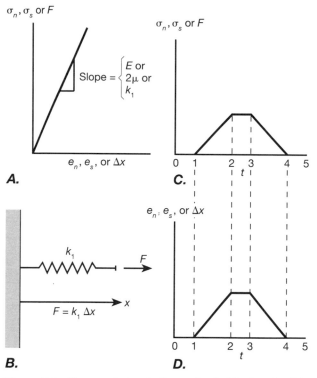

Figure 18.1 Characteristics of a linear elastic (Hookean) solid. *A.* A plot of normal stress (or force) versus extension (or displacement) for an elastic solid (or a spring) where the slope E (or k_1) is the Young's modulus (or spring constant). *B.* A mechanical analogue to elastic behavior consists of a spring with spring constant k_1 attached to a rigid wall, subjected to a force F, and displaced a distance Δx beyond its unloaded length. *C.* A stress–time history imposed on the elastic material. *D.* The extension–time behavior resulting from the stress history.

Viscous Materials

At room temperature and pressure, a rock clearly reacts to stress very differently from a material such as water. We cannot expect, therefore, that an equation for elastic behavior, which applies very well to cold rocks, would be very successful in accounting for the behavior of water. We need a different type of constitutive equation.

If a stress is applied to a fluid, it begins to flow. When the stress is removed, the flow stops, but the fluid does not return to its undeformed configuration, and the deformation is said to be **nonrecoverable.** The larger the applied stress, the faster the fluid flows, which suggests a relationship between the stress and the strain *rate*.

This type of behavior is most simply idealized as a linearly viscous, or Newtonian, constitutive equation, and the appropriate one-dimensional equation for constant-volume deformation relates the normal deviatoric stress $_\Delta\sigma_n$ to the incremental extension per unit time $\dot{\varepsilon}_n$, or the shear stress σ_s to the incremental shear strain per unit time $\dot{\varepsilon}_s$. (For a discussion of strain rates, see Box 18.1).

$$_\Delta\sigma_n = 2\eta\dot{\varepsilon}_n \qquad \sigma_s = 2\eta\dot{\varepsilon}_s \qquad (18.2)$$

The incremental extension rate is the rate of change in length divided by the instantaneous deformed length; the incremental shear strain rate is half the rate at which the tangent of the incremental shear angle changes, where the incremental shear angle is measured for two material lines instantaneously normal to one another (see Box 18.1). These strain rates are nothing more than the strains defined by the incremental strain ellipse (Section 15.4) divided by the increment of time. The proportionality constant η is called the **coefficient of viscosity** of the fluid.

The linear equation for viscous behavior is graphed in Figure 18.2A, which is similar to the graph for elastic behavior except that the abscissa is the strain *rate* rather than the strain. When the stress goes to zero, the strain rate goes to zero but the strain does not. Thus the deformation is permanent, and the amount of strain that can accumulate in such a material has no relationship whatsoever to the magnitude of the stress. Given sufficient time, any stress can produce an arbitrarily large strain.

A mechanical analogue for viscous behavior is a device called a dash pot, which consists of a fluid-filled cylinder containing a porous piston (Figure 18.2B). When a force is applied across the system, the motion of the piston is governed by the rate at which fluid can flow through the pores in the piston. The greater the force applied, the faster the piston moves. This device is familiar as the mechanism that is often used to prevent screen doors from slamming and as hydraulic shock absorbers on motor vehicles.

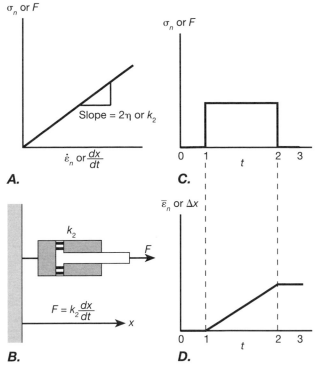

A.

B.

$F = k_2 \dfrac{dx}{dt}$

C.

D.

Figure 18.2 Characteristics of a linear viscous (Newtonian) fluid. A. A plot of stress (or force) versus incremental strain rate (or displacement rate) for a viscous material (or dash pot) has a slope η (or k_2), the coefficient of viscosity (dash pot constant). B. A mechanical analogue for viscous behavior is a dash pot consisting of a porous piston in a cylinder containing a viscous fluid, attached to a rigid support, and subjected to a force F. C. A stress–time history imposed on the fluid. D. The natural strain–time history resulting from the imposed stress.

A force suddenly applied to the dash pot at time t_1 and suddenly removed at time t_2 (Figure 18.2C) produces a linear increase of displacement between those times, and it leaves a permanent displacement when the force is removed (Figure 18.2D). Under some conditions of high temperature and pressure, rocks may deform according to a viscous constitutive equation (see Box 19.1).

Plastic Materials

Under many conditions, including common experimental conditions in the laboratory, the model of the linearly viscous fluid does not describe well the observed behavior of polycrystalline solids, such as rocks and metals. Such materials commonly undergo no permanent deformation if the applied stress is smaller than a characteristic **yield stress,** but they flow readily if it is at or slightly above the yield stress. Materials exhibiting this type of behavior are **plastic materials.**

We idealize plastic behavior mathematically by assuming that there is no deformation at all (the material

is rigid) below the yield stress and that during the deformation, the stress cannot rise above the yield stress except during acceleration of the deformation. This model describes a **perfectly plastic material** or a **rigid-plastic material.** The stress for ductile flow is a constant, and the constitutive equation is the **von Mises yield criterion** (see Figure 9.9):

$$|\sigma_s| \le K \qquad (18.3)$$

This equation requires that the magnitude of the shear stress σ_s not be greater than the yield stress K, which is a characteristic of the material. Thus for this ideal model, the stress does not determine the strain rate, and the constitutive equation is merely a limit on the possible values of the stress (Figure 18.3A). In a Mohr diagram, K is the radius of the Mohr circle at the yield point of the material.

A mechanical analogue for this type of behavior is the idealized frictional resistance to the sliding of a block

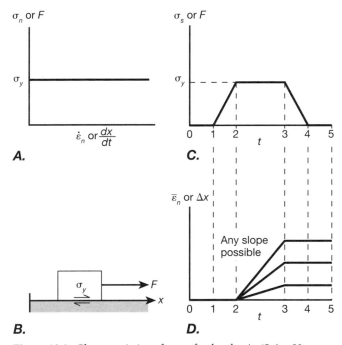

A.

B.

C.

D.

Any slope possible

Figure 18.3 Characteristics of a perfectly plastic (Saint Venant) material. A. Characteristics of a plastic material and its mechanical analogue on a plot of stress (or force) versus incremental strain rate (or velocity). B. A mechanical analogue to plastic flow is the frictional resistance to sliding of a block on a plane. No displacement occurs until the applied force exceeds the frictional resistance. Once sliding has started, the applied force cannot rise above the frictional resistance regardless of the velocity of sliding, except during acceleration. C. A stress–time history imposed on a plastic material. The maximum possible stress is the yield stress. D. The natural-strain–time history resulting from the imposed stress. Any slope on the graph is possible, because the rate of change of strain is not a function of the stress.

Box 18.1 Measures of Strain Rate

Two strain rates are commonly used in the literature: one based on the rate of change of the finite strain, the other based on the incremental strain per unit time. Unfortunately, the symbols representing the different rates are not standardized, and it is not always clear which of the two rates is intended. The reader must therefore be careful about assuming which is meant. We demonstrate the difference between the two rates below.

If we use the index (i) to indicate the initial deformed state at some time t and the index (f) to indicate the final deformed state at a small increment of time Δt later, and if we use L for the undeformed length and ℓ for the deformed length, the rate of change of the finite extension is given by

$$\dot{e}_n = \frac{de_n}{dt} = \lim_{\Delta t \to 0} \frac{e_n^{(f)} - e_n^{(i)}}{\Delta t}$$

$$\dot{e}_n = \lim_{\Delta t \to 0} \frac{1}{\Delta t}\left[\left(\frac{\ell_f - L}{L}\right) - \left(\frac{\ell_i - L}{L}\right)\right]$$

$$\dot{e}_n = \lim_{\Delta t \to 0} \frac{1}{\Delta t}\left[\frac{\ell_f - \ell_i}{L}\right]$$

$$\dot{e}_n = \frac{d\ell/dt}{L} = \frac{d}{dt}\left[\frac{\ell}{L}\right] = \dot{s}_n \qquad (18.1.1)$$

We drop the subscript i or f on ℓ when it is clear that ℓ refers to the instantaneous deformed length. Note that the rate of change of the finite extension is also the rate of change of the finite stretch s_n.

The finite shear strain rate is

$$\dot{e}_s = 0.5 \lim_{\Delta t \to 0} \frac{\tan \psi^{(f)} - \tan \psi^{(i)}}{\Delta t}$$

$$= 0.5 \lim_{\Delta t \to 0} \frac{[u^{(f)}/L] - [u^{(i)}/L]}{\Delta t} = 0.5 \lim_{\Delta t \to 0} \frac{\Delta u}{L \Delta t}$$

$$\dot{e}_s = \frac{0.5(du/dt)}{L} \qquad (18.1.2)$$

where u is the displacement normal to L associated with the shear of the line L through the angle ψ (Figure 18.1.1).

The rate of incremental extension is just the incremental extension that occurs in the time Δt.

$$\dot{\varepsilon}_n = \lim_{\Delta t \to 0} \frac{\varepsilon_n}{\Delta t} = \lim_{\Delta t \to 0} \frac{1}{\Delta t}\left[\frac{\ell_f - \ell_i}{\ell_i}\right] = \frac{d\ell/dt}{\ell} \qquad (18.1.3)$$

The rate of incremental extension $\dot{\varepsilon}_n$ is also the rate of change of the natural strain $\bar{\varepsilon}_n$ defined in Equation (15.4).

$$\frac{d\bar{\varepsilon}_n}{dt} = \frac{d}{dt}\left(\ln \frac{\ell_f}{L}\right) = \frac{L}{\ell_f}\frac{d\ell_f/dt}{L} = \frac{d\ell/dt}{\ell} = \dot{\varepsilon}_n$$

The rate of incremental shear strain is

$$\dot{\varepsilon}_s = 0.5 \lim_{\Delta t \to 0} \frac{\tan \Delta\psi}{\Delta t} = 0.5 \lim_{\Delta t \to 0} \frac{\Delta u}{\Delta t \ell} = \frac{0.5(du/dt)}{\ell} \qquad (18.1.4)$$

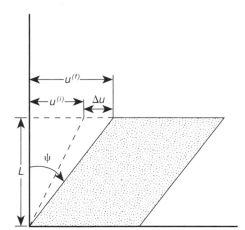

Figure 18.1.1 The finite shear strain rate in progressive simple shear is defined in terms of the increment of displacement $\Delta u = u^{(f)} - u^{(i)}$ per unit length L, per unit time, where the superscripts f and i indicate final and initial displacements.

---

on a surface (Figure 18.3B). Consider that the load F increases linearly from t_1 to the yield stress at t_2, remains constant to t_3, and then decreases linearly to t_4. The velocity of the block is zero until the applied force exceeds the force of friction, which is equivalent to the yield stress. At that point the block starts to move, but the velocity of the block cannot be determined from the magnitude of the applied force, because that force never exceeds the frictional resistance except during acceleration. Thus the force history (Figure 18.3C) produces

an undefined rate of displacement (Figure 18.3D). The stress and the strain rate behave in an analogous manner in a perfectly plastic material.

Other Continuum Models for Material Behavior

The simple elastic, plastic, and viscous models we have considered are useful, but they do not describe adequately all important material behavior. We will not

where $\Delta\psi$ is the very small incremental shear angle that measures the shear of a pair of instantaneously perpendicular lines, and Δu is a very small incremental displacement for a line of length ℓ, where $\Delta\psi$, ℓ, and Δu are defined in a similar manner to ψ, L, and $u^{(i)}$ in Figure 18.1.1.

Thus the two strain rates differ in that for the rates of finite extension and shear strain, $\dot{e}_n$ and $\dot{e}_s$, the reference length is the original length L, whereas for the rates of incremental extension and shear strain, $\dot{\varepsilon}_n$ and $\dot{\varepsilon}_s$, the reference length is the instantaneous deformed length ℓ. The different rates are related by the stretch s_n.

$$\dot{\varepsilon}_n = \frac{d\ell/dt}{\ell} = \frac{d\ell/dt}{L}\frac{L}{\ell} = \frac{\dot{e}_n}{s_n} \tag{18.1.5}$$

$$\dot{\varepsilon}_s = \frac{du/dt}{\ell} = \frac{du/dt}{L}\frac{L}{\ell} = \frac{\dot{e}_s}{s_n} \tag{18.1.6}$$

Thus at small strains, the difference between the rates is negligible because s_n is close to 1. At large strains, however, the difference becomes significant.

The strain rate tensor $\dot{\varepsilon}_{k\ell}$ is a generalization of the rates of incremental strain $\dot{\varepsilon}_n$ and $\dot{\varepsilon}_s$. The tensor components of this strain rate are the components of the incremental strain tensor $\varepsilon_{k\ell}$ for a unit increment of time. These components can be shown to be the symmetric part of the velocity gradient tensor, which is defined by

$$\dot{\varepsilon}_{k\ell} = \dot{\varepsilon}_{\ell k} = 0.5\left[\frac{\partial v_k}{\partial x_\ell} + \frac{\partial v_\ell}{\partial x_k}\right] \tag{18.1.7}$$

where k and ℓ independently can take on any of the values 1, 2, and 3, and where v_k (or v_ℓ) are the components of the velocity of material points in the deforming continuum. Because the velocity is the displacement rate, these equations are comparable to Equations (15.1.17) with the displacement replaced by the velocity. The tensor components for which k and ℓ are equal (for example, $k = \ell = 1$) are rates of incremental extension. Those for which k and ℓ are unequal (for example, $k = 1$ and $\ell = 3$) are rates of incremental shearing.

Use of the finite extension rate $\dot{e}_n$ or the finite shear strain rate $\dot{e}_s$, or in general $\dot{e}_{k\ell}$, can be justified when the mechanical behavior of the material depends on the total amount of strain accumulated in the body from the undeformed state, for in this case the undeformed state has a mechanical significance for the behavior of the material.

If the behavior of the material is independent of the total strain, then the undeformed state is mechanically indistinguishable from the deformed state, as, for example, in the deformation of a viscous fluid such as water. In these circumstances, there can be little physical meaning in a strain rate that uses some arbitrarily defined state of the material as a reference state, and it is physically more relevant to refer the strain rate at any given time to the state of the material at that time. Thus the incremental extension rate $\dot{\varepsilon}_n$ or the incremental shear strain rate $\dot{\varepsilon}_s$, or in general $\dot{\varepsilon}_{k\ell}$, is the most appropriate measure to use.

For many materials, the present behavior is affected by the strain that has accumulated within a constant span of time before the present, and the material is said to have a fading memory. The primary creep of many polycrystalline materials is an example of this behavior. Once steady-state creep is established, however, any state of strain is mechanically indistinguishable from any other state of strain, and the incremental strain rate, such as $\dot{\varepsilon}_n$, is most appropriate for describing the mechanical behavior.

Although experiments on ductile flow of rocks have concentrated mostly on the high-temperature steady-state deformation, the incremental strain rate is not universally employed, and in many cases, published reports of experimental deformation are not even explicit about which definition of strain rate is used.

delve deeply into more sophisticated models, but it is easy to gain an intuitive appreciation for some of them.

Figures 18.4 through 18.7 show four different rheologic models that result from combining the simple models in series (Figures 18.4 and 18.5) or in parallel (Figures 18.6 and 18.7). In each figure the relationships between stress and strain or between stress and strain rate are plotted in part A. The mechanical analogue is shown in part B. Part C shows a stress–time history imposed on the material that is chosen to demonstrate the major characteristics of the material behavior. Part D shows the strain–time response of the material to the imposed stress. We use the natural strain as a strain measure because it is related directly to the incremental strain rate that appears in the constitutive equation (see Box 18.1). For the elastic part of the deformation, the strains are small and the finite strain and the natural strain are almost the same.

A visco-elastic (Maxwell) solid (Figure 18.4) behaves like a combination in series of an elastic spring

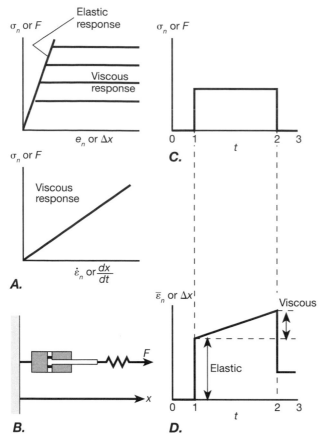

Figure 18.4 The characteristics of a visco-elastic, or Maxwell, material. *A.* The constitutive behavior of a visco-elastic material, showing the stress as a function of both natural strain and incremental strain rate. The graph of stress versus natural strain shows that the magnitude of the elastic response depends on the magnitude of the stress. The graph of stress versus incremental strain rate shows the dependence of the viscous strain rate on the stress. *B.* The mechanical analogue consists of a dash pot in series with a spring. *C.* A stress–time history imposed on the material. *D.* The natural-strain–time response to the imposed stress includes an instantaneous recoverable elastic deformation and a nonrecoverable viscous deformation.

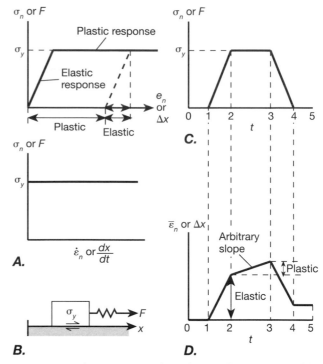

Figure 18.5 Characteristics of an elastic–plastic, or Prandtl, material. *A.* The constitutive behavior of an elastic–plastic material, showing the stress as a function of both natural strain and incremental strain rate. The graph of stress versus natural strain shows the elastic response below the yield stress. The graph of stress versus incremental strain rate shows the independence of the plastic strain rate from the stress. *B.* The mechanical analogue consists of a friction block connected in series with a spring. *C.* A stress–time history imposed on the material. *D.* The natural-strain–time history resulting from the imposed stress. Below the yield stress, the material deforms elastically. At the yield stress, plastic deformation occurs at an undetermined rate. When the stress is removed, the elastic portion of the strain recovers, leaving a permanent strain equal to the plastic portion.

and a viscous dashpot. The permanent strain accumulates viscously, and it begins as soon as a stress is applied. For high viscosities, this material behaves like an elastic material for loads of short duration but like a viscous material for long-term loads. For a constant imposed strain, the initial elastic response is gradually converted into permanent viscous deformation, and the associated stress decays with time. Visco-elastic behavior resembles that of silly putty. It is useful in modeling the response of the Earth's crust, which is observed to undergo short-term elastic deformation when subjected to a rapid loading, but which gradually flows if the load is maintained for long periods.

An elastic–plastic (Prandtl) material (Figure 18.5) also exhibits a combination of recoverable elastic strain and permanent deformation. The permanent deformation is plastic, however, and it does not begin until the yield stress is reached. Release of the stress allows the elastic deformation to disappear, but a permanent deformation remains. Elastic–plastic behavior characterizes the behavior of some crystals at high temperatures, as discussed below.

A visco-plastic (Bingham) material (Figure 18.6) displays linear viscous behavior only above a yield stress such as characterizes plastic materials. A natural example of this behavior is wet paint; it is fluid but has a small yield stress, which prevents it from running off a wall after a thin layer is a applied.

A firmo-viscous (Kelvin or Voigt) material (Figure 18.7) is elastic to the extent that the equilibrium strain

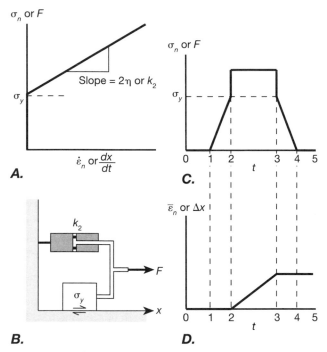

A.

B.

C.

D.

Figure 18.6 Characteristics of a visco-plastic, or Bingham, material. *A.* The graph of stress versus incremental strain rate for a visco-plastic material. Above the yield stress, the material behaves like a viscous fluid. *B.* The mechanical analogue consists of a dash pot and a friction block connected in parallel. *C.* A stress–time history imposed on the visco-plastic material. *D.* The natural-strain–time history resulting from the imposed stress.

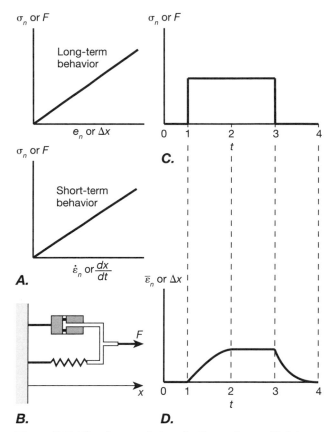

A.

B.

C.

D.

Figure 18.7 The characteristics of a firmo-viscous (Kelvin or Voight) material. *A.* The long- and short-term behaviors of a firmo-viscous material are revealed on plots of stress versus natural strain (long-term elastic dependence) and stress versus incremental strain rate (short-term viscous dependence). *B.* The mechanical analogue consists of a dash pot and a spring connected in parallel. *C.* The stress–time history imposed on the material. *D.* The natural-strain–time response to the imposed stress shows that the incremental strain rate depends both on the initial stress and on the strain.

is a linear function of the applied stress, but the strain rate is governed by a viscous response. An everyday application of the mechanical analogue is an automobile suspension system consisting of a spring connected in parallel with a dash pot (a shock absorber), which damps out the elastic oscillations of the spring.

Other models can be derived by combining the simple mechanical analogues in different configurations. The potential usefulness of any such model depends on whether its response corresponds to that observed for some particular material. All these models, with the exception of plasticity (Figures 18.3 and 18.5), have the limitation that they are superpositions of the linearly proportional responses of a spring or a dash pot to an imposed load. Such linear models have proved extremely useful in many applications in geology and engineering, but materials do not have to behave in mathematically convenient ways. Thus nonlinear elasticity theory is necessary to account for the elastic properties of rocks under very high pressures, and some of the processes of ductile flow of crystalline solids are inherently nonlinear, as we see, for example, in Section 18.4. We discuss the generalization of these equations to three dimensions in Box 18.3.

18.2 Experimental Investigation of Ductile Flow

Experimental investigations of the ductile behavior of rocks are generally conducted using an apparatus that deforms a sample of rock under high pressures and temperatures. The sample is held inside a furnace within a pressure vessel. Pressure is applied to the sample by a gas, liquid, or ductile solid. The temperature is maintained by controlling the power to the furnace. The sample is deformed by squeezing it between opposing pistons. The samples are generally small cylinders of rock ranging from several millimeters to a few centimeters long, with a diameter between one-third and one-fourth of the length. They are subjected to confined compression (Figure 8.14D) at confining pressures ranging from a few hundred to a few thousand megapascals and at temperatures ranging between room temperature and the melting temperature of the material.

The experiments generally involve the slow, continuous deformation, or **creep,** of the specimen under varied conditions. For technical and practical reasons, the lower bound for experimental strain rates is roughly 10^{-7} per second (s^{-1}), although some techniques can push that down another order of magnitude or so. At this rate, a specimen 1 cm long shortens by 10^{-7} cm each second or less than 1 percent per day. Shortening the sample by only 10 percent from 10 mm to 9 mm takes approximately 11.6 days. Many experiments are performed at 10^{-4} to 10^{-5} s^{-1}.

Experiments are performed at either constant stress (creep experiments) or constant strain rate. The so-called constant-stress experiments are often done at constant load, and the stress actually decreases slightly with the increase in cross-sectional area of the sample as it deforms. Similarly, the constant-strain-rate experiments are often done at constant displacement rate, which results in an increase in the rate of incremental strain with progressive shortening of the sample. More sophisticated experiments, however, use computer control to continually adjust the load or the displacement rate so the calculated stress or strain rate remains at a constant value.

For investigating the ductile flow of rocks, experiments at constant stress are preferable to those at constant strain rate, because the flow mechanism depends more directly on stress than on strain rate, and the material constants are more directly determinable from the experimental data (see Boxes 18.2 and 19.1). Constant-strain-rate experiments are generally simpler to perform, however, and for steady-state deformation, they give similar information (see Box 18.2).

The ductile behavior of different materials is best compared at an equal **homologous temperature,** defined by the ratio T/T_m, where T is the temperature of the material and T_m is its melting temperature, both expressed in kelvins (K). The melting temperature is a rough measure of the strength of the bonds binding a crystalline material together. Thus the behavior of many different materials tends to be the same at the same homologous temperature, even if their absolute melting temperatures are very different. For example, at the high homologous temperature of 0.95, ice and olivine display a similar type of behavior even though that temperature for ice is 259 K ($-14°C$), whereas for olivine it is approximately 2017 K ($1744°C$). The mechanical behavior of a material at different pressures is also approximately the same at the same homologous temperature if the pressure effect on the melting temperature is included in the calculation of the homologous temperature.

The results of a constant-stress experiment are often plotted on a graph of strain versus time (Figure 18.8A, B). At homologous temperatures below about 0.5, creep curves generally show continuously decreasing creep rates (Figure 18.8A). This phenomenon is called **cold working,** and the creep is **logarithmic creep** because the total strain increases with the logarithm of time.

For high-temperature creep at homologous temperatures above about 0.5, a typical creep curve has several characteristic parts (Figure 18.8B). The first part is an essentially instantaneous recoverable elastic strain that accompanies the initial, ideally instantaneous loading of the material. If the applied stress is above the yield stress, then the material begins creeping immediately upon application of the load. The creep rate is relatively high at first, but it declines steadily as the experiment proceeds. This phase of the curve is called **primary creep,** and the phenomenon of decreasing creep rate at constant stress is called **work hardening** or **strain hardening.** In this phase of the deformation, the material becomes less ductile with increasing strain.

Eventually the creep rate stabilizes at a constant value, and this phase is called **steady-state creep** or **secondary creep.** Steady-state creep is the part of the experiment that most interests geologists, because at constant stress it could continue indefinitely. Thus it presumably represents the long-term deformation processes that occur within the Earth.

In many experiments, the steady-state regime eventually gives way to **tertiary creep,** during which the strain rate accelerates and ultimately the sample fractures. Tertiary creep is most common during high-stress, low-temperature experiments and/or extensional stress experiments. In high-temperature experiments, it is most likely to result from the change in stress or strain rate imposed by the changing geometry of the sample during creep. It is therefore an artifact of the method of experimentation, rather than a fundamental characteristic of the deformation process.

On a plot of differential stress versus time, the curve for constant-strain-rate experiments can be divided into comparable portions (Figure 18.8C, D). In this case, the buildup of elastic strain is not instantaneous because of the constant strain rate. Thus the stress builds up in proportion to the accumulating elastic strain. When the stress reaches the yield point, ductile deformation begins.

In an experiment at low homologous temperatures (Figure 18.8C), the stress continues to increase indefinitely, at least to the strength limits of the sample or the apparatus. In these cases, work hardening is the dominant mode of deformation.

In experiments at high homologous temperatures (Figure 18.8D), primary creep commences at the yield

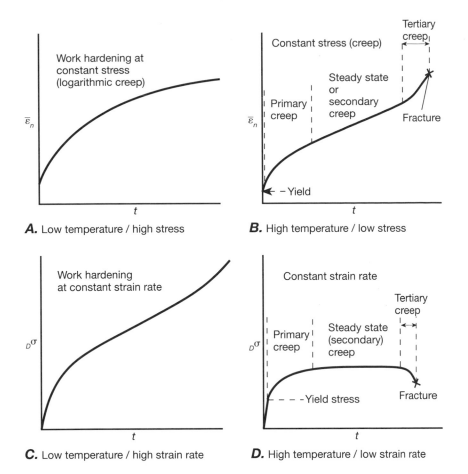

Work hardening at
constant stress
(logarithmic creep)

$\bar{\varepsilon}_n$

t

A. Low temperature / high stress

Constant stress (creep)

Tertiary
creep

Steady state
or
secondary
creep

Primary
creep

Fracture

$\bar{\varepsilon}_n$

← – Yield

t

B. High temperature / low stress

Work hardening
at constant strain rate

$_D\sigma$

t

C. Low temperature / high strain rate

Constant strain rate

Tertiary
creep

Primary
creep

Steady state
(secondary)
creep

$_D\sigma$

– – –Yield stress

Fracture

t

D. High temperature / low strain rate

Figure 18.8 Characteristic creep curves for ductile deformation of a polycrystalline material. *A.* A curve showing logarithmic creep at constant stress, which is characteristic of work hardening during a low-temperature experiment. *B.* A creep curve for a constant-stress, high-temperature experiment, showing the three different stages of creep. *C.* A creep curve for a constant-strain-rate, low-temperature experiment, showing work-hardening behavior. *D.* A creep curve for a constant-strain-rate, high-temperature experiment, showing the three different stages of creep.

point, and the rate of increase in stress gradually declines to zero. Where the stress becomes constant, the creep rate has reached a steady-state value. Steady-state creep continues for an indefinite period until the onset of tertiary creep, which is accompanied by a decline in the stress.

In practice, the yield point may be difficult to identify exactly, because there is a gradational transition from purely elastic to predominantly ductile strain, and because in creep experiments, the apparatus cannot increase the stress rapidly enough to approximate an instantaneous loading. The shape of the curves may also be affected by the practice of maintaining constant load or constant displacement rate rather than constant stress or constant strain rate.

18.3 Experiments on Cataclastic Flow and Friction

Cataclastic flow occurs in cataclastic rocks during displacement along a fault zone (see Sections 4.2 and 9.4, Figure 4.5*B*, and Table 4.1). An understanding of the process contributes to the continuing effort to under-

stand and predict earthquakes in earthquake-prone regions, such as along the San Andreas fault in California.

To investigate the rheology of this flow, experiments are performed on sliding interfaces that have been produced either by fracture or by cutting and various degrees of polishing. Finely crushed rock may be introduced between the rock interfaces to simulate fault gouge. Experiments at constant displacement rate or, less commonly, at constant applied stress determine the effects on mechanical behavior of differential stress, displacement rate, normal stress, interface surface roughness, grain size of the gouge, and thickness of the gouge layer.

Frictional effects dominate the behavior of the system. Figure 18.9 shows some characteristic results of experiments conducted at room temperature on a layer of simulated granite gouge 1 mm thick, sheared between two granite blocks. Figure 18.9*A* shows displacement-versus-time curves for creep experiments carried out at different constant applied stresses (given at the end of each curve in megapascals). The curves exhibit two patterns: Either there is a continual decrease of velocity with time, shown by the slope of the curves, or there is a continual increase leading to catastrophic unstable slip. Note that only relatively small changes in stress lead to significant changes in the creep behavior.

Box 18.2 **Experimental Determination of the Material Constants in the High-Temperature Creep Equation**

In order to define the rheological equations for a given material, we first choose an appropriate form of constitutive equation whose general properties fit those observed for the material. For polycrystalline solids during high-temperature steady-state creep, one of Equations (18.4), (18.7), (18.8), and (18.10) generally accounts well for the observations. We then use experimental data to determine the values of the constants that characterize the particular material in the constitutive equation.

Our approach depends on whether the experiments are to be constant-stress experiments or constant-strain-rate experiments. Because the latter are more common among rock deformation experiments, we describe the procedure for obtaining the material constants n, E^*, V^*, b, and K_1 with constant-strain-rate experiments.

We write the basic form of the constitutive equation as follows:

$$_D\sigma = K_1|_s\dot{\varepsilon}_n|^{1/n}/d^{b/n}\exp\left[\frac{H^*}{nRT}\right] \qquad H^* = E^* + pV^*$$

(18.2.1)

where H^* is the activation enthalpy. If we take the natural logarithm of both sides and rearrange, we can isolate different variables as the independent variable on the right side of the equation. Each of the following equations is arranged such that the terms in braces either are intrinsically constant or are held constant during a particular set of experiments, and the independent variable is in the last term on the right.

$$\ln{}_D\sigma = \frac{1}{n}\left\{-\ln K_1 + b\ln d + \left[\frac{H^*}{RT}\right]\right\} + \left[\frac{1}{n}\right]\ln|_s\dot{\varepsilon}_n|$$

(18.2.2)

$$\ln{}_D\sigma = \frac{1}{n}\left\{-\ln K_1 + b\ln d + \ln|_s\dot{\varepsilon}_n|\right\} + \left[\frac{H^*}{nR}\right]\frac{1}{T}$$

(18.2.3)

$$\ln{}_D\sigma = \frac{1}{n}\left\{-\ln K_1 + b\ln d + \ln|_s\dot{\varepsilon}_n| + \left[\frac{E^*}{RT}\right]\right\} + \left[\frac{V^*}{nRT}\right]p$$

(18.2.4)

$$\ln{}_D\sigma = \frac{1}{n}\left\{-\ln K_1 + \left[\frac{H^*}{RT}\right] + \ln|_s\dot{\varepsilon}_n|\right\} = \left[\frac{b}{n}\right]\ln d$$

(18.2.5)

The experimental technique involves holding constant all variables written between the braces so that, in effect, there is a linear relationship between $\ln{}_D\sigma$ and the one independent variable. For each type of experiment, plotting the dependent variable (left side of the equation) versus the independent variable on a graph enables us to measure the slope of the data distribution, which determines the value of the coefficient of the independent variable. The procedures are summarized in the table below.

The stress exponent n is determined directly as the inverse of the slope of Equation (18.2.2). Data for dunite from Figure 18.12 are plotted in Figure 18.2.1. The slopes for these data give $n = 2.35$ and 2.99, values which are slightly low compared to the preferred value of 3 to 3.5. In practice, however, many more experiments are required to determine the constant accurately (see, for example, the moderate-stress regime in Figure 18.14). The activation enthalpy H^* can be determined from the slope H^*/nR of Equation (18.2.3) because n is determined independently, and R is a constant. Data for dunite from Figure 18.11 are plotted in Figure 18.2.2 which, for $n = 3.0$ to 3.5, gives an activation enthalpy between 518 and 604 kJ/mole. The result depends strongly on the value of n but is consistent with the prefered value of about 540 kJ/mole. Similarly, the activation volume V^* can be determined from the slope of Equation (18.2.3) because n is determined independently, R is a known constant, and T is held at a known constant value

Variables held constant	Graph		Slope	Equation	Figure
	Ordinate	Abcissa			
T, p, d	$\ln{}_D\sigma$	$\ln|_s\dot{\varepsilon}_n|$	$1/n$	(18.2.2)	18.2.1
$_s\dot{\varepsilon}_n$, p, d	$\ln{}_D\sigma$	$1/T$	H^*/nR	(18.2.3)	18.2.2
$_s\dot{\varepsilon}_n$, T, d	$\ln{}_D\sigma$	p	V^*/nRT	(18.2.4)	18.2.3
$_s\dot{\varepsilon}_n$, T, p	$\ln{}_D\sigma$	$\ln d$	b/n	(18.2.5)	18.17B

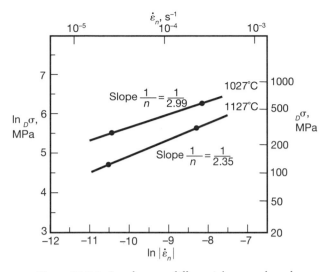

Figure 18.2.1 Steady-state differential stress plotted versus strain rate on a natural logarithmic scale for experiments on dunite (see Figure 18.12). The slopes of the plots reveal the value of the inverse stress exponent $1/n$. For an accurate determination, many more experiments than shown here are required (compare Figure 18.15).

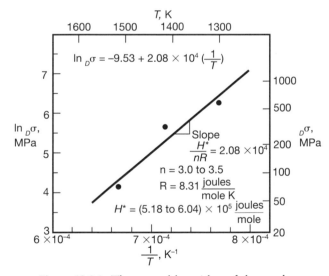

Figure 18.2.2 The natural logarithm of the steady-state differential stress plotted against inverse temperature for experiments on dunite from Figure 18.11. The slope provides a value for H^*/nR and thus determines the activation enthalpy. For an accurate determination, many more experiments than shown here are required.

for the experiment. The data for dunite from Figure 18.16 are plotted in Figure 18.2.3 and indicate that the activation volume is approximately $V^* = 2.8 \times 10^{-5}$ m³/mole. Finally, the grain size exponent b is determined from the slope of Equation (18.2.4) because n is already known. For dunite, Figure 18.17C shows the relationship that defines the slope to be $b/n = 1.21$ (Equations 18.9 and 18.10), and because n was determined to be about 1.4, we find $b = 1.7$. Knowing all the material parameters in the creep equation except K_1 enables us to determine the value of that constant from the value of the intercept on any of the graphs.

Note, however, that the determination of all these constants is dependent on the value of n. The same material constants can be determined independent of one another from experiments performed at constant stress. The constitutive equation is then rearranged so that $\ln |_s\dot{\varepsilon}_n|$ appears as the dependent variable on the left of the equation. In the table, $_D\sigma$ replaces $|_s\dot{\varepsilon}_n|$ in the column of variables held constant, and $\ln |_s\dot{\varepsilon}_n|$ is plotted on the ordinate instead of $\ln {_D\sigma}$. Equation (18.2.3), for example, becomes

$$\ln |_s\dot{\varepsilon}_n| = \{\ln K_1 - b \ln d + n \ln {_D\sigma}\} - \left[\frac{H^*}{R}\right]\frac{1}{T} \quad (18.2.6)$$

Thus the slope of a plot of $\ln |_s\dot{\varepsilon}_n|$ versus $1/T$ provides the value of H^*/R directly and avoids the additional error associated with having to determine the value of H^* on the basis of an independent determination of n Equation (18.2.3). For this reason, constant stress experiments are preferable for determining the material constants.

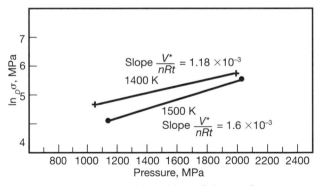

Figure 18.2.3 The natural logarithm of the steady-state differential stress plotted against the pressure for dunite (see Figure 18.16). The slope determines the value of V^*/nRT and thus provides a value for the activation volume if n is known.

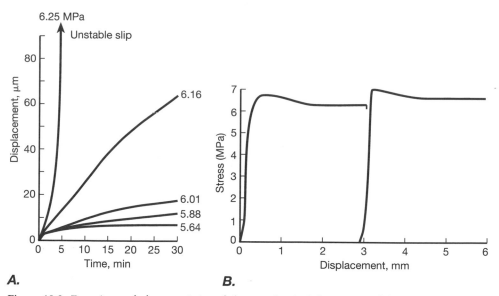

Figure 18.9 Experimental characteristics of the cataclastic deformation of fault gouge. *A.* Displacement–time record for a 1-mm-thick layer of simulated gouge of 250-μm grain size sheared between two granite blocks whose surfaces were finished with #60 abrasive. The various curves are for applied stresses ranging from 5.64 MPa to 6.25 MPa. *B.* Stress–displacement record for a constant-strain-rate experiment using the same configuration as in part *A.* At 3 mm displacement, the displacement was stopped and then restarted.

Figure 18.9*B* shows the result of a constant-displacement-rate test for the same experimental configuration. After almost 3 mm of displacement, the displacement was halted and then restarted. The behavior is very similar to that of the elastic–plastic (Prandtl) material described in Section 18.4, except the stress peaks at the yield point and then decays with further displacement to a constant residual value.

A diagnostic feature of cataclastic deformation is that the differential stress required for deformation increases markedly with increasing confining pressure. This behavior indicates a frictional mechanism in which the force of friction increases with the normal stress across the plane of sliding (see Section 9.4).

Laboratory experiments on frictional behavior indicate that the coefficient of friction on a surface has two values, one for static friction and a slightly lower one for sliding friction. During sliding, the coefficient of sliding friction decreases slightly with increasing velocity of slip; and during periods of no slip, the coefficient of static friction in fault gouge under pressure increases with time, leading to a strengthening of the shear zone.

These experimental results resemble observations of natural faulting. Stress buildup on a fault can lead to the sudden rapid displacement associated with an earthquake. This process implies a dynamic instability caused by decreasing strength with increasing displacement. On the other hand, the fact that faults with a substantial movement history still have a significant shear strength requires a static strengthening mechanism so that resistance to slip can increase during periods of no displacement.

Figure 18.10*A* shows a simple mechanical analogue that reproduces the expected behavior of a fault system. It consists of a spring that has a spring constant K and a slider that has a coefficient of static friction higher than the coefficient of sliding friction. The sliding interface of the block corresponds to the fault, and the spring corresponds to the behavior of rock surrounding the fault that transmits the stresses to the fault surface.

Assume that a constant velocity v is imposed on the free end of the spring. The force exerted on the block by the spring is resisted by the force of friction across the base of the slider. As the spring stretches, the shear stress on the base of the block builds up until it reaches the level of static friction. At that point the block begins to slide, and the coefficient of friction decreases progressively to a residual value after a displacement d_c. As the block slides, the spring relaxes, and according to Hooke's law (Figure 18.1*A*, *B*; see also Equation 18.1), the force exerted by the spring on the slider must decrease. The behavior of the system depends on the relative rates at which the spring force and the frictional resistance decrease with displacement (Figure 18.10*B*, *C*).

If the frictional resistance (the solid line in Figure 18.10*B*) decreases more rapidly with displacement than

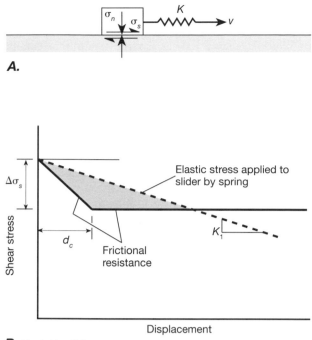

A.

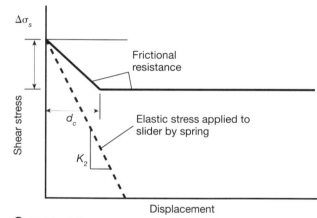

B. Unstable sliding

C. Stable sliding

Figure 18.10 An analogue for the behavior of a fault system. *A.* A mechanical analogue consisting of a spring, with spring constant *K*, connected in series to a frictional slider for which the coefficient of friction decreases from the static value to the sliding value over a finite amount of displacement. If a constant velocity is imposed on the end of the system, it can respond either by stick-slip or by stable sliding. *B.* The instability leading to stick-slip is shown on a record of stress versus displacement. During the slip event, the frictional resistance decreases more per unit of displacement than does the elastic stress, leaving the excess elastic stress available to accelerate the system. This accounts for the slip instability. *C.* If slip is by stable sliding, the elastic stress tends to decrease more per unit displacement than the frictional stress, which means that the stress on the block cannot exceed the frictional resistance. It is maintained by the constant displacement rate imposed on the system.

the force exerted by the spring (the dashed line in Figure 18.10*B*), then the difference in stress between the two lines represents the stress available to accelerate the block, and the sliding becomes unstable: An earthquake occurs. Eventually the block slides far enough that the force exerted by the spring is less than that necessary to maintain the sliding, and the block stops. Sliding cannot resume until the force exerted by the spring once again exceeds the force of static friction. The resulting intermittent behavior is reminiscent of the stick–slip motion characteristic of many active faults (see Figure 9.12).

If, on the other hand, the slope of the stress-displacement line for the spring (the dashed line in Figure 18.10*C*) is steeper than that for the frictional resistance (the solid line in Figure 18.10*C*), then sufficient force for sliding is maintained only by the displacement rate *v* imposed on the free end of the spring, and sliding is stable while stress on the system follows the solid line.

Thus an understanding of the mechanics of cataclasis and friction may hold the key to understanding the mechanical behavior of faults.

18.4 Steady-State Creep

Below depths of roughly 15 km to 20 km in the Earth, the homologous temperature generally exceeds 0.5. Deformation of rocks under these conditions is characterized by creep that can reach a steady state and therefore can accommodate large amounts of deformation. Thus it is generally assumed that the steady-state constitutive equations, or flow laws, can be used to characterize the large-strain, high-temperature ductile deformation that occurs in the Earth.

Many experiments have been performed to investigate the dependence of steady-state creep rate on stress and temperature, as well as on other parameters such as pressure, grain size, and chemical environment (see Section 18.5). For discussion, it is convenient to divide the results into low-, moderate-, and high-stress deformation regimes. The boundaries vary for different rocks, but for dunite, for example, which is a common mineral in the upper mantle, the moderate-stress range is roughly between 20 MPa and 200 MPa, and it is on the same order of magnitude for many other minerals and rocks.

We discuss the moderate-stress regime first, because this regime is the most intensively investigated and the most commonly applied to deformation in the Earth. The high-stress regime is probably not widely applicable to deformation in the Earth, although it may occur in some areas in association with faulting. The low-stress regime may well be more important than has been supposed for deformation in the Earth. Because deformation proceeds very slowly at these low stresses, however, it is difficult to investigate experimentally and therefore has been ignored by many researchers.

The Moderate-Stress Regime: Power-Law Creep

The constitutive equation that accounts for most moderate-stress, steady-state deformation is the **power-law equation,** so called because the absolute value of the steady-state rate of incremental strain[1] $|_s\dot{\varepsilon}_n|$ is related to the differential stress[2] $_D\sigma$ (Equation 8.47) raised to a power n. The equation is written to give either the strain rate as a function of the stress for constant-stress experiments or the stress as a function of the strain rate for constant-strain-rate experiments:[3]

$$|_s\dot{\varepsilon}_n| = A_1{}_D\sigma^n \exp\left[\frac{-E^*}{RT}\right]$$

$$_D\sigma = K_1|_s\dot{\varepsilon}_n|^{1/n} \exp\left[\frac{E^*}{nRT}\right] \tag{18.4}$$

where A_1 and K_1 are constants, n is the stress exponent, E^* is the activation energy per mole for the creep process, and R is the Boltzmann constant per mole (also called the gas constant). The constant A_1 and K_1 are related by

$$K_1 = \left[\frac{1}{A_1}\right]^{1/n}$$

The constants A_1, E^*, and n (or alternatively K_1, E^*, and n) are characteristic for any particular material. The experimental techniques that enable us to determine these different constants are discussed in Box 18.2. It is clear from Equation (18.4) that both the temperature and the stress have a large effect on the strain rate.

Creep is a thermally activated process, which means there is an energy barrier that inhibits the creep mechanism. The exponential term in Equations (18.4) is a characteristic of such processes, and the activation energy E^* is a measure of the energy barrier. At low temperatures, a high stress is required to overcome the energy barrier and produce a particular strain rate. At sufficiently high temperatures, however, random thermal fluctuations can provide the energy needed to surmount the energy barrier, so the strain rate can be maintained by a lower stress. Thus, in general an increase in temperature increases the strain rate for a constant stress (the first Equation 18.4) or lowers the stress required to produce a given strain rate (the second Equation 18.4). This effect is accounted for by the rapid increase, with increasing temperature, of the exponential term in Equation (18.4), and by the corresponding decrease of the exponential term in the second Equation (18.4).

Figure 18.11 shows the results of a typical series of constant-strain-rate experiments, in this case on dunite (polycrystalline olivine), at three different temperatures. The stress reaches a nearly constant magnitude with increasing strain, indicating steady-state conditions, which are the only conditions to which Equations (18.4) apply. The steady-state stress decreases markedly with increasing temperature.

The yield stress is very difficult to identify. In fact, the initial slopes of the stress–strain curves are all different, indicating that they result from a mixture of elastic strain, primary-creep strain, and the non-rigid behavior of the experimental apparatus. The yield stress of the material, therefore, is probably below the stress at which the curve turns substantially toward horizontal, which normally would be identified as the yield stress.

Figure 18.12 shows the effect of different strain rates at two different constant temperatures. Decreasing the strain rate by an order of magnitude results in a drop in the stress under these conditions by a factor of slightly more than 0.5. This behavior implies stress exponents of $n = 3$ for the 1300 K experiments (Figure 18.12A) and $n = 2.35$ for the 1400 K experiments (see Box 18.2 and Figure 18.2.1). The best fit for a much more ex-

---

[1] We use the left subscript s to indicate steady state.

[2] Remember that from its definition in Equation (8.47), the differential stress is the maximum minus the minimum principal stress, so it is always positive regardless of whether stresses are tensile or compressive. Thus we must use the absolute value for the rate of incremental strain.

[3] We use the standard notation $\exp[-x] \equiv e^{-x} = 1/e^x$ and $\exp[x] \equiv e^x$, where e is the base of the natural logarithm: $e = 2.718$. The exponentials in Equations (18.4) are the Arrhenius factors, which are characteristic of thermally activated processes. They are related to the theoretical probability that random thermal fluctuations can provide enough energy to surmount the energy barrier for a process.

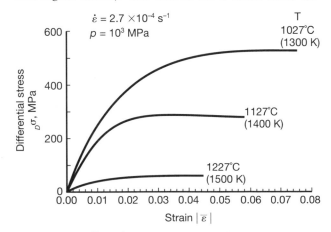

Figure 18.11 Effect of temperature on steady-state creep stress for coarse-grained dunite (average grain size $d = 1$ mm). Experiments were performed at constant strain rate, pressure, and temperature for three different temperatures.

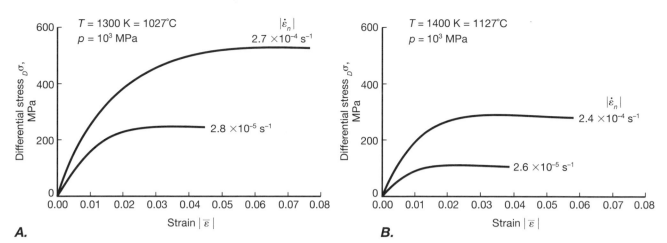

Figure 18.12 The effect of different strain rates on the steady-state stress for dunite at constant temperature and pressure. *A.* Two experiments at $T = 1300$ K, $p = 10^3$ MPa for strain rates of roughly $10^{-4}\,\text{s}^{-1}$ and $10^{-5}\,\text{s}^{-1}$. *B.* Two experiments at $T = 1400$ K, $p = 10^3$ MPa for strain rates of roughly $10^{-4}\,\text{s}^{-1}$ and $10^{-5}\,\text{s}^{-1}$.

tensive set of experiments on olivine is actually about $n = 3.5$. For many polycrystalline materials, including silicates, metals, and oxides, the stress exponents have values roughly between 3 and 5, and some can be as high as 7. Even higher values of the stress exponent are occasionally observed, but these are usually in the high-stress regime where a different form of the stress dependence is more appropriate (see below).

Comparing Figures 18.11 and 18.12 with the stress–strain curve in Figure 18.5*A* indicates that the behavior of a power-law material at a given strain rate, temperature, and pressure approaches that of the elastic–plastic

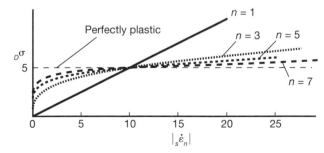

Figure 18.13 Relationship between a general power-law model for steady-state creep rate and a perfectly plastic rheology. Curves for power-law creep are plotted for stress exponents of 1, 3, 5, and 7. The constant in the equation is adjusted to give the same strain rate for each curve at a stress of 5.

model for a continuum. As the stress exponent increases, the steady-state strain rate becomes more sensitive to changes in stress. This behavior is evident when we compare Figure 18.5*A* with Figure 18.13, in which the steady-state strain rate is plotted against stress for power-law creep models with the stress exponents $n = 1, 3, 5$ and 7. As the exponent increases, the power-law behavior approaches the idealized model of perfect plasticity more closely. For real power-law materials, however, the yield point increases with increasing strain rate, and the steady-state strain rate is a determinate function of stress, which is not true for perfectly plastic materials.

A compilation of steady-state creep data for olivine from several different investigators is plotted in Figure 18.14. The strain rate is plotted as a normalized strain rate in order to eliminate the effect of the different temperatures at which experiments were performed. In the range of moderate stresses, the data are fit reasonably well by a straight line with a slope of $1/n = 0.33$ (see the second Equation 18.4), which gives $n = 3$.

The High-Stress Regime: The Exponential Creep Law

At high stresses, roughly $_D\sigma > 200$ MPa, the strain rate becomes increasingly sensitive to differential stress as the stress increases. If we attempt to explain the data with the power-law creep model (Equation 18.4), we find that the stress exponent n increases with increasing stress. The data are better fit by a different creep model called the **exponential creep law**. It has the form

$$|_s\dot{\varepsilon}_n| = A_2 \exp(\beta\,_D\sigma) \exp\left[\frac{-E^*}{RT}\right] \qquad (18.6)$$

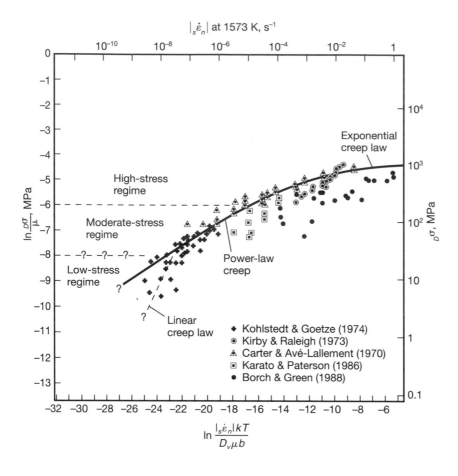

Figure 18.14 Normalized creep data for olivine showing regimes of high, moderate, and low stress. The two most recent data sets indicate slightly lower flow stresses than the older sets, probably reflecting improvements in the experimental technique. The data of Kohlstedt and Goetze are for single crystals. The line that is fit to the older data is a hyperbolic sine creep law; this creep equation has the characteristics of power-law creep in the moderate-stress regime and exponential-law creep in the high-stress regime. Plotting the non dimensional normalized strain rate $|_s\dot{\varepsilon}_n|kT/D_v\mu b$, where $D_v = D_0 \exp[-E^*/RT]$, eliminates the effect of the different temperatures of the experiments from the data plot. In this expression, k is the Boltzmann constant, D_v is the coefficient of volume self-diffusion, μ is the shear modulus, and b is the magnitude of the Burgers vector, a lattice constant. For olivine, we used $\mu = 7.91 \times 10^4$ MPa; $k = 1.38 \times 10^{-29}$ MPa m^3 K^{-1}, $D_v = (10^{-1}$ m^2 s$^{-1})$ exp $[(-5.44 \times 10^5$ J mole$^{-1})/$ $(8.31$ J mole^{-1} K$^{-1})(T$ K$)]$; and $b = 6.98 \times 10^{-10}$ m. The creep rates on the top scale are for 1573 K (1300°C), which is not necessarily the temperature at which the data were collected.

where A_2, β, and E^* are constants whose values are characteristic of different materials. In this case, the steady-state strain rate depends exponentially on the differential stress, which means that it depends on e raised to the power $(\beta _D\sigma)$. This type of creep behavior is evident from the steady-state data in the high-stress regime of Figure 18.14, where the data define a gradual curve, indicating the exponential-law behavior.

The Low-Stress Regime: Power-Law Creep with Low n

At stresses roughly below 20 MPa, the constitutive equation is similar to Equations (18.4), but in this case the stress exponent n has a value between 1 and 2, which is distinctly lower than the values observed for moderate stresses.

Figure 18.15 shows the steady-state creep behavior of the Solenhofen limestone over a range of more than two orders of magnitude in stress and at a variety of temperatures. The low-stress regime occupies most of the graph and shows power-law creep with a stress exponent of about $n = 1.7$. In the moderate-stress regime, the data are consistent with power-law creep with a stress exponent of about $n = 5$. One set of data extends into the high-stress regime and shows exponential-law creep. A similar transition to a low-stress exponent in olivine has been observed at very fine grain sizes (see the discussion of grain size dependence in the next section), and the data on the deformation of single crystals of olivine at the lowest stresses in Figure 18.14 suggest that even in single crystals there is a transition to a lower-stress exponent at the lowest stresses.

These different creep laws reflect different processes dominating at the different levels of stress. We discuss the creep mechanisms in more detail in Chapter 19.

18.5 The Effects of Pressure, Grain Size, and Chemical Environment on Steady-State Creep

Effect of Pressure

The effect of pressure on steady-state creep is relatively small at crustal depths within the Earth, so it is commonly ignored. Within the mantle, however, the effect becomes important. An increase in pressure decreases

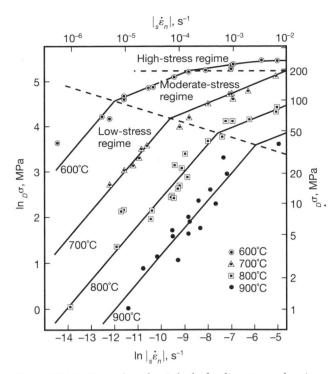

Figure 18.15 Creep data for Solenhofen limestone, showing three different creep laws in the high-, moderate-, and low-stress regimes. Data are from constant-strain-rate tests, and points represent the strain rate and the stress at 10 percent strain, which is at or close to steady state. Different lines are plotted for each temperature because the strain rate is not normalized for temperature differences as in Figure 18.14.

the steady-state strain rate at constant stress, and it increases the steady-state stress at constant strain rate.

Figure 18.16 illustrates the pressure effect with two curves for steady-state creep of olivine at constant strain rate and temperature and at two different pressures. The pressure effect is commonly accounted for by modifying the first Equation (18.4) to read

$$\left|{}_s\dot{\varepsilon}_n\right| = A_1 {}_D\sigma^n \exp\left[\frac{-(E^* + pV^*)}{RT}\right] \qquad (18.7)$$

where p is the pressure, V^* is the activation volume per mole, and $E^* + pV^* = H^*$, the activation enthalpy. V^* is generally interpreted as the volume of crystal affected by the activation process. As the pressure increases, the value of the exponential term decreases, and therefore so does the strain rate at constant stress. For olivine, for example, the activation volume is approximately 2.7×10^{-5} m^3/mole (see Figure 18.2.3 in Box 18.2). At a depth of 30 km, the pressure is roughly 10^9 N/m^2, so the pV^* term is less than 5 percent of the activation energy $E^* = 540,000$ J mole^{-1}. For a temperature of 1000 K, a constant strain rate, and a stress exponent of $n = 3.5$, the differential stress at a depth of 30 km would be only about 2.5 times that at the surface. For the same

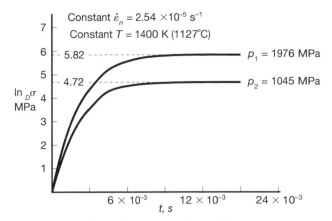

Figure 18.16 Effect of pressure on the steady-state creep stress for dunite. Experiments were run at constant strain rate, temperature, and pressure for two different pressures.

temperature and depth, for a constant differential stress, the strain rate would be about 0.04 times that at the surface (compare the variability of experimental data in Figure 18.14, for example).

Experiments have shown that the dependence of the creep rate of polycrystalline materials on temperature and pressure is predictable if the temperature is normalized by the solidus melting temperature for the material:

$$\left|{}_s\dot{\varepsilon}_n\right| = A_1 {}_D\sigma^n \exp\left[\frac{-gT_m}{T}\right] \qquad (18.8)$$

where g is a dimensionless constant characteristic of a particular material. The pressure dependence is implicit in the equation, because the melting temperature T_m depends on the pressure (it generally increases with increasing pressure). This form of the constitutive equation, in fact, seems to take account of complexities in the pressure dependence of the material behavior that are not adequately modeled by Equation (18.7).

Effect of Grain Size

When coarse-grained polycrystalline solids, such as rocks, are deformed at low homologous temperature in the moderate-to-high-stress regime, the yield stress tends to decrease slightly with increasing grain size. This effect is illustrated in Figure 18.17A for limestones and marbles, in which the room temperature yield stress varies inversely as the square root of the grain diameter. A similar effect is observed at higher temperatures in the moderate-stress regime, where during power-law creep, the steady-state flow stress for fine-grained Solenhofen limestone is considerably higher than it is for coarser-grained marbles.

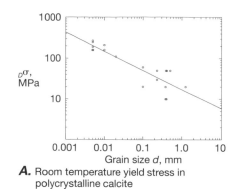

A. Room temperature yield stress in polycrystalline calcite

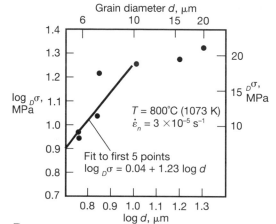

$T = 800°C$ (1073 K)
$\dot{\varepsilon}_n = 3 \times 10^{-5}$ s^{-1}

Fit to first 5 points
$\log {}_D\sigma = 0.04 + 1.23 \log d$

B. Steady-state stress in Solenhofen limestone

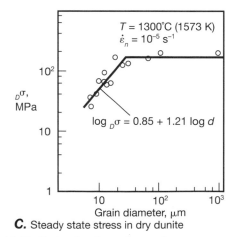

$T = 1300°C$ (1573 K)
$\dot{\varepsilon}_n = 10^{-5}$ s^{-1}

$\log {}_D\sigma = 0.85 + 1.21 \log d$

C. Steady state stress in dry dunite

Figure 18.17 Grain size dependence of creep properties. *A.* Grain size dependence of the yield stress for a variety of calcite rocks of different grain size at room temperature and 100 MPa confining pressure. *B* and *C.* Grain size dependence of the steady-state creep stress during constant-strain-rate experiments for (*B*) fine-grained Solenhofen limestone and (*C*) synthetic dunite.

The opposite effect, however, is observed for fine-grained materials in the low-stress regime, where the stress exponent is less than 2. Here the steady-state stress at a constant strain rate decreases rapidly with decreasing grain size. This effect has been observed experimentally both in fine-grained Solenhofen limestone below a mean grain diameter of 10.5 μm (Figure 18.17*B*) and in a synthetic dunite below a mean grain diameter of about 27 μm (Figure 18.17*C*). For constant temperature and strain rate, a log-linear fit to the data below these grain sizes gives

$$\log {}_D\sigma = B + C \log d \quad \text{or} \quad \log {}_D\sigma \, d^{-C} = B \quad (18.9)$$

where d is the mean grain diameter, and the constants (B, C) are equal to (0.004, 1.23) for the limestone and to (0.85, 1.21) for the dry dunite (Figure 18.17*B, C*). In order to account explicitly for the dependence on grain size in the creep equation, we must modify Equation (18.4) by adding a grain-size-dependent term.

$$|_s\dot{\varepsilon}_n| = A_2 d^{-b} {}_D\sigma^n \exp\left[\frac{-E^*}{RT}\right] \quad (18.10)$$

The second Equation (18.9) implies that the product $({}_D\sigma d^{-C})$ is a constant for constant strain rate and temperature. In Equation (18.10) if we keep the temperature and strain rate constant the product of the stress and the grain size terms ($d^{-b} {}_D\sigma^n$) must also be constant. By raising the former product to the power n and equating the result to the latter product, we conclude that the grain diameter exponent $b = Cn$. Taking $n = 1.7$ and 1.4 as appropriate for the low-stress regimes of Figures 18.15 and 18.14, respectively, therefore gives values of about $b = 2.1$ and 1.7 for the grain diameter exponent in Equation (18.10) for limestone and dry dunite. On theoretical grounds, we expect values of b between 2 and 3, and the data in Figure 18.17*B, C* are consistent with the lower end of this range. Similar values are commonly observed for many polycrystalline materials in comparable conditions.

In Box 18.2 we summarize the experimental techniques used to determine the values of the constants in constitutive equations such as (18.4), (18.7), and (18.10).

Effect of the Chemical Environment

The chemical environment of deformation has a profound effect on the rheology of rocks.

At elevated temperatures and pressures, water dissolves in very small amounts (parts per billion) in the lattices of silicates. In at least some minerals, such as quartz, feldspar, and olivine, it drastically reduces the creep strength, apparently by reducing the activation energy for creep. This so-called **water weakening** or

hydrolytic weakening is notable in quartz, which is extremely strong when it is dry but becomes relatively weak when water is dissolved in the crystal lattice. The effect is pressure-dependent because the solubility of water in the silicate lattices increases with increasing pressure. The presence of water tends also to decrease the melting temperature of a silicate, and this effect seems to account for the change in creep behavior through Equation (18.8). Because of the common occurrence of water in crustal rocks, it seems likely that it has a strong influence on deformation in fault zones and in crustal metamorphic terranes.

Silicates, of course, are all compounds of oxygen. Experiments have demonstrated that the partial pressure of oxygen, which is a measure of the amount of oxygen in the environment, strongly affects the creep properties. Figure 18.18 shows, for example, that as the partial pressure of oxygen increases in a series of constant-strain-rate experiments on quartz, the creep strength decreases dramatically. A similar dependence has been demonstrated for olivine.

The fact that the chemical environment can have important effects on the deformation properties of a mineral severely complicates the application of laboratory experiments to the Earth. We must know not only the temperature, pressure, stress, and grain size but also the chemical environment of the mineral before we can apply a creep law with confidence. It may be possible, however, to account for the chemical environment through Equation (18.8) if the effect on the melting temperature is known. In a large fraction of published creep experiments, the chemical environment has not been carefully controlled, so this influence on the results is not known. The different chemical environments associated with different experimental setups may explain part of the variation in results reported by different laboratories for comparable materials (see Figure 18.14).

Nevertheless, these deformation experiments have not been futile. They have given us good approximate data for the magnitude of creep strengths to be expected in the Earth and for the dependence of those strengths on the physical and chemical environment. We can expect that the added complications will ultimately enable us to determine even more about the conditions within the Earth and to explain with greater precision the processes that occur and the structures that we observe in deformed rocks.

18.6 Application of Experimental Rheology to Natural Deformation

Equations that describe the flow of rocks such as (18.2), (18.10), (18.3.4), (18.3.7), and (18.3.8) all show that the stress is related to the strain rate and not to the finite strain. This has important implications for interpreting structures of ductile deformation that we observe in the rocks. In deformed rocks, we generally see a record only of the finite strain, which is the sum of all the deformations the rocks have experienced. Because the strain rate is essentially the incremental strain per unit time, and because the finite strain ellipse does not provide any information about the deformation path and therefore the incremental strain ellipse, it is incorrect to presume that the structures can be directly interpreted in terms of the orientations of stress. For example, one cannot assume in general that the maximum shortening direction of the finite strain is parallel to the maximum compressive stress.

Only for coaxial deformations do the principal axes of finite strain remain parallel to those of incremental strain and of strain rate. For isotropic materials, the principal stresses are parallel to the corresponding principal strain rates, as shown by the fact that in Equations (18.3.4), (18.3.7), and (18.3.8) a particular stress com-

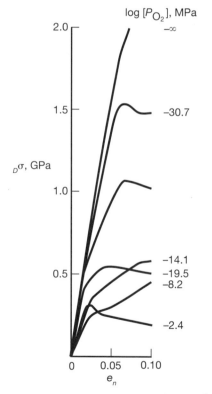

Figure 18.18 Constant-strain-rate experiments for quartz, showing the effect of the partial pressure of oxygen on the creep strength. Experiments were done at 10^{-5} s^{-1}, 800°C, and a confining pressure of 1.64 GPa (*note:* 1 GPa = 10^3 MPa). Curves are labeled with the value of log [P_{O_2}], the logarithm of the oxygen partial pressure.

Box 18.3 Constitutive Equations in Three Dimensions

Although one-dimensional constitutive equations are useful for explaining the different models of material behavior, the complete constitutive relationships must take into account all three dimensions, and accordingly the equations are more complex. We summarize here the three-dimensional constitutive relationships for isotropic elastic, viscous, plastic, and power law continua. Isotropic materials are the simplest because they have the same mechanical properties in all directions. In anisotropic materials, the response of the material is different in different directions, and the constitutive equations become even more complex.

ELASTIC BEHAVIOR

For three-dimensional elasticity, we need six equations, one relating each independent component of the stress to the strain components, or (conversely) one relating each of the independent strain components to the stress components. For such materials two elastic moduli, called the **Lamé parameters** λ and μ, are required to describe the mechanical behavior (anisotropic materials require more than two constants).

Normal-stress components   Shear-stress components

$$\sigma_{11} = \lambda e_v + 2\mu e_{11} \qquad \sigma_{12} = 2\mu e_{12}$$

$$\sigma_{22} = \lambda e_v + 2\mu e_{22} \qquad \sigma_{13} = 2\mu e_{13}$$

$$\sigma_{33} = \lambda e_v + 2\mu e_{33} \qquad \sigma_{23} = 2\mu e_{23}$$

$$(18.3.1)$$

The equations for normal-stress components are on the left, and those for shear-stress components are on the right. There are two terms in each equation for the normal stresses, because a normal stress generally produces both a change in volume and a change in shape. The first term accounts for the volume change, in which e_v is the volumetric extension ($e_v = e_{11} + e_{22} + e_{33}$; see Equations 9.6 and 15.16).

The second term accounts for the change in shape. Shear stresses, on the other hand, produce only changes in shape and therefore, are related to the strain by only one elastic constant.

The six equations can be summarized in compact notation by the single equation

$$\sigma_{k\ell} = \lambda e_v \delta_{k\ell} + 2\mu e_{k\ell} \qquad (18.3.2)$$

where Equations in (18.3.1) are obtained from Equation (18.3.2) by using different combinations of the subscript values chosen from ($k = 1$, 2, or 3) and ($\ell = 1$, 2, or 3), and where $\delta_{k\ell}$ is a tensor called the Kronecker delta, whose components equal 1 for $k = \ell$ and equal 0 for $k \neq \ell$.

The Lamé parameter μ in Equations (18.3.1 and 18.3.2) is the same as the shear modulus in Equation (18.1). Young's modulus E and Poisson's ratio v, both of which appear in Equation (9.5) and the first of which appears in Equation (18.1), are related to the Lamé parameters by

$$E = \frac{\mu(3\lambda + 2\mu)}{\lambda + \mu} \qquad v = \frac{\lambda}{2(\lambda + \mu)} \qquad (18.3.3)$$

Thus one could rewrite Equations (18.3.1) to express λ and μ in terms of E and v, and it is strictly a matter of convenience which elastic moduli are used to describe the behavior.

Although most minerals in rocks are anisotropic, the average elastic constants for the rock are isotropic if the mineral grains are randomly oriented. A rock with a preferred orientation of mineral grains such as a schist, however, is mechanically anisotropic—a fact that is often ignored for the sake of simplicity in modeling.

VISCOUS BEHAVIOR

To describe constant-volume viscous behavior in three dimensions, we must specify how each of the six independent components of the stress and of the strain rate are related. Therefore, we need six independent

ponent is determined by the corresponding strain rate component. If a material is anisotropic, however, even this relationship does not hold.

Thus only if the entire deformation history of the rock was strictly coaxial, and only if the rock is and always was isotropic could the principal axes of finite strain be parallel to those of stress. Most deformation histories, however, probably are not strictly coaxial, and most rocks probably are not isotropic. Thus in interpreting structures, it is safest to restrict the interpretation to the observable characteristics, which are the finite strains, and not to make simplified assumptions regarding the geometry of the deformation path or the mechanical properties of the material at the time of deformation in order to discuss the stress orientations.

In comparing the results of experimental deformation of rocks with the continuum models we discuss in Section 18.1, we see that when the stress exponent n is close to 1, the steady-state behavior of the material exhibits approximately a linear relationship between

equations. These six can be summarized as one equation:

$$\Delta\sigma_{k\ell} = 2\eta\dot{\varepsilon}_{k\ell} \qquad (18.3.4)$$

where $\Delta\sigma_{k\ell}$ is the deviatoric stress tensor defined by Equations (8.43) and (8.45), and $\dot{\varepsilon}_{k\ell}$ is the rate of incremental strain tensor (see Box 18.1). Here the subscripts k and ℓ both take on values of 1, 2, or 3, and all possible combinations of these subscript values produce nine equations. Three of the six equations for the shear stresses are redundant, however, because of the symmetry of the stress tensor and the strain rate tensor (see Equations 8.24 and 15.12).

The volumetric strain associated with the flow of a fluid is usually very small and therefore is generally ignored. By using the deviatoric stress in the constitutive equation, we ignore any effect of the pressure, which is appropriate if the volumetric deformation is negligible. Equation (18.3.4) is similar in form to the equation for elasticity (Equation 18.3.2) except that the volumetric strain is omitted.

PLASTIC BEHAVIOR

In three dimensions, the constitutive equation for perfectly plastic behavior is written in terms of the deviatoric stress $\Delta\sigma_{k\ell}$ because, again, the mean normal stress does not affect the behavior of the material. Thus the yield criterion is that the second invariant of the deviatoric stress tensor I_2 must be equal to a constant K^2.

$$I_2 \equiv \sum_{k=1}^{3} \sum_{\ell=1}^{3} (\Delta\sigma_{k\ell})(\Delta\sigma_{k\ell}) = K^2 \qquad (18.3.5)$$

The double summation implies summation over all combinations of values of k and ℓ. In principal coordinates, this relationship, expressed in terms of the components of the full stress tensor, can be written

$$I_2 = \tfrac{1}{3}[(\hat{\sigma}_1 - \hat{\sigma}_2)^2 + (\hat{\sigma}_2 - \hat{\sigma}_3)^2 + (\hat{\sigma}_3 - \hat{\sigma}_1)^2] = \tau_{oc}^2 \qquad (18.3.6)$$

τ_{oc} is the shear stress on the octahedral planes, which are the planes that are equally inclined to all three principal stress axes. The normal stress on those planes is $\bar{\sigma}_n$. Thus this yield criterion is simply a generalization of the requirement that the maximum shear stress be equal to a constant. In two dimensions, I_2 is directly related to the square of the radius of the Mohr circle (see Equation 8.27).

The strain rate is constrained by the requirement

$$\dot{\varepsilon}_{k\ell} = \frac{\sqrt{E_2}}{K}\sigma_{k\ell} \qquad (18.3.7)$$

where E_2 is the second invariant of the strain rate tensor $\dot{\varepsilon}_{k\ell}$ defined in a similar manner to I_2 in Equation (18.3.5). Equation (18.3.7) requires the principal axes of the strain rate to be parallel to those of the stress, but the complete strain rate is indeterminate, because E_2 is not constrained by the relationship. In fact, by forming the second invariant of both sides of the equation, we simply recover Equation (18.3.5), which is therefore implicit in Equation (18.3.7).

POWER LAW BEHAVIOR

A generalized tensor form of the power-law equation can be written in terms of the second invariant of the deviatoric stress I_2.

$$|s\dot{\varepsilon}_{k\ell}| = A_3 I_2^{0.5(n-1)}\Delta\sigma_{k\ell}\exp\left[\frac{E^*}{RT}\right] \qquad (18.3.8)$$

$$I_2 = \sum_{i=1}^{3} \sum_{j=1}^{3} (\Delta\sigma_{ij})(\Delta\sigma_{ij}) \qquad (18.3.9)$$

In the case of confined compression for which $\hat{\sigma}_2 = \hat{\sigma}_3$, $\sqrt{I_2}$ reduces to a constant times the differential stress $_D\sigma$, and in principal coordinates for $(k, \ell) = (1, 1)$, Equations (18.3.8 and 18.3.9) reduce to the first Equation (18.4).

---

stress and strain rate—that is, viscous behavior. When n has a value of 3 or more, on the other hand, the steady-state behavior shows a stress–strain-rate curve that resembles that of plastic behavior (see Figure 18.13), the yield stress in effect being exponentially dependent on the temperature.

The rheological equations discussed in the preceding sections are strictly one-dimensional. A more general description of material behavior requires that the constitutive equations be formulated for three-dimensional

stress and strain. The resulting equations are considerably more complex than the one-dimensional equations, and we discuss some of these generalizations in Box 18.3.

Most experiments investigating rock rheology have used either monomineralic rocks such as quartzite, dunite, or marble, or single crystals of particular minerals, such as quartz, olivine, calcite, or feldspar. A few experiments, however, have used polymineralic rocks such as granite (see Table 18.1). Unfortunately, there is no way to predict accurately the behavior of a polyminer-alic rock from data for its constituent minerals or the rheologic effect of a change in mineralogy. If a rock consists predominantly of one mineral, then we may assume that the rheology of the whole rock is governed by that mineral. A small percentage of a significantly weaker mineral, however, may strongly affect the ductile behavior of a rock. Our inability to predict this phenomenon fundamentally restricts our attempts to model accurately the behavior of Earth materials.

It is worth asking how the common experimental strain rates of 10^{-7} s^{-1} or higher compare with natural geologic strain rates. Consider the strain rates associated with plate tectonics. The spreading rates of modern oceanic ridges average about 5 cm per year. If we assume that this displacement is accommodated by shear distributed linearly across the asthenosphere, and that the asthenosphere is approximately 200 km thick, then it is simple to calculate the shear strain rate:

$$\frac{5 \text{ cm/yr}}{200 \text{ km}} = \frac{1.6 \times 10^{-7} \text{ cm/sec}}{2 \times 10^{7} \text{ cm}} = 0.8 \times 10^{-14} \text{ s}^{-1}$$

It is difficult to change this number by a factor of 10 without making unreasonable assumptions. Thus a strain rate of 10^{-14} s^{-1} seems a reasonable order-of-magnitude approximation for mantle deformation associated with convection and plate tectonics. Comparison of the total strains measured in deformed crustal rocks with the stratigraphically constrained times available for the deformation gives comparable orders of magnitude, although higher rates occur in faults and ductile shear zones. In general, we expect geologic deformation to proceed at strain rates between 10^{-12} s^{-1} and 10^{-15} s^{-1}.

Thus geologic strain rates are generally 5 to 10 orders of magnitude lower than the rates practical for laboratory experiments. This difference prompts us to ask whether experimentally determined rheologies are applicable to natural situations. Figure 18.15, which shows the rheology for Solenhofen limestone, illustrates the problem. If experiments had been restricted to conditions that produced creep only in the moderate-stress regime (the zone of power-law creep), then extrapolation to geologic conditions would lead to an erroneous inference of much lower strain rates at a given stress, or much higher stresses at a given strain rate. Whether the same or different rheology dominates the behavior at stresses and strain rates beyond the reach of experiment is a persistent problem that we face when trying to apply experiments to real Earth processes.

There is a method for attacking this problem indirectly. It is to understand the mechanisms by which ductile deformation occurs and to identify microscopic and submicroscopic features in the rock or in the crystal lattices that characterize the different ductile processes.

Comparing microstructures in experimentally deformed samples with those observed in naturally deformed rocks could reveal whether similar deformation mechanisms operated in the rocks and, therefore, whether the laboratory data may be extrapolated to natural conditions. We discuss this approach in more detail in Chapter 19.

Comparing observed natural deformation rates with the predictions based on laboratory-determined rheology is another possible means of checking the applicability of laboratory data. For example, the rates of uplift of the Earth's surface in response to unloading caused by the melting of the continental ice sheets from the last ice age can be determined by dating raised beaches along shorelines that were near or under the ice. These rates reflect flow in the mantle under loads that can be estimated from knowledge of the original area and thickness of the ice. Because the response to those loads is governed by the rheology of the mantle, the uplift data constrain the possible rheology and therefore offer a clue to the relevance of laboratory data. Some analyses of such data indicate that the mantle behaves as a linearly viscous fluid with a viscosity[4] of about 10^{20} Pa s. This value is high compared to that of more familiar geologic fluids such as basaltic magma, which is on the order of 3800 Pa s to 7 Pa s, for temperatures roughly between 1100°C and 1400°C, and water, which is about 10^{-3} Pa s. Given the size of the Earth, however, and the great lengths of time available, even this high a viscosity is sufficiently low to allow convective flow in the solid mantle. This conclusion is highly controversial, however—in part because it implies that most laboratory experiments on olivine, which give a stress exponent of about 3, cannot be extrapolated to mantle conditions.

Thus, although experimental data provide some of the best information available on the rheology of rocks, we must be aware that indiscriminate application of those results to the Earth can lead to error.

Experimentally determined values of n, E^*, and K_1 (the first Equation 18.4) for a variety of rock types are summarized in Table 18.1. These data are only approximate, because there is much variability among different experiments on the same material, but they give an order-of-magnitude estimate of how rocks might behave in the Earth.

The variation of mechanical properties with depth is of major interest in our efforts to understand the tectonic behavior of the Earth's lithosphere. To determine the variation of properties with depth, we must account for the increase of both temperature and pressure, and the change in mineralogic composition. The

---

[4] The SI unit of viscosity is the Pascal-second (Pa s), or (N s)/m^2; the cgs unit is the poise, or (dyne s)/cm^2. 1 Pa s = 10 poise.

Table 18.1 Examples of Material Constants for Steady-State Power-Law Flow of Selected Rocks in the Moderate-Stress Regime

Material	n	E^*, kJ mole^{-1}	K_1, MPa^{-n} s^{-1}
Rock salt	5.3	102	6.29
Marble[a] (20–100 MPa)	7.6	418	5.07×10^{-28}
Marble[a] (< 20 MPa)	4.2	427	1.98×10^{-9}
Quartzite	2.4	156	6.31×10^{-6}
Quartzite (wet)	2.3	154	2.52×10^{-4}
Granite	3.2	123	1.26×10^{-9}
Granite (wet)	1.9	137	2.0×10^{-4}
Quartz diorite	2.4	219	1.26×10^{-3}
Diabase	3.4	260	2.02×10^{-4}
Albite rock	3.9	234	2.59×10^{-6}
Anorthosite	3.2	238	3.27×10^{-4}
Dunite (dry)	3.0	540	4.0×10^{6}
Dunite (wet)[b]	3.0	420	1.9×10^{3}

Source: After Kirby (1983) and Ranalli and Muphy (1987).

[a] From Schmid et al. (1980); the applicable range of differential stress $_D\sigma$ is listed.

[b] Chopra and Paterson (1981); Karato and Paterson (1986).

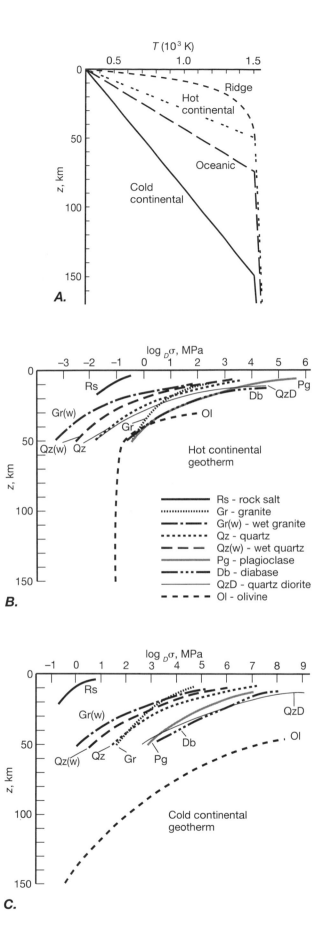

pressure effect for lithospheric depths (100 km or less) is relatively small and for a first approximation can be ignored. The temperature increase, however, is a major effect, and the inferred geothermal gradient can vary significantly from one tectonic setting to another. Figure 18.19A shows simplified geotherms for regions below midoceanic ridges, hot continental areas such as the Basin and Range province of the western United States, oceanic areas underlain by mature lithosphere (greater than 50 million years old), and cold continental areas (such as continental shield areas). The stress required to maintain a strain rate of 10^{-14} s^{-1} during power-law creep for a variety of rock types (see Table 18.1) is plotted in Figure 18.19B for the hot continental geotherm and in Figure 18.19C for the cold continental geotherm. Note that the quartz-rich rock types are generally weaker than the plagioclase-rich rock types.

At shallow depths, the low temperatures require high stresses to maintain the creep rate, but creep deformation is unlikely because fracture and cataclastic

Figure 18.19 (*Right*) Variation of power-law creep rheology with depth for various rock types listed in Table 18.1. *A.* Simplified linearized geotherms for a midocean ridge, hot continental lithosphere, oceanic lithosphere, and cold continental lithosphere. *B* and *C*. Steady-state creep stress required to maintain a creep rate of 10^{-14} s^{-1} for (*B*) the hot continental geotherm in part *A* and for (*C*) the cold continental geotherm in part *A*.

flow can occur at lower stress. The differential stress at which these brittle processes occur, however, increases with increasing pressure (see for example Figure 9.11), and the differential stress required for ductile creep decreases with increasing temperature (see Figure 18.11 and the second Equation 18.4). Because both pressure and temperature increase with increasing depth, there must be a transition region at which brittle deformation gives way to ductile deformation. This is the so-called brittle–ductile transition (see Section 9.4 and Figure 9.9). The fact that earthquakes occur no deeper than 15 to 20 km along the strike–slip San Andreas fault of California, for example (see Chapter 7), is probably because the brittle–ductile transition occurs at that depth.

The actual variation of strength with depth thus depends on the deformation mechanism as well as on the type of rock. Figures 18.20A and B show two possible distributions of crustal strength based on the assumption of Coulomb-type brittle behavior at low temperature and pressure, and power-law creep at high temperature and pressure (Table 18.1). Figure 18.20A reflects a hot continental geotherm with a quartz/granitic crust and a crust-mantle boundary (Moho) at 30 km. Figure 18.20B is for a cold continental geotherm with a 40-km thick quartz/granitic crust. Models assuming other distributions of rock type with depth would result in different distributions of crustal strength with depth, which could include distributions with two strength maxima within the crust. The most significant factors, however, are the maximum in strength at the brittle–ductile transition in the crust and a second maximum at the Moho, where the rock composition changes from quartzo-feldspathic rocks to peridotite dominated by olivine.

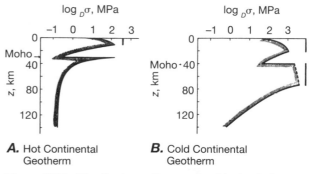

Figure 18.20 Distributions of strength with depth for two different models of the lithosphere. *A.* Model of a quartz/granite crust in a hot continental geotherm with the Moho at a depth of 30 km. *B.* Model of a quartz/granite crust in a cold continental geotherm with the Moho at 40 km.

This model differs considerably from the homogeneous elastic crustal model assumed by Hafner (see Figures 10.14 through 10.16), and to that extent, Hafner's models must be regarded as over simplifications. On the other hand, as noted in the preceding discussion, the extrapolation of laboratory data over many orders of magnitude in strain rate is potentially unreliable. In addition, these models do not take into account possible chemical stratification of the crust, changes in the mineralogy caused by metamorphic reactions, or the potential for changes in rheology caused by metamorphic reactions during deformation. Thus the models presented above must also be considered oversimplifications and approximations.

Additional Readings

Amin, K. E., A. K. Mukherjee, and J. E. Dorn. 1970. A universal law for high-temperature, diffusion-controlled transient creep. *Jr. Mech. Phys. Solids* 18: 413–426.

Borch, R. S, and H. W. Green II. 1989. Deformation of peridotite at high pressure in a new molten salt cell: Comparison of traditional and homologous temperature treatments. *Phys. of the Earth and Planet. Int.* 55: 269–276.

Carter, N. 1976. Steady-state flow of rock. *Rev. of Geophys. and Space Phys.* 14: 301–360.

Carter, N. L., and H. G. Avé-Lallement. 1970. High-temperature flow of dunite and peridotite. *Geol. Soc. Am. Bull.* 81: 2181

Carter, N. L., and S. H. Kirby. 1978. Transient creep and semi-brittle behavior of crystalline rocks. *Pure Appl. Geophys.* 116: 807–839.

Cathles, L. M., III. 1975. *The viscosity of the earth's mantle.* Princeton, N. J.: Princeton University Press.

Chopra, P. N., and M. S. Paterson. 1981. The experimental deformation of dunite. *Tectonophysics* 78: 453–473.

Dieterich, J. 1981. Constitutive properties of faults with simulated gouge, in N. L. Carter, M. Friedman, J. M. Logan, and D.W. Stearns (eds.), *Mechanical behavior of crustal rocks.* Geophysical Monograph 24. Washington, D. C.: American Geophysical Union, pp. 103–120.

Garofalo, F. 1965. *Fundamentals of creep and creep rupture of metals.* New York: Macmillan.

Green, H. W., II, and R. S. Borch. 1987. The pressure dependence of creep. *Acta Metall.* 35 (6): 1301–1305.

Griggs, D. T. 1967. Hydrolytic weakening of quartz and other silicates. *Jr. Roy. Astron. Soc.* 14: 19–31.

Johnson, A. M. 1970. *Physical processes in geology.* San Francisco: Freeman, Cooper.

Karato, S.-I., and M. S. Paterson. 1986. Rheology of synthetic olivine aggregates: Influence of grain size and water. *Jr. Geophys. Res.* 91 (B8): 8151–8176.

Kirby, S. H. 1983. Rheology of the lithosphere. *Rev. of Geophys. and Space Physics* 21: 1458–1487.

Kirby, S. H., and C. B. Raleigh. 1973. Mechanisms of high temperature, solid-state flow in minerals and ceramics and their bearing on the creep behavior of the mantle. *Tectonophysics* 19: 165

Kohlstedt, D. L., and C. Goetze. 1974; Low-stress high-temperature creep in olivine single crystals. *J. Geophys. Res.* 79: 2054.

Mukherjee, A. K., J. E. Bird, and J. E. Dorn. 1969. Experimental correlations for high-temperature creep. *Trans. ASM* 62: 155–179.

Olsson, W. A. 1974. Grain size dependence of yield stress in marble. *Jr. Geophys. Res.* 79: 4859–4862.

Ord, A., and B. E. Hobbs. 1986. In H. C. Heard and B. E. Hobbs, eds., *Mineral and rock deformation: Laboratory studies—The Paterson Volume.* Geophysical Monograph 36. Washington, D. C.: Am. Geophys. Union.

Pfiffner, O. A., and J. G. Ramsay. 1982. Constraints on geological strain rates: Arguments from finite strain states of naturally deformed rocks. *Jr. Geophys. Res.* 87: 311–321.

Ranalli, G., and D. C. Murphy. 1987. Rheological stratification of the lithosphere. *Tectonophysics* 132: 281–296.

Ross, J. V., et al. 1979. Activation volume for creep in the upper mantle. *Science* 203: 261–263.

Schmid S. M., J. N. Boland, and M. S. Paterson. 1977. Superplastic flow in fine-grained limestone. *Tectonophysics* 43: 257–291.

Schmid S. M., M. S. Paterson, and J. N. Boland. 1980. High-temperature flow and dynamic recrystallization in Carrara marble. *Tectonophysics* 65: 245–280.

Takeuchi, S., and A. S. Argon. 1976. Review: Steady-state creep of single-phase crystalline matter at high temperatures. *Jr. Mater. Sci.* 11: 1542–1566.

Tullis, J. 1979. High-temperature deformation of rocks. *Rev. of Geophys. and Space Phys.* 17: 1137–1154.

Twiss, R. J. 1976. Structural superplastic creep and linear viscosity in the Earth's mantle. *Earth and Planet. Sci. Letts.* 33: 86–100.

Microscopic Aspects of Ductile Deformation

Mechanisms and Fabrics

In previous chapters we have considered rock deformation from the continuum point of view, which assumes the rock is homogeneous and has no discontinuities in structure or in mechanical properties. In fact rocks are made of crystal grains, which are discontinuous at the grain boundaries and deform by the migration of crystal defects. With the assumption of homogeneity, therefore, we can describe only the behavior averaged over a volume that is large compared to the grain size. In this chapter, however, we explore the processes of deformation at the microscopic and submicroscopic scale.

We wish to answer several questions: What mechanisms permit solid rocks to flow? Under what conditions do these mechanisms operate? What rheology (that is, what relationship between strain rate and stress) is associated with each of these mechanisms? What microscopic and submicroscopic structures in the rock reflect the deformation mechanisms that produced them? What can we infer from these structures about the conditions of deformation?

Macroscopic flow of rocks can result from a number of different mechanisms, each of which may dominate the rheology of a rock at a particular range of physical conditions such as temperature, confining pressure, and differential stress. The conditions of normalized stress and homologous temperature at which some of the flow mechanisms dominate can be plotted on a **deformation mechanism map** or, more simply, a **defor-**mation map. Figure 19.1 shows deformation maps for quartz, calcite, and olivine. Quartz is one of the most common minerals in the crust; calcite is of major importance in sedimentary sequences, as limestone and marble; and olivine is the most abundant mineral in the upper mantle. The mechanical behavior of these minerals probably dominates much of the deformation that we observe. The heavy lines indicate conditions under which the two mechanisms in adjacent fields of the graph contribute equally. The fine lines are contours of the strain rate at which the material would deform for any specified stress and temperature. The expected range of strain rates in the Earth is shaded.

These maps are not based on the specific rheologic constants listed in Table 18.1 and are undoubtedly incorrect in detail. Evidence from natural rocks suggests that for quartz and calcite, dislocation creep may dominate at lower stresses and temperatures than these maps suggest. Moreover, the maps do not include all known deformation mechanisms, nor do they account for all conditions that rocks encounter in the natural environment. For example, hydrolytic weakening (see Section 18.5), which is not accounted for in these maps, reduces the stress and temperature at which dislocation creep can dominate the deformation process in quartz and olivine. Moreover, the maps are constructed for a particular grain size and therefore do not indicate the effect of changing grain size (see Section 18.5). Thus the maps

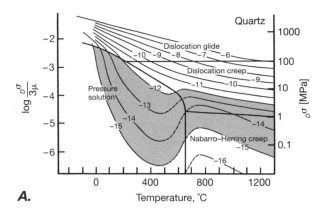

A.

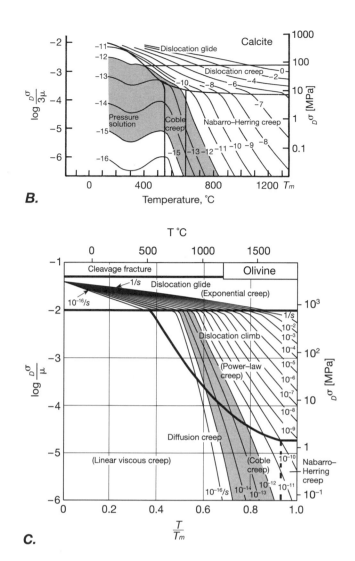

B.

C.

Figure 19.1 Deformation mechanism maps. Plot of normalized stress versus temperature, showing the fields in which different mechanisms dominate the strain rate, and the boundaries between the fields at which the strain rates from the two mechanisms are equal. Fine lines are contours of strain rate. Expectable geologic strain rates (10^{-12} s^{-1} to 10^{-15} s^{-1}) are shaded. A. Deformation map for wet quartzite with 100-μm grain size. Stress scale at right is evaluated for the shear modulus μ at 900°C. B. Deformation map for marble with 100-μm grain size. Stress scale at right is evaluated for the shear modulus μ at 500°C. C. Deformation map for olivine with 100-μm grain size. Temperature scale is homologous temperature.

should be considered at best only a qualitative illustration of the rheologic behavior of the different minerals.

The motions of linear crystal defects called dislocations, or diffusive mass-transport processes, account for most mechanisms on the deformation maps in Figure 19.1. Exponential-law creep and power-law creep result from dislocation motion in the high- and moderate-stress regimes, respectively. At lower stresses, deformation is accommodated by diffusion either along grain boundaries (solution creep and Coble creep) or through the volume of crystals (Nabarro–Herring creep). At high stresses, especially in the upper crust where temperature and pressure are low, cataclastic flow dominates the rheology, but this mechanism is not plotted on the deformation maps (see Table 4.1 and Sections 9.4, 18.3, and 19.1).

The **fabric** of a rock is defined by the geometric organization of the structures in the rock.[1] The preferred orientations of fold axes and axial surfaces, foliations, lineations, joints, veins, and similar macroscopic structures constitute the **macrofabric**. The preferred orientations of crystallographic axes of mineral grains, the preferred orientations of the principal dimensional axes of elongate or tabular grains, and the characteristics of

the shape and arrangement of crystal grains, or the **texture**, constitute the **microfabric**. On the submicroscopic scale, the fabric is defined by the **substructure**, which includes the arrangement of linear and planar imperfections in the crystal lattices. A **fabric element** is a structure that constitutes part of the fabric of a rock. Any fabric element must be a penetrative feature of a volume of rock that is large compared to the scale of the structure.

Most mechanisms of deformation produce some sort of microfabric in the rock. Accordingly, in this chapter we discuss the different deformation mechanisms, many of which determine the forms of the deformation maps in Figure 19.1, and the characteristic microscopic and submicroscopic fabrics that result from their operation. We look first at mechanisms that characteristically operate at low temperatures, then at diffusion and solution mechanisms, and finally at dislocation mechanisms and their associated microfabrics.

---

[1] Different authors use the term *fabric* in slightly different ways. We adopt a broad definition of the term.

19.1 Mechanisms of Low-Temperature Deformation

Elastic Behavior

Elastic behavior involves the interatomic or interionic forces and the potential energy of the atoms or ions in a crystal lattice. The position of any ion in a lattice is one of minimum potential energy that results from the balance of attractive and repulsive forces. For a simple crystal such as salt (NaCl), for example, the oppositely charged ions Na^+ and Cl^- are attracted to each other by long-range electrostatic forces associated with the net charge on the whole ion. The positively charged nuclei of both ions, however, are repelled from each other by a repulsive force that, because of the shielding of the electron cloud, is not effective at large distances. The equilibrium spacing of the ions in the crystal lattice[2] occurs where these two forces balance each other.

During elastic deformation, the ions are forced out of their positions of minimum potential energy. Young's modulus of elasticity is a measure of the increase in potential energy for a given applied stress. Removal of the stress causes the lattice to relax to its original position of minimum potential energy. Thus elastic deformation disappears when the stress is removed.

Friction

The physical mechanism associated with friction depends on the nature of the contact between the two surfaces. Because no surface is perfectly flat, the contact between two surfaces is supported by asperities (Figure 19.2; see also Section 14.6), and the actual contact area is much smaller than the nominal area of the interface. Thus the stress is concentrated in the asperities and is much higher than the nominal applied stress. The behavior of the asperities governs the friction.

The asperity model of a frictional interface can account for the observed behavior of friction in fault gouge (see Section 18.3). The normal stress, concentrated on the supporting asperities, causes them to creep, presumably by some ductile mechanism, thereby progressively increasing the area of contact between the surfaces. The more interlocked the surfaces become as a result of the creep of the asperities, the higher the frictional resistance to sliding. The decrease in friction with sliding and with increasing velocity of sliding can be accounted for by the shearing off of asperities and by the generation of a thin layer of gouge, both of which reduce the interlocking of the two surfaces.

Granular Flow

Granular flow involves the rolling and sliding of rigid particles past one another. It characterizes the deformation of either unlithified sedimentary layers or slurries and occurs only at low effective confining pressures.

When particles move past one another, they cannot remain in a close-packed arrangement such as is shown in Figure 19.9. As such particles are displaced, the space between them increases, so the volume of the body must increase as well. High effective confining pressure suppresses granular flow, because an increase in volume requires that work be done against the effective confining pressure. Moreover, the sliding of particles past one another involves friction, and frictional resistance to sliding increases with increasing normal stress across the sliding surface. Thus the higher the effective confining pressure, the more work is expended in dilation and frictional sliding, and the less efficient the granular flow mechanism becomes.

High pore-fluid pressure lowers the effective confining pressure by reducing the normal stress supported by grain-to-grain contacts (see Section 9.5). Thus even at considerable depths, a pore-fluid pressure approaching the minimum compressive stress permits granular flow.

Granular flow is important in soft-sediment deformation, such as the slumping of sediment layers on the ocean floor, and it is probably the mechanism of deformation in at least some parts of the accretionary prisms that accumulate above subduction zones. The effects of granular flow are difficult to detect in rock microfabrics. If the rocks are deformed (for example, folded) but the mineral grains themselves show no evidence of deformation, granular flow may have occurred. Subsequent crystalline flow and recrystallization, however, would make it impossible to recognize the earlier operation of granular flow from the microfabric.

Cataclastic Flow

Cataclastic flow is a process of deformation that involves continuous brittle fracturing of grains in a rock, with attendant frictional sliding and possibly rolling of the fractured particles past one another (see Sections 4.2, 9.4, and 18.3). Although it results in a macroscopically coherent and permanent deformation, it differs

---

[2] It is worth remembering that a mineral is simply a three-dimensional network of ions arranged such that ionic charges balance and the ions are at an equilibrium spacing. Silicate minerals fundamentally consist of SiO_4 tetrahedra that are linked in particular arrangements by sharing various numbers of oxygens of the tetrahedron. These linked structures are in turn bound together by other ions necessary to preserve charge balance. The nature and strength of the various bonds determine the resistance of the mineral to deformation.

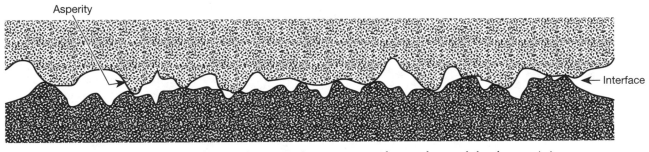

Figure 19.2 Asperities support the contact between two surfaces and control the characteristics of friction.

from granular flow in that small-scale fracturing is an integral part of the process. It occurs instead of granular flow when the work required to fracture grains is less than that required to slide or roll them past one another. Cataclastic flow therefore occurs where effective confining pressures are too high to permit pure granular flow. There exists between the two processes a continuous gradation that involves different relative amounts of fracturing, sliding, and rolling. Cataclastic flow differs rheologically from ductile flow (discussed below) in that for cataclastic flow, the stress required to maintain a given strain rate increases strongly with increasing effective confining pressure.

Cataclastic fabrics are characterized by the sharp angular shapes of the clasts and grains, by pervasive cracks, by a broad spectrum of grain and clast sizes, and commonly by an absence of any foliation (see Figure 4.5B and Table 4.1). Cataclastic flow produces a progressive decrease in the grain size and a progressive increase in the volume fraction of cataclastic material through the continual comminution of larger grains into smaller and through the fracturing and abrasion of the sides of the fault zone. For individual rock types, a correlation exists between displacement on a fault and the thickness of gouge in the fault, and in principle a correlation also exists between grain size and total strain, although such simple relationships are complicated by numerous other factors.

Pseudotachylites (see Figure 4.6 and Table 4.1) are unfoliated rocks of glass or cryptocrystalline material that are commonly associated with cataclasites and that occur as veins along some fault zones. We interpret the glass to be a quenched melt of the rock that forms locally by frictional heating when the rock is dry and slip rates are sufficiently high. These conditions are most likely to occur during fault slip events associated with earthquakes. The resulting melt intrudes through small fractures in the adjacent rock before being quenched.

Although the process by which cataclasites form seems clear, our understanding of the rheology of this process is based strictly on experimental work (see Section 18.3). Theoretical models for the rheology associated with cataclastic flow have not been developed.

19.2 Twin Gliding

Many minerals have a crystal structure that is capable of forming a twin across particular crystallographic planes. The twinned structure may be related to the original structure by a mirror reflection across the twin plane, such as in calcite, or by rotation of the original structure about an axis normal to the twin plane, such as the albite twin in plagioclase. Twins may form during either crystallization or deformation.

Twin gliding is a process whereby twins are produced from an original structure by a simple shearing parallel to the twin plane. Calcite twins commonly form in this manner, as shown in Figure 19.3A–C. Figure 19.3A shows the untwinned calcite structure; the dots represent the Ca^{2+} ions, and the short lines represent the edges of the CO_3^{2-} planes. Figure 19.3B shows the calcite structure after a simple shear deformation of the Ca^{2+} lattice parallel to the e twin plane, defined by the Miller indices $\{\bar{1}012\}$ (see footnote to Table 19.1). The resulting structure is not quite a twin because of the orientation of the CO_3^{2-} radicals. A minor rotation of the CO_3^{2-} planes, however (Figure 19.3C), brings the structure into a perfect reflection twin.

In calcite, this twin gliding occurs at a relatively low shear stress (10 MPa) on the twin plane. Twinning an entire calcite crystal grain makes possible the accumulation of a maximum shear strain of $e_s = 0.35$. Greater strains must be accommodated by another mechanism. Thus the mechanism does not give rise to a steady-state creep, so it is not represented on the deformation map for calcite (Figure 19.1B)

The ease of twinning in calcite, however, makes twin gliding a significant deformation mechanism at low temperatures in limestones and marbles. Rocks thus deformed have abundant twins, most of which form in an orientation of high incremental shear strain and presumably high shear stress. Because total strains associated with this mechanism typically are not large, the twin planes do not rotate far from the orientation in which they formed. Thus the twin planes and the slip directions of twinning can be used to infer the orientations of the principal incremental strains and, by fur-

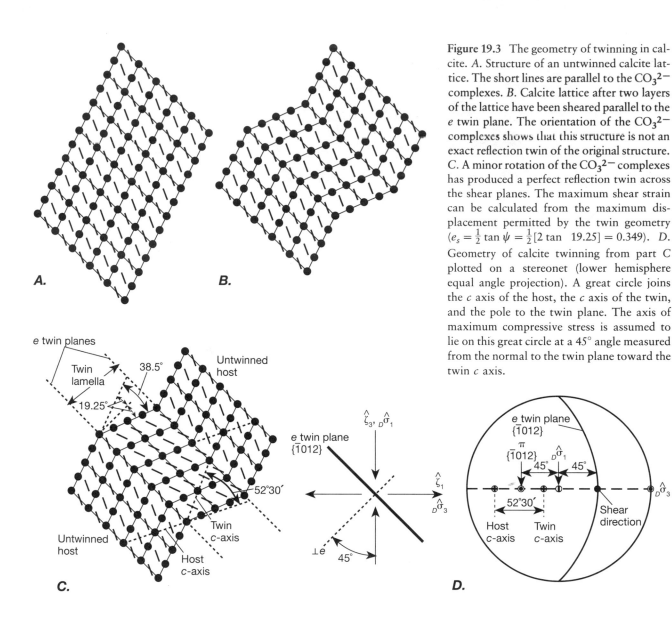

Figure 19.3 The geometry of twinning in calcite. *A.* Structure of an untwinned calcite lattice. The short lines are parallel to the CO_3^{2-} complexes. *B.* Calcite lattice after two layers of the lattice have been sheared parallel to the *e* twin plane. The orientation of the CO_3^{2-} complexes shows that this structure is not an exact reflection twin of the original structure. *C.* A minor rotation of the CO_3^{2-} complexes has produced a perfect reflection twin across the shear planes. The maximum shear strain can be calculated from the maximum displacement permitted by the twin geometry $(e_s = \frac{1}{2} \tan \psi = \frac{1}{2}[2 \tan 19.25] = 0.349)$. *D.* Geometry of calcite twinning from part *C* plotted on a stereonet (lower hemisphere equal angle projection). A great circle joins the *c* axis of the host, the *c* axis of the twin, and the pole to the twin plane. The axis of maximum compressive stress is assumed to lie on this great circle at a 45° angle measured from the normal to the twin plane toward the twin *c* axis.

ther assumption, of the principal stresses associated with deformation of the rocks (Figure 19.3*D*). The dispersion of twin plane orientations can also be used to constrain the magnitude of the differential stress (see Box 19.2).

19.3 Diffusion and Solution Creep

In **diffusion creep,** deformation takes place by the actual transfer of material from areas of high compressive stress to areas of low compressive stress. Diffusion creep may result from the diffusion of point defects through a crystal lattice, the diffusion of atoms or ions along grain boundaries, or the diffusion of dissolved com-

ponents in a fluid along the grain boundaries. Diffusion also occurs together with grain boundary sliding to greatly enhance the possible rate of strain.

Point defects in crystal lattices include both interstitials and vacancies. **Interstitials** are extra atoms that are stuffed between the lattice sites of a crystal (Figure 19.4*A*). The lattice is deformed around an interstitial because it must expand to accommodate the defect. A **vacancy** is a lattice site that is unoccupied by any atom (Figure 19.4*B*). The lattice around a vacancy is also deformed because it tends to collapse into the void. In general, an interstitial is a higher-energy defect than a vacancy, and the concentration of stable vacancies in a lattice increases with increasing temperature. As a result, vacancies are more common and more important in ductile flow.

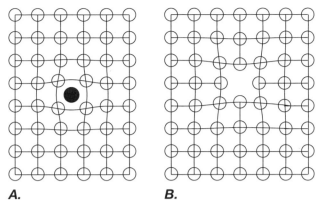

A.   **B.**

Figure 19.4 Point defects in a crystal lattice. *A.* An interstitial. The black circle represents an extra atom in the crystal, causing distortion of the lattice. *B.* A vacancy. The lattice tends to collapse in around the vacancy, thereby distorting the structure.

If an atom jumps from its lattice site into an adjacent vacant site, it fills that site but leaves a vacant site behind, in effect causing the vacancy to jump in the opposite direction from the atom. (Figure 19.5). Thus the movement of vacancies in one direction is equivalent to the movement of atoms in the opposite direction.

Nabarro–Herring Creep

The motion of vacancies is a thermally activated process, which means that vacancies make random jumps from site to site through the lattice with a frequency that depends on the temperature (see footnote 3 in Section 18.4). If differential stress is applied to a crystal, vacancies are created at the surface of the crystal where the compressive stress is a minimum (Figure 19.6*A*, *B*), and they are destroyed at the surface where the compressive stress is a maximum (Figure 19.6*A*, *C*). The resulting concentration gradient of the vacancies causes a diffusive flux: The vacancies move from the surface of low compressive stress toward the surface of high compressive stress. The flux of atoms is opposite to that of the vacancies, and a crystal grain changes shape by atoms moving away from the high-stressed face (Figure 19.6*C*) and accumulating on the low-stressed face (Figure 19.6*B*).

The deformation that results is called **Nabarro–Herring creep**, and it is characterized by a linear relationship between the strain rate and the differential stress (Box 19.1). A solid deforming by this mechanism is comparable to a Newtonian, or linearly viscous, fluid (see Section 18.1). Figure 19.1 indicates that Nabarro–Herring creep is an effective deformation mechanism

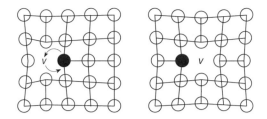

Figure 19.5 The motion of a vacancy (indicated by "v") from one lattice site to an adjacent site is opposite to the motion of the atom from the adjacent site into the vacant site. Thus material diffuses in a direction opposite to that of the vacancies.

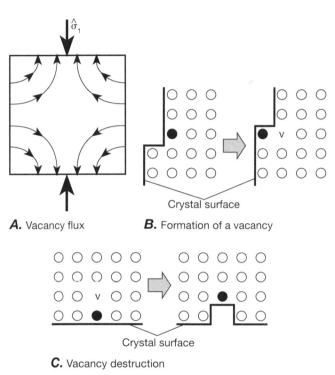

A. Vacancy flux   **B.** Formation of a vacancy

C. Vacancy destruction

Figure 19.6 Nabarro–Herring creep caused by volume diffusion of vacancies in response to a uniaxial compressive stress. *A.* Under a uniaxial compressive stress, vacancies diffuse toward the surface of highest normal stress along the indicated paths through the crystal; atoms diffuse in the opposite direction. *B.* The process of creation of a vacancy at a crystal surface of minimum compressive stress. The solid lines mark the surface of the crystal, which is never perfectly planar on the atomic scale. The solid circle marks the ion whose position changes during the process to create the vacancy (marked v). The surface thus gets gradually built out, lengthening the crystal normal to the compressive stress, and the vacancies diffuse away toward a surface of high compressive stress. *C.* The process of destruction of a vacancy at a crystal surface of maximum compressive stress. Symbols have the same meaning as in part *B*. Removal of atoms from the surface gradually results in displacement of the surface, shortening the crystal parallel to the maximum compressive stress.

Box 19.1 Rheologies Inferred from Mechanisms of Ductile Deformation

One method for trying to understand the mechanisms of ductile deformation is to make models of how the process might function and to use those models to derive an equation describing the theoretical rheology. The model, of course, is based on observation of the physical phenomenon, and one test of whether the model is appropriate is whether the resulting theoretical rheology is comparable to the observed rheology. In this box, we give descriptive sketches of various models of ductile deformation and look at the rheological equations derived from such models to see how they compare with observation.

The model for *Nabarro–Herring creep* is that deformation occurs by a flux of vacancies through the crystal lattice. Thus the strain rate depends on the vacancy density and the rate of migration. The vacancy density is dependent on temperature, and it also varies at different surfaces of the crystal grain, depending on the magnitude of the normal component of stress at the surface. Under nonhydrostatic stress, therefore, the vacancy concentration differs at differently stressed surfaces, setting up a concentration gradient down which the vacancies diffuse. The strain rate is then related to the vacancy flux, which yields, for the theoretical rheology,

$$|_s\dot{\varepsilon}_n| = \frac{6V_v^* {}_vD_0}{RT} d^{-2} {}_D\sigma \exp\left[\frac{-H_v^*}{RT}\right] \quad (19.1.1)$$

where ${}_D\sigma$ is the differential stress; d is the grain diameter; T is the absolute temperature; $H_v^* = E_v^* + pV_v^*$ is the activation enthalpy for self-diffusion through the volume, which equals the activation energy E_v^* plus the pressure p times the activation volume V_v^*; ${}_vD_0$ is the diffusion constant for self-diffusion through the volume of the crystal; and R is the Boltzmann constant for a mole of material (the "gas constant").

This equation has several interesting features. First, it has a form very close to Equation (18.10) with $n = 1$, except that in Equation (18.10) the constant A_2 does not include an inverse temperature dependence. Second, the stress exponent is 1, which means that this is a creep mechanism that produces the rheology of a Newtonian viscous fluid. Third, the strain rate depends inversely on the square of the grain diameter, which means that as the grain size increases, this mechanism rapidly becomes less efficient.

A similar approach for *Coble creep* and for *solution creep* differs mainly in that the diffusion path must be along the grain boundaries rather than through the volume. The theoretical model for Coble creep leads to the rheological equation

$$|_s\dot{\varepsilon}_n| = \frac{6V_b^* {}_bD_0\delta}{RT} d^{-3} {}_D\sigma \exp\left[\frac{-H_b^*}{RT}\right] \quad (19.1.2)$$

and a similar form results for solution creep. This equation is very similar to that for Nabarro–Herring creep with three exceptions. (1) ${}_bD_0$, H_b^*, and V_b^* are the diffusion constant, the activation enthalpy, and the activation volume, respectively, for grain boundary diffusion rather than volume diffusion. (2) The parameter δ appears in the coefficient; it is the thick-

---

only at high homologous temperatures and low stresses. In olivine, for example, it is effective only near the melting temperature (Figure 19.1C).

Coble Creep

Diffusion of atoms occurs not only through the volume of crystals but also along grain boundaries. Diffusion is more rapid along grain boundaries than through the volume, largely because the activation energy for grain boundary diffusion is roughly two-thirds that for volume diffusion. Thus, although the diffusion path for an atom or ion around the sides of a grain may be longer than the path directly through the volume, the higher rate of grain boundary diffusion can make the former mechanism the more efficient of the two.

The process is the same as for volume diffusion: Atoms diffuse away from surfaces subjected to high

compressive stress, and they accumulate on surfaces of low compressive stress. The resulting **Coble creep** also exhibits a linear stress–strain-rate relationship. Coble creep is effective at lower temperatures than Nabarro–Herring creep because of its lower activation energy. The two fields are separated by a line of constant homologous temperature (Figure 19.1).

Solution Creep

During **solution creep**, mineral grains dissolve more readily at faces under high compressive stress, ions diffuse through the fluid phase on the grain boundary, and they precipitate on surfaces of low compressive stress. Evidence for solution creep and for the mobility of components of the rock comes from the study of folds (Section 12.3), foliations (Sections 14.3 and 17.2), and lineations (Sections 14.3 and 14.6).

ness of the grain boundary. (3) The strain rate depends on the inverse third power of the grain diameter, which means that as grain size increases, the efficiency of Coble creep decreases more rapidly than that of Nabarro–Herring creep. Nevertheless, this diffusive mechanism also provides a Newtonian viscous rheology.

One model for the mechanism of dislocation creep, known as **Weertman creep,** assumes that the rate of dislocation glide is limited by the rate at which dislocations can climb. The strain rate then depends on the density of dislocations ρ, the magnitude of the Burgers vector b which defines the amount of slip associated with each dislocation, and the climb velocity of the dislocations ν_c.

$$|_s\dot{\varepsilon}_n| = \beta b \rho \nu_c \qquad (19.1.3)$$

where β is a geometric constant. Theoretically, the dislocation density should vary as the square of the differential stress. That is,

$$\rho = \alpha_D \sigma^2/(\mu b)^2 \qquad (19.1.4)$$

where α is a constant, μ is the shear modulus, and b is the magnitude of the Burgers vector (see Box 19.2). The climb velocity depends on the differential stress and on the coefficient of volume self-diffusion, which in turn is a thermally activated quantity.

$$\nu_c \propto {}_D\sigma_\nu D_0 \exp\left[\frac{-H^*}{RT}\right] \qquad (19.1.5)$$

Substituting these dependencies into Equation (19.1.3) gives a power-law relationship that has the form

$$|_s\dot{\varepsilon}_n| = \frac{\beta_0}{kT} D \sigma^3 \exp\left[\frac{-H^*}{RT}\right] \qquad (19.1.6)$$

where β_0 is a combination of other constants, and k is the Boltzmann constant. This equation is identical in form to Equation (18.7) with $n = 3$ except for the inverse temperature dependence of the first term. Other assumptions can be made that give rise to a stress exponent of $n = 4.5$, and empirical values are commonly between 3 and 5.

For Harper–Dorn creep Equation (19.1.4) does not apply, because the dislocation density apparently is independent of stress. The rheological equation that results from substituting the relation (19.1.5) into Equation (19.1.3) is a linear relationship between strain rate and stress.

A theoretical model based on kinetic theory has been used to calculate the rate of glide of dislocations through a crystal lattice when the glide rate is controlled by obstructions. The equation has the form

$$|_s\dot{\varepsilon}_n| = A_3 \exp\left[1 - \frac{D\sigma}{\sigma_0}\right] \exp\left[\frac{-E^*}{RT}\right] \qquad (19.1.7)$$

where A_3 is a constant, and σ_0 is the differential flow stress at 0 K. The dependence of the creep rate on an exponential function of the stress is comparable to the experimentally determined Equation (18.6), which suggests that the model of dislocation glide may be appropriate.

Solution creep probably is very similar to Coble creep, the principal difference being that the diffusivity is higher along fluid-filled grain boundaries than along dry grain boundaries. Like the other diffusion creep mechanisms, solution creep probably has a Newtonian viscous rheology. On the deformation mechanism maps for calcite and quartz, the rheology estimated for solution creep restricts (Figure 19.1B) or eliminates (Figure 19.1A) the field of Coble creep. Solution creep is the dominant deformation mechanism at temperatures of several hundred degrees Celsius or less, such as would be found in low-grade metamorphic rocks and at the strain rates expected for geologic deformation. This result is in accordance with the field evidence for this mechanism of deformation.

For any of these three diffusion mechanisms, increasing the grain size increases the length of the diffusive path and therefore decreases the efficiency of the deformation mechanism (Box 19.1). In high-grade meta-

morphic regions and in the mantle, it seems very likely that the mean grain size of deforming rocks is larger than the 100 μm (0.1 mm) assumed for the maps in Figure 19.1. Thus the actual range of conditions at which diffusion mechanisms operate in the Earth is probably considerably more restricted than shown. For a tenfold increase in grain size, strain rates decrease by a factor of 100 for Nabarro–Herring creep and by a factor of 1000 for Coble and solution creep. [This result derives from the dependence of the creep rate on the grain size d in Equations (19.1.1) and (19.1.2).] In such a case, the dislocation creep field would expand at the expense of the diffusion creep fields, and the Nabarro–Herring field would expand into the Coble and solution creep fields.

During any diffusive mass-transport process, grains tend to shorten parallel to the maximum compressive stress and to lengthen parallel to the minimum compressive stress (Figure 19.7). As the shape of the grains changes, the lengths of the diffusive paths increase, and

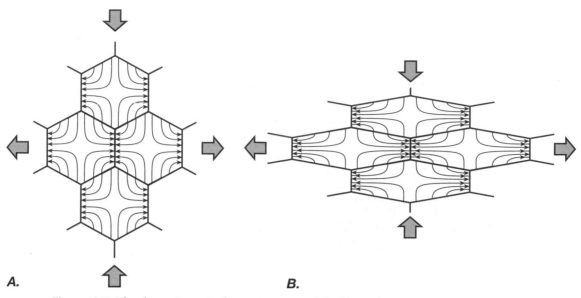

A.   **B.**

Figure 19.7 The change in grain dimension accomplished by Nabarro–Herring creep increases the lengths of the diffusive paths of the material (indicated by the curved arrows) and leads to a slowing down of the diffusion creep. A similar geometry results from grain boundary diffusion.

consequently the strain rate decreases. At high temperatures, however, grain boundary migration becomes important (see Section 19.5), and this process could maintain a roughly equant grain shape, thereby allowing diffusion creep to continue as a steady-state process.

Solution creep, of course, requires the presence of an aqueous fluid in the rocks. This is common in metamorphic rocks, but if deformation occurs under dry conditions, then solution creep is impossible, and the diffusion creep field in the deformation maps is considerably smaller.

Superplastic Creep

Grain boundaries in a polycrystalline material can also be important because sliding on the boundaries can accommodate strain. If voids are not to open along the boundaries, however, the shapes of the grains must change slightly as they slide past one another, and this deformation can be accommodated by diffusion (Figure 19.8). **Superplastic creep** is a phenomenon observed in some metals and inferred for some rocks. It results from coherent grain boundary sliding in which deformation occurs without the opening of gaps or pores between adjacent crystal grains. It is characterized by rapid strain rates at low stresses, compared to other mechanisms of ductile deformation, and by a power-law rheology with a stress exponent between 1 and 2.

Figure 19.8 shows one possible mechanism of superplastic creep for four isolated grains in a two-dimensional polycrystal. The changes in grain shape

required to make the grain boundary sliding possible are indicated by the isolated grains in the diagram. The shape changes from the solid to the dashed outline by diffusion along paths indicated by the curved arrows of the material shaded black. The straight arrows at the grain centroids indicate the direction of motion of the grains as a result of grain boundary sliding. In the diagram, grains 2 and 4 start out as neighbors, and after the grain boundary sliding, grains 1 and 3 have become neighbors. Like other diffusion creep mechanisms, this mechanism results in a linear viscous rheology. However, because the average diffusive path lengths are shorter than in Coble or Nabarro–Herring creep, and because the grains slide along their boundaries, the strain rate is theoretically about five times higher than for Nabarro–Herring or Coble creep. In this mechanism, the grains remain approximately equant, so large strains can be accommodated (Figure 19.8; compare Figure 19.7). The effect of this process on the deformation maps of Figure 19.1 is to increase the strain rate at any given temperature within the diffusion creep field and to increase the area of this field at the expense of the dislocation creep field.

Other mechanisms may operate during superplastic creep to accommodate grain boundary sliding. If the grains change shape by the motion of dislocations, which we discuss in the next sections, a power-law rheology is predicted with a stress exponent of 2. Such a relationship is commonly observed during superplastic creep in metals. It is possible for both the diffusion and the dislocation mechanisms to be active at the same time in accommodating grain boundary sliding.

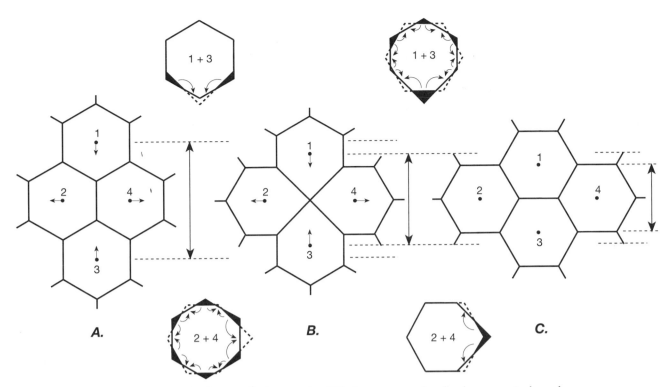

Figure 19.8 Model of superplastic creep by diffusion-accommodated coherent grain boundary sliding. The curved arrows in the isolated grains indicate the diffusive paths by which material is transferred in order to change the shape of the grains from part A to parts B and C. Solid lines show the initial grain shape, dashed lines the final shape. Black shading indicates the part of the old crystal that must diffuse to fill in the area outlined by the dashed lines. This change of shape permits coherent grain boundary sliding by which neighboring grains become separated and non-neighboring grains become neighbors: a "neighbor switching" event.

19.4 Linear Crystal Defects: The Geometry and Motion of Dislocations

The ductile deformation of a crystal causes a change in its shape but no change in its crystalline nature. We can visualize how this process might occur in a perfect crystal by considering the model shown in Figure 19.9. Deformation could proceed by the shifting of lattice points along a crystallographic glide plane, as illustrated in the progression from Figure 19.9A to B. Repeating such shifts on parallel planes in the crystal would allow accumulation of large homogeneous strains (Figure 19.9C). Each step would require a very large stress, because all atoms above the glide plane would have to be lifted at the same time out of their potential-energy wells and moved over the adjacent atoms lying below the glide plane before they could settle back into an equivalent lattice site. Because of this large stress, the theoretical yield strength of a perfect crystal is very large.

The observed yield strengths for real crystals, however, are roughly one order of magnitude smaller than theoretical strengths for perfect crystals, which suggests

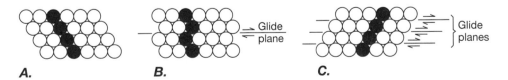

Figure 19.9 The theoretical process of slip in a perfect crystal by the sliding of an entire lattice plane one lattice distance over the adjacent lattice plane. Black circles identify the same set of atoms in each diagram. A. Low-energy close-packed array of atoms. B. Low-energy close-packed array after slip of one lattice distance along the glide plane. C. Homogeneous deformation due to sliding on all adjacent glide planes.

the presence of crystal defects. For the two-dimensional crystal in Figure 19.9, shearing only one lattice point at a time on the glide plane, instead of an entire row requires a much smaller stress. This process, however, introduces into the lattice a defect that separates sheared from unsheared parts. Such defects are linear features in three dimensions which are called **dislocations.** The propagation of a dislocation across an entire glide plane has the net effect of shearing the entire glide plane one lattice spacing as in Figure 19.9B. The motion of dislocations through a crystal lattice is probably the most important mechanism for producing ductile deformation in crystalline materials.

Dislocation Geometry

The two principal types of linear defects that occur in crystal lattices are edge dislocations and screw dislocations. Both types of dislocation mark the boundary between sheared and unsheared crystal. They are distinguished by whether the sheared part of the crystal has moved perpendicular or parallel to the dislocation line.

In the perfect crystal shown in Figure 19.10A, we produce dislocations by shearing the shaded volume along the glide plane relative to the rest of the crystal. A dislocation develops along the interior edge of each

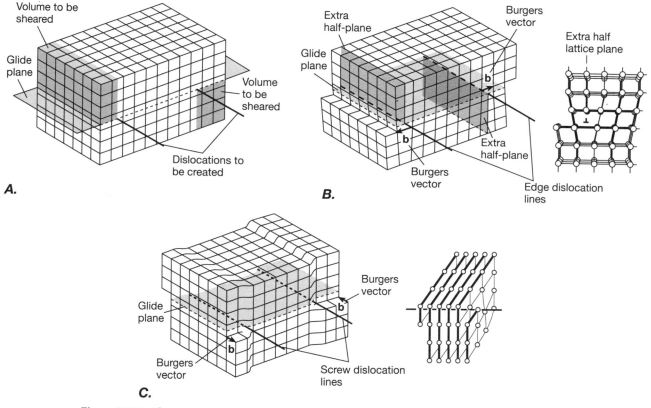

Figure 19.10 The geometry of edge dislocations and screw dislocations. The two examples of each type of dislocation in the lattice blocks of parts *B* and *C* are dislocations of opposite sign. Here **b** is the Burgers vector for each dislocation. *A.* A perfect crystal lattice into which we introduce edge dislocations (part *B*) or screw dislocations (part *C*). The dislocations will lie in the labeled glide plane along the interior edge of the shaded volumes, and each dislocation will separate the area of the glide plane along which slip has occurred from the area on which there has been no slip. *B.* Edge dislocations of opposite sign (note the orientation of the Burgers vectors) produced by shearing of the shaded volumes of crystal lattice in part *A* in a direction *perpendicular* to its interior edge. Each dislocation is at the edge of an extra half-plane of lattice points in the crystal lattice. The diagram to the right shows the crystal structure looking down the edge dislocation line. The inverted T symbol is a standard symbol for an edge dislocation with the stem representing the extra half-plane, and the cross-bar representing the glide plane. *C.* Screw dislocations of opposite sign produced by shearing the shaded volumes of the crystal lattice in part *A* one lattice dimension in a direction *parallel* to its interior edge. The diagram to the right shows that the crystal lattice planes form a continuous helical surface around the dislocation line.

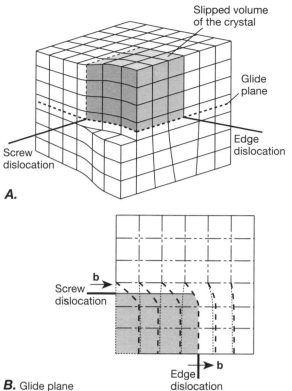

A.

B. Glide plane

Figure 19.11 Boundaries to slipped portions of a crystal lattice. *A.* The shaded volume of the crystal has slipped relative to the unshaded portion. The boundaries on the glide plane are an edge and a screw dislocation. *B.* View of the glide plane. The shaded area is the area over which slip has occurred. Where the Burgers vector is normal to the boundary of the slipped area, the boundary is an edge dislocation; where it is parallel to the boundary, the boundary is a screw dislocation. Dotted and dashed lines show lattice planes below and above the glide plane, respectively.

shaded volume, so it lies on the glide plane at the boundary between the sheared area and the unsheared area of the glide plane. To produce an edge dislocation, we shear the shaded volume by one lattice spacing *perpendicular* to its interior edge (Figure 19.10*B*). The shearing leaves an extra half-plane of atoms on one side of the glide plane at the boundary between the sheared and the unsheared lattice, and the edge of that extra half-plane is the edge dislocation. To produce a screw dislocation, we shear the shaded volume by one lattice spacing *parallel* to its interior edge (Figure 19.10*C*). The shearing leaves the initially planar arrays of lattice points shifted into a continuous spiral whose axis is the screw dislocation line.

The slip vector that is characteristic of a dislocation is called the **Burgers vector**[3] (Figure 19.10). It is parallel to the glide plane, has a length of one lattice spacing, and is perpendicular to an edge dislocation and parallel

---

[3] After the Dutch–American materials scientist J. M. Burgers.

to a screw dislocation. Both edge and screw dislocations can develop on a given glide plane with two different signs. Any counterclockwise circuit around a dislocation consisting of an equal number of lattice steps to the left, down, to the right, and up does not close. The vector from the end of the circuit to the starting point is the characteristic Burgers vector for the dislocation, and its orientation defines the sign of the dislocation. Dislocations of opposite sign on the same glide plane are shown for edge dislocations in Figure 19.10*B* and for screw dislocations in Figure 19.10*C*. Note that in each case the Burgers vectors **b** have opposite directions, although the convention for deciding which dislocation is positive and which negative is arbitrary.

Because a dislocation marks the boundary on a glide plane between slipped and unslipped portions of the crystal lattice, it cannot simply stop inside the crystal. It must either continue to the edge of the crystal or form a closed loop. A dislocation can change character from edge to screw, however, depending on its orientation. In Figure 19.11*A* the slip is limited to a rectangular corner of the glide plane. Where the slip vector—that is, the Burgers vector—is perpendicular to the boundary of the slipped region, the dislocation is an edge (Figure 19.11*B*); where it is parallel to the boundary, the dislocation is a screw.

Dislocations need not be straight lines through the lattice. Curvature of a dislocation results from offsets in the dislocation line. The offsets can be short segments of either edge or screw dislocation, called jogs and kinks, respectively. For example, in Figure 19.12*A*, the slipped region of the glide plane is bounded by a curved dislocation composed of segments of edge dislocations offset by kinks and segments of screw dislocation offset

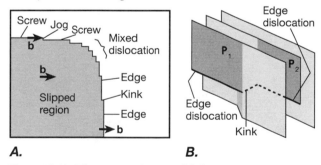

A.   **B.**

Figure 19.12 The nature of a curved dislocation. *A.* Curvature of an edge dislocation within the slip plane can be ascribed to an increasing density of kinks along the edge. At moderate angles to the Burgers vector, the dislocation line takes on mixed character, and at low angles it becomes a screw dislocation with edgelike jogs along it. *B.* Geometry of a kink in an edge dislocation. The portions of the planes that are darkly shaded and labeled P_1 and P_2 are the portions that are extra half-planes in the crystal lattice. The edge dislocation is offset by the kink. The kink has the character of a short segment of screw dislocation, as indicated by tracing a circuit around the kink segment which forms a helical path from plane P_1 to P_2.

by jogs. Where neither type of dislocation is dominant, the dislocation is of mixed character. Figure 19.12*B* shows the detailed geometry of a kink in an edge dislocation. Jogs also permit both types of dislocations to curve out of their glide planes.

In the diagrams illustrating dislocation geometry, we have used a cubic lattice. In many materials (including silicates), however, the chemical compositions and the crystal structures are much more complex. For such structures, the principles of dislocation geometry remain the same except that instead of individual atoms and planes of atoms, we must consider unit cells and planes of unit cells of the crystal structure. In complex crystals, the Burgers vector is the spacing of the unit cell rather than the interatomic spacings implied by the diagrams of simple structures.

Although dislocations are too small to be directly observable with an optical microscope, several techniques provide direct evidence for them. In some materials they can be made observable by decorating them with another material. For example, when natural olivine is held at elevated temperature in an oxidizing atmosphere for a short period of time, oxygen diffuses rapidly along the dislocations and oxidizes some of the iron in the olivine. The small lines of oxides are then visible under an ordinary petrographic microscope, revealing the dislocation structure (Figure 19.13*A*).

When a polished surface of a crystal is chemically etched, the strained material near an emerging dislocation reacts with the etchant more vigorously than the unstrained material, thereby producing a pit in the surface. Such etch pits are readily observable with a reflecting microscope (Figure 19.13*B*).

Dislocations in very thin foils of material can be imaged directly in a transmission electron microscope. The image results from the interaction of the electron beam with the strain field around a dislocation. Under ideal conditions, the characteristics of this interaction can even be used to determine the orientation of the Burgers vector of a dislocation and thus to identify its nature, but in normal electron micrographs the two types of dislocations cannot be distinguished. Figure 19.13*C* shows an electron beam image of dislocations in quartz.

The Motion of Dislocations

Ductile deformation can occur in crystal lattices by propagation of enormous numbers of dislocations through the lattice. We look now at the ways in which

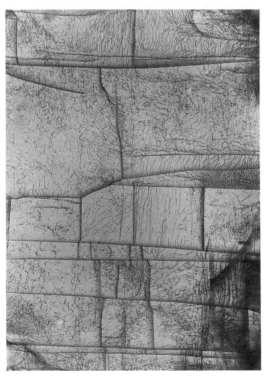

A.

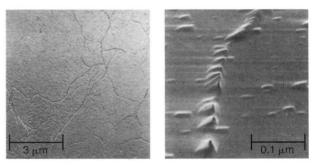

B.

C.

Figure 19.13 Images of dislocations in silicates. *A.* Dislocations in olivine decorated by oxidation and imaged under the petrographic microscope. The fine linear features within the rectangular cells are the dislocation lines. *B.* Etch pits in quartz, showing locations on the crystal surface where dislocations intersect the surface. Left: low magnification; right: high magnification. *C.* Electron microscope image of curved and tangled dislocations in quartz.

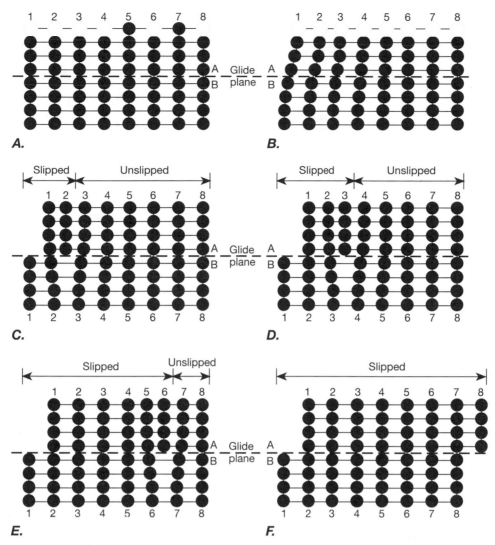

Figure 19.14 The glide of a dislocation through the crystal lattice is accomplished by switching the bonds of neighboring atoms across the glide plane. Parts *A* through *F* illustrate the process of forming an edge dislocation and propagating it through the crystal to produce one Burgers vector of offset in the crystal. Crystallographic planes in the undeformed crystal are numbered 1 through 8, and the letters A and B refer to the parts of the planes above and below the slip plane, respectively.

dislocations produce ductile flow and at some of the simple interactions that can occur between and among dislocations as they move through the crystal.

Figure 19.14 illustrates how **glide** of an edge dislocation through a crystal produces ductile deformation. Figure 19.14*A* shows a perfect lattice with one of the glide planes marked. Columns of atoms are numbered from 1 to 8, and the letters A and B indicate the parts of the column above (A) and below (B) the glide plane, respectively. A shear stress applied to the crystal (Figure 19.14*B*) breaks the bonds crossing the glide plane in columns 1 and 2 (Figure 19.14*C*); column 1A then forms a new bond with column 2B. Column 2A becomes an extra half-plane with an edge dislocation in the glide plane. Continued shear stress (Figure 19.14*D*) causes

column 3B to break from 3A and then to join with 2A, leaving 3A as the extra half-plane and moving the dislocation one lattice spacing to the right. Several more such bond switching events (Figure 19.14*E*) result in the dislocation gliding completely out of the crystal, leaving a step one Burgers vector wide on the surface of the crystal (Figure 19.14*F*). Propagation of many dislocations through the crystal from one side to the other on the set of parallel glide planes can result in a significant amount of ductile shear of the structure.

The model in Figure 19.14 is only two-dimensional, so in fact a whole row of bond switching events must accompany each step made by the dislocation. Because only one row of bonds is broken and reformed for each Burgers vector that a dislocation moves, the motion of a dislocation requires much less stress than the shearing of a whole sheet of atoms all at once. In fact, the stress required for advancing a whole dislocation line is minimized by propagating kinks or jogs down the dislocation line, thereby requiring the breaking of only one bond at a time.

A screw dislocation and an edge dislocation that have the same Burgers vector propagate in perpendicular directions under the same shear stress. For example, in Figure 19.12A the edge dislocation propagates to the right whereas the screw propagates up, leaving behind a region of crystal that has slipped one Burgers vector. Under a given stress, dislocations of opposite sign propagate in opposite directions.

A set of parallel crystallographic glide planes with an associated Burgers vector constitutes a **slip system** in a crystal. The Burgers vector defines the glide direction on the glide plane. In general, the most closely packed planes and the directions with the shortest Burgers vectors tend to be the most easily activated slip systems. Table 19.1 lists the principal slip systems of some common minerals. Whether a slip system is activated depends on the magnitude of the critical shear stress necessary to move a dislocation and on the magnitude of the shear component of the applied stress on the glide plane in the direction of slip.

Edge dislocations glide only on planes that contain both the dislocation line and the Burgers vector. They can leave their original glide plane by a process called **climb.** In order for an edge dislocation to climb, atoms must be either added to or taken away from the edge of the extra half-plane. The edge dislocation in Figure 19.15A, for example, climbs down one lattice spacing (the half-plane lengthens) if an atom from a neighboring site jumps onto the bottom of the half-plane, leaving behind a vacancy (Figure 19.15A to B) which can then diffuse away. The edge dislocation climbs up if a vacancy diffuses to the bottom of the half-plane, and the plane shortens (Figure 19.15B to A). By the process of climb, edge dislocations act as sources or sinks for vacancies in the lattice. Jogs are created in the dislocation line if different segments of the dislocation climb by different amounts, as shown by the irregular edge of the isolated extra half-plane in Figure 19.15C.

Table 19.1 Dominant Slip Systems of Some Common Minerals

Mineral	Low-temperature slip systems[a]	High-temperature slip systems[a]
Calcite[b]	$\{10\bar{1}1\}\langle\bar{1}012\rangle:\{r_1\}\langle r_1 \cap f_2\rangle$ twinning: $\{\bar{1}012\}\langle10\bar{1}1\rangle:\{e_1\}\langle e_1 \cap a_2\rangle$	$\{\bar{2}021\}\langle1\bar{1}02\rangle:\{f_1\}\langle r_2 \cap f_3\rangle$ $\{\bar{2}021\}\langle1\bar{2}10\rangle:\{f_1\}\langle r_3 \cap f_1\rangle$
Quartz[b]	$(0001)\langle11\bar{2}0\rangle:(\text{base})\langle a\rangle$	$\{10\bar{1}0\}[0001]:\{m\}[c]$ $\{10\bar{1}0\}\langle1\bar{2}10\rangle:\{m\}\langle a\rangle$ $\{10\bar{1}0\}\langle1\bar{2}13\rangle:\{m\}\langle c+a\rangle$ $\{1\bar{1}01\}\langle11\bar{2}0\rangle:\{r\}\langle a\rangle$
Micas		$(001)\langle110\rangle$ $(001)[100]$
Halite	$\{110\}\langle1\bar{1}0\rangle$	$\{110\}\langle1\bar{1}0\rangle$ $(001)\langle1\bar{1}0\rangle$
Olivine	$(100)[001]$ $\{110\}[001]$	$\{110\}[001]$ $(010)[100]$ $\{0k\ell\}[100]$

Source: Data from Nicolas (1988) and Nicolas and Poirier (1976).

[a] In labeling slip systems, we give the Miller indices of the slip plane first, followed by the components of the slip direction vector. If a specific plane and direction are indicated, they are written in parentheses and in square brackets, respectively: (plane) [direction], for instance. If a set of symmetrically equivalent slip systems is indicated, one plane of the set is written in braces and the corresponding direction is written in angle brackets: {plane}⟨direction⟩.

[b] The letter symbols following the Miller indices for the slip systems are standard abbreviations for the different planes and directions. Some directions are indicated by the intersection of two planes $[p_1 \cap p_2]$.

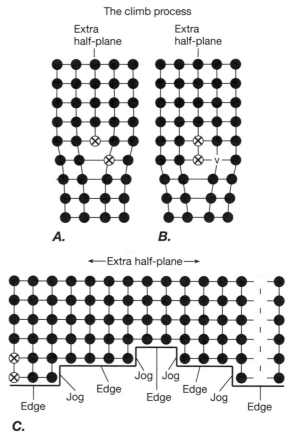

The climb process

Extra half-plane   Extra half-plane

A.   **B.**

←Extra half-plane→

Edge   Jog   Edge   Jog   Jog   Edge   Edge   Jog   Edge

C.

Figure 19.15 Climb of an edge dislocation. The dislocation climbs downward if an atom from a neighboring site jumps onto the half-plane (A to B) leaving a vacancy behind which can then move away by diffusion. The dislocation climbs upward if a vacancy diffuses to a neighboring site and then jumps onto the half plane (B to A). C. The extra half-plane associated with an edge dislocation along which there have been different amounts of climb, leaving jogs in the dislocation line. This view shows the third dimension of the extra half-plane labeled in part B.

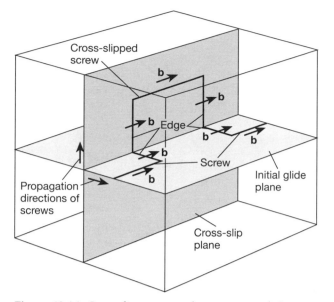

Figure 19.16 Cross-slip occurs when a screw dislocation changes slip planes, and the process leaves segments of edge dislocation that connect the cross-slipped segment with the rest of the screw dislocation. In this diagram, the left and right portions of the screw have continued gliding on the original slip plane, whereas the central portion has cross-slipped.

Screw dislocations need not glide on a single plane, because the Burgers vector and the dislocation line are parallel and therefore do not define a unique plane. The glide planes for a screw dislocation belong to a family of planes each of which must be parallel to the Burgers vector (Figure 19.16). Thus it is possible for a segment of a screw to change glide planes, which is referred to as **cross-slip**. Figure 19.16 shows that when a segment of a screw dislocation undergoes cross-slip, it leaves behind a segment of edge dislocation that connects it to the rest of the screw.

If dislocations of the same type but opposite sign come together, they annihilate one another. Two edge dislocations of opposite sign, for example, may either bound opposite ends of the same extra lattice plane (Figure 19.17A) or bound two half-planes extending in opposite directions in the crystal lattice (Figure 19.17B). When the two dislocations climb toward each other, the extra lattice plane is either eliminated (Figure 19.17A) or made into a whole lattice plane (Figure 19.17B), and no dislocations remain. Similarly, two screw dislocations of opposite sign come together and eliminate any screw offset to the lattice planes. Because the elimination of dislocations from the lattice lowers

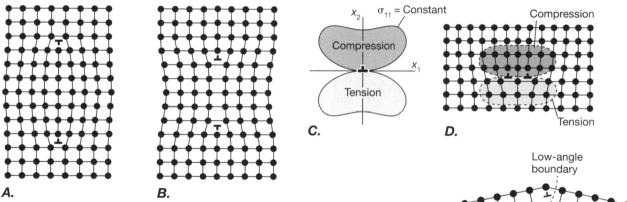

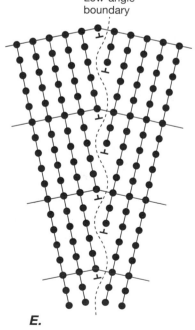

Figure 19.17 Dislocation interactions. *A* and *B*. Dislocations of opposite sign tend to attract each other and annihilate, thereby eliminating imperfections in the crystal and lowering the strain energy. In part *A*, dislocations climb toward each other and eliminate the extra lattice plane between them. In part *B*, they climb toward each other and complete the missing lattice plane between them. *C*. The form of a contour of constant stress σ_{11} about an edge dislocation. Note that one side of the dislocation is in compression and the other side is in tension. *D*. The local distortion of the crystal lattice, and thus the local strain energy per unit volume, increases as two dislocations of the same sign on the same glide plane get closer together. The equilibrium configuration corresponds to the lowest possible energy state. Thus these dislocations effectively repel each other, because the local strain energy per unit volume is lowered as the dislocations move apart. *E*. Edge dislocations tend to accumulate in a wall because it is a low-energy configuration. If two parallel dislocations of the same sign occur one above the other, the strain energy per unit volume for the lattice between the dislocations is reduced, because the compressional strain associated with one dislocation partly cancels the extensional strain associated with the other. The accumulation of a large number of dislocations in this relationship forms a dislocation wall—also called a tilt boundary—which separates two parts of a crystal lattice with slightly different orientation.

the internal strain energy of the crystal, dislocations of opposite sign tend to attract one another.

Dislocations of the same sign on the same glide plane, however, tend to repel one another. The strain about a single edge dislocation is contractional on one side and extensional on the other, leading to compressional and tensile stresses on opposite sides of the glide plane around the dislocation (Figure 19.17C). As two edge dislocations approach each other on the same glide plane, their strain fields overlap and add together (Figure 19.17D). The tendency of the lattice to adopt a low-energy configuration provides an effective force that drives the two dislocations apart.

Edge dislocations of the same sign on different parallel slip planes are attracted into **dislocation walls** (Figure 19.17E) in which the dislocations are aligned one above another. This is a low-energy configuration, because the contractional strain associated with one side of one dislocation is in part canceled out by the extensional strain associated with the adjacent dislocation. A dislocation wall is also called a **low-angle boundary** or a **tilt boundary,** because it is a boundary between two parts of the crystal lattice that are tilted at a low angle with respect to each other (Figure 19.17E). They may form the boundaries of **subgrains** (Figure 19.13A), which are small volumes within individual crystal grains that commonly differ in orientation by 1° to 2° or less from neighboring subgrains. The greater the density of dislocations in a tilt boundary, the greater the angle of tilt.

Dislocations are easily lost from a lattice by gliding or climbing out of the lattice or by annihilation within the lattice, as described above. Thus, in order for them to be able to produce an arbitrary amount of ductile deformation, there must be some mechanism for generating dislocations.

A common mechanism for generating dislocations is called a **Frank–Read source.** Assume that the segment AB of the dislocation in Figure 19.18A is pinned at both ends A and B. This might happen, for example, if the segment had climbed or cross-clipped out of its original

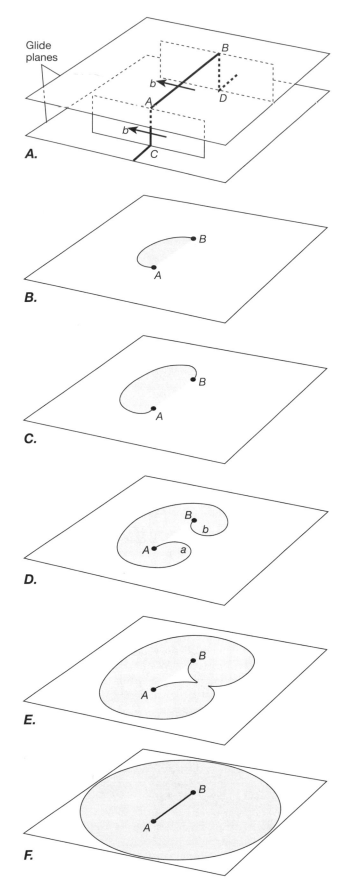

Figure 19.18 The operation of a Frank–Read source for dislocation generation. A. An edge dislocation has climbed from the upper glide plane to the lower one, leaving a segment AB on the upper plane and generating two dislocations AC and BD that are not on the glide plane and therefore cannot move easily. They pin the segment AB at its ends. B. The dislocation segment AB bows out in response to an applied shear stress. C. The dislocation pivots about the pinning points A and B. D. The parts of the dislocation in lobes a and b, which are of opposite sign, approach each other. E. The lobes a and b meet and annihilate. F. The dislocation snaps back to the position AB, leaving a dislocation loop expanding in the glide plane.

slip plane and the jogs were unable to move easily. With increasing shear stress on the material, the dislocation tends to bow out more and more from the pinning points (Figure 19.18*B*, *C*, *D*). In each figure, the shaded area shows the region of the crystal in which slip of one Burgers vector has occurred, starting from the configuration in Figure 19.18*A*. As the stress increases further, the dislocation pivots around each pinning point (Figure 19.18*C*). The two lobes of the dislocation at *a* and *b* (Figure 19.18*D*) are of opposite sign, and when they come together they annihilate (Figure 19.18*E*). By this process, all the material within the dislocation loop has slipped one Burgers vector, and a dislocation segment is left between the original pinning points *A* and *B*. The dislocation loop that has been created continues to expand in response to the applied stress (Figure 19.18*F*), and the segment *AB* can bow out again an unlimited number of times to create unlimited numbers of dislocations. The same mechanism works for both edge and screw dislocations. With many sources such as this operating throughout a crystal, unlimited amounts of ductile deformation can be produced.

19.5 Mechanisms of Dislocation Creep

Propagation of dislocations through crystal lattices is one of the most common mechanisms of ductile deformation of crystalline materials. Different processes affect the rate of dislocation motion and thus give rise to different rates of deformation in response to the same applied stress, which accounts for some of the observed differences in rheology. In this section we discuss the most important of these processes.

For a crystal to undergo an arbitrary constant-volume deformation, it is necessary for five independent slip systems to be active, a condition that is referred to as the von Mises condition, not to be confused with the plastic failure criterion named for the same person. This requirement arises because an arbitrary deformation is described by the strain tensor, which has six independent components. The constant-volume constraint provides one relationship among the three extensional strains (e_v = constant, Equation 15.16; see footnote 7, Section 15.2), leaving five of the strain components independent. Five independent slip systems are required to produce these five independent components of strain. In a polycrystal, if the strain is allowed to be inhomogeneous from one crystal grain to another but is still required to be coherent (that is, grain boundaries do not separate), then four, or even three, independent slip systems may be enough to accommodate an arbitrary strain of the material, especially if grain boundary sliding can occur.

Dislocation Glide

At low temperatures and/or high stresses, ductile deformation of crystalline solids occurs predominantly by glide of dislocations through the lattice, a process called **cold working.** Resistance to the motion of the dislocations comes in part from the lattice itself because of the necessity of breaking bonds in order for the dislocations to move. In silicates, the bonds are largely covalent and therefore strong, and the stress necessary to overcome the bond strength is high. Resistance also comes from obstacles that occur in the path of a gliding dislocation, such as other dislocations or impurities.

When several slip systems operate simultaneously, dislocations gliding on different slip systems inevitably intersect. This process introduces kinks and jogs, into the dislocation lines that impede the dislocation motion. As strain increases, the number of dislocations increases, and the complexity of interactions produces tangles of dislocations that impede dislocation motion even more. The higher the density of dislocations, the more difficult it becomes for any dislocation to move. Thus higher stress is required to force the dislocations to glide and to produce more dislocations. This process is the origin of the work-hardening behavior described in Section 18.2. Figure 19.13*C* shows an example of dislocation tangles that resulted from deformation of quartz.

If the stress is held constant during cold working, the creep rate gradually declines and may approach zero because of the work hardening. If the strain rate is kept constant, the stress increases with increasing strain until either the material fractures or a steady-state saturation stress is reached that is high enough to force the dislocations to glide out of the crystal. The steady-state condition is approached only at high strains, such as $e_n > 1$. Thus the transient creep stage is very long, and in many low-temperature experiments a steady state is not observed. Dislocations may become so closely packed that they act like Griffith cracks (see Section 9.7), or grain boundaries may open to form Griffith cracks, leading to the brittle failure of the sample before steady state can be established.

A theoretical model based on kinetic theory enables us to calculate the rate of glide of dislocations through a crystal lattice when the glide rate is controlled by obstructions. The theory predicts a strain rate dependent on an exponential function of the differential stress (Box 19.1). This rheology coincides with the experimentally observed behavior of crystalline solids in the high-stress regime.

The glide mechanism dominates the creep rate of minerals at very high stresses across a wide range of homologous temperatures (Figure 19.1). Because the temperature dependence of creep is the same for the exponential law and the power law, the boundary between the two mechanisms depends only on the normalized stress (Figure 19.1).

Recovery by Dislocation Climb

At homologous temperatures above approximately 0.5, work hardening is counteracted by **recovery processes** that permit the rearrangement of dislocations. One of the most important recovery mechanisms is climb of edge dislocations. Edge dislocations can climb toward others of opposite sign and annihilate them, they can climb into low-energy walls, tangles can become untangled through climb, and the climb of jogs on dislocations can allow the dislocations to glide more freely. Glide of the dislocations remains an important factor in ductile deformation, but climb is the rate-limiting process, because it is the slowest of the processes that enable deformation to occur.

Thus the steady-state creep observed at homologous temperatures above about 0.5 is a balance between the processes of work hardening, which tend to decrease the mobility of dislocations, and those of recovery, which tend to enhance their mobility. The mutual repelling force between dislocations, called the **backstress**, counteracts the effect of the applied stress on dislocations and thus inhibits dislocation glide. At steady state, the backstress is essentially equal to the applied stress, and new dislocations cannot be produced and cannot move unless others are eliminated, either by annihilation or by climb or cross-slip into walls or out of the crystal.

A number of different models have been proposed to explain steady-state creep on the basis of assumptions about what the rate-limiting step is in the creep or recovery process. All of these models lead to equations of the power-law type (Box 19.1 and Equation 18.4), which is consistent with the inference that dislocation processes control the ductile deformation observed in experiments.

Dynamic Recrystallization

Dynamic recrystallization or **syntectonic recrystallization** is a process by which new crystal grains form from old grains during the deformation process.[4] Two mechanisms are recognized by which dynamic recrystallization occurs: boundary migration recrystallization, and subgrain rotation recrystallization.

Boundary migration recrystallization is recrystallization by means of the migration of a grain boundary separating highly strained from unstrained crystals of the same mineral. The material in the highly strained crystal crosses the grain boundary and is added to the unstrained crystal. Thus the grain boundary migrates into the more highly strained region and leaves unstrained crystal behind it. This process is comparable to a recovery process because it lowers the dislocation density, and therefore the strain energy, of deformed grains. In this case, however, deformed grains are completely replaced by new, undeformed grains. The new grains in turn become deformed and are subsequently replaced. This process can be an important means of maintaining steady-state deformation at stresses much lower than normal recovery processes require. Figure 19.19A, B shows the effects of boundary migration recrystallization in quartz. The highly serrated grain boundaries indicate high mobility, and such boundaries can bulge rapidly out into a volume of strongly strained crystal to nucleate a new recrystallized grain.

During **subgrain rotation recrystallization**, subgrain boundaries in highly strained regions of a crystal accumulate an increasing number of dislocations, which causes a progressive rotation of the subgrain lattice with respect to the surrounding crystal. Eventually, at roughly 10° of relative rotation, the boundary becomes saturated with dislocations. Any further rotation changes the boundary into the high-angle grain boundary of a newly recrystallized grain, which can no longer be represented by a dislocation wall. Thus subgrain rotation recrystallization is not distinct from recovery but rather results from the operation of recovery mechanisms such as dislocation climb. Figure 19.19C shows an old, large grain of quartz that is filled with subgrains. The misorientation of the subgrains increases progressively from the center of the grain (region A) toward the boundary (region C), where the subgrains become indistinguishable from the new, recrystallized grains that surround the old quartz grain.

Grain boundaries are high-angle boundaries characterized by large misorientations or structural discontinuity between adjacent grains. Such boundaries consist of a patchwork of areas of partial fit and areas of no fit between the lattices of adjacent grains. High-angle boundaries are very mobile, because it is easy for atoms or ions to transfer one at a time across them. Subgrain boundaries are relatively immobile, however, because they can move only by the coordinated glide of all the dislocations in the subboundary wall.

No models of dislocation creep rigorously account for the effects of recrystallization by boundary migration

---

[4] As used here, the term *recrystallization* refers to processes that leave the chemical composition of the affected mineral grains essentially unchanged. Such recrystallization is "dynamic" if it occurs during deformation and "static" if deformation is not occurring. Metamorphic petrologists commonly use the same word to describe the result of chemical reactions during which the reactant minerals break down and new minerals form. The term **neomineralization** can also be used for the latter process. The meaning is clear from the context, although failure to recognize the difference in usage can cause confusion.

A.

50 μm

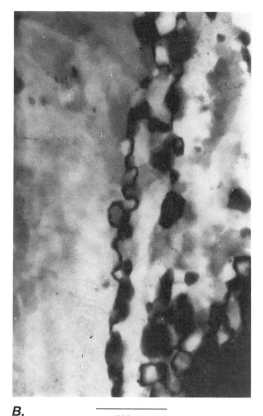

B.

500 μm

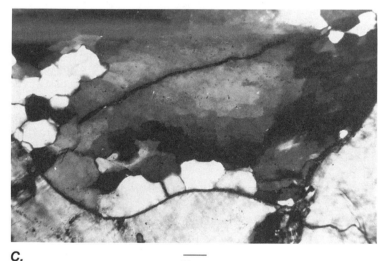

C.

50 μm

Figure 19.19 Mechanisms of dynamic recrystallization in quartz. Photomicrographs under crossed nicols. *A.* The highly serrated crystal boundaries in this polycrystalline quartz are characteristic of high mobility and migration of the grain boundaries. *B.* Boundary migration recrystallization in polycrystalline quartz begins with the migration of a high-angle boundary to form a lobe into the adjacent crystal. *C.* Subgrain rotation recrystallization. The misorientation of the subgrains, indicated by the shade of grey, increases progressively from the core of the grain toward the rim, where there is little or no distinction between recrystallized grains and highly misoriented subgrains.

or subgrain rotation. It has been determined empirically that steady-state creep during boundary migration recrystallization displays a higher creep rate than steady-state creep during recovery. In both cases, however, the material displays a power-law rheology. The increased creep rate presumably results from the higher rate of primary creep relative to steady-state creep (see Figure 18.8*B*), which must occur in each new, recrystallized grain immediately after its formation. The continual

resupplying of such new grains by dynamic recrystallization maintains a significant portion of the crystal grains in the rock in the primary-creep regime.

At the strain rates characteristic of geologic deformation, dislocation creep is probably important in the deformation of quartz at high grades of metamorphism (Figure 19.1*A*), especially if the dislocation creep field is enlarged by higher creep rates associated with recrystallization; by the presence of water, which causes hydrolytic weakening; and by larger grain sizes, which suppress diffusion creep. This is consistent with the development of the crystallographic preferred orientations that are commonly observed in quartz tectonites (see Section 19.7). The deformation map for olivine (Figure 19.1C) indicates that power-law creep probably is also a major creep mechanism in upper-mantle deformation, which is again consistent with the evidence for crystallographic preferred orientation of olivine in peridotite tectonites. Calcite apparently should not deform much by dislocation creep (Figure 19.1B), although the large grain size and strong crystallographic preferred orientation of many marbles suggest that the boundaries of the dislocation creep field in this map may be at too high a stress.

Harper–Dorn Creep

The creep rates controlled by recovery and recrystallization (described above) all have power-law rheologies. For olivine (Figure 19.1C), the field of the deformation map occupied by these mechanisms covers most of the expected range of stresses for mantle deformation at high temperature, especially considering that this field would expand at the expense of the diffusion creep field for the larger grain sizes characteristic of the mantle. Thus power-law creep may dominate in much of the upper mantle, at least. This conclusion is consistent with interpreting the observed anisotropy of seismic velocities in the upper mantle as resulting from preferred crystallographic orientation produced by dislocation creep mechanisms (see Section 19.7). It is not consistent, however, with geophysically derived inferences of a linear rheology for the entire mantle (see Section 18.6).

The resolution of this problem may lie in the existence of yet another dislocation deformation mechanism called **Harper–Dorn creep.** The exact mechanism is not understood, but it is a dislocation mechanism for which the dislocation density is constant and independent of the differential stress, which leads to a linear relationship between strain rate and stress (Box 19.1). Like diffusion creep, it is effective at low stresses, and it produces a linear viscous rheology. In contrast to diffusion creep, however, it can provide higher strain rates than other mechanisms in coarse-grained materials, roughly greater than 0.5 mm. Because of these features, it is tempting to speculate that Harper-Dorn creep may be an important creep process in the mantle. Although it has so far been definitely identified only in some metals, the similarity between other deformation mechanisms in metals and silicates suggests that it also might be applicable to the Earth. Therefore, the dislocation microstructures associated with dislocation creep (see Section 19.6 and 19.7) are not necessarily incompatible with a linear rheology, but the suggestion

is at present very speculative and in need of much research. If Harper–Dorn creep does occur in rocks, the deformation maps in Figure 19.1 will have to be significantly revised, especially in the low-stress areas.

Annealing

The microstructure of a polycrystalline material can continue to evolve at high temperature even after the differential stress is removed, because recovery and recrystallization processes continue to operate. In the absence of a differential stress, **static recovery** and **static recrystallization**—referred to collectively as **annealing**—are driven by the associated decrease in the internal strain energy per unit volume of the material.

Static recovery involves the rearrangement and annihilation of dislocations by climb and cross-slip. Static recrystallization may be subdivided into **primary** and **secondary** recrystallization. **Primary recrystallization,** like dynamic recrystallization, reduces the internal strain energy of the material by replacing deformed grains that have a high dislocation density with new grains that are essentially strain-free. Grain boundaries become straight, and for materials such as quartz and olivine that have nearly isotropic grain boundary energies, the grain boundaries at triple grain junctions tend to meet at angles of 120° (see Figure 17.3A). During **secondary recrystallization,** small crystal grains are preferentially eliminated by the exaggerated growth of a few larger grains. Such grains tend to have highly irregular boundaries, and grains that on a planar section are apparently isolated, in fact, can be part of a single grain that is highly convoluted in three dimensions. The growth of very large grains decreases the internal energy of a material because the grain boundary energy per unit volume is reduced.

Evaluation of Relative Creep Rates

Different deformation mechanisms give rise to different rheologic equations (Box 19.1), and the strain rate for each mechanism is therefore different under the same conditions. This effect is illustrated in Figure 19.20, where the logarithm of the normalized creep rate is plotted against the logarithm of the normalized stress for olivine. The rheologies for Nabarro–Herring diffusion creep (Equation 19.1.1 in Box 19.1) and for power-law creep by dislocation glide and recovery (Equation 19.1.6) are plotted by using material constants appropriate for olivine and four different grain sizes for the diffusion creep mechanism. Because the graph is a log–log plot, the slope of each line is equal to the exponent on the stress in the corresponding rheologic equation (Equations 19.1.1 and 19.1.6).

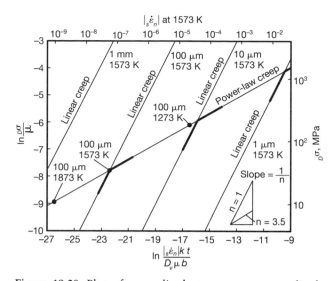

Figure 19.20 Plot of normalized stress versus normalized strain rate for olivine, showing how the strain rates for power-law creep and for diffusion creep are related. The heavy obtuse angles mark the stresses where the dominant deformation mechanism changes for grain sizes of 1 μm, 10 μm, and 100 μm at 1573 K. The heavy lines indicate the line that provides the highest strain rate under the given conditions. The solid dots indicate the change in crossover stress for a 100-μm grain size at three temperatures, each separated from the next by 300 K. The vertical scale on the right gives the differential stress, and the horizontal scale at the top gives the absolute strain rate for 1573 K (1300°C). The equations for the linear and power-law creep curves are, respectively,

$$\ln \frac{|_s\dot{\varepsilon}_n|kT}{D_v\mu b} = \left[-9.92 + \ln \frac{D_b}{D_v} \right] - 2 \ln \frac{d}{b} + \ln \frac{D\sigma}{\mu}$$

$$\ln \frac{|_s\dot{\varepsilon}_n|KT}{D_v\mu b} = 4.87 + 3.5 \ln \frac{D\sigma}{\mu}$$

The ratio D_b/D_v appears in the linear creep law because the plotted strain rate is normalized by the coefficient of volume self-diffusion D_v, whereas the appropriate quantity for normalizing the linear diffusion creep should be the coefficient of grain boundary self-diffusion D_b. This term introduces the temperature dependence of the plot for linear creep. $D_v = 0.1$ exp $[-Q_v/RT]$; $D_b = 0.1$ exp $[-Q_b/RT]$; $Q_v = 5.4 \times 10^5$ joule; $Q_b = 2.9 \times 10^5$ joule; other constants are listed in the caption to Figure 18.14.

The stress exponent is 3.5 for power-law creep and is 1 for Nabarro–Herring creep. Because of the different slopes, the lines cross. At stresses above the cross-over stress, the highest strain rate is provided by power-law creep, but below that stress, the linear Nabarro–Herring creep provides a higher strain rate. The variation of this cross-over stress with homologous temperature for a particular grain size (the heavy dots in Figure 19.20) defines the boundary on the deformation maps between the different deformation mechanisms (Figure 19.1C). In principle, both mechanisms can operate on either side

of the cross-over stress, and the total strain rate is the sum of strain rates contributed by each mechanism. Except for a small stress range immediately around the cross-over stress, however, the contribution of one mechanism or the other to the strain rate dominates the behavior of the material.

It is apparent in Figure 19.20 that the cross-over stresses for realistic mantle grain sizes of roughly 1 mm or larger are at such low strain rates that they could never be examined directly in an experiment. Only by decreasing the temperature several hundred degrees or decreasing the grain sizes to about 1 to 10 μm can the cross-over stresses be made accessible experimentally, and these conditions are unlikely to be representative of the mantle. Our inference of which mechanism actually should dominate under mantle conditions, therefore, depends strongly on our ability to identify the appropriate mechanisms and to determine accurately the rheologic equations. Only then can we extrapolate reliably to mantle conditions. The uncertainty of the appropriate values for the material constants in the equations—to say nothing of the possibility that there are other deformation mechanisms that we do not know about or understand sufficiently—means that interpretation of the appropriate rheology for the mantle is still a much-debated issue.

19.6 Microstructural Fabrics Associated with Dislocation Creep

Creep and recrystallization result in the development of characteristic fabrics in mineral grains and in rocks, including dislocation microstructures, grain textures, and crystallographic preferred orientations. These fabrics provide important clues to the nature of the deformation.

Crystals deformed by cold working, when examined in thin section between crossed polarizers, commonly exhibit a characteristic undulatory extinction that reflects deformation—imposed curvature in the lattice (Figure 19.21A). This curvature represents the excess of dislocations of one sign over those of the other. If the temperature is high enough for recovery to occur, either during or after the deformation, these dislocations may climb and be annihilated in pairs or form low-energy walls, which may be evident as subgrain boundaries or a series of tilt boundaries of the same sign (Figure 19.21B).

Deformation lamellae are features that develop in quartz (see Box 19.2) and olivine; they indicate that deformation has occurred at relatively high strain rate or high differential stress and at temperatures in the low range for dislocation creep (Figure 19.22). They appear

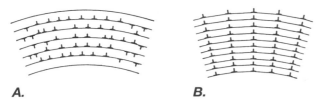

Figure 19.21 A dislocation model for undulatory extinction and polygonization. *A.* Glide planes in the crystal contain dislocations of both signs, but more of one sign than of the other, giving rise to the curvature of the lattice planes. *B.* Recovery during annealing allows dislocations of opposite sign to annihilate and dislocations of the same sign to array themselves into low-energy walls. The crystal is thereby segmented into subgrains separated by low-angle tilt walls such that each subgrain has a small misorientation with respect to its neighbors.

in thin section as thin, discontinuous planar features in a mineral grain and have a slightly different index of refraction from the surrounding crystal. They are commonly formed by pile-ups and tangles of dislocations on the active glide planes in the crystal and are evidence of a low rate of recovery.

Ribbon-shaped grains, especially of quartz, that have been severely flattened and stretched (Figure 19.23) indicate that the rocks were deformed at moderate temperature below that required for recrystallization.

Subgrains in crystals are another indication that dislocation creep mechanisms have operated. These subgrains are often visible in the petrographic microscope as a slight difference in extinction position of areas within a single crystal grain (Figure 19.19C). The misorientation of many subgrains, however, is too small to be optically visible, and techniques such as decorating

the dislocations (Figure 19.13A), etching polished crystal surfaces (Figure 19.13B), and imaging the crystal in the transmission electron microscope (Figure 19.13C) must be used.

Highly serrated grain boundaries (Figure 19.19A, B) are evidence that the grain boundaries have been mobile and therefore that the homologous temperature of deformation was relatively high. Mobility of the grain boundaries promotes dynamic recrystallization and grain growth and allows the grains to maintain a more equant morphology during deformation. If deformation ceases while temperatures are still high, annealing occurs and mobile grain boundaries tend to become straight, which is a minimum-energy configuration (Figure 17.3A). Straight grain boundaries, however, are not a unique characteristic of postdeformational annealing. They also develop during dynamic subgrain rotation recrystallization and as a result of diffusional mechanisms of deformation.

The effects of dynamic recrystallization are easiest to recognize where there has been a dramatic reduction in grain size but at least some original grains are still preserved. Boundary migration recrystallization may produce highly irregular and sutured grain boundaries (Figure 19.19B). Subgrain rotation recrystallization results in increasing misorientation of subgrains from the core to the boundary of large relict grains (Figure 19.19C).

Piezometers, or stress measures (Box 19.2), are steady-state fabric characteristics that change with the magnitude of the applied differential stress. If they are preserved in the mineral grains, we can use them to infer the magnitude of the paleostress that caused the ductile deformation. Because dislocations multiply until the backstress equals the applied stress, the dislocation

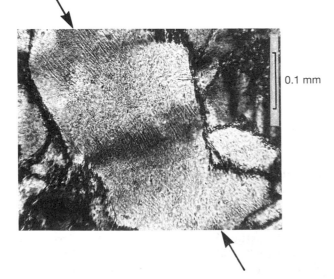

0.1 mm

Figure 19.22 Deformation lamellae in quartz seen under the optical microscope.

0.1 mm

Figure 19.23 Ribbon quartz in a strongly deformed quartzite.

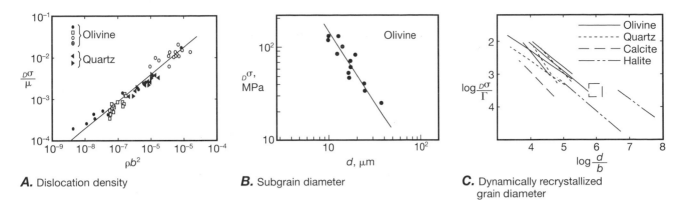

A. Dislocation density

B. Subgrain diameter

C. Dynamically recrystallized grain diameter

Figure 19.24 Dependence of stress-sensitive fabric elements on differential stress. *A.* Variation of free dislocation density with differential stress for olivine and quartz, showing the relationship given in Equation (19.2.1). *B.* Variation of the mean subgrain diameter with the differential stress at steady state, illustrating the relationship given in Equation (19.2.2). *C.* Variation of mean dynamically recrystallized grain diameter with differential stress, illustrating the relationship given in Equation (19.2.3). Data include relationships for wet and dry olivine and quartz, for calcite (box as well as dashed line), and for halite recrystallizing by migration and rotation recrystallization. The normalization modulus is $\Gamma = \mu/(1 - v)$ where μ is the shear modulus and v the Poisson ratio.

density[5] at steady-state creep is a function of the applied differential stress (Figure 19.24*A* and Box 19.2). The size of subgrains and of dynamically recrystallized grains formed during steady-state dislocation creep is also dependent on the magnitude of the applied differential stress, and the size decreases with increasing stress (Figure 19.24*B*, *C* and Box 19.2). Piezometrically determined stresses as high as 100 to 200 MPa have been found in crustal rocks associated with mylonites in ductile shear zones. At temperatures that put the deformation into the dislocation creep regime, such stresses must be associated with geologically rapid strain rates. For quartz, at temperatures as low as 400°C, the strain rate at 100 MPa differential stress is roughly 10^{-10} s^{-1} (Figure 19.1*A*).

19.7 Preferred Orientation Fabrics of Dislocation Creep

Ductile deformation by dislocation creep produces characteristic preferred orientations of mineral crystallographic axes. The pattern of preferred orientation that

develops for a particular mineral from an initially randomly oriented polycrystalline aggregate depends on the slip systems that are active in that mineral and on the geometry and magnitude of the externally applied deformation. The number and orientation of the slip systems that operate in the particular mineral may change with different conditions of temperature and/or stress, thereby producing different fabrics under the different conditions. Diffusion creep does not lead to preferred orientations, because it is not associated with any preferred plane and orientation in the crystal.

Coaxial deformations produce fabrics that are symmetric with respect to the principal axes of finite strain, whereas noncoaxial deformations tend to produce asymmetric fabrics. The strength of the preferred orientations increases with increasing strain; in some cases, however, it approaches a limiting value. Thus we should be able to deduce constraints on the geometry, magnitude, and conditions of deformation from observations of natural preferred orientation fabrics and our knowledge of the ductile behavior of particular minerals.

Formation of a Crystallographic Preferred Orientation

The formation of a crystallographic preferred orientation during ductile deformation of a rock requires some mechanism by which individual crystals rotate toward a preferred orientation as a result of slip on their particular slip systems. Two models have been proposed for this process. One, called the Taylor–Bishop–Hill theory, requires crystals to rotate such that their spin

---

[5] Dislocation density is measured in units of total dislocation length per unit volume, which reduces to units of inverse length squared. Common units are cm^{-2}. Undeformed crystals may have free dislocation densities of 10^3 to 10^5 cm^{-2}; deformed crystals may have densities up to 10^{10} cm^{-2} (Figure 19.24*A*). For the larger figure, the total length of dislocations in a cubic centimeter is comparable to the total length of blood vessels in the human body. If they were strung end to end, they would reach two and a half times around the world!

Box 19.2 **Determining the Orientation and Magnitude of Paleostresses in Deformed Rocks**

In applying our understanding of the rheology of rocks to the interpretation of structures in the Earth, we face a dilemma: The orientations and magnitudes of the stresses, as well as the strain rates associated with the observable ductile deformation, are usually impossible to determine. Without one of these pieces of information, it is impossible to use the rheological equations, and inferences of rheology are speculative or at best very approximate. Under some circumstances, however, it is possible to use the fabrics of deformed mineral grains to determine the orientations and magnitudes of the stresses that caused the deformation. Such measurements are extremely important in enabling us to check laboratory and theoretical models against actual conditions of deformation in the Earth. The orientations of calcite twin lamellae and of quartz deformation lamellae can be used to determine the orientations of the principal stress axes. Dislocation density, subgrain diameter, dynamically recrystallized grain size, and to some extent calcite twin lamellae can all be used to infer the magnitude of the paleostresses.

Calcite e twin lamellae (Figure 19.3) and quartz deformation lamellae (Figure 19.22) are planar microstructures that form in orientations of high incremental shear strain. It is assumed that these planes represent planes of high resolved shear stress and that subsequent deformation of the crystal has not reoriented the planar features significantly. Because planes of high shear stress have a restricted range of orientations relative to the principal stress axes, the preferred orientations of the planar microstructures and the crystallographic axes of the crystals in which they occur can be used to infer the original orientation of the stress axes.

Deformation twins in calcite can develop only if the shear stress is large enough to twin the calcite crystal, and if the crystal is oriented such that the resolved shear stress on the twin plane is sufficiently high and roughly parallel to the direction for twinning shear. During the twinning process, the c-axis of the twin is flipped into an orientation closer to the axis of maximum compression than the c-axis of the host grain (Figure 19.3A, C, D). By measuring the orientations of the twin plane as well as the c axes in the twin and in the host, we can use the geometry of Figure 19.3D to infer the optimum stress orientation for producing that twin. The preferred orientation of these directions determined for a large number of calcite grains defines the best estimate of the orientation of the principal stresses for the deformation.

A critical resolved shear stress of roughly 10 MPa is required to produce a twin in a calcite crystal. If a differential stress (which is twice the maximum shear stress) of 20 MPa is applied to a calcite rock, twinning can occur only in those grains oriented such that the twin plane is exactly parallel to the plane of maximum resolved shear stress and the twinning direction is exactly parallel to the direction of maximum resolved shear stress. For such a situation, the inferred $\hat{\sigma}_1$ axes for twinned grains should form a tight maximum. For higher differential stresses, grains that are not in the ideal orientation would also be able to twin, and the inferred directions for $\hat{\sigma}_1$ would spread out into a larger maximum. Thus the size of the maximum can be used to infer the approximate magnitude of the differential stress.

The method of analysis of quartz lamellae is illustrated in Figure 19.2.1A, B, C for the fabrics on the limbs of a small fold in a quartzite layer shown in Figure 19.2.1D. The orientations of the lamellae planes and of the host c axes are determined. On a stereonet plot of the data, poles to deformation lamellae tend to form two maxima which may also spread out to form a partial small-circle girdle (the top diagrams in Figure 19.2.1A, B, C). The symmetry axes of the lamella pole distribution are assumed to be parallel to the principal axes of stress.* An arrow along the great circle connecting the host c axis to the lamella pole through the acute angle tends to point toward the axis of minimum compressive stress and away from the axis of maximum compressive stress (the bottom diagrams in Figure 19.2.1A, B, C).

Figure 19.2.1D shows the folded layer. Here the arrows indicate the directions of the principal stresses inferred from the fabric diagrams in Figure 19.2.1A, B, C, and the dashed lines represent the average orientation of the lamellae. Figure 19.2.1E shows the same layer unfolded to the point where the maximum compressive stresses are all parallel. Apparently, the lamellae were formed fairly early in the deformation (see Figure 20.5A) and were later rotated by the folding. The fact that the lamellae do not represent the stresses expected later in the folding process suggests that the magnitude of the stress decreased as the folding progressed. These conclusions are consistent with the mechanical models discussed in Section 20.3, and

(continued)

---

* In fact, because the lamellae record a deformation, it is more accurate to assume that the symmetry axes are related to the incremental strain ellipse. Generally, however, it is implicitly assumed that the principal axes of stress are parallel to those of incremental strain. Although this assumption may be a reasonable approximation, it is not, strictly speaking, a necessary relationship, especially if the material is anisotropic, which rocks generally are.

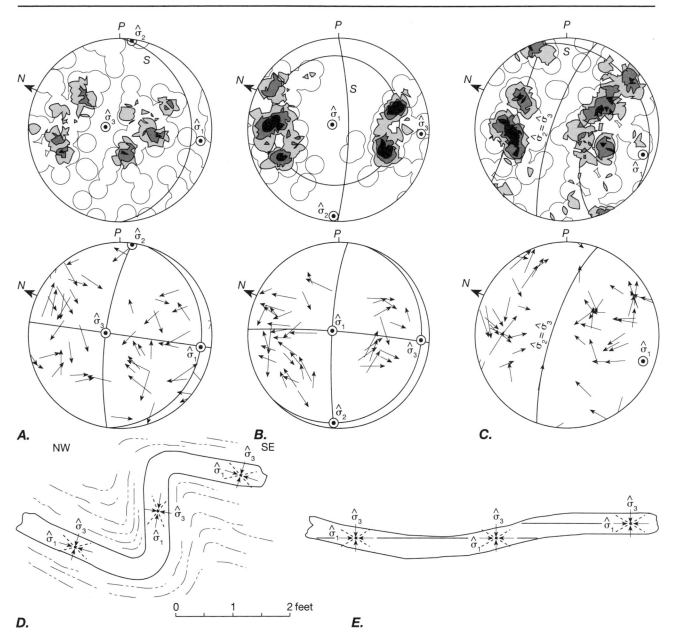

Figure 19.2.1 Principal stress axes in a minor fold deduced from quartz deformation lamellae. *A, B,* and *C. Top:* Poles to deformation lamellae with bedding orientation (great circle labeled S; N is north direction), small-circle distribution of lamella poles, and symmetry axes of the fabric (circled dots). *Bottom:* Arrow diagrams with arrows drawn along great circles connecting the optic *c*-axis (tail) with the lamella pole (head). Only arrows subtending angles between 10° and 25° are shown. Great circles are symmetry planes of the lamella fabric. The principal stress axes are inferred to be parallel to the symmetry axes of the lamella fabric, and the arrows tend to point toward the minimum compressive stress $\hat{\sigma}_3$ and away from the maximum compressive stress $\hat{\sigma}_1$. The point *P* on each stereonet is the orientation of the line normal to the cross sections in parts *D* and *E*. Lower-hemisphere equal-area projections onto the horizontal plane. *A.* 110 lamellae; contours 0.9, 2.7, 4.5, and 6.4 percent per 1 percent area. *B.* 116 lamellae; contours 0.9, 2.6, 4.4, and 6.1 percent per 1 percent area. *C.* 163 lamellae; contours 0.6, 1.8, 3.1, and 4.3 percent per 1 percent area. *D.* Profile of the fold, showing the stress axes deduced from quartz lamellae in the fold limbs. The point *P* on each stereonet in parts *A* through *C* is the direction normal to the cross section of the fold. *E.* The quartzite layer unfolded to the point at which the axes of maximum compressive stress are all parallel.

Box 19.2 (*continued*)

they indicate a complication of interpreting stress orientations inferred from fabric data: Later deformation may rotate the inferred stress axes out of their original orientations.

Where calcite twin lamellae and quartz deformation lamellae have been studied in the same rock, each method gives results consistent with the other. Accurate information on stress orientation, however, can be obtained from twin and deformation lamellae only if the initial distribution of c axes for grains in the rock is close to random, and if the total strain is limited to relatively low values—perhaps 20 percent or less.

For rocks that have deformed at steady state, the magnitude of the paleostress may be determined from three different elements of the microstructural fabric. With increasing differential stress $_D\sigma$, the free dislocation density ρ increases, and the mean diameters of subgrains d_s and of dynamically recrystallized grains d_r decrease according to the following relationships:

Fabric element	Piezometric equation	
Dislocation density	$_D\sigma = \alpha\mu b\rho^{1/2}$	(19.2.1)
Subgrain diameter	$_D\sigma = K_1\mu b d_s^{-r}$	(19.2.2)
Dynamically recrystallized grain diameter	$_D\sigma = K_2\mu b d_r^{-p}$	(19.2.3)

In these equations, μ is the elastic shear modulus, b is the magnitude of the Burgers vector, α is a constant approximately equal to 1, K_1 and K_2 are empirically determined constants, and r and p are constants generally with values $r \approx 1$, $p \approx 0.7$, although values of p as high as 1.4 have been reported. Experimental calibrations of these relationships are shown in Figures 19.23A, B, and C, respectively.

Although dislocation densities decrease rapidly during annealing at high temperature after the stress is removed, the dislocation substructure can be preserved if deformation occurs at low temperature, if cooling is rapid, or if cooling occurs under constant differential stress.

The subgrain diameter responds more slowly than the dislocation density to changes in stress, because it requires larger strains to form subgrains and because boundaries are less mobile than individual dislocations during annealing. Subgrains therefore provide a more stable piezometer.

The constants in Equation (19.2.3) for the two mechanisms of dynamic recrystallization are different. Subgrain rotation recrystallization produces smaller grains than boundary migration recrystallization, at least in calcite and halite, but the difference between the two mechanisms in silicates has not been determined, so there is some question about whether laboratory calibrations of this piezometer are based on the same recrystallization mechanism or combination of mechanisms of recrystallization that occur under natural conditions.

The highest differential paleostresses inferred for tectonic processes in the Earth by means of these different piezometers have come from mylonites in ductile shear zones. Magnitudes on the order of 100 MPa to 200 MPa are indicated from crustal shear zones such as the Moine Thrust in northern Scotland, the Arltunga complex in central Australia, metamorphic core complexes in the western Cordillera of the United States, and ultramafic bodies of mantle origin that have been thrust up on crustal rocks in orogenic belts. The application of piezometry to olivine in xenoliths brought up from depths of 100 to 200 km suggests that stresses in the mantle range from the high stresses mentioned above down to stresses on the order of a few megapascals. The higher stresses may be associated with the local process—probably diapirism—that resulted in the eruption of the lavas containing the xenoliths. The lower stresses may represent the regional flow stresses in the upper mantle. The difficulty of distinguishing steady-state microstructures from those resulting from annealing makes the determinations of the lower stress values somewhat suspect, however.

The application of these piezometers to rocks is still not very precise. Experimental calibration of the constants in the equations is not adequate for most minerals. In order to record the stress correctly, the fabrics must be the result of steady-state deformation, which is difficult to prove. The fabrics must not have altered since the steady-state deformation, which is also difficult to prove. The distinction between the constants for the two different recrystallization mechanisms has not been determined for any silicates. And there are indications that the sensitivity of the microstructures to changes in stress may decrease—perhaps to zero—at low stresses. Despite all the difficulties, the information the piezometers provide is an invaluable and unique key to understanding the conditions of deformation in the Earth.

equals that of the imposed large-scale homogeneous deformation. The model assumes that at least five independent slip systems are active in the crystals, satisfying the von Mises condition. The other model, which we refer to as the misfit minimization model, is based on determining the rotations required to minimize the geometric misfit in an aggregate of crystals that can deform on only a limited number of slip systems. Both models predict aspects of observed crystallographic preferred orientations, and the actual process of crystallographic rotation probably includes both mechanisms. We discuss the basic aspects of the two mechanisms of rotation and then describe characteristics of natural crystallographic fabrics found in olivine and in quartz.

In a deforming continuum, the **spin** at a point is the average of the instantaneous rates of rotation of all material lines of all orientations through that point, about each of the reference axes.[6] The angular velocities of the material lines in a continuum undergoing either progressive pure shear or progressive simple shear are shown in Figure 19.25A, B, respectively. The length of the arrow at the tip of each material line indicates the magnitude of the angular velocity of that line. The symmetry of these arrows in Figure 19.25A shows that for progressive pure shear, the average of these angular velocities—and therefore the spin—is zero. For progressive simple shear, however (Figure 19.25B), the spin is clearly not zero.

To describe the essential idea of the Taylor–Bishop–Hill theory, we consider the simplified model of a crystal grain of random orientation imbedded in a continuum undergoing a macroscopic progressive pure shear. We assume that the crystal has only one slip direction on one slip plane which is oriented normal to the crystallographic c axis. We also assume that the deformation in the crystal grain is homogeneous (Figure

19.25C, D). The theory then requires the spin of the crystal to be identical to that in the surrounding continuum; that is, it must be zero in this case (Figure 19.25A). Slip on the one slip system in the crystal produces a progressive simple shear such that the strain ellipse in the crystal is identical in shape to that of the macroscopic pure shear. The simple shearing, however, has a nonzero spin (Figure 19.25B). The spin in the crystal can be made zero if a compensating rigid-body angular velocity is added to the progressive simple shear (Figure 19.25E, F). This angular velocity maintains the principal axes of strain in the crystal parallel to the principal axes of macroscopic strain. In so doing, it rotates the slip plane toward a limiting orientation normal to the shortening direction and rotates the c axis toward $\hat{\zeta}_3$, the axis of maximum incremental shortening. Note that the crystallographic axes, such as the c axis, are not material lines and are not rotated by shear on the slip planes. The rate of rotation decreases toward zero as the limiting orientation is approached.

If we apply this model to all grains of a polycrystalline aggregate and ignore the constraints of the von Mises condition (see Section 19.5), we predict a symmetrically distributed maximum of c axes around the direction of maximum incremental shortening $\hat{\zeta}_3$ which means, because this example is a coaxial deformation, around the maximum finite shortening axis $\hat{s}_3$ as well. The concentration of the maximum increases with increasing strain.

Under similar conditions, a progressive simple shear of the continuum should cause the c axis of the crystal to rotate toward the perpendicular to the macroscopic shear plane. In this limiting orientation, the crystallographic shear plane is parallel to the macroscopic shear plane, and the incremental strain rate and spin are the same in the crystal as in the surrounding continuum. The rate of rotation of the crystal decreases asymptotically toward zero as it approaches the limiting orientation.

For the coherent deformation of a polycrystalline aggregate, however, crystals must satisfy the von Mises condition of having five independent slip systems. The analysis and the preferred orientation patterns consequently become more complex. The crystallographic fabrics that develop are affected by different choices for the independent slip systems, by their relative ease of operation, by the deformation path, and by any initial crystallographic fabric in the material. The basic principles described above, however, still apply for understanding the production of crystallographic preferred orientation fabrics.

In many minerals, it is unlikely that there are five independent slip systems that actually operate during ductile deformation. Having to choose five to permit an

---

[6] This property of the spin can be shown to be a consequence of its definition as the antisymmetric part of the velocity gradient tensor.

$$w_{k\ell} = 0.5 \left[\frac{\partial v_k}{\partial x_\ell} - \frac{\partial v_\ell}{\partial x_k} \right]$$

where the antisymmetric character means $w_{k\ell} = -w_{\ell k}$. Adding $\dot{\varepsilon}_{k\ell}$ from Equation (18.1.7) to $w_{k\ell}$ shows that the velocity gradient tensor $\partial v_k/\partial x_\ell$ is the sum of a symmetric part, which is the incremental strain rate $\dot{\varepsilon}_{k\ell}$, and an antisymmetric part, which is the spin $w_{k\ell}$. It can be shown further that the spin is the angular velocity of the material lines parallel to the principal axes of the incremental strain rate tensor $\dot{\varepsilon}_{k\ell}$. The definition shows that there are only three independent components to the tensor, because for $k = \ell$, $w_{k\ell} = 0$. Thus we can define an equivalent vorticity vector ω_i whose components are:

$$[\omega_1, \omega_2, \omega_3] \equiv [(w_{32} - w_{23}), (w_{13} - w_{31}), (w_{21} - w_{12})]$$

(compare footnote 9 in Section 15.4). Some authors refer to the spin tensor as the vorticity tensor and give the word *spin* a specialized definition.

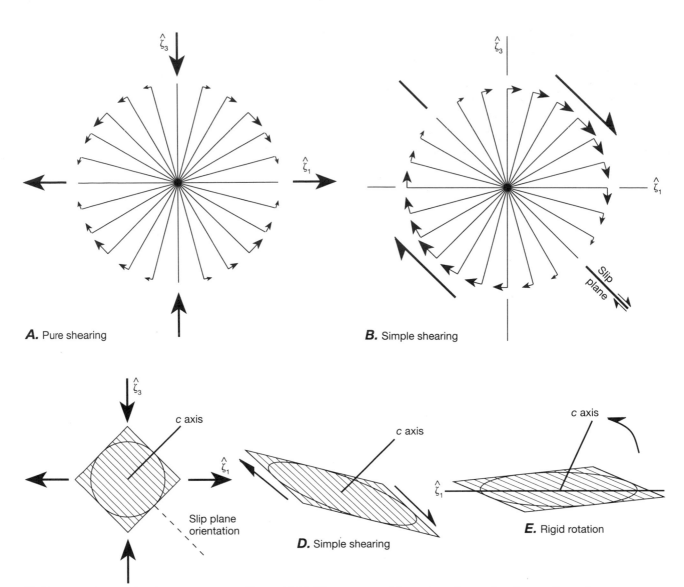

A. Pure shearing

B. Simple shearing

C. Macroscopic pure shearing

D. Simple shearing

E. Rigid rotation

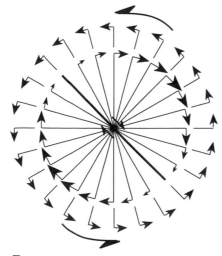

F. Simple shearing and rigid rotation rate to give zero spin

Figure 19.25 The process of formation of a preferred crystallographic orientation by dislocation glide. *A.* During progressive pure shear, the angular velocities of material lines, indicated by the arrows at the ends of the lines, are distributed symmetrically about the principal axes of incremental strain, and the spin is therefore zero. *B.* For progressive simple shear, the angular velocities of the material lines are not symmetric about the incremental strain axes, and the average of these rotational velocities—the spin—is not zero. *C.* The initial orientation of the crystal. The set of parallel solid lines indicates the orientation of the slip planes; the slip direction is in the plane of the diagram; the line normal to the slip planes shows the orientation of the optic *c* axis. *D.* Macroscopic pure shearing as in part *A* can be accommodated in the crystal only by simple shearing on the single slip system. The crystal deforms to a strain equal to the macroscopic pure shear strain. The slip does not rotate the *c* axis, but it does rotate material lines in the crystal. The spin is not zero, as shown in part *B*. *E.* The rigid-body angular velocity of the crystal reduces the spin to zero and keeps the principal axes of crystal strain parallel to the macroscopic strain axes as in part *A*. The angular velocity rotates the *c* axis toward the direction of maximum incremental shortening. *F.* The rigid-body angular velocity of the crystal grain (indicated on the outer circle) is of such a magnitude that when it is added to the angular velocities from progressive simple shear (inner circle; compare part *B*), the resulting angular velocities of the material lines are the same as for pure shearing (part *A*). That is, the spin is zero. The rigid-body angular velocity reorients the *c* axis, causing it to rotate toward the principal shortening direction.

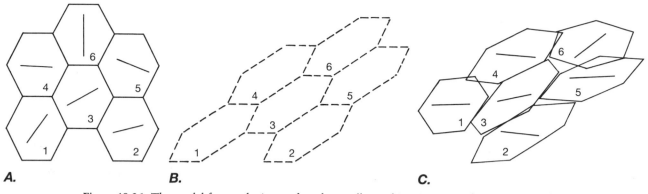

Figure 19.26 The model for producing preferred crystallographic orientations by minimizing the geometric misfit in a polycrystalline aggregate composed of crystals that have only one slip system. A. The undeformed polycrystal. Lines in each crystal show the orientation of the slip plane. The slip direction is in the plane of the diagram. B. The form the crystals would take if the deformation were homogeneous. C. The best approximation to the homogeneous form of the deformation, given that crystals can slip on only one slip system. The area of misfit plus overlap is a minimum.

arbitrary deformation of the crystals may not be realistic and may lead to preferred orientations that do not develop in nature.

Crystallographic rotations can also result from the requirement that the geometric misfit of a heterogeneously deforming polycrystal be a minimum. The essence of the misfit minimization theory is illustrated in Figure 19.26. Figure 19.26A shows an idealized undeformed polycrystal. The lines within each grain show the orientation of the single permitted slip direction. The von Mises condition is ignored. Figure 19.26B shows the polycrystal as it would look if the deformation were homogeneous. Figure 19.26C shows the closest approximation to the homogeneous deformation that can be attained, given that each crystal has only one slip system. Rotations and translations of the crystals are introduced in order to minimize the gaps and overlaps resulting from the heterogeneity of the actual crystal slip. It is these rotations that result in the reorientation of the crystal lattices and in the development of a preferred orientation. The remaining gaps and overlaps are ignored, the implicit assumption being that other deformation mechanisms (including solution and diffusion) can operate in nature to keep the deformation coherent. This model also predicts a concentration of crystallographic slip planes subparallel to the plane of flattening in coaxial deformation and subparallel to the shear plane in simple shear.

No simple model has emerged from experimental results to predict the effects of either dynamic or static recrystallization on preferred orientation fabrics. In general, the orientation of new grains produced during static recrystallization seems to be controlled by the orientation of the original grain, although the crystallographic relationship is not a simple one. Dynamic recrystallization, however, does not seem to alter preferred orientation fabrics significantly.

Olivine Fabrics

The dominant slip systems in olivine change with temperature and strain rate, as shown by the crystal diagrams and the experimental data plotted in Figure 19.27 (see also Table 19.1). At low temperatures the dominant slip system is {110}[001], which means the slip planes are those that are symmetrically equivalent to the (110) plane, and the slip direction is [001] (see the footnote to Table 19.1 for a fuller explanation of the notation for specifying slip systems). At intermediate temperatures the dominant slip system is {0kℓ}[100], where k and $ℓ$ can have any integer value. For this slip system, called **pencil glide**, any plane parallel to the [100] direction is a slip plane that has the slip direction parallel to [100]. At the highest temperatures, the dominant slip system is (010)[100].

Examples of preferred orientation fabrics for the [010] crystallographic axis in naturally deformed olivine are shown in Figure 19.28. In Figure 19.28A, the [010] axes form a maximum normal to the foliation plane but are somewhat spread out in a girdle also normal to the foliation. If the foliation is interpreted to be the plane of flattening of the finite strain ellipsoid, $\hat{s}_1$–$\hat{s}_2$, with the lineation parallel to $\hat{s}_1$, the symmetry of the [010] fabric with respect to the foliation indicates a coaxial deformation. The [010] maximum then corresponds with the shortening direction, as would be expected for dominant slip on the high-temperature single slip system (010)[100] (Figure 19.27). The girdle distribution is consistent with a contribution of slip on the multiple slip systems {0kℓ}[100], and this distribution indicates that deformation occurred under the moderate-temperature conditions that lead to the activation of multiple slip (Figure 19.28).

The [010] fabrics in Figure 19.28B and C are for unrecrystallized and recrystallized grains, respectively,

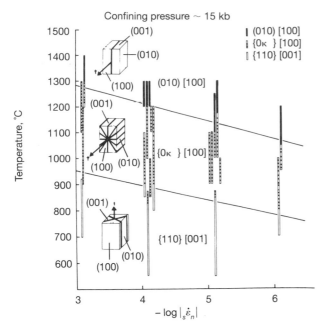

Figure 19.27 Experimentally determined slip systems that operate in olivine as a function of temperature and strain rate. The different patterns on the vertical bars indicate the conditions under which the different slip systems have been observed. Parentheses, ( ), indicate the Miller indices of particular crystallographic slip planes; braces, { }, indicate a complete set of symmetrically related crystallographic slip planes; and brackets, [ ], indicate the coordinates of particular crystallographic slip vectors. The blocks represent an olivine crystal on which are indicated the active slip planes and slip direction (arrow).

in a naturally deformed peridotite. They indicate deformation under noncoaxial conditions because the [010] maximum is inclined to the foliation, not normal to it. For unrecrystallized grains (Figure 19.28B), the fabric is consistent with slip on a single slip system (010)[100] in a simple shear deformation. The [010] maximum is diffuse but at a high angle to the inferred macroscopic shear plane, indicated by the pair of opposing arrows. The foliation plane indicates the orientation of the plane of flattening, $\hat{s}_1-\hat{s}_2$, and the acute angle between the plane of flattening and the shear plane indicates the sense of shear that produced the fabric (compare with Figure 15.15).

The angle between the shear plane and the foliation plane should decrease with increasing strain, as long as the fabric is not affected by dynamic recrystallization. This relationship can be used to infer the minimum magnitude of the shear strain. The shear strain indicated by Figure 19.28B falls roughly between the strains shown in Figure 15.15C and D. In the same rock, the newly recrystallized grains show a pattern (Figure 19.28C) similar to, but even more diffuse than, the unrecrystallized grains (Figure 19.28B). The interpretation of fabrics in this manner can provide essential information about the coaxial or noncoaxial nature of the deformation and at least a minimum estimate of the magnitude of the shear in naturally deformed rocks. Olivine fabrics of this nature have been used to infer the relative motion of fossil subduction zones as reflected in peridotites at the base of thrust faults that lie beneath blocks of mantle rocks.

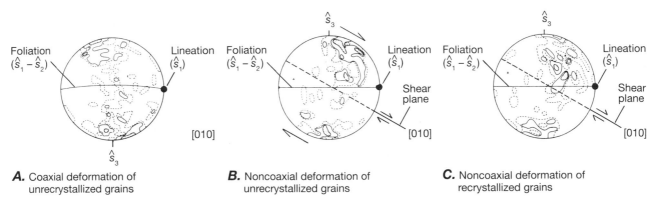

A. Coaxial deformation of unrecrystallized grains

B. Noncoaxial deformation of unrecrystallized grains

C. Noncoaxial deformation of recrystallized grains

Figure 19.28 [010] olivine fabrics from naturally deformed dunites. Lower-hemisphere, equal-area projections; contours at 1, 2, 4, and 8 percent per 0.45 percent area. A. [010] olivine fabric from 100 unrecrystallized grains from the Baldissero ultramafic body (lvrea zone, western Alps). A girdle normal to the foliation plane contains a maximum normal to the foliation. The symmetry indicates coaxial deformation. If the foliation is taken to be the $\hat{s}_1-\hat{s}_2$ plane of the finite strain ellipsoid (the plane of flattening), and the lineation is the $\hat{s}_1$ direction, then this pattern is consistent with (010)[001] slip producing the maximum and {0kℓ}[100] slip producing the girdle. The inference is that deformation was at moderate temperatures for ductile deformation in olivine. B and C. [010] olivine fabrics from the Lanzo ultramafic body (lvrea zone, western Alps). B. [010] fabric in 100 unrecrystallized olivine grains. C. [010] fabric in 100 dynamically recrystallized olivine grains. Both show a diffuse maximum inclined to the foliation. This indicates noncoaxial deformation and is consistent with dominant glide on (010)[100] during deformation having a strong component of simple shear.

Olivine fabrics produced experimentally at high temperatures during coaxial deformation are consistent with the natural fabric in Figure 19.28A. In these experiments, the unrecrystallized and recrystallized grains showed the same fabric. The models presented at the beginning of this section for the formation of preferred orientations of crystals with one slip system during progressive pure shear predict the same fabrics as these natural and experimental olivine fabrics.

Quartz Fabrics

Quartz is a more difficult mineral to understand than olivine because of its higher symmetry, because of its greater number of possible slip systems, and possibly also because the conditions in the crust under which quartz is commonly deformed span the range of conditions over which the preferred slip systems in the mineral change. The major slip planes and directions that have been identified in quartz on the basis of experimental deformation of single crystals are shown in Figure 19.29 (see also Table 19.1). At lower temperatures and higher strain rates, slip most commonly occurs on the basal plane, which is normal to the c axis, in a direction parallel to one of the a axes, $(0001)\langle 11\bar{2}0\rangle$. At higher temperatures and lower strain rates, slip occurs on one of the prism planes m parallel to an a axis or to c, $\{10\bar{1}0\}\langle 1\bar{2}10\rangle$, $[0001]$, and on the rhomb planes r parallel to the a and the $c + a$ directions $\{1\bar{1}01\}\langle 11\bar{2}0\rangle$, $\langle 11\bar{2}3\rangle$. Numerous other slip systems also have been observed, including slip on the prism

planes m parallel to $c + a$ and slip on second-order pyramidal planes $\{2\bar{1}\bar{1}1\}$ parallel to $c + a$.

In natural rocks, however, fabric data for both the c and a axes shows that there is a dominant tendency for an a axis to be oriented parallel to the direction of slip, indicating that it is the dominant slip direction. This direction is common to several slip planes, such as the basal plane, the prism planes m, and the positive and negative rhombohedral planes r and z. Thus the preferred orientation of c axes depends on the geometry of the deformation and on the particular slip planes that are active.

Figure 19.30 shows idealizations of the common types of orthorhombic quartz c-axis and a-axis fabrics, along with examples of natural c-axis fabric patterns. Diagrams A through E are superimposed on a Flinn diagram illustrating the states of strain associated with the various patterns. The ratios of the principal strains for the natural fabrics (the large dots in Figure 19.30) were determined largely from the shapes of deformed grains in the rocks. The idealized c-axis diagrams are **fabric skeletons,** which one constructs by drawing lines on contoured fabric data through the maxima and along the lines of maximum concentration. The idealized a-axis fabrics are shown as schematic contour diagrams. Although a-axis patterns provide additional information that is important for interpreting the kinematic significance of the patterns, their determination requires x-ray techniques and sophisticated computer analysis. Determining these complete fabrics is difficult and time-consuming and has been done in only a few studies.

The symmetry of the fabrics in Figure 19.30 with respect to the principal stretches indicates that the de-

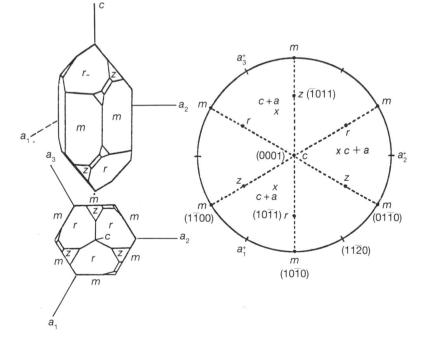

Figure 19.29 Crystal faces and directions in quartz, showing the most important slip planes and slip directions. A. Crystal morphology with crystallographic axes and faces labeled. Here m indicates the prism planes; r and z indicate the positive and negative rhombohedral planes, respectively; and the plane normal to c is the basal plane. B. Upper-hemisphere equal-angle projection of the poles to crystallographic planes (dots) and the crystallographic directions (x's) in quartz, oriented as shown in the lower of the two diagrams in part A.

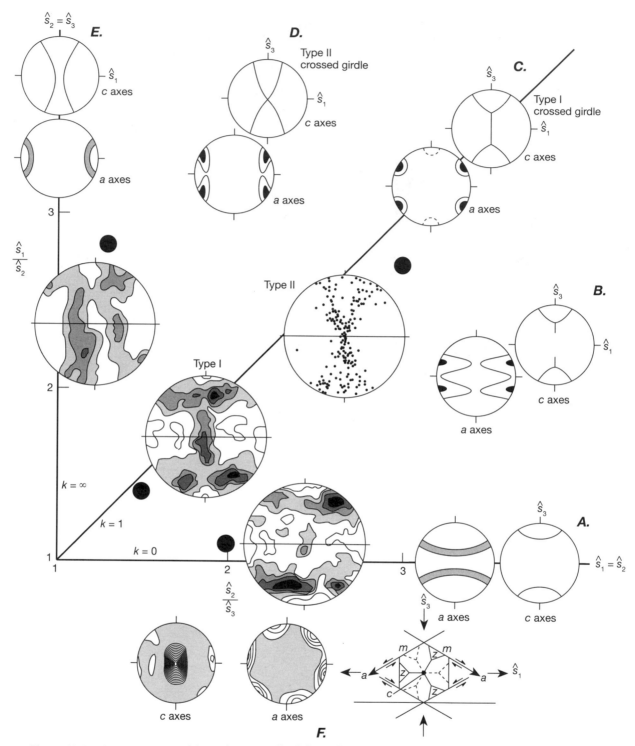

Figure 19.30 Common quartz fabrics for coaxially deformed rocks: natural *c*-axis fabrics, idealized *c*-axis fabric skeletons, and schematic contours of *a*-axis fabrics as a function of the value of *k* on a Flinn diagram. The large black dots plotted on the Flinn diagram indicate the principal stretch ratios for the natural fabrics, which have been independently determined from deformed grain shapes. In each diagram, the foliation plane is oriented E–W and vertical. A. Simple flattening ($k = 0$). B. General flattening ($0 < k < 1$). C. Plane strain ($k = 1$). The type I crossed girdle fabric consists of a small-circle girdle with a connecting partial great-circle girdle. The type II cross-girdle fabric consists of two crossing great-circle girdles. D. General extension ($1 < k < \infty$), showing a type II crossed girdle pattern transitional from plane strain to simple extension. E. Simple extension ($k = \infty$). F. Quartz *c*-axis fabric consisting of a single maximum parallel to the foliation and perpendicular to the lineation, the associated *a*-axis fabric, and the interpretation of the active slip systems. Such fabrics are common in high-grade metamorphic rocks.

formation was coaxial. For simple flattening ($k = 0$), c axes tend to lie on a small-circle girdle about the maximum shortening direction $\hat{s}_3$, and a axes also define a small-circle girdle about $\hat{s}_3$ that has a larger opening angle (Figure 19.30A). In the general flattening field ($0 < k < 1$), patterns are transitional between simple flattening and plane strain (Figure 19.30B). For plane strain ($k = 1$), the c axes define a crossed girdle pattern (Figure 19.30C). The **type I crossed girdle pattern** looks like a distorted small-circle girdle about the maximum shortening direction $\hat{s}_3$, connected by a partial great-circle girdle in the $\hat{s}_2$–$\hat{s}_3$ plane. The a axes fall predominantly into two maxima symmetric about the direction of maximum extension $\hat{s}_1$ (Figure 19.30C). Less common is the **type II crossed girdle pattern,** which approximates a pair of crossing great-circle distributions. This pattern has been associated both with plane strain (Figure 19.30C) and with a general extension (Figure 19.30D) for which the pattern represents a transition to the pattern for simple extension. In simple extension, the c axes tend to define a **cleft girdle pattern,** which consists of a small circle with a large opening angle centered on the direction of maximum extension $\hat{s}_1$ (Figure 19.30E). The associated a-axis pattern evolves toward a small-circle distribution with a small opening angle also centered about the $\hat{s}_1$ direction.

Another c-axis pattern common in high-grade metamorphic rocks is a single maximum parallel to the foliation and normal to the lineation (Figure 19.30F). The a-axis pattern mimics the distribution of the a axes in a quartz single crystal.

These fabrics consistently show that the a directions of the crystals have a strong preferred orientation symmetric relative to the principal stretches. The associated c-axis orientations then consistently place one of the principal slip planes containing the a axis in an orientation of high shear strain. The implication is that a is the dominant direction of slip and that the slip systems tend to become oriented in such a way as to accommodate the imposed deformation. Most of the orthorhombic c-axis fabric diagrams are more easily understood after we discuss the geometrically simpler case of simple shear. For Figure 19.30F, however, the preferred crystallographic orientation accommodates coaxial plane strain by slip in the a direction on two prism planes symmetrically oriented with respect to $\hat{s}_1$ and $\hat{s}_3$.

In noncoaxial deformation, the c-axis pattern becomes asymmetric with respect to the foliation in the rock—and therefore presumably with respect to the plane of flattening $\hat{s}_1$–$\hat{s}_2$. Figure 19.31 shows idealized c-axis fabric skeletons and a-axis contour diagrams, along with some examples of c-axis fabrics from naturally sheared rocks. The diagrams are plotted so that the foliation plane is the vertical great circle from left

to right across each stereogram, and the shear sense on the shear plane is dextral. The progression from part A to part E represents an increase in the angle Ω between the maximum principal axes of the incremental and the finite strain ellipses, $\hat{\zeta}_1$ and $\hat{s}_1$. As the amount of shear increases from zero to infinity, the angle Ω increases from $0°$ to $45°$, which reflects the fact that with increasing shear strain, $\hat{s}_1$ rotates toward parallelism with the shear plane.

For simple shear, the initial c-axis fabric skeleton is a type I crossed girdle pattern characteristic of coaxial plane strain (compare Figures 19.31A and 19.30C). Ω is essentially zero, and the shear plane is at $45°$ to the center of the distorted small-circle girdle. With increasing strain, c axes tend to concentrate toward a great-circle girdle roughly normal to the shear plane. The inclination of the great-circle girdle relative to the foliation, which marks the $\hat{s}_1 - \hat{s}_2$ plane, indicates the direction of relative shear on the shear plane. Thus, well-defined quartz fabrics, asymmetric with respect to the foliation, can be used to infer the sense of shearing in the rocks. The type of information, if gathered across an entire region, is useful in inferring the kinematic history.

In simple shear, the crystallographic slip direction tends to become aligned with the direction of macroscopic shear, and the crystallographic slip planes tend to become oriented parallel to the plane of simple shear. Because slip in the a crystallographic direction appears to be dominant in naturally deformed quartz, particular c-axis maxima can be associated with slip on particular slip systems in the a direction. Thus in Figure 19.32A, maximum I is associated with slip on the prism planes $\{m\}\langle a \rangle$, because with c in this orientation, one m plane is parallel to the shear plane, and the a axis in that m plane is parallel to the shear direction (Figure 19.32B). Similarly, maxima II (Figure 19.32A) are associated with slip on the rhombohedral planes $\{r\}\langle a \rangle$ and $\{z\}\langle a \rangle$ (Figure 19.32C), and maximum III (Figure 19.32A) is associated with slip on the basal planes, which is normal to the c axis, in one of the a directions—that is, $(0001)\langle a \rangle$ (Figure 19.32D).

From this association of c-axis maxima with different slip systems, we can see that the c-axis patterns that exhibit orthorhombic symmetry are consistent with a coaxial deformation accommodated on slip systems symmetrically oriented relative to the principal axes of strain. Thus, for example, from Figure 19.32A we interpret the crossed girdle patterns to indicate that coaxial deformation was accommodated by slip in two directions symmetrically oriented with respect to both $\hat{s}_1$ and $\hat{s}_3$ (Figure 19.32E).

Deformation experiments have successfully reproduced some aspects of these quartz fabrics. Single c-axis maxima and small-circle girdle distributions have been

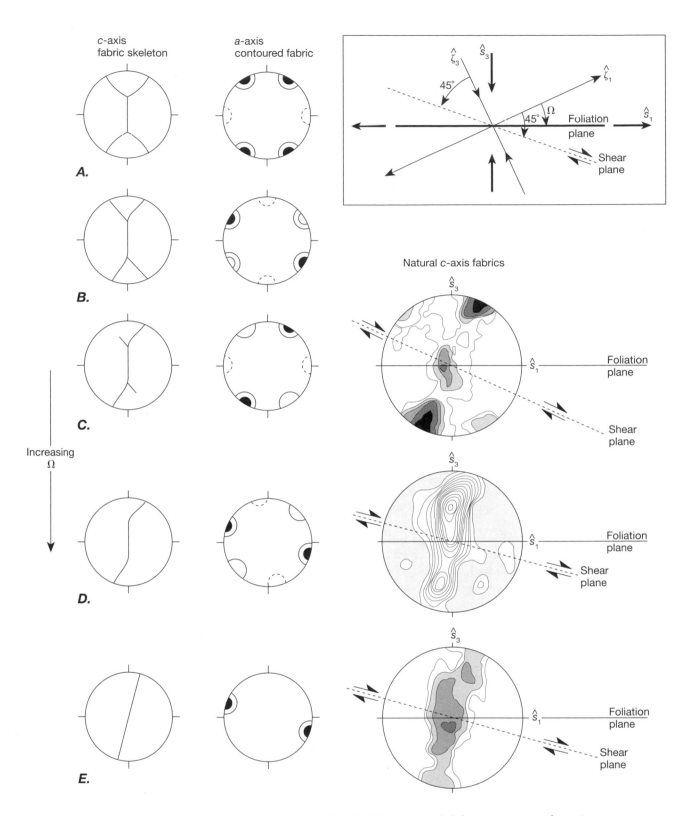

Figure 19.31 Common quartz fabrics associated with noncoaxial deformation: natural *c*-axis fabrics, idealized *c*-axis fabric skeletons, and schematic contours of *a*-axis fabrics as a function of increasing amounts of noncoaxial deformation (parts *A* through *E*), as measured by increasing values of the angle Ω between the maximum principal axes of the incremental and finite strain ellipses.

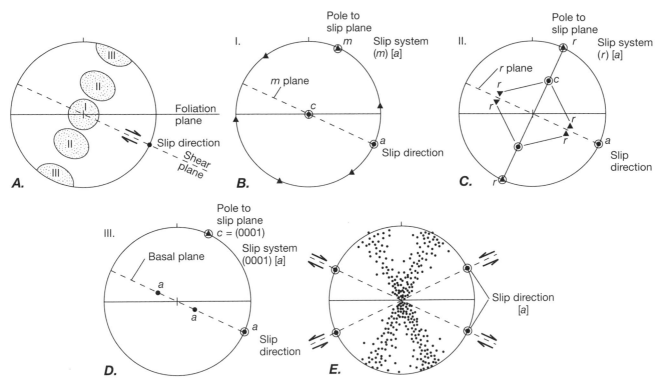

Figure 19.32 Slip systems in quartz associated with particular c-axis maxima, assuming that the active slip plane must be parallel to the shear plane and that a is the crystallographic slip direction. Crystallographic axes are shown as dots, poles to crystallographic planes as triangles. Circled points indicate directions and poles relevant to the active slip system. A. Locations of three common c-axis maxima, I, II, and III, in noncoaxial deformations, relative to the foliation and shear planes. B. Crystallographic orientation for maximum I associated with prismatic slip $(m)[a]$ (compare Figure 19.30F). C. Crystallographic orientation for maximum II associated with rhombohedral slip $(r)[a]$ and $(z)[a]$. D. Crystallographic orientation for maximum III associated with basal slip $(0001)[a]$. E. Hypothetical type II crossed-girdle c-axis pattern and the inferred slip directions and slip plane orientations associated with it. Crystallographic slip directions, commonly a-axes, tend to become aligned parallel to the circled dots (compare parts A through D).

produced experimentally under conditions of simple flattening (compare Figure 19.30A). The transition from a maximum to a small-circle girdle with progressively increasing angle occurs as temperature increases and strain rate decreases. This transition is consistent with a shift from predominantly basal plane slip parallel to a (see maximum III in Figure 19.32A, D), to mixed basal plane and prism slip both parallel to a with an increasing dominance of the prism over the basal slip system (compare maximum I in Figure 19.32A, B). Experiments in simple shear with a component of axial shortening have also succeeded in producing the asymmetric fabrics characteristic of natural noncoaxial deformation.

The fabrics that develop depend in large part on which of the slip systems require the least critical resolved shear stress in order to be activated. This critical shear stress for a particular slip system varies with temperature and with the dissolved water content of the crystal, and at best it is only sketchily known for the

different slip systems in quartz. Nevertheless, by using reasonable values for the different slip systems, investigators making theoretical calculations with the Taylor–Bishop–Hill model have produced many of the characteristics of quartz c-axis fabrics shown in Figures 19.30 and 19.31.

For simple shear, the Taylor–Bishop–Hill model and the misfit minimization model yield quite different results. The Taylor–Bishop–Hill model predicts a c-axis fabric skeleton comparable in many respects to that in Figure 19.31B and C, but the fabric does not evolve to the single girdle of Figure 19.31E. The misfit minimization model predicts the formation of a single girdle inclined to the $\hat{s}_1$–$\hat{s}_2$ plane, similar to the fabric skeleton of Figure 19.31E.

Thus although the theoretical models successfully reproduce various aspects of naturally observed quartz c-axis fabrics, they are not completely successful. We need better knowledge of the relative ease of slip on the different slip systems and of the effect of conditions such as temperature and dissolved water content, as well as more sophisticated models that incorporate the effects of both spin and geometric fit. We can anticipate that such developments will make it possible to interpret quartz fabric data more accurately and afford us greater insight into the deformation conditions under which the fabrics were produced.

Additional Readings

Ashby, M. F., and R. A. Verrall. 1977. Micromechanisms of flow and fracture, and their relevance to the rheology of the upper mantle. *Phil. Trans. Roy. Soc. Lond.* 288A: 59–95.

Ball, A., and S. White. 1977. An etching technique for revealing dislocation structure in deformed quartz grains. *Tectonophysics* 37(4): T9–T14.

Bell, T. H., and M. A. Etheridge. 1976. The deformation and recrystallization of quartz in a mylonite zone, central Australia. *Tectonophysics* 32: 235–269.

Carter, N. L. 1976. Steady state flow of rocks. *Rev. of Geophys. and Space Phys.* 14: 301–360.

Carter, N. L., and M. Friedman. 1965. Dynamic analysis of deformed quartz and calcite from the Dry Creek Ridge anticline, Montana. *Am. Jr. of Sci.* 263: 747–785.

Carter, N. L., and S. H. Kirby. 1978. Transient creep and semi-brittle behavior of crystalline rocks. *Pure and Appl. Geophys.* 116: 807–839.

Etchecopar, A., and G. Vasseur. 1987. A 3-D kinematic model of fabric development in a polycrystalline aggregate: Comparisons with experimental and natural examples. *J. Struct. Geol.* 9: 705–717.

Etheridge, M. A., and J. C. Wilkie. 1981. An assessment of dynamically recrystallized grain size as a paleopiezometer in quartz-bearing mylonite zones. *Tectonophysics* 78: 475–508.

Friedman, M. 1964. Petrofabric techniques for the determination of principal stress directions in rocks. In W. R. Judd, ed., *State of stress in the Earth's crust.* New York: American Elsevier, pp. 450–552.

Hirth, J. P., and J. Lothe. 1968. *Theory of Dislocations,* New York: McGraw-Hill.

Karato, S.-I., M. S. Paterson, and J. D. FitzGerald. 1986. Rheology of synthetic olivine aggregates: Influences of grain size and water. *Jr. Geophys. Res.* 91: 8151–8176.

Karato, S.-I. 1984. Grain size distribution and rheology of the upper mantle. *Tectonophysics* 104: 155–176.

Kirby, S. H. 1983. Rheology of the lithosphere. *Rev. of Geophys. and Space Phys.* 21: 1458–1487.

Kirby, S. H., and A. K. Kronenberg. 1987. Rheology of the lithosphere: Selected topics. *Rev. of Geophys. and Space Phys.* 25: 1219–1244.

Knipe, R. J. 1989. Deformation mechanisms—recognition from natural tectonites. *Jour. Struct. Geol.* 11(1/2): 127–146.

Kohlstedt, D. L., and M. S. Weathers. 1980. Deformation-induced microstructures, paleopiezometers, and differential stresses in deeply eroded fault zones. *Jr. Geophys. Res.* 85: 6269–6285.

Lister, G. S., and B. E. Hobbs. 1980. The simulation of fabric development during plastic deformation and its application to quartzite: The influence of deformation history. *J. Struct. Geol.* 2: 355–370.

Lister, G. S., and M. S. Paterson. 1979. The simulation of fabric development during plastic deformation and its application to quartzite: Fabric transitions. *J. Struct. Geol.* 1: 283–297.

Lister, G. S., M. S. Paterson, and B. E. Hobbs. 1978. The simulation of fabric development in plastic deformation and its application to quartzite: The model. *Tectonophysics* 45: 107–158.

Lister, G. S., and P. F. Williams, 1983. The partitioning of deformation in flowing rock masses. *Tectonophysics* 92: 1–33.

Nicolas, A., F. Boudier, and A. M. Boullier. 1973. Mechanisms of flow in naturally and experimentally deformed peridotites. *Am. Jr. Sci.* 273: 853–876.

Nicolas, A., and J. P. Poirier. 1976. *Crystalline plasticity and solid state flow in metamorphic rocks,* New York: Wiley.

Price, G. P. 1985. Preferred orientations in quartzites. In H.-R. Wenk, ed., *Preferred orientation in deformed metals and rocks: An introduction to modern texture analysis.* New York: Academic Press, pp. 385–406.

Rutter, E. H. 1976. The kinetics of rock deformation by pressure solution. *Phil. Trans. Roy. Soc. Lond.* A283: 203–219.

Schmid, S. M. 1982. Microfabric studies as indicators of deformation mechanisms and flow laws operative in mountain building. In K. J. Hsü, ed., *Mountain building processes.* New York: Academic Press, pp. 95–110.

Schmid, S. M., and M. Casey. 1986. Complete fabric analysis of some commonly observed quartz c-axis patterns. In B. E. Hobbs and H. C. Heard, eds., *Mineral and rock deformation: laboratory studies. The Paterson volume.* Geophysical Monograph 36. Am. Geophys. Union, Washington D. C., 263–286.

Twiss, R. J. 1986. Variable sensitivity piezometric equations for dislocation density and subgrain diameter and their relevance to olivine and quartz. In H. C. Heard and B. E. Hobbs, eds., *Mineral and rock deformation: laboratory studies. The Paterson volume.* Geophysical Monograph 36, Am. Geophys. Union, Washington D. C., 247–261.

Tullis, J. A. 1979. High temperature deformation of rocks and minerals. *Rev. of Geophys. and Space Phys.* 17: 1137.

Weertman, J., and J. R. Weertman. 1964. *Elementary dislocation theory.* New York: Macmillan.

Wenk, H. R., ed. 1985. *Preferred orientation in deformed metals and rocks: An introduction to modern texture analysis.* New York: Academic Press.

CHAPTER

20 Quantitative and Scale Models of Rock Deformation

A principal reason why we study the rheology of rocks under various conditions is so that we have a firm basis for formulating mechanical models of deformation. We can then test the models by comparing model predictions with actual observations. If a model reproduces the observed characteristics accurately and reliably, we accept that the model can provide insight into the natural conditions of deformation.

In Chapters 12, 14, 16, and 17, the emphasis is on kinematic models of ductile deformation for which we prescribe the motion. Such models do not attempt to account for the motions as a necessary consequence of the rheology of the material. Models that incorporate the rheology, however, can help us evaluate whether the kinematic models are mechanically possible and reasonable and can give us a deeper level of understanding of the origin of the structures.

We use both mathematical and scale models to study the mechanics of rock deformation. **Mathematical models** describe material behavior in terms of equations which we solve to find the distribution of displacement, stress, and strain throughout a deforming body. In principle, we can solve the equations either analytically or numerically. Analytic solutions are general solutions to the equations that express the values of the unknown quantities as a mathematical function of position and material properties. Thus the solution can be evaluated for a wide variety of specific conditions, and we can determine from the general solution how changes in

conditions, such as the viscosity, affect the specific solution. Analytic solutions, however, can be found only if the geometry of the body and that of the deformation are both relatively simple, and they generally apply only for small strains. Numerical solutions provide the approximate values of the unknown quantities at a specific set of points throughout the body. The solutions are not general, however; they are specific to a particular set of conditions. Thus to investigate the effect of a change in viscosity, for example, we must calculate another specific solution for the different viscosity. The advantage of numerical solutions for investigating geologic deformations, however, is that we can analyze geometrically complex bodies and large and complex deformations.

Scale models are actual physical models of parts of the Earth. The models are constructed with materials whose properties scale in such a way that their behavior over short times and small distances reproduces the behavior of rocks over long periods of time and large distances. The correct choice of the model materials requires a knowledge of scaling theory as well as appropriate compromises, because it is difficult if not impossible to scale the model correctly in every respect.

The literature on mechanical models in geology is very large. In this chapter, therefore, we can only introduce some fundamental ideas behind mathematical and scale models of rock deformation and give a few examples of each type of model. Our intent is to provide some insight into how the models are constructed and to demonstrate the value of mechanical models in helping us understand geologic structures.

20.1 Formulation of a Mathematical Model

In making a mathematical model, we wish to calculate the distribution of stress, strain, and displacement throughout a body by specifying only the mechanical properties of the body and the stresses or the displacements that are imposed on its boundaries. To solve any set of equations, we need as many equations as there are unknowns. Problems in mechanics generally require three different types of equations: constitutive equations, conservation laws (or balance equations), and boundary conditions (see Box 20.1).

Constitutive Equations

In Chapter 18 we discussed the constitutive equations that describe the behavior of elastic, viscous, plastic, and power-law materials. These equations express the stress as some function of the strain or the strain rate, or they express one of these variables as some function of the stress. For a viscous material, for example, the deviatoric stress is directly proportional to the strain rate. The constitutive equation (Equation 18.3.4) represents six independent equations (compare Equations 18.3.1 for elastic materials). There are, however, twelve independent quantities that we do not know a priori: six independent stress components $\sigma_{k\ell}$ and six independent strain rate components $\dot{\varepsilon}_{k\ell}$. Clearly, we do not have enough equations to determine all twelve unknowns from the constitutive equations alone.

Equation (18.1.7) gives us six equations that define the strain rate in terms of velocity. We can thus write the constitutive equation in terms of the three velocity components v_k instead of the six strain rate components. This gives us six constitutive equations for the nine unknowns $\sigma_{k\ell}$ and v_k but we still do not have a complete system of equations.

Conservation Laws

In general, the constitutive equations alone are not sufficient to solve a problem involving the deformation of a body. The motion of each point in a body, however, must also obey the fundamental conservation laws of physics. For the strictly mechanical problems that we are concerned with here, these conservation laws include the conservation of mass, the conservation of momentum, and the conservation of angular momentum.

The **conservation of mass** requires that mass be neither created nor destroyed, and this condition is automatically satisfied if we assume constant-volume deformation and a homogeneous mass density.

The **conservation of momentum** is Newton's second law. It states that the net force on a body is equal to the mass multiplied by the acceleration.

$$\sum_i \mathbf{F}^{(i)} = m\mathbf{a} \qquad (20.1)$$

Where $\mathbf{F}^{(i)}$ include all the possible forces acting on a body.

The **conservation of angular momentum** requires that the net torque on the body be equal to its moment of inertia multiplied by its angular acceleration. At a material point, however, the net torque must be zero, because the point has infinitesimal dimension and hence an infinitesimal moment arm. The symmetry of the stress tensor, Equations (8.24), is an expression of this requirement.

Geologic deformations are generally so slow that the acceleration is negligible. Under these circumstances, the right side of Equation (20.1) becomes zero, and Newton's second law reduces to the **equilibrium equation,** which requires that all the forces on a body be balanced. If we ignore the force of gravity, which we generally can do only for deformation on a relatively small scale (see, for example, Box 20.1), the only forces we must consider are those provided by the resistance of the material to deformation, and these are expressed by the stresses in the body. At each point in the body, the equilibrium equation then takes the form

$$\sum_{k=1}^{3} \frac{\partial \sigma_{k\ell}}{\partial x_k} = \frac{\partial \sigma_{1\ell}}{\partial x_1} + \frac{\partial \sigma_{2\ell}}{\partial x_2} + \frac{\partial \sigma_{3\ell}}{\partial x_3} = 0 \qquad (20.2)$$

Each partial derivative of a stress component has units of force per unit volume. The sum is the net force per unit volume resulting from stress components acting parallel to the coordinate axis x_ℓ, giving us three equations, for $\ell = 1$, 2, and 3, respectively. For the simple example of fluid flow through a pipe parallel to x_1, we take $\ell = 1$ in Equation (20.2), and the first term in the sum is then the pressure gradient that drives the flow.

Equation (20.2) provides us with the three additional equations to complete our system. Thus with Equations (18.3.4), (18.1.7), and (20.2), we have fifteen equations with which to solve for the fifteen unknowns $\sigma_{k\ell}$, $\dot{\varepsilon}_{k\ell}$ and v_k.

Boundary Conditions

Combining these equations gives us a set of differential equations (see Box 20.1), and solving them requires integration. The process of integration introduces arbitrary constants or functions which, because they are not specified, permit a whole set of acceptable solutions to the problem. A specific solution can be obtained only if the arbitrary terms can be evaluated, and we can do this only if we specify additional constraints for the problem called **boundary conditions.** For example, we may want to determine the stresses, velocities, and strain rates throughout a body when known stresses are applied to its boundary. Alternatively, we may want to calculate the stresses, velocities, and strain rates throughout a body that result when we impose certain velocities on the boundaries. The stresses or velocities required on the boundary constitute the boundary conditions for the problem. The solution for the unknowns within the body must be consistent with the conditions we prescribe on the boundary. Thus the constitutive equations, the balance laws, and the boundary conditions are the three elements we must have to find a solution to a problem of deformation.

Solution

If the differential equations can be solved analytically, it is possible to obtain an exact solution for the velocity as a function of position in the body. We then obtain the displacement by multiplying the velocity by an increment of time. That displacement, however, changes the geometry of the body (for example, a flat layer may have become folded), and for the new, more complicated geometry, an analytic solution is difficult if not impossible to obtain. Thus we generally cannot follow the development of a geometrically complex structure through time with analytic solutions.

With numerical techniques, however, we can obtain approximate solutions for the shape of the body and the stress distribution at the end of each successive increment of time. We use iterative techniques by which we make an initial guess at a solution for the stress and the velocity of a representative set of material points in the body and then repeatedly adjust that guess to get progressively closer to the actual solution. When the adjustments required at each iteration are small enough to be ignored, we multiply the velocity of those points by a small time increment to find the displacement and thus determine the new shape of the body. This new shape then becomes the starting point from which we solve in a similar manner for the displacement field and stress distribution at the end of the next time step. The actual smooth evolution of the system through time is thereby modeled as a series of stepwise changes. If the time increments are not too large, the complex time-dependent behavior of a system can be followed reasonably accurately.

In order to study such problems as the deformation in a collisional orogen, the driving forces of plate tectonics, or convective motions in the mantle, we must include the effects of temperature and heat transfer. The additional variables that are introduced require additional constitutive equations, as well as the balance law for the conservation of energy. The solutions quickly become very complex but, with the aid of electronic computers, not intractable.

The detailed development of the mathematics and the solutions, both analytic and numerical, are beyond the scope of this book (see Box 20.1). We give however, a general idea of the method and its applications, by discussing briefly several models for the formation of folds and a model for faulting.

Cause and Effect

Very commonly we tend to think of the stress as being the cause in a mechanical process and of the deformation as being the effect. In mathematical terms, the cause should be the independent variable, which can be varied at will, and the effect is the dependent variable, whose value is determined by the independent variable. The balance and constitutive equations, however, are completely neutral in terms of which variables are independent and which dependent. Thus the distribution of the displacements or velocities throughout a body—that is the displacement or velocity field—can be considered the cause of the distribution of stresses—that is, the stress field—just as easily as the stress field can be considered the cause of the displacement or velocity field.

Thus we must look to the boundary conditions to distinguish between cause and effect. The boundary conditions, after all, define what we choose to impose on the body. We can require the stresses on the surface of the body to be constant (stress boundary conditions), in which case the stress is the cause of the displacement or velocity field. Or we can specify the velocity of the boundaries of the body, in which case the velocity is the cause of the stress field within the body. Of course, we can also specify stress on some parts of the boundary and velocity on the other parts (mixed boundary conditions), in which case the cause is neither all one nor all the other.

The next question is, what type of boundary condition best represents geologic situations? Unfortunately, there is no simple answer. In many cases, the boundaries of a model do not correspond to a boundary in the Earth on which either the stress or the velocity is necessarily constant. Under such circumstances, the boundary conditions that are required for solutions to the mathematical model must be viewed as deviations from probable conditions in the Earth. In some cases, the assumption of velocity or displacement boundary conditions seems to provide a simpler explanation of observed structures than the assumption of stress boundary conditions, implying that deformation or de-

Box 20.1 **Mathematical Formulation of the Problem of Viscous Deformation**

We develop here the equations from which we calculate the mechanical behavior of a body undergoing viscous, constant-volume deformation.

The viscous constitutive equation (Equation 18.3.4) is written in terms of the deviatoric stress $_\Delta\sigma_{k\ell}$, which we define in terms of the stress $\sigma_{k\ell}$ and the mean normal stress $\bar\sigma_n$ by

$$_\Delta\sigma_{k\ell} = \sigma_{k\ell} - \bar\sigma_n\,\delta_{k\ell} \qquad (20.1.1)$$

where $\delta_{k\ell}$, called the Kronecker delta, is defined to be a tensor whose components equal 1 for $k = \ell$ and 0 for $k \neq \ell$, and where $k = 1, 2$, and 3 and $\ell = 1, 2$, and 3. Equation (20.1.1) is equivalent to Equation 8.43.

The viscous constitutive equation (Equation 18.3.4) also is written in terms of the strain rate components $\dot\varepsilon_{k\ell}$. We can reduce the number of unknowns from these six independent components to three velocity components v_k by using Equation (18.1.7). Making this substitution and using Equation (20.1.1), we can write the constitutive equation for constant-volume viscous deformation as

$$\sigma_{k\ell} = \bar\sigma_n\,\delta_{k\ell} + \eta\left[\frac{\partial v_k}{\partial x_\ell} + \frac{\partial v_\ell}{\partial x_k}\right] \qquad (20.1.2)$$

This relationship gives us six equations for the six independent components of the stress $\sigma_{k\ell}$, but the three velocity components v_k are still unknowns.

For a constant-volume, finite deformation, the volumetric extension e_v must be zero, from Equations (15.5) and (15.18). The same expression must apply to the incremental volumetric extension ε_v. Combining Equations (15.5), (15.16), and (15.18), and writing the result in terms of the principal incremental extensions, we get

$$\varepsilon_v = (\hat\varepsilon_1 + 1)(\hat\varepsilon_2 + 1)(\hat\varepsilon_3 + 1) - 1 = 0 \qquad (20.1.3)$$

Because the principal incremental extensions are very small ($\hat\varepsilon_k \ll 1$), products of two or more incremental extensions are negligible. Thus carrying out the multiplication in Equation (20.1.3) and dropping second- and third-order terms gives

$$\varepsilon_v = \hat\varepsilon_1 + \hat\varepsilon_2 + \hat\varepsilon_3 = \varepsilon_{11} + \varepsilon_{22} + \varepsilon_{33} = 0 \qquad (20.1.4)$$

where we can write the second relationship because the sum is a scalar invariant of the incremental strain tensor. Taking the time derivative of the second Equation (20.1.4), and using Equation (18.1.7) with $k = \ell$ gives the condition for constant-volume deformation in terms of the velocity gradient components.

$$\dot\varepsilon_v = \frac{\partial v_1}{\partial x_1} + \frac{\partial v_2}{\partial x_2} + \frac{\partial v_3}{\partial x_3} = \sum_{k=1}^{3}\frac{\partial v_k}{\partial x_k} = 0 \qquad (20.1.5)$$

This gives us one more equation for the three velocity components, but we still do not have a complete system of equations.

A complete system of equations must include the conservation of momentum, Newton's second law, which requires that the net force equal the mass multiplied by the acceleration (Equation 20.1). This con-

---

formation rate is the cause. An example is the origin of gash fracture orientations, which we explained in terms of the stress in Section 10.7 (Figure 10.11) and in terms of the displacements in Section 17.4 (Figure 17.18). The latter provides a simpler model for the observed variety of gash fracture orientations.

Thus in interpreting structures, it is important not to be prejudiced toward one mode of thinking, even if intuitively it seems somehow more "natural." Intuition is often very useful, but uninformed intuition can lead one astray.

20.2 Analytic Solution for the Viscous Buckling of a Competent Layer in an Incompetent Matrix

As an example of an analytic mechanical model for the formation of a geologic structure, we consider a model of viscous buckling of a layer that has a thickness h and a high viscosity η, imbedded in an infinitely thick viscous matrix that has a lower viscosity η_0 (Box 20.1). The measure of viscosity is one example of a precise method of defining the competence; a relatively high-viscosity material is competent, and a relatively low-viscosity material is incompetent.

The geometry of the problem is set out in Figure 20.1, where the initial deflection is exaggerated for the sake of clarity. We wish to calculate how the layer will behave if a known force **P** is applied parallel to the layer and if the resistance **K** of the matrix is proportional to the rate of deflection of the layer. For the boundary conditions, we specify that the deflection caused by deformation is zero along the entire length of the layer before the deformation begins, and it must always be zero at $x_1 = 0$ and $x_1 = L$, where L is the unknown wavelength we want to determine. Thus the layer is pinned at 0 and L, and these points are nodes of the folding. In addition, we ignore any possible longitudinal shortening of the layer in response to the force **P**.

The solution to this model gives the magnitude of the displacement u normal to the layer for an initial

dition, applied at each point in a deformable body, can be shown to be

$$\sum_{k=1}^{3} \frac{\partial \sigma_{k\ell}}{\partial x_k} + \rho g_\ell = \rho a_\ell \qquad (20.1.6)$$

which is called the **equation of motion**. Because ℓ can independently take on any of the values 1, 2, or 3, we have a set of three equations. Each term has units of force per unit volume. The right side of the equation is the inertial force term—that is, the mass per unit volume ρ multiplied by the acceleration a_ℓ. The left side of the equation includes the applied forces per unit volume. The second term on the left side is the force of gravity per unit volume, where g_ℓ is the acceleration due to gravity. The terms in the sum of the derivatives of the stress components are the forces per unit volume arising from the resistance of the material to deformation; they are discussed in the main text (Equation 20.2).

If the acceleration is negligible and we ignore the force of gravity, we obtain Equation (20.2) for the equation of equilibrium. For many geologic problems, these are acceptable approximations. For large-scale deformations, however, such as the gravitational collapse of overthickened crust, the diapiric rise of salt domes or gneiss domes, and mantle convection, the gravitational force cannot be ignored. We use the simplest equation (Equation 20.2), however, to illustrate the formulation of a mechanical model.

We can combine the constitutive equation (Equation 20.1.2) with the equation of equilibrium (Equation 20.2) by taking the appropriate derivatives of Equation (20.1.2), adding them, and setting the result equal to zero as required by Equation (20.2). We obtain

$$\sum_{k=1}^{3} \frac{\partial \sigma_{k\ell}}{\partial x_k} = 0 = \frac{\partial \bar{\sigma}_n}{\partial x_\ell} + \eta \sum_{k=1}^{3} \left[\frac{\partial^2 v_k}{\partial x_\ell \, \partial x_k} + \frac{\partial^2 v_\ell}{\partial x_k^2} \right] \qquad (20.1.7)$$

The first term under the summation on the right is zero because of Equation (20.1.5). Thus the final differential equations that we have to solve are

$$\frac{\partial \bar{\sigma}_n}{\partial x_\ell} = -\eta \sum_{k=1}^{3} \frac{\partial^2 v_\ell}{\partial x_k^2} \qquad \sum_{k=1}^{3} \frac{\partial v_k}{\partial x_k} = 0 \qquad (20.1.8)$$

where the second equation is Equation (20.1.5). The first equation represents three equations for $\ell = 1, 2,$ and 3, respectively. Note that although we combined the constitutive equation with the equation of equilibrium to eliminate the stress components, we still have the mean normal stress as an unknown, along with the three velocity components v_k. These equations provide a set of four differential equations to solve for the four unknowns.

Integrating the equation introduces constants of integration that we evaluate by prescribing appropriate boundary conditions on the surfaces of the body. By solving for the velocity and mean normal stress as a function of position in the body, we can then determine the stress from Equation (20.1.2).

---

wavelength L_0 that has an initial displacement δ_0. The solution is a function of the time t, of a, which is a function that depends on the viscosities of the layer and the medium, of the initial wavelength L_0, and of the distance x along the layer.

$$u = \delta_0 \exp [t/a] \sin (2\pi x/L_0) \qquad (20.3)$$

The time dependence applies only for the small increment of deformation before the initial geometry of the layer is appreciably changed. The dominant wavelength L of the folds that develop is the one that grows the fastest, and it is therefore the one for which the function a is a minimum. It can be shown that this condition leads to Biot's relationship for L:

$$L = 2\pi h \left[\frac{\eta}{6\eta_0} \right]^{1/3} \qquad (20.4)$$

Thus the fold wavelength is directly proportional to the thickness h of the competent layer, but it depends only on the cube root of the layer-to-matrix viscosity ratio.

This solution is applicable only to small-amplitude folds for which the folding angle is less than about 20°

or 30° and for which the wavelength of the fold λ and the length along the layer L for the same wavelength are approximately the same. With large amounts of shortening, the wavelength decreases as the folding angle increases, and the difference between λ and layer L increases.

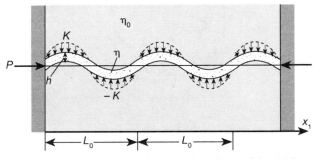

Figure 20.1 Model for the theoretical analysis of the folding of viscous materials. A competent viscous layer of viscosity η and thickness h is compressed parallel to the layer by a force **P**. It is imbedded in an incompetent viscous matrix of viscosity η_0, which resists the folding of the layer with a force **K**. L_0 is the wavelength of the initial deflection in the layer.

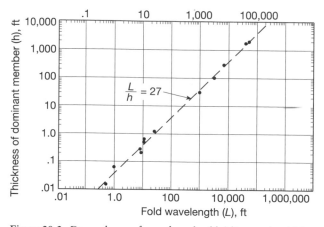

Figure 20.2 Dependence of wavelength of folding on the thickness of the folded layer determined from natural folds. The line is for a wavelength-to-thickness ratio $L/h = 27$.

Despite this limitation, the relationship of Equation (20.4) seems to hold in nature. Figure 20.2 shows the relationship between layer thickness and fold wavelength measured from natural folds. It shows that the ratio of wavelength to layer thickness, L/h, is essentially a constant. This result is consistent with Equation (20.4) if the variation in the viscosity ratio is small.

In principle, it should be possible to measure the wavelength of a train of folds and the thickness of the folded layer and thus determine the ratio of the viscosities of the layer and the medium. For all the folds plotted in Figure 20.2, for example, the wavelength-to-thickness ratio L/h is approximately 27, implying that the corresponding ratio of the layer viscosity to the matrix viscosity, η/η_0, is approximately 476. The wave-

length, however, is not a very sensitive function of the viscosity ratio. Thus if L/h increases only a small amount to 27.5, we would infer from Equation (20.4) that the viscosity ratio would increase to about 503. The result is also not reliable because the folds plotted in Figure 20.2 are multilayer folds, whereas the analysis was done for single layers. Ptygmatic folds would be preferable for this type of analysis, because they develop in isolated competent layers in an incompetent matrix and thus closely resemble the geometry of the theoretical model.

This model of folding implicitly makes one major assumption: that the layers of rock actually deform as a viscous material. Our discussion of rock rheology in Chapter 18, however, reveals that this assumption may not be valid. Moreover, the predicted form of the folds would be no different if we had assumed the rocks were elastic materials, because the constitutive equations for viscous fluids and elastic solids are very similar in form. Thus an analysis of the bending of a stiff elastic layer in a soft elastic matrix gives a differential equation identical in form to that for the viscous layer. The relationship among wavelength, thickness, and the ratio of the elastic moduli of the layer and the matrix is

$$L = 2\pi h \left[\frac{B}{6B_0} \right]^{1/3} \qquad B = \frac{E}{1 - v^2} \qquad (20.5)$$

where B and B_0 are the elastic moduli of the layer and the matrix, respectively, E is Young's modulus, and v is the Poisson ratio and h is the layer thickness.

Figure 20.3A illustrates this relationship for elastic materials with three rubber strips, one twice the thickness of the other two. The wavelength for the thicker strip is very close to twice the wavelength for the strips that are half as thick, as predicted by Equation (20.5).

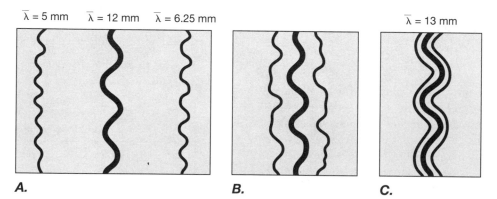

A. **B.** **C.**

Figure 20.3 Folding of elastic rubber strips imbedded in a softer elastic gelatin matrix. A. Widely separated strips 0.4 mm thick (two strips) and 0.8 mm thick (center strip). The average wavelength in the thicker strip ($\lambda \cong 12$ mm) is approximately twice the average wavelength in the strips that are half as thick ($\lambda \cong 5.6$ mm). B. With less separation, the folding of adjacent strips begins to interfere. C. With still less separation, the multilayer behaves as a single layer, forming folds with a wavelength larger than that for any of the individual strips but smaller than that for a single strip having a thickness equal to the combined thicknesses of all the strips.

Figure 20.3B, C shows the effects of folding of a multilayer in which the competent layers are so close together that the folding patterns interfere and the wavelength of the folding is modified. The wavelength of the multilayer folds in Figure 20.3C is equivalent to that for a single layer whose thickness is greater than any individual layer in the multilayer but less than the combined thicknesses of all three layers. Experiments like these serve to test theoretical predictions and provide an intuitive grasp of the factors that affect the form of folds.

Because Equation (20.5) is identical in form to Equation (20.4), it is impossible to distinguish elastic from viscous effects on the basis of the form of folds alone. Despite this ambiguity, the theory provides insight into the process of fold formation and into the probable influence of initial geometry and material properties on the final fold geometry.

20.3 Numerical Models of Viscous and Power-Law Buckling

In this section, we consider numerical models principally of the buckling of a competent layer in an incompetent matrix, using both a viscous and a power-law rheology (see also Section 12.1 and Figure 12.5). Because we want to calculate the development of the model through time into the realm of large-scale deflections of the layer, we cannot use the general analytic solution (Equation 20.3) for the viscous rheology, which applies only to small deflections, and we cannot even obtain an analytic solution for the power-law rheology. Thus we must use numerical solutions and therefore must find a new solution for each different set of material properties that we want to investigate.

Viscous Rheology and Folding

Figure 20.4 shows the geometry of the model. An initial irregularity of wavelength L and amplitude A_0, chosen to give initial dips between $1°$ and $2°$, is built into the layer of thickness h. Because of the symmetry of the geometry, calculations need only be performed for points within the area $ABCD$. The boundary conditions require that the lines AB and CD both approach the stationary axis $x = 0$ at a constant velocity and that there be no shear stresses on BC. The time increments for each iteration are chosen such that the average extensional strain for each increment parallel to the initial layer is $e_n = -0.02$.

The geometric irregularities included in the model layer approximate natural irregularities at a wavelength near the theoretically dominant one (Equation 20.4).

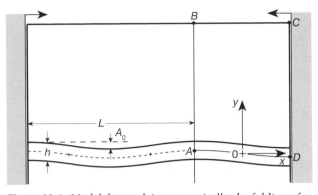

Figure 20.4 Model for studying numerically the folding of a viscous layer in a less viscous medium. A layer of thickness h has an initial irregularity of wavelength L and amplitude A_0 built into it. Because of the geometric symmetry of the model, calculations need only be done for the section of the model outlined by points $ABCD$. The boundary conditions require that the planes AB and DC approach the origin $x = 0$ at the same constant rate.

Irregularities are inherent in natural systems, and unless they are introduced into models, the perfection of the model geometry causes the system to be artificially stable such that folds would never form.

Figure 20.5 shows one model for the development of a fold for which $L/h = 12$ and the viscosity ratio $\eta/\eta_0 = 42.1$. The figure illustrates the fold geometry and the variations across the structure and through time in the orientation of the axes of maximum compressive stress $\hat{\sigma}_1$. The model produces a class 1B fold. Note that the folding angle ϕ increases much more rapidly for early increments of shortening than it does for later increments. In the hinge zones, the maximum compressive stress is parallel to the layer on the concave sides of folds where layer-parallel shortening occurs, and it is roughly perpendicular to the layer on the convex sides where layer-parallel elongation occurs. In the limbs, the maximum compressive stress tends to rotate with the limbs until limb dips become steep, at which point it returns toward its original orientation and tends to be at high angles to the bedding. The orientations of $\hat{\sigma}_1$ within the layer can be compared with those of Figure 10.12C, which were observed experimentally in an elastic bar of gelatin with no surrounding medium.

The magnitudes of the stresses also vary across the fold and throughout the course of the deformation. On the convex side of the hinge zone, $\hat{\sigma}_1$ drops to very small values early in the deformation and remains small throughout. On the concave side of the hinge zone, the compressive stresses are high throughout the folding process. Within the limbs, $\hat{\sigma}_1$ starts off at high values but decreases by a factor of 10 by the time the folding angle approaches $180°$. This change reflects the fact that the competent layer bears a large proportion of the force

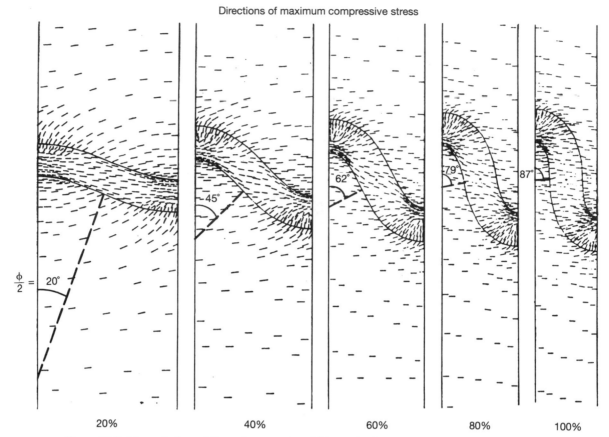

20%   40%   60%   80%   100%

Figure 20.5 Numerical modeling of progressive folding of a layer in a matrix, where the layer is 42.1 times more viscous than the matrix, a high competence contrast. The wavelength-to-thickness ratio is $L/h = 12$. The short lines indicate the orientation of $\hat{\sigma}_1$, the maximum compressive stress. For each frame, the half folding angle $\phi/2$ is indicated, and the overall shortening is shown as a natural strain at the bottom of each frame.

applied to the system when the layer is parallel to the shortening direction, but its strengthening effect decreases as the limbs rotate to higher angles. (We reached this same conclusion in Box 19.2 from studying the quartz deformation lamellae in a folded layer [compare Figure 20.5A with Figure 19.2.1E].)

The orientations of the principal stretches can also be calculated from the numerical models of folding. Figure 20.6 shows the distribution of $\hat{s}_1$ axes throughout the system for the fold model shown in Figure 20.5. The maximum principal stretch axes $\hat{s}_1$ form a convergent fan in the competent layer. Within the incompetent matrix, the $\hat{s}_1$ axes form a divergent fan on the convex side or outside of the hinge in the layer (areas labeled O) and are parallel to the axial surface or slightly convergent on the concave side or inside of the folded layer (areas labeled I).

Figure 20.7 shows the distribution of minimum compressive stress axes $\hat{\sigma}_3$, for the fold in the last frame of Figure 20.5. Note that the principal axes of stress (Figure 20.7) and finite strain (Figure 20.6B) are not

necessarily parallel. It is instructive to compare the orientations of $\hat{\sigma}_3$ axes (Figure 20.7) and of $\hat{s}_1$ axes (Figure 20.6B) with those of foliations from natural folds of a competent material such as sandstone in an incompetent matrix such as shale (Figure 13.13). The main differences between the $\hat{\sigma}_3$ and $\hat{s}_1$ orientations are the pattern in the hinge zone and the degree of refraction of the axes at the limbs. Overall, the pattern shown by the maximum principal stretch axes $\hat{s}_1$ bears a closer resemblance to the natural foliation pattern than does the pattern shown by the minimum compressive stress axes $\hat{\sigma}_3$. This similarity lends credence to the hypothesis that foliations subparallel to the axial surface mark the orientation of the plane of flattening $\hat{s}_1$–$\hat{s}_2$ of the finite strain ellipsoid. It also suggests that the refraction of foliations across lithologic boundaries reflects differences in the mechanical properties of the adjacent layers, which give rise to the different amounts and orientations of strain across the lithologic contact.

A change in the viscosity contrast has a profound effect on the geometry of folds and on the orientation

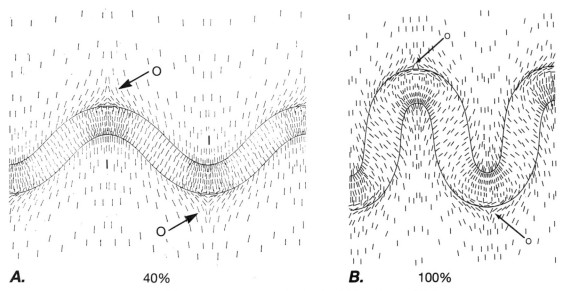

A.   40%   **B.**   100%

Figure 20.6 Orientations of $\hat{s}_1$, the maximum stretch, for the folds in the model shown in Figure 20.5 for natural strains of (A) 40 percent overall shortening and (B) 100 percent overall shortening. The $\hat{s}_1$ directions are divergent on the convex sides of the folded layer (regions marked O) and are parallel to slightly convergent on the concave sides of the folded layer (regions marked I). Within the layer, the $\hat{s}_1$ directions are convergent in the limbs and, in the hinge zone, may be convergent (O) or may vary from divergent to convergent across the layer (I).

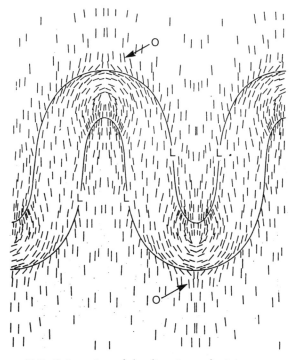

Figure 20.7 Orientation of the directions of minimum compressive stress $\hat{\sigma}_3$ in the fold shown in the last frame of Figure 20.5. The pattern of orientations is similar to that of $\hat{s}_1$ in Figure 20.6B, except that the divergent pattern is not so pronounced on the convex side of the folded layer (regions labeled L), and there is little refraction of orientations across the layer boundaries in the limbs (regions marked I).

of $\hat{s}_1$. Figure 20.8 shows the fold produced by a model for which $L/h = 9$ and the viscosity ratio is 17.5. In this model, homogeneous shortening and thickening of the layer occurred along with the folding, especially during the early stages of folding. The result is a fold of class 1C geometry (see Section 11.3 and Figures 11.19, 11.20 and 12.12). The refraction of the orientations of $\hat{s}_1$ at the layer boundary is considerably less than in the first model for folds that have similar limb dips (compare Figure 20.6 with Figure 20.8). This model also demonstrates that buckle folding need not produce layer-parallel extensional strains on the convex side of the hinge zone in the layer. Note that in Figure 20.8B, the concave side of each fold develops into a cusp, and the convex side forms a lobe comparable to a fold mullion (Figure 13.22A). The cusps point into the material that was more competent at the time of deformation. Because the relative competence for the same two rock types can actually change with changing conditions of deformation (see Figure 20.9 and the discussion that follows), the cusps can be a useful criterion for determining relative competence in naturally deformed rocks.

Although it is not certain whether the assumption of linear viscosity for the model materials accurately reflects the rheology of real rocks, the correspondence between the calculated structures and observations in naturally deformed rocks suggests that the assumption may not be bad at least as a first approximation.

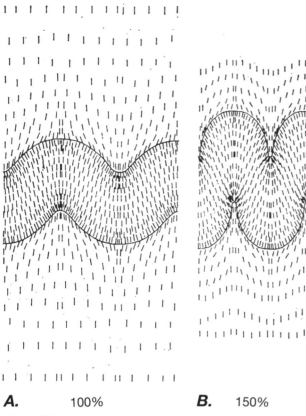

A. 100%   **B.** 150%

Figure 20.8 Numerical modeling of progressive folding of a layer in a matrix, where the ratio of the layer viscosity to the matrix viscosity is 17.5, a low competence contrast. The wavelength-to-thickness ratio is $L/h = 9$. Short lines show the orientations of $\hat{s}_1$. The shortening measured as natural strain is 100 percent in part A and 150 percent in part B.

Power-Law Rheology and Folding

The assumption of a power-law rheology for models that have the geometry shown in Figure 20.4 provides an alternative approximation to the rheology of real rocks, at least for some conditions of ductile deformation. This model assumes experimentally determined power-law rheologies of wet quartzite for the layer, and of marble for the matrix, as defined by:

Wet quartzite $\quad |_s\dot{\varepsilon}_n| = 1.77 \times 10^{-2} \exp\left[\dfrac{-2.30 \times 10^5}{RT}\right]_D\sigma^{2.8}$

$$(20.6)$$

Marble $\quad\quad |_s\dot{\varepsilon}_n| = 1.20 \times 10^{-4} \exp\left[\dfrac{-2.59 \times 10^5}{RT}\right]_D\sigma^{8.3}$

$$(20.7)$$

where the stresses are in megapascals, the activation energies are in joules per mole, and the strain rates are in inverse seconds (s^{-1}). (Note that the constants used

here are different from those listed in Table 18.1. This reflects the differences that occur in various experimental evaluations of these constants.)

The materials that obey a power-law rheology are not characterized by a single viscosity. Nevertheless, it is convenient to express the strain rate under any particular set of conditions in terms of an **effective viscosity,** which is defined as the ratio of the shear stress to the shear strain rate. In terms of the differential stress and the axial strain rate, the effective viscosity η_{ef} is

$$\eta_{ef} = \frac{D\sigma}{3|\dot{\varepsilon}_n|} \qquad (20.8)$$

The effective viscosity is dependent on the stress and the temperature, as can be seen by substituting for the strain rate in Equation (20.8) from either Equation (20.6) or Equation (20.7). For a power-law material at a constant stress and temperature, the effective viscosity equals that of a linearly viscous material flowing at the same rate under the same stress. It is a mechanically exact measure of the competence of the different rocks under consideration, but in this case it is important to realize that the competence increases with decreasing stress as well as with decreasing temperature.

In Figure 20.9 we plot the effective viscosity ratio η_{qtz}/η_{mbl} for wet quartzite and marble, as a function of temperature, for a strain rate of 10^{-14} s^{-1}. At temperatures below about 550°C, $(\eta_{qtz}/\eta_{mbl}) \geq 1$, and the quartzite has the higher effective viscosity (the greater

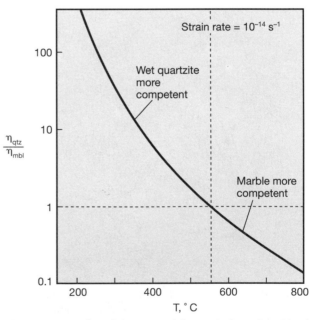

Figure 20.9 A plot of the ratio of the equivalent viscosity of wet quartzite to that of marble (η_{qtz}/η_{mbl}) as a function of temperature at a strain rate of 10^{-14} s^{-1}. The relative competence reverses at a temperature of about 550°C (dashed lines).

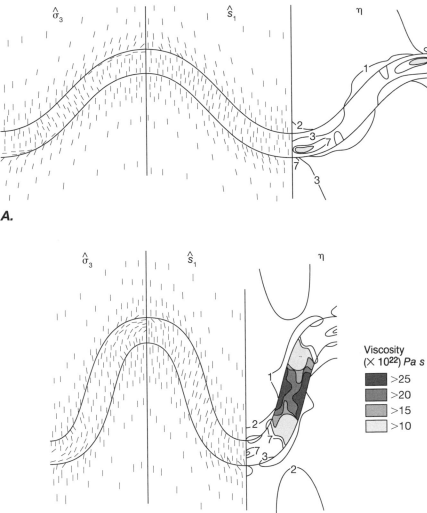

A.

B.

Figure 20.10 Numerically calculated folds deforming with power-law rheologies for a wet quartzite layer in a marble matrix at 375°C. Initial shortening is at a rate of 10^{-14} s^{-1}. In the left panel, short lines are parallel to $\hat{\sigma}_3$; in the center panel, they parallel $\hat{s}_1$. The right panel is a contour plot of the magnitude of the equivalent viscosity η. Shortening measured as natural strains are 40 percent in part A and 80 percent in part B.

Viscosity
($\times 10^{22}$) *Pa s*

■	>25
▨	>20
▦	>15
□	>10

competence). Because marble has a higher activation energy, however, the marble becomes the more competent material above that temperature. This behavior has interesting implications for the possible development of structures in metamorphic rocks, as we discuss below.

Figure 20.10 shows a plot of the $\hat{\sigma}_3$ axes, the $\hat{s}_1$ axes, and the equivalent viscosities for one numerical experiment at a temperature of 375°C, which gives an apparent viscosity ratio from Figure 20.9 of $\eta_{\text{qtz}}/\eta_{\text{mbl}} = 10$ at $\dot{\varepsilon}_n = 10^{-14}$ s^{-1}. The initial dip of the limbs is 10°, and the initial strain rate is 10^{-14} s^{-1}. Time increments were chosen to give increments of natural strain of $\bar{\varepsilon}_n = 0.05$.

The fold formed is a class 1B fold. The $\hat{s}_1$ axes form a divergent fan in the marble on the convex side of the hinge zone and a convergent fan within the quartzite layer. The difference between the $\hat{\sigma}_3$ and the $\hat{s}_1$ orientations is similar to the case of viscous folding (compare Figures 20.10B with 20.6B and 20.7). The effective viscosity increases significantly as the stress decreases, so it does not necessarily remain constant

around the structure or through time (Figure 20.10). Compare the folds in Figure 20.10 with those in Figures 20.5 through 20.7, also of class 1B geometry, which formed in linearly viscous materials. The geometry is very much the same, even though the effective viscosity ratio for the power-law materials is one-fourth the viscosity ratio for the viscous materials. For the power-law rheologies, the angular variation in the fans of $\hat{s}_1$ is not so large as for the linear rheology—a difference particularly apparent at high strains. The differences between the two models are fairly subtle, however, and it would be difficult to infer a rheology for a natural fold from the geometry of the fold and the orientation of $\hat{s}_1$ axes.

Because of the temperature dependence of the rheologies and of the effective viscosity ratio, an increase in the temperature of metamorphism could change the mechanical behavior of the system, thereby affecting the geometry of the folds that develop. As an example, Figure 20.11 shows the the model of the quartz layer in the marble matrix subjected to a 20 percent shortening at 375°C (top diagram), followed by continued short-

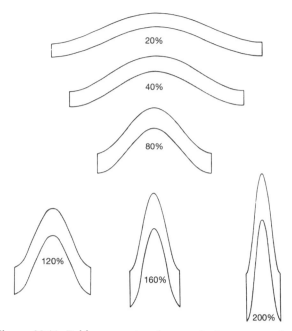

Figure 20.11 Fold geometries that result from power-law rheologies for a quartzite layer in a marble matrix shortened to 20 percent natural strain at 375°C ($\eta_{qtz}/\eta_{mbl} = 10$) and then shortened up to 200 percent natural strain at 550°C ($\eta_{qtz}/\eta_{mbl} = 1$). Numbers indicate the natural strain in percent overall shortening.

ening up to 200 percent at 550°C. At the higher temperature, the effective viscosity ratio of quartzite to marble is 1, indicating that the mechanical difference between these two materials is very small. Thus this example is a well-defined mechanical model equivalent to the kinematic model discussed in Section 12.4 (Figure 12.12), for which folding was followed by homogeneous flattening normal to the axial surface. The initial class 1B fold (top diagram in Figure 20.11) is flattened into class 1C geometry, which, beyond about 160 percent shortening, closely resembles a class 2 geometry.

Figure 20.12 shows the effect of different temperatures on the same model. In this case, all folds initially formed by shortening to 40 percent at 375°C ($\eta_{qtz}/\eta_{mbl} = 10$) and then were shortened to a total of 100 percent at different temperatures and thus different effective viscosity ratios. With increasing temperature from 450°C to 550°C, the geometry of the folds is class 1C, with characteristics ranging from nearly class 1B to nearly class 2. In this temperature range, the fold geometry reflects the behavior of the more competent quartzite layer. Above 550°C, however, the quartzite layer becomes less competent than the marble and ceases to control the geometry of the folding. In these cases,

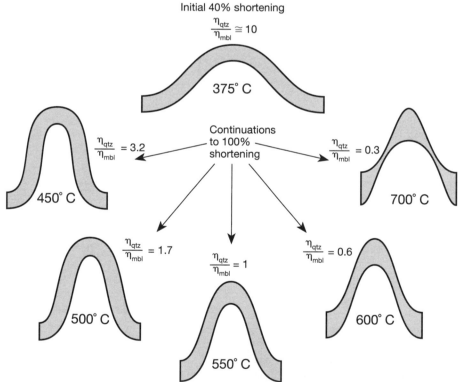

Figure 20.12 Power-law rheology folding for a wet quartzite layer in a marble matrix. Folds were deformed to 40 percent overall shortening at 375°C ($\eta_{qtz}/\eta_{mbl} = 10$) and shortened further to 100 percent overall shortening (measured as natural strain) at the temperature and effective viscosity ratio indicated for each fold. Note the change in fold style from class 1 in the folds for which $\eta_{qtz}/\eta_{mbl} = 1$ to class 3 folds, for which the quartzite layer becomes less competent than the marble matrix ($\eta_{qtz}/\eta_{mbl} < 1$).

we see the quartzite layer develop class 3 folds, with the curvature at the hinges dominated by the behavior of the marble. This fold geometry resembles that of the incompetent layer in Figure 11.22B, C, 12.17, and 16.6.

These numerical fold models give a sound mechanical basis for the relationship between rheology and fold geometry (Section 12.5). The degree to which they represent natural conditions, however, is uncertain. Linear viscosity may be an unrealistic rheology for real rocks although at very low stresses it might be reasonable for solution folding. This mechanism, however, is not being modeled in these examples. On the other hand, power-law rheologies derived experimentally at high strain rates may not be appropriate for modeling real rocks at geologically realistic strain rates, which are many orders of magnitude smaller. The rheology adopted for marble in Equation (20.7), for example, was determined from high-stress experiments on coarse-grained Yule marble. Data from Carrara marble are quite different (Table 18.1). The possibility of a transition to lower-stress exponents at low stress, as demonstrated for the fine-grained Solenhofen limestone (Figure 18.15), also was not considered.

The differences in behavior predicted by the viscous and power-law rheologies, however, are not large. Accordingly, the models seem to provide insight into the geometries of natural folds. These results also indicate that assuming a linear viscous rheology is probably a reasonable first-order approximation.

The models described so far generally produce subrounded to rounded folds (Figure 11.16), but they do not seem to account for chevron folds with sharp hinges and planar limbs. Such folds are characteristic of rock sequences from a wide variety of conditions, ranging from unconsolidated sediments to high-grade metamorphic rocks. The rocks in which chevron folds develop, however, all have in common a strong planar anisotropy, which may control the geometry of developing folds. Alternatively, it is possible that a "strain softening" mechanism, such as recrystallization or solution creep, might cause the stress required for steady-state creep to *decrease* with increasing strain. In this case, because strain is largest in the hinge area, the material there would tend to become progressively less resistant to deformation than elsewhere, and this in turn would cause an even greater concentration of strain. The result would be a fold in which almost all the bending was concentrated in a small, sharp hinge area, leaving the limbs almost planar and undeformed.

Comparison with Kinematic Models

In these numerical models, the kinematic behavior of the layer during folding is determined by the constitutive equations, the balance laws, the boundary conditions, the material properties, and the temperature. We can make a qualitative comparison of the distribution of $\hat{s}_1$ orientations predicted by these models (Figures 20.6, 20.8, and 20.10) with those derived from the kinematic models, for which we imposed a preconceived kinematic behavior (Figures 16.1 and 16.2). In making the comparison, we must recognize that the numerical models provide a higher resolution of the strain distribution than is shown for the kinematic models.

Volume is conserved in the numerical models, so the kinematic models for volume-loss folding (Figures 16.1D and 16.2D) are not pertinent to the comparison. The boundary conditions imposed on the numerical models lead to the buckling mode of folding, and therefore a comparison is relevant only with the fold models in Figure 16.2B and C. These kinematic fold models compare in folding angle ϕ most closely with the numerical models in part B of Figures 20.6, 20.8, and 20.10. Of these, all are class 1B except for the folds in Figure 20.8, which are class 1C. The class 1C geometry is attributable to the low-competence contrast between the layer and matrix which leads to a large amount of layer shortening and thickening in the early stages of folding, and to nearly homogeneous flattening during later stages of folding. Thus the most relevant comparisons are between the numerical models of Figures 20.6B and 20.10B and the kinematic models of Figure 16.2B, C.

The strain distribution in the numerical model based on Newtonian rheology (Figure 20.6B) has aspects similar to both orthogonal flexure (Figure 16.2B) and flexural shear (Figure 16.2C). In particular, the layer-parallel orientation of $\hat{s}_1$ on the convex side of the hinge zone is similar to that resulting from orthogonal flexure but quite distinct from the orientations for flexural shear. In the limbs of the numerical model, however, the angle between $\hat{s}_1$ and the bedding is not as high as in orthogonal flexure, but instead is more akin to the angle in the limbs of the flexural shear folds (Figure 16.2C).

The strain distribution in the numerical model based on the power law rheology is also best interpreted as a mixture of orthogonal flexure and flexural shear, despite the fact that in the numerical model $\hat{s}_1$ is not parallel to the layer in the convex side of the hinge zone. This difference implies that there is a smaller amount of layer-parallel lengthening in this region compared with orthogonal flexure, even though the minimum principal stress $\hat{\sigma}_3$ is parallel to the layer there (Figure 20.10B). In the formation of these folds, the differential stress is smaller on the convex side of the hinge zone because the layer parallel compression is reduced during lengthening of the layer. For a power-law rheology, the lower differential stress in turn means that the effective viscosity is higher than in other parts of the fold (see the contours of effective viscosity in Figure 20.10A).

Thus the strain rate is lower and the total accumulated strain is smaller. The power-law rheology therefore accounts for the difference in the hinge zone between these folds and orthogonal flexure folds. In the limbs, the orientation of the $\hat{s}_1$ axes for the power law model is closer to that for flexural shear than for orthogonal flexure.

Thus, based on the results of numerical modeling, we conclude that folding is probably not well represented by either pure orthogonal flexure or pure flexural shear. This result should come as no great surprise, because our kinematic models are simple postulated end-members of possible kinematic behaviors, and natural processes generally do not conform to our ideals of simple behavior.

20.4 Plastic Slip-Line Field Theory and Faulting

Faults in the Earth are characterized by the fact that material moves parallel to the fault and that the velocity of motion is discontinuous across the fault. These characteristics are reproduced by a rigid-plastic, isotropic, homogeneous material undergoing plane strain. For these conditions, it can be shown that the constitutive equations for plasticity (Equation 18.3.5 in Box 18.3) and the equation of equilibrium (Equation 20.2) provide a hyperbolic differential equation whose solution gives rise to two families of lines called slip lines, which by convention are designated α and β (Figures 20.13A and 20.14). The mathematical development of the theory is beyond the scope of this book (it can be found in a variety of references[1]). The results, however, are not difficult to understand. Each line in one family is orthogonal to all the lines in the other family, and each line is everywhere tangent to the maximum shear stress and the maximum shear strain rate in the material. Thus the maximum and minimum principal stresses ($\hat{\sigma}_1$ and $\hat{\sigma}_3$) and the principal axes of the incremental strain ellipse ($\hat{\zeta}_3$ and $\hat{\zeta}_1$) bisect the right angles between any two intersecting slip lines (Figure 20.13A).

The components of displacement and velocity parallel and normal to the slip lines must obey a few simple rules. Extensional strain rate parallel to any of the slip lines must be zero, and this requires the component of displacement or velocity parallel to a slip line to be constant along that slip line (Figure 20.13B). It also means that the component of displacement or velocity normal to any slip line must be constant across the slip line, because the slip lines are mutually orthogonal (Fig-

---

[1] See, for example, Johnson et al. (1970) and Odé (1960).

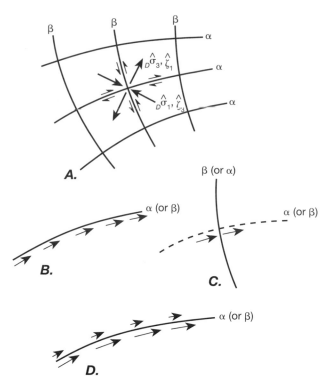

Figure 20.13 Characteristics of plastic slip-line fields. *A.* Slip lines form two sets of curves —α and β, respectively—that are everywhere mutually orthogonal. Each line is tangent to the orientation of maximum resolved shear stress and maximum resolved incremental shear strain, so the principal axes of both quantities bisect the angles between the two sets of slip lines. *B.* Along any one slip line, the tangential component of the velocity and displacement must be constant. *C.* The component of velocity and displacement normal to a slip line must be the same on both sides of the slip line. *D.* The component of velocity and displacement parallel to a slip line can be discontinous across the slip line, so the slip line can behave like a fault.

ure 20.13C). The components of displacement and of velocity parallel to a slip line, however, may be different on opposite sides of that slip line without violating these restrictions (Figure 20.13D), but any such discontinuity must be the same along the entire length of that particular slip line. It is this last property that makes the slip lines relevant to the interpretation of faults.

The geometry of the slip lines depends on the geometry of the deforming system. As an example, we consider the slip lines resulting from a rigid punch impinging on the edge of a plastic layer (Figure 20.14). In Figure 20.14A, the plastic material is constrained between two rigid boundaries symmetrically placed on either side of the punch. The heavily shaded area is a region of "dead" material that is fixed to the punch and does not deform. In Figure 20.14B, the punch impinges on the straight edge of a plastic layer that is unconstrained on the sides. The slip lines are the lines along

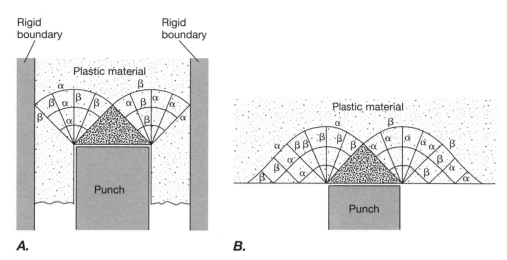

Figure 20.14 Geometry of plastic slip-line fields in an isotropic, homogeneous, rigid-plastic material deformed by a rigid punch. The stippled area represents the plastic material. The heavily stippled triangular area is the "dead zone" in which no deformation occurs. *A.* The plastic material is constrained by rigid boundaries symmetrically placed on either side of the punch. *B.* The punch impinges on the straight edge of the plastic material, which is unconstrained on either side of the punch.

which the material shears during the deformation, and because discontinuities in the tangential component of velocity are permitted across the lines, they have been used to interpret patterns of strike-slip faulting in the Earth's crust that result from tectonic collisions between a less ductile continental block, the punch, and a more ductile one, the plastic material.

20.5 The Theory of Scale Models

The formulation of an analytic or a numerical model of a geologic system becomes more complex and difficult as the system becomes more realistic. For many geologic situations, it is useful to make a scale model whose behavior on a scale of meters and hours is equivalent to what occurs in the natural system, or **prototype,** on a scale of kilometers and millions of years. The main problem is to know how to construct a model such that its behavior reliably represents that of the prototype.

In order to describe completely the mechanical relationship between a scale model and the prototype, we need the three independent scale factors for length (λ), time (τ), and mass (μ). We define them by

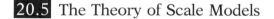

$$\lambda = \frac{\ell_m}{\ell_p} \qquad \tau = \frac{t_m}{t_p} \qquad \mu = \frac{m_m}{m_p} \qquad (20.9)$$

where subscripts m and p refer to the model and prototype, respectively, and where ℓ is length, t is time, and m is the mass of material in geometrically similar volumes of the model and prototype. The prototype and

model are **geometrically similar** if all linear dimensions of the model are λ times the equivalent dimension in the prototype. They are **kinematically similar** if the time required for the model to undergo a change in size, shape, or position is τ times the time required for the prototype to undergo a geometrically similar change.

The scaling of all the mechanical quantities of interest can be derived from these three basic scale factors, and these mechanical scale factors restrict the properties of the scale model. Table 20.1 lists the different mechanical quantities, their units, and the scale factors derived for them.

In Chapter 8 we distinguish between two kinds of forces, body forces and surface forces. Of the body forces on mechanical systems, we need consider only the inertial force (F_i), which is the force associated with acceleration, and the gravitational force (F_g). We ignore others, such as electrostatic and magnetic forces. Surface forces arise from the resistance to deformation, and the relationship is expressed by a constitutive equation such as one of those discussed in Chapter 18.

Conservation of momentum requires that the sum of all body and surface forces acting on a point be zero (see Equation 20.1). Scaling laws require that the geometry and time span of a deformation must scale appropriately from the prototype to the model. Thus the response to each type of scaled force in the model must be geometrically and kinematically similar to the prototype's response to natural forces. These requirements are satisfied if each of the various forces in the model is related to the corresponding one in the prototype by the same scale factor, in which case the model and the prototype are **dynamically similar.**

Table 20.1 Scale Factors for Selected Variables in Mechanics

Quantity	Symbol	Units	Ratios	Scale factor
Area	A	ℓ^2	$\dfrac{A_m}{A_p} = \dfrac{(\ell_m)^2}{(\ell_p)^2}$	λ^2
Volume	V	ℓ^3	$\dfrac{V_m}{V_p} = \dfrac{(\ell_m)^3}{(\ell_p)^3}$	λ^3
Density	ρ	$\dfrac{m}{V}$	$P = \dfrac{\rho_m}{\rho_p} = \dfrac{m_m}{m_p}\dfrac{(\ell_p)^3}{(\ell_m)^3}$	$P = \dfrac{\mu}{\lambda^3}$
Velocity	v	$\dfrac{\ell}{t}$	$\dfrac{v_m}{v_p} = \dfrac{\ell_m/t_m}{\ell_p/t_p} = \dfrac{\ell_m}{\ell_p}\dfrac{t_p}{t_m}$	$\dfrac{\lambda}{\tau}$
Acceleration	a	$\dfrac{\ell}{t^2}$	$\dfrac{a_m}{a_p} = \dfrac{\ell_m/t_m^2}{\ell_p/t_p^2} = \dfrac{\ell_m}{\ell_p}\dfrac{t_p^2}{t_m^2}$	$\dfrac{\lambda}{\tau^2}$
Force	F	$\dfrac{m\ell}{t^2}$	$\dfrac{F_m}{F_p} = \dfrac{m_m(\ell_m/t_m^2)}{m_p(\ell_p/t_p^2)} = \dfrac{m_m\ell_m t_p^2}{m_p\ell_p t_m^2}$	$\dfrac{\mu\lambda}{\tau^2}$
Stress	σ	$\dfrac{F}{A}$	$\Sigma = \dfrac{\sigma_m}{\sigma_p} = \dfrac{F_m/A_m}{F_p/A_p} = \dfrac{F_m\ell_p^2}{F_p\ell_m^2}$	$\Sigma = \dfrac{\mu}{\lambda\tau^2}$
Viscosity	η	$\dfrac{Ft}{A}$	$\dfrac{\eta_m}{\eta_p} = \dfrac{\sigma_s^{(m)}/\dot\varepsilon_s^{(m)}}{\sigma_s^{(p)}/\dot\varepsilon_s^{(p)}} = \dfrac{\sigma_s^{(m)}\dot\varepsilon_s^{(m)}}{\sigma_s^{(p)}\dot\varepsilon_s^{(p)}}$	$\Sigma\tau = \dfrac{\mu}{\lambda\tau}$

Scale factors for different material constants can be derived from the appropriate constitutive equation. For viscous fluids, for example, Equations (18.2) and (18.3.4) show that the viscosity coefficient is given by the shear stress divided by the shear strain rate. Because strain is a dimensionless quantity, the unit of strain rate is inverse time. From these relationships we derive the scale factor for viscosity shown in Table 20.1.

In some circumstances, it may be convenient to replace one of the fundamental scale factors λ, τ, and μ with a different one defined in terms of the one it replaces. For example, we will see in the next section that it is convenient to consider the scale factor for stress Σ as one of the independent scale factors in place of μ. We can never specify a priori the values of more than three of the scale factors, and those three must be independent. The remaining scale factors are then determined, thus constraining the characteristics of the scale model.

Although the construction of scale models may seem straightforward in principle, numerous problems complicate the process. It is impossible to devise a model in which all factors can be scaled correctly, and compromises and approximations are necessary. The greater the number of physical phenomena to be modeled, the more difficult it is to scale all factors correctly.

In order to have an accurate mechanical scale model, it is necessary to know the rheology of the prototype rocks. All too often, this fundamental feature must simply be guessed. Even if we can devise an appropriate scale model, we still must find a usable material that satisfies the scaling requirements.

Any model is of finite size and therefore must include boundary conditions that do not necessarily reflect the prototype geologic environment. Boundary conditions for any restricted part of the Earth are commonly difficult or impossible to define precisely in any case. Thus, although the technique of scale modeling can provide some fascinating insights into the behavior of geologic systems, it is not without drawbacks, and we must keep its limitations in mind as we interpret model behavior.

In the following sections, we analyze the results of scale model experiments of different geologic situations. Examples of other models illustrated before include Figures 9.7, 9.8, 10.12, 12.30, 20.3, and 22.30.

20.6 Scale Models of Folding

Many investigators have made scale models of folding by using plasticine, silicone putty, and similar materials. As with any model, the important question is whether we are justified in interpreting the model structures as equivalent to those observed in rocks. We can use the theory of scale models to answer this question, and we illustrate the analysis by considering below a model of fold development in the Jura Mountains.

We have already mentioned the Jura in the chapters on strike-slip faults and on folds (see the maps in Figures 6.20, and 7.11 and the detailed cross section in Figure 11.2B). The structure of the mountains consists of a

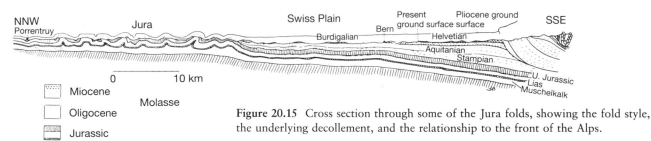

0 _____ 10 km

☐ Miocene

Molasse

☐ Oligocene

▦ Jurassic

☐ Keuper and Muschelkalk

▨ Buntsandstein and Paleozoic Basement

Figure 20.15 Cross section through some of the Jura folds, showing the fold style, the underlying decollement, and the relationship to the front of the Alps.

series of large folds developed in sedimentary rocks lying north of the main Alpine orogenic zone (Figure 20.15). The folds are in general of class 1 geometry, and they are underlain by a décollement.

The width of the Jura is about 30 km, or 3×10^6 cm. A reasonable model would have a corresponding cross-sectional length of 30 cm. Thus the geometric scale factor is

$$\lambda = \frac{\ell_m}{\ell_p} = \frac{30 \text{ cm}}{3 \times 10^6 \text{cm}} = 10^{-5} \qquad (20.10)$$

The time required for the Jura folds to form was approximately one million years. Suppose we want to be able to run a model experiment within 9 hours, or approximately 10^{-3} years (9 hr ÷ 24 hr/da ÷ 365 days/yr = 1.03×10^{-3} yr). The scale factor for time is

$$\tau = \frac{t_m}{t_p} = \frac{10^{-3} \text{ yr}}{10^6 \text{ yr}} = 10^{-9} \qquad (20.11)$$

Finally, the model materials we are likely to be able to use do not have a very great range in density. As a first approximation, we can assume that the ratio of model density to prototype density is approximately 1, whereby

$$P = \frac{\rho_m}{\rho_p} \cong 1 = \frac{\mu}{\lambda^3} \qquad \mu \cong \lambda^3 \qquad (20.12)$$

We thus have values for the three independent scale factors and can derive values of all the other mechanical scale factors. We first inquire what the scale factor for forces should be. Using Equations (20.10) through (20.12) in the scale factor for force (Table 20.1), we find that

$$\frac{F_m}{F_p} = \frac{\mu\lambda}{\tau^2} = \frac{\lambda^4}{\tau^2} = \frac{10^{-20}}{10^{-18}} = 10^{-2} \qquad (20.13)$$

Thus each type of force on the model must be 0.01 times the corresponding force on the prototype.

This requirement immediately presents a problem, because both the prototype and the model are subject to the same gravitational force, which implies that

the scale factor for force should be 1. There are two ways to deal with this contradiction. First, if, compared to surface forces, gravitational force is not a significant factor in fold development, then we can ignore the gravitational forces and the constraint they impose on the force scale factor. In fact, we adopt this approximation in the analytic and the numerical models for folding discussed in Sections 20.2 and 20.3. Second, if, compared to surface forces, gravitational force *is* the dominant factor, as it must be in any convective process such as the rise of diapiric structures and gneiss domes or in isostatic adjustments, then it is possible to scale the body force up in the model by doing the scaled experiment in a centrifuge. We discuss this possibility further in the next section.

For the Jura model, we assume that gravity plays a negligible role in the development of the folds. Furthermore, because the deformation proceeds very slowly, accelerations are very small, and therefore inertial forces are also negligible. With these assumptions, mass does not appear explicitly in the balance equation (Equation 20.2). Assuming constant-volume deformation, the density is not a factor in the constitutive relationships either. For this reason, the scale factors for mass or density are not directly pertinent to our problem, and we may choose a model material of any convenient density without affecting the correlation between model and prototype.

Thus we may choose the stress ratio Σ as a third independent scale factor. In order to model the natural behavior of rocks, however, we must know what that behavior is. Experiments on ductile deformation of limestone show that it behaves as a power-law material at high stress but that at low stress and fine grain size, its rheology approaches that of a viscous fluid (Figure 18.15). None of these experiments, however, evaluate the rheology of solution creep, which was significant in the Jura folding, according to observation of the deformed rocks and which in principle could increase the range of stresses and grain sizes for which limestone would deform viscously (see Figure 19.1B).

Let us assume, therefore, that the rocks deformed viscously and that the viscosity was between 10^{12} and 10^{16} Pa s (see Section 18.6 footnote 4). Using the scale factor for viscosity (Table 20.1), the scale factor for time (Equation 20.11), and this range of values for the prototype viscosity, we find that

$$\frac{\eta_m}{\eta_p} = \Sigma\tau \tag{20.14}$$

$$\eta_m = 10^{-9}\,\eta_p\Sigma = 10^3\Sigma \text{ to } 10^7\Sigma \tag{20.15}$$

Equation (20.15) shows that it is possible to maintain dynamic similarity for the model over a large range of model viscosities simply because it is easy to vary the stresses applied to the model—and thus the stress scale factor—over a wide range. This result indicates that for this model in particular, and for fold models in general for which gravitational forces are insignificant, the viscosity of the particular material used to model the rock deformation is not important. For as long as sufficient stress can be applied to deform the material at the appropriate rate, dynamic similarity can be maintained.

It is convenient to choose a model material that flows on time scales of hours but not on time scales of minutes. Such a property permits relatively large deformations within reasonable lengths of time at easily attained stresses. At the same time, such materials are easy to handle during construction of the model, and after deformation it is possible to cut the model apart for examination without destroying the geometry of the experiment. An appropriate viscosity range is between 10^4 and 10^7 Pa s, and the materials that are approximately visco-elastic and have viscosities within this range include silicone putty and stitching wax.

The rocks in the Jura include shale interbeds between the dominant limestone layers. The rheologic properties of shale have not been well studied, but field evidence indicates that shales are considerably less competent than limestones. The models shown in Figures 20.16 and 20.17 were constructed from layers of stitching wax separated by layers of grease to approximate the properties of the interlayered limestone and shale.

Figure 20.16 shows the results of one model experiment on the Jura Mountains. The details of the experimental setup are shown in Figure 20.16A. The interlayered sequence of stitching wax and grease was compressed by the gravitationally induced collapse and spreading of a large block of stitching wax. The resulting deformation (Figure 20.16B) bears a striking resemblance to the deformation depicted in the cross section of the Jura shown in Figure 20.15.

At the elevated temperatures characteristic of high grades of metamorphism, the style of folds commonly observed indicates that large contrasts in the competence of different rock types are minimized or eliminated. The model in Figure 20.17 was designed to reproduce this style of deformation. It consists of a rigid block adjacent to a multilayer of warm and softened stitching wax. The block and the wax multilayer are overlain by a thin covering of interlayered stitching wax and grease. Deformation of the warm wax was driven by the advancing rigid block, which caused the wax to shorten and thicken and then to spread laterally as a result of gravitational collapse. The resulting structure is a recumbent anticline that has spread out underneath the covering layers. The lower limb of the fold is highly attenuated, and the style of folding has much in common with major recumbent folds that are observed in the metamorphic cores of many orogenic belts (see, for example, Figure 11.1). The

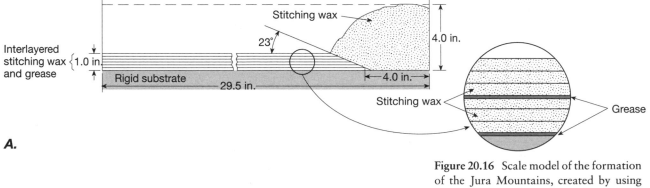

Figure 20.16 Scale model of the formation of the Jura Mountains, created by using interlayered stitching wax and grease. A. Details of the experimental setup, showing the sequence of layers of stitching wax and grease and the block of stitching wax that compressed the layered sequence as it flowed out and flattened. B. The result of one of the deformation experiments, showing folding comparable to that observed in the cross section of the Jura (see Figure 20.15).

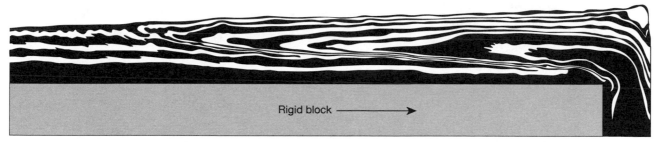

Figure 20.17 Stitching wax model of the formation of a recumbent anticline. A short, thick multilayer of warmed stitching wax with no interlayers of grease was compressed by a rigid block representing colder, more competent material. Overlying the warm wax and rigid block are several thin layers of wax with grease interbeds. Deformation is driven by the motion of the block into the warmed wax and by gravitational spreading of the resulting uplift.

advancing fold has also shortened and folded the layers that lie ahead of it.

This experiment does not properly model orogenic deformation, however. The model does not fulfill all the requirements for dynamic similarity with a prototype orogenic zone, because the gravitational forces, on which the deformation depends, are not correctly scaled. Moreover, the boundary conditions represented by the motion of the rigid block are unlikely to duplicate the conditions actually associated with orogeny. The results of this experiment are intriguing because of their similarity to real geologic structures, but they do not provide a reliable model of deformation in the Earth because the model is not correctly scaled.

20.7 Scale Models of Gravity-driven Deformation

In many model experiments, both the model and the prototype are deformed under the same gravitational acceleration. Thus, using the scale factors for acceleration from Table 20.1, we must have

$$\frac{g_m}{g_p} = \frac{\lambda}{\tau^2} = 1 \qquad \tau = \lambda^{1/2} \qquad (20.16)$$

where g_m and g_p are the gravitational accelerations for model and prototype, respectively.

We consider first the modeling of brittle deformation in the Earth's crust, which is governed by the Coulomb fracture criterion (Equation 9.8). Because time is not a variable in the Coulomb fracture criterion, the scale factor τ is not important for the scale model. Of the two material constants in the fracture criterion, the coefficient of internal friction μ_c is dimensionless and thus should be the same in both prototype and model (we introduce a subscript c for the coefficient of internal friction to distinguish it from the scale factor for mass).

The cohesion c_0 has dimensions of stress and must be scaled accordingly. For a deformation driven by gravitational forces, such as extensional normal faulting, we use the stress scale factor from Table 20.1 with the second Equation (20.16) to find

$$\Sigma = \frac{\sigma_m}{\sigma_p} = \frac{\mu}{\lambda\tau^2} = \frac{\mu}{\lambda^2} = \frac{\mu}{\lambda^3}\lambda = P\lambda \qquad (20.17)$$

where we used the scale factor for density from Table 20.1. Thus if the density scale factor is roughly 1, the strength of the materials must scale with the length. Cohesions for rocks are generally less than 50 MPa, and many are more than an order of magnitude less. Thus taking $\lambda = 10^{-5}$ from Equation (20.10), we see that even to model the strongest rocks requires a model material with negligible cohesion. The brittle deformation of the Earth's crust is therefore commonly modeled by using dry sand!

Figure 20.18 shows one experiment in which layers of dry sand were deformed to 50 percent extension by the uniform stretching of a rubber substrate. The results show horst and graben structure bounded by conjugate normal faults, listric normal faults bounding tilted fault blocks, and domains extending across several fault blocks in which bedding is tilted uniformly in one direction—all features that are found in the Basin and Range province (compare Figures 5.4, 5.6, 5.12 and 5.13).

If both time and the force of gravity are significant factors in the formation of a structure, as is the case for gravity-driven viscous deformation, special problems arise in the development of a scale model. If we choose a length scale factor reasonable for modeling geologic systems,—say $\lambda = 10^{-5}$—Equation (20.16) requires that the time scale factor be 3.16×10^{-3}. Thus a geologic event that occurs over a period of 10^6 years would require 3160 years to occur in a dynamically similar scale model. Clearly it is not practical to do such experiments.

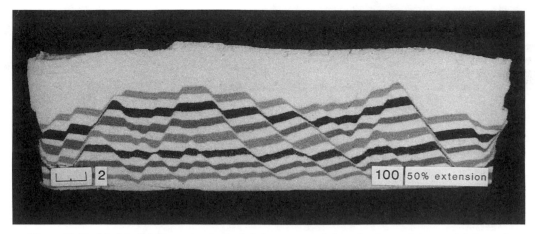

Figure 20.18 Model of deformation in the brittle crust, made by using layered dry sand deformed to 50 percent extension by uniform stretching of a rubber substrate. Note horst and graben structure, listric normal faults, and tilted fault blocks.

The use of a centrifuge to substitute inertial forces in the model for gravitational forces in the prototype can provide a way around this problem. Deformation in the prototype is so slow that inertial body forces are very much smaller than the gravitational body forces and therefore can be ignored. If we place the model in a centrifuge where accelerations equivalent to several thousand times the acceleration of gravity can be attained, then for the model, the gravitational body forces are very much smaller than the inertial body forces and can therefore be ignored. If one or both quantities in the ratio defining a scale factor are unimportant for the behavior of the system, then scaling that quantity from prototype to model is immaterial, and the constraints imposed by the scale factor for that particular type of force can be ignored. From this line of argument, then, we can ignore the constraint implied by Equation (20.16) for centrifuged models, and we may treat λ and τ as independent scale factors.

Assuming the rocks behave viscously, the significant constraint is that the body forces and the viscous forces must be scaled by the same scale factor. The body force per unit mass is provided by the mechanical acceleration a_m in the model and by the gravitational acceleration g_p in the prototype. Thus, using the superscripts (v), (i), and (g) to identify viscous, inertial, and gravitational forces, respectively, we must have

$$\frac{F_m^{(v)}}{F_p^{(v)}} = \frac{F_m^{(i)}}{F_p^{(g)}} \tag{20.18}$$

$$\frac{\eta_m(\ell_m)^2/t_m}{\eta_p(\ell_p)^2/t_p} = \frac{[\rho_m(\ell_m)^3]a_m}{[\rho_p(\ell_p)^3]g_p}$$

$$\frac{\eta_m}{\eta_p} = P\lambda\tau\frac{a_m}{g_p} = 5 \times 10^{-12} \tag{20.19}$$

where the three independent scale factors are the density scale factor P, which we take to be 0.5 (an appropriate

value if silicone putty is the model material for rocks), and λ and τ, whose values are 10^{-5} and 10^{-9} from Equations (20.10) and (20.11), respectively. The ratio of mechanical to gravitational acceleration is not a formal scale factor, because it is a ratio of two different quantities. Its value is determined, by considering the capabilities of the centrifuge, to be approximately 10^3. If we assume the prototype viscosities in an orogenic zone to be in the range of 10^{14} to 10^{18} Pa s, then the result from Equation (20.19) gives

$$10^3 \leq \eta_m \leq 10^7 \text{ Pa s} \tag{20.20}$$

Salt is commonly deposited in thick layers of evaporite. It is a low-density substance (2.2×10^3 kg/m^3) with a very low strength at room temperature (see Table 18.1 and Figure 18.19B, C). The sediments that often overlie the salt, such as sandstone, shale, and limestone, have densities significantly greater than the salt (2.3 to 2.8×10^3 kg/m^3). Thus when thick salt layers occur under a sufficiently thick overburden, the salt tends to move slowly up, forming diapirs (see Section 12.10). Figure 12.33 illustrates a variety of shapes that are observed in salt diapirs (see also Figure 12.34).

Considerable work has been done modeling diapiric intrusions in a centrifuge by using silicon putties, modeling clay, and similar materials, and many features observed in nature have been successfully reproduced. Such models show that the shear between the rising diapir and the surrounding material induces a torroidal flow along the margins of the diapir (Figure 20.18). The effective viscosity ratio $m = \eta_{(hi\,\rho)}/\eta_{(lo\,\rho)}$ between the high-density and low-density materials has a strong effect on the shape of the diapir, because it governs the location of the torroidal flow. If $m \ll 1$, the diapiric material is much more viscous, the torroidal flow is concentrated in the high-density material, and the diapir has a fairly columnar shape (Figure 20.19A). If the vis-

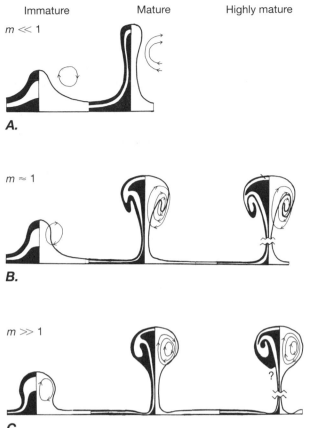

Figure 20.19 Styles of diapiric structures revealed by model experiments. Diapir shape and structure are a function of the ratio *m* of the viscosity of the high-density material to the viscosity of the low-density material, and of the maturity of diapiric development. The structure and shape are determined largely by the torroidal flow that develops around the diapir as a result of drag on the diapir as it intrudes the overburden. *A.* If the viscosity of the diapir is much greater than that of the surroundings, the torriodal flow occurs mainly in the high-density material, and columnar diapiric stocks develop. *B.* If the viscosities of diapir and surroundings are approximately equal, the torroidal flow lines cross the diapir boundary, and mushroom-shaped diapirs develop. *C.* If the viscosity of the diapir is much less than the surroundings, the torroidal flow is restricted to the interior of the diapir, and bulb-shaped diapirs develop.

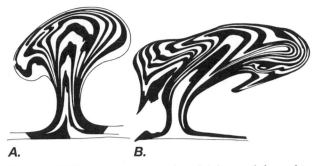

Figure 20.20 Internal structure of model diapirs deformed in a centrifuge, showing the folding of initially planar layers in the low-density material. *A.* Asymmetric diapir with a bulb shape on the right and a mushroom shape on the left. *B.* Tilted, highly mature, mushroom-shaped diapir.

cosities are approximately equal, the torroidal flow lines cross the margin between the diapir and the surrounding material, deforming the margin to yield a mushroom-shaped diapir (Figure 20.19*B*). If the diapir has a much lower viscosity than the surrounding material, the torroidal flow is concentrated within the diapir, and it adopts a bulb shape (Figure 20.19*C*). The mushroom shape can also develop as a result of lateral spreading of the diapir if it reaches a level above which it cannot rise (see Figure 20.21). Figure 20.20 shows some of the complexity of internal structure that develops from initially horizontal layering in centrifuged silicone putty diapirs. These models compare favorably with the diagrams in Figures 12.33 and 12.34, which show the structure of naturally occurring salt diapirs.

Gneiss domes are a common feature of many high-grade metamorphic belts, and they have been interpreted as the result of the diapiric rise of lower-density gneisses into a higher-density cover (Figure 12.35). Centrifuged models designed to simulate orogenic zones (Figure 20.21) bear a striking resemblance to the structures associated with mantled gneiss domes. These studies support the belief that gravity-driven deformation is a significant factor in the development of major structures in orogenic zones.

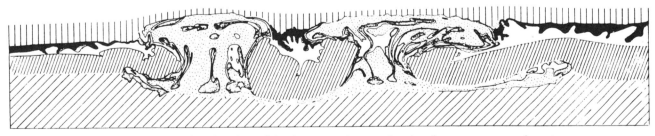

Figure 20.21 Centrifuged scale models of an orogenic zone, showing the consequences of gravity-driven deformation. Different patterns of decoration show materials having different densities and viscosities.

Additional Readings

Backofen, W. A. 1972. *Deformation processing*. Reading, Mass.: Addison-Wesley.

Balk, R. 1949. Structure of Grand Saline salt dome, Van Zandt County, Texas. *Amer. Ass. Petrol. Geol. Bull.* 33: 1791–1829.

Biot, M. A. 1961. Theory of folding of stratified viscoelastic media and its implications in tectonics and orogenesis. *Geol. Soc. Am. Bull.* 72: 1595–1620.

Bucher, W. H. 1956. Role of gravity in orogenesis. *Geol. Soc. Am. Bull.* 67: 1295–1318.

Currie, J. B., H. W. Patnode, and R. P. Trump. 1962. Development of folds in sedimentary strata. *Geol. Soc. Am. Bull.* 73: 655–673.

Dennis, J. G., and R. Häll. 1978. Jura-type platform folds: A centrifuge experiment. *Tectonophysics* 45: T15–T25.

Dieterich, J. H. 1969. Origin of cleavage in folded rocks. *Am. Jr. Sci.* 267: 155–165.

Dieterich, J. H., and Carter, N. L. 1969. Stress-history of folding. *Am. Jr. Sci.* 267: 129–154.

Hubbert, M. K. 1937. Theory of scale models as applied to the study of geologic structures. *Geol. Soc. Am. Bull.* 48: 1459.

Jackson, M. P. A., and C. J. Talbot. 1989. Anatomy of mushroom-shaped diapirs. *J. Struct. Geol.* 11(1/2): 211–230.

Johnson, A. M. 1970. *Physical processes in geology*. San Francisco: Freeman, Cooper.

McClay, K. R., and P. G. Ellis. 1987. Analogue models of extensional fault geometries. In M. P. Coward, J. F. Dewey, and P. L. Hancock, eds., *Continental extensional tectonics*. Geol. Soc. Lond. Special Pub. 28. Oxford, England: Blackwell Scientific Publications, pp. 109–125.

Odé, H. 1960. Faulting as a velocity discontinuity. In D. T. Griggs and J. Handin, eds., Rock deformation. Geol. Soc. Am. Memoir 79.

Parrish, D. K., A. L. Krivz, and N. L. Carter. 1976. Finite-element folds of similar geometry. *Tectonophysics* 32: 183–207.

Ramberg, H. 1981. *Gravity, deformation, and the earth's crust*. London: Academic Press.

AT THIS POINT, we have described the basic types of structures observed on the Earth, and we have considered models for the mechanics of rock deformation and the mechanisms by which rocks deform. These topics provide a basis from which to consider the tectonics of the Earth, the deformation of the Earth's crust on a regional to global scale, and its history through geologic time.

We discuss first the major tectonic features of the Earth (Chapter 21) and then the tectonics of orogenic (mountain) belts (Chapter 22). These latter features have traditionally been the subject of much attention in structural geology and tectonics, because it is predominantly in these belts that the structures discussed in the foregoing chapters are developed.

Principal Tectonic Features of the Earth

The major tectonic features of the Earth and the types of structures they exhibit provide the key to understanding large-scale dynamic processes in the Earth. And studying the small-scale structures associated with major features provides insight into their large-scale significance.

In this chapter we discuss first oceanic and then continental features. Figure 21.1 summarizes the age of the oceans and the broad outlines of continental geology. Although the oceans occupy by far the majority of the Earth's surface area, oceanic crust is substantially younger than continental crust. Relatively young oceanic crust near active spreading centers occupies a larger area than the oldest crust, most of which has disappeared down subduction zones (Figure 21.1). Thus most oceanic crust has a relatively simple history that reflects only the youngest events of the Earth's tectonics.

Continental crust, on the other hand, ranges in age from 0 to at least 3.96 billion years and displays a more complex geology (compare Figure 1.4). Thus the geologic age used to characterize the oceanic crust in Figure 21.1 is not so helpful for our discussion of the continents as a subdivision according to broad tectonic features, specifically Precambrian shields, interior lowlands, orogenic belts, and continental rifts and margins. Although subordinate in area to oceanic crust, the continental regions are better exposed, better studied, and better

known. The greater age range of rocks in continents means that for times prior to formation of the oldest oceanic crust, only the continents preserve any information on the processes that were active on or within the Earth. Interpretation of many ancient continental features, however, is hampered by the effects of a long and complex history during which the oldest records may have been obscured, destroyed, or buried.

21.1 Ocean Basins

Vast areas of the oceanic bottom are flat or nearly so. The oceanic crust underlying these areas is remarkably uniform in thickness and composition. Ranging in thickness from 3 to 10 km and averaging 5 km thick, oceanic crust is thin compared to continental crust. It consists predominantly of igneous rocks of basaltic composition.

These nearly flat areas of the ocean bottom include the ridges[1] as well as abyssal plains. Scattered throughout the ocean basins are plateaus of anomalously thick

---

[1] Although the midocean ridge system represents one of the most important topographic and tectonic features on Earth, the average slope of its flanks is generally less than one or two degrees.

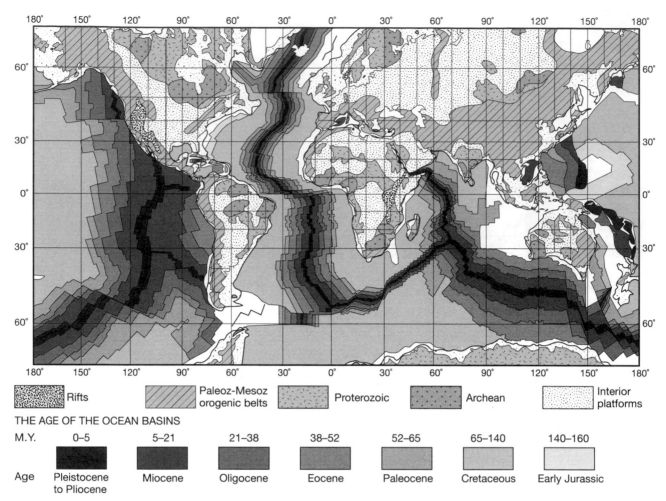

THE AGE OF THE OCEAN BASINS						
M.Y.	0–5	5–21	21–38	38–52	52–65	65–140	140–160

| Age | Pleistocene to Pliocene | Miocene | Oligocene | Eocene | Paleocene | Cretaceous | Early Jurassic |

Rifts   **Paleoz-Mesoz orogenic belts**   **Proterozoic**   **Archean**   **Interior platforms**

Figure 21.1 Generalized world map showing major features of continental crust, as well as age of oceanic crust. Continental features include Precambrian shields with Archean and Proterozoic areas, interior lowlands, orogenic belts, rifts, and margins.

crust, island arc-trench systems, and aseismic ridges of relatively thick crust (Figure 21.2).

Gravity measurements over the oceans indicate that generally the free-air anomaly is nearly zero (compare Section 2.6). Thus, for the most part, ocean basins are in isostatic equilibrium, and differences in elevation reflect differences in density or thickness of the underlying crust and/or mantle.

An average layered model for the oceanic crust, shown in Figure 21.3A, is based on P-wave seismic velocity (V_p) measurements. The lithologic interpretation of these layers results from direct sampling of the oceanic crust and from comparison with on-land exposures of rock sequences thought to represent old oceanic crust.

The uppermost layer, layer 1, has a seismic P-wave velocity (V_p) of 3 to 5 km/sec and is interpreted as unconsolidated sediment of pelagic, hemi-pelagic, or turbiditic origin.[2] In layer 2, commonly subdivided into layers 2A, 2B, and 2C, V_p ranges from 5 to 6 km/sec. It is interpreted as consisting largely of submarine basaltic extrusive and shallow intrusive rocks. Layer 2A has a relatively low velocity that rapidly increases with depth; layer 2B is a layer of relatively constant velocity; layer 2C again displays a rapid increase of seismic velocity with depth. Layer 3, with its subdivisions 3A and 3B and a P-wave velocity ranging from 6 to 7.5 km/sec,

---

[2] Pelagic sediments are derived by settling of suspended material throughout the ocean water column. The material comes either from wind-borne dust from land or from shells of microscopic animals and plants. Hemi-pelagic sediments contain significant continental or volcanic material. Turbidite is a sediment formed by sediment-laden bottom currents and generally derived from a continent or island source.

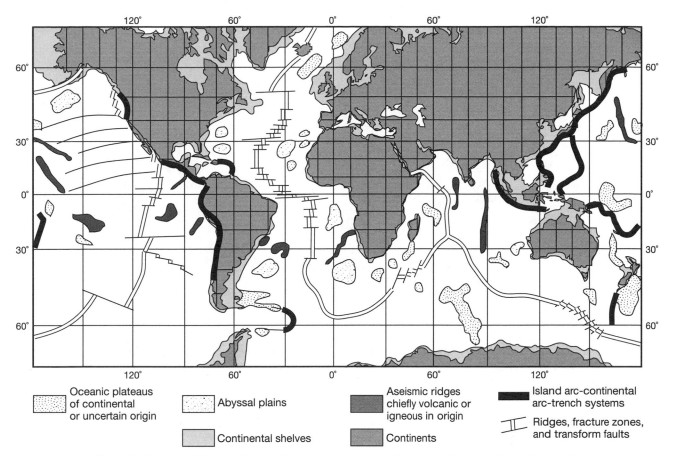

Figure 21.2 Generalized world map showing major oceanic features: ridges, transform faults and fracture zones, oceanic plateaus, aseismic ridges, and island or continental arc–trench systems.

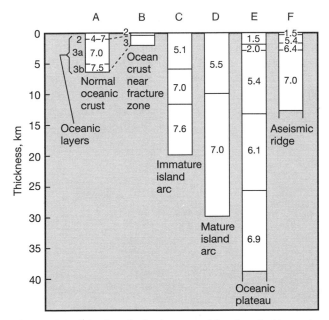

Figure 21.3 Seismic velocity layer models of typical oceanic crust and other oceanic features. P wave velocities are indicated for the different layers.

is thought to represent mafic–ultramafic[3] plutonic rocks and/or serpentinized mantle peridotite. Layer 3 subdivisions may reflect varying quantities of olivine in plutonic rocks.

For descriptive purposes, we divide the principal features of oceanic crust into those features characteristic of plate margins and those characteristic of plate interiors.

Features of Oceanic Plate Margins

Divergent plate margins are topographically high regions that characteristically occur in the middle of the

---

[3] Mafic and ultramafic are terms used to indicate the composition of the rocks in question. "Mafic" rocks have one or more Fe-Mg–bearing mineral, such as amphibole, pyroxene, or olivine. "Ultramafic" rocks consist chiefly of Fe-Mg–bearing minerals. Rocks of basaltic or gabbroic composition are mafic, whereas peridotites and serpentinites are ultramafic.

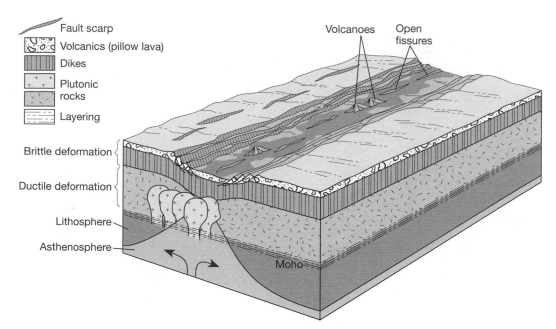

Figure 21.4 Block diagram illustrating schematically the principal features of a midoceanic divergent plate margin. Extensional (normal fault) structures at the surface pass downward into a zone of magmatic intrusion and ductile stretching. The lithosphere thickens away from the plate margin. Not to scale.

ocean basins (except for the eastern Pacific Ocean and the northwestern Indian Ocean). These **midoceanic ridges** form a continuous, world-girdling topographic swell approximately 40,000 km long, 2.5 km high above the abyssal floors of the ocean basins on either side, and 1000 to 3000 km wide. Structures on ridges are predominantly active normal faults, as revealed by morphology and first-motion studies of earthquakes (Figure 21.4). The faulting is consistent with extension perpendicular to the trend of the ridge and parallel to the inferred relative plate motion.

Transform fault boundaries in the oceans are the seismically active portions of **fracture zones**—great rectilinear fracture systems within the oceanic crust. They

are characterized by pronounced differential topographic relief, sharp ridge and trough topography, steeply dipping faults, and deformed oceanic rocks (Figure 21.5). They range in length up to 10,000 km. Although they often are fairly narrow features, some transform faults are up to 100 km or more in width. The seismically inactive portions of these fracture zones represent fossil transform faults. First-motion studies of the earthquakes along the active transform fault portions indicate strike-slip faults with characteristic horizontal relative motion *opposite* in sense to that of the apparent offset of the ridge crest. The thickness of oceanic crust near fracture zones and transform faults tends to be less than average (Figure 21.3*B*).

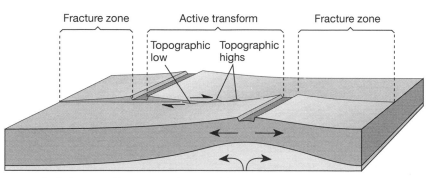

Figure 21.5 Block diagram illustrating schematically a conservative, or transform fault, boundary in oceanic crust offsetting a divergent margin (ridge). The structures of each offset portion of the ridge are as shown in Figure 21.4. Not to scale.

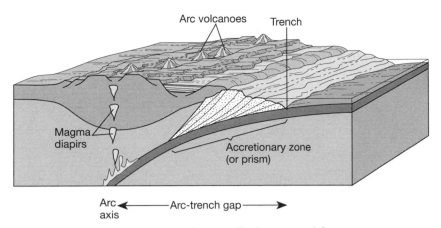

Figure 21.6 Block diagram illustrating schematically the principal features in an intra-oceanic convergent plate margin, or subduction zone. One plate descends beneath another along a marginal zone of thrust faults. Partial melting of down-going crust produces blobs of magma that rise and become volcanoes. Not to scale.

Convergent plate margins in the oceans exhibit chains of volcanic islands accompanied by parallel trenches, which are the deepest parts of the ocean basins. These **island arc–deep sea trench** pairs generally are arrayed in a series of arcs that join at cusps and extend for thousands of kilometers. The volcanic islands are spaced approximately 80 km apart and rise above submerged ridges that tend to be a few hundred kilometers wide. Trenches are up to 12 km deep and approximately 100 km wide. Systems of active thrust faults usually characterize the landward side of trenches, whereas island arcs and regions behind the arcs exhibit active normal faults (Figure 21.6). Trenches are associated with pronounced negative Bouguer gravity anomalies, which indicates a marked deficiency in mass below the sea floor.

The crust of island arc regions is an average of 25 km thick—considerably thicker than that of normal oceanic crust. It is rather variable, however, thinning abruptly to oceanic thicknesses on either side of the arc axis (Figure 21.3C, D). Younger, immature arcs tend to have a thinner crust (Figure 21.3C) than older, mature ones (Figure 21.3D).

Features of Oceanic Plate Interiors

Away from plate margins, the deepest regions of the ocean are vast areas of very flat ocean floor, the **abyssal plains.** These plains represent areas of normal oceanic crust covered by sediments of turbidity current and pelagic origin.

Broad elevated regions, or **oceanic plateaus,** have a variety of origins. Some are apparently continental rocks, others inactive volcanic arcs. The origin of still others is unclear. They range in area from a few hundred to many thousands of square kilometers, and they stand 1 to 4 km above the normal ocean floor. Crustal thickness is generally of continental rather than typically oceanic dimensions (Figure 21.3E).

Linear ridges characterized by high-elevation, anomalously thick oceanic crust, and by a general lack of associated seismic activity, are called **aseismic ridges.** Their lack of seismic activity and their more limited dimensions (see Figure 21.2) set them apart from the midoceanic ridges. In most cases they represent linear constructional ridges formed by chains of basaltic volcanoes. The Hawaiian Islands–Emperor Seamount chain, extending northwest from Hawaii to Midway Island and thence north to the Kamchatka trench (Figure 21.2) is the most famous example of this type of crustal feature. The crustal thickness of aseismic ridges is considerably greater than that of normal oceanic crust and is comparable to that of island arcs (Figure 21.3F).

21.2 Structure of Continental Crust

Because it is older and has experienced more tectonic history, continental crust is more complex in structure than oceanic crust. Continental crust is thicker and less dense than oceanic crust and has lower seismic velocity. Figure 21.7, an idealized cross section of North America, shows features typical of continental crust.

The average thickness of continental crust is about 35 km. There is considerable deviation from that average, however, depending on location and tectonic setting. The crust tends to be thickest (up to 70 km or more) under mountainous regions, about average under sedimentary platforms, and relatively thin under rifts

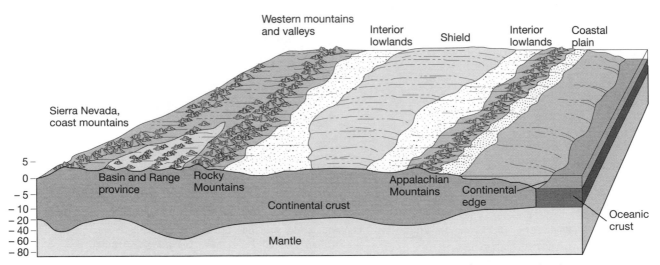

Figure 21.7 Generalized block diagram of the North American continent, showing variation in crustal thickness in diverse tectonic provinces. Note the change in vertical exaggeration at about 10 km.

such as the Basin and Range province, Precambrian shields, and marginal areas. The crustal seismic velocity tends to increase with depth, but velocity inversions are reported from some regions, such as the Basin and Range province.

Clues to the nature of the continental crust at depth come from direct examination of rocks formed at depth and exposed by uplift and erosion and from geophysical sounding of the crust. Exposures of the shallow crustal rocks show that faulting, folding, metamorphism, and igneous intrusion are characteristic of orogenic activity. High-grade metamorphic rocks of the lower crust may be exposed in very deeply eroded terranes. These rocks contain very complex structure, such as the folded and

refolded interlayers of pyroxene granulite and granitic gneiss from Greenland. Such structures are relatively common in many Precambrian shields and in deeply eroded central zones of Phanerozoic orogenic belts. It seems reasonable to infer that they are common in the lower crust everywhere.

Correlation of field data with the seismic velocity structure of the continental crust suggests the generalized petrologic and seismic model shown in Figure 21.8. The upper levels of the crust, beneath the sedimentary cover, consist of metasedimentary and metavolcanic rocks intruded here and there by granitic rocks. The middle levels are composed largely of migmatite, a silicic rock that was partially melted and strongly deformed

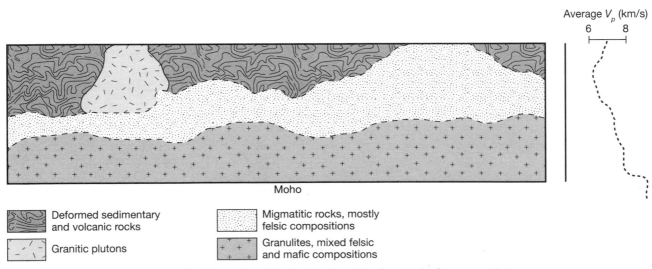

	Deformed sedimentary and volcanic rocks			Migmatitic rocks, mostly felsic compositions
	Granitic plutons			Granulites, mixed felsic and mafic compositions

Figure 21.8 Generalized crustal model showing lateral and vertical inhomogeneities, to account for observations in deeply eroded regions and for observed seismic variations. The diagram to the right shows idealized average P-wave velocities (V_p) through the crust.

during metamorphism and that is characterized by a slightly lower seismic velocity than the upper crust. The lower levels of the crust consist of highly folded rocks that commonly exhibit granulite facies[4] or metamorphism, intruded by mafic and silicic plutonic rocks. These rocks have slightly higher seismic velocity than the upper crust.

This model of the crust includes great lateral and vertical heterogeneity, which is consistent with the available information. A great deal of detailed exploration of the continental crust is taking place around the world, and we can expect much refinement of our knowledge and change in our concepts in the years ahead.

21.3 Precambrian Shields

All continents exhibit large areas where Precambrian rocks greater than 600 million years old (Ma) are exposed at the surface (Figure 21.1). These regions commonly form topographically rolling uplands that stand higher than the lowlands surrounding them, thus giving rise to the name Precambrian *shield*.

We subdivide Precambrian shields into Archean and Proterozoic terranes on the basis of the age of the rocks. (*Archean* comes from the Greek word *archi*, which means "beginning." *Proterozoic* comes from the Greek words *proteron*, which means "before," and *zoe*, which means "life," an allusion to geologists' original impression that these rocks bore no fossils.) This subdivision turns out to have tectonic significance that is of worldwide utility. Most Archean rocks are greater than 2500 million years old (Ma), and Proterozoic rocks range in age from 570 to approximately 2500 Ma. Archean regions display evidence of greater crustal instability or mobility than Proterozoic regions. The tectonic distinction is not universal or abrupt, however, and the transition from one tectonic style to the other varies by a few hundred million years or so from place to place.

Archean Terranes

Rocks in Archean terranes are divisible on the basis of their metamorphic grade into **high-grade gneissic regions,** which exhibit amphibolite or granulite facies of

metamorphism,[5] and **greenstone belts,** which are characterized by rocks at greenschist or lower grades of metamorphism. Both types are characteristically intruded by younger granitic plutons. The part of the Kaapvaal craton of southern Africa that is shown in Figure 21.9A displays the typical division into greenstone belt, gneiss, and granitic rocks (Figure 21.9B).

High-grade gneisses form the bulk of Archean regions. They consist largely of quartzo-feldspathic gneisses derived by metamorphism of felsic igneous rocks, but they also contain subordinate metasedimentary rocks, including metamorphosed quartzites, volcanogenic sediments, iron formations, and carbonate rocks (Figure 21.9C). Deformed mafic–ultramafic complexes form the rest of the gneissic regions.

The high-grade gneissic regions are complexly mixed on a scale of tens to hundreds of kilometers with lower-grade greenstone belts (Figure 21.9B) that contain mafic to silicic volcanic rocks and shallow intrusive bodies, volcanogenic sediments of similar composition, and subordinate flows and shallow sills of olivine-rich magmas (Figure 21.9D).

Three tectonic and structural features are common to all Archean terranes. First, most rocks are highly deformed and display more than one generation of folds. (Figure 21.9D). The most obvious structural features are upright folds; less obvious are refolded low-angle faults and recumbent folds. Figure 21.9D shows a complex pattern of folding in the Barberton Mountain Land of Swaziland and South Africa; the overall pattern of the belt is reminiscent of a type 2 interference structure (compare Figure 12.31B). Figure 21.10 shows a map and cross section of a complex region of types 1 and 3 (Figure 12.31A, C) interference folds in a mafic–ultramafic complex—the Fiskenaesset complex—and surrounding gneiss and amphibolite in southwestern Greenland.

Second, the contacts between greenstone belts and high-grade gneissic areas are complex. In some places, the contacts are shear zones that mask the original relationship. Elsewhere, greenstone rocks are deposited on older gneissic basement. In still other areas, gneissic granitic rocks intrude rocks of the greenstone belt.

Third, the sedimentary rock types fall into one of two broad categories: Either they are immature volcanogenic sediments that are characteristic of the greenstone belts and of parts of the gneissic terranes, or they are a quartzite–carbonate–iron formation assemblage

---

[4] The term *metamorphic facies* refers to a distinctive assemblage of metamorphic minerals that are characteristic of certain conditions of pressure and temperature. The granulite facies of metamorphism is generally characterized by an assemblage of garnet, pyroxene, and feldspar. It usually indicates high temperature (above 650°C) and high pressure (above 500 MPa).

[5] The greenschist facies is characterized by the presence of chlorite and actinolite. The amphibolite facies typically includes hornblende and may or may not include aluminosilicate minerals, garnet, and the like. Pressure and temperature conditions for greenschist facies are approximately 400–500°C and 200–500 MPa; for amphibolite facies they are about 500–650°C and 200–500 MPa.

A.

B.

C.

D.

Figure 21.9 Generalized maps of a typical Archean crustal region (a portion of the Kaapvaal craton, southern Africa), showing gneissic terranes with metasedimentary units, granitic rocks, and a greenstone belts. *A.* Regional map of Kaapvaal craton, showing areas of granite, gneiss, and the Barberton greenstone belt. *B.* Detailed map of part of the Kaapvaal craton. *C.* The Mankayane inlier, showing infolds in gneiss of metasedimentary and meta-igneous rocks consisting of a mafic–ultramafic unit structurally overlain by a sedimentary unit of metaquartzite, quartzo-feldspathic, pelitic, and calcareous schist, and iron formation. *D.* Detailed map of Barberton greenstone belt, showing internal structure. (See Figure 21.4*A* for location.)

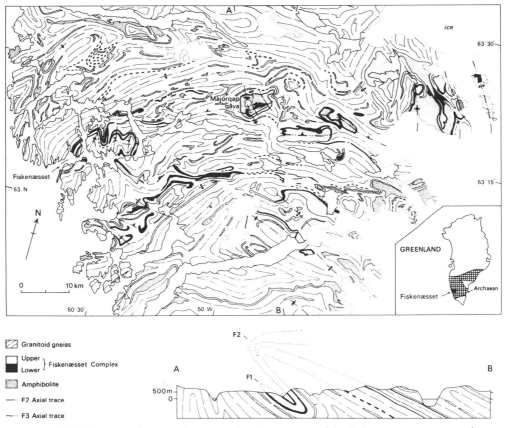

Figure 21.10 Generalized map and cross section of a portion of the Fiskenaesset region, southern Greenland, showing refolded folds. The Fiskenaesset complex is a mafic–ultramafic stratiform sequence; the lower unit is peridotite, the upper unit gabbroic.

associated in many areas with multiply deformed mafic–ultramafic layered igneous complexes and found only in gneissic terranes (Figure 21.9C).

The study of Archean tectonics has really just begun, for numerous large-scale detailed maps and sufficiently precise radiometric dating techniques have become available only since about 1975. The worldwide presence of these characteristic sedimentary and structural associations implies that similar sedimentary and tectonic conditions occurred globally during Archean time and that these conditions differed markedly from those that characterized Phanerozoic time. In particular, the widespread metamorphism and the presence of ultramafic magmatic rocks indicate higher temperatures in the Earth during Archean times. The petrology of the ultramafic magmas imply that they formed by about 50 percent melting of a mantle source at temperatures of approximately 1500°C. Theoretical heat-budget calculations suggest that the rate of increase of temperature with depth in the Earth (the geothermal gradient) was approximately 2 or 3 times the present one.

Proterozoic Terranes

Proterozoic terranes display both slightly deformed stable regions and highly deformed mobile areas, in contrast to the ubiquitous evidence for mobility that Archean terranes display. Regions of the Earth's crust that have achieved tectonic stability, called **cratons** (from the Greek word *kratos*, which means "power"), first appear in Proterozoic time. In these areas, vast deposits of weakly deformed, unmetamorphosed Proterozoic sediments typically overlie a basement of deeply eroded, deformed, and metamorphosed Archean rocks.

Proterozoic cratonic sediments display evidence of relatively stable tectonic environments; mature sediments such as quartzites and quartz–pebble conglomerates are common in areally extensive stratigraphic units. Quartzites are frequently intercalated with abundant iron formations composed of interstratified iron-rich oxides, iron carbonates, and iron silicates. These predominantly undeformed cratonic sediments are deposited on preexisting older basement. They host many

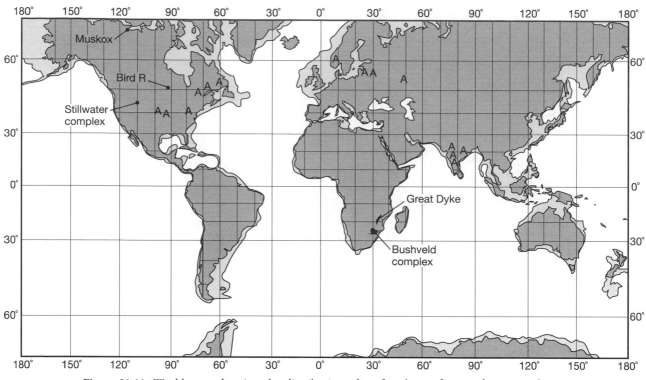

Figure 21.11 World map showing the distribution of mafic–ultramafic stratiform complexes (block dots and blobs) and of anorthosite complexes (indicated by A), all of Proterozoic age.

vast Precambrian placer gold and uranium deposits, as well as most of the world's deposits of iron ore. These sedimentary sequences extend over large distances.

Proterozoic deformed belts are of two general types. Some display multiply deformed regions rich in volcanic rocks and reminiscent of Archean terranes, as well as many Phanerozoic (post-Precambrian) volcanic-rich orogenic belts. Others exhibit thick sedimentary sequences deposited in linear troughs, presumably along ancient continental margins, and subsequently deformed to form linear fold and thrust belts similar to those of the Phanerozoic orogenic belts.

Proterozoic igneous rocks also display distinctive differences when compared with older and younger terranes. Archean regions are often cut by regionally extensive Proterozoic regional dike swarms of basaltic composition.

A number of Proterozoic dike systems are associated with extensive mafic–ultramafic stratiform complexes, such as the Muskox and Bird River complexes in Canada, the Stillwater complex in the United States, and the Bushveld complex in south Africa (Figure 21.11). These complexes are essentially undeformed masses of layered igneous rocks hundreds to tens of thousands of square kilometers in area. They resemble the layered igneous complexes of the Archean but differ in that they are only weakly deformed. Figure 21.12 shows an ex-

ample of such a feature, the Bushveld complex. The vertical columnar sections from three widely separated locations in the complex illustrate the amazingly continuous nature of the different distinctive layers in the complex. The dark layers in the columnar sections in Figure 21.12 represent individual layers of chromite within a sequence of gabbroic cumulate rocks that can be traced for tens of kilometers.

Large, intrusive massifs of anorthosite[6] are another distinctive igneous–metamorphic rock suite that appeared in late Proterozoic time (1000–2000 m.y. B. P.) (indicated by the symbol A in Figure 21.11). In some cases, these rocks are clearly igneous in origin, but in other cases, deformation and recrystallization have so modified the primary texture and structures of the rock that their origin is difficult to decipher.

Associated with many Proterozoic fold and thrust belts are a series of smaller, linear, sediment-filled grabens called **aulacogens** that generally strike at high angles to the trend of the deformed belts. (The term is derived from the Greek word *aulax,* which means "furrow.") Those aulacogens that have been mapped in the

---

[6] An igneous rock composed almost completely of plagioclase feldspar.

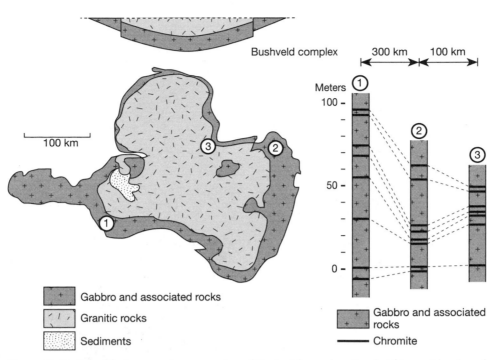

Figure 21.12 Simplified map and cross section of Bushveld complex, South Africa, with selected stratigraphic sections. Dark lines are chromite layers. Note the similarity in the sections over great distances.

North American continent are shown in Figure 21.13*A*. Careful mapping of several aulacogens shows that their sediments are correlated with the thick sediments of the deformed belt, as well as with thinner, undeformed platform sediments on either side (Figure 21.13*B*). The sediments commonly are undeformed or only slightly folded, with fold axes trending parallel to the axis of the trough.

Thus tectonic conditions during the Proterozoic apparently differed from those in the Archean. Widespread undeformed platform sequences indicate the existence in Proterozoic times of large, stable continental regions. Regional dike swarms and linear sediment troughs, such as aulacogens, indicate that these regions were capable of undergoing brittle extension. In this respect, Proterozoic tectonics more closely resembled processes that operated in the Phanerozoic than those that operated in the Archean. Some workers have even suggested that the Archean–Proterozoic transition is the single most important tectonic event in all of Earth history. What caused this transition, and what does it indicate? Why was the Archean so different from later times? How did the Proterozoic differ from later times? What was the nature of global tectonics during Proterozoic and Archean times? Did either bear any resemblance to the Phanerozoic plate tectonics? Addressing these provocative questions is beyond the scope of this book.

Here, however, we shall discuss in more detail the Phanerozoic tectonic processes, for which the evidence is so much clearer.

21.4 Phanerozoic Regions

As we consider Phanerozoic (Cambrian and younger) features of the Earth, the available evidence increases vastly, and we can obtain a much more detailed picture of the structural characteristics of the younger parts of continents than we can of the Precambrian areas. We shall briefly describe features of continental platforms, of orogenic belts, of continental rifts, and finally of modern continental margins.

Continental Platforms

All continents contain regions of interior lowlands and cratonic platforms where relatively thin sequences of sedimentary rocks overlie Precambrian rocks that are sub-surface continuations of the shields. With minor exceptions, these sedimentary rocks are flat-lying and are composed of lithologic units that are continuous over vast areas larger than the Precambrian shields themselves. For the most part, they are plains that stand

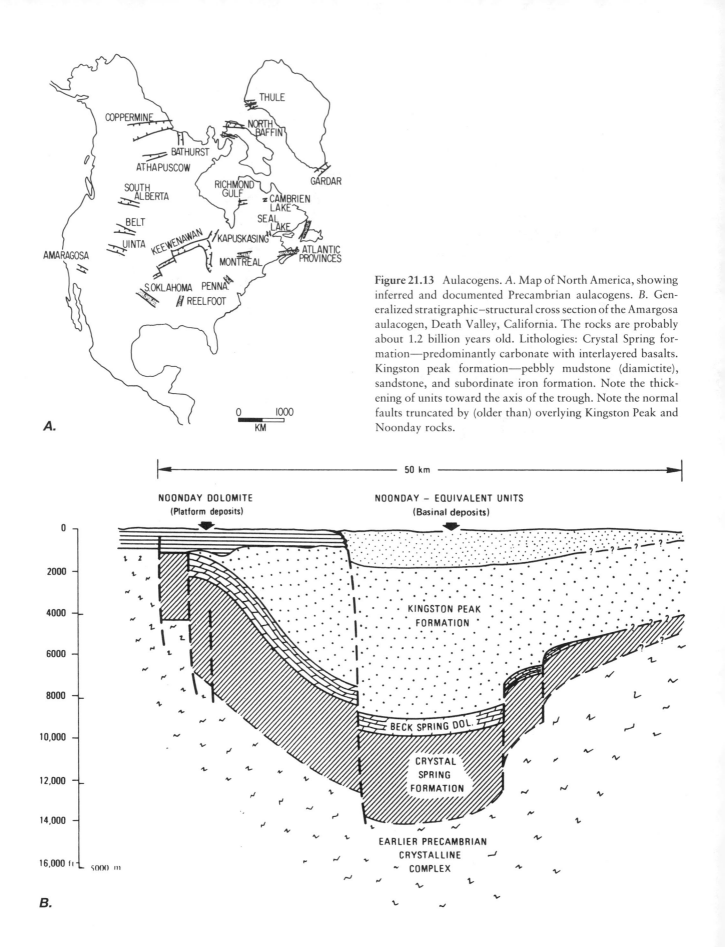

Figure 21.13 Aulacogens. *A.* Map of North America, showing inferred and documented Precambrian aulacogens. *B.* Generalized stratigraphic–structural cross section of the Amargosa aulacogen, Death Valley, California. The rocks are probably about 1.2 billion years old. Lithologies: Crystal Spring formation—predominantly carbonate with interlayered basalts. Kingston peak formation—pebbly mudstone (diamictite), sandstone, and subordinate iron formation. Note the thickening of units toward the axis of the trough. Note the normal faults truncated by (older than) overlying Kingston Peak and Noonday rocks.

a few hundred meters above sea level. To a structural geologist these regions are relatively monotonous. Yet the economic wealth of these regions in coal, petroleum, mineral deposits, and agricultural resources is such that a large body of knowledge about them has accumulated, giving rise to what the late American structural geologist P. B. King called "the science of gently dipping strata."

Most interior platform sedimentary sequences begin with middle Cambrian or younger deposits; lower Cambrian or older Phanerozoic rocks are generally found only at the edges of the platforms. Throughout much of the world, and especially in North America, the contact with the underlying Precambrian shield rocks is a profound unconformity, usually called a great unconformity, that marks a worldwide transgression of the sea over older continental interiors. In most places, this unconformity represents a time gap of tens to hundreds of millions of years.

Most platform sediments are marine and represent deposition in epeiric seas (the Greek word *epiros* means "continent"). A major exception is the platform sequence of much of Gondwanaland,[7] which is mostly nonmarine in origin. The marine sediments record major periods of transgression and regression throughout the Phanerozoic, which in turn reflect major fluctuations in the level of the oceans relative to that of the continent (Figure 21.14) The main structural characteristic of the platforms is a group of cratonic basins separated by intervening domes or arches. Figure 21.15 shows the distribution of cratonic basins throughout the world and identifies the basins and arches of North America. Many of these features exhibit evidence of intermittent vertical movement of the crust lasting over ten to hundreds of millions of years. Arches served as sources of sediment during some stratigraphic intervals and in others were covered, but with thinner stratigraphic sequences than the surrounding platforms. Basins may contain thicker sequences deposited in deeper water. In times of general regression, these basins show evidence of restricted circulation and even desiccation.

North America provides numerous good examples of these features (Figure 21.15). The Transcontinental Arch is a region that stood high relative to the surrounding area through most of Paleozoic time. The sedimentary facies of certain stratigraphic intervals shows that at some times it actually was emergent in an otherwise flooded continental region. Conversely, the Michigan and Illinois basins are areas that were relatively depressed features through most of the Paleozoic. During times of high sea level, sediments in these basins were of deeper-water origin and were thicker than sed-

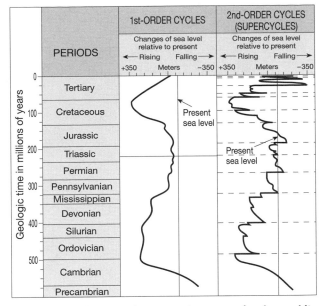

Figure 21.14 Transgression regression curves for the world's continents. These curves show major long-term changes in sea level (first-order cycles) and more detailed changes (second-order cycles).

iments on the surrounding platforms. During times of low sea level, basin sediments record evidence of restricted circulation. During some regressive periods, evaporite deposits developed.

The existence of domes and basins on the continental platform and the reflection of fluctuations of sea level in platform sediments have been known for decades. In light of our present understanding of plate tectonics and its operation during part or all of Phanerozoic time, two questions come to mind: What tectonic processes have caused these domes and basins to form, and in response to what plate tectonic processes?

Orogenic Belts

Orogenic[8] belts are one of the most prominent tectonic features of continents, and they have been the primary focus of work in structural geology for the past century. These belts are characteristically formed of thick sequences of shallow-water sandstones, limestones, and shales deposited on continental crust, and oceanic

---

[7] The supercontinent formed by India, Africa, Australia, and Antarctica.

[8] The term *orogenic* is derived from the Greek words *oros,* which means "mountain," and *genesis,* which means "origin, birth." We use the term to refer to areas that are major belts of pervasive deformation. The term *mobile belt* also means approximately the same thing; it denotes regions that have been tectonically mobile. *Mountain belt* is a geomorphic term that refers to areas of high and rugged topography. Most mountain belts are also orogenic belts, so the two terms are often used interchangeably. Not all orogenic belts, however, are mountainous.

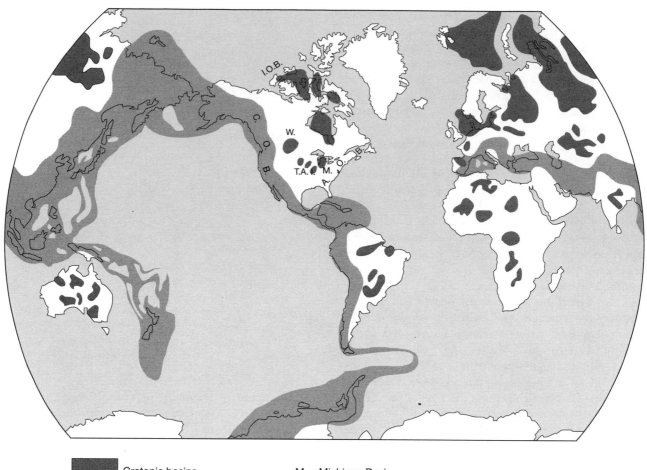

	Cratonic basins
	Mesozoic-Cenozoic orogenic belts

T.A. – Transontinental Arch

I. – Illinois Basin

M. – Michigan Basin

W. – Western Canada / Williston Basin

I.O.B. – Innuitian Orogenic Belt

C.O.B. – Cordilleran Orogenic Belt

A.O.B. – Appalachian Orogenic Belt

Figure 21.15 Map of the world, showing the distribution of cratonic basins. On North America, basins and arches of the interior platform are identified.

deposits characterized by deep-water turbidites and pelagic sediments, commonly with volcaniclastic sediments and volcanic rocks. Orogenic belts have typically been deformed and metamorphosed to varying degrees and intruded by plutonic rocks, chiefly of granitic affinity.

Structurally, most orogenic belts display a crude bilateral symmetry that is manifest in a linear central area of thick deformed and metamorphosed sedimentary and/or volcanic accumulations bordered on either side by undeformed regions, either oceanic or continental. In the past, much significance was attached to the symmetric nature of orogenic belts. Recent work, however, has demonstrated that the symmetry is more apparent than real, because in many cases the structures on the two sides of the center are of different ages.

The application of the plate tectonic model to the study of orogenic belts has revolutionized ideas on how orogenic belts form. We now believe that orogenic belts form at convergent margins as a result of the collision of two continents or that of a continent and an island arc or other thick crust of oceanic origin. Different types of mountain belts form, depending on the nature of the colliding blocks and on which side overrides the other.

Thus the tectonic history of an orogenic belt may record some aspects of the history of plate tectonic activity. By studying the tectonic history of young orogenic belts, we can discover the relationship between orogenic structures and associated plate tectonic activity. Similar structures in inactive or older orogenic belts can then be used to infer the existence of similar plate tectonic activity in the geologic past.

Because of the importance of orogenic belts for understanding large-scale tectonic processes and because of the importance of minor structures in working out the origin of orogenic belts, we devote the following chapter to a detailed description of the structural characteristics of such belts.

Continental Rifts

Continental rifts are areas marked by abundant normal faulting, by shallow earthquake activity, and by mountainous topography. The North American Basin and Range province is an example, as we noted in Chapter 5. This region exhibits north-trending grabens and horsts extending over an area approximately 100 to 600 km from east to west and 2000 km from north to south (Figure 5.9). In such regions, the continental crust is undergoing extension that has often, in the geologic record, preceded the breakup of continents and the formation of new ocean basins.

Modern Continental Margins

The margins of the present continents are apparently marked by a relatively sharp transition from continental crust to oceanic crust that is poorly exposed and difficult to resolve with common exploration geophysical techniques. Seismic refraction, which utilizes layered models, cannot be applied where the layers are discontinuous, as at margins of continents. Only recently has it become possible to penetrate the thick marginal sedimentary sequences with seismic reflection techniques and produce images of the transition from continental to oceanic crust. Consequently, the structure of continental margins is still poorly known.

Four types of continental margins, however, based on their tectonic environment, are recognizable (Figure 21.16): passive, or Atlantic-style, margins; convergent, or Andean-style, margins; transform, or California-style, margins; and back-arc, or Japan Sea-style, margins. The geographic name sometimes used to refer to

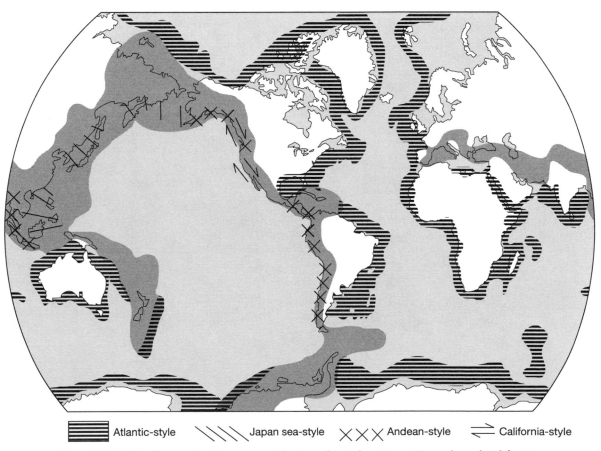

Figure 21.16 World map showing present Atlantic-style, Andean, Japan-Sea-style, and California-style continental margins.

each style is taken from a region where it is characteristically developed.

Passive margins, or **rifted margins,** or **Atlantic-style margins,** are present on both sides of the Atlantic as well as around the Indian and Arctic Oceans and around Antarctica. They are created as continents rift apart to form new ocean basins. They initiate at a divergent plate boundary, but as spreading proceeds and the ocean basin widens, they end up in a midplate position (Figure 21.17).

Passive margins include a coastal plain and a submarine topographic shelf of variable width, generally underlain by a thick (10–15 km) sequence of shallow-water mature clastic or biogenic sediments. Along some margins, an outer ridge is present in the thick sedimentary sequence, generally at the point where the shelf passes into a steeper topographic slope toward the ocean basin. A relatively thick (roughly 10 km) sequence of sediments is often present along the continental rise and slope (Figure 21.17). Normal faults, including growth faults, are the most characteristic structural features found in the sediments along these margins.

Convergent margins, or **Andean-style margins,** are present where consuming plate boundaries are located along a continental margin. They exhibit an abrupt topographic change from a deep sea trench offshore to a high belt of mountains within 100 to 200 km of the coast. Continental shelves tend to be narrow or absent. The mountains along these margins are characterized by a chain of active stratovolcanoes of principally an-

desitic composition (Figure 21.18). Active deformation results in thrust complexes near the trench, high-angle normal faults near the volcanic axis, and either normal or thrust faults between the volcanic axis and the continent.

Transform margins, or **California-style margins,** are also characterized by sharp topographic differences between ocean and continent. They are marked by active strike-slip faulting, sharp local topographic relief, a poorly developed shelf, irregular ridge-and-basin topography, and many deep sedimentary basins. Figure 21.19 shows schematically the development of such topography by strike-slip displacement on two faults along an irregular continental margin. As the faults move, they progressively displace portions of the continent from each other (Figure 21.19*B*), thereby producing an alternation in places of narrow ocean basins and continental fragments. The Pacific margin of the United States is a typical example, and the many faults of the San Andreas fault system have produced a ridge-and-basin topography off southern California (see Figures 7.2 and 7.13).

Back-arc margins, or **Japan Sea-style margins,** are composite margins consisting of a passive Atlantic-style margin separated by a narrow oceanic region from an active island arc. The Japan Sea is a narrow ocean between the passive east coast of Asia and the active volcanic arc of Japan. Both the passive and the active margins of the composite margin exhibit the features of the individual margins that we have noted (Figure 21.20).

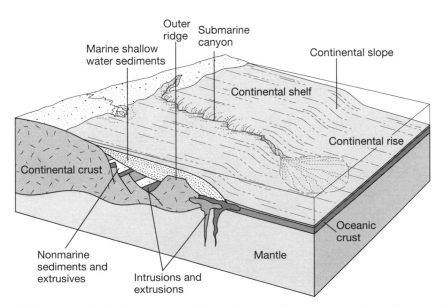

Figure 21.17 Generalized block diagram of a passive, or Atlantic-style, continental margin. Not to scale.

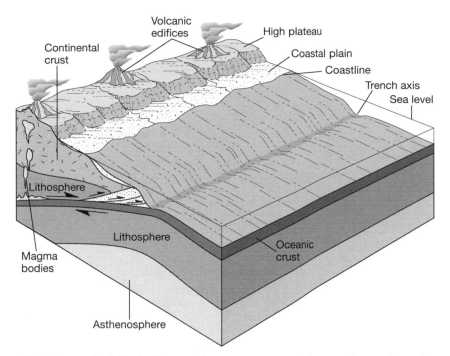

Figure 21.18 Generalized block diagram of a convergent, or Andean–style, continental margin. Not to scale. Note the similarity to Figure 6.19.

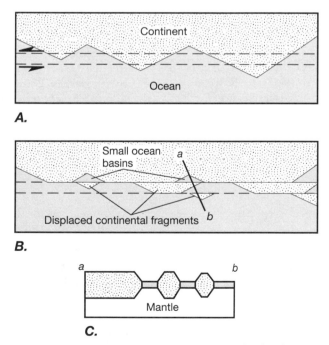

Figure 21.19 Generalized maps and cross section illustrating the development of a California-style, or transform, continental margin. Not to scale. *A.* Irregular continental margin and a two-fault strike-slip system. *B.* After motion on both faults of the system, portions of the continent are displaced to new positions. *C.* Cross section *ab,* showing ridge-and-basin structure.

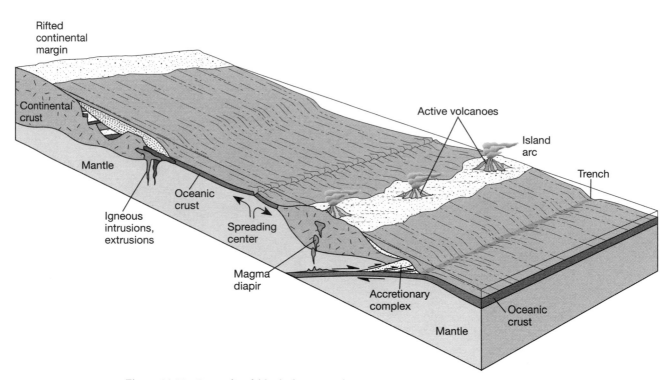

Figure 21.20 Generalized block diagram of a Japan-Sea-style margin. Not to scale.

The labels on the figure are: Rifted continental margin; Continental crust; Mantle; Igneous intrusions, extrusions; Oceanic crust; Spreading center; Magma diapir; Accretionary complex; Active volcanoes; Island arc; Trench; Oceanic crust; Mantle.

Additional Readings

Burchfiel, B. C. 1983. The continental crust. In R. Siever, ed., *The dynamic earth.*® *Scientific American* (September): 114.

Francheteau, J. 1983. The oceanic crust. In R. Siever, ed., *The dynamic earth*,® *Scientific American* (September): 130.

Hoffman, P. 1988. United plates of America. *Ann. Rev. Earth and Planet. Sci.* 16: 543–603.

Kröner, A., and Greiling, eds. *Precambrian tectonics illustrated*. Stuttgart: Schweizerbartsche.

National Academy of Sciences–National Research Council. 1980. *Continental tectonics*. Washington: National Academy of Sciences.

Nisbet, E. G. 1987. *The young earth: An introduction to Archaean geology*. Boston: Allen and Unwin.

Uyeda, S. 1978. *The new view of the earth*. New York: W. F. Freeman.

Windley, B. F. 1984. *The evolving continents*. New York: Wiley.

CHAPTER

22 Anatomy of Orogenic Belts

Orogenic belts coincide, not accidentally, with some of the Earth's great mountain chains. Since time immemorial, these features have inspired poets and philosophers as much as geologists. Only recently, however, have we come to recognize orogenic belts to be the geologic record of plate tectonic activity. This activity includes subduction of one plate beneath another, as well as collisions between crustal masses—such as two continents, a continent and an island arc, or a continent and an oceanic plateau. Because all oceanic crust older than about 200 Ma (early Jurassic) has been subducted, orogenic belts are the prime repository of information about plate tectonic interactions for the first 95 percent of Earth history. Studying these features gives us an opportunity to decipher part of the tectonic history of the Earth that is preserved nowhere else.

In this chapter we discuss the structural characteristics of the major parts of orogenic belts, illustrating them with examples from the major mountain belts of the world, chiefly the North American Cordillera, the Appalachian–Caledonide, and the Alpine–Himalayan systems (see Figure 22.1).

Figure 22.2 shows generalized maps of the North American Cordillera, the Alpine–Iranian portion of the Alpine–Himalayan orogen, and the Appalachian–Caledonide orogen on a predrift reconstruction. Of these three orogens, only the latter is inactive. It formed by collision of Africa and Europe with North America in Paleozoic time and was fragmented into its separate parts by subsequent opening of the Atlantic Ocean in Mesozoic–Cenozoic time. The Cordillera and the Alpine–Himalayan orogens include plate margins that are still active. The major plates involved in the North American orogen are the North American, Pacific, and Juan de Fuca plates; those involved in the Alpine–Himalayan orogens include the Eurasian, African, Arabian, and Indo-Australian plates. Other, smaller plates are involved in each orogen.

From the maps of Figure 22.2 it should be clear that no single map or cross section can provide a universal model of an orogenic belt. Nevertheless, these and other orogens have a number of features in common: a rough bilateral structural symmetry; a foreland or undeformed plate on either side; outer foredeeps; fold and thrust belts; sutures marked by ophiolitic rocks; a slate belt; and an internal crystalline core zone of metamorphosed and deformed sedimentary and volcanic rocks, mafic–ultramafic complexes, and granitic plutons. Thus we can discuss these belts in terms of the features they share.

Similarly, although the cross section of any portion of one of these orogens will differ from that of another

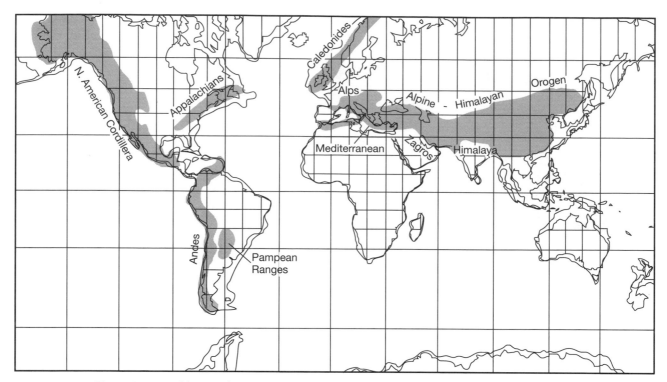

Figure 22.1 World map, showing the location of the orogenic belts referred to in this chapter: North American Cordilleran, Appalachian–Caledonide, Alpine–Himalayan, and Andean.

part or of another orogen, the structures displayed have enough in common for us to discuss them in terms of a composite cross section, as shown in Figure 22.3.

22.1 The Outer Foredeep or Foreland Basin

Between the main orogenic belt and the undeformed continental platform commonly lies a thick series of clastic sediments derived from a rising source area in the adjacent mountains. These sediments were deposited in a **foredeep,** or a **foreland basin** (a **molasse basin** in the terminology of Alpine geology; Figures 22.2 and 22.3), and they reach thicknesses of as much as 8 to 10 km near the mountain front. Generally the coarseness of the basin fill decreases away from the mountain front. Conglomerates pass into sandstones and shales, which in turn may pass into carbonate marine shelf sediments (Figure 22.4).

Environmental indicators from different basins suggest two possible modes of basin development. Along the eastern side of the Cordilleran belt in the western United States, for example, the Cretaceous stratigraphy begins with deep-water sediments that pass upward into shallow-water deposits, suggesting that the basin

formed relatively abruptly and was gradually filled up. This gradual shallowing of the basin is indicated by the expansion of the shallow-water sandstones and continental deposits, through time, from west to east across the basin (Figure 22.4). By contrast, the Devonian foredeep trough of the northern Appalachians, exposed in New York, formed slowly enough for sedimentation to keep pace with subsidence.

In some basins the clasts in the conglomerates and sandstones reveal an **unroofing sequence,** in which stratigraphically younger deposits contain debris from successively deeper levels in the mountains. Such a sequence reflects the progressive uplift and erosion, or unroofing, of the adjacent mountains.

The amount of deformation of the foredeep rocks is generally slight, indicating that either most were deposited after the main phase of deformation in the interior of the orogenic belt or that they were far enough away not to feel its effect. Near the mountain front, however, the foredeep deposits commonly are deformed by folds and thrust faults. The folds are generally open, multilayer, class 1B folds that have angles of 90° or less. The wavelengths, which often exceed 1 km, are controlled by the thick competent layers in the stratigraphy. Folding increases in intensity toward the mountain front, as evidenced by higher fold angles, larger aspect ratios, a change toward multilayer class 1C geometry, and (in some instances) inclined, possibly overturned

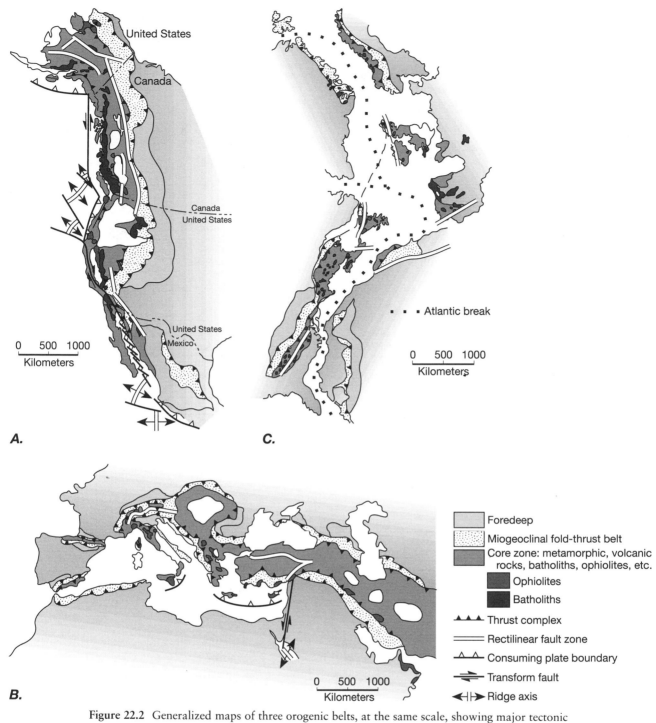

Figure 22.2 Generalized maps of three orogenic belts, at the same scale, showing major tectonic features to be compared with model cross section. Note also the locations of other figures. *A.* Generalized map of North American Cordillera. *B.* Generalized map of the Alpine–Iranian, or western, segment of the Alpine–Himalayan orogen. *C.* Generalized map of the Appalachian–Caledonide orogenic belt (including the West African orogen), on a predrift reconstruction of the continents around the Atlantic Ocean.

folds with a vergence toward the stable platform. Allochthonous masses of the sediments are found in a few places in the foredeep basins, emplaced presumably by gravity sliding of surficial sheets of sediment.

In some regions, such as the central Rocky Mountains of the United States (Figure 22.2*A*) and the Pampean ranges of the southern Andes (Figure 22.1), low to moderately dipping thrust faults bring basement to

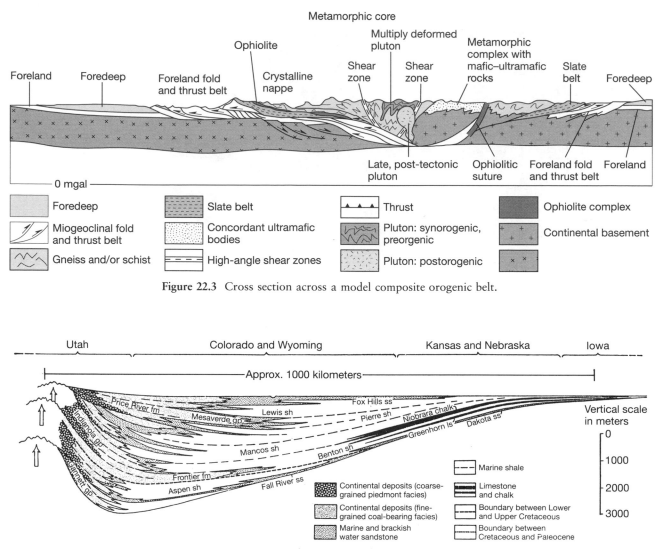

Figure 22.3 Cross section across a model composite orogenic belt.

Figure 22.4 Diagrammatic cross section of a Cretaceous foredeep on the east side of the Cordillera from Utah to Iowa. Such basins are also called molasse basins and foreland basins. Note the thickening and coarsening of sediments toward the western source area.

the surface over shelf or outer foredeep deposits (Figure 22.2*A*, *B*). Seismic reflection over one such uplift, the Wind River Mountains of Wyoming, indicates that the thrust penetrates to the base of the continental crust.

22.2 The Foreland Fold and Thrust Belt

Behind the foredeep basins, toward the center of the orogenic belt, lies a foreland fold and thrust belt. This belt consists predominantly of folded and thrust-faulted miogeoclinal sedimentary rocks that have been pushed away from the orogenic core and out over the stable foreland (Figures 22.2 and 22.3). The foredeep deposits are overthrust by the miogeoclinal rocks at the front of

the fold and thrust belt, and some foredeep deposits may be incorporated into the hanging wall blocks of some of the thrusts. The miogeoclinal sediments generally thicken toward the core of the orogenic belt.

Typically examples of these structural regions include the Appalachian Valley and Ridge province (see Figures 6.11*A* and 6.12*A*), the Cordilleran overthrust belt, especially north of the Basin and Range province (see Figures 6.11*B* and 6.12*B*), the Jura mountains north of the Alps (see Figures 11.2*B* and 20.15), the southern Himalayas, and the Zagros mountains of Iran.

We have already described the general structural features of foreland fold and thrust belts (Section 6.3). The fundamental characteristic is the presence of a sole fault, which separates the deformed rocks of the thrust sheet from the underlying undeformed basement and rises through the stratigraphic section toward the foreland to give the thrust sheet a wedge-shaped geometry. Above the sole fault, there may be several décollements, all of which ultimately are branches off the main sole fault and, like the sole fault, tend to rise through the stratigraphic section toward the foreland. Each fault characteristically adopts a ramp-flat geometry in which it cuts steeply up through competent layers such as sandstone or limestone and forms bedding-paralle faults in incompetent layers such as shale, gypsum, or salt. Duplex structures are common (see Figures 6.15 and 6.16).

Movement on such faults accommodates shortening and thickening of the thrust wedge and creates fault-ramp folds, which may tighten to accommodate further deformation. Other folds may form above a flat décollement to accommodate shortening and thickening of the thrust wedge. Some thrust faults may develop when folding becomes too tight to accommodate more shortening, and faults cut up from the décollement through the steep or overturned limb of a fold (see Figure 6.10A). The dominant fold style is class 1B to 1C (see Figure 11.19), which of geometric necessity must be associated with a décollement (Section 11.3 and Figure 11.23). Many of these folds are asymmetric, and in most cases they have a vergence away from the orogenic core (Section 11.3 and Figure 11.13D). The fold wavelength is related to the thickness of the folded competent layers (see Figures 20.2 and 20.3).

In many fold and thrust belts, age relationships consistently show that thrusts and folds near the orogenic core and those shallower in the thrust stack are older than those near the foreland and those deeper in the thrust stack. This decrease in age of deformation toward the foreland is sometimes called a **prograding deformation** (see Section 6.4). However, **out-of-sequence thrusts**—new thrust faults that form behind the frontal thrust—are not uncommon. The simple model of thrust wedges (Section 10.11 and Box 10.2) requires that a thrust wedge maintain a critical taper, which implies a prograding deformation as the thrust wedge grows. The thrust wedge must also thicken in order to prograde, however; the thickening can take place by folding or by out-of-sequence thrusting.

Faults that form early in the deformation are commonly folded when a new fault propagates out under the older fault, and folds develop above the décollement formed by the younger fault. This folding of the older faults makes continued slip on them increasingly difficult, and they eventually become inactive. Such deformed faults may subsequently be cut by later out-of-sequence faults.

In some cases, imbricate thrust faults dominate the deformation of the thrust wedge, as in the southern Appalachian Valley and Ridge province (Figures 6.11A and 6.12A). In other cases, the formation of folds characterizes the deformation, as in the northern Valley and Ridge.

Most major sole faults remain above the strong crystalline basement and within the miogeoclinal sedimentary sequence, which characteristically contains abundant layers of weak rocks such as shales, gypsiferous layers, or salt. This style of deformation, where the basement remains undeformed by the thrusting, is known as **thin-skinned tectonics.**

In the inner or rearward parts of foreland fold and thrust belts, however, fault slices of crystalline rocks become incorporated into the thrust sheets (Figure 22.3). In some cases—such as the western Alps, for example (Figure 22.5)—these rocks were the basement on which the miogeoclinal rocks were deposited; they are called **external massifs** (Figure 22.5A). In other cases the crystalline rocks bear no obvious relationship to the sediments of the thin-skinned belt. Regardless of their origin, these basement crystalline rocks are variably deformed, and in some cases they contain mylonite zones and folds that reflect faults and folds in the associated cover. Such blocks could be easily incorporated into the thrust sheets if earlier normal faults associated with rifting were reactivated as thrust faults.

Observations of major thrust sheets indicate that they have a wedge shape (Section 10.11). If major thrust systems are effectively branches off a subduction zone that cut through the sediments of a down-going passive continental margin, the polarity of the system is determined by the subduction zone, and the thrust fault must rise through the stratigraphy in the direction of the foreland (as shown in Figure 6.12 for example). Moreover, the sedimentary sections involved in foreland fold and thrust belts are miogeoclinal sections that become thinner with increasing distance from the orogenic core. Thus there is a gentle but significant upward slope to the basement of the sedimentary section from the hinterland toward the foreland. The slope is accentuated by, and in part attributable to, the rifted and thinned character of the continental basement at a passive continental margin (Figures 5.9 and 21.17). Where these sections are deformed, they ubiquitously exhibit transport from the margins toward the foreland, and the basement slope helps direct the thrust faults upward in this direction.

Most rocks in foreland fold and thrust belts are unmetamorphosed. In some regions, however, low-grade metamorphism is present. Clay minerals, for example, are recrystallized to chlorites and micas, coal is anthracite grade, or the magnetic vectors of the rocks are reset. Slaty cleavage is characteristic of argillaceous sediments and may be cut by a second generation of spaced foliation formed by solution during deformation. In spite of these metamorphic effects, fossils are fairly

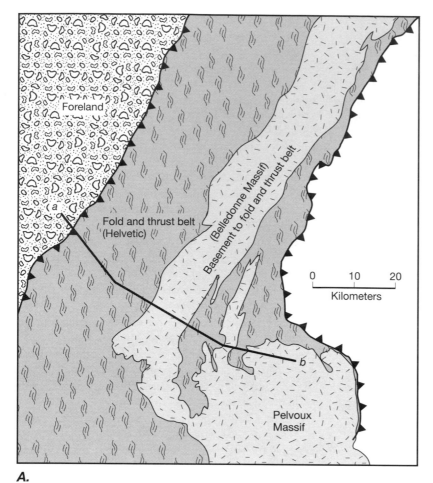

A.

Figure 22.5 External massifs of the western Alps, resulting from involvement of basement rock in fold and thrust belt. *A.* Generalized map of the Belledonne and Pelvoux massifs, France. *B.* Cross section along line *ab*. *C.* Index map, showing location of *A*.

Foreland

Fold and thrust belt (Helvetic)

(Belledonne Massif)

Basement to fold and thrust belt

0   10   20

Kilometers

Pelvoux Massif

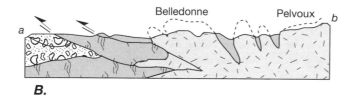

Belledonne   Pelvoux

a   *b*

B.

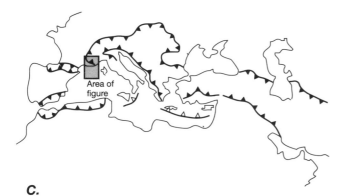

Area of figure

C.

common, the stratigraphy is relatively easy to work out, and the rocks are amenable to correlation from one thrust block to another.

Moving from the front of the fold and thrust belt further toward the interior of the orogenic belt, several changes take place. The character of the sediments involved in the deformation changes from shallow-water miogeoclinal sediments to deeper-water eugeoclinal sediments. In some areas, called **slate belts,** the rocks are characterized by a monotonous dominance of relatively unfossiliferous shales and slates. The monotony of the stratigraphy, the lack of fossils, and the poor exposure of the easily eroded shales and slates make stratigraphic analysis and correlations difficult and imprecise. The rocks in these regions apparently were laid down off the edges of continental margins as continental rise or abyssal deposits, or in an off-shore volcanic environment. It is possible to distinguish between these different provenances by stratigraphic and petrologic analysis of the sediments.

Because slate belts are closer to the orogenic core, the grade of metamorphism increases, reaching high-zeolite or low-greenschist facies. Ductile deformation becomes increasingly prominent, and the style of folding becomes dominated by class 1C to class 2 folds, reflecting a more ductile sedimentary pile and a decrease in the dominance of competent layers on the folding of the sedimentary section (Section 12.5). In places, multiple generations of folding are present. Folds become more inclined, even recumbent, and they can form huge fold nappes. This change in fold style is attributable in part to the higher temperature indicated by the increase in metamorphic grade, and in part to a change in lithology from predominantly sandstone and limestone to the more ductile shales and slates.

As the rocks are more highly deformed and more recrystallized, they tend to exhibit pervasive continuous foliation such as slaty cleavage and phyllitic foliation, which in places may be overprinted by later spaced foliations. Faults are present, but in many cases the lack of distinctive markers or piercing points makes them difficult to interpret.

The slate belt of Wales, Britain, is one of the best-studied examples (Figure 22.6). There a thick sequence of late Precambrian and lower Paleozoic deep-sea sediments and associated volcanic rocks and mélange displays a pervasive sequence of upright faults and penetrative cleavage with down-dip extension. Thrust faults are recognized in a few areas, but rarely can they be traced for long distances. As shown in the cross

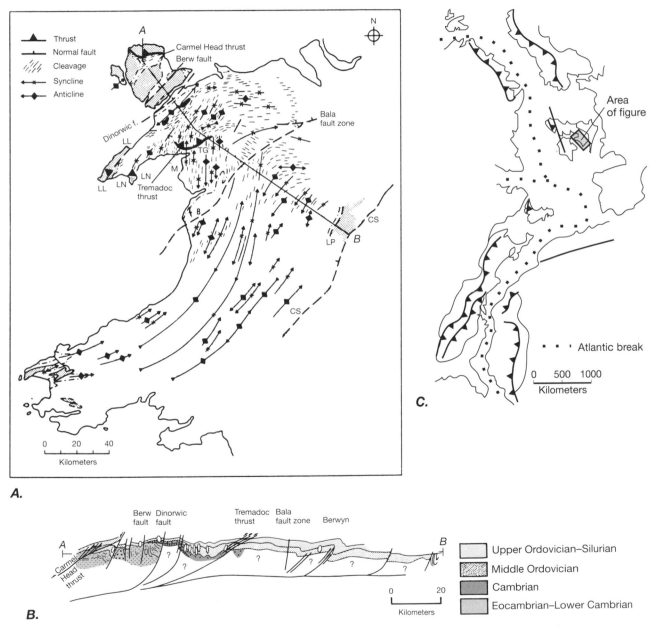

Figure 22.6 The slate belt of Wales, an example of an orogenic slate belt. See Figure 22.2C for location. *A.* Map. *B.* Cross section. *C.* Index map, showing location of *A.*

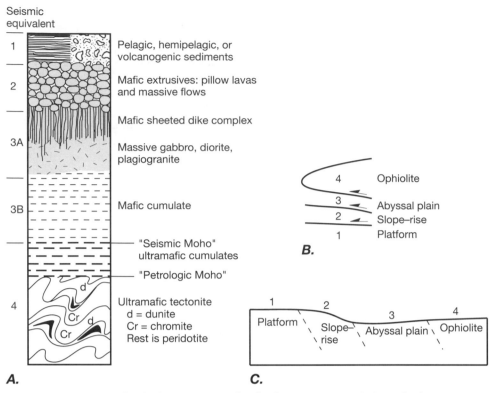

Figure 22.7 Features of ophiolites. *A.* Generalized columnar section of idea ophiolite sequence, showing principal lithologic units, and inferred correlation with oceanic seismic model. "Seismic Moho" is the boundary between seismic crust and mantle. "Petrologic Moho" is the boundary between mantle tectonite below and igneous rocks above. *B.* Diagram illustrating typical tectonic stacking of thrust sheets of different provenance beneath ophiolites. *C.* Diagram illustrating palinspastic restoration of the thrust sheets of *B* to their original relative position.

section, Figure 22.6*B*, the structures are thought to be part of a series of imbricate thrusts above a basal décollement, but the question marks on the section emphasize the uncertainty in this interpretation.

Most orogenic belts contain ophiolites. Where best preserved, these complexes are pseudostratiform sequences of ultramafic and mafic rocks, from bottom to top, of peridotite tectonite, overlain in turn by a plutonic complex of layered ultramafic–mafic cumulate igneous rocks and massive gabbro and related rocks, a dike complex, and extrusive rocks, overlain by pelagic, hemipelagic, or volcanigenic sediments (see Figure 22.7*A*). Thicknesses of well-preserved complexes resemble those of the oceanic crust (see Figure 21.3), and it is possible to make a tentative correlation between units of an ophiolitic sequence and those of the oceanic crust, as indicated in Figure 22.7*A*. Such ophiolites are thought to represent fragments of the oceanic crust and mantle, formed in an oceanic environment (midoceanic or back-arc spreading center) and preserved on continents.

Ophiolites occur predominantly in two ways. Small tectonic slices or blocks of incomplete sequences are present in many accretionary prisms formed by sub-duction of an oceanic plate. Complete ophiolitic sequences commonly are present in large subhorizontal thrust sheets which are hundreds of kilometers in dimension. Belts of many ophiolites of similar ages, such as in the Alpine–Iranian orogenic belt (Figure 22.2), may extend for thousands of kilometers. Individual ophiolite thrust sheets commonly overlie a tectonic complex of thrust slices of platform, slope–rise, and abyssal sediments, as illustrated in Figure 22.7*B*. Palinspastic restoration of these tectonic slices (Figure 22.7*C*) suggests that successively higher thrust sheets originated from positions progressively further from the foreland, with the ophiolite complex representing the highest and most oceanic thrust sheet.

The structural relationships of these ophiolite complexes indicate that they represent thrust sheets of oceanic crust and mantle, original rooted in the oceanic mantle. Many geologists believe that these well-preserved ophiolitic thrust complexes are the result of a continental margin on a down-going plate colliding with a subduction zone. If this idea is correct, then the presence of such ophiolites indicates not only continental margin thrusting but also an unknown amount of intra-oceanic thrusting. Thus they mark a zone of possibly vast but uncertain displacement. Because unknown but presumably substantial amounts of oceanic lithosphere and overlying sediment have disappeared downward beneath these thrusts, it is intrinsically impossible to balance cross sections across such zones, unlike structures in foreland fold and thrust belts (Section 4.5). Although ophiolitic thrust sheets have received relatively less attention than the more familiar and easily analyzed fold and thrust complexes, they may be more important in the interpretation of the history of an orogenic belt.

22.3 The Crystalline Core Zone

The crystalline center or "core" of an orogenic belt contains metamorphic and plutonic rocks that have deformed extensively by ductile flow. The resulting structures include large thrust or fold nappes and complex multiple-deformation features. The core is invariably thrust out over the rocks of the foreland fold and thrust belt (Figure 22.3). Some of the nappes are very large (100,000 to 250,000 km^2).

Multiple generations of folding in core zones produce a variety of fold inteference structures. A structural sequence that frequently emerges from the geometric analysis of such areas includes one or more generations of recumbent isoclinal folds, refolded by a generation of upright, more open folds, and finally deformed by a generation either of smaller-scale kink and chevron folds or of ductile shear zones (Figure 22.8).

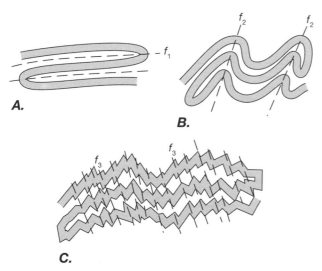

A.

B.

C.

Figure 22.8 Diagrammatic cross sections illustrating progressive sequence of folding proceeding from (A) first-generation isoclinal folds through (B) more upright second-generation folds superposed on the earlier deformation to (C) a third-generation kinking superposed on all earlier foldings.

In rocks that preserve both fine-scale layering and large-scale stratigraphy, small-scale (high-order) folds mimic the orientations and styles of the larger (lower-order) folds. Detailed studies of high-order folds in critical outcrops, therefore, can be used to infer the basic geometry of regional deformation. In plutonic igneous rocks where no original layering is present, the deformed rocks become foliated, and mylonite zones of high-ductile-shear strain define boundaries of areas where deformation has been less intense.

The rocks in core zones of mountain belts are of diverse origins. In the rest of this section, we shall consider the principal components of most orogenic core zones.

Sedimentary Rocks and Their Basement

In some cases, such as in some Alpine Penninic nappes (Figures 22.5A and 22.9), the crystalline rocks represent metamorphosed deep-water sedimentary rocks and their thinned continental crystalline basement. In such regions, former basement rocks commonly form the cores of nappe structures and are surrounded by an envelope or sheath of metasedimentary rocks. Figure 22.9 shows an example of such structures: the Adularia, Tambo, and Suretta nappes of the Penninic zone and their metasedimentary cover.

In the deeper structural levels of an orogenic belt, metamorphic temperatures can approach or even exceed the granite solidus. The resulting highly ductile rocks (Figure 22.10A) are gravitationally unstable, and they rise diapirically (Figure 22.10B), forming huge **mantled gneiss domes** that may contain a core of intrusive granite, as well as gneiss mantled by a metasedimentary envelope (Figure 22.10C).[1]

Volcanic and Igneous Rocks and Associated Sediments

Orogenic core zones generally contain large areas of rocks characterized by a lack of pronounced or continuous layered stratigraphy and by an abundance of intrusive rocks. Metamorphism is commonly intense in these rocks, reaching amphibolite or even granulite conditions. They thus tend to form vast areas of massive

---

[1] Gneiss domes appear to be gradational into domes formed by multiple folding. Indeed, it is possible in some cases that little or no piercement—that is, diapiric activity—of the gneiss through the overlying rock has taken place. The contact between the gneiss and metasedimentary rocks generally is a ductile shear zone, however.

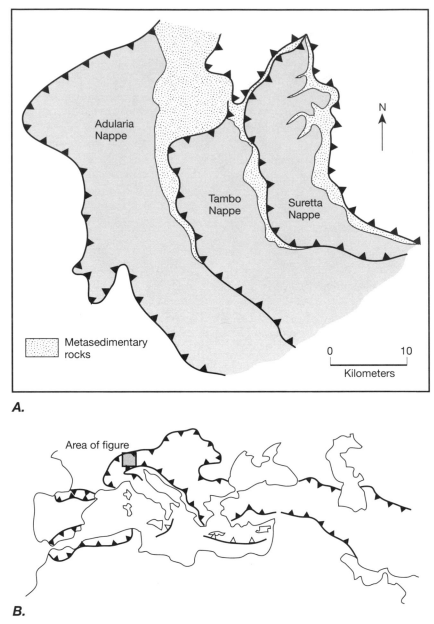

A.

B.

Figure 22.9 *A.* Generalized map of part of the Penninic (core) zone of the Alps, showing map view of three major subhorizontal crystalline nappes (the Adularia, Tambo, and Suretta nappes) separated by thin septa of metasedimentary rocks. Compare with figure 22.16. *B.* Index map showing location of *A.*

or banded amphibolite or granulite in which the original stratigraphic and intrusive relationships become next to impossible to determine.

The Appalachians in New England and southern Canada provide some especially well-documented examples of these rocks. As shown in Figure 22.11, the rocks there include metasedimentary rocks, metavolcanic rocks, and a number of gneiss domes. The rocks display complex multiple folded structures that generally exhibit patterns and numbers of deformational phases similar to those outlined for the miogeoclinal or shelf sequences discussed above.

Such metavolcanic rocks were originally interpreted as "eugeosynclinal" sequences, but they are now understood to be a juxtaposition of continental rise–slope deposits, with volcanic rocks and volcanogenic sediments that formed in continental volcanic arcs, oceanic arcs, or midoceanic volcanic complexes. The juxtaposition is a consequence of collisions at one or more subduction zones.

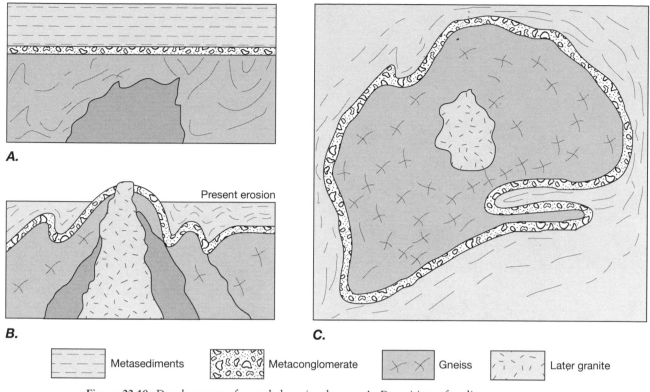

A.

Present erosion

B.

C.

Metasediments   Metaconglomerate   Gneiss   Later granite

Figure 22.10 Development of mantled gneiss domes. *A.* Deposition of sedimentary sequence unconformably on metamorphosed sediments and intrusive rocks. *B.* Deformation of entire sequence in part *A* during a new deformation, followed by intrusion of a new granitic body. *C.* Generalized map of the resulting gneiss dome after exposure by erosion.

Metamorphosed Ophiolitic Sequences

Ophiolite belts are a feature of many orogenic core zones, and in many cases they have participated in the regional deformation and metamorphism. When this has happened, the pseudostratigraphy of the ophiolite complexes develops complex structures including large-scale recumbent or multiply refolded folds. At high grades of metamorphism and large amounts of deformation, pillow lavas become massive greenschists or amphibolites; mafic dikes and plutonic complexes become massive or banded amphibolites; peridotites become serpentinized at lower grades of metamorphism and at higher grades are dehydrated back into peridotite, with the original mantle fabric overprinted or even obliterated by the later deformation.

Lower Continental Crust and Mantle

In some regions of collisional orogenic belts where overthrusting and subsequent uplift have been extreme, highly metamorphosed quartz-feldspathic gneisses are found overlying peridotite. The Ivrea zone of the Alps is a good example of this situation (Figure 22.12). These regions may represent the contact between the lower continental crust and the underlying mantle—the original Moho. Like ophiolitic peridotites, both the peridotites and the gneisses of continental assemblages typically display an old metamorphic assemblage and deformation fabric overprinted by a younger one. Some subcontinental peridotites contain irregular layers and veinlets of gabbroic rocks that pass into bodies or dikes, suggesting that partial melting has occurred. The pressure–temperature relationships inferred from the gabbro and peridotite mineralogy suggest that this melting took place during emplacement of the complex.

*Gneissic Terranes with Abundant
Ultramafic Bodies*

Some orogenic core zones include terranes characterized by amphibolite or granulite-facies gneisses and schists containing numerous small, discontinuous ultramafic bodies, most less than 1 km long, consisting of fresh peridotite, pyroxenite, or dunite or their serpentinized

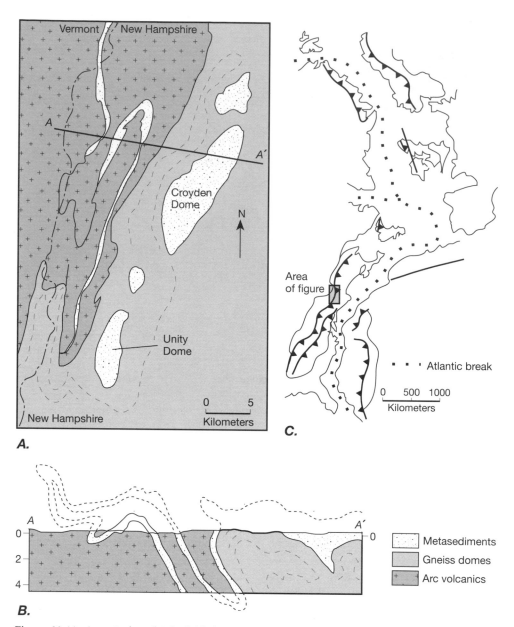

Figure 22.11 A typical multiply folded eugeoclinal core zone, the northern Appalachians in Vermont. Of particular interest is the metavolcanic sequence, possibly an early Paleozoic island arc complex. A. Map. B. Cross section. C. Index map showing location of A.

equivalents. The southern Appalachians provide an especially good example of such bodies (Figure 22.13), but similar terranes are present in the Alps and the Caledonides. Characteristically, the bodies are elongate parallel to the regional structural grain and display a regionally concordant internal fabric.

The interpretation of these terranes is one of the great unsolved tectonic problems of mountain belts. We understand neither the mechanism by which the ultramafic rocks are incorporated into the metamorphic terranes nor the protolith of the metamorphic rocks. (The term protolith is derived from the Greek words *protos,*

which means "first," and *lithos,* which means "rock.") We can suggest four possible origins for these enigmatic regions. They may be exposures of deep crustal levels where fragments of the subcontinental or subisland arc mantle was somehow incorporated into the crust during deformation; they may be remnants of ophiolitic or mantle slabs that have been completely disrupted after emplacement by extreme deformation and metamorphism that produced the gneissic terranes; they may be metamorphosed melange terranes; or they may be mafic and ultramafic igneous rocks intruded into continental margin sediments during rifting.

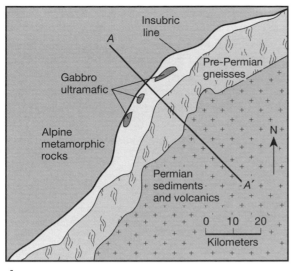

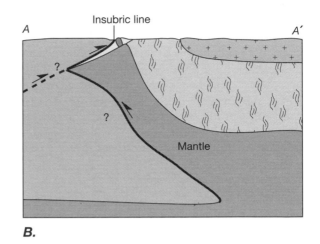

A.

B.

N

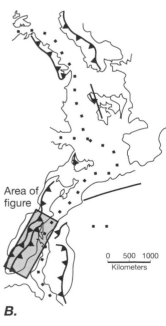

C.

Figure 22.12 Exposure of lower continental crust and mantle, lvrea zone, southern Alps. *A.* Generalized map. *B.* Cross section based on geology and geophysics, showing mantle thrust over northern continental edge, as well as back thrust in opposite direction. *C.* Index map showing location of *A.*

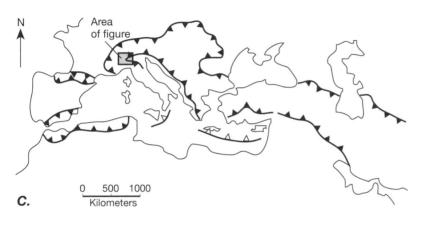

A.

▦ Metavolcanic rocks   • Concordant ultramafic body

Figure 22.13 Concordant ultramafic bodies in the crystalline core zone of the southern Appalachians. *B.* Index map showing location of *A.*

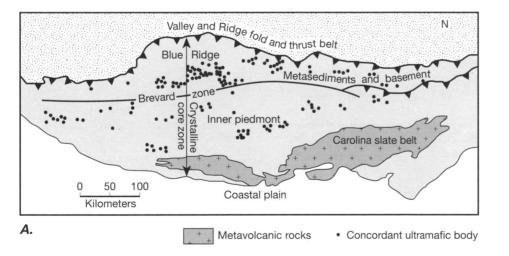

B.

Granitic Batholiths

Granitic batholiths are large areas of plutonic igneous rocks, generally of dioritic to granitic composition, that occupy vast areas in many mountain belts, in some cases dominating the area of the orogenic core. The North American Cordillera provides a good example, where vast areas of batholithic rocks extend discontinuously from northern Alaska to northern Mexico (Figure 22.2A). Most batholiths are not a single body but rather comprise tens to hundreds of individual plutons, each a few tens to hundreds of square kilometers in area. They exhibit wide differences in rock type, in degree of deformation, and in apparent depth of emplacement.

Granitic rocks are either I-type and S-type granites, where the I and S denote derivation by partial melting of an originally igneous or sedimentary source, respectively. I-type granites are characteristically hornblende–

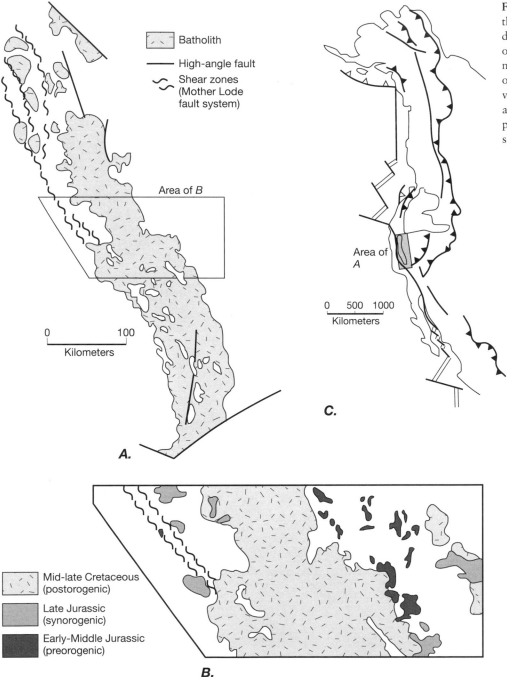

Figure 22.14 Batholiths in the North American Cordillera. *A.* Generalized map of Sierra Nevada, California. *B.* More detailed map of part of central Sierra Nevada, showing radiometric age relationships of granitic plutonic rock. *C.* Index map showing location of *A.*

Batholith

High-angle fault

Shear zones (Mother Lode fault system)

Area of *B*

0   100
Kilometers

A.

Area of *A*

0   500   1000
Kilometers

C.

Mid-late Cretaceous (postorogenic)

Late Jurassic (synorogenic)

Early-Middle Jurassic (preorogenic)

B.

biotite quartz diorites in composition. They are thought to have formed by partial melting of a hydrous mantle or of a previously crystallized igneous rock. S-type granites are richer in potassium, and they characteristically contain both biotite and muscovite and, more rarely, garnet. They may be derived from the partial melting of sedimentary rocks.

The timing of batholithic activity relative to deformation in the orogen also is variable. In a number of mountain belts, the age of granitic rocks overlaps with periods of deformation. Older granitic plutons commonly exhibit evidence of this deformation (such as the development of foliation and/or folds parallel to regional trends), whereas younger intrusive rocks do not. Thus they can be considered preorogenic if they intruded prior to the main deformation, synorogenic if they intruded during deformation, and postorogenic if they intruded after deformation.

The dominant rock type in batholiths is different in different mountain ranges. Quartz diorite or granodiorite is dominant in the western U.S. Cordillera and the Andes, whereas granite is the most common in the Appalachians and Caledonides. The Alpine belt displays very few granitic rocks. The composition of the granitic rock displays a crude correlation with the type of country rock and the timing of intrusion. Both preorogenic batholiths and those intruding oceanic sedimentary or volcanic rocks tend to be poorer in potassium than postorogenic batholiths and those invading continental rocks.

The Sierra Nevada of California and Nevada provide a good example of the variability of batholithic history. They have an intrusive history ranging in age from approximately 70 Ma to 270 Ma (Figures 22.14*A*, *B*). Some plutons in the Sierra Nevada are deformed; others are not. Granitic rocks in the Sierra Nevada cluster into three age groups: early-mid Jurassic, late Jurassic, and mid-late Cretaceous (Figure 27.14*B*). Because the principal deformation in this region was mid-late Jurassic (175 to 140 Ma), these groups generally correspond to preorogenic, synorogenic, and postorogenic intrusives.

We can gain some insight into the variations observed in granitic batholithic terranes from plate tectonic considerations. Preorogenic I-type granitic bodies may be the product of normal consuming margin igneous activity that occurs before the collision in which the granitic bodies are deformed. The granites are sodium-rich and are associated with volcanic rocks of similar composition that are also deformed in the later collision. The synorogenic and postorogenic S-type granites may result from partial melting of the lower part of the continental crust thickened by formation of a mountain root during collision. They are potassium-rich—and possibly even aluminum-rich—granitic rocks. Postoro-

genic alkalic granites form after all orogenic phases have ceased and may reflect the early stages of an episode of subsequent continental rifting.

22.4 The Deep Structure of Core Zones

The core zones of orogenic belts generally overlie the roots of mountain belts where the crust is thickest. Thus their deep structure is related to the formation of mountain roots. What is the structure of these core zones at depth? How far down do the structures that we observe at the surface extend?

In many orogenic belts viewed in cross section, the major folds are recumbent, and faults dip at a low angle. In places, however, these features are steeper in dip. Folds are upright or vertical to steeply reclined, thrust faults are nearly vertical, pronounced down-dip lineations occur, and in places, ductile strains become very large. Some of these zones are also the sites of major shear zones, as we noted in the previous section. Generally, structures and tectonic units are not continuous across the region.

Recent work in the Alps has shown that the steep-dip region is in fact a limb of a huge second-generation fold that deforms originally subhorizontal nappes and thrust sheets (Figure 22.15; see Figure 11.1*A* for a comparable older interpretation of the Alps). Because these folds have a vergence that is opposite to the vergence of the nappe structures, they are referred to as **back-folds**. Such regions formerly were viewed as the source area, or root zone, of large-scale nappes and folds.

Back-fold structures may develop as a consequence of collision and subsequent change in direction of dip of a subduction zone, as illustrated schematically in Figure 22.16. Alternatively, they may form as a result of continued shortening during isostatic uplift.

The depth to which surface structures descend seems to vary from one mountain belt to another. In the Alps, seismic reflection and refraction evidence suggests that the structures involved in the back fold extend to deep levels. In contrast, seismic, reflection work in the Southern Appalachians by the Consortium for Continental Reflection Profiling (COCORP) suggests that the crystalline core zone is allochthonous and tectonically overlies a series of flat-lying reflectors that may be the little-deformed equivalents to the sediments of the Valley and Ridge province (Figure 22.17; see also Figure 22.2*C*). This result implies that the entire deformed belt may be allochthonous and may be displaced hundreds of kilometers over an autochthonous continental basement. Similar seismic results from elsewhere along this orogenic belt suggest that the entire core zone is allochthonous along much of its length. Such a thrust feature

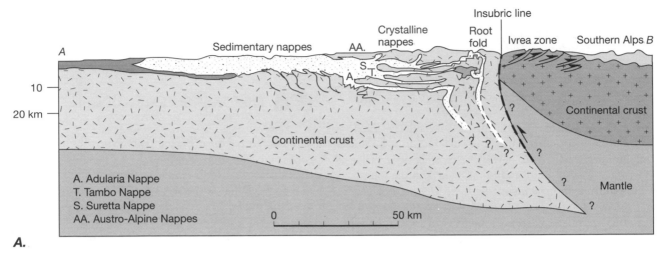

A.

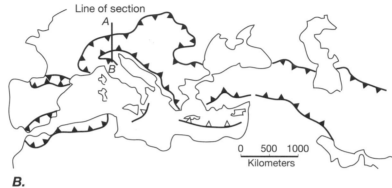

B.

Figure 22.15 Cross section of the Swiss Alps, showing recumbent nappes and root fold in the crystalline core zone of the Alps. Note off-set in Moho, comparable with but not identical to that of Figure 22.13. Compare also with Figure 22.10, which is located just west of this cross section. *B.* Index map showing location of *A.*

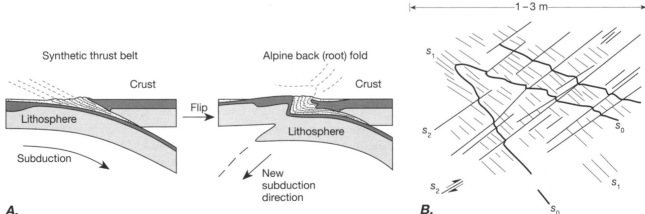

A.   **B.**

Figure 22.16 Relationship between thrust belts, back folds, and subduction. *A.* Model for formation of multiply deformed structures such as the Alpine back fold by formation of synthetic thrust faults during collision of continents and subsequent back folding after a "flip" of subduction polarity. *B.* Outcrop scale structure from within the root zone, showing two episodes of deformation, marked by cleavage and kinkbands S_2, throught to represent deformation during synthetic thrusting and back folding, respectively.

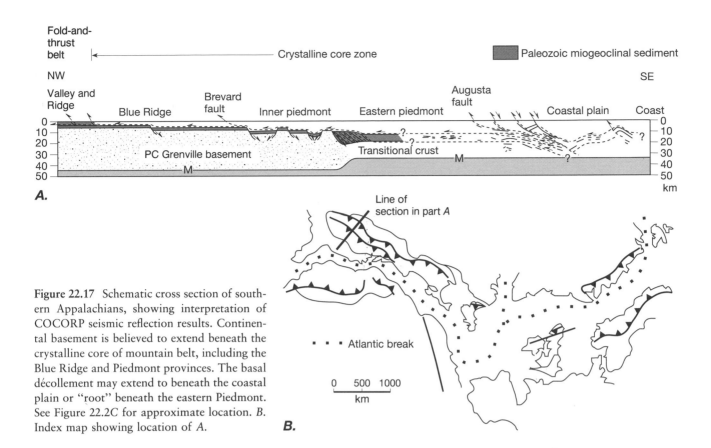

Fold-and-thrust belt |← —————————— Crystalline core zone ——————————— | ▨ Paleozoic miogeoclinal sediment

A.

Figure 22.17 Schematic cross section of southern Appalachians, showing interpretation of COCORP seismic reflection results. Continental basement is believed to extend beneath the crystalline core of mountain belt, including the Blue Ridge and Piedmont provinces. The basal décollement may extend to beneath the coastal plain or "root" beneath the eastern Piedmont. See Figure 22.2C for approximate location. *B.* Index map showing location of *A.*

B.

involving an entire orogenic belt has profound implications for the nature of the movements and forces that caused it.

22.5 High-Angle Fault Zones

High-angle fault zones, which in general are moderately to steeply dipping and transect all the other features, are present in nearly every mountain belt (see rectilinear faults in Figure 22.2, for example). They vary from a few hundred meters to 10 km in width, although most are relatively narrow and are marked by a band of well-developed mylonite. They typically extend for tens or hundreds of kilometers roughly parallel to the axis of the deformed belt, forming a major structural boundary along the orogenic belt.

Some shear zones offset geologic features identifiable on both sides of the fault, so estimates of the displacement are possible. In such regions, both dip-slip and strike-slip faults may be present. The lineations observed in such zones tend to be approximately parallel to the direction of demonstrable displacement.

Other fault zones separate regions of distinctly different geologic history. For example, the dominant metamorphic ages on the two sides of the fault may be

radically different, or the geologic or paleogeographic histories may differ. In such cases, no correlation can be made across the fault, and such zones probably represent **sutures** (remnants of formerly existing oceanic regions that have disappeared).

Rocks within such shear zones are highly recrystallized and even mylonitic, and they often possess a different metamorphic grade, generally lower, but in places higher, than the surrounding region. In most places a variety of diverse lithologies are present in discontinuous lenses. Mineral lineations are common, as are minor folds, with their axes parallel to the mineral lineations. These lineations can be horizontal or steep, depending on the geometry of deformation within the fault zone in question.

The timing of the deformation in these faults is variable. Some zones are late faults that sharply transect the preexisting structures. Others, however, are gradational into the structure of the surrounding rocks and are correspondingly difficult to interpret. Near the fault zones, fold axial surfaces tend to be deflected from the regional attitude into attitudes parallel to the zones.

Most well-documented zones display a complex history of repeated movement with several different senses. Three examples serve to illustrate this characteristic. The Brevard zone of the southern Appalachians (Figures 22.13 and 22.17) extends for more than 500 km along strike. It appears to have an earlier dip-slip history, then a subsequent dextral strike-slip motion. The Insubric line of the Alps (Figures 22.5*A*, 22.12 and 22.15) also offers a good example of one of these zones. It is a steep fault zone that separates the Ivrea zone from the Penninic zone of the Alps. It also appears to have an early dip-slip history with a subsequent dextral strike slip, so the displacement history is complex and the cross section varies along its length. The Mother Lode fault system of the Sierra Nevada, California (Figure 22.14), is another example of such faults. These steeply east-dipping faults in part represent the traces of highly deformed thrust faults that display predominantly west-over-east dip-slip motion and that have been folded and reactivated during an episode of east-over-west back-folding, followed by recent dextral strike-slip and normal faulting. Both the Mother Lode fault system and the Insubric line represent the remnant of a suture in their respective mountain system.

22.6 Metamorphism and Tectonics

The rocks in the central portions of all orogenic belts are metamorphic rocks. The distribution of metamorphic zones in most deformed belts is roughly symmetrical, so the highest-grade rocks underlie the central portions of the belt, and unmetamorphosed rocks are found on the flanks. Although a comprehensive discussion of the subject of metamorphism is beyond the scope of this book, a few points are of structural and tectonic interest.

The series of metamorphic assemblages developed in metamorphic zones are a reflection of the temperature–depth profile at the time of metamorphism (Figure 22.18) On the basis of the different mineral assemblages, we can distinguish metamorphism that has occurred under conditions of low pressure and high temperature (Buchan type), normal pressure and temperature (Barrovian type), and high pressure and low temperature (blueschist type). The temperature–depth profiles implied by these types of metamorphism result from different tectonic conditions.

Buchan metamorphism implies that temperatures are elevated above a normal geothermal gradient. This condition develops in contact aureoles around shallow, level igneous intrusions in a volcanic arc environment. The metamorphism may be preorogenic, in which case it may be overprinted by subsequent synorogenic Barrovian metamorphism, or it may be postorogenic, in which case it overprints all earlier metamorphic phases.

Blueschist metamorphism implies that temperatures in the Earth are depressed significantly below their normal values. This situation occurs in subduction

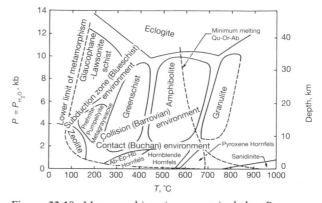

Figure 22.18 Metamorphism in mountain belts. Pressure–temperature diagram, showing major metamorphic facies, and the conditions of high pressure–low-temperature (blueschist) metamorphism characteristic of a subduction zone environment, intermediate-pressure-and-temperature (Barrovian) metamorphism characteristic of normal temperature gradients in a collisional environment, and low-pressure–high-temperature (Buchan) metamorphism in a shallow contact aureole environment.

zones where cold, shallow rocks are carried to great depths faster than normal temperatures can be reestablished. It tends to occur relatively early in the orogeny and is rare in most mountain belts. Where present, it commonly is overprinted by younger Barrovian metamorphism. Buchan (low-pressure–high-temperature) and blueschist (high-pressure–low-temperature) metamorphism where present together may constitute what is called a paired-metamorphic belt.

Barrovian metamorphism reflects a normal geothermal gradient and is widespread in all mountain belts. For this reason it is also known as classic regional metamorphism.

The Alps provide a good example of the relationships between the various metamorphic types (Figure 22.19). The blueschist metamorphism of 70–85 Ma age is overprinted in places by the 15–25 Ma Barrovian regional metamorphism. Elsewhere, only Barrovian metamorphism is reflected in the rocks. Evidence of any previous metamorphic event has been completely obliterated.

In areas of good exposure, one can get an idea of the shape of isograd surfaces in three dimensions. Where they are undeformed or only mildly deformed, they appear to be gently curved surfaces that intersect the Earth's surface at small angles. For example, Figure 22.20 shows a map of the structure of part of the northern Appalachians (Figure 22.20*A*) and of the Barrovian metamorphic zones in the same region (Figure 22.20*B*). In Figure 22.20*A* the upper, central, and lower nappes form a subhorizontal stack along which there is a north-trending antiform–synform pair and a concentration of

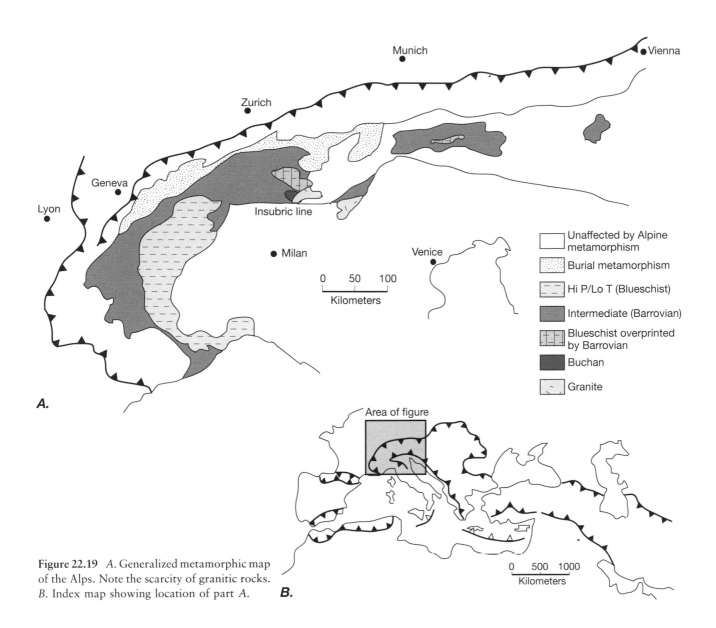

Munich

Zurich

Vienna

Geneva

Lyon

Insubric line

Milan

Venice

| 0 | 50 | 100 |

Kilometers

Unaffected by Alpine metamorphism

Burial metamorphism

Hi P/Lo T (Blueschist)

Intermediate (Barrovian)

Blueschist overprinted by Barrovian

Buchan

Granite

A.

Area of figure

| 0 | 500 | 1000 |

Kilometers

B.

Figure 22.19 *A. Generalized metamorphic map of the Alps. Note the scarcity of granitic rocks. B. Index map showing location of part A.*

gneiss domes. These folds in turn are warped into a series of culminations and depressions about west-northwest–trending axes, giving rise to type 1 and type 2 interference patterns that are particularly evident in the boundary between the upper and central nappes. The grade of metamorphism generally decreases from the lower to the upper nappes (Figure 22.20B), and the metamorphic isograds are gently folded about the same two directions as the nappes, giving the interference patterns on the isograd map (Figure 22.20B). Thus, although the isograds do cut across the nappe boundaries, they share in the two gentle foldings of the tectonic units, suggesting that deformation of the nappes ceased before or during peak metamorphism and that subsequently the nappes and isograds together were gently folded.

In other regions, metamorphic zones clearly have been displaced or even inverted. In the central Himalaya,

for example (Figure 22.21), the metamorphic grade increases from the chlorite zone continuously up through to the sillimanite zone with progressively higher positions in the structure. As shown in Figure 22.22, such relations could indicate a primary inversion of isotherms such as would develop at a subduction zone above a down-going cold slab, (Figure 21.22A) or it could indicate a tectonic inversion of the isotherms following metamorphism, as would occur on the inverted limb of a recumbent fold (Figure 22.22B).

The interpretation of metamorphic zones in terms of plate tectonics is complex because the rates of down-warping and uplift of the rocks are comparable to the rate at which they can heat up or cool off. The rate of erosion must be considerably less than the rate of tectonic thickening in order to account for the thickened crust in orogenic belts such as the Zagros and Himalaya–Tibet areas. Thus a given volume of rock might

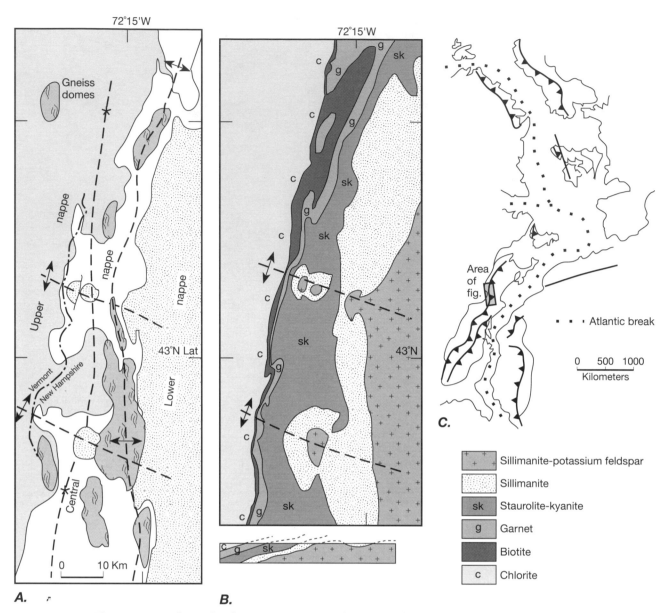

Figure 22.20 Relationship between structure and metamorphism, an example from the New England Appalachians. See Figure 22.2C for location. *A.* Tectonic map showing the distribution of major nappes and gneiss domes. *B.* Map and schematic cross section of mineral isograds of Barrovian metamorphism. Note the general increase of grade with tectonic level. *C.* Index map showing area of *A.*

Legend:

Sillimanite-potassium feldspar

Sillimanite

sk Staurolite-kyanite

g Garnet

Biotite

c Chlorite

be expected to be buried quickly and then exhumed more slowly. The temperatures achieved during the process depend on the heat flux into the rock and the rate of uplift and denudation.

Figure 22.23 schematically illustrates different burial and uplift histories for a volume of rock as a set of pressure–temperature paths along which time increases nonlinearly in the direction of the arrows—so-called **PTt paths.** A rock starting out at the surface (upper left-hand corner) is rapidly buried along path A to its maximum depth at B. If it is uplifted immediately, it might follow a path such as C. If the rock is uplifted

more slowly, it might follow path D or E. In case E, considerable time elapses before uplift, so the temperature in the rock approaches the steady-state geotherm.

Note that in all cases, the maximum temperature is not reached at the maximum pressure experienced by the rock. Thus the peak temperatures of metamorphism must occur after the end of the collisional process. If the mineral assemblages most likely to be preserved are those formed at peak temperatures of metamorphism, the metamorphic events recorded in the rocks must be postcollisional events. Numerical calculations for models such as this applied to the Alps suggest that the

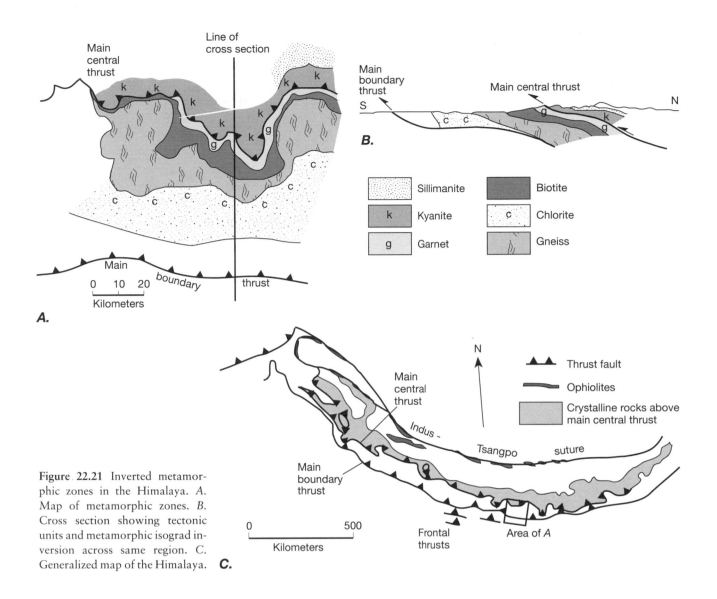

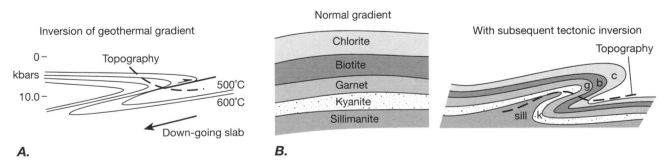

Figure 22.21 Inverted metamorphic zones in the Himalaya. *A.* Map of metamorphic zones. *B.* Cross section showing tectonic units and metamorphic isograd inversion across same region. *C.* Generalized map of the Himalaya.

time lag between the end of the collision and the peak of metamorphism may be 10 Ma or more.

Metamorphic mineral assemblages exhibited by a rock vary depending on the PTt path and the relative rate of chemical reaction. Thus path A-B-C might give rise to surface exposure of blueschist metamorphism,

path A-B-D to blueschist overprinted by Barrovian metamorphism, and A-B-E to Barrovian metamorphism.

The radiometric age that a rock exhibits is a measure of the time since the radiometric clock within the rock was isolated. Thus it is the time at which the rock

Figure 22.22 Possible origin of inverted metamorphic zones. *A.* Inversion of isotherms during thrusting, showing possible temperature distribution during metamorphism. *B.* Development of a recumbent nappe of metamorphic rocks after metamorphism.

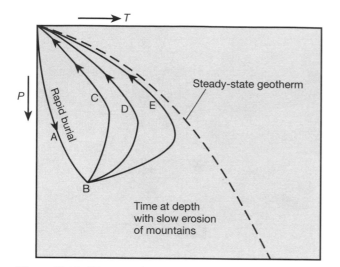

Figure 22.23 Diagram showing various pressure–temperature–time (PTt) trajectories of rocks undergoing metamorphism. See the text for discussion.

cooled through the "closure temperature" for a particular radioactive element and its decay products. That temperature is different for different decay schemes. For example, the "closure temperature" for the K–Ar decay scheme is lower than that for the U-Pb or the Rb-Sr scheme, so K–Ar dates are invariably slightly to greatly younger than the others. Thus discordant ages from different decay schemes in the same rock reflect the rock's thermal history.

22.7 Minor Structures and Strain in the Interpretation of Orogenic Zones

Our efforts to understand the origin of orogenic cores and their relationship to plate tectonic events prompt several questions: In what directions were the rocks transported during the deformation? What is the dis-

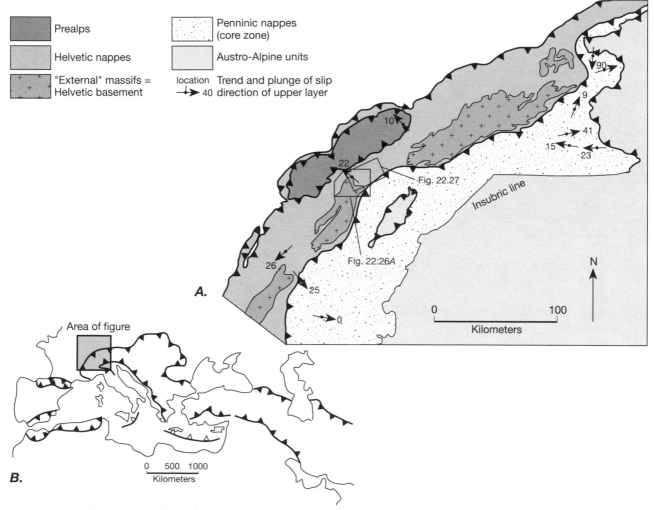

Figure 22.24 Generalized tectonic map of the Alps, showing the orientation of the slip lines deduced by Hansen's method from various parts of the orogenic belt. Arrows point in the direction of movement of the upper layers. The reason for some arrows pointing in toward the core is not understood. Note the locations of Figures 22.26 and 22.27. *B*. Index map showing location of *A*.

tribution of strain through the rocks? What large-scale pattern of flow gave rise to the observed structures and strain distributions, and how can these features be explained in terms of tectonic processes? These questions bear on the way that orogenic cores evolve, and answering them may ultimately help us understand how the evolution of the core zone and plate kinematics are correlated.

In this section we present examples of the application of the analysis of minor structures, outlined in previous chapters, to orogenic core zones. Our focus here is on kinematic analysis of folds (discussed in Sections 12.2, 12.5, 12.7, and 12.8); of foliations (Sections 16.3 and 16.4); of mineral fibers (Sections 14.6 and 16.5); of strain (Section 17.1), and of crystallographic preferred orientations (Section 19.7). We cannot aim here to give a comprehensive account of the worldwide significance of such analyses—a formidable if not impossible task. Rather it is to give some idea of how kinematic analyses may help us understand the formation of a given individual orogenic belt.

Kinematic Analysis of Folds

Application of the Hansen method of determining the slip direction (Section 12.8) to deformed rocks in the Alps shows that in the outer areas of the thrust nappes, the shear direction is transverse to the orogenic belt, as is generally expected. In the central region of the orogenic core, however, the shear direction tends to be parallel to the axis of the orogenic belt (Figure 22.24). Similar results are found in the Norwegian Caledonides and the northern Appalachians. These longitudinal shear directions could be accounted for by flow models in which localized collision results in lateral flow away from the collision zone. Such models have been proposed for the Himalaya–Tibet region.

Kinematic Interpretation of Foliations

Because foliations are generally parallel to the plane of flattening of the finite strain ellipsoid (Section 16.6), they can be used to infer a shear sense in a fault zone if the orientation of the shear zone is also known. The intersection of the foliation plane and the shear plane is a line approximately perpendicular to the direction of shear, and the acute angle between the foliation and the shear plane points in the direction of relative motion of the material on the opposite side of the shear plane (Figure 22.25A).

This relationship is confirmed by studies of the foliation developed in sediments in southern Alaska that have been deformed in the accretionary prism above the subduction zone (Figure 22.25B). Earthquake focal mechanisms and Pacific plate reconstructions determine the relative plate motion independently, and it indeed lies perpendicular to the intersection of the foliation with the thrust plane.

Strain Analysis

The regional distribution of strain in the Morcles nappe of the Swiss Alps southeast of the Lake of Geneva is shown in Figure 22.26. The map (Figure 22.26A) shows the distribution of horizontal planes of the strain ellipsoids and in a few places of discrepancy $\hat{s}_1$–$\hat{s}_2$ axes with $\hat{s}_1$ correctly oriented. Note that $\hat{s}_1$ is commonly perpendicular to the fold axes and parallel to the direction of displacement of the nappe, although in places it is parallel to the fold axes and perpendicular to the direction of displacement. Examination of the cross section (Figure 22.26B) shows that the strains are highest near the base of the nappe where the $\hat{s}_1$ axis is oriented at small angles to the subhorizontal thrust. Higher in the nappe, the strain magnitude decreases progressively, and the orientation of the $\hat{s}_1$ axes becomes steeper.

The deformation history, all of which is recorded in the final finite strain ellipsoid, includes a component of initial flattening associated with sedimentary compaction, followed by multilayer buckle folding during which the limestones behaved as the competent members, itself followed by inhomogeneous simple shearing associated with the emplacement of the nappe. Thus the finite strain ellipsoids do not record just the process of nappe emplacement.

Kinematic Interpretation of Mineral Fibers

Mineral fibers in oriented overgrowths on pyrite concretions have been used to infer the extension history over a region of the Swiss Alps that overlaps the previous figure (Figure 22.27). The lines are everywhere parallel to axes of maximum incremental extension $\hat{\zeta}_1$, and the length of any segment of the line is proportional to the magnitude of the extension in that direction. Note that the lines do *not* indicate the direction and amount of *displacement* of material points in the rocks. The triangular dots indicate the location of the measurements and of the youngest end of the extension history.

The Morcles nappe is shown in both Figures 22.26 and 22.27. In the west, lower in the nappe, the extension history recorded by the fibers is simple, and the maximum extension direction $\hat{s}_1$ of the finite strain ellipsoids is approximately parallel to the fiber extension directions. In the east, however, which is structurally high in the nappe, the fiber extension histories are curved,

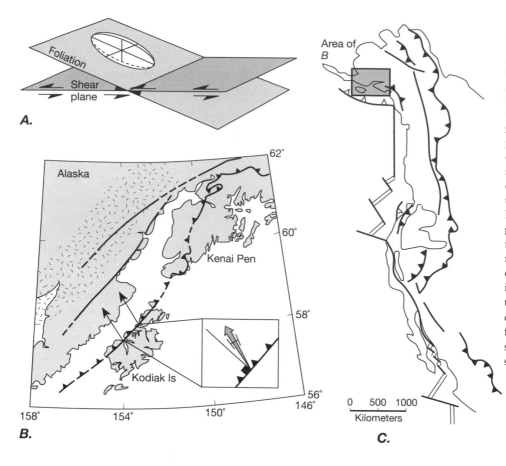

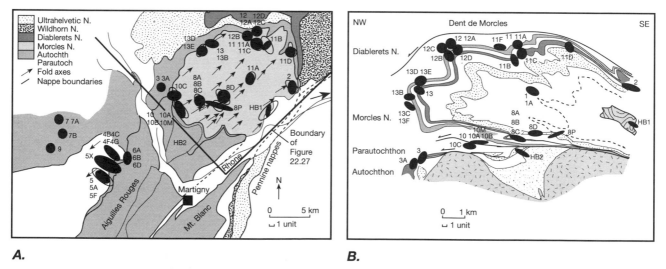

Figure 22.25 Relationship between foliation and shear plane. *A.* The foliation and the shear plane intersect in a line perpendicular to the diagram and perpendicular to the direction of shearing. The acute angles of the intersection (shaded) point in the direction of motion of the material on the opposite side of the shear plane. *B.* An example of the application of this relationship in the field, Kodiak Island, Alaska. The generalized map shows arrows indicating the relative plate motion deduced from analysis of Pacific plate motions. The inset shows that relative motion (thick arrow) and the deduced shear direction from the foliation-shear plane relationship (thin line). *C.* Index map showing location of *B.*

Figure 22.26 Distribution of strain in the Morcles Nappe in the Alps of western Switzerland. See Figure 22.24 for location. *A.* Geologic map of the Morcles Nappe and its surroundings, showing horizontal sections through the strain ellipsoid (solid black ellipses) and the $\hat{s}_1 - \hat{s}_2$ section of the finite strain ellipsoids with s_1 oriented parallel to its correct bearing (open ellipses). *B.* Composite down-plunge projection through the Morcles Nappe, showing the distribution of the $\hat{s}_1 - \hat{s}_3$ section through the finite strain ellipsoids.

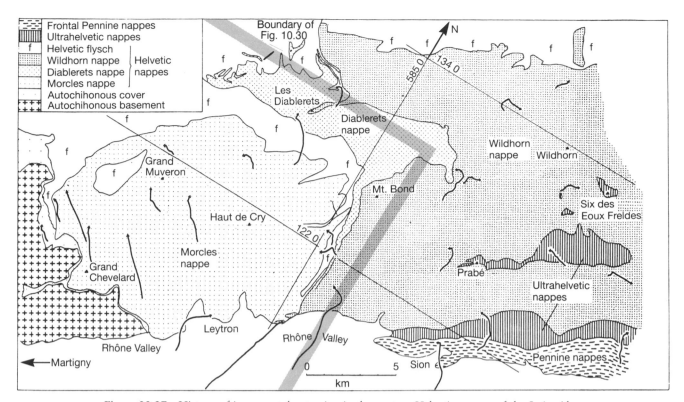

Figure 22.27 History of incremental extension in the western Helvetic nappes of the Swiss Alps, as deduced from fibrous overgrowths on pyrite crystals. Lines are parallel to the directions of maximum incremental extension $\hat{z}_1$, and line lengths in any given direction indicate magnitudes of the extension in that direction. The lines do not indicate the *displacement* of material points in the rock. Triangular dots are at the youngest end of the line and are plotted at the location where the data were measured. See Figure 22.24 for location.

and the orientation of $\hat{s}_1$ is less regular. The strain ellipsoids, of course, probably record more of the deformation than the mineral fibers. Nevertheless, the information deduced from the fibers about the history of the strain accumulation provides more stringent constraints on any model that attempts to account for the emplacement of these masses of rock.

In the lower and central parts of the Morcles nappe, the extension directions were relatively constant to the northwest throughout the period recorded by the fiber growth. A notable feature of much of the Wildhorn nappe and of the upper parts of the Morcles nappe is the change from roughly north–south extension approximately normal to the orogenic core zone early in the deformation to significant components of east–west extension roughly parallel to the core zone later in the deformation. This consistent pattern over a large area suggests a fundamental change in the geometry of the deformation. Such details of deformation history have not yet been incorporated, however, into a unified model of the emplacement of these nappes during orogeny.

Kinematic Analysis of Crystallographic Preferred Orientations

Along the basal thrusts of ophiolites, the ultramafic rocks or underlying metamorphic rocks exhibit a foliation oriented approximately parallel to the thrust surface. Fabrics of olivine and orthopyroxene in the ultramafic rocks and of quartz in the underlying metamorphic rocks in some cases develop an orientation that can be related to sense of shear (see Section 19.7). If such basal thrust contacts represent fossil plate boundaries, as suggested above, these fabrics may indicate the relative plate motions along the boundaries.

22.8 Models of Orogenic Deformation

One of our objectives in conducting a structural study of an area is to understand the relationship between local small-scale structures and large-scale tectonic processes, up to and including plate tectonics. The significance of the local structures can be deduced only from the integration of detailed studies over large areas, which requires much time-consuming field work and analysis. Despite decades of study, however, a general model for orogenic deformation has not emerged, and we can only describe pieces of a puzzle that have not yet been assembled into a complete picture. This difficulty in relating structural geology to tectonics remains a major fascinating problem of orogenic belts.

It is easiest to interpret results from areas of good exposure. The models of regional tectonics derived from such areas can then provide the framework for interpreting the structure of regions that are less well exposed. Because the Alps and the Caledonide mountains have recently been scraped clean by continental glaciation, exposure is exceptionally good, and much of the work on these problems has come from studies in these areas.

Simple Patterns of Ductile Flow

The pattern of flow of rock during ductile deformation determines the distribution of strain in the rocks and the types and orientations of the structures that form. It is best described by the **streamlines** of the flow, which are everywhere tangent to the velocity vectors of material points. We consider three idealizations of the patterns of flow that could occur: convergent flow, divergent flow, and shear flow. The formation of structures in the rocks is the result of inhomogeneities in the flow, which are not considered explicitly in these models.

In convergent flow (Figure 22.28A) all streamlines converge in the downstream direction, and velocity must increase in that direction if the material maintains a constant volume. Any volume of rock is subjected to a coaxial deformation, and the strain path plots in the constrictional strain field of the Flinn diagram (Figure 15.19). The $\hat{s}_1$ direction of the finite strain ellipsoid is oriented essentially parallel to the streamlines (Figure 22.28A). We therefore expect fold hinges and stretching lineations formed during the deformation to be parallel to the streamlines. Because in general, material lines rotate toward the $\hat{s}_1$ direction during a coaxial deformation (see Figure 15.15), we also expect older lineations defined by material lines to rotate toward parallelism with the streamlines.

In divergent flow, all streamlines diverge from one another downstream, and the material velocity decreases in the downstream direction (Figure 22.28B). The deformation is again coaxial, but the strain path in this case lies in the flattening strain field of the Flinn diagram (Figure 15.19), and the $\hat{s}_3$ direction of the finite strain ellipsoid is essentially parallel to the streamlines. Stretching lineations and fold axes formed during the deformation are perpendicular to the streamlines, and lineations defined by material lines are rotated toward

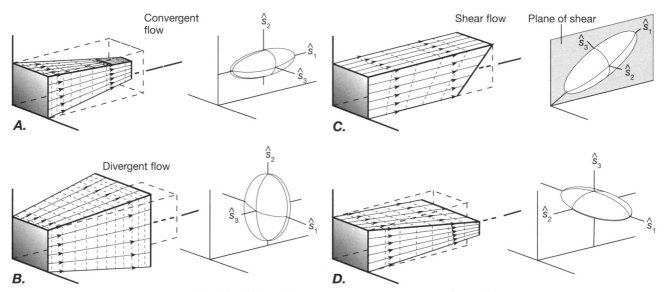

Figure 22.28 Streamlines for different flow geometries. A. Convergent flow. The streamlines all converge in the downstream direction, and velocity increases downstream. B. Divergent flow. The streamlines all diverge in the downstream direction, and velocity decreases downstream. C. Shear flow. The streamlines are all parallel, and velocity does not change in the downstream direction. The velocity of the upper surface, however, is the highest, and of the lower surface the lowest. D. Combined convergent and divergent flow.

the $\hat{s}_1$ direction and therefore toward being perpendicular to the streamlines.

In shear flow, the streamlines are parallel, and the velocity of the material does not change in the downstream direction. The velocity does change, however, in a direction perpendicular to the streamlines (Figure 22.28C). The deformation is noncoaxial. Progressive simple shear is one example of such a shear flow (Figure 15.13B). It illustrates the fact that the $\hat{s}_1$ direction and material lines rotate progressively toward the stream lines. In principle, however, neither $\hat{s}_1$ nor rotated material lines ever become exactly parallel with the streamlines.

We may imagine these simple types of flow to be combined to give more complex types of flows. For example, the streamlines may converge in one plane but diverge in a plane normal to the first (Figure 22.28D). Moreover, a shear flow may be combined with any of the flows that involve convergent or divergent flow.

The orientation of the streamlines is equivalent to what is commonly called the **direction of tectonic transport**. The foregoing discussion shows that there is no simple relationship between streamlines and the principal axes of finite strain, the fold axes, or other material lineations. If the regional distribution of principal strain axes can be determined, however (Section 22.7), they constrain the possible flow patterns (Figure 22.28). Alternatively, local inhomogeneities in the flow may produce particular structures from which it is possible to deduce the orientation of the streamlines (Section 22.7), although any component of rigid translation cannot be recorded by structures in the rock.

Models of Nappe Emplacement

In Chapter 10 we discussed the problem of the emplacement of thrust sheets. We paid particular attention to foreland fold and thrust belts in which high pore-fluid pressure plays an important role in reducing frictional resistance at the base of the thrust wedge. We also mentioned the possibility that friction may be an irrelevant mechanism to invoke if the thrust sheet deforms in a ductile manner. It is to this possibility that we now turn our attention.

We consider three simple models that have been proposed to explain the emplacement of ductile fold or thrust nappes: gravity glide, a horizontal compression, and gravitational collapse. Each of these models predicts a different distribution of the principal finite strain axes, so in principle we could distinguish the different mechanisms of emplacement.

The critical differences show up on a cross section through the thrust sheet parallel to the streamlines. In Figure 22.29 we represent a portion of such a cross

section in two dimensions by a rectangular block resting on a base. The gravity glide model of nappe motion assumes that a nappe may be emplaced by gravitational forces that cause the nappe to glide down a gently inclined base by ductile shearing within the nappe (Figure 22.29B). In the simplest case, the nappe neither shortens nor extends parallel to the movement direction. The resulting flow is an example of inhomogeneous progressive simple shear flow (compare Figure 22.28C). The base of the nappe is a zone of intense shear strain. With progressively shallower depths in the nappe, the shear strain decreases, reaching zero near the surface (Figure 22.29B). Thus at the top of the nappe, the $\hat{s}_1$ axis of the finite strain ellipse is oriented at 45° to the shear plane (Figure 22.29B; see also Figure 15.15A). With increasing depth in the nappe, the orientation of the $\hat{s}_1$ axis rotates increasingly toward lower angles with the shear plane, reaching the minimum angle at the base of the nappe (see Figure 15.15D).

The second model assumes the nappe is emplaced by a horizontal compression applied to the rear of the nappe, the so-called push from behind (Figure 22.29C). The result is a ductile shearing over the base, similar

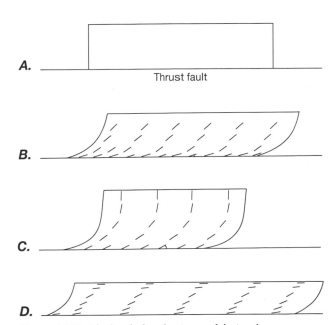

Figure 22.29 Idealized distributions of $\hat{s}_1$ in thrust nappes deformed according to different possible mechanisms of emplacement. *A.* A portion of a nappe before deformation. *B.* Gravity glide. The nappe deforms by simple shear parallel to the basal fault. *C.* Horizontal compression, or a "push from behind." The nappe deforms by a combination of simple shear and shortening parallel to the basal fault. *D.* Gravitational collapse. The nappe deforms by a combination of simple shear and extension parallel to the basal fault.

to that described above for the first model, on which is superimposed a shortening of the nappe parallel to the streamlines. For simplicity we assume the shortening is homogeneous through the nappe. The flow is then a combination of shear flow (Figure 22.28C) and divergent flow (Figure 22.28B). Thus at the top of the nappe where the shear strain is zero, the shortening causes the $\hat{s}_1$ axes to be vertical. With increasing depth approaching the basal shear zone, the shortening component of the strain added to the simple shear causes the $\hat{s}_1$ axes to be oriented at a higher angle to the shear plane than for the case of simple shear alone.

The third model assumes the nappe is emplaced by a process of gravitational collapse (Figure 22.29D), which involves the ductile spreading and thinning of the nappe in a manner analogous to the flow of a continental ice sheet. In this model, for both the nappe and the ice sheet, the flow is driven by the tendency to flatten the slope of the top surface of the body. We assume for simplicity that the flattening of the nappe is homogeneous within the illustrated portion of the cross section. The resulting deformation is a combination of shear flow (Figure 22.28C) and convergent flow (Figure 22.28A). At the top of the thrust sheet where the shear strain is zero, the flattening of the nappe causes the $\hat{s}_1$ axes to be horizontal and parallel to the streamlines (Figure 22.29D; see also Figure 15.13A). Where the shear strain is non-zero, the extensional strain tends to rotate the $\hat{s}_1$ axes toward lower angles with the shear plane than is the case for simple shear alone. The net result is a sigmoidal pattern of $\hat{s}_1$ orientations with the smallest angles between $\hat{s}_1$ and the shear plane near the top and bottom of the nappe, and with the highest angles near the center (Figure 22.29D).

Figure 22.30 shows a plasticine model of gravitational collapse that includes vertical thinning, horizontal extension, and shear along the base, with a complex rolling under of the top of the nappe at the front. Such a process could explain the major fold noses and inverted limbs of fold nappes, and it could account for the superposed crenulation cleavage found in the inverted limb of the Morcles nappe.

These models, of course, are highly simplified, particularly in the ignoring of end effects, in the assumed geometry of the nappes, and in the assumed deformation that the nappes undergo. Nevertheless, they provide a basis on which to begin to build an interpretation of field data. The strain in the Morcles nappe, for example (Figures 22.26 and 22.27), is reasonably consistent with the models for gravity glide or possibly gravitational collapse, but not with the push-from-behind model (Figure 22.29). Nevertheless, the inhomogeneity of the strain and the complicated history of the deformation (Section 22.7) make comparison with such simple models very imprecise.

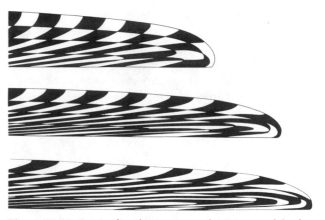

Figure 22.30 Strain distribution in a plasticine model of a nappe undergoing gravitational collapse with shear along the base. Note that the top of the nappe at the front gets rolled under the advancing front of the nappe, a process that would explain the development of recumbent folds at the fronts of nappes.

More complex models of nappe formation have been investigated via scale-model experiments. One hypothesis proposes that the emplacement of some nappes is driven by convective overturn of crustal rocks in an orogen. Models of the process produced by centrifuging layered blocks of different puttylike materials to induce density-driven flow (Section 20.7) show striking similarities to cross sections of some deformed orogens (such as Figures 22.12 and 22.16). Although the similarity does not prove the hypothesis, it is sufficient to indicate that this model must also be considered as a possible way of interpreting the field data.

Plate Tectonic Models of Foreland Fold and Thrust Belts

Structural evidence from active décollement-style fold and thrust belts in the southwest Pacific and from the Zagros mountains in Iran indicate that they develop during the subduction of a passive continental marginal and are synthetic to the direction of subduction (Figure 6.18C). This model implies that orogeny and fold and thrust belt development should be an episodic process that occurs only when continental crust arrives at a subduction zone on a down-going plate. Secondary back-folding and/or back-thrusting may result from isostatic rise of the thickened crust during continued convergence, or from a "flip" in the direction of subduction, following collision (Figure 22.17A).

This model is by no means universally accepted as a general explanation of foreland fold and thrust belts. A major difficulty is the presence in the Andes of a fold and thrust belt antithetic to the present subduction of the Nasca plate. An orogenic belt such as the Andes could record an ancient collision of a continent with oceanic island arcs lying above a subduction zone dipping away from the continental margin. Indeed such a model has been invoked to explain some features of the western North American Cordillera. Such a model has not been invoked for the Andes, however. It is also possible that such fold and thrust belts are the result of gravitational collapse of crust thickened above a subduction zone (see Figure 6.19C).

Plate Tectonic Models of Orogenic Core Zones

The core zones of many collisional orogens show evidence of several generations of deformation, which commonly includes an early isoclinal folding and subsequent upright and/or kink folding (Section 22.3 and Figures 22.8, 22.15 and 22.16). Attempts to relate such structural features to plate tectonics must assume that the deformation is a consequence of the relative plate motions, which may or may not be true.

Numerous proposals have been made to account for the different generations of deformation, although the data are not adequate to support any of them clearly. The multiple generations might correspond to shearing during subduction, followed by shortening and thickening during collision and then by isostatic collapse of the orogenic welt. Alternatively; they could be related simply to changes in plate motion during subduction and removal of the rocks in question from the active part of the accretionary zone at the plate margin.

Because generally there is no evidence that minerals defining foliations have recrystallized after deformation, we infer that the deformation was more or less synchronous with the peak of metamorphism, which probably postdated any collision by a few tens of millions of years (Section 22.6). Thus it remains unclear whether any of the structures observed in the core zones reflect original subduction directions; rather, they may result from internal deformation within the collision zone during isostatic adjustment and gravity collapse of the orogenically thickened crust.

Despite such ideas and decades of assiduous study of orogenic core zones by hundreds of geologists, it has been extraordinarily difficult to establish any precise associations between particular structures and tectonic events or to make any general models that account for the observed structural characteristics in terms of plate tectonics. Perhaps as absolute dating techniques improve, it will become possible to relate precisely the age of formation of a single structure or fabric-forming event to the inferred relative plate motion for the same time, but at present the problem remains unresolved.

22.9 The "Wilson Cycle" and Plate Tectonics

Observations concerning the development of orogenic belts that have accumulated over the past century or so suggested a pattern that was termed the orogenic cycle.[2] Although the advent of the plate tectonic theory has swept away many of the old concepts, any new model must account for the same observations. The characteristics of an "orogenic cycle" include

1. Accumulation, in separate areas, of thick deposits of both shallow-water (miogeoclinal) and deep-water (eugeoclinal) marine sediments, the latter in association with intrusions or extrusions of mafic or intermediate magmatic rocks.
2. Commencement of deformation in the foreland fold and thrust belt, together with the emplacement of ophiolitic rocks and the subsequent isostatic rise of the ophiolite and deformed sediments beneath it.
3. Continued deformation in the fold and thrust belt, and metamorphism, deformation, and intrusion of granitic batholiths in the core zone, together with deposition of synorogenic sediments.
4. Further isostatic rise of the orogenic region, and deposition and partial deformation of postorogenic continental sediments in the outer foredeep (the "molasse" of Alpine geology).
5. Block faulting, development of fault-bounded basins, and intrusion of scattered alkalic dikes or intrusive bodies.

In 1966, just as the idea of sea floor spreading was gaining widespread acceptance, the Canadian tectonicist J. T. Wilson proposed that the Atlantic Ocean had closed and then reopened. This early idea spawned the concept of the **Wilson cycle**, which includes the process of orogeny in a "cycle" consisting of continental rifting and the formation of an ocean basin, followed by subduction and the closing of the ocean basin, and then ending in collision and orogeny followed by reopening, and so on.

The idea of the Wilson cycle can accommodate the observations of events leading to orogeny that we have outlined. Figure 22.31 shows schematically one possible, but by no means unique, scenario that might give rise to the orogenic events outlined above.

1. The rifting of a continent and the opening of a new ocean basin produces gradually subsiding passive

---

[2] The term *cycle* is used rather loosely. Rarely, if ever, is there evidence of exactly the same sequence of events repeating itself in the same orogen.

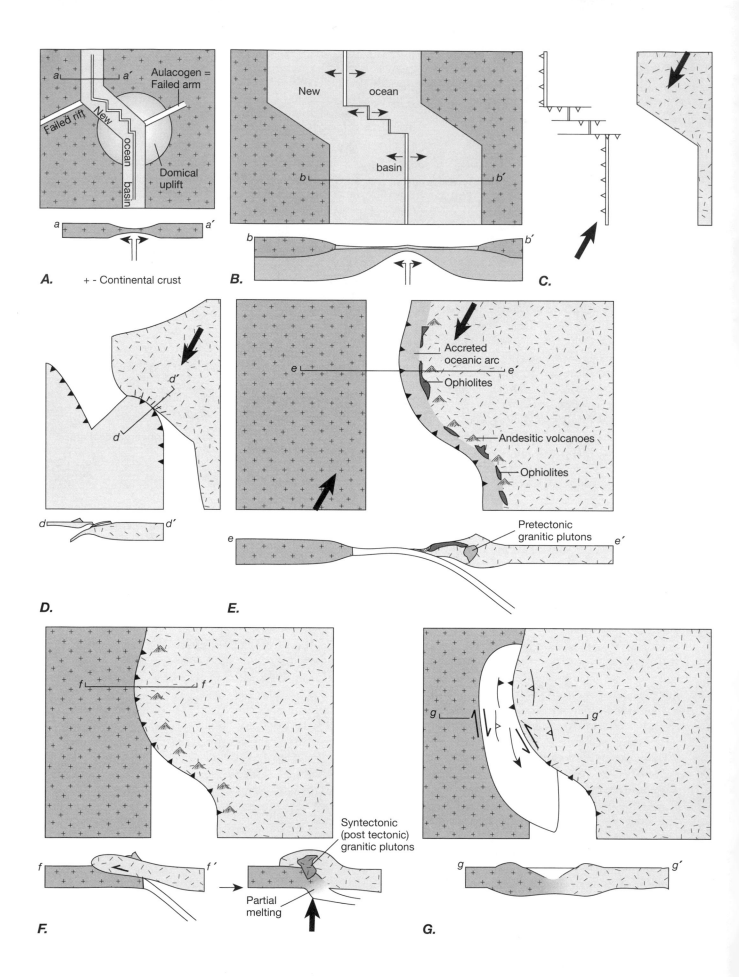

A.   + - Continental crust

B.

C.

D.

E.

    Accreted
    oceanic arc

    Ophiolites

    Andesitic volcanoes

    Ophiolites

Pretectonic
granitic plutons

F.

Syntectonic
(post tectonic)
granitic plutons

Partial
melting

G.

Figure 22.31 (*Opposite page*) Diagrammatic sketch maps and cross sections illustrating possible development of a plate tectonic Wilson cycle to account for traditional observations of mountain belts. *A.* Rifting of continental margin and the formation of domical uplifts and aulacogens along failed arms. *B.* Development of mature Atlantic-style ocean with passive margins. *C.* Change of relative plate motion, commencement of closure of ocean, and development of subduction zones along preexisting fault zones in oceans along fracture zones, ridge-transform fault intersections, or ridge parallel faults. *D.* Collision of hypothetical intra-oceanic arc system with continental margin and emplacement of ophiolites. *E.* "Flip" of subduction direction, development of Andean-style margin on continent to right, cessation of subduction on continent to left, and continued convergence with the other continent. *F.* Collision of two continents with Andean- and Atlantic-style continental margins, respectively, and formation of mountain root. *G.* Adjustment of continental margin collision zone by strike-slip movement and/or renewed rifting.

continental margins on which thick deposits of shallow-water (miogeoclinal) sediments accumulate, while offshore the ocean basin is formed by basaltic volcanism and deposition of deep-water sediments in the abyssal plains and continental rises (Figure 22.31A, B). Eventually, with a shift in the pattern of plate motions, the spreading pattern changes and a subduction zone develops in the ocean basin, probably along preexisting fractures such as transform faults, oceanic fracture zones, or ridge-parallel faults (Figure 22.31C). The eugeoclinal suite of deposits is completed by island arc volcanic and volcanogenic sedimentary rocks.

2. A passive continental margin on the down-going plate collides with the subduction zone (Figure 22.31D), emplacing a piece of oceanic crust and island arc crust (an ophiolite) onto the continental margin, juxtaposing the eugeoclinal and miogeoclinal suites, and initiating the formation of a foreland fold and thrust belt in the sediments of the passive margin. This collision probably is not synchronous along the entire margin but rather migrates along strike with time, depending on the relative geometries of the subduction zone and the continental margin.

3. Following the first collision, the polarity of the subduction zone flips, producing a continental arc or Andean-style continental margin along the now-deformed former Atlantic-style margin (Figure 22.31E) and initiating subduction of the remainder of the ocean basin. The consuming continental margin depicted in Figure 22.31E *need not* be a part of the original rifted continent, because the convergent relative velocities need not be the exact opposite of the original divergent relative velocities. Preorogenic plutons intrude the deformed continental margin. Deep basins in the trench, along the continental margin, or deep-sea fans derived from the continental margin in the narrowing ocean constitute the deep-water "orogenic sediments." Eventually the second continental margin arrives at the subduction zone and begins a continent–continent collision (Figures 22.31F, G) that sutures the formerly separate

pieces of continent. All the rocks in the suture region are deformed, the thrusting of the continental crust over continental crust creates a deep root, and partial melting at the base of the root produces late-stage intrusive or extrusive rocks. Strike-slip faulting takes place as the irregular edges of the continents adjust to each other.

4. Isostatic rise of the collision zone creates a source for postorogenic sediments and deforms earlier sediments. Strike-slip faults continue as the zone accommodates further postcollisional convergence (Figure 22.31G).

5. Finally, the suture zone may be torn apart by another rifting event, which causes normal faulting and intrusion and extrusion of alkalic to basaltic magmas.

Although the Wilson cycle as outlined above is based largely on Appalachian–Caledonide tectonic history, recent investigations have shown that history to be much more complex. The cycle is no more than a very broad generalization that does not provide a great deal of insight into the details of orogeny. It is highly unlikely, moreover, that two rifted continental margins will come back together in exactly the same place from which they rifted. Rather, given the changes in plate motion observed as a normal consequence of multiplate tectonic evolution, it is much more likely that two rifted continents will reapproach each other along different parts of the margin and that a misfit will occur. Indeed it is probable that many continental collisions have occurred between two continental margins that were never close to one another before. Misfits along colliding continental margins will adjust by lateral motion of crustal blocks, possibly by plastic slip-line processes (Figure 20.14). Thus we can expect to find that strike-slip motion in orogenic regions is as important as the more obvious contractional motion. Finally, the Wilson cycle does not account for the agglomeration of exotic terranes, which the examination of several orogenic belts reveals to be a common feature of orogenies.

22.10 Terrane Analysis

Many, if not most, of the world's orogenic belts include a composite of distinct terranes that originate not only from the continent(s) or arc(s) involved in the final collision but also from areas that are clearly "exotic" to the main crustal blocks or whose relationship to those blocks is "suspect." We call such terranes **exotic** or **suspect terranes.**

A terrane is thus an area surrounded by sutures[3] and characterized by rocks having a stratigraphy, petrology, or paleolatitude that is distinctly different from that of neighboring terranes or continents. In a collisional mountain belt, it is a remnant of crust that has had a different history from the subducted oceanic crust or from the main crustal blocks that have collided. Some mountain belts, such as the North American Cordillera, are characterized by collisions not between two continents but between a single continent and numerous exotic terranes. Some of these terranes apparently are similar to the areas of anomalous oceanic crust discussed in Section 21.1.

We need a different kind of analysis to understand the role of these exotic or "suspect" terranes in a mountain belt, and their recognition has begun to shed new light on some of the history of these complex regions. The object of terrane analysis is to work out when adjacent terranes were apart and when they came together, as revealed by a detailed comparison of the geologic histories. Figure 22.32 illustrates schematically the method of analysis with a hypothetical map (Figure 22.32A) and a stratigraphic–tectonic diagram (Figure 22.32B). Four terranes, numbered 1 through 4, are separated from each other and from continents A and D by sutures. By plotting the stratigraphy from each terrane in a column, with adjacent columns for adjacent terranes, it is possible to determine when the terranes began a common history and thereby to constrain when they collided, or "docked." The time of collision or docking of two terranes or of a terrane and a continent is given by the maximum age either of cross-cutting intrusives or of sediments unconformably overlying both terranes. Apparent polar wander (APW) paths (see Section 2.6) of two separate terranes should be different until the docking, after which they exhibit a single APW path.

Terranes 3 and 4 contain rocks of Carboniferous to Cenozoic age and Silurian to Cenozoic age, respectively. They share a common history from Jurassic time onward, as indicated by a date on the pluton (crosses) that intrudes the suture. In mid-Jurassic time, they collided

---
[3] Defined in Section 22.5 as zones along which oceans have disappeared.

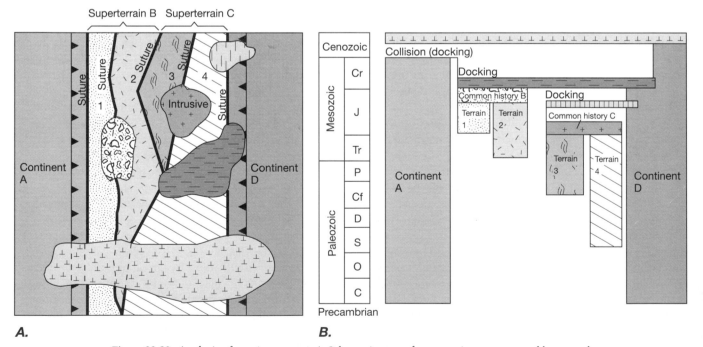

Figure 22.32 Analysis of exotic terranes. *A.* Schematic map of two continents separated by several exotic terranes. *B.* Stratigraphic–tectonic diagram illustrating the ages of rocks in individual terranes and the ages of "docking" and of common histories. See the text for discussion.

with continent D, as indicated by the sedimentary unit (vertical-ruled) that overlies that suture. Rocks in terrane 1 are Jurassic through Cenzoic in age, and rocks in terrane 2 are Triassic through Cenzoic in age. The diagram shows that they share a common history beginning in late Jurassic, when a rock unit (irregular blob pattern) was deposited across the suture between them. Terranes 1 and 2 docked or collided with the already amalgamated terranes 3 and 4 and Continent D in early Cretaceous time, as indicated by the sedimentary unit overlying all the sutures between terranes 2-3, 2-4, 3-4, and 4-D (horizontal-ruled). Continent A collided with the composite of the terranes and continent D in Cenozoic time, when the sedimentary deposits covering all the previously accreted terranes began also to cover continent A.

The recognition of exotic terranes in mountain belts around the world requires a major modification of the Wilson cycle and of our ideas about how mountain belts develop. Given that areas of anomalously thick oceanic crust can develop in the ocean basins (Section 21.1), the process of subduction must sweep them into the subduction zone ahead of any continent that rides on the same plate. Before collision, these terranes have their own individual geologic history, but after collision, they become part of the overriding crustal block. Thus the argument about whether orogeny is an on-going or a highly episodic event may be settled in favor of both sides; that is, the docking of exotic terranes may be a continual process during subduction, but the collision of major continents is only episodic. Thus the motions of the major crustal blocks in a collision may represent only a part of the activity that constructs an orogenic belt, and many other relative plate motions may be represented by the numerous sutures among different exotic terranes.

Additional Readings

Best, M. G. 1982. *Igneous and metamorphic petrology.* New York: W. H. Freeman.

Boudier, F., A., Nicolas, and J.L., Bouchez. 1982. Kinematics of oceanic thrusting and subduction from basal sections of ophiolites. *Nature* 296: 825–828.

Cook, F. A., et al. 1981. COCORP seismic profiling of the Appalachian orogen beneath the coastal plain of Georgia. *G.S.A. Bull.* 92: 739–748.

Hansen, E.C. 1971. *Strain facies.* New York: Springer Verlag. A. L. Harris, C. H. Holland, and B. E. Leake, eds., *The Caledonides of the British Isles reviewed.* Geological Society of London special publication 8, pp. 187–199.

Hatcher, R. D., Jr., Williams, H., and Zietz, I. 1982. Contributions to the tectonics and geophysics of mountain chains. *G.S.A. Memoir* 158: 225.

Hatcher, R. D., Jr., and Williams, R. T. 1986. Mechanical model for single thrust sheets. I: Taxonomy of crystalline thrust sheets and their relationships to the mechanical behavior of orogenic belts. *G.S.A. Bull.* 97: 975–985.

Hsü, K. J. 1982. *Mountain building processes.* New York: Academic Press.

King, P. B. 1977. *Evolution of North America.* Princeton, N.J.: Princeton University Press. 2nd edition.

Le Fort, P. 1975. Himalaya: The collided range, present knowledge of the continental arc. *Am. J. Sci.* 275A: 1–44.

Moores, E. M., ed. 1990. *Shaping the Earth: Tectonics of continents and oceans.* New York: W. H. Freeman.

Ramsay, J. G. 1963. Structures stratigraphy, and metamorphism in the western Alps. *Proc. Geol. Assoc.* 74: 389–391.

Rodgers, J. 1970. *Tectonics of the Appalachians.* New York: Wiley.

Roeder, D. H. 1973. Subduction and orogeny. *J. Geophys. Research* 78: 5005–5024.

Zen E-An, W. S, White, J. B., Hadley, and J. B., Thompson, Jr., eds. *Structural studies of the Appalachians.* II. Northern and Maritime. New York: Wiley.

Epilogue

In this book, we have examined structures in rocks ranging from the submicroscopic scale of dislocations to the regional scale of orogenic belts. We have examined fractures, faults, folds, foliations, lineations, crystallographic preferred orientations, and dislocation structures. We have discussed kinematic models for the formation of these structures, as well as continuum mechanical models involving strain and rheology. In discussing these different models, we have tried to show that the better we understand deformation structures in rocks and the way they form, the more we can infer about the conditions and processes that occur within the Earth. We have also attempted to show how all scales of investigation, and all techniques by which we can better understand the origin of structures, help provide a unified picture of deformation in the Earth, especially in orogenic belts, where much of the deformation of the continental crust is concentrated.

In the final analysis, however, the study is not complete, because apart from not understanding thoroughly the origin of many structures, we are left with a host of questions about what ultimately causes the deformation of the Earth's crust. Why is so much of it concentrated in narrow linear belts on the continents? And how has the type of deformation that has occurred evolved through the Earth's history?

In effect, what we have done in this book is to investigate the small-scale deformation structures and deformation processes and connect them up in scale to regional belts of deformation. What remains now is to investigate the large-scale deformation processes in the Earth and connect them down in scale to the regional belts of deformation. This is the realm of tectonics, and it is the subject of our next book.

It is at the scale of regional deformation and orogenic belts that a continuum exists from structural geology to tectonics. Different lines of evidence constrain the interpretation of deformation in orogenic belts from the analysis of small-scale structures and from the large scale plate motions. The two lines must be consistent with a single model for the development of regional deformation, so both are necessary aspects of our study.

The geologic record in the ocean basins, however, permits the direct inference of plate tectonic motions for only the past 200 million years at most. Beyond that, we must rely on the geologic record in continental rocks. Thus the connection between structural geology and tectonics takes on another dimension, for the analysis of structures is a major source of information about the large-scale tectonic processes that occurred prior to 200 Ma. From the evidence of the past 200 million years, we must try to understand the relationship between the small- to regional-scale structures on the one hand and the large-scale tectonic processes on the other. Then when we look at older orogenic belts, we can infer the large-scale processes from the evidence provided by the smaller-scale structures. At present, our ability to accomplish this is by no means complete, but we are making progress.

Thus the investigation of structural and tectonic processes should go hand in hand. Structural investigations are incomplete unless they can be placed in the context of tectonic processes. Conversely, our inference of tectonic processes must be consistent with our interpretations of structures in the affected rocks for deformations younger than about 200 million years and in part must be guided by those interpretations for older deformations.

We encourage students to pursue this fascinating subject into the large-scale realm of tectonics, the subject of our next book. We rely heavily on material presented in this volume, but we approach if from the point of view of large-scale processes, seeking ultimately to obtain an integrated view on all scales of the deformation of our planet.

The gradually increasing data base on the other terrestrial planets of our solar system challenges us further to apply our understanding of the Earth to explaining the similarities and differences that we observe on Mercury, Venus, the Moon, Mars, and the moons of the outer giant planets. The observation of worlds other than our own gives us the opportunity to test our ideas about our planet and to refine them to fit in the still more general context of the solar system.

Sources of Illustrations

Fig. 1.1 After Willie 1975. *Scientific American.*

Fig. 1.2 After Uyeda 1978.

Fig. 2.5 From R.R. Shrock. 1948. *Sequence in layered rocks.* New York: McGraw-Hill.

Fig. 2.6 Modified after Blatt, Middleton, Murray 1980.

Fig. 2.8 After Shrock 1948.

Fig. 2.10*A* From M.G. Best. 1982. *Igneous and metamorphic petrology.* New York: W.H. Freeman, Fig. 5-34b, p. 176.

Fig. 2.12 After R.P. Nichelsen and V.D. Hough. 1967. Jointing in the Appalachian plateau of Pennsylvania. *Geol. Soc. Am. Bull.* 78:609–630, Fig. 2.

Fig. 2.14 From J. Suppe. *Principles of structural geology.* Englewood Cliffs, N.J.: Prentice-Hall, Fig. 2.9.

Fig. 2.15 From Lindseth 1982.

Fig. 2.16 After Sheriff 1978, p. 210.

Fig. 2.18 After Sheriff 1978.

Fig. 2.19 After F. Press and F. Siever. 1986. *Earth.* New York: W.H. Freeman, p. 425; ref. McElhinny 1973.

Fig. 2.1.2 Redrawn after Press and Siever 1986, p. 414.

Fig. 2.3.1 After Lindseth 1982.

Fig. 3.1 After Kulander, Barton, and Dean 1979.

Fig. 3.2*B* T. Engelder. 1985. Loading paths to joint propagation during a tectonic cycle: An example from the Appalachian Plateau, U.S.A. *Jour. Struct. Geol.* 7:459–476.

Fig. 3.3 R.A. Hodgson. 1961a. Regional study of jointing in Comb Ridge-Mavarre Mountain area, Arizona and Utah. *Amer. Assoc. Petrol. Geol. Bull.* 45:1–38, Figs. 12, 16, 17.

Fig. 3.4 From S. Stanley. 1986. *Earth and life through time.* New York: W.H. Freeman, p. 164.

Fig. 3.5 Photo by N.K. Hubert, in Twidale 1982.

Fig. 3.8 Photo courtesy of Richard Sibson.

Fig. 3.9 After Nichelsen and Hough 1967.

Fig. 3.10 In part after Hodgson 1961a.

Fig. 3.11 Data sets A through D from Ladeira and Price 1981; data set E from McQuillan 1973, reported in Ladeira and Price 1981.

Fig. 3.12 From Engelder and Geiser 1980.

Fig. 3.13 D. Bahat. 1979. Theoretical considerations on mechanical parameters of joint surfaces based on studies of ceramics. *Geol. Mag.* 116:81–166, plate 1B, plate 4.

Fig. 3.14*A* After R.A. Hodgson. 1961b. Classification of structures on joint surfaces. *Am. Jour. Sci.* 259:493–502. Fig. 1; Kulander, Barton, and Dean 1979, Fig. 77.

Fig. 3.14 *B,C* After Kulander and Dean 1985, Fig. 10.

Fig. 3.15 After Kulander, Barton, and Dean 1979.

Fig. 3.16 From D.W. Stearns. 1968. Certain aspects of fractures in naturally deformed rocks. In R.E. Riecker, ed., *NSF advanced science seminar in rock mechanics for college teachers of structural geology.* Bedford, Mass.: Terrestrial Sciences Laboratory, Air Force Cambridge Research Laboratories, pp. 97–118.

Fig. 3.17 After Price 1966; Stearns 1968.

Fig. 4.4 After Sibson 1977.

Fig. 4.13 Courtesy of J. Shelton.

Fig. 4.16 After Petit 1987.

Fig. 4.17*C,E,F,G* After C. Simpson and S.M. Schmidt. 1983. An evaluation of criteria to deduce the sense of movement in sheared rocks. *Geo. Soc. Am. Bull.* 94:1281–1288, Fig. 5.

Fig. 4.17*D* After Simpson 1986.

Fig. 4.18 After W. Addicott. 1967. In W.R. Dickinson and A. Grantz, eds., *Problems of the San Andreas fault system.* Stanford Univ. Pubs. Geol. Sci.

Fig. 4.26 Woodcock and Fischer 1986.

Fig. 4.31*B* Redrawn after Platker. 1965. Tectonic deformation associated with the 1964 Alaska earthquake. *Science* 148:1685.

Fig. 5.1*A* Photograph by N. Lindsley-Griffin.

Fig. 5.1*B* Photograph by J. Stewart.

Fig. 5.2 After M.P. Billings. 1972. *Structural geology,* 3d. Englewood Cliffs, N.J.: Prentice-Hall.

Fig. 5.3 From Wintershalle in Wernicke and Burchfiel 1982.

Fig. 5.6 From I. Effimoff and A.R. Pinezich. 1986. Tertiary structural development of selected basins: Basin and Range province, northeastern Nevada. In L. Mayer, ed., *Geol. Soc. Am. Sp. Paper* 208.

Fig. 5.7*A* After E.A. Wentlandt. 1951. Hawkins field, Wood County, Texas. Austin, Univ. of Texas Pub., No. 153–158.

Fig. 5.7*B* After E. Cloos. 1968. *Am. Assoc. Petrol. Geol. Bull.* 52:420–444.

Fig. 5.8*B* After Press and Siever 1986.

Fig. 5.9 Modified after Uyeda 1977.

Fig. 5.10 After P.J. Coney. 1980. Review of Cordilleran metamorphic core complexes. *Geol. Soc. Am. Mem.* 153; S.L. Wust. 1986. Regional correlation of extension directions in Cordilleran metamorphic core complexes. *Geology* 14(10):828–830, Fig. 1; and J.H. Stewart. 1978. Basin-range structure in western North America, a review. In R.B. Smith and G.P. Easton, eds., Cenozoic tectonics and regional geophysics of the western Cordillera. *Geol. Soc. Am. Mem.* 152:1–32.

Fig. 5.11 Modified after G.S. Lister, M.A. Etheridge, P.A. Symonds. 1986. Detachment faulting and the evolution of passive continental margins. *Geology* 14:246–250.

Fig. 5.12 After J.H. Stewart. 1978. Basin-range structure in western North America: A review. In R.B. Smith and G.P. Eaton, eds., Cenozoic tectonics and regional geophysics of the western Cordillera. *Geol. Soc. Am. Mem.* 152:1–32.

Fig. 5.13*A,B* G.A. Davis, J.L. Anderson, E.G. Frost, and T.J. Shackelford. 1980. Mylonitization and detachment faulting in the Whipple-Buckskin-Rawhide Mountains terrane, southeastern California and western Arizona. *Geol. Soc. Am. Mem.* 153:79–129.

Fig. 5.14 From D.M. Worrall and S. Snelson. 1989. Evolution of the northern Gulf of Mexico, with emphasis on Cenozoic growth faulting and the role of salt. In A.W. Bally and A.R. Palmer, eds., *The geology of North America: An overview.* DNAG The Geology of North America, Vol. A. Boulder, Colo.: Geological Society of America, pp. 97–138.

Fig. 5.15 After C.H. Bruce. 1973. Pressured shale and related sediment deformation: Mechanism for development of regional contemporaneous faults. *Amer. Assoc. Petrol. Geol. Bull.* 57:878–886.

Fig. 5.17 In part after Hamblin 1965.

Fig. 5.18 After Wernicke and Burchfiel 1982.

Fig. 5.19 After Wernicke and Burchfiel 1982.

Fig. 5.20 From A.D. Gibbs. 1984. Structural evolution of extensional basin margins. *Jour. Geol. Soc. Lond.* 141:609–620.

Fig. 5.21 From Lister and Davis 1989.

Fig. 6.1 Photos courtesy of R.J. Varga.

Fig. 6.3 From S.B. Smithson et al. 1978. Nature of the Wind River thrust, Wyoming, from COCORP deep-reflection data and from gravity data. *Geology* 6(11):648–652, Fig. 5.

Fig. 6.5 From C.D.A. Dahlstrom. 1970. Structural geology in the eastern margin of the Canadian Rocky mountains. *Canad. Petrol. Geol. Bull.* 18:332–406.

Fig. 6.6 Dahlstrom 1970.

Fig. 6.7*A* From Boyer and Elliot 1982.

Fig. 6.7*C* From S. Mitra. 1988. Three-dimensional geometry and kinematic evolution of the Pine Mountain thrust system, southern Appalachians. *Geol. Soc. Am. Bull.* 100:72–95.

Fig. 6.8 After Dahlstrom 1970.

Fig. 6.11*A* After Harris and Bayer 1979.

Fig. 6.11*B* After R.D. Price and R.D. Hatcher, Jr. 1983. Tectonic significance of similarities in the evolution of the Alabama-Pennsylvania Appalachians and the Alberta-British Columbia Canadian Cordillera. *Geol. Soc. Am. Mem.* 158:149–160.

Fig. 6.11C After A. Gansser. 1981. Himalaya and overview. In H.K. Gupta and F.M. Delaney, eds., Zagros, Hindu Kush, Himalaya, Geodynamic evolution. *Geodynamics series* 3:215–242.

Fig. 6.12A D. Davis, J. Suppe, and F.A. Dahlen. 1983. Mechanics of fold-and-thrust belts and accretionary wedges. *Jour. Geophys. Res.* 88:1153–1172; after Roeder et al. 1978.

Fig. 6.12B Davis et al. 1983; after A.W. Bally, P.L. Gordy, and G.A. Stewart. 1966. Structure, seismic data and orogenic evolution of southern Canadian Rockies. *Canadian Petrol. Geol. Bull.* 14:337–381.

Fig. 6.14 After Boyer and Elliot 1982.

Fig. 6.15 From Boyer and Elliot 1982.

Fig. 6.16 From Dahlstrom 1970.

Fig. 6.17 After Boyer and Elliot 1982.

Fig. 6.18 After Boyer and Elliot 1982.

Fig. 6.20 Modified from Ernst 1973.

Fig. 7.1A From J.S. Shelton. 1966. *Geology illustrated.* New York: W.H. Freeman, Fig. 237.

Fig. 7.1B ERTS Photo (8) No. 14-490-40625-A0 from P. Molnar and P. Tapponnier. 1975. Cenozoic tectonics of Asia: Effects of a continental collision. *Science* 189:419–426.

Fig. 7.2A After J.C. Crowell. 1979. The San Andreas fault through time. *Bull. Geol. Soc. London* 133:292–302, Fig. 2.

Fig. 7.2B After P. Tapponnier and P. Molnar. 1977. Active faulting and tectonics in China. *Jour. Geophys. Res.* 82:2905–2930.

Fig. 7.4 Modified after Sylvester 1988.

Fig. 7.6B After Woodcock and Fischer 1986, Fig. 12a.

Fig. 7.7B After Woodcock and Fischer 1986, Fig. 12b.

Fig. 7.8 After T.P. Harding. 1985. Seismic characteristics and identification of negative flower structures, positive structures, and positive structural inversion. *Am. Assoc. Petrol. Geol. Bull.* 69:582–700.

Fig. 7.9 Landsat image 1584-03070-5; from P. Tapponnier and P. Molnar. 1979. Active faulting and Cenozoic tectonics of the Tien Shan, Mongolia, and Baykal regions. *Jour. Geophys. Res.* 84:3425–3459.

Fig. 7.10E After R. Freund. 1974. Kinematics of transform and transcurrent faults. *Tectonophysics* 21:93–134.

Fig. 7.11 After H. Laubscher. 1972. Some overall aspects of Jura dynamics. *Amer. Jour. Sci.* 272:296.

Fig. 7.12 After J.S. Tchalenko and N.N. Ambraseys. 1970. Structural analysis of the Dasht-e Bayly (Iran) earthquake fractures. *Geol. Soc. Am. Bull.* 81:41–60.

Fig. 7.13 After B.P. Luyendyk, M.J. Kamerling, and R.

Terres. 1980. Geometric model for Neogene crustal rotations in southern California. *Geol. Soc. Am. Bull. Part 1* 91:211–217.

Fig. 7.14 From the Geologic map of Pakistan, with interpretation after Lawrence and Yeats 1978.

Fig. 7.15 After Freund 1974.

Fig. 7.17 After Luyendyk et al. 1980.

Fig. 9.5A Data from J. Handin. 1966. Handbook of physical constants. *Geol. Soc. Am. Mem.* 97. Edited by S.P. Clark, Jr. Table 11-3.

Fig. 9.7 After Hoeppener et al. 1969; referenced in Freund 1974.

Fig. 9.8 After Freund 1974.

Fig. 9.10 From M.S. Paterson. 1978. *Experimental rock deformation—The brittle field.* New York: Springer Verlag, Fig. 48.

Fig. 9.11 J. Byerlee. 1975. The fracture strength and frictional strength of Weber sandstone. *Int. Jour. Rock Mechanics Mining Sci.* 12:1–4.

Fig. 9.12C After K. Sieh. 1982. Late Halocene displacement history along the south-central reach of the San Andreas fault. Ph.D. diss., Stanford University.

Fig. 9.14. F. Donath. 1961. Experimental study of shear failure in anisotropic rocks. *Geol. Soc. Am. Bull.* 72:985–990.

Fig. 9.21A After Z.T. Bieniawski. 1967. Mechanism of brittle fracture of rock: Parts I, II, and III. *Int. Jour. Rock Mech. Min. Sci.* 4:395–430.

Fig. 9.21B Modified from Johnson 1970, Fig. 9.20.

Fig. 10.1B After R.H. Merrill. 1968. Measurement of in situ stress and strain in rock. Rock Mechanics Seminar, NSF advanced science seminar in rock mechanics for college teachers of structural geology, Boston College, 1967. Terrestrial Science Laboratory, Air Force Cambridge Research Laboratory, Bedford, Mass., pp. 151–199.

Fig. 10.2B From Zoback et al. 1980.

Fig. 10.3 After McGarr and Gay 1978.

Fig. 10.4 After McGarr and Gay 1978.

Fig. 10.5B,C After McGarr and Gay 1978.

Fig. 10.6A From Richardson et al. 1979.

Fig. 10.6B From Zoback and Zoback 1980.

Fig. 10.12 After Currie et al. 1962.

Fig. 10.15 After Hafner 1951.

Fig. 10.16 After Hafner 1951.

Fig. 10.17 From Twiss et al. 1991.

Fig. 10.19 After Davis et al. 1983.

Fig. 10.20 From Davis et al. 1983.

Fig. 10.1.1 After Engelder 1985.

Fig. 11.1A J. Debelmas, A. Escher, and R. Trumpy. 1983. Profiles through the western Alps. In N. Rast and F.M. Delany, eds., Profiles of orogenic belts, *AGU-GSA Geodynamics series* 10:83–97.

Fig. 11.1B Haller collection of photographs from East Greenland, Geologisk Museum, Copenhagen, Denmark.

Fig. 11.2A Synthetic aperature side-looking radar image courtesy of the U.S. Geological Survey, U.S. Dept. of the Interior.

Fig. 11.2B After A. Buxtorf. 1916. Prognosen und Befund beim Hauensteinbasis-und Grenchenbergtunnel und die Bedeutune der letzteren fürdie Geologie des Juragebirges. *Verh. Naturforsch. Ges. Basel* 27:185–254.

Fig. 11.12 Graph after Fleuty 1964.

Fig. 11.15 After Twiss 1988.

Fig. 11.16 After Twiss 1988.

Fig. 11.17 After Twiss 1988.

Fig. 11.19 After Ramsay 1967.

Fig. 11.20 After Ramsay 1967.

Fig. 11.26 Photo courtesy of R.J. Varga.

Fig. 12.4B Photo courtesy of R.J. Varga.

Fig. 12.10 After Groshong 1975.

Fig. 12.11 Groshong 1975.

Fig. 12.15 Modified after Donath and Parker 1964.

Fig. 12.20 After Weiss 1980.

Fig. 12.21 After Ramsay 1967.

Fig. 12.22 From Patterson and Weiss 1966.

Fig. 12.24 After Suppe 1983.

Fig. 12.25 From Suppe 1983.

Fig. 12.26 From Suppe 1983.

Fig. 12.27 From Suppe 1985.

Fig. 12.30 From Skjernaa 1975.

Fig. 12.31 Modified from Ramsay 1967.

Fig. 12.32A From L. Skjernaa. 1989. Tubular folds and sheath folds: Definitions and conceptual models of their development, with examples from the Grapesvare area, northern Sweden. *Jour. Struct. Geol.* 11(6):689–703.

Fig. 12.33 After M.P.A. Jackson and C.J. Talbot. 1989. Anatomy of mushroom-shaped diapirs. *J. Struct. Geol.* 11:211–230.

Fig. 12.34A From C.J. Talbot and M.P.A. Jackson. 1987. Internal kinematics of salt diapirs. *Am. Assoc. Petrol. Geol. Bull.* 71(9):1068–1093, Fig. 13.

Fig. 12.34B After W.R. Meulberger, P.S. Clabaugh, and M.L. Hightower. 1962. Palestine and Grand Saline salt domes, eastern Texas. In E.H. Rainwater, and R.P. Zingula, eds., *Geology of the Gulf Coast and central Texas guidebook of excursions.* Houston Geol. Soc., Fig. 5; and W.R. Meulberger. 1968. Internal structures and mode of uplift of Texas and Louisiana salt domes. *Special Paper 88.* Geol. Soc. Am. 359–364, Fig. 1.

Fig. 12.34C After Jackson and Talbot 1989.

Fig. 12.35 From J.B. Thompson, Jr. et al. 1968. Nappes and gneiss domes in west-central New England. In E-An Zen et al., eds., *Studies of Appalachian geology: Northern and maritime.* New York: Interscience Publishers, pp. 203–218.

Fig. 13.1 Modified from Powell 1979.

Fig. 13.2A Courtesy of N. Lindsley-Griffin.

Fig. 13.2B Courtesy of M. St. Orge.

Fig. 13.3 From Powell 1979.

Fig. 13.4 From Powell 1979.

Fig. 13.5B Thin section courtesy of G. Protzman; photo by E. Rodman.

Fig. 13.6 Alvarez et al. 1978.

Fig. 13.7 Gray 1978.

Fig. 13.9A Glen 1982.

Fig. 13.9B Photo courtesy of J. Treagus.

Fig. 13.9C Gray 1977.

Fig. 13.11 Weber 1981.

Fig. 13.12A Photo courtesy of E. Rodman.

Fig. 13.13B Photo courtesy of Peter Stringer.

Fig. 13.14B Photo courtesy of R.J. Varga.

Fig. 13.16A C. Simpson and S. Schmid. 1983. An evaluation of criteria to deduce the sense of movement in sheared rocks. *Geol. Soc. Am. Bull.* 94:1281–1288.

Fig. 13.17 Turner and Weiss 1963.

Fig. 13.19A Photo courtesy of Martin G. Miller.

Fig. 13.19B Photo courtesy of Carol Simpson.

Fig. 13.20B Photo courtesy of R. J. Varga.

Fig. 13.22A B.E. Hobbs et al. 1976. *An outline of structural geology.* New York: Wiley.

Fig. 13.22B L.E. Weiss. 1972. *The minor structures of deformed rocks: A photographic atlas.* New York: Springer Verlag, Fig. 65A.

Fig. 13.26B A. Etchecopar and J. Malavielle. 1987. Computer models of pressure shadows: A method for strain measurement and shear sense determination. *J. Struct. Geol.* 9:667–677.

Fig. 14.2 After Ramsay 1976.

Fig. 14.3A After J. G. Ramsay. 1967. *Folding and fracturing of rocks.* New York: McGraw-Hill, Fig. 3-44.

Fig. 14.5 After Ramsay 1976.

Fig. 14.6 After Means 1987.

Fig. 14.7 After Durney and Ramsay 1973; Ramsay and Huber 1983.

Fig. 14.9 Ramsay and Huber 1983.

Fig. 15.7A Sample courtesy of G. Protzman; photo by E. Rodman.

Fig. 15.7B Sample courtesy of John Christie; photo by Mary Graziose.

Fig. 15.8 Structure diagrams after Ramsay 1967.

Fig. 15.19 After B.E. Hobbs, W.D. Means, and P.F. Williams. 1976. *An outline of structural geology.* New York: Wiley.

Fig. 16.3C After Ramsay 1967.

Fig. 16.4 From Hobbs 1971.

Fig. 16.5 After Hobbs 1971.

Fig. 16.7A,B From Alvarez et al. 1978.

Fig. 16.7C From Alvarez et al. 1976, Fig. 3.

Fig. 16.8 After Ramsay and Huber 1983.

Fig. 16.13 After Williams and Schoneveld 1981.

Fig. 16.17 From Gray 1979.

Fig. 17.2A,C From Mitra 1978.

Fig. 17.2B From S. Mitra. 1976. A quantitative study of deformation mechanisms and finite strain in quartzites. *Contrib. Mineral. Petrol.* 59:203–226.

Fig. 17.3 After Ramsay and Huber 1983.

Fig. 17.4 From D.M. Ragan. 1985. *Structural geology: An introduction to geometrical techniques*, 3d ed. New York: Wiley, Fig. X10.2.

Fig. 17.5 After Wellman 1962.

Fig. 17.6 R_f-ϕ curves from Lisle 1985.

Fig. 17.7 R_f-ϕ curves from Lisle 1985.

Fig. 17.8 Modified from Ramsay and Huber 1983, Fig. 5.11.

Fig. 17.9 From Wright and Platt 1982.

Fig. 17.10 From Hundleston and Holst 1984.

Fig. 17.11A After E.T. Cleaves, J. Edwards, Jr., and J.D. Glaser. 1968. Geologic map of Maryland. Maryland Geological Survey.

Fig. 17.11B After Cloos 1947, 1971.

Fig. 17.12 After Hobbs 1971.

Fig. 17.13 After Hobbs 1971.

Fig. 17.14A,B After Wood 1973.

Fig. 17.14 After Ramsay 1967.

Fig. 17.15 From Ramsay and Huber 1983.

Fig. 17.16A D.K. Mukhopadhyay and B.W. Haimanot. 1989. Geometric analysis and significance of mesoscopic shear zones in the Precambrian gneisses around the Kolar schist belt, south India. *J. Struct. Geol.* 11(5):569–581.

Fig. 17.16B From Ramsay and Graham 1970.

Fig. 17.17 Ramsay 1980, Fig. 3.

Fig. 17.18 Modified from Ramsay and Huber 1983.

Fig. 17.19 After Durney and Ramsay 1973, Fig. 15.

Fig. 18.9 From Dieterich 1981.

Fig. 18.10 After J.H. Dieterich. 1979. Modeling of rock friction 1. Experimental results and constitutive equations. *Jour. Geophys. Res.* 84(B5):2161–2168.

Fig. 18.11 From Borch and Green 1989.

Fig. 18.12 Data from Borch and Green 1989.

Fig. 18.14 Modified after Twiss 1976, Fig. 1.

Fig. 18.15 From Schmid et al. 1977.

Fig. 18.16 From Borch and Green 1987.

Fig. 18.17A From Olsson 1974.

Fig. 18.17B From Schmid et al. 1977.

Fig. 18.17C From Karato and Paterson 1986.

Fig. 18.18 From Ord and Hobbs 1986.

Fig. 18.19 From Ranalli and Murphy 1987.

Fig. 18.20 From Ranalli and Murphy 1987.

Fig. 18.2.1 From Borch and Green 1989.

Fig. 18.2.2 From Borch and Green 1989.

Fig. 18.2.3 From Borch and Green 1989.

Fig. 19.1A,B Rutter 1976.

Fig. 19.1C Ashby and Verrall 1977.

Fig. 19.8 After M.F. Ashby and R.A. Verrall. 1973. *Acta Metall.* 21.

Fig. 19.10B Right diagram after Hirth and Lothe 1968.

Fig. 19.10C Right diagram after Weertman and Weertman 1964.

Fig. 19.13A Courtesy of Jin Zhen-Ming.

Fig. 19.13B Ball and White 1977.

Fig. 19.13C R.J. Twiss. 1976. Some planar deformation features, slip systems, and submicroscopic features in synthetic quartz. *Jour. Geol.* 84:701–724.

Fig. 19.14 After A. Spry. 1969. *Metamorphic textures.* Oxford: Pergamon Press.

Fig. 19.17C From Hirth and Lothe 1968.

Fig. 19.19A Kohlstedt and Weathers 1980.

Fig. 19.19B,C Bell and Etheridge 1976.

Fig. 19.20 Plot based on results of Karato et al. 1986.

Fig. 19.21 From Spry 1969, Fig. 22.

Fig. 19.22 A.C. McLaren amd B.E. Hobbs. 1972. Transmission electron microscope investigation of some naturally deformed quartzites. In *Flow and fractures of rocks*, The Griggs Volume. Am. Geophys. Union Monograph 16. Washington, D.C.: Am. Geophys. Union, pp. 55–66.

Fig. 19.23 Kohlstedt and Weathers 1980.

Fig. 19.24*A* Kohlstedt and Weathers 1980.

Fig. 19.24*B* S.-I. Karato, M. Toriumi, and T. Fujii. 1980. Dynamic recrystallization of olivine single crystals during high-temperature creep. *Geophys. Res. Letts.* 7(9):649–652.

Fig. 19.24*C* Schmid 1982.

Fig. 19.26 From Ethecopar and Vasseur 1987.

Fig. 19.27 From N. L. Carter and H. Avé Lallement. 1970. High-temperature flow of dunite peridotite. *Geol. Soc. Am. Bull.* 81:2181–2202.

Fig. 19.28*A,B* From Nicolas and Poirier 1976.

Fig. 19.29 Nicolas and Poirier 1976.

Fig. 19.30 After Schmid and Casey 1986. Natural fabric diagrams from Price 1985, samples (A) 9, (C) 46, 62, (E) 72; (F) Schmid and Casey 1986, sample Gran 133.

Fig. 19.31 Natural fabrics in (C) and (E) from Price 1985; (B) from Schmid and Casey 1986; fabric skeletons and schematic *a*-axis fabrics from Schmid and Casey 1986.

Fig. 19.32 After Schmid and Casey 1986.

Fig. 19.2.1 After W.H. Scott, E.C. Hansen, and R.J. Twiss. 1965. Stress analysis of quartz deformation lamellae in a minor fold. *Am. Jour. Sci.* 263:729–746.

Fig. 20.2 From Currie et al. 1962.

Fig. 20.3 Currie et al. 1962.

Fig. 20.4 From Dieterich and Carter 1969.

Fig. 20.5 From Dieterich and Carter 1969.

Fig. 20.6 From Dieterich 1969.

Fig. 20.7 From Dieterich 1969.

Fig. 20.8 From Dieterich 1969.

Fig. 20.9 From Parrish et al. 1976.

Fig. 20.10 Parrish et al. 1976.

Fig. 20.11 After Parrish et al. 1976.

Fig. 20.12 After Parrish et al. 1976.

Fig. 20.13 After Backofen 1972.

Fig. 20.14 After Backofen 1972.

Fig. 20.15 From Dennis and Häll 1978.

Fig. 20.16 After Bucher 1956.

Fig. 20.17 After Bucher 1956.

Fig. 20.18 From McClay and Ellis 1987.

Fig. 20.19 From Jackson and Talbot 1989.

Fig. 20.20 From Jackson and Talbot 1989.

Fig. 20.21 Ramberg 1981.

Fig. 21.1 Modified after Anonymous. 1950. Der Bau der Erde Gotha, *Justus Perthes* 1:140,000,000; and Stanley 1986.

Fig. 21.2 Modified after A.W. Bally. 1980. A.G.U. Dynamic series 1, p. 4; and Uyeda 1978.

Fig. 21.8 Modified after Smithson et al. 1979, p. 263.

Fig. 21.9*C* Modified after M.P.A. Jackson. 1984. Archean structural styles in ancient gneiss complex of Swaziland, southern Africa. In A. Kröner and R. Greiling, eds., *Precambrian tectonics illustrated.* Stuttgart: Schweizerbart'sche, pp. 1–18.

Fig. 21.9*D* After K. Anhausser. In Kröner and Greiling 1984.

Fig. 21.10 After J.S. Meyers. 1984. Archean tectonics in the Fiskenaesset region of southwest Greenland. In Kröner and Greiling 1984.

Fig. 21.11 Modified after R.L. Stanton. 1972. *Ore petrology.* New York: McGraw-Hill, p. 374.

Fig. 21.12 Modified after Stanton 1972, pp. 313, 316.

Fig. 21.13 Modified after K. Burke. 1981. In *Continental tectonics,* NAS-NRC.

Fig. 21.14 Modified after S. Cloetingh. 1986. Intraplate stresses: A new tectonic mechanism for fluctuations of relative sea level. *Geology* 14:617–620; and P.R. Vail, R.M. Michum, Jr., and S. Thompson. 1977. *AAPG Mem.* 19:84.

Fig. 21.15*A* Modified after A. Bally et al. 1979. *Continental Margins,* NAS-NRC.

Fig. 21.15*B* Modified after M. Kay. 1951. North American geosynclines. *Geol. Soc. Amer. Mem.*

Fig. 21.16 Modified after Bally et al. 1979, p. 78.

Fig. 22.2*A* Modified after King 1977.

Fig. 22.2*B* Modified after J.F. Dewey. 1977. Suture zone complexities, a review. *Tectonophysics* 40:53–67.

Fig. 22.2*C* Modified after H. Williams. 1984. Miogeoclines and suspect terranes of the Caledonian-Appalachian orogen: Tectonic patterns in the North Atlantic region. *Canadian Jour. of Earth Sciences* 21:887–901.

Fig. 22.3 Generalized after Hatcher and Williams 1986.

Fig. 22.4 After King 1977.

Fig. 22.5 Modified after Ramsay 1963.

Fig. 22.6 Modified after Coward and Siddans 1979. Index map after Williams 1984.

Fig. 22.9 After A. Spicher. 1980. Tektonische Karte der Schweiz, Schweizer Geologisches Kommission.

Fig. 22.10 Redrawn after P. Eskola. 1949. *Origin of mantled gneiss domes.*

Fig. 22.11 Modified after J.B. Thompson et al. 1968. Nappes and gneiss domes in west-central New England. In E-An Zen et al., eds., *Structural studies of the Appalachians.*

Fig. 22.12B After A. Zingg and R. Schmid. 1979. Multidisciplinary research on the Ivrea zone. *Schweizerische Mineralogische und Petrografische Mitteilung* 59:189–197.

Fig. 22.13 After K.C. Misra and F.B. Keller. 1978. Ultramafic bodies in the southern Appalachians. *Am. Jour. Sci.* 278:389–419.

Fig. 22.14 After P.C. Bateman. 1981. In W.G. Ernst, ed., *The geotectonic evolution of California.* Englewood Cliffs, N.J.: Prentice-Hall.

Fig. 22.15 After H. Laubscher. 1982. Detachment, shear, and compression in the central Alps. *Geol. Soc. Am. Mem.* 158:191–213; H. Miller, St. Mueller, and G. Perrier. 1983. Structure and dynamics of the Alps: A geophysical inventory. In H. Berckhemer and Kim Hsu, eds., Alpine-Mediterranean geodynamics. *GSU-AGU geodynamics series* 7:175–204.

Fig. 22.16 After Roeder 1973.

Fig. 22.17 After Cook et al. 1981.

Fig. 22.18 Modified after F.J. Turner. 1968. *Metamorphic petrology: Mineralogical and field aspects.* New York: McGraw-Hill, Fig. 8-6.

Fig. 22.19 Redrawn from M. Frey, J.C. Hunziker, W. Frank, et al. 1974. Alpine metamorphism of the Alps, a review. *Schweizerische Mineralogische und Petrografische Mitteilung* 54:247–290.

Fig. 22.20B After Thompson et al., 1968, pp. 212–213.

Fig. 22.21A After LeForte 1975.

Fig. 22.21B A. Gansser. 1974. The ophiolitic mélange, a worldwide problem on Tethyan examples. *Eclogae geolog. Helv.* 67:479–508.

Fig. 22.21C After A. Gansser. 1981. The geodynamic history of the Himalaya. *A.M. GSA-AGU Geodynamics series* 3:111–122.

Fig. 22.22 After LeForte 1975.

Fig. 22.23 After P.C. England and A.B. Thompson. 1984. Pressure-temperature-time paths of regional metamorphism. I. Heat transfer during the evolution of regions of thickened continental crust. *J. Petrol.* 25:894–928.

Fig. 22.24A After E.C. Hansen, W.H. Scott, and R.S. Stanley. 1967. Reconnaissance of slip line orientations in parts of three mountain chains. *Year Book,* vol. 65. Washington, D.C.: Carnegie Institute, pp. 406–410.

Fig. 22.25B After J.C. Moore et al. 1978. Orientation of underthrusting during latest Cretaceous and earliest Tertiary time, Kodiak Islands, Alaska. *Geology* 6. Fig. 1, p. 209.

Fig. 22.26 From A.W.B. Siddans. 1983. Finite strain patterns in some Alpine nappes. *J. Struct. Geol.* 5 (3/4):444–445.

Fig. 22.27 After D.W. Durney and J.G. Ramsay. 1973. Incremental strain measured by syntectonic crystal growths. In K.A. DeJong and R. Scholter, eds., *Gravity and tectonics.* New York: Wiley, p. 67.

Fig. 22.28 After E.C. Hansen. 1971. *Strain facies.* New York: Springer Verlag.

Fig. 22.29 After D.J. Sanderson. 1982. Models of strain variation in nappes and thrust sheets: A review. *Tectonophysics* 88(3-4):201–233.

Fig. 22.30 After O. Merle. 1986. Patterns of stretch trajectories and strain rates within spreading-gliding nappes. *Tectonophysics* 124:211–222.

Index

Note: Page numbers in *italics* indicate illustrations; those followed by t indicate tables and n indicate footnotes.

Fold(s), (Cont.)
    drag, 57–58, *58*, 75, 77, 254
    enveloping surfaces of, 224
    fault-bend, *99*, 100
        multilayer, 251–254, *252*, *253*
    fault-propagation, 102, *103*
        multilayer, *253*, 253–254
    fault-ramp, *99*, 100
    flank of, 220, *221*
    flattening of, superposed homogeneous, 243–245
    flexural, 239–241, *240*
        bending and, 314–315, *316*
        buckling and, 315–318, *317*
        orthogonal, 240, *240*
    flexural-shear, 240–241, *241*
        multilayer, 245–248
    flexural-slip, multilayer, 246, *246*
    flow, 241–242, *242*, *243*
    foreland, 102, 468–469, 469–471. *See also* Foreland
        fold and thrust belts
    formation of, strain and, 301, *302*
    fractures and, 48–50, *49*, 200–202, *201*, 202t
    generations of, 255
        interpretation of, 258–259
    geologic significance of, 217–220
    geometry of, 220–224, *220–224*
        foliation and, 269–272, *270*, *271*
    harmonic, 232, *233*
    hinge line of, *60*, 61, 220, *221*, 222, *222*
    inclined, 224, 225, *225*
    inflection line of, 220, *220*
    interference patterns for, 255–258, *257*, *258*
    kinematic models of, 238–261
        vs. mathematical models, 435–436
        in orogenic belts, 487, *488*
    kink, 236, *236*
        formation of, *248*, 248–249, *249*, 250–251
        layered, 250–251
    layered. *See* Folded layer
    limb of, 220, *221*
    lineations and, 280–281, 314–321
    mathematical models of. *See* Mathematical models
    median surface of, 224
    multilayered. *See* Folded multilayer
    neck, 276, *276*
    neutral surface of, 240
    noncylindrical, 221
    normal faults and, *76*, 76–77
    order of, 234–235, *235*
    in orogenic belts, 217, *218*
        kinematic analysis of, 487, *488*
    orthogonal flexure, 240, *240*
    overturned, 225, *225*
    parallel, 231, *231*, 235
    parasitic, 235
    passive-flow, 241–242, *242*, *243*
    passive-shear, 241–242, *242*, *243*, *318*, 318–319
        multilayer, 245–248
    perfect, 228–230
    profile of, 221, *222*

    ptygmatic, 237, *237*
        in a multilayer, *247*, 247–248
    reclined, 225, *225*
    recumbent, 224, *225*
    s, 227, *227*
    scale models of, 438–441, *439–441*
    scale of, 217, *218*, 224, *224*
    sheath, 62, *63*
    similar, 236
    solution, 242–243, *243*, *244*
    strain and, 314–320, *316–321*
        in folds, *347–350*, 348–352
    stress trajectories and, 201, *201*
    strike-slip faults and, 116, *116*, 122, *122*
    style of, 226t, 228t, 228–230, *229*, *230*
        common, 235–237
    superposed, geometry of, *255–258*, 255–259
    superposed homogenous flattening and, *319*, 319–320,
        320
    symmetrical, 224, 227, *227*–228
    synform, 220
    thrust faults and, 102, *103*, 110. *See also* Fold and
        thrust belts
    tightness of, 228, 228t, *229*, *230*
    trough line of, 221, *221*
    upright, 224, *225*
    vergence of, 110, 227–228
    volume-loss, 242–243, *243*, *244*
    wavelength of, 224
    z, 227, *227*
Fold and thrust belts, 102–103, *104*, *105*, 217, *219*
    foreland, 468–473, *470–472*
    geometric and kinematic models of, 108–109, *109*
    plate tectonic models of, 492–493
    strike-slip faults and, 122, *122*
Folded angular unconformity, 18–20, *19*
Folded layer
    axial trace thickness of, 231, *231*, 232, 232t
    bending of, 314–315, *316*
    buckling of, 315–318, *317*
    chevron folding of, 250–251, *251*
    dip isogon variation in, 231, *231*, 232t
    flexural folding of, 239–241, *240*, *241*
    flexure of, 239–241
    geometry of, 222–224, *223*, *224*
    homogeneous flattening of, 243–245, *244*, *245*
    inflection surface of, 223, *223*
    kinematic models of, 238–245
    kink folding of, 250–251
    orthogonal thickness of, 231, *231*, 232, 232t
    passive-shear folding of, 241–242, *242*
    Ramsey's classification of, *231*, 231–232, *232*, 232t
    relative curvature of, 231, *231*, 232t
    strain and, 301, *302*
    stratigraphic up-direction in, bedding-foliation
        relationship and, *271*, 271–272
    style of, *231*, 231–232, *232*, 232t
    volume-loss folding of, 242–243, *243*, *244*
Folded multilayer
    class 1A, 246